The Engineering Guide to LEED— New Construction

About the Author

Liv Haselbach is currently a faculty member in the Department of Civil & Environmental Engineering at the University of South Carolina. She is a licensed engineer and a U.S. Green Building Council LEED® accredited professional with 30 years of experience in the field. The author of numerous articles for industry journals, Dr. Haselbach was the founding owner of a civil/environmental engineering firm specializing in land development and regulatory compliance.

The Engineering Guide to LEED— New Construction

Sustainable Construction for Engineers

Liv Haselbach

New York Chicago San Francisco
Lisbon London Madrid Mexico City
Milan New Delhi San Juan
Seoul Singapore Sydney Toronto

The McGraw·Hill Companies

Library of Congress Cataloging-in-Publication Data

Haselbach, Liv.
   The engineering guide to LEED-new construction : sustainable construction
for engineers / Liv Haselbach.
        p.   cm.
   Includes bibliographical references and index.
   ISBN 978-0-07-148993-5 (alk. paper)
   1. Sustainable buildings—Design and construction.   2. Leadership in
Energy and Environmental Design Green Building Rating System.   I. Title.
TH880.H37 2008
690—dc22

                                                                    2008002073

McGraw-Hill books are available at special quantity discounts to use as premiums and sales
promotions, or for use in corporate training programs. To contact a representative please visit
the Contact Us pages at www.mhprofessional.com.

The Engineering Guide to LEED–New Construction

1 2 3 4 5 6 7 8 9 0 DOC/DOC 0 1 4 3 2 1 0 9 8

ISBN 978-0-07-148993-5
MHID 0-07-148993-2

 This book is printed on recycled, acid-free paper made from 100% postconsumer
waste.

Sponsoring Editor	**Proofreader**
Cary Sullivan	Honey Paul
Acquisitions Coordinator	**Indexer**
Rebecca Behrens	Jeff Evans
Editorial Supervisor	**Production Supervisor**
David E. Fogarty	Pamela A. Pelton
Project Manager	**Composition**
Harleen Chopra, International Typesetting and Composition	International Typesetting and Composition
Copy Editor	**Art Director, Cover**
Patti Scott	Jeff Weeks

Contents at a Glance

Contents

Preface

*T*he *Engineering Guide to LEED–New Construction* is intended as a reference or a textbook to aid in the understanding and application of green building design concepts for the engineering and development community. It focuses on the U.S. Green Building Council (USGBC) Leadership in Energy and Environmental Design® (LEED) rating system as an example format for sustainable vertical construction.

Sustainability has many definitions. The definition that has been generally accepted in the context of human beings building and living in a more "sustainable" world was initially developed at the World Commission on Environment and Development (WCED) in 1987. It is simply this: "Sustainable Development is development that meets the needs of the present without compromising the ability of future generations to meet their own needs."

Sustainable Construction is a subset of sustainability, which focuses more on the built environment, both during the construction phase and during the operational life cycle of the facility. However, both sustainability and sustainable construction are subject to interpretation and are very difficult to define. What may seem sustainable in one culture or per one set of values, may not appear sustainable to another. Likewise, the concepts that may be viewed as more important for sustainability by some people, may not be as important to others.

A very common term used for a major focus area in sustainable construction is *Green Building*. Green Building may not really represent true sustainable construction in some people's opinion, but it is an attempt to approach sustainability in a format that fits readily into our current culture. It is a movement that tries to put some of the concepts of sustainability into the construction or renovation of our buildings and facilities.

In like manner, even though this book has sustainable construction in its title, it by no means represents a fully comprehensive method to construct sustainably. This book is intended to be an introduction to some of the major concepts that are being accepted in methodologies to introduce sustainability into construction practices in the United States. It specifically covers the concepts that are currently being promoted by a rating system developed by the USGBC. The system is entitled LEED and has grown out of energy-saving efforts in the United States.

This book focuses more on the "Environmental" aspects of LEED. The "Energy" aspects have been fairly well developed and might require another volume or two to adequately explain in greater detail. In addition, this book expands on some of the environmental issues that are focused on in LEED and gives some direction into means to accomplish the goals, or gives more detailed background information on the environmental systems and impacts that many of the LEED subcategories are based on. In this way, an engineer or other professional may better grasp the intent of the proposed sustainability methods. A better understanding may result in better and more comprehensive, or perhaps alternative designs to obtain the goals. This book is meant to be a guide for all professionals working on sustainable construction in the United States.

The book is divided into three different types of sections. The first section, which consists of Chap. 1, gives an introduction to both sustainable construction and the USGBC LEED–New Construction rating system. The second section, Chaps. 2 through 7, goes into detail on many of the prerequisites and credits used by the USGBC for certification through its rating system for new construction and major renovations. The last section, Chaps. 8, 9, and 10, consists of several distinct parts broadening out from the specific rating system items with emphasis on application to various sectors of particular interest to many groups, management, the military, and low-impact development (LID). Chapter 8 gives some overviews of construction management and organization for helping a *green* professional make a sustainable construction project successful. Chapter 9 gives a view of how the Federal Government and one of its largest departments, the Department of Defense (DoD), have evolved as major contributors to the sustainable development movement in the United States. This chapter starts with a broad overview of how the DoD is involved in the movement and gradually details it down to the example of how the DoD is starting to address the improving of indoor air quality. Chapter 10 gives a broad overview of how this rating system may effectively intersect with another growing movement in the sustainable construction arena in the United States, LID, which focuses on the outdoor impacts of construction, particularly with respect to stormwater.

The Engineering Guide to LEED–New Construction can be used either as a textbook, or a reference. The exercises in the chapters help to develop sustainable skills and understanding for the students, while also allowing them to research the new principles and guidances that may evolve after publication of the book. I hope that others find as great a satisfaction and enjoyment in becoming more *green* as I have over the past few years.

Acknowledgments

I would like to thank the people who have helped educate me on various aspects of green building, politely listened to me talk excitedly about sustainability, and have helped me research or review some of the specific items in the text over the past few years. Of particular note are Major Steve Bruner and Lee D. Pearce, both in the U.S. Army Corps of Engineers; Delisa Clark of USC Campus Planning and Construction; Michael Koman in the USC Housing Department; Dr. M. Hanif Chaudhry, Chair of the USC Civil and Environmental Engineering Department; Dr. Lynn Odom; Chris Estes of Estes Design Inc.; Heramb Haldankar in the USC Civil and Environmental Engineering Department; and all the students who have taken my Sustainable Construction for Engineers course. I am very appreciative for the financial support I received to help in developing this course in Sustainable Construction for Engineers in 2003 from the Sustainable University Initiative.

I am really grateful for the help in editing from my friends, Nancy Elizabeth Bjorklind Bove, Susan Marie Cornett, and Joyce Striebel; and from my beautiful daughters, Candace Elizabeth Brakewood and Heidi Marie Brakewood. Finally, I would like to thank my husband, Mike Navarro, for encouraging me to write this book, for always being so enthusiastic about it, and mostly for his constant love.

CHAPTER 1

Introduction

1.1 An Introduction to Sustainable Development

There are many definitions of *sustainability* and *sustainable development*. What is sustainable for one group may not be as sustainable for another. The most accepted worldwide definition of *sustainable development* is "development that meets the needs of the present without compromising the ability of future generations to meet their own needs" (Brundtland Commission, 1987). However, in an effort to address some of the sustainability ideals and goals in the United States, a subset of sustainability referred to as *sustainable construction* and its subcomponent called *green building* are emerging. They offer a look at sustainability in terms of a smaller scope, such as typical building or construction in an area of the developed world, and in the shorter term, perhaps the life of a building.

Sustainable construction is any construction, while green building focuses on vertical construction. There is also a movement developing to address horizontal construction (transportation and utility corridors). The horizontal construction program related to roads in the United States is usually referred to as *green highways*, although there are also publications that refer to some of the designs as *sustainable streets*. The Green Highway Partnership is an organization which has the U.S. Environmental Protection Agency (EPA) and the Federal Highway Administration (FHWA) among its partners. This partnership began in the Washington, D.C., area and is expanding nationally. Another initiative relating mainly to the outdoors and horizontal construction is the Sustainable Sites Initiative (SSI), which is a collaboration between the American Society of Landscape Architects (ASLA), the Lady Bird Johnson Wildflower Center (The Wildflower Center), and the United States Botanic Garden (USBG) among others. More information can be found at their website: http://www.sustainablesites.org/.

Sustainable construction is an international concern. Although this book focuses on the process in the United States, great efforts have been made to develop sustainable policies and practices throughout the world. The sustainable practices range from procurement or supply chain practices through wastewater reuse practices in the operational phases. Good solutions and practices for sustainable construction may not be the same in all areas of the world and for all societies. Therefore, the rating systems developed and the techniques used will differ. Energy efficiency is a high priority in many countries, particularly in countries with cold winters. There has been much publication and research into improving energy efficiencies internationally both for new construction and as retrofits in existing buildings.

By looking at the green building movement in the United States, it is obvious that the green rating system as developed by the U.S. Green Building Council (USGBC) has had wide acceptance to date and is becoming more widespread. This system is referred to as LEED®, which stands for Leadership in Energy and Environmental Design.* Since LEED grew out of energy program initiatives, it is already developed in focus areas of both architects and mechanical engineers. Sometimes the construction that focuses on energy efficiency and also water efficiency is referred to as *high-performance* building.

Sustainable construction research and applications are still in their infancy. There is a great need for research, education, and case studies from applications to further develop a more sustainable future in development and construction. Even the definitions of what is or what is not sustainable need to be researched and further evaluated. For instance, certain people may believe that keeping as much as possible of the natural environment pristine is one of the most important goals of sustainability, whereas others may believe that an improved food supply for humans is more important. Thus, not only do the principles behind sustainability differ from group to group, or may be in contrast or seeming incompatibility with one another, but also the ranking or value judgments of the importance of various factors to sustainability are very difficult to develop and find consensus on.

To facilitate the further development and implementation of the green building system in the United States, engineers and other professionals must be educated in the rating systems used, and in the parameters and principles that have been established in the use of these rating systems. The intention of the author is to develop a text that educators can use to teach engineers and professionals about sustainable development, particularly the LEED system. The emphasis is on the development of skills to facilitate the use of this system, as well as guidance on potential additional research avenues to further improve the green building movement, with stress on environmental aspects, as future developments and needs arise.

The author likes to explain that a sustainable environmental goal will focus on both how construction impacts the environment and how environmental decisions impact development. It is a two-way road, and things must be looked at from both directions. In addition, other impacts in the entire life cycle of the facilities or practices need to be looked at. What may seem like an environmentally sound practice may in fact not turn out quite that way when input into an anthropogenic world. Two examples which follow are the sustainability of stormwater ponds and the use of certain additives in gasoline.

A common form of stormwater management is the use of a retention or a detention pond for storage and possibly infiltration of additional runoff caused by an increase in paved or roofed areas. Therefore in an approved site plan set, a pond may be called for. However, it has been shown that in many cases the specified ponds are not installed, much less maintained to sustain a more manageable stormwater system. Why? The reasons are not known, but perhaps the contractor or owner does not understand the environmental impact that the lack of a pond may cause; installation of one may not seem important.

Then there is the use of MTBE (methyl-tertiary-butyl ether) as an additive in gasoline to improve air quality, which was a common practice in the 1990s. It was considered to be a sound environmental practice with respect to air issues. However, there were some unexpected consequences. Traditionally in the United States, gasoline

---

*LEED is a registered trademark of the U.S. Green Building Council.

is made at a refinery, is transported and distributed through a network of petroleum facilities, and ends up at the retail outlet, the gasoline station. Gasoline is usually stored in underground tanks, for many reasons including safety, at this end resale site. Underground tanks are currently fairly well regulated and monitored in the United States, but that was not always the case. In many sites, gasoline has leaked into the ground, and the product sits on or in the groundwater. Most constituents in gasoline are nonpolar organics which do not dissolve readily in polar water, but might prefer to adsorb to organic material in the soil. So even though the groundwater may slowly flow off-site, most of the gasoline contaminants remain closer to the site. However, MTBE is an ether, and ethers are more polar than many gasoline constituents such as benzene. Polar organics tend to be more soluble in water and may have a greater tendency to travel with the groundwater gradient. Therefore, there are many places where some gasoline leakage in the soils at a site did not initially cause an off-site concern until the MTBE was added and moved more rapidly off-site. In this way a practice that is an environmental benefit for one system may not be as sustainable in another.

There are several other green rating systems for vertical construction in the United States, but the LEED system and the alternative Green Globes system are by far the most prominent. Green Globes is distributed by the Green Building Initiative (GBI) and is based on a system developed in Europe and later also used in Canada. The European version is referred to as the Building Research Establishment's Environmental Assessment Method (BREEAM). This text is based on the LEED system, but this in no way implies that the other systems are not useful and viable. The LEED rating system was chosen as the focus of this text as it is currently more widespread in the United States. It also does not allow for a self-certification method, which makes it more restrictive and possibly more difficult to complete, but at the same time also gives greater control and consistency to green building.

The text has a heavy focus on some of the civil and environmental aspects of the rating system, since the author is first targeting this engineering community. However, it also addresses in some fashion other issues in the rating system. One of the reasons for its being more all encompassing in the topics reviewed is to educate the civil and environmental engineers about the other credits and criteria so that they can effectively work with other disciplines in a cooperative fashion to better implement sustainable construction practices.

There is a need for further involvement from the civil and environmental engineering community in the development of the LEED rating system. The evaluation of best practices and goals for many of these aspects is still in its infancy. It is here that much research and teaching may be needed to further develop the rating system effectively in a timely manner. There has been a substantial amount of green building at the University of South Carolina since 2003, and a few of the categories of sustainable construction in which additional research, the development of best practices, and optimization are needed include the following:

1. Construction and demolition (C&D) debris recycling and reuse. There is a need to optimize construction practices to facilitate C&D debris recycling in an economic fashion and to develop the recycling and reuse infrastructure in many areas of the United States to support these practices. Figure 1.1.1 shows a construction debris waste container yard at a LEED registered project in Columbia, S.C.

FIGURE 1.1.1   Construction debris container yard at a LEED registered project in 2005 at the University of South Carolina in Columbia, S.C. (*Photograph taken by Steve Bruner.*)

2. Design-procurement-construction process integration and optimization.

3. Construction management processes.

4. Stormwater management and low-impact development (LID).

There is a need for research and development of best management practices and integrated management practices with respect to the nonpoint source type of pollutants. One area includes new infrastructure and materials with respect to stormwater management such as pervious pavements. An example of a multipurpose landscape amphitheater and stormwater management feature at a LEED certified project in Columbia, S.C., is shown in Fig. 1.1.2. Additional information on LID can be found in Chap. 10.

The text is also intended to be used by other engineers and professionals such as mechanical engineers, architects, planners, community leaders, and construction managers, as it does present the overall holistic approach of the LEED rating system and can facilitate the understanding of many of the credits and criteria outside the traditional purview of these disciplines. Again, the intention is to let all the professionals and affected parties learn to "talk the talk" of the other team members and understand the viewpoints and engineering decisions in an interdisciplinary fashion. It is very important that all interested parties, including the community, be involved in the sustainable construction process. Criteria that are important to many participants can be incorporated into the design and implementation of green construction. One example concerns the maintenance of the "historical" feel in a community. Green does not mean the buildings and facilities have to look different. Figure 1.1.3 depicts a structure built

FIGURE 1.1.2 Landscaped amphitheater also functions as a stormwater management feature at the LEED Certified West Quad Housing Complex at the University of South Carolina. (*Photograph taken in September 2006.*)

in 1939 at the University of South Carolina, and it shows how the new "green" building in Fig. 1.1.2 can still fit in well with the other historic buildings at this university.

There are many other excellent references to consult in the understanding and educating of others about sustainable construction and green building. Several of these are also focused on the USGBC LEED. However, this book is different from many of the other references that cover the LEED rating system in that it presents the information and criteria in a fashion that is useful for then doing mathematical and design exercises for further understanding and familiarity with the criteria and parameters in use. Several excellent references on sustainable development, industrial ecology (terminology referring to sustainable development from a more industrial viewpoint), sustainable construction, and green building are listed in the References section of this chapter. There are also many website resources from which sustainable education materials can be accessed. One of note is the ImpEE (Improving Engineering Education) Project website developed by the University of Cambridge in the United Kingdom, and another is the Green Design Institute at Carnegie Mellon.

There are a plethora of current publications in journals based on how universities and other institutions are implementing sustainability into the curricula. For instance, *The International Journal of Engineering Education* focused Part I of its vol. 23, no. 2, 2007 issue on "Educating Students in Sustainable Engineering."

One very important reference which should be highlighted here is the LEED NC-2.2 Reference Guide, as developed by the USGBC. This text in sustainable construction is intended to supplement the subject Reference Guide. The Reference Guide, just like building codes and legislation, presents the parameters and suggested practices for

Figure 1.1.3 Preston College Building at the University of South Carolina, built in 1939. Similar architecture to the new green West Quad Facility opened in 2004 (as shown in Fig. 1.1.2).

implementation of the rating system that other entities can adopt to facilitate building in a more sustainable fashion. The Reference Guide also expands upon the environmental, societal, and economic principles used in the determination of the various prerequisites and credits, and presents many excellent examples, case studies, and references which can be used to aid in the implementation of the LEED system. This text takes the basic parameters and principles of the subject Reference Guide, particularly with respect to many of the civil and environmental design issues, and further expands them into equations and teaching formats which may be helpful in a classroom situation. It is with gratitude that the author acknowledges the USGBC in allowing her to use most of the USGBC rating system verbiage for each prerequisite and credit in this text and many of the tables, definitions, and other information from the Reference Guide.

LEED-NC 2.2 is the current (2007) version of the USGBC rating system for New Construction and Major Renovations. The first edition of LEED-NC version 2.2 was released in October 2005. It replaced LEED-NC version 2.1 as the current rating system for newly registered projects. Since that time, there have been several corrections and addenda to version 2.2, and a listing of these can be found as errata available for download to the general public from the USGBC website. This text is based on the LEED-NC version 2.2 released in October 2005 and as amended through the second edition dated September 2006 and errata as posted in the spring of 2007. Projects that were started under the LEED-NC version 2.1 may remain under that version, although all new projects will be expected to adhere to version 2.2 or future versions as they are officially adopted. There are substantial differences between versions 2.2 and 2.1 for many of the subcategories, and it is expected that the next version also will contain substantial changes to many of the credits. Practitioners and educators should ensure that these changes are incorporated in future projects and class curricula.

This text focuses on the LEED-NC rating system, which is the premier and most heavily used green rating system in the United States at this time. LEED-NC is for new construction and major renovations for mainly commercial, institutional, industrial, and large residential (four stories or more) projects. The USGBC and several other organizations have also developed, or are in the process of developing, rating systems for other types of projects. One other organization of note is the National Association of Home Builders (NAHB), which through the GBI is offering its voluntary Model Green Home Building Guidelines.

Some of the other LEED rating systems are listed in Table 1.1.1.

		Status as of May 2007	Comments
LEED-EB	Existing Building operations and maintenance	Approved 10/22/04 Current Version 2.0	Both for later recertification of buildings previously certified and for existing building renovations
LEED-CI	Commercial Interiors	Approved 11/17/04 Updated November 2004 Current Version 2.0	For commercial tenant space in a building which may or may not have the core and shell certified
LEED-CS	Core and Shell	Approved July 2006 Current Version 2.0	For the core and shell of a building which may or may not have the commercial interior tenant space certified
LEED-H	Homes	Pilot started in 2005 Pilot Version 1.11a released 2/1/07 Launch expected 2007	
LEED-ND	Neighborhood Development	Pilot expected start in 2007	Integration of smart growth, urbanism, and green building for neighborhood design
LEED for Schools	Schools (K–12)	Launched in April 2007	Based on LEED-NC, but required for K–12 academic buildings. May be used for K–12 nonacademic buildings
LEED for Retail	Retail construction*	Draft Version 2.0 April 2007	Two rating systems. One is based on NC 2.2. Second is based on CI 2.0
LEED for Healthcare	Healthcare facilities*	Under development	
LEED for Labs	Facilities with laboratories*	Under development	

*Can also be viewed as a part of LEED-NC (or CI for retail) with suggested guidances as listed in Table 1.2.1.

TABLE 1.1.1   Other LEED Rating Systems

The USGBC is currently working on revamping its premier rating system (LEED-NC) to version 3.0. It is unclear whether the various other rating systems will be incorporated directly into version 3.0 or whether they will remain stand-alone systems. Either way, one of the goals in the revamping is to make the various rating systems for the numerous types of applications more consistent and interchangeable.

Finally, note that sustainability is not just an environmental concept. Many are quick to point out that sustainability focuses on a triple bottom line: economic, environmental, and societal. Many of the concepts relating to future economic and societal concerns are dependent on current resource and environmental management and best practices. Hawken, Lovins, and Lovins further expand on this in their book *Natural Capitalism, Creating the Next Industrial Revolution.*

1.2 Introduction the USGBC LEED-NC Rating System

LEED-NC is applicable to a large portion of the most expensive vertical construction projects in the United States. It basically covers most commercial, institutional, and industrial projects and includes residential construction of facilities of four or more stories.

Guidances

There are many projects which have unique criteria that may not fit well into the credit intents and requirements as listed in LEED-NC 2.2. To facilitate LEED certification and sustainable practices for certain categories of projects, the USGBC has or is developing guidances specific to many special subsets of project application types. They are referred to as application guides. Table 1.2.1 is a listing of several of the guidances being developed.

These guidances serve many purposes. They serve to aid in modifying or embellishing criteria for specific credits when it may be difficult to apply the credit directly as written for a particular type of project, and they also identify some items that may be used for a particular project type for one of the first four Innovation & Design credits. Table 1.2.1 can also be found as Table 7.0.1 in Chap. 7. Chapter 9 is a special section on the U.S. military and sustainable construction and explores in greater detail how one of the guidances, LEED for Lodging, may relate to LEED-NC, with particular emphasis on indoor air quality. For any project that deals with any of these special applications, the guidances should be used in conjunction with the LEED-NC 2.2 rating system.

Application	Status as of September 2006
LEED for Multiple Buildings/Campuses	Guide available
LEED for Lodging (<4 stories)	Developed by USAF but not yet voted on for adoption by USGBC membership
LEED for Healthcare	Under development
LEED for Labs	Under development

TABLE 1.2.1 Application Guidances

How LEED-NC Is Set Up

LEED-NC contains both credits and prerequisites. All eight prerequisites are mandatory for each project. None of the credits (currently up to 69 points from these credits) are mandatory at this time. (The carbon commitment and other future goals may make some credits mandatory for certain projects in the future.) The credits are worth a certain number of points, and a combination of credit points adds up to a certain level of certification. The current levels of certification and the number of points associated with each level are as listed:

- Certified: 26 to 32 points
- Silver: 33 to 38 points
- Gold: 39 to 51 points
- Platinum: 52 to 69 points

LEED-NC is subdivided into six categories for which there are prerequisites, subcategories, and credits representing possible points. Usually each credit has one associated point, but two of the Energy and Atmosphere credits are multipoint. The six categories and their associated numbers of prerequisites, subcategories, and possible points (excluding exemplary performance points associated with that category) are as noted:

- **Sustainable Sites** (SS)
  - Prerequisites: 1
  - Subcategories: 8
  - Possible points: 14

- **Water Efficiency** (WE)
  - Prerequisites: none
  - Subcategories: 3
  - Possible points: 5

- **Energy and Atmosphere** (EA)
  - Prerequisites: 3
  - Subcategories: 6
  - Possible points: 17

- **Materials and Resources** (MR)
  - Prerequisites: 1
  - Subcategories: 7
  - Possible points: 13

- **Indoor Environmental Quality** (EQ)
  - Prerequisites: 2
  - Subcategories: 8
  - Possible points: 15

- **Innovation and Design Processes** (ID)
  - Prerequisites: None
  - Subcategories: 2
  - Possible points: 5

LEED-NC Documents

Information is provided from the USGBC on several different levels. The main level of information provided for LEED-NC 2.2 is the governing document as approved by the USGBC for the rating system. It is entitled *LEED-NC Green Building Rating System for New Construction & Major Renovations Version 2.2 for Public Use and Display*. It is available free of charge to the general public for download and public display and can be found on the USGBC website (www.usgbc.org). The verbiage from various sections of this document is included verbatim throughout this text. The exact wording is important to read to adequately interpret the intent and expanse of the requirements listed. Some other items provided to the general public for free from the USGBC website are the errata sheets (which include corrections to both the governing document and the Reference Guide as described in the following paragraph) and various fact sheets and sample items, including sample templates. Templates are the online format used by the USGBC for credit submittals and calculations. The USGBC also provides a project checklist for overall summary credit and prerequisite tracking.

As mentioned earlier, the USGBC provides a reference guide for the LEED-NC 2.2 rating system. This reference guide is supplemental to the governing document and is available for purchase. It gives example calculations, case studies, and additional references, and it expands upon the intent and definitions necessary to better understand the prerequisite and credits for certification. Members of the USGBC and people associated with registered projects can also log on to the system and view the Credit Interpretation Rulings (CIRs), which are replies to Credit Interpretation requests. The Credit Interpretation Rulings are further explained in the following sections and allow the members or projects to keep up with future interpretations and enhancements to the rating system. Project personnel can also use the template system directly online to continually track credit and prerequisite performance.

Credit Formats

The information given in the credit write-ups for the various credits is set in a standard format. The format in version 2.2 is slightly different from that of version 2.1. The formats used in the governing document (Public Display Portion) and the Reference Guide in version 2.2 are also slightly different and are summarized as follows:

- In Public Display Portion: Intent
  Requirements
  (No longer lists submittals to allow for greater flexibility)
  Potential Technologies and Strategies
- In Reference Guide: Intent
  Requirements
  Summary of Referenced Standards
  Approach and Implementation
  Calculations (Yes or No)
  Exemplary Performance Point (Yes or No)
  Submittal Documentation

Registration, Certification, Membership, and Accreditation

There are two main definitions of various status categories with respect to building projects and the USGBC LEED process: registration and certification. There are currently fees associated with both registration and certification.

Registration with the USGBC is completed at the inception of a project to begin the certification process. The project team then receives access to many of the management features and templates available on the USGBC website to aid in keeping track of the project. The project is usually registered at the current version of the rating system, and the project is tracked at this level.

Certification at one of the certification levels listed previously is intended for that project after it is complete. Certification is requested through an application process. Certification is received at the version of the rating system for which the project is registered. Figures 1.2.1 and 1.2.2 are photographs of the first private building in South Carolina that received certification under the USGBC LEED-NC rating system.

Figure 1.2.1 Cox and Dinkins Engineers and Surveyors in Columbia, S.C: First private building to receive certification under the USGBC LEED-NC rating system in South Carolina. (*Photograph taken in June 2007.*)

Figure 1.2.2 Cox and Dinkins Engineers and Surveyors in Columbia, S.C. USGBC LEED-NC rating system certification plaque. (*Photograph taken in June 2007.*)

Formerly, the application for certification was one step near the end of a project. In LEED-NC 2.2 this application process has been optionally divided into two phases. Several of the design phase credits can be applied for prior to completion of the project, with the remaining applied for at the end of the construction phase. Each project is allowed one design phase review. This aids in expediting paperwork and understanding final interpretations of the applicable credits.

There are some additional definitions of various status categories with respect to people and the USGBC LEED process:

Membership to the USGBC is by organization, and then individuals within that organization can become members under the umbrella of their organization (organizations pay dues, individuals do not). A person can also be a **member of a chapter** such as the USGBC-SC (South Carolina) chapter (individuals usually pay dues to a chapter). Students and recent graduates may also become members of the Emerging Green Builders (EGB), a special membership category usually associated with local chapters for students and young professionals.

Accreditation is the mechanism by which individuals can become LEED-Accredited Professionals (LEED-AP). This is a process developed by the USGBC to determine whether a person can be recognized as a professional with respect to its rating system. It is an exam-based accreditation process.

Project Checklist and Templates

Development of effective project management and green concept integration into construction projects will be a continual challenge as the green movement expands in the United States. Some simplified suggestions for improvement and implementation will be discussed in later chapters. In addition, substantial efforts are under way to research optimization methods for implementation of the green process. There are many tools and suggestions developed by the USGBC to facilitate becoming green, which can be found in example format on the USGBC website (www.usgbc.org). Two of the main tools are the project checklist and the templates. The project checklist is used to keep track of overall project success with respect to green construction, and the templates are used to keep track of individual prerequisites and credits during the project and to submit information to the USGBC during the certification application process.

At the onset of a project after registration, the USGBC supplies the LEED-NC Registered Project Checklist to the project team. Table 1.2.2 shows a blank LEED-NC 2.2 checklist as downloaded from the USGBC website in May 2007.

Templates, or letter templates as they were frequently referred to in LEED-NC 2.1, are provided for use by the project team after project registration. They are online and provide an interactive format for many of the calculations needed for credit verification. They are used for summary calculations and as a tool to track the necessary documentation for submittals for each prerequisite and credit applied for. Figure 1.2.3 is a summary figure of the format for part of a submittal for a prerequisite that requires drawing submittals and narratives explaining the project adherence to the requirements. It is based on the example template available for download from the USGBC website for the Sustainable Sites prerequisite one (SSp1) for Construction Activity Pollution Prevention. Note that it refers to "credit" requirements, even though it is a prerequisite and does not earn points, but this is purely a matter of keeping the template formats consistent. Figure 1.2.4 is a summary figure of the format for the first portion of Option 1 for the Energy and Atmosphere credit number 1 (EAc1), Optimize Energy Performance. It is depicted to show a template where actual numbers and data are input to the USGBC. The actual online template process is interactive and will provide appropriate calculations from some of the data input.

What to Do If Things Are Not Clear for Your Project

Credit Interpretation Rulings (CIRs) represent the format that the USGBC has chosen to continually aid projects with interpretations of the rating system for their specific circumstances. Requests for interpretation of credits are submitted to the USGBC, and the USGBC replies with Credit Interpretation Rulings. It is a formal process that lists the rulings of the interpretations online for use by other projects to allow for both consistency from project to project and expediency in having the explanation and interpretations readily available without constant revisions to the governing documents. The CIR process is very important to the rating system's success and certification. It is expected that project team members are very familiar with this process. In the LEED-NC 2.1 Reference Guide, the CIR process was summarized in the following four steps:

1. Consult the Reference Guide.

2. Review the Reference Guide and self-evaluate.

3. Review the CIR web page for previously logged CIRs.

4. If still unanswered, submit a CIR with an online form.

LEED for New Construction v2.2 Registered Project Checklist (p.1)				
Project Name and Address:				
Yes	?	No		
			Sustainable Sites	**14 Points**
Y			Prereq 1 — **Construction Activity Pollution Prevention**	Required
			Credit 1 — **Site Selection**	1
			Credit 2 — **Development Density & Community Connectivity**	1
			Credit 3 — **Brownfield Redevelopment**	1
			Credit 4.1 — **Alternative Transportation**, Public Transportation Access	1
			Credit 4.2 — **AlternativeTransportation**, Bicycle Storage & Changing Rooms	1
			Credit 4.3 — **Alternative Transportation**, Low-Emitting & Fuel-Efficient Vehicles	1
			Credit 4.4 — **Alternative Transportation**, Parking Capacity	1
			Credit 5.1 — **Site Development,** Protect of Restore Habitat	1
			Credit 5.2 — **Site Development,** Maximize Open Space	1
			Credit 6.1 — **Stormwater Design,** Quantity Control	1
			Credit 6.2 — **Stormwater Design,** Quality Control	1
			Credit 7.1 — **Heat Island Effect,** Non-Roof	1
			Credit 7.2 — **Heat Island Effect,** Roof	1
			Credit 8 — **Light Pollution Reduction**	1
			Water Efficiency	**5 Points**
			Credit 1.1 — **Water Efficient Landscaping**, Reduce by 50%	1
			Credit 1.2 — **Water Efficient Landscaping**, No Potable Use or No Irrigation	1
			Credit 2 — **Innovative Wastewater Technologies**	1
			Credit 3.1 — **Water Use Reduction**, 20% Reduction	1
			Credit 3.2 — **Water Use Reduction**, 30% Reduction	1
			Energy & Atmosphere	**17 Points**
Y			Prereq 1 — **Fundamental Commissioning of the Building Energy Systems**	Required
Y			Prereq 2 — **Minimum Energy Performance**	Required
Y			Prereq 3 — **Fundamental Refrigerant Management**	Required
			Credit 1 — **Optimize Energy Performance**	1 to 10
			10.5% New Buildings or 3.5% Existing Building Renovations	1
			14% New Buildings or 7% Existing Building Renovations	2
			17.5% New Buildings or 10.5% Existing Building Renovations	3
			21% New Buildings or 14% Existing Building Renovations	4
			24.5% New Buildings or 17.5% Existing Building Renovations	5
			28% New Buildings or 21% Existing Building Renovations	6
			31.5% New Buildings or 24.5% Existing Building Renovations	7
			35% New Buildings or 28% Existing Building Renovations	8
			38.5% New Buildings or 31.5% Existing Building Renovations	9
			42% New Buildings or 35% Existing Building Renovations	10
			Credit 2 — **On-Site Renewable Energy**	1 to 3
			2.5% Renewable Energy	1
			7.5% Renewable Energy	2
			12.5% Renewable Energy	3
			Credit 3 — **Enhanced Commissioning**	1
			Credit 4 — **Enhanced Refrigerant Management**	1
			Credit 5 — **Measurement & Verification**	1
			Credit 6 — **Green Power**	1

TABLE 1.2.2   Blank LEED-NC 2.2 Project Checklist (Sheet 1)

The LEED-NC 2.2 Reference Guide summarizes the CIR process in three steps following the project team's being unable to adequately answer a question based on its interpretation of the Reference Guide. The three steps are as follows:

1. Review the CIR web page for previously logged CIRs. Note that some of the CIRs for other rating systems or versions may not be applicable.

2. If no applicable CIR exists, then submit a CIR via the CIR web page. This web page has guidelines for how a CIR should be submitted, with particular

LEED for New Construction v2.2 Registered Project Checklist (p.2)				
			Materials & Resources	**13 Points**
Y			Prereq 1	**Storage & Collection of Recyclables**	Required
			Credit 1.1	**Building Reuse**, Maintain 75% of Existing Walls, Floors & Roof	1
			Credit 1.2	**Building Reuse**, Maintain 100% of Existing Walls, Floors & Roof	1
			Credit 1.3	**Building Reuse**, Maintain 50% of Interior Non-Structural Elements	1
			Credit 2.1	**Construction Waste Management**, Divert 50% from Disposal	1
			Credit 2.2	**Construction Waste Management**, Divert 75% from Disposal	1
			Credit 3.1	**Materials Reuse**, 5%	1
			Credit 3.2	**Materials Reuse**, 10%	1
			Credit 4.1	**Recycled Content**, 10% (post-consumer + ½ Pre-consumer)	1
			Credit 4.2	**Recycled Content**, 20% (post-consumer + ½ Pre-consumer)	1
			Credit 5.1	**Regional Materials**, 10% Extracted, Processed & Manufactured Regionally	1
			Credit 5.2	**Regional Materials**, 20% Extracted, Processed & Manufactured Regionally	1
			Credit 6	**Rapidly Renewable Materials**	1
			Credit 7	**Certified Wood**	1
				Indoor Environmental Quality	**15 Points**
Y			Prereq 1	**Minimum IAQ Performance**	Required
Y			Prereq 2	**Environmental Tobacco Smoke (ETS) Control**	Required
			Credit 1	**Outdoor Air Delivery Monitoring**	1
			Credit 2	**Increased Ventilation**	1
			Credit 3.1	**Construction IAQ Management Plan**, During Construction	1
			Credit 3.2	**Construction IAQ Management Plan**, Before Occupancy	1
			Credit 4.1	**Low-Emitting Materials**, Adhesives & Sealants	1
			Credit 4.2	**Low-Emitting Materials**, Paints & Coatings	1
			Credit 4.3	**Low-Emitting Materials**, Carpet Systems	1
			Credit 4.4	**Low-Emitting Materials**, Composite Wood & Agrifiber Products	1
			Credit 5	**Indoor Chemical & Pollutant Source Control**	1
			Credit 6.1	**Controllability of Systems**, Lighting	1
			Credit 6.2	**Controllability of Systems**, Thermal Comfort	1
			Credit 7.1	**Thermal Comfort**, Design	1
			Credit 7.2	**Thermal Comfort**, Verification	1
			Credit 8.1	**Daylight & Views**, Daylight 75% of Spaces	1
			Credit 8.2	**Daylight & Views**, Views for 90% of Spaces	1
				Innovation & Design Process	**5 Points**
			Credit 1.1	**Innovation in Design**: Provide Specific Title	1
			Credit 1.2	**Innovation in Design**: Provide Specific Title	1
			Credit 1.3	**Innovation in Design**: Provide Specific Title	1
			Credit 1.4	**Innovation in Design**: Provide Specific Title	1
			Credit 2	**LEED® Accredited Professional**	1
				Project Totals(pre-certification estimates)	**69 Points**
			Certified: 26-32 points, **Silver:** 33-38 points, **Gold:** 39-51 points, **Platinum:** 52-69 points		

TABLE 1.2.2   Blank LEED-NC 2.2 Project Checklist (Sheet 2)

information on what should be in the request. The main focus is on the intent of the credit. The CIR will eventually be posted on the CIR web page and is not intended to include a long description of a particular project, but rather a more overall question to be interpreted for application for the project and other similar project circumstances.

3. The USGBC will rule on the CIR either via e-mail or by posting on the CIR web page.

As mentioned earlier under LEED-NC documents, the USGBC has a process to periodically post errata to both the governing document (Public Display Portion) and the Reference Guide, which are also periodically included in new editions of the versions.

LEED-NC 2.2 Submittal Template
SS Prerequisite 1: Construction Activity Pollution Prevention

(Responsible Individual)   (Company Name)
I, _____, from_____
Verify that the information provided below is accurate to the best of my knowledge.
_____

CREDIT COMPLIANCE
Please select the appropriate compliance path

Option 1: The Erosion and Sedimentation Control Plan (ESC) conforms to the 2003 EPA Construction General Permit, which out lines the provisions necessary to comply with Phase I and Phase II of the National Pollutant Discharge Elimination System (NPDES) program. _____

OR

Option 2: The ESC Plan follows local erosion and sedimentation control standards and codes, which are more stringent than the NPDES program requirements. _____

SUPPORTING DOCUMENTATION
The noted project drawing(s) have been uploaded. The drawing(s) shows the erosion and sedimentation control measures implemented on the site.

Sheet Description Log
Please include sheet name, sheet number and file name for each uploaded, refereneced drawing (e.g. A-101, Site Plan, siteplan.pdf)
_____
_____
_____

_____ I have provided the appropriate supporting documentation in the document upload section of LEED Online. Please refer to the above sheets.

NARRATIVE (Required)
Provide a narrative to describe the Erosion and Sedimentation control measures implemented on the project. If local standard has been followed, please provide specific information to demonstrate that the local standard is equal to or more stringent than the referenced NPDES program.
_____
_____

NARRATIVE (Optional)
Please provide any additional comments or notes regarding special circumstances or considerations regarding the project's credit approach.
_____
_____ The project is seeking point(s) for this
credit using an alternate compliance approach.The compliance approach including references to any applicable Credit Interpretation Rulings is fully documented in the narrative above.

Project Name:

FIGURE 1.2.3 Portion of example template from the USGBC website for a prerequisite which requires drawings and narrative for submittal (SSp1).

LEED-NC 2.2 Submittal Template
EA Credit 1: Optimize Energy Performance

(Responsible Individual)   (Company Name)
I, _____ , from_____
Verify that the information provided below is accurate to the best of my knowledge.
_____

CREDIT COMPLIANCE
(Please complete the color coded criteria(s) based on the option path selected)

Please select the appropriate compliance path option
___ Option 1 (Pg2): Performance Rating Method, ASHRAE 90.1-2004 Appendix G or equivalent (upto 10 points possible)
___ Option 2 (Pg14): ASHRAE Advanced Energy Design Guide for Small Office Buildings 2004 (4 points)
___ Option 3 (Pg14): Advanced Buildings Benchmark™ Version 1.1, Basic Criteria & Prescriptive Measures (1 point)

OPTION 1: PERFORMANCE RATING METHOD
___ I Confirm that the energy simulation software used for this project has all capabilities described in EITHER section 'G2 Simulation General Requirements' in Appendix G of A SHRAE 90.1-2004 OR the analogous section of the alternative qualifying code used.

___ I Confirm that the baseline building and proposed building in this project's energy simulation runs use the assumptions and modeling methodology described in EITHER Appendix G of ASHRAE 90.1-2004 OR the analogous section of the alternative qualifying energy code used.

Complete the following sections to document compliance using Option1:
        Section 1.1 – General Information
        Section 1.2 – Space Summary…..:…

………….Section 1.8 – Performance Rating Method Compliance Report

Section 1.1 – General Information
   …………

Section 1.2 – Space Summary
Provide the space summary for your project
(click "CLEAR" to clear the contents of any row. All numeric entries must be entered as whole numbers without commas):

Table 1.2 – Space Summary

Building Use (Occupancy Type)	Conditioned Area (sf)	Unconditioned Area (sf)	Total Area (sf)	
				CLEAR
				CLEAR
				CLEAR
				CLEAR
				CLEAR
				CLEAR
				CLEAR
Total				

FIGURE 1.2.4 Portion of example template from the USGBC website for a credit which requires data and narrative for submittal (EAc1: Option 1). Shaded areas in the table are for user input of data. Blank areas in the table will be filled in automatically.

1.3 New and Future Developments

Organizations

An important part of specifications and design for construction is the use of established standards. Many organizations are recognized as established sources for many standards and have procedures for accrediting, developing, reviewing, and revising standards. The organization that accredits organizations as a standards developer in the United States is the American National Standards Institute (ANSI). On November 27, 2006, ANSI accredited the U.S. Green Building Council as an official Standards Developing Organization. Some of these organizations which help develop or accredit standards as mentioned in this text include the following:

AIA   American Institute of Architects.

ANSI   The American National Standards Institute accredits standards for products, services, processes, and systems in the United States; accredits organizations that perform certifications; and coordinates U.S. standards with international standards.

ASHRAE   The American Society of Heating, Refrigerating and Air-Conditioning Engineers, Inc., publishes standards in the areas of HVAC and refrigeration.

ASTM   ASTM International was originally formed in 1898 as the American Society for Testing and Materials. It develops technical standards for materials, products, systems, and services.

IESNA   The Illuminating Engineering Society of North America is recognized as an authority on lighting and illumination standards in the United States.

ISO   The International Organization for Standardization (Organisation Internationale de Normalisation) is the largest international developer of standards. ANSI is the U.S. voting member body for ISO.

Standard 189

In 2006, the USGBC, ASHRAE, IESNA, and AIA joined together to develop Standard 189, a new minimum standard for high-performance green building. The standard is being led by the newly developed ASHRAE Standard Project Committee 189 (SPC 189). The standard was initially proposed in January 2006 and as a standard under development is referred to as SPC 189P. The preliminary title, purpose, and scope as revised through November 8, 2006 read as follows:

> **Standard for the Design of High-Performance, Green Buildings Except Low-Rise Residential Buildings**
>
> **1 Purpose:**
>
> The purpose of this standard is to provide minimum requirements for the design of high-performance, green buildings to:
> (a) Balance environmental responsibility, resource efficiency, occupant comfort and well being, and community sensitivity, and
> (b) Support the goal of the development that meets the needs of the present without compromising the ability of future generations to meet their own needs.

2 Scope:

2.1 This standard provides minimum criteria that:

(a) Apply to new buildings and major renovation projects (new portions of buildings and their systems): a building or group of buildings, including on-site energy conversion or electric-generating facilities, which utilize a single submittal for a construction permit or which are within the boundary of a contiguous area under single ownership

(b) Address sustainable sites, water use efficiency, energy efficiency, the building's impact on the atmosphere, materials and resources, and indoor environmental quality (IEQ).

2.2 The provisions of this standard do not apply to:

(a) single-family house, multi-family structures of three stories or fewer above grade, manufactured houses (mobile homes) and manufactured houses (modular).

(b) buildings that do not use either electricity or fossil fuel.

2.3 This standard shall not be used to circumvent any safety, health or environmental requirements.

Computer-Aided Design

There has been much interest in incorporating environmental assessments as well as material selection options and evaluations based on life-cycle information and built performance into the computerized early design stages of projects. One result is the development of an *LCADesign* program in Australia. In November 2006, the USGBC and Autodesk announced that they had formed a relationship to further the use of technology and the development of sustainable design. Autodesk is one of the largest companies in the world developing computer-aided design software. Its premier product is Autocad and is used on many projects in the United States.

The Carbon Commitment

Global climate change is a major issue of concern in the world. Some details of mechanisms and suspected causes of this phenomenon are treated in subsequent chapters. The USGBC has made a commitment to focus on emphasizing strategies and green building goals which would help lower the emission of carbon dioxide into the atmosphere. Carbon dioxide is recognized as a gas with a global warming potential (GWP) and is thought to be a major contributor to current changes in the atmosphere. Carbon dioxide levels in the atmosphere have increased in recent decades, as has the use of fossil fuels for energy. Fossil fuels are carbon-based, and their use is thought to impact the global carbon cycle, with more carbon being released to the atmosphere (in the form of carbon dioxide) than sequestered in the crust, ocean, or flora (usually in dissolved, fossil fuel, mineral rock, or vegetation form). Fuels that cannot be readily reproduced in our human-generation time frames, such as fossil fuels, are referred to as *nonrenewable* while energy sources such as wind or wood which can be regrown are referred to as *renewable*. Even though the burning of wood releases carbon into the air in the form of carbon dioxide, it is considered renewable with respect to the carbon cycle, since new trees recycle carbon from the atmosphere during their growth. Renewable energy sources that are still a part of the carbon cycle such as corn-based ethanol or wood-burning are typically referred to as *carbon-neutral*. Other energy sources such as wind harvesting which do not use carbon-based chemicals as part of the energy transfer mechanism are referred to by the author as *carbon-free*.

In late 2006, the USGBC announced some very progressive goals for reducing carbon emissions. These include a proposed 2007 goal of all newly registered projects to strive to attain a minimum of 2 credits for LEED credit EAc1 (Energy and Atmosphere credit 1, *Optimize Energy Performance*), and to reduce carbon emissions by at least 50 percent (the reduction percentage is higher for the higher levels of certification). Optimizing energy performance as a part of the carbon commitment will reduce energy use, and as most energy use in the United States is based on fossil fuels, this should result in decreased dependence on fossil fuels and decreased carbon emissions from these nonrenewable fuels. The goals also include a commitment that the USGBC itself, in its operations, be carbon-neutral by 2008.

Miscellaneous USGBC Initiatives

There are several other official initiatives at the USGBC to improve the development of the LEED rating system. Four of the major initiatives as listed in July 2007 are Harmonizing and Aligning LEED Rating System, Integrated Committee Structure, Technical Development, and Regular Update Cycle for LEED. The intention of Harmonizing and Aligning LEED Rating System is to try to align the various USGBC rating systems so that they are easier to use and have consistency between similar credits. The proposed Integrated Committee Structure will recategorize the committees from being LEED rating system specific and will instead focus on Technical, Market, and Certification issues. The Technical Development initiative will help focus on some of the environmental issues of concern, add in greater regional variability, and incorporate more technical analyses such as life-cycle assessment. Finally, the proposed regular update cycle for LEED is intended to help incorporate market and technological developments into the rating system on a consistent basis that allows for improved input from the community and stakeholders.

Other environmental concerns that are currently not a part of LEED-NC 2.2 are also being incorporated into future rating systems; one example is sound pollution and acoustics. In the USGBC LEED for Schools rating system released in 2007, there is both a prerequisite and a credit in the Indoor Environmental Quality category relating to acoustics. The green building rating system is constantly being improved and built upon, and as noted previously, the movement is responding to this perpetual need for changes and enhancements by reestablishing the mechanisms by which changes are incorporated and distributed.

References

ASHRAE (2006), http://spc189.ashraepcs.org/, website for Standard Project Committee SPC 189, accessed December 11, 2006.

Bon, R., and K. Hutchinson (2000), "Sustainable Construction: Some Economic Challenges," *Building Research and Information*, 28(5/6): 310–314.

Brundtland Commission (1987), *Our Common Future*, Report by the Brundtland Commission [formally the World Commission on Environment and Development (WCED)], Oxford University Press.

du Plessis, C. (2001), "Sustainability and Sustainable Construction: The African Context," *Building Research and Information*, 29(5): 374–380.

Engel-Yan, J., et al. (2005), "Toward Sustainable Neighbourhoods: The Need to Consider Infrastructure Interactions," *Canadian Journal of Civil Engineering*, 32: 45–57.

Fedrizzi, R. (2006), "A Look Back at Greenbuild," *Buildings.com Greener Facilities Newsletter*, November, vol. 4, Issue 11, Stamats Business Media, Cedar Rapids, Iowa.

FMLink Group (2006), "Autodesk and USGBC Partner on Technology to Help Make Building Industry Greener," *Facilities Management News*, November 20, 2006.

Forman, R. T. T., et al. (2003), *Road Ecology: Science and Solutions*, Island Press, Washington, D.C.

Graedel, T. E., and B. R. Allenby (2002), *Industrial Ecology*, 2d ed., Prentice-Hall, New York.

Green Design Institute (2007), http://www.ce.cmu.edu/GreenDesign/, website at Carnegie Mellon University, accessed May 10, 2007.

Green Highways Partnership (2007), www.greenhighways.org, website accessed May 9, 2007.

Haselbach, L. M., and C. M. Fiori (2006), "Construction and the Environment: Research Foci for a Sustainable Future," *Journal of Green Building*, Winter, 1(1).

Haselbach, L. M., S. R. Loew, and M. E. Meadows (2005), "Compliance Rates for Stormwater Detention Facility Installation," *ASCE Journal of Infrastructure Systems*, March, 11(1).

Hawken, P. (1993), *The Ecology of Commerce, A Declaration of Sustainability*, HarperBusiness, New York.

Hawken, P., A. Lovins, and L. H. Lovins (1999), *Natural Capitalism, Creating the Next Industrial Revolution*, Little, Brown, Boston, Ma.

Horman, M. J., et al. (2006), "Delivering Green Buildings: Process Improvements for Sustainable Construction," *Journal of Green Building*, Winter, 1(1).

IESNA (2006), http://www.iesna.org/, Illuminating Engineering Society of North America website, accessed December 12, 2006.

ImpEE (2007), http://www-g.eng.cam.ac.uk/impee/, Improving Engineering Education Project website, accessed May 10, 2007, University of Cambridge, United Kingdom.

ISO (2006), http://www.iso.org/iso/en/ISOOnline.frontpage, International Organization for Standardization website, accessed December 12, 2006.

Jefferson, C., J. Rowe, and C. Brebbia (eds.) (2001), *The Sustainable Street: The Environmental, Human and Economic Aspects of Street Design and Management*, WIT Press, Southampton, United Kingdom.

Jia, H., et al. (2005), "Research on Wastewater Reuse Planning in Beijing Central Region," *Water Science & Technology*, 51: 195–202.

Katz, L., and J. Sutherland (Guest eds.) (2007), "Educating Students in Sustainable Engineering," *The International Journal of Engineering Education*, pt. I, 23(2).

Kibert, C. J. (2005), *Sustainable Construction; Green Building Design and Delivery*, Wiley, New York.

Lomberg, B. (2001), *The Skeptical Environmentalist: Measuring the Real State of the World*, Cambridge University Press, Cambridge, United Kingdom.

Mulder, Karel (ed.) (2006), *Sustainable Development for Engineers, A Handbook and Resource Guide*, Greenleaf Publishing, Sheffield, United Kingdom.

Ofori, G. (2000), "Greening the Construction Supply Chain in Singapore," *European Journal of Purchasing and Supply Management*, 6: 195–206.

Passa, J., and D. Rompf (2007), "Energy Efficient Sustainable Schools in Canada South," *Journal of Green Building*, Spring, 2(2): 14–30.

Pearce, A. R., J. R. DuBose, and S. J. Bosch (2007), "Green Building Policy Options for the Public Sector," *Journal of Green Building*, Winter, 2(1).

Pearce, D. W., G. D. Atkinson, and W. R. Dubourg (1994), "The Economics of Sustainable Development," *Annual Review of Energy and the Environment*, 19: 457–474.

Sahely, H. R., C. A. Kennedy, and B. J. Adams (2005), "Developing Sustainability Criteria for Urban Infrastructure Systems," *Canadian Journal of Civil Engineering*, 32: 72–85.

Shen, L., et al. (2005), "A Computer-Based Scoring Method for Measuring the Environmental Performance of Construction Activities," *Automation in Construction*, 14: 297–309.

Seongwon, S., S. Tucker, and P. Newton (2007), "Automated Material Selection and Environmental Assessment in the Context of 3D Building Modelling," *Journal of Green Building*, Spring, 2(2): 51–61.

Sick, F., and A. Kerschberger (2007), "Innovative Low Energy Renovation of an Office Building: Concept and Simulation," *Journal of Green Building*, Spring, 2(2): 31–41.

SSI (2007), http://www.sustainablesites.org/, Sustainable Sites Initiative website, accessed July 1, 2007.

Straube, J., and C. Schumacher (2007), "Interior Insulation Retrofits of Load-Bearing Masonry Walls in Cold Climates," *Journal of Green Building*, Spring, 2(2): 42–50.

Tam, V. W. Y., and K. N. Le (2007), "Predicting Environmental Performance of Construction Projects by Using Least-Squares Fitting Method and Robust Method," *Journal of Green Building*, Winter, 2(1).

USAF, USGBC, and Paladino and Company, Inc. (2001), *Application Guide for Lodging Using the LEED Green Building Rating System*, U.S. Air Force Center for Environmental Excellence, U.S. Green Building Council, Washington, D.C.

USGBC (2003), *LEED-NC for New Construction, Reference Guide*, Version 2.1, 2d ed., May, U.S. Green Building Council, Washington, D.C.

USGBC (2004–2005), *LEED-CI Green Building Rating System for Commercial Interiors;* Version 2.0, November 2004, updated December 2005, U.S. Green Building Council, Washington, D.C.

USGBC (2004–2005), *LEED-EB Green Building Rating System for Existing Buildings, Upgrades, Operations and Maintenance*, Version 2, October 2004, updated July 2005, U.S. Green Building Council, Washington, D.C.

USGBC (2005), *LEED-NC Application Guide for Multiple Buildings and On-Campus Building Projects, for Use with the LEED-NC Green Building Rating System Versions 2.1 and 2.2*, October, U.S. Green Building Council, Washington, D.C.

USGBC (2005–2007), *LEED-NC for New Construction, Reference Guide*, Version 2.2, 1st ed., U.S. Green Building Council, Washington, D.C., October 2005 with errata posted through Spring 2007.

USGBC (2006), *LEED Green Building Rating System for Core & Shell Development;* Version 2.0, July, U.S. Green Building Council, Washington, D.C.

USGBC (2006), "USGBC Unveils 8 Climate Actions," News Release at Press Conference, 2 p.m. November 15, 2006, Denver, Colo., US Greenbuild Conference, as accessed December 12, 2006, http://www.fypower.org/pdf/USGBC_GHGPlan.pdf.

USGBC (2006), www.usgbc.org, U.S. Green Building Council website, accessed September 2006.

USGBC (2007), *LEED for Homes Program Pilot Rating System*, Version 2.11a, January, U.S. Green Building Council, Washington, D.C.

USGBC (2007), *LEED for Neighborhood Development Rating System*, Pilot, Congress for the New Urbanism, Natural Resources Defense Council, U.S. Green Building Council, Washington, D.C.

USGBC (2007), *LEED for Retail—New Construction and Major Renovations*, Pilot Version 2, April, U.S. Green Building Council, Washington, D.C.

USGBC (2007), *LEED for Schools for New Construction and Major Renovations*, Approved 2007 Version, April, U.S. Green Building Council, Washington, D.C.

Vanegas, J. A. (ed.) (2004), *Sustainable Engineering Practice: An Introduction*, American Society of Civil Engineers, Reston, Va.

Wikipedia (2006), http://en.wikipedia.org/wiki/ANS I, ANSI information accessed December 12, 2006.

Wikipedia (2006), http://en.wikipedia.org/wiki/ASTM, ASTM information accessed December 12, 2006.

Xiaohua, W., and Z. Feng (2003), "Energy Consumption with Sustainable Development in Developing Country: A Case in Jiangsu, China," *Energy Policy*, 31: 1679–1684.

Yingxin, Z., and L. Borong (2004), "Sustainable Housing and Urban Construction in China," *Energy and Buildings*, 36: 1287 1297.

Exercises

1. What is the current version of LEED-NC and when was it adopted?

2. Download the errata for LEED-NC 2.2. Separate errata into two categories, errata to the rating system and errata to the Reference Guide only. Incorporate these errata into appropriate sections of this text.

3. Update Table 1.1.1 with the current status of the rating systems listed, and add other or modified rating systems from the USGBC website www.usgbc.org.

4. Update Table 1.2.1 as to the current status of the guidances. Are there additional guidances mentioned on the USGBC website?

5. What are the categories in the BREEAM rating system?

6. What is the current status of the Green Globes rating system? What are the categories in this system?

CHAPTER 2

LEED Sustainable Sites

The Sustainable Sites category of the U.S. Green Building Council (USGBC) rating system for new construction version 2.2 (LEED-NC 2.2) consists of one prerequisite and eight credit subcategories which together may earn a possible 14 points (credits), not including exemplary performance points. The notation format for the prerequisites and credits is, for example, SSp1 for Sustainable Sites prerequisite 1 and SSc1 for Sustainable Sites credit 1. Also, in this chapter, to facilitate easier cross-referencing between this text and the USGBC rating system, the second digit in a section heading, equation, table, or figure number represents the credit subcategory number for sections that deal directly with a USGBC LEED-NC 2.2 credit subcategory.

The Sustainable Sites (SS) portion deals with issues outside of the building, including some of the building exterior, the land that is being developed, and the surrounding community. It is not the only category that includes factors outside of the building, but comprises those exterior issues that do not fit readily into the other categories and those that have a direct impact on the local community or microenvironment. The intent of the Sustainable Sites portion is summarized in LEED-NC 2.2 as follows:

Project teams undertaking building projects should be cognizant of the inherent impacts of development on the following:

- land consumption
- ecosystems
- natural resources
- energy use

Preference should be given to buildings with high performance attributes in locations that enhance existing neighborhoods, transportation networks and urban infrastructures. During initial project scoping, preference should be given to sites and land use plans that preserve natural ecosystem functions and enhance the health of the surrounding community.

In summary, the Sustainable Sites category encourages revitalizing and using existing infrastructure, and impacting the environment and natural resources as little as possible in the local area (community) near or within the proposed development.

The prerequisite and 14 credits available in the Sustainable Sites category are as follows:

- **SS Prerequisite 1** (SSp1): *Construction Activity Pollution Prevention* (previously referred to as *Erosion and Sedimentation Control* in LEED-NC 2.1)
- **SS Credit 1** (SSc1): *Site Selection*

- **SS Credit 2** (SSc2): *Development Density and Community Connectivity* (previously referred to as *Urban Redevelopment* in LEED-NC 2.1) (EB)
- **SS Credit 3** (SSc3): *Brownfield Redevelopment*
- **SS Credit 4.1** (SSc4.1): *Alternative Transportation: Public Transportation Access* (EB)
- **SS Credit 4.2** (SSc4.2): *Alternative Transportation: Bicycle Storage and Changing Rooms*
- **SS Credit 4.3** (SSc4.3): *Alternative Transportation: Low-Emitting and Fuel-Efficient Vehicles* (previously referred to as *Alternative Transportation, Alternative Fuel Vehicles* in LEED NC 2.1)
- **SS Credit 4.4** (SSc4.4): *Alternative Transportation: Parking Capacity*
- **SS Credit 5.1** (SSc5.1): *Site Development: Protect or Restore Habitat* (previously referred to as *Reduced Site Disturbance: Protect or Restore Open Space* in LEED-NC 2.1)
- **SS Credit 5.2** (SSc5.2): *Site Development: Maximize Open Space* (previously referred to as *Reduced Site Disturbance: Development Footprint* in LEED-NC 2.1)
- **SS Credit 6.1** (SSc6.1): *Stormwater Management: Quantity Control* (previously referred to as *Stormwater Management: Rate and Quantity* in LEED-NC 2.1)
- **SS Credit 6.2** (SSc6.2): *Stormwater Management: Quality Control* (previously referred to as *Stormwater Management: Treatment* in LEED-NC 2.1)
- **SS Credit 7.1** (SSc7.1): *Heat Island Effect: Non-Roof*
- **SS Credit 7.2** (SSc7.2): *Heat Island Effect: Roof*
- **SS Credit 8** (SSc8): *Light Pollution Reduction*

LEED-NC 2.2 in the Reference Guide provides a table which summarizes some of these prerequisite and credit characteristics and specifically notes the following:

- A significant change from LEED version 2.1
- The main phase of the project for the submittal document preparation (design or construction)
- The project team member(s) who have the most decision-making responsibility as to whether this credit may or may not be achieved (owner, design team, or contractor).

The submittals for each credit or prerequisite have either been sectioned into Design Submittals and/or Construction Submittals by the USGBC. The prerequisites or credits which are noted as Construction Submittal credits may have major actions going on through the construction phase which, unless documented through that time, may not be able to be verified from preconstruction documents or the built project. In addition, some have activities that are typically contractor-specified, and therefore the construction phase decisions may impact the credit substantially. There are three items in the Sustainable Sites category, which are designated by the USGBC as Construction Submittals: SSp1 (Construction Activity Pollution Prevention), SSc5.1 (Site Development: Protect or Restore Habitat), and SSc7.1 (Heat Island Effect: Non-Roof). These have been so noted in Table 2.0.1.

Two of the credits listed previously have the icon EB noted after their title (SSc2 and SSc4.1). This icon is a tool to help those who wish to proceed with the continuing LEED-EB certification (Existing Building) after certification of LEED-NC (New Construction

Credit	Construction Submittal	EB Icon	EP Point Availability	EP Point Criterion	Location of Site Boundary or Area in Calculations
SSp1: Construction Activity Pollution Prevention	*	NA	NA		Yes
SSc1: Site Selection			No		Yes
SSc2: Development Density and Community Connectivity		Yes	Yes For Option 1 Only	Site density is Double the area density or the area density is Doubled for Double the density radius	Yes
SSc3: Brownfield Redevelopment			No		No
SSc4.1: Alternative Transportation: Public Transportation Access		Yes		Overall Plan (1 pt) and for SSc4.1 (1 pt) Double Lines and Minimum Total 200 Transit Rides Daily	Yes
SSc4.2: Alternative Transportation: Bicycle Storage and Changing Rooms					No
SSc4.3: Alternative Transportation: Low-Emitting and Fuel-Efficient Vehicles			Yes (2)		No
SSc4.4: Alternative Transportation: Parking Capacity					No
SSc5.1: Site Development: Protect or Restore Habitat	*		Yes	75%	Yes
SSc5.2: Site Development: Maximize Open Space			Yes	Double	Yes
SSc6.1: Stormwater Management: Quantity Control			No		Yes
SSc6.2: Stormwater Management: Quality Control			No		Yes
SSc7.1: Heat Island Effect: Non-Roof	*		Yes	100%	No
SSc7.2: Heat Island Effect: Roof			Yes	100% Green	No
SSc8: Light Pollution Reduction			No		Yes

TABLE 2.0.1   SS EB Icon, Exemplary Performance (EP) Point Availability, and Miscellaneous

or Major Renovation) is obtained. Those credits noted with this icon are usually significantly more cost-effective to implement during the construction of the building than later during its operation. They are also shown in Table 2.0.1.

Exemplary performance (EP) points under the Innovation and Design category are available which relate to several items in the Sustainable Sites category. One is available for the SSc4 Alternative Transportation subcategory as a whole. Six additional point options are also available with one point related directly to six of the individual credits. These are available for exceeding each of the respective credit criteria to a minimum level and meeting other criteria as noted in the credit descriptions for SSc2, SSc4.1, SSc5.1, SSc5.2, SSc7.1, and SSc7.2. There are no EP points available for the other subcategories or credits in the Sustainable Sites category. The items available for EP points in this category are shown in Table 2.0.1 with a column summarizing the point achievement criteria. Note that only a total of four EP points are available in total for a project, and they may be from any of the noted EP point options in any of the first five main LEED categories or may be other special point items (see Chap. 7).

Table 2.0.1 also has a column that notes the importance of the site boundary and site area in the credit calculations or verification. These are variables that should be determined early in a project as they impact many of the credits. Some of the credits are easier to obtain if the site area is less, while others are easier to obtain if the site area is greater. Determination of the site area may sometimes be flexible, particularly for campus locations, but it should be reasonable and must be consistent throughout the LEED process, as well as in agreement with other nearby project submissions. It is mentioned again in Chap. 8 as an important variable to analyze early and set in a project.

Land Area Definitions

Many of the Sustainable Sites credits relate to the site land area and the areas within the sites that are developed or not developed in certain ways (see Figure 2.0.1). To understand this better, it is helpful to define the various portions of the site. Definitions as given by LEED-NC 2.2 for many of these site areas are given in App. B.

Let the following symbols represent the various areas of a site:

A_T Total area of the lot (also referred to as the total project site area)

BF   Building footprint—the planar projection of the built structures onto the land

DF   Development footprint—total building and hardscape areas

EQR   Total roof areas that are covered by equipment and/or solar appurtenances and other appurtenances

HS   Hardscape: All nonbuilding manmade hard surface on the lot such as parking areas, walks, drives, and patios. Hardpacked areas such as gravel parking areas are included. Pools, fountains, and similar water-covered areas are included if they are on impervious surfaces which can be exposed when the water is emptied.

LSMAN   Manmade or graded landscaped area totals

LSNAT   Natural landscaped area totals

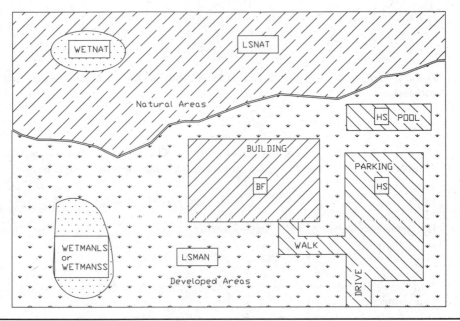

Figure 2.0.1 Various basic land areas for the Sustainable Sites calculations.

OS	Open space—all natural areas and landscaped areas, also wet areas with vegetated low slopes
PD	Previously developed—all areas that have been built upon, landscaped, or graded (not including areas that may have been farmed in centuries past and have returned to natural vegetation)
TR	Total roof—area defined as the sum of the horizontal projections of the roofs on the site plan less the sum of the roofed areas that hold equipment, solar energy panels, or other appurtenances (BF − EQR)
WETMAN	Total manmade wetland, ponds, or pool areas
WETMANLS	Total manmade wetland, ponds, or pools with vegetated low slopes (<25 percent) leading to its edges. This does not include pools and fountains with impervious bottoms as these are part of the hardscape
WETMANSS	Total manmade wetland, ponds, or pools with any steep (>25 percent) or unvegetated slopes leading to its edges. This does not include pools and fountains with impervious bottoms as these are part of the hardscape
WETNAT	Total natural wetlands, ponds, stream, or pool areas

The various areas relate to each other in the following ways. The total manmade wet areas are the sum of those with low-sloped edges and those which have some steep slopes.

$$\text{WETMAN} = \text{WETMANLS} + \text{WETMANSS} \qquad (2.0.1)$$

The total site area is the sum of all the independent areas.

$$A_T = \text{LSNAT} + \text{WETNAT} + \text{LSMAN} + \text{WETMAN} + \text{BF} + \text{HS} \qquad (2.0.2)$$

The *development footprint* (see SS Credit 1) is defined as the areas with hard or hardpacked surfaces.

$$DF = BF + HS \qquad (2.0.3)$$

Open space, as defined in SS Credit 5.2, is the combination of all natural and landscaped areas and low-sloped wet areas except for those considered to be hardscape:

$$OS = \text{LSNAT} + \text{WETNAT} + \text{WETMANLS} + \text{LSMAN} \qquad (2.0.4)$$

The previously developed area, as defined in SS Credit 5.1, is everything that is not natural:

$$PD = A_T - \text{LSNAT} - \text{WETNAT} \qquad (2.0.5)$$

And the total roofed (TR) area is the planar projection of the roofed areas (BF) less the roof equipment areas:

$$TR = BF - EQR \qquad (2.0.6)$$

Areas of natural ledge have not been specifically addressed; but if in the natural areas, these would most likely be included in the open space and excluded from the previously developed areas. If areas of natural ledge are in the developed areas, then they would most likely be included in the hardscape summation. Each case would be subject to interpretation.

Sustainable Sites Prerequisites

SS Prerequisite 1: *Construction Activity Pollution Prevention*

USGBC Rating System
This prerequisite is intended to control erosion and pollutant transport off-site during and after construction to minimize their impact on the surrounding areas and on water and air quality. LEED NC-2.2 lists the Intent, Requirements, and Potential Technologies and Strategies for this credit as follows:

Intent
Reduce pollution from construction activities by controlling soil erosion, waterway sedimentation and airborne dust generation.

Requirements
Create and implement an Erosion and Sedimentation Control (ESC) Plan for all construction activities associated with the project. The ESC Plan shall conform to the erosion and sedimentation requirements of the 2003 EPA Construction General Permit OR local erosion and sedimentation control standards and codes, whichever is more stringent. The Plan shall describe the measures implemented to accomplish the following objectives:

- Prevent loss of soil during construction by stormwater runoff and/or wind erosion, including protecting topsoil by stockpiling for reuse.
- Prevent sedimentation of storm sewer or receiving streams.
- Prevent polluting the air with dust and particulate matter.

The Construction General Permit (CGP) outlines the provisions necessary to comply with Phase I and Phase II of the National Pollutant Discharge Elimination System (NPDES) program. While the CGP only applies to construction sites greater than 1 acre, the requirements are applied to all projects for the purposes of this prerequisite. Information on the EPA CGP is available at http://cfpub.epa.gov/npdes/stormwater/cgp.cfm.

Potential Technologies and Strategies

Create an Erosion and Sedimentation Control Plan during the design phase of the project. Consider employing strategies such as temporary and permanent seeding, mulching, earth dikes, silt fencing, sediment traps and sediment basins.

Calculations and Considerations

This strategy for construction phases of projects was commonly referred to as SEC (soil and erosion control) and now is more commonly known as ESC (erosion and sedimentation control). The current LEED-NC 2.2 requirement is based on the USEPA Document No. EPA 832R92005, Chap. 3, which can be downloaded from the EPA Office of Water at www.epa.gov/OW. It separates the practices into two overlapping categories, one for the prevention of erosion (stabilization) and the other for the containment or control of eroded sediment to other parts of the site or off-site. Some of the common stabilization methods used include the installation of geotextiles, temporary or permanent seeding or mulching on exposed soils, and installation of a construction vehicle drive. Typical sediment control measures include installation of silt fencing downslope of graded areas, installation of silt barriers around catch basins and other inlets to storm sewers or waterway access points, and installation of sedimentation ponds. The methods used are sometimes referred to as structural for actual items installed or nonstructural for other types of practices, such as scheduling grading activities during a drier season.

The steps for compliance for this credit include having and implementing an ESC Plan. Many of the structural and nonstructural measures used for ESC will be part of the project drawings. These project drawings will usually show the location of the methods on the site plans, list a material schedule for the measures, and include a time schedule for implementation, maintenance, and, if appropriate, removal of the control measure. The ESC Plan and this prerequisite adherence also apply to the demolition phase of the project and to the full extent of the construction, even areas outside of the LEED boundary.

The submittals for the credit include copies of the applicable project drawings and specifications addressing ESC, a statement as to whether the project adheres to the national EPA criteria or a stricter local standard, and a narrative outlining the ESC methods implemented. When a local standard is used, it must be demonstrated in the narrative that it is at least as strict as the NPDES standard. There are no required calculations for this credit other than those that might be necessary to verify that the ESC Plan adheres to the minimum standards required. The submittal is part of the LEED Construction Submittal, as most of the items are a part of the construction phase and adherence to the requirements should be documented.

Many states now have certification programs for ESC professionals, and a listing of these can be found at www.cpesc.net. This is a good resource for a review of common terminology and measures used for ESC in the United States. Additional information on erosion and sediment control can be found in Chap. 10.

2.1 SS Credit 1: *Site Selection*

USGBC Rating System

The intention of this credit is to minimize the impact of building construction on natural resources both on-site and nearby. LEED-NC 2.2 lists the Intent, Requirements, and Potential Technologies and Strategies for this credit as follows:

Intent

Avoid development of inappropriate sites and reduce the environmental impact from the location of a building on a site.

Requirement

Do not develop buildings, hardscape, roads or parking areas on portions of sites that meet any one of the following criteria:

- Prime farmland as defined by the United States Department of Agriculture in the United States Code of Federal Regulations, Title 7, Volume 6, Parts 400 to 699, Section 657.5 (citation 7CFR657.5).
- Previously undeveloped land whose elevation is lower than 5 feet above the elevation of the 100-year flood as defined by the Federal Emergency Management Agency (FEMA).
- Land which is specifically identified as habitat for any species on Federal or State threatened or endangered lists.
- *Land* within 100 feet of any wetlands as defined by United States Code of Federal Regulations 40 CFR, Parts 230-233 and Part 22, and isolated wetlands or areas of special concern identified by state or local rule, OR within setback distances from wetlands prescribed in state or local regulations, as defined by local or state rule or law, whichever is more stringent.
- Previously undeveloped land that is within 50 feet of a water body, defined as seas, lakes, rivers, streams, tributaries which support or could support fish, recreation or industrial use, consistent with the terminology of the Clean Water Act.
- Land which prior to acquisition for the project was public parkland, unless land of equal or greater value as parkland is accepted in trade by the public landowner (Park Authority projects are exempt).

Potential Technologies and Strategies

During the site selection process, give preference to those sites that do not include sensitive site elements and restrictive land types. Select a suitable building location and design the building with the minimal footprint to minimize site disruption of those environmentally sensitive areas as identified.

Calculations and Considerations

This credit tries to avoid disturbing pristine or special natural resources by selectively locating buildings and the development footprint away from these resources or developing on previously developed areas of the site. To obtain this credit, all six of the criteria must be met. Two important definitions to understand are those of development footprint and previously undeveloped land. *Development footprint* as defined by LEEDNC-2.2 can be found in App. B, as well as the definition of *previously developed sites,* which are assumed to be the opposite of *previously undeveloped land.* The differences between the two are as follows: the development footprint, as defined, does not include landscaped areas, whereas previously developed sites as defined include areas graded or altered by direct human activities which may include landscaped areas.

There are no calculations for this credit other than as appropriate to verify distances, elevations, and setbacks. Confirmation that all six criteria are met must be submitted. The second and fifth criteria are applicable only if the land in question was previously undeveloped, but again this must be shown. There is room for "special circumstances" or "nonstandard compliance paths" in this credit, but these must be described.

Also of note is the first criterion which refers to prime farmland. Not all farmland or potential farmland is prime. The U.S. Department of Agriculture (USDA) definition of *prime agricultural land* as stated in the applicable Code of Federal Regulations (CFR) part is as follows:

> Prime farmland is land that has the best combination of physical and chemical characteristics for producing food, feed, forage, fiber, and oilseed crops, **and is also available for these uses** (the land could be cropland, pastureland, rangeland, forest land, or other land, but not urban built-up land or water). It has the soil quality, growing season, and moisture supply needed to economically produce sustained high yields of crops when treated and managed, including water management, according to acceptable farming methods. In general, prime farmlands have an adequate and dependable water supply from precipitation or irrigation, a favorable temperature and growing season, acceptable acidity or alkalinity, acceptable salt and sodium content, and few or no rocks. They are permeable to water and air. Prime farmlands are not excessively erodible or saturated with water for a long period of time, and they either do not flood frequently or are protected from flooding. Examples of soils that qualify as prime farmland are Palouse silt loam, 0 to 7 percent slopes; Brookston silty clay loam, drained; and Tama silty clay loam, 0 to 5 percent slopes.

The second criterion relates to designated floodplain elevations. The 100-year floodplain is usually as designated on the FEMA Flood Insurance Rate Maps (FIRMs). A 100-year flood is not one that occurs every 100 years, but rather is a flood with a 1 percent probability whose magnitude or greater might occur every year. If the FEMA flood lines are not available, then the Army Corps of Engineers or other recognized authority maps may be substituted. There are some exceptions to building in a floodplain for projects which have some redevelopment in the floodplain area. Another exception is that areas can be "elevated" above the 5-ft minimum if both the 100-year floodplain is not impacted and there are balanced cuts and fills in these areas to not impact storage volumes. In LEED-NC 2.1 a credit interpretation ruling (CIR) allowed for the definition of the facilities which needed to be at least 5 ft above the elevation of the 100-year floodplain to be only those included in the building footprint. As the version 2.2 verbiage is more lenient than that of version 2.1 (version 2.1 did not specifically exempt previously developed land), it is reasonable to assume that this definition is also applicable to version 2.2.

As with the floodplain criterion, already existing facilities that are located within the setbacks as defined for wetlands can remain. Maintenance and minor improvements on these nonconforming existing facilities would normally be appropriate, whereas major renovations or additions would not. Each facility should be examined individually for conformance and intent.

Wetlands are defined by the Code of Federal Regulations 40 CFR Parts 230–233 and Part 22: "Wetlands consist of areas that are inundated or saturated by surface or ground water at a frequency and duration sufficient to support, and that under normal circumstances do support, a prevalence of vegetation typically adapted for life in saturated soil conditions."

The definition of *water body* also needs clarification. Per the LEED errata posted by the USGBC in the spring of 2007, small manmade ponds, such as those used in stormwater retention, fire suppression, and recreation are excluded, whereas manmade wetlands and other water bodies created to restore natural habitat and ecological systems are not exempt.

Special Circumstances and Exemplary Performance
There is no EP point available for this credit.

2.2   SS Credit 2: *Development Density and Community Connectivity*

USGBC Rating System
The intention of this credit is to infill or revitalize urban areas that already have supportive utility, transportation, and service infrastructure, or to increase services near denser residential communities close to other supporting services such as would be found in an urban or dense suburban area. This credit has an EB notation. LEED-NC 2.2 lists the Intent, Requirements, and Potential Technologies and Strategies for this credit as follows:

Intent
Channel development to urban areas with existing infrastructure, protect greenfields and preserve habitat and natural resources.

Requirements
OPTION 1- DEVELOPMENT DENSITY

Construct or renovate building on a previously developed site AND in a community with a minimum density of 60,000 square feet per acre net. (Note: density calculation must include the area of the project being built and is based on a typical two-story downtown development.)

OR

OPTION 2- COMMUNITY CONNECTIVITY

Construct or renovate building on a previously developed site AND within ½ mile of a residential zone or neighborhood with an average density of 10 units per acre net AND within ½ mile of at least 10 Basic Services AND with pedestrian access between the building and the services.

   Basic Services include, but are not limited to: 1) Bank; 2) Place of Worship; 3) Convenience Store; 4) Day Care; 5) Cleaners; 6) Fire Station; 7) Beauty 'shop'; 8) Hardware; 9) Laundry; 10) Library; 11) Medical/Dental; 12) Senior Care Facility; 13) Park; 14) Pharmacy; 15) Post Office; 16) Restaurant; 17) School; 18) Supermarket; 19) Theater; 20) Community Center; 21) Fitness Center; 22) Museum. Proximity is determined by drawing a ½ mile radius around the main building entrance on a site map and counting the services within the radius.

Potential Technologies and Strategies
During the site selection process, give preference to urban sites with pedestrian access to a variety of services.

Calculations and Considerations

Option 1 The submittal requirements for Option 1 of this credit have several calculations associated with them. The minimum development density must be met *both* for the site in question (the development density of the site DD_{site}) and for the surrounding area (the development density of the surrounding area DD_{area}) that is within the density boundary (DB) as defined by the density radius (DR). The following set of symbols is used in the equations, with GFA referring to gross floor area as defined through the errata as posted by the USGBC in the spring of 2007 (see the definitions in App. B):

A_T	Total land area of the project lot or site
A_{Ti}	Total land area of a lot i
DD_{site}	Development density of the site [usually given in square foot per acre (ft²/acre)]
DD_{area}	Development density of the area within the density boundary as determined by the density radius for the density boundary
DDD_{area}	Development density of double the surrounding area as determined by the density radius for double (DRD) the density boundary area
DR	Density radius for the density boundary area (usually given in feet)
DRD	Density radius for double the density boundary area (usually given in feet)
GFA_i	Gross floor area of other building i (square feet)
GFA_p	Gross floor area of the subject project (square feet)

One point can be earned for Option 1 if the inequalities in Eqs. (2.2.1) and (2.2.2) are true, given the definition of the density radius in Eq. (2.2.3):

$$DD_{site} = (GFA_p/A_T) \geq 60,000 \text{ ft}^2/\text{acre} \tag{2.2.1}$$

$$DD_{area} = \frac{\sum GFA_i}{\sum A_{Ti}} \geq 60,000 \text{ ft}^2/\text{acre} \qquad \text{for all applicable buildings and lots within DR} \tag{2.2.2}$$

$$DR = 3 \times \sqrt{A_T} \tag{2.2.3}$$

Based on a CIR dated September 22, 2006, an exemplary performance point can be earned over and above Option 1 for a project which meets one of the two following requirements after all the requirements of Option 1 are met:

- The project itself must have a density of at least double that of the average density within the calculated area [see Eq. (2.2.4)].

or

- The average density within an area twice as large as that for the base credit achievement must be at least 120,000 ft²/acre. [To double the area, use Eq. (2.2.3), but double the property area first.]

Therefore, to obtain an EP point for SSc2 in addition to the point earned for Option 1, one of the following two options must be valid. Either the inequality in Eq. (2.2.4) must be achieved given the development density of the area as defined in Eq. (2.2.2)

$$DD_{site} = (GFA_p / A_T) \geq 2 = DD_{area} \qquad (2.2.4)$$

or both Eqs. (2.2.5) and (2.2.6) must hold. These are based on the two additional variables the development density of double the surrounding area (DDD_{area}) as defined by the density radius for double the area (DRD).

$$DD_{area} = \frac{\sum GFA_i}{\sum A_{T_i}} \geq 120,000 \ ft^2/acre \qquad \begin{array}{l} \text{for all applicable buildings} \\ \text{and lots within DR} \end{array} \qquad (2.2.5)$$

$$DR = 3 \times \sqrt{2 \times A_T} \qquad (2.2.6)$$

There are several additional definitions of the variables involved that must be explained. The following definitions are per LEED-NC version 2.2 when noted in quotation marks:

Project site area   "For projects that are part of a larger property (such as a campus), define the project area as that which is defined in the project's scope. The project area must be defined consistently throughout LEED documentation." This will typically be represented as A_T for most of the credit equations.

Density radius   The density radius (DR) is determined by Eq. (2.2.3). A circle of a radius equal to the density radius is then drawn on an area map with its center near the center of the project site.

Density boundary   This is the area within the circle as determined by drawing the density radius. "For each property within the density boundary and for those properties that intersect the density boundary, create a table with the building square footage (GFAs) and site area of each property. Include all properties in the density calculations except for undeveloped public areas such as parks and water bodies. Do not include public roads and right-of-way areas." Do not include the subject project, but do include site areas of vacant lots except as previously noted. These GFAs and site areas are used in the summations in the numerator and denominator, respectively, of Eq. (2.2.2). In general the GFA of a building is all the interior space taken from outer wall to outer wall and summed over all the floors, not including unfinished attic/dormer space; and the GFA has also been defined in the errata posted by the USGBC in the spring of 2007 to include up to two stories of parking, but not single-story surface parking [definitions are given in App. B as *site area* and *square footage* of a building (GFA)].

As can be seen in Eq. (2.2.2), the development density of the surrounding area DD_{area} is calculated as the sum of all the GFAs divided by the sum of all the applicable site areas. A common mistake is to take the average of individual development densities for the properties within the density boundary. This type of average would not be appropriately weighted for the sizes of each property and facility and is not correct. The submittals to the USGBC would not allow this error to be made, but it may be made in early estimates not using the templates and is therefore emphasized here.

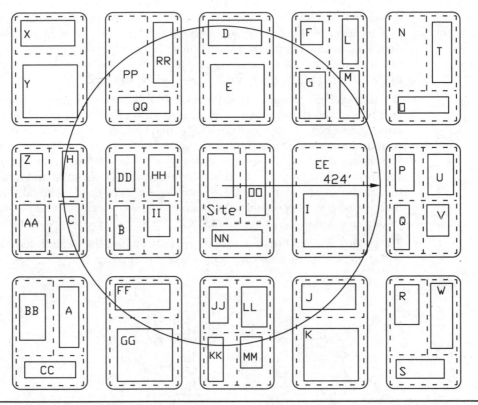

FIGURE 2.2.1   Density boundary area map.

All the listed equations and tabulations and a copy of the area map with the density boundary drawn must be submitted along with the template for validation of Option 1 of this credit. Figure 2.2.1 shows an example density boundary area map for this credit.

Some exceptions do exist for the numbers in Eqs. (2.2.1) through (2.2.3). For instance, let us say that the circle drawn intersects a very small portion of a large lot that is not well developed. This small exception at the edge of the density boundary may be the cause for an exception to the summation of area densities, especially if the density is appropriate in other areas outside the circle. This might still meet the intent of the credit. The site itself may sometimes not meet the full density criterion, but if perhaps taken together with other neighboring campus facilities with more than adequate densities, it may still meet the intent. Existing single-family housing and historic sites may not have to be included in the calculations as they are usually considered to be valuable assets to a community. Stadiums may also be exempt if they exist in pedestrian-friendly urban areas. In addition, future development may be used in the calculations if there is sufficient documentation that these types of facilities will be built.

Option 2   The submittal requirements for Option 2 of credit SSc2 include the following:

- A site vicinity map showing the parcel and building in question (project site) with a ½-mi-radius circle drawn from the main entrance, or the drawing scale noted on the map. The location of all services within this circle shall be shown. The location of the residential neighborhoods shall also be shown with the dense residential neighborhood of at least 10 units per acre noted. Many types of maps may be used including aerial photographs, sketches, and assessor maps. An example in a very urban area is given in Fig. 2.2.2.

- A listing of all the community services within this circle including business name and type. Even though there may be multiples of many of the basic services within this circle, only one of each type may be used for this credit, except for restaurants, for which multiple restaurants may count for 2 of the minimum 10 services. List only those services for which there is pedestrian access to the site. Do not list any that are blocked by highways, waterways, and other areas that hinder access. Again, future development may be used in the calculations if there is sufficient documentation that these types of facilities will be built. It is uncertain whether the project in question can be included as one of the 10.

- The project site and building area in square feet. [See the definitions of *site area* and *square footage* of a building (GFA) given in the App. B.]

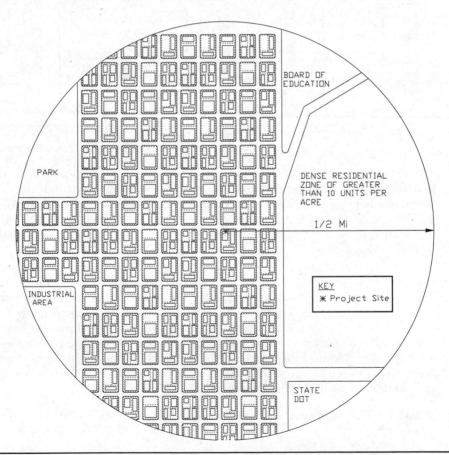

FIGURE 2.2.2 Area map for community connectivity. Locations of all applicable services must also be shown.

Special Circumstances and Exemplary Performance

If there are special circumstances for either of the options, a narrative must be submitted describing these circumstances.

The criteria which must be met for exemplary performance for Option 1 were listed under the Option 1 section. There is no EP point available for SS Credit 2 Option 2.

2.3 SS Credit 3: *Brownfield Redevelopment*

The intent of this credit is twofold: to reduce developing on undeveloped land and to appropriately redevelop those sites listed as brownfields. The site is first evaluated via a site assessment to determine if there are complications by environmental contamination. Brownfields can be designated for both "real" and "perceived" contamination. A risk assessment is also used to determine if and what remediation may be needed as applicable for proposed development. Remediation means the removal of hazardous materials from the groundwater or the soils. Typical remediation methods include in situ (in place) or ex situ (off-site) technologies. Sometimes the contaminated soils or waters remain on-site and are remediated with facilitated on-site processes such as "pump and treat" or are left in place and allowed to be treated by natural attenuation methods if their presence is not expected to be a liability if left in place for an extended period. They can also be encapsulated if appropriate. The ex situ processes involve many types of options but must be carefully manifested for shipment off-site. It is possible to become liable for pollution on another site if the waste material is sent off-site and not properly remediated or disposed of. With in situ processes, the owner must carefully consider the liability of contaminants potentially migrating off-site. With ex situ processes, the owner must carefully consider the proper disposal costs and the potential for off-site liabilities.

The two applicable federal laws are CERCLA (Comprehensive Environmental Response, Compensation and Liability Act) and RCRA (Resource Conservation and Recovery Act). CERCLA, commonly referred to as Superfund, is a program to help offset the costs of remediation. RCRA is the act that was originally intended to promote recycling, but was expanded and now also governs the handling of hazardous waste.

USGBC Rating System

The Intent, Requirements, Potential Technologies and Strategies, and Summary of Referenced Standards as given in LEED-NC 2.2 are as follows:

Intent

Rehabilitate damaged sites where development is complicated by environmental contamination, reducing pressure on undeveloped land.

Requirements

Develop on a site documented as contaminated (by means of an ASTM E1903-97 Phase II Environmental Site Assessment or a local Voluntary Cleanup Program) OR on a site defined as a brownfield by a local, state or federal government agency.

Potential Technologies and Strategies

During the site selection process, give preference to brownfield sites. Identify tax incentives and property cost savings. Coordinate site development plans with remediation activity, as appropriate.

Summary of Referenced Standards

ASTM E1903-97 Phase II Environmental Site Assessment, ASTM International, www.astm. org. This guide covers a framework for employing good commercial and customary practices in conducting a Phase II environmental site assessment of a parcel of commercial property. It covers the potential presence of a range of contaminants that are within the scope of CERCLA, as well as petroleum products.

EPA Brownfields Definition
EPA Sustainable Redevelopment of Brownfields Program, www.epa.gov/brownfields. With certain legal exclusions and additions, the term "brownfield site" means real property, the expansion, redevelopment, or reuse of which may be complicated by the presence or potential presence of a hazardous substance, pollutant, or contaminant (source: Public Law 107-118, H.R. 2869 – "Small Business Liability Relief and Brownfields Revitalization Act"). See the web site for additional information and resources.

Calculations and Considerations
The required submittals include confirmation that the project site was established as being contaminated either by an ASTM E1903-97 Phase II Environmental Site Assessment or by a local, state, or federal government agency, and a detailed narrative which describes the contamination and the remediation actions being taken. A method or means to monitor the cleanup effort should also be described. There is no need to submit the entire Phase II document; the executive summary is sufficient. No minimum level of contamination is specified. The contamination can be widespread or localized such as surface asbestos contamination or confined leakage from an underground storage tank (UST).

Special Circumstances and Exemplary Performance
There is no EP point available for credit SSc3.

2.4 SS Credit Subcategory 4: *Alternative Transportation*
This credit subcategory has a total of four subsections, each worth one credit point. The main intention of all four Alternative Transportation credits is to reduce our dependency on single-occupancy vehicles (SOVs), particularly those which use gasoline and petroleum diesel fuels. The goals are to reduce the need for building more road infrastructure, reduce motor vehicle pollution, decrease the dependency on foreign oil, and reduce the vast amount of paved surfaces found in urban areas that may contribute to nonpoint source pollution from stormwater runoff and the urban heat island effect. The four subsections (credits) are as follows:

- **SS Credit 4.1** (SSc4.1): *Alternative Transportation: Public Transportation Access* (EB)
- **SS Credit 4.2** (SSc4.2): *Alternative Transportation: Bicycle Storage and Changing Rooms*
- **SS Credit 4.3** (SSc4.3): *Alternative Transportation: Low-Emitting and Fuel-Efficient Vehicles* (previously referred to as *Alternative Transportation, Alternative Fuel Vehicles* in LEED NC 2.1)
- **SS Credit 4.4** (SSc4.4): *Alternative Transportation: Parking Capacity*

There are two possible exemplary performance points available under Alternative Transportation. The EP point available in the Innovation and Design category for Alternative Transportation as per the LEED-NC 2.2 first edition may be awarded by "instituting a comprehensive transportation management plan that demonstrates a quantifiable reduction in personal automobile use through the implementation of multiple alternative options." It is related to the overall subcategory, and not to each individual credit. In addition, another EP point can be earned according to the following language from a USGBC CIR dated September 11, 2006:

> Based on evidence that locations with higher transit density can achieve substantially and a quantifiably higher environmental benefit, meeting the following threshold qualifies a project for exemplary performance Innovation Credit. This follows the Center for Clean Air Policy's finding that average transit ridership increases by 0.5% for every 1.0% increase in growth of transit service levels, which leads to the conclusion that quadrupling transit service generally doubles transit ridership.
>
> To accomplish this quadrupling of service and doubling of ridership, at a minimum:
>
> - Locate the project within ½ mile of at least two existing commuter rail, light rail, or subway lines, OR locate project within ¼ mile of at least two or more stops for four or more public or campus bus lines usable by building occupants;
>
> AND
>
> - Frequency of service must be such that at least 200 transit rides per day are available in total at these stops. A combination of rail and bus is allowable. This strategy is based on the assumption that the threshold of the base credit would provide, in most cases, at least 50 transit rides per day (half-hourly service 24 hours per day or more frequent service for less than 24 hours per day). If, on average, transit ridership increases by 0.5 percent for every 1.0% increase in transit service, then quadrupling the number of rides available would, on average, double the transit ridership. (4×50 rides = 200 rides). Include a transit schedule and map within your LEED certification submittal.

A summary of the credit criteria and exemplary performance criteria for the entire Alternative Transportation subcategory is given in Table 2.4.1.

SS Credit 4.1: *Alternative Transportation—Public Transportation Access*

USGBC Rating System

The first Alternative Transportation credit, 4.1, provides for access to public transportation as an alternative to SOVs. This credit is noted with the icon EB as it may be very costly to provide these services at a site where they are not already available. The applicable forms of public transport are commuter-type rail service and bus lines. LEED-NC 2.2 lists the Intent, Requirements, and Potential Technologies and Strategies for this credit as follows:

Intent
Reduce pollution and land development impacts from automobile use.

Requirements
Locate project within ½ mile of an existing—or planned and funded—commuter rail, light rail or subway station

OR

Locate project within ¼ mile of one or more stops for two or more public or campus bus lines usable by building occupants.

Potential Technologies and Strategies
Perform a transportation survey of future building occupants to identify transportation needs. Site the building near mass transit.

Alternative Transportation Credit Section	Use	Option and/or EP	Minimum Criteria
Entire Category	Any	EP	Comprehensive Transportation Management Plan
4.1: *Public Transportation Access*	Any	1	<½ mi to a commuter rail, light rail, or subway station
4.1: *Public Transportation Access*	Any	2	<¼ mi to 1 or more stops for 2 public or campus bus lines usable by occupants
4.1: *Public Transportation Access*	Any	1-EP	<½ mi to 2 commuter rail, light rail or subway stations and 200 transit rides/day (rail and/or bus)
4.1: *Public Transportation Access*	Any	2-EP	<¼ mi to 1 or more stops for 4 public or campus bus lines usable by occupants and 200 transit rides/day (rail and/or bus)
4.2: *Bicycle Storage and Changing Rooms*	Commercial or institutional	—	Showers = $0.005 \times$ FTE and BR = $0.05 \times$ (Maximum PBU_j)
4.2: *Bicycle Storage and Changing Rooms*	Residential uses	—	BSF = $0.15 \times$ (DO)
4.3: *Low-Emitting and Fuel-Efficient Vehicles (LEFEVs) or Alternative Fuel Vehicles (AFVs)*	Commercial or institutional usually	1	LEFEVP1 = $0.03 \times$ FTE (both the number of LEFEV vehicles provided and the number of LEFEV preferred parking spaces provided)
4.3: *Low-Emitting and Fuel-Efficient Vehicles or Alternative Fuel Vehicles*	Any	2	LEFEVP2 = $0.05 \times$ TP (the number of LEFEV preferred parking spaces provided)
4.3: *Low-Emitting and Fuel-Efficient Vehicles or Alternative Fuel Vehicles*	Any	3	AFV = $0.03 \times$ TP (the AFV dispensing capability)
4.4: *Parking Capacity*	Nonresidential	1	Do not exceed minimum parking and CVPP1 = $0.05 \times$ TP (Van/carpool preferred parking provided)
4.4: *Parking Capacity*	Nonresidential	2	If TP < $0.05 \times$ FTE, CVPP2 = $0.05 \times$ TP (Van/carpool preferred parking provided)
4.4: *Parking Capacity*	Residential	3	Do not exceed minimum parking and have ride-sharing programs/infrastructure
4.4: *Parking Capacity*	Any	4	No new parking

TABLE 2.4.1 Summary of Alternative Transportation Credits and Exemplary Performance (EP) Points

Calculations and Considerations

To confirm adherence to this credit, a vicinity map should be provided showing the project location and the location of the applicable rail/subway station or the applicable bus stops. The ½-mi or ¼-mi distances required, respectively, are usually not a straight line from the property site, but are rather an appropriate pedestrian pathway from the site to the public transportation station or stop. These pathways must be shown on the vicinity map. The submittal also must include a listing of the stations/stops and the pedestrian distances traveled to get there from the site. The rail or subway station must be for use by commuters. The bus lines may or may not be from the same company or agency; the important thing is that they are separate lines and usable by the building occupants. Figure 2.4.1 shows an example area map which fulfills both options, although only one is needed for the credit.

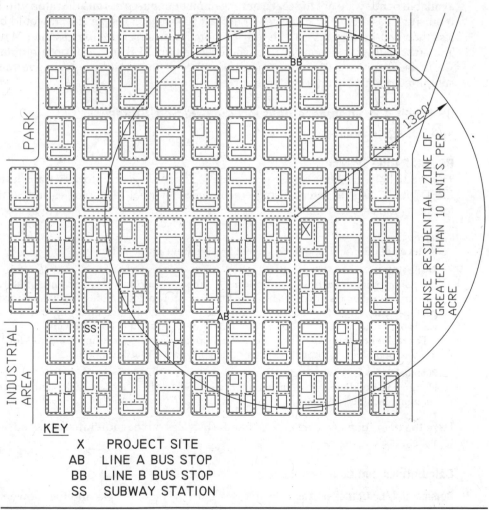

KEY

X	PROJECT SITE
AB	LINE A BUS STOP
BB	LINE B BUS STOP
SS	SUBWAY STATION

FIGURE 2.4.1 Area map for alternative public transportation.

In the case of Option 1, note that if a train station is not available, but is *both* planned and funded, then this will count. If something is funded, then there is a high likelihood that it will actually be built in the near future.

In either case, a permanent, private shuttle may also be established to the bus or train stops to potentially qualify if the minimum distance requirements cannot be met. Details, accessibility, and availability should be described.

Special Circumstances and Exemplary Performance

As noted previously, an exemplary performance point based on SSc4.1 was approved in a Credit Ruling dated September 11, 2006. To obtain this additional point, there must be double the number of either rail or bus lines within the requisite distances and there must be a combined frequency of service such that at least 200 transit rides per day are available at these stops; a combination of rail and bus is allowable. Transit rides available per day are not the seats, but the number of buses or rails that stop within the requisite areas daily. For instance, for at least 50 transit rides per day, it would be just less than half-hourly service 24 h/day or more frequent service for less than 24 h/day. Transit schedules and maps must be included in the LEED certification submittal.

If there are special circumstances for either of the options, a narrative must be submitted describing these circumstances.

SS Credit 4.2: *Alternative Transportation—Bicycle Storage and Changing Rooms*

USGBC Rating System

This credit helps facilitate bicycle transportation as an alternative to SOV transportation. LEED-NC 2.2 lists the Intent, Requirements, and Potential Technologies and Strategies for this credit as:

Intent
Reduce pollution and land development impacts from automobile use.

Requirements
For **commercial or institutional buildings,** provide secure bicycle racks and/or storage (within 200 yards of a building entrance) for 5% or more of all building users (measured at peak periods), AND provide shower and changing facilities in the building or within 200 yards of a building entrance, for 0.5% of Full-Time Equivalent (FTE) occupants.

For **residential buildings**, provide covered storage facilities for securing bicycles for 15% or more of building occupants in lieu of changing/shower facilities.

Potential Technologies and Strategies
Design the building with transportation amenities such as bicycle racks and shower/changing facilities.

Here the definition of *secure* means that the bicycles can be individually locked such as to a rack.

Calculations and Considerations

Commercial/Institutional Uses   In all cases, both the bicycle racks and the showers must be provided free of charge. Several calculations need to be made to determine the number of bicycle rack or storage spaces and the number of changing/shower facilities.

For projects that are located on a campus, the changing/shower facilities need not be in the project in question, but can be in other buildings on campus as long as they are located within 200 yards of the entrance to the project building. Given the following definitions:

FTE   Full-time equivalent occupant (during the typically busiest part of a day)

FTE_j Full-time equivalent occupant during shift j

$FTE_{j,i}$ Full-time equivalent occupancy of employee i during shift j

$FTE_{j,i}$ is equal to 1 for a full-time employee and is equal to the normal hours worked (less than 8) divided by 8 for a part-time employee. Obviously, this would need to be modified if shifts are different from the standard 8 h.

The FTE is then determined from the following set of equations:

$$FTE_{j,i} = (\text{worker } i \text{ hours})/8\,h \qquad \text{where } 0 < FTE_{j,i} \le 1 \tag{2.4.1}$$

$$FTE_j = \Sigma\, FTE_{j,i} \qquad \text{for all employees in shift } j \tag{2.4.2}$$

$$FTE = \text{maximum } (FTE_j) \tag{2.4.3}$$

This FTE is used to determine the number of changing/shower facilities (Showers) which must be provided for any occupant wishing to bicycle to the location:

$$\text{Showers} = 0.005 \times FTE \tag{2.4.4}$$

The number of showers must always be a whole number, and any fractional results must be rounded up to the next-higher whole number.

The calculations get more complex for the number of secure bicycle racks and/or storage spaces that must be provided (bicycle rack spaces) since many facilities service not only the regular employees but also many others who may frequent the facility, such as clients, customers, or students in an educational facility; and for many industrial applications there are shift overlaps. These additional users are referred to as *transient occupants*. Together the FTE_j and the transient occupants TO_j can be used to determine the peak occupancy.

There may be some confusion as to the definition of the peak occupancy for transient occupants. The LEED-NC 2.2 reads as follows: "Estimate the transient occupants such as students, visitors and customers for the **peak period** for the facility." This does not mean to give maximum occupancy, but rather the typical transient occupancy during a busy part of the day. This may be difficult in the cases of facilities with very erratic attendance such as stadiums and concert halls. For instance, a typical baseball game is probably attended by a much smaller crowd than the maximum seating capacity. Using the maximum capacity based on the dozen or so times a year when there are tournaments or special games does not necessarily constitute the intent of this credit, just as popular concerts do not constitute typical occupancies for an auditorium. A reasonable evaluation might be based on the number of games with average attendance and perhaps the number of practice days with typical occupancies during those periods such as team members, trainers, spectators, and other staff. Then an averaging of these values based on frequencies should be attempted to give reasonable numbers for transient occupancies.

If the required number of bicycle rack spaces is defined as BR, the peak transient occupancy during shift or period j as TO_j, and the peak building users during shift j as PBU_j, then BR can be calculated by using the following two equations:

$$PBU_j = FTE_j + TO_j \tag{2.4.5}$$

$$BR = 0.05 \times (\text{maximum } PBU_j) \tag{2.4.6}$$

The number of bicycle rack spaces (BR) must always be a whole number, and any fractional calculations must be rounded up to the next-higher whole number. Figure 2.4.2 depicts an outdoor bicycle rack at a LEED certified building.

Submittals for obtaining this credit for commercial/institutional uses must include the calculations plus a map showing that the bicycle and changing/shower facilities are within 200 yd of the entrance to the project building (or within the building).

Residential Uses   The secure bicycle spaces for residential projects must also be covered to protect the bicycles from the weather and from theft. They must be provided for a minimum of 15 percent of the building occupants. The building designer should declare the design occupancy (DO) for the buildings. Note that a design occupancy is not the same as a maximum occupancy as may be established by a fire code. The submittal should show that there are a minimum of secure *and* covered bicycle storage facility (BSF) spaces for at least 15 percent of this design occupancy as shown in Eq. (2.4.7). Again, the number of bicycle storage spaces must be a whole number. Additional changing/shower facilities are not required as they are already available in the residences.

$$BSF = 0.15 \times DO \tag{2.4.7}$$

FIGURE 2.4.2   Bicycle rack outside of the West Quad Learning Center at the University of South Carolina in September 2006.

Special Circumstances and Exemplary Performance

There is not an exemplary performance point tied individually to SSc4.2; however, there is an overall point that can be earned for a transportation plan as previously mentioned.

SS Credit 4.3: *Alternative Transportation—Low-Emitting and Fuel-Efficient Vehicles (LEFEVs)[and also Alternative Fuel Vehicles(AFVs)]*

Air quality and the global climate may be affected by emissions from motor vehicles and the combustion of fossil fuels. This credit seeks to reduce these impacts by promoting motor vehicles which have lower emissions either due to advanced or alternative designs or due to high fuel efficiency. However, full life-cycle analyses should be considered when changing to alternative fuels, as there may be other environment issues relating to the manufacture or use of the fuels.

USGBC Rating System

LEED-NC 2.2 lists the Intent, Requirements, and Potential Technologies and Strategies for this credit as follows:

Intent

Reduce pollution and land development impacts from automobile use.

Requirements

OPTION 1

Provide low-emitting and fuel-efficient vehicles for 3% of Full-Time Equivalent (FTE) occupants AND provide preferred parking for these vehicles.

OR

OPTION 2

Provide preferred parking for low-emitting and fuel-efficient vehicles for 5% of the total vehicle parking capacity of the site.

OR

OPTION 3

Install alternative-fuel refueling stations for 3% of the total vehicle parking capacity of the site (liquid or gaseous fueling facilities must be separately ventilated or located outdoors).

For the purpose of this credit, low-emitting and fuel-efficient vehicles are defined as vehicles that are either classified as Zero Emission Vehicles (ZEV) by the California Air Resources Board (CARB) or have achieved a minimum green score of 40 on the American Council for an Energy Efficient Economy (ACEEE) annual vehicle rating guide.

"Preferred parking" refers to the parking spots that are closest to the main entrance of the project (exclusive of spaces designated for handicapped) or parking passes provided at a discounted price.

Potential Technologies and Strategies

Provide transportation amenities such as alternative-fuel refueling stations. Consider sharing the costs and benefits of refueling stations with neighbors.

Calculations and Considerations

This credit actually covers two different sets of vehicles: one is the low-emitting and fuel-efficient vehicle (LEFEV) category for which Options 1 and 2 apply. The other is the alternative-fuel vehicle (AFV) category for which Option 3 applies. There is some overlap between the two sets, and they are further defined in the following sections.

Options 1 and 2   For Option 1 determine the FTE by Eqs. (2.4.1) through (2.4.3), and then determine both the minimum number of low-emitting and fuel-efficient vehicles to provide and the minimum number of preferred parking spaces designated for these types of vehicles (LEFEVP1) by Eq. (2.4.8).

$$LEFEVP1 = 0.03 \times FTE \qquad (2.4.8)$$

The number of low-emitting and fuel-efficient vehicles and preferred parking spaces (LEFEVP1) must always be a whole number, and any fractional results must be rounded up to the next-higher whole number.

The submittal for Option 1 should include these LEFEVP1 preferred parking calculations and a site plan showing the number and location of LEFEVP1 preferred parking spaces. The submittal should also include a listing with the number, make, model, manufacturer, and ZEV or ACEEE vehicle score of each low-emitting and fuel-efficient vehicle provided for the FTE occupants.

For the Option 2 submittal, confirm the number of total parking spaces provided for the project (TP), confirm that the minimum number of preferred parking spaces for low-emitting and fuel-efficient vehicles (LEFEVP2) is at least 5 percent of this total, and show the location(s) of the preferred parking for these low-emitting and fuel-efficient vehicles on an area map. The number of LEFEVP2 preferred parking spaces is given by Eq. (2.4.9). It must always be a whole number, and any fractional calculations must be rounded up to the next-higher whole number.

$$LEFEVP2 = 0.05 \times TP \qquad (2.4.9)$$

Options 1 and 2 may be much more viable than it might seem from reading the credit description, as the typical person may assume that the low-emitting and fuel-efficient vehicles are mainly alternative-fuel vehicles, and prior to 2007 not many people in the United States owned such vehicles. However, this category also includes many gasoline-fueled vehicles. Nearly all hybrids fall into this category, but so do many of the smaller, more fuel-efficient cars on the market. The current year's listings and information on how to purchase a book with ratings for prior years are available at www.greenercars. com. For instance, in 2006 in addition to many hybrids, vehicles with ratings above 40 on the ACEEE rating included the following models: Toyota Corolla, Hyundai Accent, Kia Rio/Rio 5, Honda Civic, Pontiac Vibe, Chevrolet Cobalt, and Saturn Ion. Since many people commute in the more energy-efficient vehicles, there might be a great demand for preferred parking spaces, and more importantly, the 5 percent requirement for the preferred parking spaces may be easy to fulfill without leaving parking spaces unoccupied. Similarly, providing fleet vehicles with these vehicles may be more economical than initially thought.

Option 3   For SS Credit 4.3 Option 3, the submittal should confirm the number of total parking spaces provided. In addition, according to the version 2.2 Reference Guide, there should also be a calculation determining that the number of refueling dispensing locations is at least 3 percent of this number. However, this may not be the proper calculation for all situations, and it may be more dependent on the alternative fuels being dispensed. It is not the intent of the credit to have dozens of refueling stations, but rather to provide alternative fuel for at least 3 percent of the vehicles which could be parked. A dispensing location for some alternative fuels can service many vehicles in a reasonable time and is usually regarded in the industry as the actual nozzle or dispensing device. A standard dispensing machine may service more than one vehicle simultaneously and may service many over a fairly short time, with some exceptions where recharging takes a long time. To take these variations into account, the submittals listed in the LEED-NC 2.2 Reference Guide include a plan with the location(s) of the alternative fueling stations, the fuel type being dispensed, the number of dispensing locations, and the fueling capacity for each for an 8-h period. It is the opinion of the author that the submittal should then also include verification that these fueling stations can adequately service the requisite number of vehicles easily within a reasonable time frame that is convenient for the vehicle users. Let AFV be this minimum number of alternative-fuel vehicles which can be adequately serviced and, as previously defined, TP is the total parking provided, then Option 3 will be obtained if Eq. (2.4.10) is valid.

$$AFV = 0.03 \times TP \hspace{3cm} (2.4.10)$$

The LEED-NC 2.2 Reference Guide provides the following definition: "Alternative Fuel Vehicles are vehicles that use low-polluting, non-gasoline fuels such as electricity, hydrogen, propane or compressed natural gas, liquid natural gas, methanol and ethanol. Efficient gas-electric hybrid vehicles are included in this group for LEED purposes." However, this is probably an older definition from previous versions of LEED that is not fully intended to be used for Option 3 in this version. It is the opinion of the author that it is not the intent of Option 3 to provide gasoline fueling for hybrids, but to instead provide some alternative fuel. In addition, biodiesel is an alternative fuel that is being accepted for use in many diesel engines. Biodiesel is a blend of vegetable/animal diesel, and petroleum-based diesel can be used in most diesel engines with no modifications up to about 20 percent biodiesel based (B20). Lower percentages of petroleum-based diesel in the mix (up to 0 percent in B100) may need some modified parts in vehicles such as different material hoses. This alternative has not been directly addressed in this credit, but the author believes that it may have the opportunity for acceptance but would need verification based on a CIR.

Special Circumstances and Exemplary Performance
In summary, Options 1 and 2 deal with low-emitting and fuel-efficient vehicles while Option 3 deals with alternative-fuel vehicles. It is the opinion of the author that low-emitting and fuel-efficient vehicles should include alternative-fuel vehicles, but that alternative-fuel vehicles do not include many of the low-emitting and fuel-efficient vehicles.

If there are special circumstances or nonstandard compliance paths taken for either of the options, a narrative must be submitted describing these circumstances. There is not an exemplary performance point tied individually to SSc4.3; however, there is an overall point that can be earned for a transportation plan as previously mentioned.

SS Credit 4.4: *Alternative Transportation—Parking Capacity*

USGBC Rating System
The intention of this credit is to limit the sizes of parking facilities and the number of vehicle trip generations to the site. *Trip generation* is a term used in transportation engineering to represent the amount of vehicular traffic generated in a network by an activity at a location.

The Institute of Transportation Engineers (ITE) publishes a manual that gives representative values for different site uses and street volumes. This reduction from the estimated trip generation for typical, usually single-occupancy, vehicles is accomplished by encouraging the use of carpooling or other shared-ride methods. LEED-NC 2.2 lists the Intent, Requirements, and Potential Technologies and Strategies for this credit as follows:

Intent

Reduce pollution and land development impacts from single occupancy vehicle use.

Requirements

OPTION 1—NON-RESIDENTIAL

Size parking capacity to not exceed minimum local zoning requirements AND provide preferred parking for carpools or vanpools for 5% of the total provided parking spaces.

OR

OPTION 2— NON-RESIDENTIAL

For projects that provide parking for less than 5% of FTE building occupants:

Provide preferred parking for carpools or vanpools for 5% of the total provided parking spaces.

OR

OPTION 3—RESIDENTIAL

Size parking capacity to not exceed minimum local zoning requirements, AND, provide infrastructure and support programs to facilitate shared vehicle usage such as carpool drop-off areas, designated parking for vanpools, or car-share services, ride boards, and shuttle services to mass transit.

OR

OPTION 4—ALL

Provide no new parking.

NOTES:

"Preferred parking" refers to the parking spots that are closest to the main entrance of the project (exclusive of spaces designated for handicapped) or parking passes provided at a discounted price.

When parking minimums are not defined by relevant local zoning requirements, or when there are no local zoning requirements, either:

A) Meet the requirements of Portland, Oregon, Zoning Code: Title 33, Chapter 33.266 (Parking and Loading)

OR, if this standard is not appropriate for the building type,

B) Install 25% less parking than the building type's average listed in the Institute of Transportation Engineers' Parking Generation study, 3rd Edition.

Potential Technologies and Strategies

Minimize parking lot/garage size. Consider sharing parking facilities with adjacent buildings. Consider alternatives that will limit the use of single occupancy vehicles.

Calculations and Considerations

The intention of this credit is to reduce the use of SOVs, which has other potential environmental benefits in addition to those just listed.

Option 1 and Option 3 for SSc4.4 require that the total number of parking spaces (TP) does not exceed the required zoning minimum. Most zoning codes have a methodology for determining the minimum number of parking spaces that should be provided for different uses and different size facilities. The submittals should include the verbiage of the applicable zoning code and the supporting calculations to determine the minimum parking required by local zoning code. The two options for this credit mandate that the maximum total parking provided (TP) be equal to or lower than the minimum as set by the local code, whether the local code has a different maximum parking calculation or not. (Sometimes the parking allowed can be lower than the minimum as set in the zoning code, if the project has been granted a variance by the regulatory authority due to special or exceptional circumstances.) In the past, most zoning codes did not have a calculation for maximum allowed parking, and the parking installed would frequently far exceed the zoning minimum. Recently, many zoning codes have been modified to include both a minimum and maximum allowable number of parking spaces, mainly to address the environmental concerns arising from the typical impervious surfaces that are used to build parking facilities and their impacts on stormwater runoff and the urban heat island effect. As mentioned in the errata as posted in the spring of 2007 by the USGBC, there are two alternatives to determining the total parking allowed if a minimum is not set by local code or zoning. In these cases the total parking (TP) should not exceed the requirements of the Portland, Oregon, Zoning Code: Title 33, Chapter 33.266 (Parking and Loading) or, if the Portland, Ore., standard is not appropriate for the building type, then the total parking should be at least 25 percent less than the building type's average listed in the document by the Institute of Transportation Engineers (ITE) entitled *Parking Generation, Third Edition*, Washington, D.C., 2004.

For example, in one municipality, a sit-down restaurant use may require a minimum of one-half parking space for each customer seat (most people go to restaurants in groups) plus one for each employee in a shift. The maximum may allow for an increase of 25 percent over this, which is reasonable based on customers who may wish to wait for service. Other zoning codes may have the formula for a similar use based on gross floor area or customer floor area. To obtain zoning approval, the designers must meet this minimum requirement unless they have a valid reason to ask for a variance. If a variance is granted from the minimum by the local municipal planning organization (MPO), then this should be included in the submittal documents. An example of a valid reason for a lower than minimum total parking may be the availability of shared parking off-site.

Option 1 Option 1 for credit SSc4.4 is for nonresidential construction and has two requirements. The first is that the total number of parking spaces (TP) does not exceed the required zoning minimum as previously defined. The second requirement in Option 1 for credit SSc4.4 is to designate the minimum number of preferred parking spaces for carpools or vanpools (CVPP1) for 5 percent of the total provided parking spaces as given in Eq. (2.4.11).

$$CVPP1 = 0.05 \times TP \tag{2.4.11}$$

The minimum number of preferred parking spaces for carpools or vanpools (CVPP1) must always be a whole number, and any fractional calculations must be rounded up to the next-higher whole number.

The submittals for Option 1 of credit SSc4.4 should include the zoning calculations for minimum required parking to obtain the total parking (or the appropriate alternative method based on the Portland, Ore., zoning code or the ITE report) and the calculation of the minimum number of preferred parking spaces for carpools or vanpools (CVPP1). A plan should be submitted showing both the total number of parking spaces (TP) and the location of at least the minimum number of preferred parking spaces for carpools or vanpools (CVPP1) as appropriately marked in relation to the other parking spaces and the entrance.

Option 2   Option 2 of credit SSc4.4 is similar to Option 1 but is for the special case where very little parking is provided, perhaps representing projects in areas with well-established mass transit or other transportation and parking opportunities. The submittals should first establish that the total parking provided (TP) is less than 5 percent of the FTE building occupants and then provide the calculation for the minimum number of preferred parking spaces for carpools or vanpools (CVPP2) as given in Eq. (2.4.12) and an area map showing their location.

$$CVPP2 = 0.05 \times TP \qquad (2.4.12)$$

The minimum number of preferred parking spaces for carpools or vanpools (CVPP2) must always be a whole number, and any fractional calculations must be rounded up to the next-higher whole number.

Option 3   Option 3 of credit SSc4.4 is for residential uses and mandates not exceeding the local zoning parking requirements and also promoting programs to reduce SOV use such as ride sharing. The submittal should include the verbiage of the applicable zoning code and the supporting calculations to determine the minimum parking required by local zoning code. This credit mandates that the maximum total parking provided (TP) equal the minimum as set by the local code and not greater, whether the local code has a different maximum parking calculation or not. As mentioned previously, there are two alternatives given for determining the total parking allowed (TP) if parking is not addressed by local zoning. The submittal should also provide a description of the infrastructure and the programs that are in place or going to be provided to not just promote, but also support ride sharing for the residents.

Option 4   Option 4 of credit SSc4.4 mandates that no new parking be provided. The submittal should include the previously existing site parking plan and the new parking plan indicating the total parking (TP) provided on each.

Special Circumstances and Exemplary Performance
If there are special circumstances or nonstandard compliance paths taken for either of the options, a narrative must be submitted describing these circumstances. There is not an exemplary performance point tied individually to SSc4.4; however, there is an overall point that can be earned for a transportation plan as previously mentioned. If there are special circumstances or nonstandard compliance paths taken for either of the options, a narrative must be submitted describing these circumstances.

2.5 SS Credit Subcategory 5: *Site Development*

SS Credit 5.1: *Site Development—Protect or Restore Habitat*
The intent of this credit is to maintain as much of the native or adapted biological species as possible in a region. There are many reasons for this. Obviously, native and

adapted species would survive well in the local climate without much additional irrigation and maintenance. There is also the desire for biodiversity in this world. Maintaining native vegetation in all regions is important both for plant biodiversity and for the local fauna to survive. It is desirable to maintain habitat for many biologically diverse organisms, not just for ecological reasons, but also because it has been found that a world with rich biodiversity benefits humans in the form of potential medicines and other products.

USGBC Rating System

LEED-NC 2.2 lists the Intent, Requirements, and Potential Technologies and Strategies for this credit as follows:

Intent

Conserve existing natural areas and restore damaged areas to provide habitat and promote biodiversity.

Requirements

On **greenfield sites**, limit all site disturbance to 40 feet beyond the building perimeter; 10 feet beyond surface walkways, patios, surface parking and utilities less than 12 inches in diameter; 15 feet beyond primary roadway curbs and main utility branch trenches; and 25 feet beyond constructed areas with permeable surfaces (such as pervious paving areas, stormwater detention facilities and playing fields) that require additional staging areas in order to limit compaction in the constructed area.

OR

On **previously developed or graded sites**, restore or protect a minimum of 50% of the site area (excluding the building footprint) with native or adapted vegetation. Native/adapted plants are plants indigenous to a locality or cultivars of native plants that are adapted to the local climate and are not considered invasive species or noxious weeds. Projects earning SS Credit 2 and using vegetated roof surfaces may apply the vegetated roof surfaces to this calculation if the plants meet the definition of native/adapted.

   Greenfield sites are those that are not developed or graded and remain in a natural state. Previously Developed Sites are those that previously contained buildings, roadways, parking lots or were graded or altered by direct human activities.

Potential Technologies and Strategies

On greenfield sites, perform a site survey to identify site elements and adopt a maser plan for development of the project site. Carefully site the building to minimize disruption to existing ecosystems and design the building to minimize its footprint. Strategies include stacking the building program, tuck-under parking and sharing facilities with neighbors. Establish clearly marked construction boundaries to minimize disturbance of the existing site and restore previously degraded areas to their natural state. For previously developed sites, utilize local and regional governmental agencies, consultants, educational facilities, and native plant societies as resources for the selection of appropriate native or adapted plant materials. Prohibit plant materials listed as invasive or noxious weed species. Native/adapted plants require minimal or no irrigation following establishment, do not require active maintenance such as mowing or chemical inputs such as fertilizers, pesticides or herbicides, and provide habitat value and promote biodiversity through avoidance of monoculture plantings.

Calculations and Considerations

The LEED definitions for *building footprint, development footprint, native* (or *indigenous*) *plants, adapted* (or *introduced*) *plants, invasive plants,* and *previously developed sites* are given in App. B, and many of the symbols used to represent them are given in the Land Area Definitions section at the beginning of this chapter.

The submittal is part of the LEED Construction Submittal since limiting construction disturbance and protecting native areas are an important part of the construction phase. It should be well documented throughout the construction phase that the limits of construction (disturbance) either do not go beyond the maximum setbacks in the first option or, for previously developed sites, do not disturb existing natural areas. For both cases, greenfield sites or previously developed sites, the submittals should include both an existing conditions map and a project site area map with the site area given and the land areas of the project building footprint, developed portions, and undeveloped and/ or restored portions noted. A narrative should also be included that describes the approach taken to obtain this credit with any special circumstances noted.

Greenfield Sites   For the greenfield sites, a grading plan should also be submitted with the designated limits of disturbance highlighted. Note that site disturbance includes any earthwork and clearing of vegetation, even if replanted afterward. Figure 2.5.1 is an example sketch of a greenfield site development and limits of disturbance plan.

Previously Developed Sites   The criteria for sites which were previously developed are based upon the building footprint, not the development footprint. The building footprint is the plan view of the site area covered by the buildings and structures, including overhangs. Shade structures do not have to be included if they are not part of the building. For sites that were previously developed, the total of the areas that have

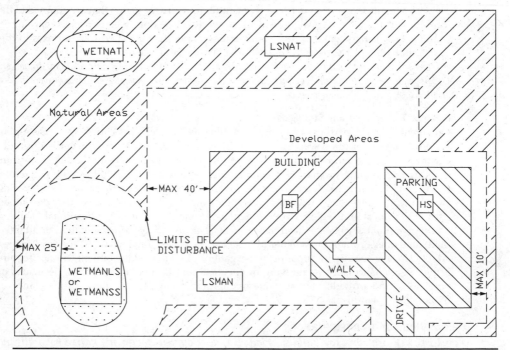

Figure 2.5.1   Example maximum setbacks for disturbance at a former greenfield site.

been restored to native/adaptive vegetation should be submitted along with a landscaping plan and schedule for the restored and protected areas. Any areas that were previously graded or landscaped count as previously developed, as do active agricultural uses. Fallow areas are sometimes considered undeveloped. Sports fields do not count as native/adaptive vegetation. (Note that sports fields do count as open space for SSc5.2.) When choosing native or adaptive vegetation, the distinction for appropriate species is very localized. Even the definition of *invasive* is local as some species are considered invasive in some areas of the country and are not considered invasive in others. One example is the ice plant, a ground cover. It is used extensively in California and is considered invasive there, but not in many other states.

Note that projects earning SS Credit 2 (Development Density and Community Connectivity) and having vegetated roof surfaces may use these vegetated roof areas in the calculation of the total of the areas restored or protected if these vegetated roof areas meet the native/adapted species criteria. Otherwise, the vegetated roof surfaces are not included. Raised planters were not specifically addressed in LEED-NC 2.2, but it is assumed that they too may count for urban areas, earning SSc2 if appropriately planted, but not for other projects. By raised planters, the author is referring to decorative planters that are substantially raised from the ground surface, not curbed landscape areas. Curbed landscape areas should count in all cases if appropriately planted.

The requirements for previously developed sites can be summarized using the following additional definitions and sets of equations and are summarized in Table 2.5.1.

LSNAP   Represents all the previously disturbed areas on the ground that are replanted with native and adapted plants

VNAPR   Represents all the roofed areas that are planted with native and adapted plants

One point is earned for most previously developed sites if the following is true:

$$\text{WETNAT} + \text{LSNAT} + \text{LSNAP} \geq 0.5 \times (A_T - \text{BF}) \qquad (2.5.1)$$

or for the previously developed urban sites where SSc2 is also earned, the following is true:

$$\text{WETNAT} + \text{LSNAT} + \text{LSNAP} + \text{VNAPR} \geq 0.5 \times (A_T - \text{BF}) \qquad (2.5.2)$$

Special Circumstances and Exemplary Performance

The previously developed sites may be awarded one point for exemplary performance if the restored/protected areas total more than 75 percent of the project area, excluding the building footprint.

Criteria for	Most Sites	Urban Sites that also Earn SSc2
Credit point	WETNAT + LSNAT + LSNAP $\geq 0.5(A_T - \text{BF})$	WETNAT + LSNAT + LSNAP + VNAPR $\geq 0.5(A_T - \text{BF})$
EP point	WETNAT + LSNAT + LSNAP $\geq 0.75(A_T - \text{BF})$	WETNAT + LSNAT + LSNAP + VNAPR $\geq 0.75(A_T - \text{BF})$

TABLE 2.5.1   Summary of Protected/Restored Habitat Areas for Previously Developed Sites

Figure 2.5.2 Landscaped area with native and adapted plants at Cox and Dinkins, Engineers and Surveyors, Columbia, S.C. (*Photograph taken June 2007.*)

An additional exemplary performance point is earned for most previously developed sites if

$$\text{WETNAT} + \text{LSNAT} + \text{LSNAP} \geq 0.75 \times (A_T - \text{BF}) \qquad (2.5.3)$$

or for the previously developed urban sites where SSc2 is also earned, the following is true:

$$\text{WETNAT} + \text{LSNAT} + \text{LSNAP} + \text{VNAPR} \geq 0.75 \times (A_T - \text{BF}) \qquad (2.5.4)$$

Figure 2.5.2 shows some examples of native and adapted plants that were used in a landscape bed in Columbia, S.C.

SS Credit 5.2: *Site Development—Maximize Open Space*

USGBC Rating System
LEED-NC 2.2 lists the Intent, Requirements, and Potential Technologies and Strategies for this credit as follows:

Intent
Provide a high ratio of open space to development footprint to promote biodiversity.

Requirements

OPTION 1:

Reduce the development footprint (defined as the total area of the building footprint, hardscape, access roads and parking) and/or provide vegetated open space within the project boundary to exceed the local zoning's open-space requirement for the site by 25%.

OR

OPTION 2:

For areas with no local zoning requirements (e.g., some university campuses and military bases), provide vegetated open-space area adjacent to the building that is equal to the building footprint.

OR

OPTION 3:

Where a zoning ordinance exists, but there is no requirement for open space (zero), provide vegetated open space equal to 20% of the project's site area.

ALL OPTIONS:

- For projects located in urban areas that earn SS Credit 2, vegetated roof areas can contribute to credit compliance.
- For projects located in urban areas that earn SS Credit 2, pedestrian-oriented hardscape areas can contribute to credit compliance. For such projects, a minimum of 25% of the open space counted must be vegetated.
- Wetlands or naturally designed ponds may count as open space if the side slope gradients average 1:4 (vertical : horizontal) or less and are vegetated.

Potential Technologies and Strategies

Perform a site survey to identify site elements and adopt a master plan for development of the project site. Select a suitable building location and design the building with a minimal footprint to minimize site disruption. Strategies include stacking the building program, tuck-under parking and sharing facilities with neighbors to maximize open space on the site.

Calculations and Considerations

The LEED definitions for *building footprint, development footprint, native* (or *indigenous*) *plants, adapted* (or *introduced*) *plants, invasive plants, open space area,* and *previously developed sites* are given in App. B. Note the exceptions for SS Credit 5.2, and also as applicable if SS Credit 2 is obtained, in the definition for *open-space area* as given in LEED-NC 2.2 with the errata as posted in the spring of 2007:

> Open Space Area is as defined by local zoning requirements. If local zoning requirements do not clearly define open space, it is defined for the purposes of LEED calculations as the property area minus the development footprint; and it must be vegetated and pervious, with exceptions only as noted in the credit requirements section. For projects located in urban areas that earn SS Credit 2, open space also includes non-vehicular, pedestrian-oriented hardscape spaces.

The exceptions are with regard to vegetation and for all projects allow for wetlands or naturally designed ponds to be included in open space if they have vegetated sides leading down to them with low slopes (<25 percent). In addition, vegetated playing fields may be included as open space. There is still some question as to whether sports fields within stadiums count as open space. These special exceptions should be reviewed through the USGBC CIR procedure.

For projects located in urban areas that *also* earn SS Credit 2, open space may include vegetated roof areas and nonvehicular, pedestrian-oriented hardscape spaces (with the caveat that at least 25 percent of the open space must be vegetated). Examples of these pedestrian-oriented spaces include items such as pocket parks, accessible roof decks, plazas, and courtyards.

For all the options, the submittal should include the total site area and the site and landscape plans which highlight the open spaces, both those vegetated and, when allowed, those not vegetated with the associated areas noted. For Option 1, also provide the calculations for the open space as required by local code; and for Option 2, also provide the area of the building footprint. As with SSc5.1, raised planters were not specifically addressed in LEED-NC 2.2, but it is assumed that they too may count for urban areas earning SSc2, but not for other projects. By *raised planters*, the author is referring to decorative planters that are substantially raised from the ground surface, not curbed landscape areas. Curbed landscape areas should count in all cases.

One of the suggested strategies is to stack the building program. This refers to the architectural decision to try to build up, with facilities stacked upon each other, versus building in a more sprawled outward design. Obviously, this would help reduce the building and thus the development footprint.

The requirements to fulfill SSc5.2 for the three options can be related to equations which are based on the land area definitions as listed in the beginning of this chapter and the three additional definitions as listed here:

OSZONE   Oszone represents the percent of the total area required to be open space by local zoning code, if such a requirement exists.

VR   VR is the vegetated roof area (see also SSc7.2).

PED25LS   PED25LS represents the special pedestrian areas with a minimum of 25 percent landscaping.

Then each option has the following sets of equations:

Option 1   One credit is earned on most sites if

$$OS \geq 1.25(OSZONE/100) \times A_T \tag{2.5.5}$$

Or, on urban sites where SSc2 is also earned,

$$OS + VR + PED25LS \geq 1.25(OSZONE/100) \times A_T \tag{2.5.6}$$

Option 2   One credit is earned on most sites if

$$OS \geq BF \text{ and is contiguous to it} \tag{2.5.7}$$

Or, on urban sites where SSc2 is also earned,

$$OS + VR + PED25LS \geq BF \tag{2.5.8}$$

Option 3   One credit is earned on most sites if

$$OS \geq 0.2A_T \tag{2.5.9}$$

Or, on urban sites where SSc2 is also earned,

$$OS + VR + PED25LS \geq 0.2A_T \qquad (2.5.10)$$

Special Circumstances and Exemplary Performance

If there are special circumstances or considerations for any of the options, a narrative must be submitted describing these circumstances. A point may be given for exemplary performance in the Innovation in Design category if it can be demonstrated that any of the three option requirements for open space are doubled. This means the following for each option:

- For Option 1, exceeding the local zoning's open-space requirement for the site by 50 percent
- For Option 2, providing vegetated open-space areas adjacent to the building that are equal to double the building footprint when the project has no local zoning requirement
- For Option 3, providing vegetated open space equal to 40 percent of the project site area when no open space (zero) is required by local zoning code

For the exemplary performance point, each option has the following sets of equations which are also summarized in Table 2.5.2 along with the regular credit criteria equations.

Option 1 Exemplary Performance   One credit is earned on most sites if

$$OS \geq 1.5(OSZONE/100)A_T \qquad (2.5.11)$$

or, on urban sites where SSc2 is also earned,

$$OS + VR + PED25LS \geq 1.5(OSZONE/100)A_T \qquad (2.5.12)$$

Criteria for	Condition	Most Sites	Urban Sites that also Earn SSc2
Option 1 credit point	OSZONE in Zoning Code	$OS \geq 1.25(OSZONE/100) A_T$	$OS + VR + PED25LS \geq 1.25(OSZONE/100) A_T$
Option 1 EP point	OSZONE in Zoning Code	$OS \geq 1.5(OSZONE/100) A_T$	$OS + VR + PED25LS \geq 1.5(OSZONE/100) A_T$
Option 2 credit point	No Zoning Code	$OS \geq BF$	$OS + VR + PED25LS \geq BF$
Option 2 EP point	No Zoning Code	$OS \geq 2BF$	$OS + VR + PED25LS \geq 2BF$
Option 2 credit point	OSZONE not in Zoning Code	$OS \geq 0.2 A_T$	$OS + VR + PED25LS \geq 0.2 A_T$
Option 2 EP point	OSZONE not in Zoning Code	$OS \geq 0.4 A_T$	$OS + VR + PED25LS \geq 0.4 A_T$

TABLE 2.5.2   Summary of Open-Space Criteria Equations

Option 2 Exemplary Performance   One credit is earned on most sites if

$$OS \geq 2BF \text{ and is contiguous to it} \tag{2.5.13}$$

or, on urban sites where SSc2 is also earned.

$$OS + V.R + PED25LS \geq 2BF \tag{2.5.14}$$

Option 3 Exemplary Performance   One credit is earned on most sites if

$$OS \geq 0.4A_T \tag{2.5.15}$$

or, on urban sites where SSc2 is also earned,

$$OS + VR + PED25LS \geq 0.4A_T \tag{2.5.16}$$

2.6   SS Credit Subcategory 6: *Stormwater Management*

SS Credit 6.1: *Stormwater Management—Quantity Control*

USGBC Rating System
LEED-NC 2.2 lists the Intent, Requirements, and Potential Technologies and Strategies for this credit as follows:

Intent
Limit disruption of natural hydrology by reducing impervious cover, increasing on-site infiltration; and managing stormwater runoff.

Requirements
OPTION 1— EXISTING IMPERVIOUSNESS IS LESS THAN OR EQUAL TO 50%

Implement a stormwater management plan that prevents the post-development peak discharge rate and quantity from exceeding the pre-development peak discharge rate and quantity for the one- and two-year, 24-hour design storms.

OR

Implement a stormwater management plan that protects receiving stream channels from excessive erosion by implementing a stream channel protection strategy and quantity control strategies.

OR

OPTION 2—EXISTING IMPERVIOUSNESS IS GREATER THAN 50%

Implement a stormwater management plan that results in a 25% decrease in the volume of stormwater runoff from the two-year, 24-hour design storm.

Potential Technologies and Strategies
Design the project site to maintain natural stormwater flows by promoting infiltration. Specify vegetated roofs, pervious paving, and other measures to minimize impervious surfaces. Reuse stormwater volumes generated for non-potable uses such as landscape irrigation, toilet and urinal flushing and custodial uses.

Calculations and Considerations

To understand this credit and determine potential technologies and strategies, the first thing that needs to be understood is what the term *imperviousness* means. Ordinarily in land development and watershed analyses, the features on the land may be divided into two characteristic types of surfaces: pervious or impervious surfaces. Impervious surfaces are typically roofed and paved areas where the roof or pavement material does not allow any significant infiltration of water into the earth below. Pervious surfaces are everything else. Frequently, it is required to determine the percent pervious or percent impervious areas for a project. In that case, the percent impervious is simply the land areas covered by characteristically impervious surfaces, divided by the total project site area A_T, given in percent. If A_{imp} is the total land areas with impervious surfaces and %Imp is the percent impervious, then

$$\%\text{Imp} = 100 A_{imp} / A_T \tag{2.6.1}$$

However, many surfaces are more or less pervious than others. For instance, hard-packed clay or a soccer field surface may be less pervious than a sandy beach. In addition, the concern here is more with the percent of rainfall that "runs off," not the percent that infiltrates. There are other mechanisms by which rainfall may exit an area other than infiltration and runoff, including evaporation and evapotranspiration. So even if an asphalt shingle roof may be considered an impervious surface, it does not shed all the rainfall as runoff, since some is absorbed or stored in crannies on the surface and eventually evaporated. Therefore, to estimate runoff totals from a development, a simple system was developed to estimate the fraction of rainfall that runs off different types of surfaces. It is referred to as the *rational method*. It has been used for more than half a century to estimate runoff rates from small sites (usually much less than 100 acres).

Rational Method, Percent Imperviousness, and Percent Impervious Many types of surfaces, based on either the material of the surface or use of the surface and other characteristics such as slope, have been given a typical rational method runoff coefficient C_i. This coefficient represents the fractional percentage of a rainfall volume from a 2- to 10-year frequency storm that is estimated to result in runoff from that particular surface. To estimate the overall runoff from these frequency storms for the site in question, the overall runoff coefficient C is calculated as the land-area-weighted average of the individual area coefficients C_i for each individual land area A_i. This overall runoff coefficient represents the estimated fraction of rainfall that will run off a site for a typical 2- to 10-year frequency storm event.

$$C = \frac{\sum C_i A_i}{\sum A_i} \tag{2.6.2}$$

There are many state and local agencies such as the North Carolina Department of Environment and Natural Resources that are also adopting the concept of percent imperviousness along with the variable percent impervious. In version 2.1 of LEED-NC, the definition of imperviousness is given as the rational method coefficient in percent, and the adopting agencies have similarly based definitions. Therefore, percent imperviousness (%Impness) can usually be calculated as

$$\%\text{Impness} = 100 C \tag{2.6.3}$$

To determine the percent imperviousness, the following steps should be taken:

- Identify the different surface types on the site.
- Calculate the total area for each of these surface types A_i, using the site drawing.
- Use Table 2.6.1 or other sources of the rational runoff coefficient such as textbooks or manufacturers' information to assign the individual runoff coefficient C_i for each type of surface.
- Finally use Eqs. (2.6.2) and (2.6.3) to determine the percent imperviousness.

The values in Table 2.6.1 have been taken from LEED-NC 2.1 with the pervious pavement grids approximated from values given there, and the pervious concrete value as determined in a laboratory study at the University of South Carolina (Valavala et al., 2006). Any surface that totally captures the rainfall and prevents it from runoff such as a retention basin or area that has the rainfall directed into a drywell or other infiltration system also has a runoff coefficient of zero unless the volume of the structure is exceeded.

Surface Type	Typical Runoff Coefficient
Pavement, asphalt	0.95
Pavement, concrete	0.95
Pavement, brick	0.85
Pervious pavement, plastic grid with grass	0.20
Pervious pavement, concrete grid with grass	0.60
Pervious pavement, concrete paver grid with gravel	0.85
Pervious concrete (0–1% slope)	~0
Turf, flat (0–1% slope)	0.25
Turf, average (1–3% slope)	0.35
Turf, hilly (3–10% slope)	0.40
Turf, steep (>10% slope)	0.45
Roofs, conventional	0.95
Roof, garden roof (<4-in substrate)	0.50
Roof, garden roof (4- to 8-in substrate)	0.30
Roof, garden roof (9- to 20-in substrate)	0.20
Roof, garden roof (>20-in substrate)	0.10
Vegetation, flat (0–1% slope)	0.10
Vegetation, average (1–3% slope)	0.20
Vegetation, hilly (3–10% slope)	0.25
Vegetation, steep (>10% slope)	0.30

TABLE 2.6.1 Typical Runoff Coefficients (Two-Year Storms)

Concept	Symbol	Governing Equation	Subsidiary Equation
Percent impervious	%Imp	$\%Imp = 100\, A_{imp}/A_T$	—
Percent imperviousness	%Impness	$\%Impness = 100C$	$C = \dfrac{\sum C_i A_i}{\sum A_i}$

TABLE 2.6.2 Summary of Percent Impervious and Percent Imperviousness Equations

The equations used to determine the two main concepts of percent impervious and percent imperviousness are summarized in Table 2.6.2. Percent impervious is used frequently in other calculations related to green building and sustainable development. Percent imperviousness is used mainly as a concept for this credit, although it is based on the rational method which is used widely in hydrological modeling.

Stormwater Management Plans and Hydrologic Models　When the imperviousness of the existing site is determined, then the new design for the proposed site must meet the listed requirement options in the credit based on whether the existing imperviousness is greater than 50 percent. If it is greater than 50 percent, then Option 2 applies and the credit is obtained only if the new design reduces the existing amount of runoff from the 2-year, 24-h design storm, in total volume only, according to the verbiage in the credit. If the existing site has an imperviousness equal to or less than 50 percent, then either of the requirements in Option 1 applies. Note that the first requirement of Option 1 applies to both the peak rate *and* the total volume of runoff from both the one-year and two-year storms.

Since the two-year event is greater than the one-year event, it might seem reasonable to assume that calculations on the two-year storms are sufficient for Option 1. However, this is not necessarily true. Statistically, one-year storms are more frequent than two-year storms, and a two-year storm tends to have a higher peak rate and a higher total volume than its equivalent duration one-year storm. However, most current stormwater codes and regulations now have requirements on controlling the runoff from 2-year and 10-year storms, but not usually 1-year storms, and this may result in changes to the outflow from the 1-year storm. One of the most common forms of stormwater management is the stormwater detention pond. The rate is usually controlled by collecting the water from the storm in a pond and releasing it through an outlet structure at a rate less than the peak requirement (two-year usually). Sometimes this is referred to as *release rate rules*. Statistically this flow rate under natural or preexisting conditions would occur only once every two years. However, now with the pond and altered hydrological state on the site, there is the opportunity for this rate to be released much more frequently. The pond can fill to similar levels for many of the smaller storms, and the outlet control is only required to keep the outflow less than the two-year rate. There is a chance, based on pond and outlet structure design, that this equivalent peak two-year rate may now occur for many smaller storms and perhaps occur on a frequency an order of magnitude greater than preexisting. In other words, this two-year outflow rate may occur a dozen or more times in two years. This not only makes the downstream impacts much more frequent, but also gives the system less time to stabilize with vegetation in between major storms. Therefore, this credit requires that the controls also restrict the outflow for smaller one-year storms to the one-year rate. There is still a chance that storms with a frequency of less than a year will now have the larger impacts of the one-year storms based on rate, but at least the impacts of potentially increasing the frequency of the two-year have been drastically reduced.

The two restrictions in Options 1 and 2 on postdevelopment runoff can be attained with a combination of many strategies. Surfaces can be chosen with lower runoff coefficients. Similarly options to detain and infiltrate or use the rainfall such as retention basins, rainwater harvesting for irrigation or interior uses, or underground infiltration systems can be used to reduce the postdevelopment runoff. The stormwater control methods should be chosen based on environmental, societal, and economic criteria, in addition to local restrictions or preferences on many of the options. These methods and options are frequently referred to as *best management practices* (BMPs). There is also a relatively new and emerging stormwater management concept referred to as low-impact development (LID) which uses a subset of these BMPs, many for multiple purposes. These are sometimes also referred to as *integrated management practices* (IMPs), and the main goal is to mimic the natural hydrologic cycle as much as possible on the site. Usually, the more the site mimics the natural hydrological cycle, the fewer the opportunities for anomalies that can disrupt downstream flows and cause nonpoint source pollution. Chapter 10 deals with other stormwater management options such as LID that may help eradicate some of the potential problems of designing to only to a few specific storm rates and volumes.

The stormwater runoff calculations for Option 2 and the first part of Option 1 should be performed using accepted hydrological models for estimating runoff. The second part of Option 1 requires that the stream protection measures be described in detail and that similar accepted runoff calculations be performed for the predevelopment and postdevelopment runoff rates and quantities. It must be demonstrated that the postdevelopment runoff rates and quantities will be below critical values for the receiving waterways after the measures are in place. No specific frequency storm is given for this requirement. It is assumed that at a minimum, both a one-year and a two-year frequency storm should be used. Again, more details on methods can be found in Chap. 10.

Stream Channel Protection   Many stream protection strategies are now using a form of control at the headwaters instead of protection within the channel. There are two main ways to control at the headwaters: either by controlling the hydrology upstream or by modifying the hydraulics at the stream headwaters.

Controlling the hydrology upstream is similar to other on-site stormwater BMPs. Usually retention and infiltration is the preferred strategy. An example is the incorporation of infiltration gardens on projects (bioretention without a subdrain) which may significantly reduce the cumulative runoff based on typical two-year storm-current regulated detention requirements. A good reference for this is the North Carolina Department of Environment and Natural Resources, Division of Water Quality, *Stormwater Best Management Practices Manual*, July 2007.

Controlling the hydraulics at the headwaters is usually by some form of energy dissipation at the culvert outlets. Other than increasing the imperviousness of the watershed, eroding headwater tributaries caused by culvert scour have typically been an observed contributor to the initiation of stream departures from stable stream equilibrium. Energy dissipation techniques are varied and include such items as channels for level spreaders, plunge pools, and rip rap aprons.

Special Considerations and Exemplary Performance

There is no EP point available as related to this credit.

SS Credit 6.2: *Stormwater Management—Quality Control*

USGBC Rating System

LEED-NC 2.2 lists the Intent, Requirements, and Potential Technologies and Strategies for this credit as follows:

Intent

Reduce or eliminate water pollution by reducing impervious cover, increasing on-site infiltration, eliminating sources of contaminants, and removing pollutants from stormwater runoff.

Requirements

Implement a stormwater management plan that reduces impervious cover, promotes infiltration, and captures and treats the stormwater runoff from 90% of the average annual rainfall using acceptable best management practices (BMPs).

BMPs used to treat runoff must be capable of removing 80 percent of the average annual post-development total suspended solids (TSS) load based on existing monitoring reports. BMPs are considered to meet these criteria if:

- they are designed in accordance with standards and specifications from a state or local program that has adopted these performance standards,

or

- there exists in-field performance monitoring data demonstrating compliance with the criteria. Data must conform to accepted protocol (e.g., Technology Acceptance Reciprocity Partnership (TARP), Washington State Department of Ecology) for BMP monitoring.

Potential Technologies and Strategies

Use alternative surfaces (e.g., vegetated roofs, pervious pavement or grid pavers) and nonstructural techniques (e.g., rain gardens, vegetated swales, disconnected imperviousness, rainwater recycling) to reduce imperviousness and promote infiltration, thereby reducing pollutant loadings.

Use sustainable design strategies (e.g., Low Impact Development, Environmentally Sensitive Designs) to design integrated natural and mechanical treatment systems such as constructed wetlands, vegetated filters, and open channels to treat stormwater runoff.

Calculations and Considerations

In LEED-NC 2.2 the requirements for this credit have been simplified, are easily understood, and are different for different climate zones in the United States. The climate zone criteria for this differentiation are based on the average annual rainfall rates, and the zones have been divided into three watershed categories: humid, semiarid, and arid. By the LEED-NC 2.2 definition, *humid* watersheds receive on average at least 40 in of rainfall per year, *semiarid* watersheds receive more than 20 and less than 40 in, and *arid* zones receive less than 20 in of rain annually.

The volume criteria for treating runoff for quality control for this credit are based on runoff from 90 percent of the average annual rainfall. For the purposes of this credit, 90 percent of the average annual rainfall may be estimated in a specific precipitation event as up to 1 in of rainfall for humid areas, 0.75 in of rainfall for semiarid areas, and 0.5 in of rainfall for arid zones. For all events less than the maximum volume criteria as so specified, all the associated runoff would be treated prior to discharge from the site; and for all storm events greater than the maximum specified, the runoff associated with this maximum criterion would need to be treated.

As far as the treatment requirement goes, it may be assumed according to this credit that all rain infiltrated on-site is 100 percent treated and any runoff that is discharged from the site from the specified maximum rainfall amount *must* be treated to attain an *80 percent TSS* removal. The removal method must be one that is either designed by acceptable state or local BMP performance standards or supported by appropriate in-field monitoring data. It is also assumed that any rainwater appropriately reused on-site does not contribute to the TSS load. However, it should be ascertained that the BMPs used adhere to any local and state health and environmental codes. There are many devices, such as rainwater harvesting systems, that have strict health code requirements for items such as length of time in the storage container, minimum treatment required, or differentiation from potable water systems and the "appearance" of potable water systems.

Treating for total phosphorus is no longer a requirement for this credit as it was in earlier versions of LEED. This seems reasonable to the author for a combination of reasons. First, a lot of the phosphorus is absorbed in, or part of, the suspended solids and will be removed by the TSS process. Second, there are usually two main nutrients that may promote eutrophication or other environmental problems in receiving waters: nitrogen and phosphorus. In general, phosphorus is the limiting nutrient for freshwater systems, and nitrogen is the limiting nutrient for saltwater bodies. Therefore, phosphorus is not always considered an important pollutant in many receiving waters. Many watersheds where phosphorus is the limiting nutrient already have discharge criteria limiting the load of phosphorus and are thus already protected.

A summary of the steps for complying with this credit is as follows:

1. Determine whether your project is in a humid, semiarid, or arid region.

2. Determine the minimum volume of water that the system must be designed to handle with a combination of on-site infiltration, appropriate on-site use, or treatment for 80 percent TSS removal prior to discharge.
   a. If in a humid region, multiply the site area by 1 in to determine this minimum volume of runoff that must be infiltrated, used appropriately on-site, or treated prior to discharge.
   b. If in a semiarid region, multiply the site area by 0.75 in to determine this minimum volume of runoff that must be infiltrated, used appropriately on-site, or treated prior to discharge.
   c. If in an arid region, multiply the site area by 0.5 in to determine this minimum volume of runoff that must be infiltrated, used appropriately on-site, or treated prior to discharge.

3. If WQ_{VH}, WQ_{VS}, and WQ_{VA} are defined as the water quality volumes for humid, semiarid, and arid climates, respectively, then the total rainfall which must be infiltrated on-site, used on-site, or treated prior to discharge off-site can be calculated as follows:

$$WQ_{VH} = (0.083 \text{ ft})A_T \qquad (2.6.4)$$

$$WQ_{VS} = (0.063 \text{ ft})A_T \qquad (2.6.5)$$

$$WQ_{VA} = (0.042 \text{ ft})A_T \qquad (2.6.6)$$

	Annual Precipitation (in)	90 percent of Annual Rainfall Estimated per Event (in)	Symbol for Volume to Be "Handled" per Event[*]	Volume to Be "Handled" per Event[*]
Humid	>40	1 over total site	WQ_{VH}	$(0.083 \text{ ft})A_T$
Semiarid	20–40	0.75 over total site	WQ_{VS}	$(0.063 \text{ ft})A_T$
Arid	<20	0.5 over total site	WQ_{VA}	$(0.042 \text{ ft})A_T$

[*]Infiltrated on-site, used on-site, or treated prior to discharge.

TABLE 2.6.3   Regional Precipitation Area Definitions and Water Quality Volumes

These water quality volumes and respective rainfall definitions by area are summarized in Table 2.6.3.

Best management practices (BMPs) are normally separated into two main categories: nonstructural and structural. For the purposes of LEED, nonstructural BMPs include such items as pervious pavements, swales, and disconnection of impervious areas to promote natural infiltration of rainwater into the soils (although others might consider swales and pervious pavements as structural controls). The structural controls include ponds, rainwater harvesting cisterns, constructed wetlands, and treatment devices in the stormwater collection system. Other agencies and organizations segregate the BMPs differently. More information on this topic can be found in Chap. 10.

For any of the methods used that involve infiltration, soil types and soil infiltration rates must be provided to determine if the soils can handle the infiltration. This credit is based on a set volume and not on a peak rate, so a mass balance under dynamic conditions may not be easily shown. However, if a minimum storage volume equal to the volume needed to be treated is maintained within the infiltration system, then there should be time for sufficient infiltration in most soils. If sufficient infiltration is designed to occur as a dynamic combination of rates and storage, then it is suggested by the author that a peak rate of rainwater inflow from both a 1-year and a 2-year 24-h design storm be used for the appropriate mass balances to be consistent with SSc6.1 or a method used that is based on a typical annual storm with the time of concentration as determined for the site. The rainfall rates used should be justified in the calculations and submittals. Likewise, for structural controls, if the volume of the control cannot handle the entire rainfall volume for treatment, then it is suggested by the author that a peak rate of rainwater inflow from both a one-year and a two-year design storm be used for the appropriate dynamic mass balances or a method appropriately related to the site time of concentration.

In all cases, for submittal, a list must be provided of all the BMPs used including a description of each, its pollutant removal capability, and the percent of rainfall that will be handled by the BMP for the site. If there are special circumstances or considerations for this credit, a narrative must be submitted describing these circumstances. Additional information on BMPs and stormwater options can be found in Chap. 10. Figure 2.6.1 shows an example site where constructed wetlands are used for both stormwater quantity and stormwater quality control.

Special Circumstances and Exemplary Performance

There is no EP point available as related to this credit.

FIGURE 2.6.1 Two interconnected constructed wetlands at a West Quad Residential Facility at the University of South Carolina. (*Photograph taken September 2006.*)

2.7 SS Credit Subcategory 7: *Heat Island Effect*

SS Credit 7.1: *Heat Island Effect—Non-Roof*

The urban heat island effect is a phenomenon where daily temperatures are higher in urban areas than in the surrounding suburban and rural areas. It is thought to be caused by many factors some of which are an increase in paved and roofed surfaces that collect warmth from solar radiation, a decrease in foliage which can keep temperatures cooler by evapotranspiration, and an increased energy use in buildings that also warms the ambient air. This effect has many negative impacts including poorer air quality, an increase in energy demand for cooling, an increase in water demand, a reduction in building material durability, and an increase in human health concerns. Therefore, measures and construction practices which reduce the urban heat island effect are encouraged.

USGBC Rating System

LEED-NC 2.2 lists the Intent, Requirements, and Potential Technologies and Strategies for this credit as follows:

Intent

Reduce heat islands (thermal gradient differences between developed and undeveloped areas) to minimize impact on microclimate and human and wildlife habitat.

Requirements

OPTION 1

Provide any combination of the following strategies for 50% of the site hardscape (including roads, sidewalks, courtyards and parking lots):

Shade (within 5 years of occupancy)

Paving materials with a Solar Reflectance Index (SRI) of at least 29

Open grid pavement system

OR

OPTION 2

Place a minimum of 50% of parking spaces under cover (defined as underground, under deck, under roof, or under a building). Any roof used to shade or cover parking must have an SRI of at least 29.

Potential Technologies and Strategies

Shade constructed surfaces on the site with landscape features and utilize high-reflectance materials for hardscape. Consider replacing constructed surfaces (i.e., roofs, roads, sidewalks, etc.) with vegetated surfaces such as vegetated roofs and open grid paving or specify high-albedo materials to reduce the heat absorption.

Calculations and Considerations

Option 1   Option 1 provides for reduced heat impacts from at least 50 percent of the site non-roof hardscape. This includes all parking areas (spaces, aisles, and other paved areas associated with parking areas), walks, roads, patios, and drives. Even water features are included in the hardscape if they are contained over impervious surfaces which can be exposed when dry. HS is defined as the sum of these non-roof hardscape areas. There are some special cases for which it is difficult to determine if a "hard" surface is a roof or non-roof or considered neither. One example is open stadium seating. These special exceptions should be reviewed through the USGBC CIR procedure.

It is very important to understand the criteria for each of the three pathways to do this. They are shaded, high SRI, or open grid pavement systems.

- *Shading.* Shading can be effected with vegetation, or if tree planting is not available, then architectural shading devices may be substituted. Roofed areas do not count in the hardscape calculation. The vegetation used must be native or adaptive trees, large shrubs, and noninvasive vines (on trellises or other supportive structures as needed). Deciduous trees are many times preferred, especially in northern climates, as they allow for benefits from solar heating of the pavement in the winter months when their leaves are gone, such as reduced icing. Variable S is defined as the effective area of the hardscape that is shaded. It is the arithmetic mean of shade coverage as calculated at 10 a.m., noon, and 3 p.m. on the summer solstice. Although not stated in LEED-NC 2.2, the time is probably what is accepted as the local time (daylight or standard) at the location.

- *SRI.* Variable R is defined as the area of the hardscape that has an SRI of at least 29. The definition of the SRI can be found in App. B. It represents a combination of high reflectance and high emissivity. Both emittance and reflectance should be calculated or determined according to the appropriate ASTM (American Society for Testing and Materials) standard as listed in the definition. These values can then be inserted into the LEED submittal template worksheet, and the appropriate SRI will be calculated. Some typical SRI values as given in LEED-NC 2.2 are listed in Table 2.7.1. Colored concretes or surface coatings on pavements can be used to increase the values for this credit. Note that the reflectance value given is solar reflectance. There is also a commonly used parameter called *visible*

Material	Emissivity	Reflectance	SRI
Typical new gray concrete	0.9	0.35	35
Typical weathered* gray concrete	0.9	0.20	19
Typical new white concrete	0.9	0.70	86
Typical weathered* white concrete	0.9	0.40	45
New asphalt	0.9	0.05	0
Weathered asphalt	0.9	0.10	6

*Reflectance of surfaces can be maintained with cleaning. Typical pressure washing of cementitious materials can restore reflectance close to the original value. Weathered values are based on no cleaning.

TABLE 2.7.1   Some Typical SRI Values of Pavements

reflectance. The solar reflectance represents a wider range of wavelengths than the visible reflectance and is often slightly less than the visible reflectance for a specific material. It is important to make sure that the reflectance quoted by a manufacturer is the solar reflectance. If not, then the solar reflectance must be gotten elsewhere. Figures 2.7.1 and 2.7.2 show example photos of weathered concrete and weathered asphalt pavements, and weathered and new concrete pavement surfaces, respectively.

- *Open grid pavement systems.* Variable O is defined as the area of the hardscape which is covered in open grid pavement systems which are at least 50 percent pervious with vegetation in the open cells.

FIGURE 2.7.1   Example of weathered concrete and asphalt pavement surfaces in Columbia, SC. The pavements are approximately 15 years old.

FIGURE 2.7.2   Example of weathered and new concrete pavement surfaces in Columbia, S.C. The weathered surface is approximately 15 years old, and the new surface is less than a month old.

The three pathways for reducing the heat island effect by keeping at least 50 percent of the non-roof hardscape as shaded, high SRI, or open grid cannot be superimposed on one another. In other words, an area that is both shaded and high SRI cannot be counted twice. There is some discretion with the shading value as it is an arithmetic mean of various areas determined by the angle of the sun, and if there is any question as to whether the appropriate areas are used, then these should be well documented. The easiest way to perform the calculations is to give modified definitions for R and O such that

R_{mod} R_{mod} is the sum of the areas with high SRI excluding those areas that are shaded.

O_{mod} O_{mod} is the sum of the areas with open grid pavement systems that are not shaded or do not have high SRI values.

By using these definitions, compliance with Option 1 can be determined if the following inequality is true:

$$HS/2 \le (S + R_{mod} + O_{mod}) \tag{2.7.1}$$

Option 2   The calculations for Option 2 are simpler. First determine the total number of parking spaces (TP) for the project and then count the number of these total spaces that are under cover (CP) for which the top roofing material above them has an SRI of at least 29. Under cover includes underground, under the building, and under shade structures such as canopies. If more than one-half of the parking spaces are under cover as so defined, then the credit is met. Note that this option only refers to the parking spaces themselves and does not mention anything about the parking aisles or other paved areas associated with these spaces. The equation for compliance with Option 2 is

$$CP \geq TP/2 \qquad\qquad (2.7.2)$$

Special Circumstances and Exemplary Performance

Submittals for both options should include the calculations for each option. In addition, for Option 1 the submittals should include site drawings showing all the non-roof hardscape areas with their heat island–related characteristics (shading, SRI, or open grid) and areas noted. The surface SRIs given may either be the actual for each paving material or the default from Table 2.7.1. For Option 2, the submittals should include plans with the number of total parking spaces and parking spaces undercover with the appropriate roofed SRI. Option 2 is a great example of how design team members need to work together to obtain the LEED credits. The parking layout may sometimes be determined by others than the architectural roofing designer. Only together will this option be met.

Concrete is a common material used for many surface treatments and other structural components of the built facilities. Many concretes are made with a combination of Portland cement and fly ash (up to 25 percent). This use of fly ash is promoted as an environmental benefit as it recycles waste from another process which might otherwise be landfilled, and it also promotes a reduction in the use of energy and the production of carbon dioxide from the cement-making process. However, concrete with fly ash is usually darker than concrete made with only Portland cement. Therefore, it may be beneficial to use concrete for surface treatments that either have no fly ash or are coated to lighten the surface color and promote reflectivity, whereas it may be good to use concrete with fly ash for other site applications to aid in earning other credits.

If there are special circumstances or considerations for this credit, a narrative must be submitted describing these circumstances.

An EP point may be awarded if it can be demonstrated that either of the options is met with a 100 percent requirement. For Option 1 this means that *all* the non-roof hardscape surfaces have been constructed with high-SRI materials, shaded within 5 years, or open grid paving systems as previously defined. For Option 2 this means that *all* on-site parking spaces are located under cover with high SRI values on the roofs above them.

A summary of the equations for credit compliance and awarding of an EP point for either option is presented in Table 2.7.2.

Criteria for	Option 1	Option 2
Credit point	$HS/2 \leq (S + R_{mod} + O_{mod})$ (At least one-half the hardscape is specially adapted to reduce the heat island effect.)	$CP \geq TP/2$ (At least one-half of the parking spaces are covered with specially adapted roofs.)
EP point	$HS = (S + R_{mod} + O_{mod})$ (All the hardscape is specially adapted to reduce the heat island effect.)	$CP = TP$ (All the parking spaces are covered with specially adapted roofs.)

TABLE 2.7.2   Summary Criteria Equations for Heat Island Effect: Non-Roof

SS Credit 7.2: *Heat Island Effect—Roof*

USGBC Rating System

LEED-NC 2.2 lists the Intent, Requirements, and Potential Technologies and Strategies for this credit as follows:

Intent

Reduce heat islands (thermal gradient differences between developed and undeveloped areas) to minimize impact on microclimate and human and wildlife habitat.

Requirements

OPTION 1

Use roofing materials having a Solar Reflectance Index (SRI) equal to or greater than:

- 78 for low-sloped roofs (slope ≤ 2V:12H)
- 29 for steep-sloped roofs (slope > 2V:12H)

for a minimum of 75% of the roof surface.

OR

OPTION 2

Install a vegetated roof for at least 50% of the roof area.

OR

OPTION 3

Install high albedo and vegetated roof surfaces that, in combination meet the following criteria:

$$(\text{Area of SRI Roof}/0.75) + (\text{Area of Vegetated Roof}/0.5) \geq \text{Total Roof Area}.$$

Potential Technologies and Strategies

Consider installing high-albedo and vegetated roofs to reduce heat absorption. SRI is calculated according to ASTM E 1980. Reflectance is measured according to ASTM E 903, ASTM E 1918, or ASTM C 1549. Emittance is measured according to ASTM E 408 or ASTM C 1371. Default values will be available in the LEED-NC v2.2 Reference Guide. Product information is available from the Cool Roof Rating Council website, at www.coolroofs.org.

Calculations and Considerations

Note that this credit references solar reflectance which is different from visible reflectance, as solar reflectance covers a wider range of wavelengths than visible reflectance does. Therefore, any reflectance information from manufacturers should be checked to ensure that the proper reflectance is used in the SRI calculation. Table 2.7.3 gives some example SRIs for various roofing materials which have been taken from the referenced table in the LEED-NC 2.2 Reference Guide. Roofing materials with the higher SRI values are sometimes referred to as *cool* roofs.

Also notice the big difference in the SRI requirement for low-slope versus steep-slope roofs. Flat roofs have a *much* stricter requirement than sloped roofs. Therefore, alternatives with some sloped roof surfaces may be preferable for obtaining the credit. Also note that cool roofs may not be as energy efficient in the colder months due to the impact of high emittance and low absorption on heating costs; however, on average, the

Example SRI Values for Generic Material	Solar Reflectance	Infrared Emittance	Temperature Rise (°F)	SRI
Gray EPDM	0.23	0.87	68	21
Gray asphalt shingle	0.22	0.91	67	22
Unpainted cement tile	0.25	0.90	65	25
White granular surface bitumen	0.26	0.92	63	28
Red clay tile	0.33	0.90	58	36
Light gravel on built-up roof	0.34	0.90	57	37
Aluminum coating	0.61	0.25	48	50
White-coated gravel on built-up roof	0.65	0.90	28	79
White coating on metal roof	0.67	0.85	28	82
White EPDM	0.69	0.87	25	84
White cement tile	0.73	0.90	21	90
White coating—1 coat, 8 mils	0.80	0.91	14	100
PVC white	0.83	0.92	11	104
White coating—2 coats, 20 mils	0.85	0.91	9	107

Source: LBNL Cool Roofing Materials Database (http://eetd,lbl.gov/CoolRoofs). These values are for reference only and are not for use as substitutes for actual manufacturer data.

TABLE 2.7.3   Some Typical SRI Values for Roofing Materials

annual overall energy use may still be reduced. In addition, structures in very northern climates tend to be highly insulated to keep warmth in during the summer and may not have any air conditioning for summer months. Thus darker roofs in the summer would tend to overheat the buildings.

The equation for Option 3 is actually applicable for all three options. It is the only one necessary to use as the limits of this equation for not using any of the appropriately rated SRI roofing material in Option 1 default to the requirement in Option 2; and conversely, not having any vegetated roof area defaults to the requirement in Option 1. Therefore, the following write-up will only address the equation covering Option 3.

The first definition that must be understood is what is meant by *roof area*. As in land areas used in all the other credit calculations, roof areas are assumed to be the horizontal planar surface as would be projected on a site plan, not the actual slanted surface. [This horizontal planar surface is the building footprint (BF) as previously defined.] Therefore a slanted shingled roof with an area as defined in this credit of 100 ft² would actually need well more than 100 squares of shingles to be covered. In addition, any roofed areas that hold equipment, solar energy panels, or other appurtenances (EQR as previously defined) are not included in the calculations. Using the site plan, the total roof surface area of the project (TR) would equal the sum of the horizontal projections of the roofs on the site plan (BF) less the sum of the roofed areas that hold equipment, solar energy panels, or other appurtenances (EQR). That is,

$$TR = BF - EQR \qquad (2.7.3)$$

In addition, the following variables should be determined from the associated areas on the site plan:

VR   Sum of the vegetated roofed areas

RRL   Sum of the low-slope roofed areas with an *areal weighted average* SRI greater than 78

RRS   Sum of the steep-slope roofed areas with an *areal weighted average* SRI greater than 29

Note that for either type of roofed surfaces, it is not the sum of the areas with the minimum SRI, but rather a sum of areas whose areal weighted average SRI meets the minimum criteria. For example, for two steep-sloped roof areas, each of equal area, one with an SRI of 28 and the other with an SRI of 30, both areas could be counted with an average SRI of 29.

Using these definitions, the project would be eligible for this credit if the following inequality held true.

$$TR \le \frac{RRL + RRS}{0.75} + \frac{VR}{0.5} \qquad (2.7.4)$$

The first half of the right side of Eq. (2.7.3) corresponds to the requirements of Option 1, and the second part of the right side of Eq. (2.7.3) corresponds to the requirements for Option 2.

Special Circumstances and Exemplary Performance

There are some special cases for which it is difficult to determine if a hard surface is a roof or non-roof or considered neither. One example is open stadium seating. These special exceptions should be reviewed through the USGBC CIR procedure. If there are other special circumstances or considerations for this credit, a narrative must be submitted describing these circumstances.

An EP point may be awarded if it can be demonstrated that Option 2 is met with a 100 percent vegetated requirement. This means that *all* the roof areas excluding mechanical equipment, solar panels, skylights, and other similar appurtenances comprise green (vegetated) roof systems.

$$VR = TR \qquad (2.7.5)$$

A summary of the criteria equations for SSc7.2 is given in Table 2.7.4.

Many websites are being developed that give alternatives for green roofs. In addition, ASTM International is developing an international standard, WK14283, "Guide for Green Roof Systems," which will aid in understanding concepts and

Criteria for	Option 1	Option 2	Option 3
Credit point	Given TR = BF − EQR $$TR \leq \frac{RRL + RRS}{0.75}$$ (At least 75% of roofed areas meet set SRI criteria.)	Given TR = BF − EQR $$TR \leq \frac{VR}{0.5}$$ (At least one-half of the roofed areas are vegetated.)	Given TR = BF − EQR, $$TR \leq \frac{RRL + RRS}{0.75} + \frac{VR}{0.5}$$ (Combination of Options 1 and 2)
EP point	n/a	Given TR = BF − EQR VR = TR (All roofs are vegetated.)	n/a

TABLE 2.7.4   Summary Criteria Equations for Heat Island Effect: Roof

terminology in green roof design. Figure 2.7.3 shows the front of a building which has a partially vegetated roof. The roof is used for outdoor receptions as well as a means to control the urban heat island effect.

Typically green roofs are divided into three categories: intensive, semi-intensive, and extensive. Extensive green roofs usually have a small substrate depth (3 to 6 in) for the vegetation to grow in and typically are planted with grasses and sedums which require no irrigation. These roofs do not require high load-bearing structures for support. Intensive green roofs, on the other hand, usually require high load-bearing support structures, can include many types of vegetation such as trees, and have a much deeper

FIGURE 2.7.3   Front of the Learning Center at the West Quad at the University of South Carolina. The roof is partially vegetated and has walkways and architectural shading structures for pedestrian access and use.

FIGURE 2.7.4   Experiment to investigate species that grow well without irrigation on a roof in central South Carolina.

substrate (soil) depth, usually more than a foot. Semi-intensive roofs are somewhere in between. Figure 2.7.4 shows an experiment on plants to test for species that might grow best without irrigation on an extensive roof in central South Carolina.

2.8 SS Credit 8: *Light Pollution Reduction*

USGBC Rating System
LEED-NC 2.2 lists the Intent, Requirements, and Potential Technologies and Strategies for this credit as follows:

Intent
Minimize light trespass from the building and site, reduce sky-glow to increase night sky access, improve nighttime visibility through glare reduction, and reduce development impact on nocturnal environments.

Requirements
FOR INTERIOR LIGHT

The angle of maximum candela from each interior luminaire as located in the building shall intersect opaque building interior surfaces and not exit out through the windows.

OR

All non-emergency interior light shall be automatically controlled to turn off during non-business hours. Provide manual override capability for after hours use.

AND

FOR EXTERIOR LIGHTING

Only light areas as required for safety and comfort. Do not exceed 80% of the lighting power densities for exterior areas and 50% for building facades and landscape features as defined in ASHRAE/IESNA Standard 90.1-2004, Exterior Lighting Section, without amendments.

(And)

All projects shall be classified under one of the following zones, as defined in IESNA RP-33, and shall follow all of the requirements as listed for that specific zone:

LZ1 – Dark (Park and Rural Settings)

Design exterior lighting so that all site and building mounted luminaires produce a maximum initial illuminance value no greater than 0.01 horizontal and vertical footcandles at the site boundary and beyond. Document that 0% of the total initial designed fixture lumens are emitted at an angle of 90 degrees or higher from nadir (straight down).

LZ2 – Low (Residential Areas)

Design exterior lighting so that all site and building mounted luminaires produce a maximum initial illuminance value no greater than 0.10 horizontal and vertical footcandles at the site boundary and no greater than 0.01 horizontal footcandles 10 feet beyond the site boundary. Document that no more than 2% of the total initial designed fixture lumens are emitted at an angle of 90 degrees or higher from nadir (straight down). For site boundaries that abut public rights-of-way, light trespass requirements may be met relative to the curb line instead of the site boundary.

LZ3 – Medium (Commercial/Industrial, High-Density Residential)

Design exterior light so that all site and building mounted luminaires produce a maximum initial illuminance value no great than 0.20 horizontal and vertical footcandles at the site boundary and no greater than 0.01 horizontal footcandles 15 feet beyond the site. Document that no more than 5% of the total initial designed fixture lumens are emitted at an angle of 90 degrees or higher from nadir (straight down). For site boundaries that abut public rights-of-way, light trespass requirements may be met relative to the curb line instead of the site boundary.

LZ4 – High (Major City Centers, Entertainment Districts)

Design exterior lighting so that all site and building mounted luminaires produce a maximum initial illuminance value no greater than 0.60 horizontal and vertical footcandles at the site boundary and no greater than 0.01 horizontal footcandles 15 feet beyond the site. Document that no more than 10% of the total initial designed site lumens are emitted at an angle of 90 degrees or higher from nadir (straight down). For site boundaries that abut public rights-of-way, light trespass requirements may be met relative to the curb line instead of the site boundary.

Potential Technologies and Strategies

Adopt site lighting criteria to maintain safe light levels while avoiding off-site lighting and night sky pollution. Minimize site lighting where possible and model the site lighting using a computer model. Technologies to reduce light pollution include full cutoff luminaires, low-reflectance surfaces and low-angle spotlights.

Calculations and Considerations

To obtain a point for this credit, one of the two interior lighting requirement options and *all* the applicable exterior lighting requirements must be met. The interior lighting requirement options are the same regardless of the use of the site or the location of the site. The exterior lighting requirements vary depending upon site use, use density, and location. The exterior lighting requirements are divided into two main portions: the first limits the exterior lighting density, and the second limits the light distribution, particularly beyond the site, either over the property lines or upward into the heavens. Note that in all cases, it is important to try to interpret between existing standards and design for the lowest possible light levels while addressing safety, security, access, way finding, identification, and aesthetics.

Interior Lighting Requirement   This requirement is very simple and can be met in two different ways. Interior lights should not shine out directly through windows, or if they do, then all interior lighting other than emergency lighting shall be on automatic controls (with manual overrides) to be turned off during nonbusiness hours. There are many building uses, such as retail businesses, where the second option might pose a potential security risk, so the first option may be preferred.

If the first route is used, then to verify that the first option for interior lighting is obtained, it is necessary to get illumination information from the manufacturer on all interior fixtures which may light any windows in the building. This information should include a lighting intensity photometric table or figure which shows the angle of maximum lighting intensity. Interior building plans should be provided with the location of each fixture provided and the angle of maximum candela traced from this fixture until it proves that it intersects an opaque wall and not a window in all exterior directions.

Exterior Lighting Requirements   There are two main parts, each with several criteria, to the exterior lighting requirements that *must* be met. The first main part of the exterior lighting requirements is based upon the ASHRAE/IESNA Standard 90.1-2004, Energy Standard for Buildings Except Low-Rise Residential—Lighting Section 9, Exterior Lighting Section, Table 9.4.5. Note that this standard does not apply to any low-rise residential uses including single-family homes. Therefore, this part is subject to interpretation for low-rise residential facilities (three habitable stories or fewer above grade); nor does this part apply to manufactured houses and buildings that do not use either electricity or fossil fuel. In addition, this does not apply to equipment or portions of building systems that use energy primarily for industrial, manufacturing, or commercial processes. Any project with these items should be examined individually by way of the USGBC CIR process and/or guidances as are being developed.

The second main part of the exterior lighting requirements limits the light distribution, particularly beyond the site, and differs according to the location of the site and the typical zoning uses in the neighborhood.

Exterior Lighting Part 1:   For all construction other than low-rise residential uses and as further exempted, the following percent of the ASHRAE/IESNA Standard 90.1-2004, Exterior Lighting Section, without amendments, Table 9.4.5 lighting power densities (LPDs) shall not be exceeded:

- 50 percent for building facades
- 50 percent for landscape features
- 80 percent for all other exterior areas

The criteria in Table 9.4.5 of the ASHRAE/IESNA Standard 90.1 are given in units of watts per square foot, watts per lineal foot, or watts per unit relating to a specific type of site feature, and are all referred to as *lighting power densities* (LPDs). The criteria are divided into two categories: those that can be traded (flexibility on the areas and exact light locations) and those that cannot be traded. Those that can be traded apply to outside areas in general. The nontradable allowances are in addition to the tradable allowances. For instance, if there is an uncovered parking lot next to a building façade, then this parking lot is allowed the parking lot allowance (80 percent of the value for the Uncovered Parking Lot LPD as given in Table 9.4.5 of the ASHRAE/IESNA Standard 90.1), plus the special building facade light fixtures (totaling, at maximum,

50 percent of the Building Facades LPD as given in Table 9.4.5 of the ASHRAE/IESNA Standard 90.1).

Examples of these LPDs from the ASHRAE/IESNA Standard 90.1 are given in Table 2.8.1. All the wattages for the fixtures in each area should be added up and divided by the appropriate unit: square feet, lineal feet, or each specific item (i.e., drive-up window),

	Application	Lighting Power Densities (LPDs)
Tradable surfaces	Uncovered parking areas	
	- Parking lots and drives	0.15 W/ft^2
	Building grounds	
	- Walkways less than 10 ft wide	1.0 W/lin ft
	- Walkways 10 ft or greater	0.2 W/ft^2
	- Plaza areas	0.2 W/ft^2
	- Special feature areas	0.2 W/ft^2
	- Stairways	1.0 W/ft^2
	Building entrances and exits	
	- Main entries	30 W/lin ft of door width
	- Other doors	20 W/lin ft of door width
	Canopies and overhangs	
	- Canopies (freestanding and attached and overhangs)	1.25 W/ft^2
	Outdoor sales	
	- Open areas (including vehicle sales lots)	0.5 W/ft^2
	- Street frontage for vehicle sales lots in addition to open-area allowance	20 W/lin ft
Nontradable surfaces These are special allowances that are in addition to the allowances in the tradable section.	Building facades	0.2 W/ft^2 or 5.0 W/lin ft
	ATMs and night depositories	270 W per location plus 90 W per additional ATM
	Guarded gatehouses or entrances	1.25 W/ft^2 of uncovered areas (covered areas are included in canopies)
	Loading areas for police, fire, EMS, or other emergency vehicle	0.5 W/ft^2 of uncovered areas (covered areas are included in canopies)
	Drive-up windows at fast food restaurants	400 W per drive-through
	Parking near 24-h retail entrances	800 W per main entry

Source: ASHRAE/IESNA Std 90.1-2004, Table 9.4.5, Energy Standard for Buildings Except Low-Rise Residential Buildings, Chapter 9: Lighting, ©American Society of Heating, Refrigerating and Air-Conditioning Engineers, Inc.

TABLE 2.8.1   Lighting Power Densities

and then calculations should confirm that each does not exceed either the 50 percent or 80 percent requirement of the values in Table 2.8.1. Again, there are some exemptions listed. Do not include solar power lighting or other lighting that does not use electricity or fossil fuels, and this does not apply to equipment or portions of building systems that use energy primarily for industrial, manufacturing, or commercial processes. Or at least, this is what the author has interpreted from the LEED-NC 2.2 Reference Guide.

Exterior Lighting Part 2: This requirement is in addition to one of the interior lighting requirement options and the exterior lighting requirement part 1, previously described. It is intended to limit the lighting distribution beyond the site, both horizontally and vertically. Since different uses have different lighting needs, the project site will have different light distribution requirements according to where it is and how it is used. There are four applicable groupings, and they are referred to as the *outdoor lighting zones* as given in IESNA RP-33 and as defined by the International Dark-Sky Association (IDA) and are given in Table 2.8.2.

The IDA is the International Dark-Sky Association and can be found at www.darksky.org. The Illuminating Engineering Society of North America (IESNA) *Recommended Practice Manual: Lighting for Exterior Environments* (RP-33) is the referenced standard for the

Zone	Ambient Illumination	Typical Examples
LZ0	Very dark	"Critical dark environments, such as especially sensitive wildlife preserves, parks, and major astronomical observatories"
LZ1	Dark	"Developed areas in state and national parks, recreation areas, wetlands and wildlife preserves; developed areas in natural settings; areas near astronomical observatories; sensitive night environments; zoos; areas where residents have expressed the desire to conserve natural illumination levels"
LZ2	Low	"Rural areas, low-density urban neighborhoods and districts, residential historic districts. This zone is intended to be the default condition if a zone has not been established"
LZ3	Medium	"Medium to high-density urban neighborhoods and districts, shopping and commercial districts, industrial parks and districts. This zone is intended to be the default condition for commercial and industrial districts in urban areas"
LZ4	High	"Reserved for very limited applications such as major city centers, urban districts with especially high security requirements, thematic attractions and entertainment districts, regional malls, and major auto sales districts"

TABLE 2.8.2 Outdoor Lighting Zone Definitions from IDA Website (http://www.darksky.org/ordsregs/l-zones.html) as Accessed May 10, 2008

definition of the lighting zones. LEED-NC 2.1 gives the 1999 version as the standard, while LEED-NC 2.2 does not reference the updated year. However, the LEED-NC 2.2 Reference Guide specifically defines the four lighting zones and the definitions as so described from the IDA website are given in Table 2.8.2. Note that Table 2.8.2 also gives examples for very dark lighting zones. These are not really applicable to new construction as they represent undeveloped or specially developed (observatory) areas only.

Each of the four lighting zones has four specific lighting distribution requirements that must be met. These limit the horizontal lighting intensity at the project property line, the vertical lighting intensity above the property line, the maximum amount of lighting intensity at a specific distance from the property edge, and the maximum amount of exterior lighting that can be directed with any angle which may go upward. These four lighting distribution requirements for each of the lighting zones are summarized in Table 2.8.3. Note that for three of the zones, the site property line can be moved out to the curb line on portions of the lot that abut public rights-of-way.

Submittals for this part of the exterior lighting requirement must include site plans that show both the maximum horizontal and vertical illuminances at the site boundaries as well as values of the maximum horizontal illuminances extending out the stated distance beyond the boundary. The horizontal lighting intensity plans are typically referred to as *isoluxes* or *isofootcandle* plans and give either contours or spot values of luxes or footcandles. Vertical lighting distributions at all the property lines (or curb lines) should also be provided. Isoluxes show the additive effects of all the lighting fixtures in the area, and they cannot be based solely on each individual fixture.

The submittal for the fourth item in this part of the exterior lighting requirements needs to consider *all the exterior lighting fixtures on the site.* It is based on the initial designed fixture lighting capacity (lumens) for each, as most lighting fixtures have a reduction in lighting intensity with age. Determination of whether the requirement has been met, can be performed by following these three steps:

Zone	Maximum Initial Horizontal Illuminance Value at Site Boundary (fc)	Maximum Initial Vertical Illuminance Value at Site Boundary (fc)	Number of Feet Beyond Site Boundary That a Maximum Initial Horizontal Illuminance of 0.01 fc Is Allowed	Maximum Allowed Percent of Total Initial Designed Fixture Lumens Emitted at an Angle of 90° or Higher from Nadir (MIDAL)
LZ1-dark	0.01	0.01	0	0
LZ2-low	*0.10	*0.10	10	2
LZ3-medium	*0.20	*0.20	15	5
LZ4-high	*0.60	*0.60	15	10

*For site boundaries that abut public rights-of-way, light trespass requirements may be met relative to the curb line instead of the site boundary.

TABLE 2.8.3 Outdoor Lighting Zone Specific Exterior Lighting Requirements

- The total initial designed lumens for all the exterior site fixtures should be added. This will be referred to as the *total initial designed lumens* (TIDL).

- In addition, any of the fixtures that have any light aimed at an angle which breaks above a horizontal plane with the fixture should have the associated lumens that shine in this direction summed over the site. This will be referred to as the *total initial designed angled lumens* (TIDAL).

- If the maximum allowed percent of total initial designed fixture lumens emitted at an angle of 90 degrees or higher from nadir is referred to as the **maximum initial designed angled lumens** (MIDAL), then this requirement can be met if Eq. (2.8.1) is valid. The applicable values for MIDAL for each zone are given in Table 2.8.3. The exception to Eq. (2.8.1) is in cases where no exterior lighting is installed; then the requirement is automatically met.

$$100(\text{TIDAL}/\text{TIDL}) \le \text{MIDAL} \qquad (2.8.1)$$

A strategy for meeting this fourth item in part 2 of the exterior lighting requirements may be to specify and install appropriate shielding and cutoffs on the exterior fixtures.

Special Circumstances and Exemplary Performance

Figure 2.8.1 gives a summary of the interior and exterior requirements necessary for fulfilling this credit.

If there are special circumstances or considerations for this credit, a narrative must be submitted describing these circumstances. There is no EP point available as related to this credit.

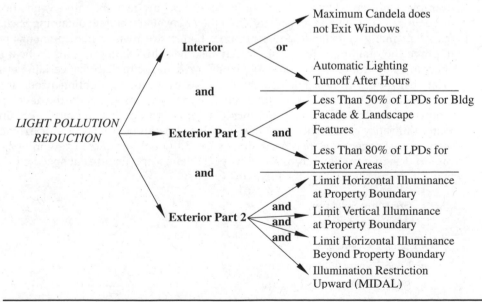

FIGURE 2.8.1 Summary of light pollution reduction credit.

2.9 Discussion and Overview

Sustainable Site credits may sometimes be mutually exclusive, and some of them may never be applicable to a project or site. For example, if none of the available sites are brownfields, then there is no possibility of obtaining SS Credit 3. However, some seemingly opposing credits may actually be achievable for the same site. For instance, SS Credit 1 encourages spaciousness by keeping certain special areas undeveloped while SS Credit 2 encourages denseness for more efficient use of infrastructure. However, many of the urban areas applicable for SS Credit 2 already meet the six criteria in SS Credit 1 as the area has already been developed and these special resources no longer, or never did, exist. What may be more difficult is to find a situation where both SS Credits 2 and 5 can be met. It may seem very difficult to promote dense urban development and yet create open space, but it is not impossible, especially when the parameters of the project allow for multistory structures. In addition, both SSc5.1 and SSc5.2 have modified requirements in dense urban environments to encourage open space but in a different manner as may be more readily doable in this environment.

It would be very difficult to meet SSc2 in the United States for a traditional retail grocery store, as U.S. customers usually use large shopping carts throughout the structure for shopping for their foodstuffs which are not made for stairwells and escalators. However, in other countries such as Switzerland and Norway, there are multistory food stores in urban areas, some with elevators for the transport of the carts. Although this practice is an unlikely one in the current culture of the United States, a retail food store could be very easily built with other businesses or offices on higher stories and the credit be attained. Why would we want to obtain both of these types of credits (SSc2 and SSc5) in dense urban areas? Dense urban areas and open space are not mutually exclusive, and the combination of both provides for access to infrastructure and a better environment for the human inhabitants.

Some of the credits should be evaluated together with respect to cost for the credit, as many of the design changes may facilitate obtaining other credits. A very good example is presented by the SSc5 credits which encourage more native vegetation and more open space. If these are sought, then they may also facilitate obtaining SSc6.1 and SSc6.2 as many vegetated spaces promote better stormwater management and can improve the quality of the runoff. It may also help in keeping the urban heat island effect to a minimum. If these specially landscaped areas are placed at certain perimeter locations, they can also aid in obtaining SSc8 by reducing lighting needs in these locations and help obtain SSc1, Site Selection, if the locations are where the development footprint is restricted. In addition, these same design options can aid in obtaining a Water Efficiency credit which encourages reduced use of potable water for irrigation as many native vegetation areas need little or no additional irrigation. In summary, taking care to design a preliminary plan with special landscape features at specific locations may at a minimum impact the following credits:

- SSc1 Site Selection
- SSc5.1 Habitat
- SSc5.2 Open Space
- SSc6.1 Stormwater Quantity
- SSc6.2 Stormwater Quality
- SSc7.1 Heat Island (non-roof)

- SSc8 Light Pollution
- WEc1.1 Water Efficient Landscaping (50 percent)
- WEc1.2 Water Efficient Landscaping (100 percent)

It was discussed earlier that site area and site boundaries are variables that are important for several credits which should be established early in the design process and used consistently for credit applications. Some other important items to establish early in the design phase are occupancies. These include employees and transient occupancies for commercial uses and design occupancies for residential facilities. These occupancies are not the maximum occupancies that are used for evaluation by fire marshals and other safety purposes. These ones are the expected normal occupancies of the facilities.

References

ACEEE (2007), http://www.aceee.org/, American Council for an Energy Efficient Economy website accessed May 14, 2007.

ACEEE (2007), www.greenercars.com, American Council for an Energy Efficient Economy website for green car listings accessed May 14, 2007.

Akan, A. O., and R. J. Houghtalen (2003), *Urban Hydrology, Hydraulics and Stormwater Quality, Engineering Applications and Computer Modeling,* John Wiley & Sons, Hoboken, Nj.

Allan, C J., and C. J. Estes (2005), "A Morphological and Economic Examination of Plunge Pools as Energy Dissipaters in Urban Stream Channels," *Journal of the American Water Resources Association* (JAWRA), Paper No. 03065.

ASHRAE (2004), ASHRAE 90.1-2004: Energy Standard for Buildings Except Low-Rise Residential Buildings, American Society of Heating, Refrigerating and Air-Conditioning Engineers, Inc., Atlanta, Ga.

ASTM (2002), Standard Guide for Environmental Site Assessments: Phase II Environmental Site Assessment Process, E1903-97, accessed May 14, 2007, http://www.astm.org/cgib-in/SoftCart.exe/DATABASE.CART/REDLINE_PAGES/E1903.htm?E+mystore.

ASTM (2007) "Sustainability Subcommittee Launches Development of Proposed Green Roof Guide," Technical News Section of *ASTM Standardization News*, July, 35(7): 14–15.

Atlanta Regional Commission and Georgia Department of Natural Resources (2001), *Georgia Stormwater Management Manual,* vol. 2: *Technical Handbook.*

CARB (2007), http://www.arb.ca.gov/homepage.htm, California Air Resources Board website accessed May 14, 2007.

Carbone, G., and D. Yow (2006), "The Urban Heat Island and Local Temperature Variations in Orlando Florida," *Southeastern Geographe, r* 46: 297–321.

CPESC (2007), http://www.cpesc.net/, Certified Professional in Erosion and Sediment Control website accessed May 14, 2007.

Davis, A. P. (2005) "Green Engineering Principles, Promote Low Impact Development," *ES&T*, August 15, pp. 338A–344A.

Davis, A. P., and R., H. McCuen (2005), *Stormwater Management for Smart Growth,* Springer, New York.

EPA (2007), http://cfpub.epa.gov/npdes/index.cfm, EPA website on NPDES accessed May 14, 2007.

EPA (2007), http://cfpub.epa.gov/npdes/stormwater/cgp.cfm, EPA website on NPDES Construction General Permit (CGP) accessed May 14, 2007.

EPA (2007), http://www.epa.gov/air/caa/, EPA website on the Clean Air Act accessed May 14, 2007.

EPA (2007), http://www.epa.gov/brownfields/, EPA website on Brownfields accessed May 14, 2007.

EPA (2007), http://www.epa.gov/heatisland, EPA website on Heat Island Effect updated through January 16, 2007, accessed on July 12, 2007.

EPA (2007), http://www.epa.gov/owow/wetlands/regs/, EPA website on Federal Regulations governing wetlands accessed May 14, 2007.

Ehrlich, P. R,. and E. O. Wilson (1991), "Biodiversity Studies: Science and Policy," *Science*, 253(August 16): 758–761.

FEMA (2007), http://gis1.msc.fema.gov/Website/newstore/viewer.htm, Federal Emergency Management Agency website regarding FIRM (Flood Insurance Rate Maps) accessed May 14, 2007.

Fay, J. A., and D. S. Golomb (2003), *Energy and the Environment*, Oxford University Press, New York.

Gelt, J. (2006), "Urban Heat Island-Higher Temperatures and Increased Water Use," *Arizona Water Resource*, September–October, 15: 1–12.

Green Roofs (2007), http://www.greenroofs.net/, Green Roofs for Healthy Cities website accessed July 12, 2007.

Gregory, R. L., and C. E. Arnold (1932), "Runoff—Rational Runoff Formulas," *Tranactions of the American Society of Civil Engineers*, 96: 1038.

Haselbach, L. M., S. Valavala, and F. Montes (2006), "Permeability Predictions for Sand Clogged Portland Cement Pervious Concrete Pavement Systems," *Journal of Environmental Management*, October, 81(1): 42–49.

Heat Island Group (2000), http://eetd.lbl.gov/HeatIsland/, Lawrence Berkeley National Laboratory (LBNL) Heat Island Group, August 30, 2000, website accessed July 12, 2007.

ITE (2003). *Trip Generation*, 7th ed., Institute of Transportation Engineers, Washington, D.C.

ITE (2004), *Parking Generation*, 3d ed., Institute of Transportation Engineers, Washington, D.C

Moran, A. C., W. F. Hunt, and G. D. Jennings (2004), "A North Carolina Field Study to Evaluate Green Roof Quantity, Runoff Quality, and Plant Growth," in *Proceedings of Green Roofs for Healthy Cities Conference*, Portland, Ore.

NCDENR, DLR (2006), *Erosion and Sediment Control Planning and Design Manual*, North Carolina Department of Environment and Natural Resources, Division of Land Resources, Raleigh.

NCDENR, DWQ (2007), *Stormwater Best Management Practices Manual*, July, North Carolina Department of Environment and Natural Resources, Division of Water Quality, Raleigh.

NCHRP (2006), *Evaluation of Best Management Practices for Highway Runoff Control*, National Cooperative Highway Research Program Report 565, Transportation Research Board (TRB), Washington, D.C.

Prince George's County, Department of Environmental Resources (1999), *Low-Impact Development Design Strategies: An Integrated Design Approach*, Prince George's County, Md.

SCDHEC (2003), *South Carolina Stormwater Management and Sediment Control Handbook for Land Disturbance Activities,* August, Accessed July 6, 2007,South Carolina Department of Health and Environmental Control, Bureau of Water, Office of Ocean and Coastal Resource Management. http://www.scdhec.gov/environment/ocrm/pubs/docs/swmanual.pdf

Schueler, T. R. (1987), *Controlling Urban Runoff: A Practical Manual for Planning and Designing Urban BMPs,* Department of Environmental Programs, Metropolitan Washington Council of Governments.

USCFR (2003), *Code of Federal Regulations, Title 7, Volume 6,* revised as of January 1, 2003, http://a257.g.akamaitech.net/7/257/2422/14mar20010800/edocket.access.gpo.gov/cfr_2003/7CFR657.5.htm, Government Printing Office website accessed May 14, 2007.

USGBC (2003), *LEED-NC for New Construction, Reference Guide,* Version 2.1, 2d ed., May, U.S. Green Building Council, Washington, D.C.

USGBC (2005–2007), *LEED-NC for New Construction, Reference Guide;* Version 2.2, 1st ed., U.S. Green Building Council, Washington, D.C., October 2005 with errata posted through Spring 2007.

Valavala, S., F. Montes, and L. M. Haselbach (2006), "Area Rated Rational Runoff Coefficient Values for Portland Cement Pervious Concrete Pavement," *ASCE Journal of Hydrologic Engineering,* American Society of Civil Engineers, Reston, Va., May-June, 11(3): 257–260.

Zheng, J., H. Nanbakhsh, and M. Scholz (2006), "Case Study: Design and Operation of Sustainable Urban Infiltration Ponds Treating Storm Runoff," *ASCE Journal of Urban Planning and Development,* March, 132(1): 36–41.

Exercises

1. Put together a CPM schedule for items for ESC for a project. (See Chap. 8 for CPM schedules.)

2. Make a site plan with ESC for a project.

3. You are designing a building on 0.4 acre. It will be three stories high, and each story will have about 12,000 ft^2.

  A: Determine the development density of the proposed building.

  B: Determine the density radius (DR) from this building to be used in the areal development density determination.

  C: Using the map in Fig. 2.2.1, please sketch in this density radius.

  D: Using the data in Table 2.E.1, calculate the average areal Development Density around this proposed building. Only use the data that is appropriate.

  E: Based on the building's, and the average area's Development Densities, is this project eligible for Credit SSc2 of LEED-NC 2.2?

4. You are designing a building on 0.5 acre. It will be four stories high, and each story will have about 11,000 ft^2.

  A: Determine the development density of the proposed building.

  B: Determine the density radius from this building to be used in the areal development density determination.

  C: Using the map in Fig. 2.2.1, sketch in this density radius.

Building	Bldg. Floor Area (ft²)	Site area (acres)	Building	Bldg. Floor Area (ft²)	Site Area (acres)
A	33,000	0.4	W	19,000	0.6
B	87,400	1.6	X	6,000	0.3
C	6,000	0.3	Y	5,000	0.3
D	27,900	0.3	Z	4,300	0.2
E	66,000	1.2	AA	33,000	0.4
F	14,000	1.4	BB	87,400	1.6
G	12,900	0.2	CC	6,000	0.3
H	6,000	0.1	DD	27,900	0.3
I	14,000	0.2	EE	Park	0.5
J	29,000	0.4	FF	14,000	1.4
K	17,000	0.3	GG	12,900	0.2
L	9,000	0.3	HH	6,000	0.1
M	24,000	0.6	II	14,000	0.2
N	28,700	0.3	JJ	29,000	0.4
O	6,700	0.2	KK	17,000	0.3
P	39,900	0.4	LL	9,000	0.3
Q	344,000	2.5	MM	24,000	0.6
R	91,300	1.9	NN	33,000	0.4
S	22,000	0.3	OO	66,000	1.2
T	33,000	0.5	PP	Vacant	1.1
U	42,000	0.5	QQ	9,000	0.3
V	14,000	0.8	RR	24,000	0.6

TABLE 2.E.1   Development Density Data for Exercises

D: Using the data in Table 2.E.1, calculate the average areal development density around this proposed building. Only use the data that are appropriate.

E: Based on the building's and the average area's development densities, is this project eligible for Credit SSc2 of LEED-NC 2.2?

5. You are designing a building on 0.3 acre. It will be three stories high, and each story will have about 10,000 ft².

A: Determine the development density of the proposed building.

B: Determine the density radius from this building to be used in the areal development density determination.

C: Using the map in Fig. 2.2.1, sketch in this density radius.

D: Using the data in Table 2.E.1, calculate the average areal development density around this proposed building. Only use the data that are appropriate.

E: Based on the building's and the average area's development densities, is this project eligible for Credit SSc2 of LEED-NC 2.2?

6. You are planning to build a textile factory. You plan to have 50 full-time workers and 20 half-time workers on the day shift, 30 full-time workers and 2 half-time workers on the evening shift, and 20 full-time workers on the night shift.

A: Determine the maximum shift FTE.

B: To get SS Credit 4.2, for how many bicycles must you provide secure storage and how many showers must you provide?

C: If the proposed parking by local regulations is 75 spaces, how many of these spaces must be designated for only carpool/vanpool to get SS Credit 4.4?

D: What would be required to obtain SS Credit 4.3?

7. You are planning to build a college hall. There are 55 offices for faculty and support staff, plus a daytime maintenance crew of four employees. There are 10 classrooms. Six of the classrooms typically have a maximum capacity of 30 per fire regulations, and four of the classrooms have a maximum capacity of 80 people. There are four computer laboratories, each with 20 computers. There are 10 research laboratories with attached offices for two graduate students each. To do the following calculations, first list your assumptions.

A: Determine the maximum shift FTE.

B: To get SS Credit 4.2, for how many bicycles must you provide secure storage and how many showers must you provide?

C: If the proposed parking by local regulations is 75 spaces, how many of these spaces must be designated for only carpool/vanpool to get SS Credit 4.4?

D: What would be required to obtain SS Credit 4.3?

8. You are a consultant for a major fast food restaurant chain which plans to totally rebuild many of its restaurants in the Columbia, S.C., area. Using the data in Table 2.E.2, calculate the LEED percent imperviousness for an existing fast food restaurant site in the northeast section of town. Which option do you need to follow to go for SS Credit 6.1?

9. The fast food chain decides to raze and rebuild another fast food restaurant in the central part of the city which was built similar to the design in Exercise 8, except the concrete drive-through lane is made of pervious concrete and there is an additional 0.10 acre of wooded land on the lot. What is the percent imperviousness of this site?

Surface	Rational Runoff Coefficient	Area
Roofed	0.95	2,000 ft²
Wooded	0.25	0.25 acre
Landscape Beds	0.4	4,500 ft²
Asphalt Parking and Drives	0.9	10,500 ft²
Grass Picnic Area	0.45	900 ft²
Concrete Drive-Through Lane	0.9	700 ft²

TABLE 2.E.2 Data for Stormwater Exercise

10. You want to analyze what must be done to obtain SS Credit 5.1 with the lot in Exercise 8. How much area (if any) would need to be restored to obtain this credit, and what areas would you suggest that you use for it? Draw a quick sketch of the site. (Use your imagination.)

11. You want to analyze what must be done to obtain SS Credit 5.1 with the lot in Exercise 9. How much area (if any) would need to be restored to obtain this credit, and what areas would you suggest that you use for it? Draw a quick sketch of the site. (Use your imagination.)

12. You want to analyze what must be done to obtain SS Credit 5.2 with the lot in Exercise 8. The local code requires a minimum of 20 percent open space. Do you meet the LEED open-space requirement with the existing layout? If not, how much additional area would need to be dedicated to open space to obtain this credit, and what areas would you suggest that you use for it? Draw a quick sketch of the site. (Use your imagination.)

13. You want to analyze what must be done to obtain SS Credit 5.2 with the lot in Exercise 8. The local zoning requirements do not address open space. Do you meet the LEED open-space requirement with the existing layout? If not, how much additional area would need to be dedicated to open space to obtain this credit, and what areas would you suggest that you use for it? Draw a quick sketch of the site. (Use your imagination.)

14. You want to analyze what must be done to obtain SS Credit 5.2 with the lot in Exercise 8. The local code requires a minimum of 40 percent open space. Do you meet the LEED open-space requirement with the existing layout? If not, how much additional area would need to be dedicated to open space to obtain this credit, and what areas would you suggest that you use for it? Draw a quick sketch of the site. (Use your imagination.)

15. You wish to get credit for the heat island effect by using a light concrete paving material for part of the paved surfaces on a lot. If the parking areas cover 20,000 ft^2 and there is an additional 4000 ft^2 of walks and drives, what minimum area of these paved surfaces must be the lighter paving material?

16. If you instead provide tree shading for 2000 ft^2 of the parking area in Exercise 15, what area in square feet, other than the shaded areas, must have the higher albedo surface to still be eligible for this credit?

17. A site has 20,000 ft^2 of land covered by buildings and hardscape. It also has 21,000 ft^2 of land covered by landscaping. Assume that the average rational runoff coefficients for the development footprint and the landscaped areas are 0.95 and 0.30, respectively. What is the percent imperviousness? What is the percent impervious? Which option is the appropriate one in SSc6.1? Why?

18. Two buildings are being built on a lot. One is a 2000 ft^2 tarred flat roof storage building. The other is an office building complex with a 4:12 pitch (33 percent slope) with a white coating on a metal roof. The actual sloped area of the roof on the office complex is 6400 ft^2. Will you be eligible for SSc7.2?

19. Go outside your classroom building and itemize all the light fixtures that you see. Incorporate these into the categories in Table 2.8.1. Sum the wattages for each category in Table 2.8.1. If wattages are unknown, then use estimated wattages or those suggested by your instructor.

20. Go to your favorite fast food restaurant. Itemize all the exterior light fixtures that you see. Incorporate these into the categories in Table 2.8.1. Sum the wattages for each category in Table 2.8.1. If wattages are unknown, then use estimated wattages or those suggested by your instructor.

21. Go to your bank. Itemize all the exterior light fixtures that you see. Incorporate these into the categories in Table 2.8.1. Sum the wattages for each category in Table 2.8.1. If wattages are unknown, then use estimated wattages or those suggested by your instructor.

CHAPTER 3

LEED Water Efficiency

The Water Efficiency category of the U.S. Green Building Council (USGBC) rating system for new construction version 2.2 (LEED-NC 2.2) consists of three credit subcategories which together may earn a possible five points (credits) maximum, not including exemplary performance (EP) points. The notation format for the credits is, for example, WEc1.1 for Water Efficiency credit 1.1. Also, in this chapter, to facilitate easier cross-referencing between this text and the USGBC rating system, the second digit in a section heading, equation, table, or figure number represents the credit subcategory number for sections that deal directly with a USGBC LEED-NC 2.2 credit subcategory.

The Water Efficiency (WE) portion deals with issues that reduce the use of potable water at the site and the discharge of wastewater from the site. This will help limit the amounts of freshwater drawn from our water bodies and aquifers, and then treated for distribution and use, which strain our freshwater supplies and our water infrastructure. It will also serve to reduce the wastewater volumes discharged to these receiving bodies. Although several of the credits in Water Efficiency might include a capital cost investment, all provide a means to potentially reduce the associated utility costs over the life of the building.

The three subcategories are Water Efficient Landscaping (WE credits 1.1 and 1.2), Innovative Wastewater Technologies (WE credit 2), and Water Use Reduction (WE credits 3.1 and 3.2). Basically, the first credit subcategory limits the use of potable water for outdoor uses, the second limits the production or discharge of wastewaters (black), and the third limits the use of potable water for indoor uses.

The five credits available in the Water Efficiency category are as follows:

- **WE Credit 1.1** (WEc1.1): *Water Efficient Landscaping: Reduce by 50%* (previously referred to as *Water Efficient Landscaping: 50% Reduction* in LEED-NC 2.1)

- **WE Credit 1.2** (WEc1.2): *Water Efficient Landscaping: No Potable Water Use or No Irrigation*

- **WE Credit 2** (WEc2): *Innovative Wastewater Technologies*

- **WE Credit 3.1** (WEc3.1): *Water Use Reduction: 20% Reduction*

- **WE Credit 3.2** (WEc3.2): *Water Use Reduction: 30% Reduction*

As in Chap. 2, the submittals for each credit or prerequisite have been sectioned into either Design Submittals and/or Construction Submittals by the USGBC. Even though the entire LEED process encompasses all phases of a project, the prerequisites or credits that are noted as Construction Submittal credits have major actions going on through the construction phase which, unless documented through that time, may not be able to be verified from preconstruction documents or the built project. All the credits in the

Water Efficiency category are listed as Design Submittal in the LEED-NC 2.2 Reference Guide and therefore are not noted as Construction Submittals in Table 3.0.1.

None of the credits in the Water Efficiency category have the icon EB noted after their title. This icon is a tool to help those who wish to proceed with the continuing LEED-EB certification (Existing Building) after certification of LEED-NC (New Construction or Major Renovation) is obtained. Those credits noted with this icon are usually significantly more cost-effective to implement during the construction of the building than later during its operation. This is also noted in Table 3.0.1.

There are EP points under the Innovation and Design category available which relate to two items in the Water Efficiency category. These are available for exceeding each of the respective credit criteria to a minimum level as noted in the credit descriptions for WEc2 and WEc3.1. These are noted in Table 3.0.1.

Two of the subcategories in the Water Efficiency category are unique in that the credit points are sequential and additive. The first subcategory (WEc1.1 and WEc1.2) gives a point for a 50 percent reduction in potable water or natural surface water or groundwater for irrigation uses, and an additional point for WEc1.2 if this reduction is 100 percent or if there is no irrigation. There are no EP points available for the first subcategory in Water Efficiency (WEc1.1 and WEc1.2) for additional reductions in potable water or natural surface or groundwater use for irrigation, as WEc1.2 already represents a 100 percent reduction.

The other subcategory that gives sequentially additive points is Water Use Reduction (WEc3.1 and WEc3.2). Here, the first point is awarded for a 20 percent reduction in indoor

Credit	Construction Submittal	EB Icon	Exemplary Performance and Other ID Point Availability	Exemplary Performance and Other ID Point Criterion	Location of Site Boundary and Site Area Important in Calculations
WEc1.1: Water Efficient Landscaping: Reduce by 50%		No	No		Yes
WEc1.2: Water Efficient Landscaping: No Potable Water Use or No Irrigation		No	No		Yes
WEc2: Innovative Wastewater Technologies		No	Yes (1)	1 point for 100%	Perhaps Option 2
WEc3.1: Water Use Reduction: 20%		No		1 point for 40% EPAct and/or	No
WEc3.2: Water Use Reduction: 30%		No	Yes (2)	1 point for 10% non-EPAct	No

TABLE 3.0.1 WE EB Icon, Exemplary Performance Point Availability, and Miscellaneous

potable water use (WEc3.1), and an additional point is available if this reduction reaches 30 percent (WEc3.2). There is an EP point under the Innovation and Design (ID) category available which relates to this subcategory in the Water Efficiency category. This is available if the reduction in indoor potable water use is at least 40 percent, and the point is in addition to the two from WEc3.1 and WEc3.2. The Water Use Reduction subcategory can also be the basis for an additional ID credit. It is not called an exemplary performance point per se, since it is not exceeding the percentages in the credit, but it is given for indoor water use reductions for water fixtures that are not specifically covered in WEc3.1 and WEc32, although the basis for the calculations is dependent on the design indoor water rates for the fixtures as regulated in WEc3.1 and WEc3.2. It is based on a reduction in other fixture water uses (non-EPAct fixtures) equal to or greater than 10 percent of the design water use rate for the fixtures (EPAct fixtures) as calculated for WEc3.1 and WEc3.2. EPAct stands for the Energy Policy Act of 1992. This is a federal act that addresses energy and water use in commercial, institutional, and residential facilities.

The second subcategory in the Water Efficiency category also has an EP point associated with it for achievement of a 100 percent reduction of sewage conveyance or treatment on-site in addition to one point for a 50 percent reduction.

The items available for EP points in this category are shown in Table 3.0.1 with a column summarizing the point achievement criteria. Note that only a total of four exemplary performance points are available for a project, and they may be from any of the noted exemplary performance point options in any of the first five main LEED categories or may be other special point items.

Table 3.0.1 also has a column that notes the importance of the site boundary and site area in the credit calculations or verification. As mentioned in Sustainable Sites, these are variables that should be determined early in a project as they impact many of the credits. Some of the credits are easier to obtain if the site area is less, while others are easier to obtain if the site area is greater. Determination of the site area may sometimes be flexible, particularly for campus locations, but it should be reasonable and must be consistent throughout the LEED process. It must also be in agreement with other nearby project submissions. The site area is mentioned again in Chap. 8 as an important variable to analyze early and set in a project.

3.1 WE Credit Subcategory 2: Water Efficient Landscaping

The intention of these credits is to reduce the use of potable water or natural surface waters or natural groundwater for outdoor irrigation. WEc1.1 is worth one point, and WEc1.2 is worth one point *in addition to* the point for WEc1.1.

Potable water is water that is assumed to be acceptable for human consumption. Strategies include using more drought-tolerant plants in landscape beds near pavements and buildings, keeping native vegetation when possible, and using recycled waters for irrigation. Figure 3.1.1 shows a landscape bed where several drought-tolerant species were used next to a drive and parking area.

WE Credit 1.1: *Water Efficient Landscaping—Reduce by 50%*

WE Credit 1.2: *Water Efficient Landscaping—No Potable Water Use or No Irrigation*

Figure 3.1.1 Landscape bed in a subdivision in Columbia, S.C., where several drought-tolerant species reduce irrigation needs. (*Photograph taken July 14, 2007.*)

USGBC Rating System WE Credit 1.1

LEED-NC 2.2 lists the Intent, Requirements, and Potential Technologies and Strategies for WEc1.1 as follows:

Intent

Limit or eliminate the use of potable water, or other natural surface or subsurface water resources available on or near the project site, for landscape irrigation.

Requirements

Reduce potable water consumption for irrigation by 50% from a calculated mid-summer baseline case. Reductions shall be attributed to any combination of the following items:

- Plant species factor
- Irrigation efficiency
- Use of captured rainwater
- Use of recycled wastewater
- Use of water treated and conveyed by a public agency specifically for nonpotable uses

Potential Technologies and Strategies

Perform a soil/climate analysis to determine appropriate plant material and design the landscape with native or adapted plants to reduce or eliminate irrigation requirements. Where irrigation is required, use high-efficiency equipment and/or climate-based controllers.

USGBC Rating System WE Credit 1.2

LEED-NC 2.2 lists the Intent, Requirements, and Potential Technologies and Strategies for WEc1.2 as follows:

Intent

Eliminate the use of potable water, or other natural surface or subsurface water resources available on or near the project site, for landscape irrigation.

Requirements

Achieve WE Credit 1.1 and use only captured rainwater, recycled wastewater, recycled greywater, or water treated and conveyed by a public agency specifically for non-potable uses for irrigation.

OR

Install landscaping that does not require permanent irrigation systems. Temporary irrigation systems used for plant establishment are allowed only if removed within one year of installation.

Potential Technologies and Strategies

Perform a soil/climate analysis to determine appropriate landscape types and design the landscape with indigenous plants to reduce or eliminate irrigation requirements. Consider using stormwater, greywater, and/or condensate water for irrigation.

Calculations and Considerations WE Credits 1.1 and 1.2

Both credit points are automatically earned if there is no irrigation on the site. Otherwise, these two credits usually require a series of calculations. For WEc1.1, the calculations must show that the design water use of potable water is less than 50 percent of a baseline irrigation case. For WEc1.2, in addition to having none of the irrigation water potable, or coming from natural surface or groundwater sources near the site, the calculations must also show that the design water use of *any* water for irrigation is less than 50 percent of a baseline irrigation case.

The calculations are done in two main steps. First a calculation for a baseline (typical) landscape scenario is performed and then calculations for the design case are performed. These are compared to determine the points available. The calculations are used to estimate the total water applied (TWA) for each of these two cases. Note that the total landscaped areas for both cases must be the same, and it is not appropriate to assume that the baseline is an entire site with only high water use such as turfgrass. The main calculation format is reviewed in the following section.

Calculating the Total Water Applied The TWA is determined using a series of landscape water usage factors and a reference evapotranspiration rate to first determine the evapotranspiration (ET) rates for both cases (baseline and design). These factors and the reference rate are as follows:

k_s The species factor (an indicator of water usage by vegetation species)

k_d The density factor (an indicator of water usage by how densely the vegetation is planted or how substantially the foliage covers the area)

k_{mc} The microclimate factor (an indicator of water usage with respect to the location of the vegetation on the site)

ET_0 The reference evapotranspiration rate ET_0 is specific to areas of the country and represents a typical evapotranspiration rate based on the local climate and a

	Factors	k_s low	k_s avg.	k_s high	k_d low	k_d avg.	k_d high	k_{mc} low	k_{mc} avg.	k_{mc} high	ET_o
Vegetation Type	Trees	0.2	0.5	0.9	0.5	1.0	1.3	0.5	1.0	1.4	Not Dependent on Vegetation Type
	Shrubs	0.2	0.5	0.7	0.5	1.0	1.1	0.5	1.0	1.3	
	Ground covers	0.2	0.5	0.7	0.5	1.0	1.1	0.5	1.0	1.2	
	Mixed: Trees/shrubs/ ground covers	0.2	0.5	0.9	0.6	1.1	1.3	0.5	1.0	1.4	
	Turfgrass	0.6	0.7	0.8	0.6	1.0	1.0	0.8	1.0	1.2	
Baseline		Conventional Design Practice			Conventional Design Practice			Same as Design			Same as Design

TABLE 3.1.1 Landscape Factors and ET_o

reference plant (grass or alfalfa). It is usually also based on conditions in that area for the month of July since often the greatest irrigation demand is in July. More information can be obtained from the American Society of Civil Engineers (ASCE) and the Irrigation Association (IA). Recently, a standardized reference evapotranspiration equation for the United States has been developed.

Each of the factors is divided into three main categories—low, average, and high water use—and further subdivided into five subcategories on the overall vegetation type or combination. The table for typical landscape factors (Table 3.1.1) is based on WE Credit 1 Table 1 from LEED-NC 2.2.

The species factor should be specific to the design species for the design case (k_{sD}), where, for instance, more succulent plants have low water use, etc. The species factor for the baseline case (k_{sB}) should be set to average values representative of conventional design practices.

The density factor is indicative of the number of plants planted and the total leaf area. More plants require more water. Some rules of thumb are based on ground shading and are shown in Table 3.1.2, but these cover only a few cases. The density factor for the design case (k_{dD}) should be set to that representative of the design plant spacings and densities. The density factor for the baseline case (k_{dB}) should be set to average values representative of conventional design practices.

The microclimate factor is based on where the landscape areas are with respect to other features on the site as well as site orientation. Some typical conditions that differentiate between low, average, and high as given in LEED-NC 2.2 are in Table 3.1.3. The same microclimate factors are used in the design and in the baseline calculations.

Vegetation Type	**Low**	**Average**	**High**
Trees	<60	60–100	Layered
Shrubs	<90	90–100	Layered
Ground covers	<90	90–100	Layered

TABLE 3.1.2 Typical Density Factor Categories Based on Percent Ground Shading

Low	Shaded areas and areas protected from wind (e.g., on north sides of buildings, courtyards, areas under wide building overhangs, and north sides of slopes)
Average	Evapotranspiration rate is unaffected by buildings, pavements, reflective surfaces, and slopes
High	Landscape area near heat-absorbing and reflective surfaces or landscape area exposed to particularly windy conditions (e.g., near parking lots, on west sides of buildings, on west or south sides of slopes, median, and wind tunnel effect areas)

TABLE 3.1.3 Typical Microclimate Factor Categories

The reference evapotranspiration rate ET_0 for a project depends on the local climate. It is the same for both the design case and the baseline case. It can be found from irrigation references or websites and is usually given in inches per day. (It may also be given in inches, which usually means inches per the month of July.) Some typical values are given in Table 3.1.4. Table 3.1.4 is taken from the *Landscape Irrigation Scheduling and Water Management Draft:* Table 1-3: Mid-Summer Daily Reference ET_0 (Grass Reference) Based on Climate for Midday Mid-Summer Air Temperature and Relative Humidity on the IA (Irrigation Association) website, January 2006 (www.irrigation.org/about_et_list.htm). The values are based on typical July daily high temperatures (degrees Fahrenheit) and associated relative humidities (RHs).

These factors and ET_0 are used to determine the baseline and the design evapotranspiration rates for each landscape area (area i) on the site, ETB_{Li} and ETD_{Li}.

$$ETB_{Li} = (k_{sB})(k_{dB})(k_{mc})(ET_0) \qquad \text{for landscape area } i \qquad (3.1.1)$$

$$ETD_{Li} = (k_{sD})(k_{dD})(k_{mc})(ET_0) \qquad \text{for landscape area } i \qquad (3.1.2)$$

There are two other factors that influence irrigation and are included in the calculations. The first is the irrigation efficiency (IE) factor, which is representative of the type of irrigation system used. A drip irrigation system usually has an irrigation

Climate	Definition	ET_0 (Max.)(in/day)
Cool humid	<70°F >50% RH	0.10–0.15
Cool dry	<70°F <50% RH	0.15–0.20
Warm humid	70–90°F >50% RH	0.15–0.20
Warm dry	70–90°F <50% RH	0.20–0.25
Hot humid	>90°F >50% RH	0.20–0.30
Hot dry	>90°F <50% RH	0.30–0.45

TABLE 3.1.4 Reference ET_0 Values

	Irrigation System	Irrigation Efficiency (IE)	Irrigation Controller	Controller Efficiency (CE)
Types	Drip irrigation	0.90	None	1
	Sprinkler	0.65	Per manufacturer	$0 \ll CE < 1$
Baseline	Conventional irrigation	0.65	None	1

TABLE 3.1.5   Other Irrigation Factors: Some Representative Values

factor of 0.90 and is much more efficient at delivering the water more directly to the plants as needed than a typical sprinkler system which has an IE of around 0.65. For the baseline case, a "conventional irrigation" system type of the region should be used (*conventional irrigation* is defined as per LEED-NC 2.2 in App. B). The second factor, the controller efficiency (CE), has to do with using special sensors to control the sprinkler use based on conditions such as soil moisture. If no special sensors are used, then CE is equal to 1; otherwise the factor used should be supported by the manufacturer's literature and is less than 1, but significantly greater than 0. Some typical or representative values are given in Table 3.1.5.

The baseline and the design evapotranspiration rates for each landscape area A_i on the site (ETB_{Li} and ETD_{Li}) together with the baseline and design irrigation efficiencies (IEB_i and IED_i), and the controller efficiencies (CE_i) for each landscape area i, are used to calculate the baseline and design total water applied ($TWAB_i$ and $TWAD_i$) for each landscape area, and the baseline and design total water applied (TWAB and TWAD) for the project.

$$TWAB_i = (A_i)(ETB_{Li})(CE_i/IE_i) \quad \text{for landscape area } i \tag{3.1.3}$$

$$TWAD_i = (A_i)(ETD_{Li})(CE_i/IE_i) \quad \text{for landscape area } i \tag{3.1.4}$$

$$TWAB = \sum_i TWAB_i \tag{3.1.5}$$

$$TWAD = \sum_i TWAD_i \tag{3.1.6}$$

Credit Determinations   Now that the total water needs for the landscaping have been determined, the amount of potable water (or other special waters that are recommended to not be used for irrigation for environmental considerations) that will be applied in the design case (TPWA) can be determined from the total water applied for the design case (TWAD) less the reuse water (RW). The author has interpreted *reuse water* to be any waters from rainwater harvesting, reused graywaters, and municipal supplied "reuse" waters applied for irrigation that are not for potable uses. Other waters from the site such as pond water or nonpotable groundwater may not be subtracted out. They are considered to be important for other environmental reasons on the site. However, even with this clarification there are many definitions of graywaters. The WEc2 subcategory section on Innovative Wastewater Technologies gives some further insight into the definition of graywaters.

$$TPWA = TWAD - RW \tag{3.1.7}$$

Note that the units must all cancel to give a unit for the TWAB and TWAD in volume per month or hour. Typically, the ET rates are given in inches per hour or month, the areas are given in square feet, and the other factors are dimensionless. In these cases, the conversion factor to obtain TWAB or TWAD in gallons per unit time is 0.6233 gal/(ft$^2 \cdot$ in), or inversely, 1.604 ft$^2 \cdot$ in/gal.

The values for TWAB, TWAD, and TWPA can be used for the credit verifications. WEc1.1 may be earned if the following inequality is true:

$$\frac{TPWA}{TWAB} \leq 0.50$$
(3.1.8)

An additional point can be given for WEc1.2 if the following two equations are both true:

$$\frac{TWAD}{TWAB} \leq 0.50$$
(3.1.9)

and

$$TPWA = 0$$
(3.1.10)

Special Circumstances and Exemplary Performance WE Credits 1.1 and 1.2

As mentioned previously, if there is no irrigation, then there are no calculations necessary to obtain both credits. If there is any irrigation, even if all is from reuse water, then calculations usually must be performed. There are some circumstances where it may be unclear whether the water used for watering is "irrigation" water per the intent of the credit or "process" water. One example is for the main field in a baseball stadium, where watering the main playing field may be considered irrigation or may be considered as a special water application (process water). Typical playing fields are irrigated, but main sports venue fields are sometimes looked at differently and have different requirements. These special exceptions should be reviewed through the CIR (Credit Interpretation Ruling) procedure.

Note that there are many special health codes governing the use or storage of reuse waters. They vary by local or state jurisdiction. The designers, owners, and managers should become familiar with any special health-related requirements for use of reuse waters so that appropriate protective measures or treatments can be included in the design and operations. The submittals must also include narratives describing the various nonpotable water sources, local codes for their use, and descriptions of any special treatment of these waters and other related features.

An interesting item to note about these credits is that the calculated baseline total water applied for irrigation (TWAB) is based upon conventional practices in the region where the project is built. The author assumes that this implies that the conventional practices are those as of around 2005, when the LEED-NC 2.2 Reference Guide was adopted and there were not many facilities percentagewise built to the LEED standards. As the "green" movement expands, the author hopes that the new "conventional" practices will become greener and greener. The author suggests that the designers date their calculations, so that in future years, these earlier baseline criteria can possibly be used as the future baseline, even if LEED and other environmental water-saving programs become the standard or conventional practice in the area. Otherwise these credits will become harder to obtain and may be disregarded, which is not the intention of LEED. There have recently been significant strides made to adopt standards and develop protocols to evaluate products and processes for water efficiency. The Irrigation Association has recently started the Smart Water Applications Technologies (SWAT) program. Much of this ongoing work is being incorporated into the EPA's Water Sense program.

If there are special circumstances for these credits, a narrative must be submitted describing these circumstances. There is no exemplary performance point available for these credits.

3.2 WE Credit 2: *Innovative Wastewater Technologies*

USGBC Rating System
LEED-NC 2.2 lists the Intent, Requirements, and Potential Technologies and Strategies for WEc2 as follows:

Intent
Reduce generation of wastewater and potable water demand, while increasing the local aquifer recharge.

Requirements
OPTION 1

Reduce potable water use for building sewage conveyance by 50% through the use of water conserving fixtures (water closets, urinals) or non-potable water (captured rainwater, recycled greywater, and on-site or municipally treated wastewater).

OR

OPTION 2

Treat 50% of wastewater on-site to tertiary standards. Treated water must be infiltrated or used on-site.

Potential Technologies and Strategies
Specify high-efficiency fixtures and dry fixtures such as composting toilet systems and non-water using urinals to reduce wastewater volumes. Consider reusing stormwater or greywater for sewage conveyance or on-site wastewater treatment systems (mechanical and/or natural). Options for on-site wastewater treatment include packaged biological nutrient removal systems, constructed wetlands, and high-efficiency filtration systems.

Calculations and Considerations
First, the intent of these two options should be further explored as there are many definitions for *sewage* and *wastewater*. For the purposes of this credit, the volumes used for calculating either option should be based on "blackwater" volumes. As a default, blackwater is the water from toilets (water closets) and urinals. Some jurisdictions also include kitchen sinks and other wastewater sources. If these other sources are included, then they should be specifically noted in the calculations and the local or state definitions provided in the submittal. In fact, it is usually easier to obtain this credit if *blackwater* is defined as the wastewater from toilets and urinals only. In some senses, this is justified for Option 1 as kitchen sinks need to use potable water. Therefore, for the calculations as outlined in the following sections for this credit and for WEc3.1 and WEc3.2, it will be assumed that blackwater is the wastewater from water closets and urinals only. Again, any variances from this should be specifically outlined and noted in the submittals.

Fixture Type	Gender	Variable Acronym	Daily Uses per Person			
			FTE* Uses	Transient: Student/ Visitor	Transient: Retail Customer	Resident
Water closet use	Male	WCUM$_i$	1	0.1	0.1	5
Water closet use	Female	WCUF$_i$	3	0.5	0.2	5
Urinal use	Male	UUM$_i$	2	0.4	0.1	n/a
Urinal use	Female	UUF$_i$	0	0	0	n/a

*The students as listed in TOW represent transient students, such as those attending a college where they may take only a handful of classes in one building on certain days. Students in primary schools and other schools where they remain in the same building every "work" day for the entire school day are actually similar to full-time employees and should be treated as such. This is not specifically addressed in LEED-NC 2.2 and should therefore be noted in the calculations and the submittals.

†These resident daily usage rates represent residents who on average also spend part of the majority of their days elsewhere, such as at school or work. It is recommended that for residential facilities where there are usually 24-h occupancies, such as nursing homes, the resident rates for water closets be, at a minimum, 8, which is a sum of 3 (for FTE) and 5 (for residences). This is not specifically addressed in LEED-NC 2.2 and should therefore be noted in the calculations and the submittals.

TABLE 3.2.1 Default Typical Daily Blackwater Fixture Use by Gender

The calculations are based on the usage of the building with both a baseline case and a design case. The baseline for this credit is always based on the same usage of the building as the design case, but with water usage rates as established as the baseline in the Energy Policy Act (EPAct) of 1992.

The number of occupants should be differentiated by gender, by employee occupancies, and by transient occupancies. There are numerous variations of the occupancies based on various usages, but the typical assumptions for blackwater generation are as listed in Tables 3.2.1 and 3.2.2. These values have been taken from the LEED-NC 2.2 Reference

Specific Fixture Type	Gallons Generated per Use [gal per flush (gpf)] (WCR$_i$ or UR$_i$)
Baseline water closet	1.6
Low-flow water closet	1.1
Dual flush and low-flow water closet	0.8
Composting toilet	0
Baseline urinal	1.0
Low-flow urinal	0.5
Nonwater urinal	0

TABLE 3.2.2 Typical Blackwater Generation by Fixture Type

Guide. They refer to the daily usage for a full-time employee equivalent (FTE) on her or his shift, for the special definition of transient occupancy for water usage (TOW) which is the summation of *each and every* transient occupant in the building during a day (the transient uses have been further subdivided into student/visitor uses and retail customer uses) and for residential design occupancies. For residential uses these may be higher for certain uses, particularly for residential facilities where there are many children or retirees, as they may be home for more hours during the day. Obviously, for special residential uses such as retirement or nursing homes, the numbers for daily uses would need to be increased appropriately. Otherwise, the default values in Table 3.2.1 should be used.

A special note is needed for some of the blackwater generation rates listed in Table 3.2.2. Composting toilets and nonwater urinals do not use water for regular usage. They are not currently customary fixtures in the United States. Therefore, they have different maintenance and waste collection needs than more common water closets and urinals. The designers and owners should be very cognizant of the different maintenance needs, and it is recommended that they determine a way to meet these special needs early in the design phase of a project so that there are no additional concerns, costs, or changes later in the project. Figure 3.2.1 is a photo of a waterless urinal.

To perform the calculations, the non-gender-specific occupancies should be established as previously defined in Chap. 2. Occupancies must be consistent throughout

FIGURE 3.2.1 Waterless urinal installed in the LEED certified public health building at the University of South Carolina, *Columbia, S.C. (Photograph taken July 25, 2007.)*

a LEED submittal for a project. The calculations for WEc2 Option 1 are different for commercial and institutional uses than for residential uses and are summarized in the following sections.

Option 1: Commercial and Institutional Uses   In a similar fashion to the calculations for occupancies in Chap. 2, the following definitions are given:

FTE_j Full-time equivalent building occupant during shift j

$FTE_{j,i}$ Full-time equivalent building occupancy of employee *i* during shift *j*. This is equal to 1 for a full-time employee and is equal to the normal hours worked (less than 8) divided by 8 for a part-time employee. Obviously, this will need to be modified if shifts are different from the standard 8 h.

The FTE_j is then determined by using the following set of equations:

$$FTE_{j,i} = (\text{worker } i \text{ h})/8 \text{ h} \qquad \text{where } 0 < FTE_{j,i} \le 1 \qquad (2.4.1)$$

$$FTE_j = \sum FTE_{j,i} \qquad \text{for all employees in shift } j \qquad (2.4.2)$$

Then the full-time employee equivalents need to be subdivided by gender. For each shift the numbers for male and female employees may not be equal. Let $FTEM_j$ and $FTEF_j$ be the male full-time employee equivalent and the female fulltime employee equivalent for shift *j*, respectively. To be consistent throughout the LEED submittal, the following equality must hold for all shifts *j*:

$$FTE_j = FTEM_j + FTEF_j \qquad \text{for all shifts } j \qquad (3.2.1)$$

In SS credit 4.2, the transient occupancies are estimated as average transient occupancies at any time over a shift, not as the total number of transient occupants who may go in or out of the building during the shift. This is done to estimate the peak number of bicycle rack spaces needed at any time. For example, if the building is a college building with faculty and staff offices and student classrooms, then in SSc4.2 the FTE_j portion of the calculations is based on the staff and faculty in the building during the typical workday, and the transient population is the average occupancy of the classrooms and study areas over this typical 8-h class time, not the total number of students who come in and out. The students may be in other campus buildings for other classes throughout the day. However, in WE credits 3.1 and 3.2, the LEED-NC 2.2 Reference Guide bases transient uses on the *total number of transients*, not the average over the shifts. To be consistent with these other water credits, let us define TOWM and TOWF to be the estimated total number of male and female transients, respectively, during a day. (TOW would be the summation of TOWM and TOWF.)

As mentioned previously, LEED-NC 2.2 recommends that the typical daily fixture usage rates for water closets and urinals be separately analyzed for the different genders, building usages, and occupancies. Let $WCUM_i$ and $WCUF_i$ be the male water closet and female water closet usage rates, respectively, for various types of buildings or occupancies. Let UUM_i and UUF_i be the male and female urinal usage rates, respectively, for various building types and occupancies. Table 3.2.1 gives some typical default values for these daily fixture usage rates. It is important to also analyze the building usage on an annual basis. Let ND_{FTE} and ND_{TOW} be the number of days in a year that each type of occupancy (FTE or TOW) uses the building. ND_{FTE} is usually 260 days for office buildings and 365 days for residential or retail. For college buildings, ND_{FTE} may be 260 for staff and faculty, but ND_{TOW} may be much less for student occupancies. Summing the fixture usage rates times the occupancies over the various shifts as applicable for the entire year gives the total number of times water closets and urinals are typically used in a year (TWCU and TUU, respectively):

$$TWCU = ND_{FTE} (\Sigma (FTEM_j \times WCUM_i) + \Sigma (FTEF_j \times WCUF_i))$$
$$+ ND_{TOW}[(TOWM \times WCUM_i) + (TOWF \times WCUF_i)]$$
over all shifts j and for each type of use i \hfill (3.2.2)

$$TUU = ND_{FTE} (\Sigma (FTEM_j \times UUM_i) + (\Sigma (FTEF_j \times UUF_i))$$
$$+ ND_{TOW} [(TOWM \times UUM_i) + (TOWF \times UUF_i)]$$
over all shifts j and for each type of use i \hfill (3.2.3)

Now both the baseline and the design blackwater annual generation rates can be determined. The blackwater baseline (BWB) generation rate, in gallons per year, can be calculated using the fixture flow rate numbers in Table 3.2.2 as follows:

$$BWB = (1.6 \times TWCU) + (1.0 \times TUU) \qquad gal/yr \hfill (3.2.4)$$

To calculate the blackwater design (BWD) generation rate, the factors in Eq. (3.2.4) are replaced with those appropriate for the alternate fixture type i used such as WCR_i (the fixture rate for water closet type i) and UR_i (the fixture rate for urinal type i) from Table 3.2.2:

$$BWD = (WCR_i \times TWCU) + (UR_i \times TUU) \qquad gal/yr \hfill (3.2.5)$$

Note that for Option 1 the only differences between the base blackwater generation rate and the design blackwater generation rate are the fixture flow rates. However, Option 1 of WE credit 3.2 is based on a reduction in the wastewater generated, a reduction in potable water used for blackwater fixtures, or a combination of both. Therefore, the annual usage of other water sources, such as rainwater or graywater, for these fixtures can be subtracted from the total design blackwater generation rate. The designer should provide calculations to estimate the rate of these alternative waters available for use on an annual basis. For instance, rainwater availability is seasonal, so the totals available during the rainy seasons may be the only opportunity for potable water reduction.

Note also that there are local and state health and other regulatory codes that cover the use, storage, and possibly treatment of nonpotable waters in buildings. In addition, there is a large variation in the definitions used from locality to locality and from state to state on what constitutes graywater, similar to with what is included in blackwater. The designers, owners, and managers should become familiar with the appropriate local definitions and any special health-related requirements for the use of reuse waters or rainwaters so that any special protective measures or treatments can be included in the design and operations. The designer should include the appropriate local definitions of graywaters and other reuse waters, as well as the associated costs and systems in the design, and include a narrative of them in the submittal. Table 3.2.3 lists some of the various state interpretations of what may be included in graywater as taken from the Appendix 3: US State Regulations Compiled in 1999 on the www.weblife.org website, humanure section.

State	Shower, Bath, Bathroom Sink Wastewaters	Kitchen Sink, Dishwater Wastewaters	Laundry Washing Machine Wastewaters
Alabama	Yes	Yes*	Yes
Alaska	Yes	Yes	Yes
Arizona	Yes	Yes*	Yes
California	Yes	—	Yes
Colorado	Yes	Yes	Yes
Connecticut	Yes	Yes	Yes
Florida	Yes	—	Yes
Georgia	Yes	—	Yes
Hawaii	Yes	Yes	Yes
Idaho	Yes	—	Yes*
Kentucky	Yes	Yes*	Yes
Maine	Yes	Yes*	Yes
Maryland	Yes	—	Yes
Massachusetts	Yes	Yes*	Yes
Michigan	Yes	Yes	Yes
Minnesota	Yes	Yes	Yes
Missouri	Yes	—	Yes
Nebraska	Yes	—	Yes
Nevada	Yes	—	Yes
New Jersey	Yes	—	Yes
New Mexico	Yes	—	Yes
New York	Yes	Yes	Yes
Oregon	Yes	Yes	Yes
Rhode Island	Yes	Yes*	Yes
South Dakota	Yes	Yes	Yes
Texas	Yes	—	Yes
Utah	Yes	—	Yes
Washington	Yes	Yes	Yes

*With some exclusions.

Source: The information is from *Appendix 3: US State Regulations* compiled in 1999 by J. C. Jenkins on the www.weblife.org website, humanure section, as interpreted by Wu et al. (2007) with Permission from College Publishing.

TABLE 3.2.3   Waters Considered to Be Graywaters by Various States

Let GW be the annual graywater or other alternative nonpotable water source usage in the wastewater fixtures; then WE credit 2 is obtained via Option 1 if the following is true:

$$\frac{BWD - GW}{BWB} \leq 0.50 \tag{3.2.6}$$

Option 1: Residential Uses   The building designer should declare the design occupancy (DO) for the buildings. Note that design occupancies are not the same as maximum occupancies. In addition, usually residential facilities do not have urinals and do have equal gender distributions as noted in Table 3.2.1. Equation (3.2.2) can be used to estimate the total water closet usage. In this case, the transient occupancies are zero, there are no urinals, and there are no shifts. Using a water closet rate of 5 uses per day as given in Table 3.2.1, and substituting DO for FTE, Eq. (3.2.2) simplifies to Eq. (3.2.7) such that the estimated total annual number of water closet usages is

$$TWCU = 365 \times 5 \times DO \qquad \text{For residential uses} \tag{3.2.7}$$

Equations (3.2.4) through (3.2.6) can again be used for the Option 1 calculations relating to these residential uses. In cases where there will be no alternative nonpotable water used for the water closets, the calculations can be further reduced to the following simple equation for obtaining Option 1 WE credit 2:

$$\frac{WCR_i}{1.6} \leq 0.50 \qquad \begin{array}{l}\text{for } WCR_i \text{ in gallons per flush, for typical mixed-gender} \\ \text{residences with only potable water to water closets}\end{array} \tag{3.2.8}$$

Option 2   Option 2 of WE credit 2 is for providing on-site wastewater treatment that is treated to tertiary standards and infiltrated and/or reused on-site. The calculations use the design wastewater generation rate (BWD) as calculated in Eq. (3.2.5) (or as applicable for residential facilities). Let TBW be the annual volume of blackwater treated and infiltrated and/or reused on-site. Option 2 of WE credit 2 may then be obtained if the following is true:

$$\frac{TBW}{BWD} \geq 0.50 \tag{3.2.9}$$

The calculations for Option 2 do not include nonpotable waters used for flushing toilets in the numerator unless these nonpotable waters are part of the treated and reused blackwater generated (i.e., part of the TBW). In addition to the calculations, the submittals must include a narrative and additional information on how, and how much of, the blackwaters are treated, infiltrated, and/or reused on-site.

Special Circumstances and Exemplary Performance WE Credits 3.1 and 3.2

As mentioned previously, if the local jurisdiction defines blackwater to include other wastewaters, then the volumes of these wastewaters generated should be included in the total baseline and design blackwater generation rates (BWB and BWD) for both options. This should be appropriately noted in the submittals. If there are special circumstances for this credit, a narrative must be submitted describing these circumstances.

There is an EP point available for this credit for each option based on the following criteria:

- Option 1: There is no potable water used for sewage conveyance (BWD = GW).
- Option 2: 100 percent of these wastewaters are treated to tertiary standards and infiltrated or used on-site (TBW = BWD).

3.3 WE Credit Subcategory 3: *Water Use Reduction*

The intention of these credits is to reduce the use of potable water for indoor use. WEc3.1 is worth one point and WEc3.2 is one point *in addition to* the point for WEc3.1.

WE Credit 3.1: *Water Use Reduction—20% Reduction*

WE Credit 3.2: *Water Use Reduction—30% Reduction*

USGBC Rating System WE Credit 3.1

LEED-NC 2.2 lists the Intent, Requirements, and Potential Technologies and Strategies for WEc3.1 as follows:

Intent

Maximize water efficiency within buildings to reduce the burden on municipal water supply and wastewater systems.

Requirements

Employ strategies that in aggregate use 20% less water than the water use baseline calculated for the building (not including irrigation) after meeting the Energy Policy Act of 1992 fixture performance requirements. Calculations are based on estimated occupant usage and shall include only the following fixtures (as applicable to the building): water closets, urinals, lavatory faucets, showers and kitchen sinks.

Potential Technologies and Strategies

Use high-efficiency fixtures, dry fixtures such as composting toilet systems and non-water using urinals, and occupant sensors to reduce the potable water demand. Consider reuse of stormwater and greywater for non-potable applications such as toilet and urinal flushing and custodial uses.

USGBC Rating System WE Credit 3.2

LEED-NC 2.2 lists the Intent, Requirements, and Potential Technologies and Strategies for WEc3.2 as follows:

Intent

Maximize water efficiency within buildings to reduce the burden on municipal water supply and wastewater systems.

Requirements

Employ strategies that in aggregate use 30% less water than the water use baseline calculated for the building (not including irrigation) after meeting the Energy Policy Act of 1992 fixture performance requirements. Calculations are based on estimated occupant usage and shall include only the following fixtures (as applicable to the building): water closets, urinals, lavatory faucets, showers and kitchen sinks.

Potential Technologies and Strategies

Use high-efficiency fixtures, dry fixtures such as composting toilets and waterless urinals, and occupant sensors to reduce the potable water demand. Consider reuse of stormwater and greywater for non-potable applications such as toilet and urinal flushing, mechanical systems and custodial uses.

Calculations and Considerations WE Credits 3.1 and 3.2

The calculations for WE credits 3.1 and 3.2 are very similar to the calculations for Option 1 of WE credit 2, with three additional types of fixtures added: lavatory faucets, showers, and kitchen sinks. Bathtubs are not addressed, but the author assumes that all residential facilities that have bathtubs have them with a shower head, and the calculations are based on the shower head flow. The intention of the credit is not to reduce the volume of water as needed for proper bathing or food preparation and cleanup, but to limit the wasted water from unused faucet flow. Also, not addressed in these credits are dishwashers and clothes washers, although these may be considered in the ID credit calculations.

Many water usage calculations for other purposes are based upon the number of fixtures in a building. The calculations for WE credits 3.1 and 3.2 are not based on the number of fixtures, but rather on the number of users, as was done in the WE credit 2 calculations and will also differentiate between the number of male and female users in case they are not equal. However, in most cases, the default assumption is that the facility is gender-neutral, with one-half of the users male and the one-half female.

Commercial and Institutional Uses   The total annual water closet and urinal usage rates (in number of times used per year) are again estimated using the FTE estimates as developed in SSc4.2, defining the following:

FTE_j Full-time equivalent building occupant during shift j.

$FTE_{j,i}$ Full-time equivalent building occupancy of employee i during shift j. This is equal to 1 for a full-time employee and is equal to the normal hours worked (less than 8) divided by 8 for a part-time employee. Obviously, this will need to be modified if shifts are different from the standard 8 h.

The FTE_j is then determined using the following set of equations:

$$FTE_{j,i} = (\text{worker } i \text{ h})/8 \text{ h} \qquad \text{where } 0 < FTE_{j,i} \leq 1 \qquad (2.4.1)$$

$$FTE_j = \sum FTE_{j,i} \qquad \text{for all employees in shift } j \qquad (2.4.2)$$

Then the full-time employee equivalents need to be subdivided by gender. For each shift the numbers for male and female employees may not be equal. Let $FTEM_j$ and $FTEF_j$ be the male full-time employee equivalent and the female full-time employee equivalent for shift j, respectively. To be consistent throughout the LEED submittal, the following equality must hold for all shifts j:

$$FTE_j = FTEM_j + FTEF_j \qquad \text{for all shifts } j \qquad (3.2.1)$$

In SS credit 4.2, the transient occupancies are estimated as average transient occupancies at any one time over a shift, not as the total number of transient occupants who may go in or out of the building during the shift. This is done to estimate the peak number of bicycle rack spaces needed at any time. However, for WE credits 3.1 and 3.2, the LEED-NC 2.2 Reference Guide bases transient uses on the *total number of transients*, not the average over the shifts. To be consistent with this, let us define TOWM and TOWF to be the estimated total number of male and female transients, respectively, during a day. TOW, as previously described, would be the summation of TOWM and TOWF.

As mentioned previously, LEED-NC 2.2 recommends that the typical daily fixture usage rates for water closets and urinals be separately analyzed for the different genders, building usages, and occupancies. Let $WCUM_i$ and $WCUF_i$ be the male water closet and female water closet usage rates, respectively, for various types of buildings or occupancies. Let UUM_i and UUF_i be the male and female urinal usage rates, respectively, for various building types and occupancies. Table 3.3.1 gives some typical default values for these daily fixture usage rates. It is important to also analyze the building usage on an annual basis. Let ND_{FTE} and ND_{TOW} be the number of days in a year that each type of occupancy (FTE or TOW) uses the building. Usually ND_{FTE} is 260 days for

Fixture Type	Gender	Daily Uses per Person				
		Variable Acronym	FTE* Uses	TOW: Student/ Visitor	TOW: Retail Customer	Resident†
Water closet use	Male	$WCUM_i$	1	0.1	0.1	5
Water closet use	Female	$WCUF_i$	3	0.5	0.2	5
Urinal use	Male	UUM_i	2	0.4	0.1	n/a
Urinal use	Female	UUF_i	0	0	0	n/a
Lavatory faucet (duration 15 s)	n/a	LFU_i	3	0.5	0.2	5
Lavatory faucet with autocontrol (duration 12 s)	n/a	LFU_i	3	0.5	0.2	5
Shower (duration 300 s)	n/a	SU_i	0.1	0	0	1
Kitchen sink (duration 15 s)	n/a	KSU_i	1	0	0	4

*The students as listed in TOW represent transient students, such as those attending a college where they may take only a handful of classes in one building on certain days. Students in primary schools and other schools where they remain in the same building every "work" day for the entire school day are actually similar to full-time employees and should be treated as such. This is not specifically addressed in LEED-NC 2.2 and should therefore be noted in the calculations and the submittals.

†These resident daily usage rates represent residents who on average also spend part of many of their days elsewhere, such as at school or work. It is recommended that for residential facilities where there are usually 24-h occupancies, such as nursing homes, the resident rates for water closets and lavatories be, at a minimum, 8, which is a sum of 3 (for full-time employees) and 5 (for residences). This is not specifically addressed in LEED-NC 2.2 and should therefore be noted in the calculations and the submittals.

TABLE 3.3.1   Default Typical Daily Fixture Use by Gender

office buildings and 365 days for residential or retail. For college buildings, ND_{FTE} may be 260 days for staff and faculty, but ND_{TOW} may be much less for student occupancies. Just as in WEc2, summing the fixture usage rates times the occupancies over the various shifts as applicable for the entire year gives the total number of times water closets and urinals are typically used in a year (TWCU and TUU, respectively). These were previously given in Eqs. (3.2.2) and (3.2.3) and are reiterated here:

$$TWCU = ND_{FTE} \times (\Sigma \ (FTEM_j \times WCUM_i) + (\Sigma \ (FTEF_j \times WCUF_i))$$
$$+ ND_{TOW} \times [(TOWM \times WCUM_i) + (TOWF \times WCUF_i)]$$
over all shifts j and for each type of use i $\qquad$ (3.2.2)

$$TUU = ND_{FTE} \times (\Sigma \ (FTEM_j \times UUM_i) + (\Sigma \ (FTEF_j \times UUF_i))$$
$$+ ND_{TOW} \times [(TOWM \times UUM_i) + (TOWF \times UUF_i)]$$
over all shifts j and for each type of use i $\qquad$ (3.2.3)

Lavatory faucets, showers, and kitchen sink usage rates are usually not considered to be gender-specific. It is typically assumed that the lavatory faucet is used for every use of a water closet and/or urinal, which combined should be the same for either male or female occupants. The LEED-NC 2.2 Reference Guide also gives default baseline values for lavatory, shower, and kitchen sink (separated for residential and nonresidential uses) which can be found with the previously listed blackwater fixture daily use rates in Table 3.3.1. These values are based on EPAct 1992.

Using these variables, the equations for total lavatory faucet use (TLFU), total kitchen sink use (TKSU), and total shower use (TSU) are

$$TLFU = [ND_{FTE} \ \Sigma \ (FTE_j \times LFU_i)] + [ND_{TOW} \ (TOW \times LFU_i)]$$
over all shifts j and for each type of use i $\qquad$ (3.3.1)

$$TKSU = [ND_{FTE} \ \Sigma \ (FTE_j \times KSU_i)] + [ND_{TOW} \ (TOW \times KSU_i)]$$
over all shifts j and for each type of use i $\qquad$ (3.3.2)

$$TSU = [ND_{FTE} \ \Sigma \ (FTE_j \times SU_i)] + [ND_{TOW} \ (TOW \times SU_i)]$$
over all shifts j and for each type of use i $\qquad$ (3.3.3)

The equations for water volume usage based on lavatory faucet, shower, and kitchen sink uses have an additional variable that was not included in the equations for blackwater generation. Unlike toilets and urinals, where the uses are calculated on a volume per flush or use, sink and shower volumes per use are based on both flow rates from the fixture and a time or duration that the fixture is kept turned on. The applicable durations are given in Table 3.3.1. Note that Table 3.3.1 gives two different durations for lavatory faucets, depending on whether there is, or is not, an autocontrol for shutoff. The baseline EPAct 1992 (conventional) flow rates and water-efficient flow rates from the LEED-NC 2.2 Reference Guide for lavatory faucets, kitchen sinks, and showers (LFR_i, KSR_i, and SR_i, respectively) are given in Table 3.3.2, as are the previously given values for the water closets and urinals from Table 3.2.2.

A special note is again needed for some of the blackwater generation rates in Table 3.3.2. Composting toilets and nonwater urinals do not use water for regular usage. They are not currently customary fixtures in the United States. Therefore, they have different

Specific Fixture Type	Gallons Generated per Use [Gallons per flush (gpf)] (WCR$_i$ or UR$_i$)	Gallons Generated per Minute (gpm) (LFR$_i$, KSR$_i$, or SR$_i$)
Baseline water closet	1.6	n/a
Low-flow water closet	1.1	n/a
Dual flush and low-flow water closet	0.8	n/a
Composting toilet	0	n/a
Baseline urinal	1.0	n/a
Low-flow urinal	0.5	n/a
Nonwater urinal	0	n/a
Conventional lavatory faucet	n/a	2.5
Low-flow lavatory faucet	n/a	1.8
Ultralow-flow lavatory faucet	n/a	0.5
Kitchen sink	n/a	2.5
Low-flow kitchen sink	n/a	1.8
Shower	n/a	2.5
Low-flow shower	n/a	1.8

*The project team may add other fixture types to this list as applicable.

TABLE 3.3.2 Typical Flow Rates by Fixture Type*

maintenance and waste collection needs than more common water closets and urinals. The designers and owners should be very cognizant of the different maintenance needs, and it is recommended that they determine a way to meet these special needs early in the design phase of a project so that there are no additional concerns, costs, or changes later in the project.

Now both the baseline and the design annual water volume usage rates for the fixtures applicable to WE credits 3.1 and 3.2 can be determined. The applicable indoor water usage baseline (IWUB), in gallons per year, can be calculated using the durations in Table 3.3.1 and the fixture flow rate numbers in Table 3.3.2 as follows:

$$IWUB_{no\text{-}auto} = (1.6 \times TWCU) + (1.0 \times TUU) + [2.5(0.25 \text{ min}) \times TLFU]$$
$$+ [2.5(0.25 \text{ min}) \times TKSU] + [2.5(5 \text{ min}) \times TSU]$$

for lavatory faucets *without* autocontrols, in gallons per year   (3.3.4)

or

$$IWUB_{auto} = (1.6 \times TWCU) + (1.0 \times TUU) + [2.5(0.2 \text{ min}) \times TLFU]$$
$$+ [2.5(0.25 \text{ min}) \times TKSU] + [2.5(5 \text{ min}) \times TSU]$$

for lavatory faucets *with* autocontrols, in gallons per year   (3.3.5)

To calculate the applicable design indoor water usage (IWUD), the factors in Eqs. (3.3.4) and (3.3.5) are replaced with those appropriate for the alternate fixture type i used such as LFR_i, KSR_i, and SR_i from Table 3.3.2:

$$IWUD_{no\text{-}auto} = (WCR_i \times TWCU) + (UR_i \times TUU) + [LFR_i (0.25 \text{ min}) \times TLFU]$$
$$+ [KSR_i (0.25 \text{ min}) \times TKSU] + [SR_i (5 \text{ min}) \times TSU]$$

for lavatory faucets *without* autocontrols, in gallons per year   (3.3.6)

or

$$IWUD_{auto} = (WCR_i \times TWCU) + (UR_i \times TUU) + [LFR_i (0.2 \text{ min}) \times TLFU]$$
$$+ [KSR_i (0.25 \text{ min}) \times TKSU] + [SR_i (5 \text{ min}) \times TSU]$$

for lavatory faucets *with* autocontrols, in gallons per year   (3.3.7)

Note that the only differences between the base water usage rate and the design water usage rate are the fixture flow rates. If autocontrols are used in the design case, then autocontrols are assumed in the baseline case. This is different from LEED-NC 2.1, and designs which meet the version 2.1 criterion might not meet the version 2.2 criterion.

As in WE credit 3.2, a reduction in potable water used for blackwater fixtures can be from a reduction in the flows, the alternate use of nonpotable water in these fixtures, or a combination of both. Therefore, the annual usage of other water sources, such as rainwater or graywater, for these fixtures can be subtracted from the total design water usage rate. The designer should provide calculations to estimate the rate of these alternative waters available for use on an annual basis. For instance, rainwater availability is seasonal, so the totals available during the rainy seasons may be the only opportunity for potable water reduction. Note also that there are local and state health and other regulatory codes that cover the use, storage, and possibly treatment of nonpotable waters in buildings. The designers, owners, and managers should become familiar with any special health-related requirements for use of reuse waters or rainwaters so that any special protective measures or treatments can be included in the design and operations. The designer should include the associated costs and systems in the design and include a narrative of them in the submittal. Let GW be the annual graywater or other alternative water source usage in the wastewater fixtures; then WE credit 3.1 is obtained if the following is true:

$$\frac{IWUD - GW}{IWUB} \leq 0.80 \tag{3.3.8}$$

In like manner, a point for WE credit 3.2 is obtained, in addition to the point for WE credit 3.1, if the following is also true:

$$\frac{IWUD - GW}{IWUB} \leq 0.70 \tag{3.3.9}$$

Residential Uses   The building designer should declare the design occupancy (DO) for the buildings. Usually, residential facilities do not have urinals, and they typically have equal gender distributions. Equation (3.2.2) can again be used to estimate the total water closet usage. In this case, the transient occupancies are zero, there are no urinals, and there are no shifts. By using a water closet rate of 5 uses per day as given in Table 3.2.1, and substituting DO for FTE, Eq. (3.2.2) simplifies to Eq. (3.3.10) such that the estimated total annual number of water closet usages per year is

$$TWCU = 365 \times DO \times 5 \qquad \text{for residential uses in number of uses per year} \qquad (3.3.10)$$

In like manner, Eqs. (3.3.1) to (3.3.3) can be modified for annual lavatory, kitchen sink ,and shower usage rates in number of uses per year to

$$TLFU = 365 \times DO \times 5 \qquad \text{for residential uses in number of uses per year} \qquad (3.3.11)$$

$$TKSU = 365 \times DO \times 4 \qquad \text{for residential uses in number of uses per year} \qquad (3.3.12)$$

$$TSU = 365 \times DO \qquad \text{for residential uses in number of uses per year} \qquad (3.3.13)$$

Similar to the commercial and institutional derivation, now both the baseline and the design annual water volume usage rates for the fixtures applicable to WE credits 3.1 and 3.2 can be determined. The applicable indoor water usage baseline (IWUB), in gallons per year, can be calculated using the durations in Table 3.3.1 and the fixture flow rate numbers in Table 3.3.2:

$$
\begin{aligned}
IWUB_{no\text{-}auto} = {} & (1.6 \times TWCU) + [2.5(0.25 \text{ min}) \times TLFU] \\
& + [2.5(0.25 \text{ min}) \times TKSU] + [2.5(5 \text{ min}) \times TSU] \\
& \text{for residences with lavatory faucets } \textit{without} \text{ autocontrols, gal/yr} \qquad (3.3.14)
\end{aligned}
$$

or

$$
\begin{aligned}
IWUB_{auto} = {} & (1.6 \times TWCU) + [2.5(0.2 \text{ min}) \times TLFU] \\
& + [2.5(0.25 \text{ min}) \times TKSU] + [2.5(5 \text{ min}) \times TSU] \\
& \text{for residences with lavatory faucets } \textit{with} \text{ autocontrols, gal/yr} \qquad (3.3.15)
\end{aligned}
$$

To calculate the applicable design indoor water usage (IWUB), the factors in Eqs. (3.3.14) and (3.3.15) are replaced with those appropriate for the alternate fixture type i used such as LFR_i, KSR_i, and SR_i from Table 3.3.2:

$$
\begin{aligned}
IWUD_{no\text{-}auto} = {} & (WCR_i \times TWCU) + [LFR_i(0.25 \text{ min}) \times TLFU] \\
& + [KSR_i(0.25 \text{ min}) \times TKSU] + [SR_i(5 \text{ min}) \times TSU] \\
& \text{for residences with lavatory faucets } \textit{without} \text{ autocontrols, gal/yr} \qquad (3.3.16)
\end{aligned}
$$

or

$$
\begin{aligned}
IWUD_{auto} = {} & (WCR_i \times TWCU) + [LFR_i(0.2 \text{ min}) \times TLFU] \\
& + [KSR_i(0.25 \text{ min}) \times TKSU] + [SR_i(5 \text{ min}) \times TSU] \\
& \text{for residences with lavatory faucets } \textit{with} \text{ autocontrols, gal/yr} \qquad (3.3.17)
\end{aligned}
$$

Note that the only differences between the base water usage rate and the design water usage rate are the fixture flow rates. If autocontrols are used in the design case, then autocontrols are assumed in the baseline case (although autocontrols are rare in most residential facilities). This is different from LEED-NC 2.1.

Similar to the cases for commercial/institutional buildings, a reduction in potable water used for blackwater fixtures can be due to a reduction in the flows, the alternate use of nonpotable water in these fixtures, or a combination of both. Therefore, Eqs. (3.3.8) and (3.3.9) can be used to determine if WE credits 3.1 and 3.2 are obtained. However, there will rarely be a case in which reuse waters or rainwaters (GW) are used in water closets for residential uses. Thus, WE credit 3.1 is obtained when there are no reuse waters or rainwaters used for typical indoor fixtures if the following is true:

$$\frac{\text{IWUD}}{\text{IWUB}} \le 0.80 \qquad \begin{array}{l}\text{for no reuse waters or rainwaters (GW)} \\ \text{used in indoor fixtures}\end{array} \qquad (3.3.18)$$

In like manner, a point for WE credit 3.2 is obtained, in addition to the point for WE credit 3.1, if the following is also true:

$$\frac{\text{IWUD}}{\text{IWUB}} \le 0.70 \qquad \begin{array}{l}\text{for no reuse waters or rainwaters (GW)} \\ \text{used in indoor fixtures}\end{array} \qquad (3.3.19)$$

Special Circumstances and Exemplary Performance

If there are special circumstances for these credits, a narrative must be submitted describing these circumstances.

There are in fact two different Innovation and Design (ID) points which can be obtained related to the WE credit 3 subcategory. One is an EP point, and the other is based on non-EPAct regulated fixtures.

The EP point available in the Innovation and Design (ID) category for Water Use Reduction as per LEED-NC 2.2 may be awarded by decreasing the water usages as defined in WEc3.1 (20 percent) and WEc3.2 (30 percent) by 40 percent. It is related to the overall subcategory and brings the total points available for overall indoor potable water use reduction of the EPAct regulated fixtures as listed in Table 3.3.1 to three if the following inequality is true:

$$\frac{\text{IWUD} - \text{GW}}{\text{IWUB}} \le 0.60 \qquad (3.3.20)$$

Project teams are also encouraged to earn an ID credit for reduction of water usage in non-EPAct regulated and/or process water (non-EPAct) consuming fixtures. These might include dishwashers and clothes washers for residential or applicable retail and commercial uses, or cooling towers for applicable uses. The criterion is based on a reduction from these changes to non-EPAct regulated or process water consuming fixtures, which is at least equal to 10 percent of the total design regulated water use (10 percent of IWUD). Therefore, for the applicable non-EPAct fixtures involved, the baseline water usage (non-EPActB) and the design water usage (non-EPActD) should be calculated. An additional ID credit may be obtained if the following is true:

$$\frac{\text{non-EPActB} - \text{non-EPActD}}{\text{IWUD}} \ge 0.10 \qquad (3.3.21)$$

This brings the total number of points available based on this subcategory for indoor water use reduction to four.

3.4 Discussion and Overview

As previously mentioned, the exemplary performance point available in the Innovation and Design (ID) category for Water Use Reduction as per LEED-NC 2.2 may be awarded by decreasing the water usages as defined in WEc3.1 (20 percent) and WEc3.2 (30 percent) by 40 percent. It is related to the overall subcategory and brings the total points available for overall indoor potable water use reduction to three. An additional ID point is available if other, non-EPAct regulated, indoor water uses are decreased by an amount equal to at least 10 percent of the design EPAct regulated water uses. This brings the total points available for overall indoor potable water use reduction to four. In addition, WEc2, *Innovative Wastewater Technologies,* allows for an additional one point if the reduction in potable water use aids in reducing "black" wastewater by 50 percent (or the treatment option), with another additional Innovation and Design point available if no potable water is used for black wastewater (or for the treatment option, if all the blackwaters are appropriately treated and infiltrated or reused on-site).

Therefore WEc2 and WEc3 together account for six total available points. Then if potable water is not used for irrigation and both WEc1.1 and WEc1.2 are attained, the total points available for a project with reduced potable water usage can go to a maximum of eight. This is a large percentage of the points needed for certification and can also result in large reductions in water-related operating costs. Furthermore, if some of the alternative water sources include rainwater, these activities can facilitate obtaining one or both of the stormwater credits (SSc6.1 and SSc6.2).

Care should be taken to evaluate the site area for the project for both the WEc1 credits and many of the Sustainable Sites subcategories. Maximizing naturally vegetated open spaces and designing the project around them can aid in stormwater management, reducing irrigation needs, lowering urban heat island impacts, and encouraging ecological diversity and sustainability. In addition, full-time, transient, and design occupancies are important in the calculations for the WEc2, WEc3, and SSc4 subcategories. These should be established early in the project phase and used consistently throughout.

References

Allen, R. G., et al. (Eds.) (2005), *Standardized Reference Evapotranspiration Equation*, The American Society of Civil Engineers, Reston, Va.

EPA ORD (2002), *Onsite Wastewater Treatment Systems Manual,* U.S. Environmental Protection Agency, Office of Research and Development, website accessed July 10, 2007, http://www.epa.gov/ORD/NRMRL/Pubs/625R00008/625R00008.htm.

EPA OW (2007), http://www.epa.gov/water/, U.S. Environmental Protection Agency, Office of Water, website accessed July 10, 2007.

EPA OW (2007), http://www.epa.gov/watersense/, U.S. Environmental Protection Agency, Office of Water, Water Sense; Efficiency Made Easy, website accessed July 10, 2007.

EPA OW (2007), "National Efficiency Standards and Specifications for Residential and Commercial Water-Using Fixtures and Appliances," U.S. Environmental Protection Agency, Office of Water, Water Sense Document, http://www.epa.gov/watersense/docs/matrix508.pdf, accessed July 10, 2007.

IA (2006), www.irrigation.org/about_et_list.htm, evapotranspiration rates on the Irrigation Association website, accessed January 2006.

IA (2007), http://www.irrigation.org/SIM/default.aspx?pg=consumer_resources. htm&id=234, website for irrigation resources, Irrigation Association, accessed June 14, 2007.

Jenkins, J. C. (1999), http://www.weblife.org/humanure/, website accessed July 26, 2007 with information from *The Humanure Handbook.* Jenkins Publishing, Grove City, Pa.

Poremba, S. M. (2007), "Smart Solutions," *Water Efficiency, The Journal for Water Conservation,* July/August, 2(4), Forester Communications, Santa Barbara, Calif.

Rainbarrel Guide (2007), http://www.rainbarrelguide.com/, Rainbarrel Guide webpage accessed July 12, 2007.

Rainbird (2007), http://www.rainbird.com/, website for Rain Bird Corporation accessed July 10, 2007.

RMI (2007), http://www.rmi.org/, Rocky Mountain Institute webpage, accessed July 10, 2007.

USGBC (2003), *LEED-NC for New Construction, Reference Guide,* Version 2.1, 2d ed., May, U.S. Green Building Council, Washington, D.C.

USGBC (2005–2007), *LEED-NC for New Construction, Reference Guide,* Version 2.2, 1st ed., U.S. Green Building Council, Washington, D.C., October 2005 with errata posted through Spring 2007.

Wu, X., B. Akinci, and C. I. Davidson (2007), "Modeling Graywater in Residences: Using Shower Effluent in the Toilet Reservoir," *Journal of Green Building,* Spring, 2(2): 111–120.

Exercises

1. You plan to plant 24,000 ft^2 of area with vegetation on a site in Columbia, S.C.
   A. You plan to use a typical sprinkler system. What is your expected TPWA for the following combination of plantings?
      - 8000 ft^2 will be north of the planned building and will be mixed tree and shrub species with average water needs, densely planted.
      - 8000 ft^2 will be average density, low-water-use lawn on a sloped area west of the building.
      - The remaining will be a shrubbed landscape bed, with species of average water needs and spaced around normally.
   B. You then decide to try to get WE credit 1.1 by keeping all the areas natural with the existing woods that require no irrigation except for the lawned area mentioned. Will you meet the credit requirement with only these changes?

2. You plan to plant 24,000 ft^2 of area with vegetation on a site in East Hampton on Long Island, N.Y.
   A. You plan to use a typical sprinkler system. What is your expected TPWA for the following combination of plantings?
      - 8000 ft^2 will be north of the planned building and will be mixed tree and shrub species with average water needs, densely planted.
      - 8000 ft^2 will be average density, low-water-use lawn on a sloped area west of the building.
      - The remaining will be a shrubbed landscape bed, with species of average water needs and spaced around normally.
   B. You then decide to try to get WE credit 1.1 by keeping all the areas natural with the existing woods that require no irrigation except for the lawned area mentioned. Will you meet the credit requirement with only these changes?

3. You plan to plant 20,000 ft^2 of area with vegetation on a site in San Francisco.
   A. You plan to use a typical sprinkler system. What is your expected TPWA for the following plantings?

  - 6000 ft^2 will be north of the planned building and will be mixed tree and shrub species with average water needs, densely planted.
  - 6000 ft^2 will be average density, low-water-use lawn on a sloped area west of the building.
  - The remaining will be a shrubbed landscape bed, with species of high water needs and the plants spaced around normally.

  B. You then decide to try to get WE credit 1.1 by using low-water-need plants and a drip irrigation system in the landscape bed. You can also extend the modified landscape bed into the lawn area. How much of the lawn area would need to be converted to obtain this credit?

4. Your client really would like an expansive, thick lawn in front of her new office building and would like to obtain at least one WEc1 point by using rainwater harvesting. The building is in Austin, Tex. The building could be oriented with the front facade facing either north or west. You are deciding between a low-water and an average water-need lawn. Determine how much rainwater harvesting is needed to irrigate the lawn with a sprinkler system per square foot for the four possible options for the month of July.

5. You are building an office building for 150 men and 100 women. Shower facilities will be provided for those who exercise before work or during lunch. You do not want to use automatic controls on the lavatory faucets.

  A. Calculate a baseline water usage.
  B. To be more green, you decide to go with the low-flow water closets, waterless urinals, and low-flow lavatory sinks. Do calculations to see if you get WE credits 2, 3.1, and 3.2.

6. You are building an office building for 150 men and 100 women. Shower facilities will be provided for those who exercise before work or during lunch. You decide to use automatic controls on the lavatory faucets.

  A. Calculate a baseline water usage.
  B. To be more green, you decide to go with the low-flow water closets, waterless urinals, and low-flow lavatory sinks. Do calculations to see if you get WE credits 2, 3.1, and 3.2.

7. You are building an apartment complex with 80 three-bedroom units near an elementary school. The facility will include a playground and pool complex. The designers estimate that there will be on average 3.5 occupants in each apartment.

  A. Calculate a baseline water usage.
  B. To be more green, you decide to go with the low-flow water closets, low-flow showers, and low-flow lavatory sinks. Do calculations to see if you get WE credits 2, 3.1, and 3.2.

8. You are building an apartment complex with 80 three-bedroom units near the downtown office district. The designers estimate that there will be on average 2.5 occupants in each apartment.

  A. Calculate a baseline water usage.
  B. To be more green, you decide to go with low-flow water closets, low-flow showers, and low-flow kitchen sinks. Do calculations to see if you get WE credits 2, 3.1, and 3.2.

9. You are building an industrial building for a mixed workforce. The day shift will have about 150 employees, and about 20 employees will be needed to work both the swing and the graveyard shifts to keep the machines running. Shower facilities will be provided. You originally do not want to use automatic controls on the lavatory faucets.

  A. Calculate a baseline water usage.
  B. You have heard about the new dual flush and low-flow water closets and decide to use them in the women's rooms. You also decide to go with low-flow urinals. Which WE credits might you earn?
  C. What minimum changes would need to be made to the specifications for the lavatories to earn the exemplary performance point for WEc3?

10. You are building an industrial building for a mixed workforce. The day shift will have about 200 employees, and about 20 employees will be needed to work both the swing and the graveyard shifts to keep the machines running. Shower facilities will be provided. You originally want to use automatic controls on the lavatory faucets.

   A. Calculate a baseline water usage.

   B. Your design will be using low-flow water closets, low-flow urinals, and low-flow lavatories. Which WE credits might you earn?

   C. You have some options to use some special controls on the process water for the plant. How much would the process water need to be reduced to obtain the Innovation and Design credit for non-EPAct fixtures?

11. Rewrite Eqs. (3.3.18) and (3.3.19) for residential uses if there are reuse waters (GW) used. Comment on the health and other protective measures which might need to be taken if these reuse waters are used.

CHAPTER 4

LEED Energy and Atmosphere

The Energy and Atmosphere category of the U.S. Green Building Council (USGBC) rating system for new construction version 2.2 (LEED-NC 2.2) consists of three prerequisite subcategories and six credit subcategories that together may earn a possible 17 points (credits) maximum, not including exemplary performance points. The notation format for the prerequisites and credits is, for example, EAp1 for Energy and Atmosphere prerequisite 1 and EAc1 for Energy and Atmosphere credit 1. To facilitate cross-referencing between this text and the USGBC rating system, the second digit in a section heading, equation, table, or figure number represents the credit subcategory number for sections that deal directly with a USGBC LEED-NC 2.2 credit subcategory.

The Energy and Atmosphere (EA) portion deals with practices and policies that reduce the use of energy at the site, reduce the use of nonrenewable energy both at the site and at the energy source, and reduce the impact on the global climate, atmosphere, and environment from both activities at the site and energy sources off-site. Many of the subcategories in the Sustainable Sites category also deal with practices that may have an impact on the climate, but usually on a local scale.

The reason why Energy and Atmosphere are combined is that a significant portion of the air pollution and global climate impacts come from energy sources. Therefore reducing or changing these energy sources has a large impact on the atmosphere, particularly on a more regional or global scale. For example, fossil-based energy sources, such as fuel oil and coal used in the production of energy, may also release many pollutants to the air and release additional carbon dioxide into the atmosphere. Carbon dioxide is considered to be a greenhouse gas, and it is considered to be nonrenewable to use these types of fuels, both from an energy source perspective and from a global carbon cycle perspective. Carbon dioxide emissions can be from the production of materials and other practices related to construction, but since most of the energy used in the United States is fossil fuel–based, reducing energy usage is advantageous in a resource and economic sense as well as in an atmospheric and air pollution prevention sense. To be inclusive, other credits related to regional or global climate issues are also included in the Energy and Atmosphere category, even though they are not specific only to energy issues. Also, one of the credits for enhanced commissioning is not limited solely to energy systems, but since a large proportion of the systems being commissioned are energy-related, the entire enhanced commissioning subcategory is included under Energy and Atmosphere.

The three prerequisite subcategories are EA Prerequisite 1: *Fundamental Commissioning of the Building Energy Systems* (EAp1), EA Prerequisite 2: *Minimum Energy Performance* (EAp2), and EA Prerequisite 3: *Fundamental Refrigerant Management* (EAp3). As is typical for prerequisites, there are no points associated with any of these three.

The points for the credits in the Energy and Atmosphere category are not individually related to a specific credit under a subcategory as was done in the Sustainable Sites and Water Efficiency categories. Instead, the six credit subcategories are each referred to as credits, and two out of the six may be worth more than one point. The six credit subcategories (a.k.a. credits) are EA Credit 1: *Optimize Energy Performance* (EAc1), for which a total of 10 points is available; EA Credit 2: *On-Site Renewable Energy* (EAc2), for which a total of 3 points is available. EA Credit 3: *Enhanced Commissioning* (EAc3), EA Credit 4: *Enhanced Refrigerant Management* (EAc4), EA Credit 5: *Measurement and Verification* (EAc5), and EA Credit 6: *Green Power* (EAc6) all have one point available for each credit. These points do not include exemplary performance (EP) points, which will be discussed later.

The prerequisites and credits in the Energy and Atmosphere category can be grouped into the following four overall sets of goals:

- EAp1 and EAc3 provide for commissioning, a design and construction service that aids in improving various system efficiencies and reliabilities at a site.

- EAp2, EAc1, and EAc5 help optimize the energy performance and reduce energy usage at the site.

- EAp3 and EAc4 reduce the amount of pollutants at the site that are related to the ozone hole in the stratosphere. These pollutants are ranked by their ozone-depleting potential (ODP). EAc4 takes into consideration those ODP pollutants that might also contribute to an index referred to as the *global warming potential* (GWP).

- EAc2 and EAc6 promote the use of renewable or "green" energy sources. EAc2 promotes its use on-site, and EAc6 promotes its use off-site.

The three prerequisites and six credit subcategories in the Energy and Atmosphere category are as follows:

- **EA Prerequisite 1** (EAp1): *Fundamental Commissioning of the Building Energy Systems* (previously referred to as *Fundamental Building Systems Commissioning* in LEED-NC 2.1)

- **EA Prerequisite 2** (EAp2): *Minimum Energy Performance*

- **EA Prerequisite 3** (EAp3): *Fundamental Refrigerant Management* (previously referred to as *CFC Reduction in HVAC&R Equipment* in LEED-NC 2.1)

- **EA Credit 1** (EAc1): *Optimize Energy Performance* (valued up to 10 points) (EB)

- **EA Credit 2** (EAc2): *On-Site Renewable Energy* (previously referred to as *Renewable Energy* in LEED-NC 2.1 and valued up to 3 points)

- **EA Credit 3** (EAc3): *Enhanced Commissioning* (previously referred to as *Additional Commissioning* in LEED-NC 2.1) (EB)

- **EA Credit 4** (EAc4): *Enhanced Refrigerant Management* (previously referred to as *Ozone Depletion* in LEED-NC 2.1)

- **EA Credit 5** (EAc5): *Measurement and Verification* (EB)
- **EA Credit 6** (EAc6): *Green Power*

The LEED-NC 2.2 Reference Guide provides a table that summarizes some of these prerequisite and credit characteristics. Specifically, the following three items are summarized:

- A significant change from LEED version 2.1
- The main phase of the project for the submittal document preparation (Design or Construction)
- The project team member(s) who have the greatest decision-making responsibility as to whether this credit may or may not be achieved (owner, design team, or contractor)

Note that the submittals for each credit or prerequisite have been sectioned into either Design Submittals and/or Construction Submittals. Even though the entire LEED process encompasses all phases of a project, the prerequisites or credits that are noted as Construction Submittal credits usually have major actions going on through the construction phase that, unless documented through that time, may not be able to be verified from preconstruction documents or the built project. Sometimes these credits have activities associated with them that cannot or need not be performed until the beginning of, or subsequent to, the construction phase. EAp1, EAc3, EAc5, and EAc6 have been noted in the construction submittal column in Table 4.0.1.

Table 4.0.1 also lists the maximum available points, not including EP or Innovation and Design (ID) points for the associated credit subcategory. This has been noted since each separate credit title does not necessarily go directly with only one point in the Energy and Atmosphere category.

Three of the credit subcategories listed in Table 4.0.1 have the icon EB noted after their title (EAc1, EAc3, and EAc5). This icon is a tool to help those who wish to proceed with the continuing LEED-EB certification (Existing Building) after certification of LEED-NC (New Construction or Major Renovation) is obtained. Those credits noted with this icon are usually significantly more cost-effective to implement during the construction of the building than later during its operation. They are also shown in Table 4.0.1.

There are EP points under the Innovation and Design category available that relate to several items in other categories. These are available for exceeding each of the respective credit criteria to a minimum level as noted in the respective credit descriptions. There is only one EP point available for EA credit 6, green power. This is also noted in Table 4.0.1.

Two of the variables which must be determined for each project and which must be kept consistent throughout the credits in a LEED submittal are the size and location of the project site. The location of the site boundary is particularly important to credit EAc2, since the locations of renewable energy sources on-site are the basis of this credit. If renewable sources are located off-site, but used on-site, then they may only be applied to EAc6. There may be exceptions to this for special circumstance, such as campus settings. Also note that there is no rule that project areas are always contiguous, so it might be possible to include portions of some campus renewable energy sources in the project; however, this might then impact other credits, such as SSc2. Therefore, there might be a need for narratives describing any such variations and how the project still

Prerequisite or Credit	Construction Submittal	EB Icon	Maximum Credit Point Total (Not Including Exemplary Performance or ID points)	Exemplary Performance and other ID Point Availability	Exemplary Performance Point Criterion	Location of the Site Boundary and the Site Area Important in the Calculations	Related Prerequisite
EAp1: Fundamental Commissioning of the Building Energy Systems	*	n/a	n/a				n/a
EAp2: Minimum Energy Performance		n/a	n/a				n/a
EAp3: Fundamental Refrigerant Management		n/a	n/a				n/a
EAc1: Optimize Energy Performance		Yes	10	No	n/a	No	EAp2
EAc2: On-Site Renewable Energy		No	3	No	n/a	Yes	n/a
EAc3: Enhanced Commissioning	*	Yes	1	No	n/a	No	EAp1
EAc4: Enhanced Refrigerant Management		No	1	No	n/a	No	EAp3
EAc5: Measurement and Verification	*	Yes	1	No	n/a	No	n/a
EAc6: Green Power	*	No	1	Yes	Doubling the amount or the contract length	Not usually	n/a

TABLE 4.0.1   EA EB Icon, Exemplary Performance Point Availability, and Miscellaneous

meets the intent of the credit. Usually, however, separate campus facilities for renewable energy sources are counted as green power for EAc6. Any renewable energy that is counted for EAc6 cannot also be counted toward EAc2. The importance of the LEED boundary to EAc2 is also noted in Table 4.0.1.

Three of the credit subcategories are actually optimizations or enhancements of EA prerequisites. The final column in Table 4.0.1 lists the prerequisites that are in this way related to the credits.

In summary, most of the Energy and Atmosphere credits address problems associated with energy use and production by a combination of methods. They aim to reduce energy usage and to use forms of energy that have a less harmful impact on the environment. Although several of the credits in the Energy and Atmosphere category might include a capital cost investment, many provide a means to potentially reduce the associated utility costs over the life of the building. The three prerequisite subcategories *must* be adhered to and verified for each project.

Energy and Atmosphere Prerequisites

EA Prerequisite 1: *Fundamental Commissioning of the Building Energy Systems*

This prerequisite is intended to aid in the efficiency and verification of the energy systems in a building.

USGBC Rating System

LEED-NC 2.2 lists the Intent, Benefits of Commissioning, Requirements, Commissioned Systems, and Potential Technologies and Strategies for this credit as follows:

Intent

Verify that the building's energy related systems are installed, calibrated and perform according to the owner's project requirements, basis of design, and construction documents.

Benefits of Commissioning

Benefits of commissioning include reduced energy use, lower operating costs, reduced contractor callbacks, better building documentation, improved occupant productivity, and verification that the systems perform in accordance with the owner's project requirements.

Requirements

The following commissioning process activities shall be completed by the commissioning team, in accordance with the LEED-NC 2.2 Reference Guide.

1) Designate an individual as the Commissioning Authority (CxA) to lead, review and oversee the completion of the commissioning process activities.
   a) The CxA shall have documented commissioning authority experience in at least two building projects.
   b) The individual serving as the CxA shall be independent of the project's design and construction management, though they may be employees of the firms providing those services. The CxA may be a qualified employee or consultant of the Owner.
   c) The CxA shall report results, findings and recommendations directly to the Owner.
   d) For projects smaller than 50,000 gross square feet, the CxA may include qualified persons on the design or construction teams who have the required experience.

2) The owner shall document the Owner's Project Requirements (OPR). The design team shall develop the Basis of Design (BOD). The CxA shall review these documents for clarity and completeness. The Owner and design team shall be responsible for updates to their respective documents.

3) Develop and incorporate commissioning requirements into the construction documents.

4) Develop and implement a commissioning plan.

5) Verify the installation and performance of the systems to be commissioned.

6) Complete a summary commissioning report. LEED for New Construction Version 2.2.

Commissioned Systems

Commissioning process activities shall be completed for the following energy related systems, at a minimum:

- Heating, ventilating, air conditioning and refrigeration (HVAC&R) systems (mechanical and passive) and associated controls
- Lighting and daylighting controls
- Domestic hot water systems
- Renewable energy systems (wind, solar etc.)

Potential Technologies and Strategies

Owners are encouraged to seek out qualified individuals to lead the commissioning process. Qualified individuals are identified as those who possess a high level of experience in the following areas:

- Energy systems design, installation and operation
- Commissioning planning and process management
- Hands-on field experience with energy systems performance, interaction, start up, balancing, testing, troubleshooting, operation, and maintenance procedures
- Energy systems automation control knowledge

Owners are encouraged to consider including water-using systems, building envelope systems, and other systems in the scope of the commissioning plan as appropriate. The building envelope is an important component of a facility which impacts energy consumption, occupant comfort and indoor air quality. While it is not required to be commissioned by LEED, an owner can receive significant financial savings and reduced risk of poor indoor air quality by including building envelope commissioning.

The LEED-NC 2.2 Reference Guide provides guidance on the rigor expected for this prerequisite for the following:

- Owner's project requirements
- Basis of design
- Commissioning plan
- Commissioning specification
- Performance verification documentation
- Commissioning report

Calculations and Considerations

There are no referenced standards for this prerequisite, but the narrative given by the USGBC is very detailed as to the expectations. Previously in LEED-NC 2.1, this prerequisite also included some systems other than energy systems, such as water systems. These other systems are no longer in the prerequisite, but many are included in EAc3: *Enhanced Commissioning*. This prerequisite has a detailed list of overall project-related documentation that must be provided and gives a detailed overview of the project as previously listed in the bulleted list. There are two main steps early on in this commissioning process:

- A commissioning authority (CxA) is designated.
- The two documents called the Owner's Project Requirements and the Basis of Design are developed and reviewed by the CxA. (Usually these documents are developed by the owner and the design team with the aid of the commissioning authority.)

As the design phase of the project proceeds, in addition to periodically updating the noted documents, the following two steps are performed:

- Commissioning requirements are incorporated into the construction documents.
- A commissioning plan is developed and implemented.

Various members of the project team assume responsibility for different portions of the commissioning plan, with the general contractor assuming responsibility for most of the requirements outlined in the construction documents. Finally, during and after construction the following two main items are done:

- The installation and performance of the listed commissioned systems are verified.
- A summary commissioning report is completed.

The CxA should have experience with at least two projects of similar magnitude and complexity as the project in question. The USGBC expects this commissioning authority to be familiar with these steps. Further details can be found in the LEED-NC 2.2 Reference Guide and will not be included here. However, it is very important that the commissioning steps be well documented and verified during the construction phase. They are a part of the construction phase submittal. The following list is a good outline from the LEED-NC 2.2 Reference Guide of the items that should be covered for the contractor in the construction documents so that his or her construction engineers and managers can adequately fulfill their portion of the fundamental commissioning (Cx) requirements:

- Commissioning team involvement
- Contractor's responsibilities
- Submittals and submittal review procedures for Cx process/systems
- Operations and maintenance documentation, system manuals
- Meetings
- Construction verification procedures
- Start-up plan development and implementation/Functional performance testing/ Acceptance and closeout
- Training
- Warranty review site visit

There is an additional commissioning opportunity that can earn a credit point, EAc3: *Enhanced Commissioning*. A summary table of the additional steps in the commissioning process that are needed for enhanced commissioning as compared to fundamental commissioning has been included in the LEED-NC 2.2 Reference Guide and is included here as Table 4.0.2.

Various definitions have also been included in the LEED-NC 2.2 Reference Guide. These are presented in App. B and cover the following terms: *basis of design* (BOD), *commissioning* (Cx), *commissioning plan, commissioning report, commission specification, commissioning team, installation inspection, owner's project requirements* (OPR), and *systems performance testing*.

Tasks	Responsibilities	
	For EAp1 Only	**For EAp1 and EAc3**
Designate commissioning authority (CxA)	Owner or project team	Owner or project team
Document Owner's Project Requirements	Owner	Owner
Develop Basis of Design	Design team	Design team
Incorporate commissioning requirements into the construction documents	Project team or CxA	Project team or CxA
Conduct commissioning design review prior to midconstruction documents	n/a	CxA
Develop and implement a commissioning plan	Project team or CxA	Project team or CxA
Review contractor submittals applicable to systems being commissioned	n/a	CxA
Verify the installation and performance of commissioned systems	CxA	CxA
Develop a systems manual for the commissioned systems	n/a	Project team and CxA
Verify that the requirements for training are completed	n/a	Project team and CxA
Complete a summary commissioning report	CxA	CxA
Review building operation within 10 months after substantial completion	n/a	CxA

TABLE 4.0.2   Tasks and Primary Responsibilities for Commissioning per LEED-NC 2.2 Reference Guide

EA Prerequisite 2: *Minimum Energy Performance*

The intention of this prerequisite is to establish and require a minimum energy efficiency in the building and many of its systems.

USGBC Rating System

LEED-NC 2.2 lists the Intent, Requirements, and Potential Technologies and Strategies for EAp2 as follows:

Intent

Establish the minimum level of energy efficiency for the proposed building and systems.

Requirements

Design the building project to comply with both

- The mandatory provisions (Sections 5.4, 6.4, 7.4, 8.4, 9.4 and 10.4) of ASHRAE/IESNA Standard 90.1-2004 (without amendments);

And

- The prescriptive requirements (Sections 5.5, 6.5, 7.5 and 9.5) or performance requirements (Section 11) of ASHRAE/IESNA Standard 90.1-2004 (without amendments).

Potential Technologies and Strategies

Design the building envelope, HVAC, lighting, and other systems to maximize energy performance. The ASHRAE 90.1-2004 User's Manual contains worksheets that can be used to document compliance with this prerequisite. For projects pursuing points under EA Credit 1, the computer simulation model may be used to confirm satisfaction of this prerequisite. If a local code has demonstrated quantitative and textual equivalence following, at a minimum, the U.S. Department of Energy standard process for commercial energy code determination, then it may be used to satisfy this prerequisite in lieu of ASHRAE 90.1-2004. Details on the DOE process for commercial energy code determination can be found at www.energycodes. gov/implement/determinations_com.stm.

Calculations and Considerations

This prerequisite uses the ASHRAE/IESNA 90.1-2004: Energy Standard for Buildings except Low-Rise Residential. This standard was sponsored by the American Society of Heating, Refrigerating and Air-Conditioning Engineers (ASHRAE) and the Illuminating Engineering Society of North America (IESNA) under an American National Standards Institute (ANSI) process. It sets minimum energy requirements and energy efficiency requirements. It does not apply to buildings that use neither electricity nor fossil fuel.

This prerequisite requires that certain mandatory provisions in the energy standard be met. In addition, there are other items that may be met by either adhering to some prescriptive requirements in the standard or alternatively meeting the intent of these prescriptive requirements by meeting certain overall design-based performance criteria. Prescriptive requirements in standards generally refer to actions that are specifically met, whereas, with design-based performance criteria, there may be many design options that will fulfill a similar goal. For instance, a prescriptive requirement might state that all windows must be double-paned, whereas a performance requirement might state that the overall energy loss from the windows must not exceed a certain value.

The general energy-related components of the building that are covered by this prerequisite have been summarized by the LEED-NC 2.2 Reference Guide and are listed in Table 4.0.3.

A review or summary of the extensive calculations needed for the energy performance is not intended to be a part of this book. There are many energy models used regularly by energy experts to determine the energy requirements of the building and system. However, there are a few items that the LEED engineer may find useful in understanding this prerequisite and its implication to other credits. Six of these are listed as follows:

- There are certain mandatory provisions that must be met in each section listed in Table 4.0.3. In addition, as previously mentioned, either certain prescriptive requirements or certain performance requirements must be met. There are two performance-based options. One performance-based option is from Section 11 of ASHRAE/IESNA Standard 90.1-2004 (The Energy Cost Budget Option), which, if used for energy analysis over the straight prescriptive method, allows for some exceedance of certain prescriptive requirements as long as there are energy cost savings in other prescribed areas that make up for the deviations. This modeling option cannot be used if EA credit 1, which gives credits for energy savings beyond this prerequisite, is sought via Option 1.

Energy-Related Component	
Section 5	Building envelope (including semiheated spaces such as warehouses)
Section 6	Heating, ventilating, and air conditioning (including parking garage ventilation, freeze protection, exhaust air energy recovery, and condenser heat recovery for service water heating)
Section 7	Service water heating (including swimming pools)
Section 8	Power (including all building power distribution systems)
Section 9	Lighting (including lighting for exit signs, building exterior, grounds, and parking garage)
Section 10	Other equipment (including all permanently wired electric motors)

Source: Courtesy LEED-NC 2.2 Reference Guide.

TABLE 4.0.3   ASHRAE/IESNA 90.1-2004 Components Applicable to EAp2 and EAc1

- If EAc1 Option 1 is sought, the alternative energy simulation documentation referred to as the Performance Rating Method option in App. G of this same standard shall be used to document these items of variance from the prescriptive requirements. EAc1 can also be obtained via two prescriptive options. If these two prescriptive options for EAc1 are sought, then it is assumed that the Energy Cost Budget Option can alternatively be used as simulation for EAp2. However, this has not been directly addressed in the LEED documentation.

- The United States is segregated into eight climate zones, and there are different prescriptive requirements for many sections in each of these zones.

- The building envelope requirements in Section 5 of Table 4.0.3 cover many mandatory items such as insulation, doors, and windows. If the prescriptive method is used, then there are maximum allowed window and skylight areas (50 percent of the gross wall area for windows and 5 percent of the roof area for skylights). This is important to note as windows and skylights impact other credits in the indoor Environmental Quality (EQ) category of LEED. Otherwise, either the cost budget method or the performance rating method can be used, and the performance requirements have greater flexibility.

- There are exceptions to Section 6 (HVAC) for projects served by existing HVAC systems such as central energy plants. For example, existing systems that service the project, but are outside of it, do not have to conform to all the requirements of this prerequisite.

- Section 9 covers lighting and includes both indoor and exterior lighting. The reduction in exterior lighting from this prerequisite is also a part of SSc8.

Table 4.0.4 is taken from the LEED-NC 2.2 Reference Guide and gives a good summary of some of the documentation for the various paths chosen for this prerequisite. Note that most LEED-NC projects do seek the EA credit 1, and therefore, only the first and fourth items in this table are applicable.

Mandatory measures (all projects)	• Building Envelope Compliance Documentation (Part I)—Mandatory Provisions Checklist • HVAC Compliance Documentation (Part II)—Mandatory Provisions Checklist • Service Water Heating Compliance Documentation (Part I)—Mandatory Provisions Checklist • Lighting Compliance Documentation (Part I)—Mandatory Provisions Checklist
Prescriptive requirements (projects using prescriptive compliance approach)	• Building Envelope Compliance Documentation (Part II) • HVAC Compliance Documentation—Part I for buildings <25,000 ft² using the simplified approach or Part III for all other buildings • Service Water Heating Compliance Documentation
Performance requirements (projects using performance compliance approach and when EA Credit 1 Option 1 is not being sought)	• Energy Cost Budget Compliance Report
Performance requirements (projects using performance compliance approach and when EA Credit 1 Option 1 is being sought)	• Performance Rating Report • Table documenting energy-related features included in the design, and including all energy features that differ between the baseline design and proposed design models

Source: Courtesy LEED-NC 2.2 Reference Guide.

TABLE 4.0.4 EAp2 Compliance Forms Based on ASHRAE Standard 90.1-2004

EA Prerequisite 3: *Fundamental Refrigerant Management*

The intent of this prerequisite is to help maintaining the ozone levels in the stratosphere. The chemical and physical makeup of the stratosphere is different from the atmosphere in the troposphere where we live and breathe. The stratosphere is located above the troposphere and is where many jets fly. Over the past century, many chemicals were developed as effective means for refrigeration and fire suppression. When these chemicals leaked or were expelled into the troposphere, the effects seemed to be fairly benign, and therefore, they became very commonly used. However, scientists have recently shown that these same chemicals react differently when they enter an atmospheric environment similar to that found in the stratosphere. There, many of these same chemical compounds apparently serve as catalysts that destroy the amount of ozone (O_3) in the stratosphere, converting it into oxygen molecules (O_2). In the troposphere, we usually regard O_2 as a "good" molecule, since this is the form of oxygen that we use to breathe, and we usually regard O_3 as a "bad" molecule, as it can cause air pollution problems and adversely impact health. However, in the stratosphere, the O_3 molecule advantageously aids in filtering out some of the ultraviolet radiation from the sun that might otherwise reach the surface of the earth and negatively impact health and property. As is well known, places in the stratosphere where the concentrations of O_3 have been reduced are commonly referred to as "ozone holes." Therefore, there is an effort to internationally reduce the use of certain refrigerants and fire suppression chemicals that might have the greatest negative impact on the amount of ozone in the stratosphere.

EA prerequisite 3 deals with a group of common refrigerants collectively referred to as chlorofluorocarbons (CFCs). They are organic carbon-based compounds in which the hydrogen atoms have been replaced by chlorine and fluorine atoms. The chlorine atoms are thought to be some of the chemicals responsible for problems in ozone depletion in the stratosphere. There are alternative compounds which can be used. Many of these are discussed under EA credit 4. Many of the CFCs also go by the DuPont trademark name *Freon*. The intent of this prerequisite is to ban or limit the use of CFCs in new projects and major renovations.

USGBC Rating System

LEED-NC 2.2 lists the Intent, Requirements, and Potential Technologies and Strategies for EAp3 as follows:

Intent
Reduce ozone depletion.

Requirements
Zero use of CFC-based refrigerants in new base building HVAC&R systems. When reusing existing base building HVAC equipment, complete a comprehensive CFC phase out conversion prior to project completion. Phase-out plans extending beyond the project completion date will be considered on their merits.

Potential Technologies and Strategies
When reusing existing HVAC systems, conduct an inventory to identify equipment that uses CFC refrigerants and provide a replacement schedule for these refrigerants. For new buildings, specify new HVAC equipment in the base building that uses no CFC refrigerants.

Calculations and Considerations

This prerequisite is quite simple. There will be no CFCs used for air conditioning or refrigeration systems in new buildings, and they must be taken out of existing buildings. For existing buildings, there must be a detailed phase-out plan. The only exceptions are for chilled water systems coming from other existing systems, in which case either there has to be a phase-out of the CFCs over a 5-year period or, in the case where this is impossible, there must be a third-party audit proving that this is not economically feasible (with leakages reduced prior to phase-out to only 5 percent for either case and several additional criteria for the latter case).

4.1 EA Credit 1: *Optimize Energy Performance*

The intention of this credit is to give credit for minimizing energy usage on-site, and also to encourage minimizing the use of off-site power and other off-site energy sources.

USGBC Rating System

LEED-NC 2.2 lists the Intent, Requirements, and Potential Technologies and Strategies for EAc1 as follows:

Intent
Achieve increasing levels of energy performance above the baseline in the prerequisite standard to reduce environmental and economic impacts associated with excessive energy use.

Requirements   Select one of the three compliance path options described. Project teams documenting achievement using any of the three options are assumed to be in compliance with EA Prerequisite 2.

OPTION 1 — WHOLE BUILDING ENERGY SIMULATION (1–10 Points)

Demonstrate a percentage improvement in the proposed building performance rating compared to the baseline building performance rating per ASHRAE/IESNA Standard 90.1-2004 (without amendments) by a whole building project simulation using the Building Performance Rating Method in Appendix G of the Standard. The minimum energy cost savings percentage for each point threshold is as follows:

New Buildings	Existing Building Renovations	Points
10.5%	3.5%	1
14%	7%	2
17.5%	10.5%	3
21%	14%	4
24.5%	17.5%	5
28%	21%	6
31.5%	24.5%	7
35%	28%	8
38.5%	31.5%	9
42%	35%	10

Appendix G of Standard 90.1-2004 requires that the energy analysis done for the Building Performance Rating Method include ALL of the energy costs within and associated with the building project. To achieve points using this credit, the proposed design—

- Must comply with the mandatory provisions (Sections 5.4, 6.4, 7.4, 8.4, 9.4 and 10.4) in Standard 90.1-2004 (without amendments);

- Must include all the energy costs within and associated with the building project; and

- Must be compared against a baseline building that complies with Appendix G to Standard 90.1-2004 (without amendments). The default process energy cost is 25% of the total energy cost for the baseline building. For buildings where the process energy cost is less than 25% of the baseline building energy cost, the LEED submittal must include supporting documentation substantiating that process energy inputs are appropriate. For the purpose of this analysis, process energy is considered to include, but is not limited to, office and general miscellaneous equipment, computers, elevators and escalators, kitchen cooking and refrigeration, laundry washing and drying, lighting exempt from the lighting power allowance (e.g. lighting integral to medical equipment) and other (e.g. waterfall pumps). Regulated (nonprocess) energy includes lighting (such as for the interior, parking garage, surface parking, facade, or building grounds, except as previously noted), HVAC (such as for space heating, space cooling, fans, pumps, toilet exhaust, parking garage ventilation, kitchen hood exhaust, etc.), and service water heating for domestic or space heating purposes. For EA Credit 1, process loads shall be identical for both the baseline building performance rating and for the proposed building performance rating. However, project teams may follow the Exceptional Calculation Method (ASHRAE 90.1-2004 G2.5) to document measures that reduce process loads. Documentation of process load energy savings shall include a list of the assumptions made for both the base and proposed design, and theoretical or empirical information supporting these assumptions.

OR

OPTION 2 — PRESCRIPTIVE COMPLIANCE PATH (4 Points)

Comply with the prescriptive measures of the ASHRAE Advanced Energy Design Guide for Small Office Buildings 2004. The following restrictions apply:

- Buildings must be under 20,000 square feet
- Buildings must be office occupancy
- Project teams must fully comply with all applicable criteria as established in the Advanced Energy Design Guide for the climate zone in which the building is located

OR

OPTION 3 — PRESCRIPTIVE COMPLIANCE PATH (1 Point)

Comply with the Basic Criteria and Prescriptive Measures of the Advanced Buildings Benchmark Version 1.1 with the exception of the following sections: 1.7 Monitoring and Trend-logging, 1.11 Indoor Air Quality, and 1.14 Networked Computer Monitor Control. The following restrictions apply:

- Project teams must fully comply with all applicable criteria as established in Advanced Buildings Benchmark for the climate zone in which the building is located.

Potential Technologies and Strategies

Design the building envelope and systems to maximize energy performance. Use a computer simulation model to assess the energy performance and identify the most cost effective energy efficiency measures. Quantify energy performance as compared to a baseline building. If a local code has demonstrated quantitative and textual equivalence following, at a minimum, the U.S. Department of Energy standard process for commercial energy code determination, then the results of that analysis may be used to correlate local code performance with ASHRAE 90.1-2004. Details on the DOE process for commercial energy code determination can be found at www.energycodes.gov/implement/determinations_com.stm.

Calculations and Considerations

EA credit 1 can be worth 1 to 10 points depending on the option chosen and the energy performance obtained. There are three options. Option 1 is based on a performance model approach that simulates the whole building. Option 1 is very calculation-intensive, and from 1 to 10 points can be obtained. The three overall strategies that will aid in obtaining many of these 10 available points are to reduce energy demand, harvest free energy, and improve the efficiency of the energy systems.

Reducing demand includes items such as occupancy sensors so that many of the systems operate only when needed, and using colors or building orientation to decrease potential loads. Figure 4.1.1 shows architectural shading features installed on a building. These help reduce the air conditioning load.

Harvesting free energy includes such items as taking advantage of natural lighting, as with light tubes, light shelves, building orientation, and clerestories. Examples of some of these are also shown in Chap. 6 where natural lighting is encouraged for indoor environmental quality, in addition to energy reduction.

Improving efficiencies refers to individual items such as more efficient equipment and higher-performance lighting. It also includes looking at the overall systems. For example, a heating system could have more effectively sized ductwork, or distributed smaller heaters could be used to reduce line losses. Figure 4.1.2 depicts a Trane TRAQ

FIGURE 4.1.1 Architectural shading features on the LEED certified Public Health Building at the University of South Carolina, Columbia, S.C. (*Photograph taken July 25, 2007.*)

FIGURE 4.1.2 Trane TRAQ damper which balances air intake for both indoor air quality and energy performance. (*Photograph Courtesy of Trane.*)

Figure 4.1.3 Trane energy wheel for pretreating incoming air to the HVAC system. (*Photograph Courtesy of Trane.*)

damper with built-in, temperature-compensated flow measurement. This damper allows the building controls to bring in the right amount of outside air to meet indoor air quality standards while eliminating excess outside air intake, which can save on energy. Figure 4.1.3 depicts a Trane energy wheel. This wheel is part of an HVAC system and is used to reduce energy demand. It can recover energy from the conditioned exhaust air and uses it to pretreat the incoming outside air. Another developing concept in improving efficiencies is the idea of thermal energy storage (TES) where cooling energy is transferred into storage via mechanisms such as chilled water or ice during off-peak hours and is then used during peak hours. This technique reduces the demand on the equipment and the energy network. These are just a few of the examples of how energy efficiencies can be improved.

Options 2 and 3 are prescriptive-based options. They are requirements where certain specific actions must be taken. They are the opposite of design-based performance requirements, where many different actions may result in the same level of performance. (The definition of prescriptive requirements is further explained in EAp2.) Option 2 is only for small office buildings and is worth four points if the standard prescriptive requirements are met. Option 3 is for any type of building. It is worth one point if the associated prescriptive requirements are met. There are no calculations for Option 2 or Option 3. Details of the prescriptive measures are not a part of this book but can be found in the LEED-NC 2.2 Reference Guide.

Option 1 The performance modeling procedures and various standard energy models used for Option 1 are not a detailed part of this book. However, there are a few items relating to the approach and calculations that the project team and engineers other than the energy modeler may find useful in acquiring a general understanding of LEED energy credits, as well as how this credit can interrelate with other credits. Some of these items are as follows:

- Both an annual baseline and an annual design energy model are established. They are based on annual energy costs and do not include any building construction or *first costs* (design, material, or installation costs) for the systems. The dollar values are based on either the local energy utility rates or, if these are not available, the 2003 default state average rates as compiled by the USGBC in the LEED-NC 2.2 Reference Guide from lists of the 2003 Commercial Sector Average Energy Costs by State (CSAECS) on the U.S. Department of Energy (DOE) website, www.eia. doe.gov. $CSAECSE_i$ and $CSAECSF_i$ represent the electrical and natural gas costs, respectively, for state *i*. These values are given in Table 4.1.1.

- The baseline and design models must be based on the ASHRAE/IESNA Standard 90.1-2004 App. G, Building Performance Rating Method. This is the method used to quantify exceedance of Standard 90.1. The energy cost budget (ECB) method in Section 11 of this same standard was accepted as a rating method in earlier versions of LEED but is no longer accepted.

- The numerical results of the baseline annual energy model are referred to as the baseline building performance (BBP), and the design annual energy costs as determined by this model are referred to as the proposed design energy model (PDEM). Unlike in earlier versions of LEED, the energy model values include both regulated and process (nonregulated) loads. The default is to assume that the process loads are approximately 25 percent of the baseline regulated loads. If this is not the case, then information should be provided to verify the difference. Since these additional loads are included in the calculations, the percent energy reductions for each point value have been reduced in LEED-NC 2.2 as compared to earlier versions.

- Both models (BBP and PDEM) are based on all the energy loads including process (or nonregulated) loads as defined specifically for LEED-NC 2.2. These types of loads have been summarized in the Requirements section. The types of loads that are included in these two categories have changed from version 2.1 of LEED-NC to version 2.2. Usually only regulated loads are reduced by the energy performance methods in the design. If some applicable process loads are reduced in the design, then a special procedure in the ASHRAE/IESNA Standard 90.1-2004 App. G (G2.5) called the *exceptional calculation method* (ECM) can be used to document and quantify these savings into the design model (PDEM).

- The potential energy savings represented as the difference between the base model (BBP) and the proposed design (PDEM) can be realized in many different ways. These may include reduced loads, more efficient equipment, and the recovery of potential energy losses by various methods, such as water preheating from waste streams, and the use of some natural sources, such as daylighting. For the purposes of the calculations presented in the following section, all these as just listed are included in the determination of the annual PDEM costs. If the energy model used cannot adequately include some of these recovery savings, then these can be calculated separately by using the ECM method and then incorporated into the PDEM.

- The savings from on-site renewable energy sources that produce power are calculated separately and are referred to as the *renewable energy cost* (REC). They are defined in greater detail in EAc2. The anticipated annual costs of off-site energy usage as represented by the PDEM can be offset by the on-site renewable

	Electrical Rate ($/kWh) (default value of CSAECSE$_i$)*	Natural Gas ($/kBtu) (default value of CSAECSF$_i$)*
Alabama	0.0682	0.00938
Alaska	0.1646	0.00355
Arizona	0.0670	0.00758
Arkansas	0.0526	0.00668
California	0.1171	0.00843
Colorado	0.0597	0.00476
Connecticut	0.0900	0.01101
Delaware	0.0693	0.00840
District of Columbia	0.0645	0.01266
Florida	0.0678	0.01083
Georgia	0.0669	0.00957
Hawaii	0.1502	0.01926
Idaho	0.0601	0.00612
Illinois	0.0758	0.00794
Indiana	0.0585	0.00844
Iowa	0.0602	0.00750
Kansas	0.0611	0.00753
Kentucky	0.0520	0.00760
Louisiana	0.0664	0.00861
Maine	0.1019	0.01086
Maryland	0.0659	0.00807
Massachusetts	0.0848	0.01071
Michigan	0.0701	0.00631
Minnesota	0.0546	0.00778
Mississippi	0.0721	n/a
Missouri	0.0505	0.00796
Montana	0.0601	0.00623
Nebraska	0.0500	0.00698
Nevada	0.0955	0.00723
New Hampshire	0.0973	0.00917
New Jersey	0.0835	0.00835
New Mexico	0.0737	0.00659

TABLE 4.1.1   Default Energy Costs per State/District *i* [DOE Energy Information Administration 2003 Commercial Sector Average Energy Costs by State (CSAECS)]

	Electrical Rate ($/kWh) (default value of CSAECSE$_j$)*	Natural Gas ($/kBtu) (default value of CSAECSF$_j$)*
New York	0.1113	0.00895
North Carolina	0.0641	0.00863
North Dakota	0.0547	0.00682
Ohio	0.0723	0.00789
Oklahoma	0.0571	0.00755
Oregon	0.0657	0.00775
Pennsylvania	0.0819	0.00898
Rhode Island	0.0834	0.00964
South Carolina	0.0652	0.00992
South Dakota	0.0605	0.00963
Tennessee	0.0631	0.00832
Texas	0.0695	0.00757
Utah	0.0538	0.00539
Vermont	0.1087	0.00778
Virginia	0.0572	0.00920
Washington	0.0624	0.00669
West Virginia	0.0545	0.00734
Wisconsin	0.0645	0.00822
Wyoming	0.0548	0.00469

*Use these values for CSAECSE$_i$ or CSAECSF$_i$ if current local values are not available.

TABLE 4.1.1   (*Continued*)

energy source cost savings (REC). The on-site renewable energy usage values (RECs) can be subtracted from the PDEM as if the power they provide to the project were free. As in the PDEM calculations, the actual capital costs of the REC are not included. The REC savings are virtual savings based on the typical off-site utility rates.

- The baseline building is different from the proposed building in many ways. The LEED-NC 2.2 Reference Guide lists many of these differences in detail. Some of them include the following:
    1. The proposed design is modeled with the actual building orientation on the lot. The baseline design is the average of this orientation and the other three orthogonal orientations.

2. Many building schedules, such as lighting or occupancy schedules, can differ between the two models if the differences can be defended. (See Chap. 8 for more information on schedules with respect to design and construction.)
3. The baseline building is based on a specific assembly for floors, walls, and roofs as outlined in App. G of the standard.
4. The total exterior wall area occupied by vertical fenestrations in the base building must be the same percent of the total exterior wall area as in the design case, unless the design case percentage is greater than 40 percent and then the base building is restricted to 40 percent.
5. The vertical fenestrations in the baseline case are distributed uniformly on all sides.
6. The baseline building can be based on a different HVAC system, but again, this variation must be defensible.

In summary, Option 1 is calculation-intensive. There are a total of five simulations that need to be performed, including one of the design case and four of the baseline case, one in each of the four orthogonal orientations. The four baseline case orientations are then averaged to obtain the baseline.

The annual energy percent cost savings (AECS) can be calculated based on a *proposed building performance* (PBP) and the *base building performance* (BBP). The use of on-site renewable energy as calculated as the REC may increase the points obtained from Option 1 of this credit, as the REC is used as a reduction in the calculation of the *proposed building performance* (PBP), offsetting this amount of energy use as if it were free. The PBP is calculated as follows:

$$PBP = PDEM - REC \qquad (4.1.1)$$

The annual energy percent cost savings (AECS) as used in the point determinations for EAc1 can then be calculated from either of the following two equations:

$$AECS = 100\left(1 - \frac{PBP}{BBP}\right) \qquad (4.1.2)$$

or alternatively

$$AECS = 100\left(1 - \frac{PDEM}{BBP} + \frac{REC}{BBP}\right) \qquad (4.1.3)$$

Equation (4.1.3) shows the positive impact that using on-site renewable energy has on the point potential for EAc1, when using Option 1.

LEED-NC 2.2 includes both regulated and process loads in assessing Option 1. Since much of the energy efficiency focus is on regulated loads, it is also useful to be able to segregate out these two parts of the energy loads for both the base building and the proposed design. This is readily done by defining BBPR and BBPP as the regulated BBP load and the process BBP load, respectively, and by defining PDEMR and PDEMP as the regulated PDEM load and the process PDEM load, respectively. By definition,

$$BBP = BBPR + BBPP \qquad (4.1.4)$$

and

$$PDEM = PDEMR + PDEMP \qquad (4.1.5)$$

It is usually easier to assess the regulated loads (BBPR and PDEMR) in the evaluations, as these are the ones most appropriately considered for energy efficiency measures. In most cases, the base process load is considered to be 25 percent of the base and does not change from the base to the proposed design. When this is the case, Eqs. (4.1.4) and (4.1.5) can be simplified to the following:

$$BBP = BBPR/0.75 \qquad \text{base process is 25\% of base load} \qquad (4.1.6)$$

and

$$PDEM = PDEMR + BBPR/3 \qquad \begin{array}{l}\text{base process is 25\% of base load and} \\ \text{does not change in design}\end{array} \qquad (4.1.7)$$

Substituting this into Eq. (4.1.3) results in the following simplification based only on regulated loads:

$$AECS = 75\left(1 - \frac{PDEMR}{BBPR} + \frac{REC}{BBPR}\right) \qquad \begin{array}{l}\text{base process is 25\% of base load} \\ \text{and does not change in design}\end{array} \qquad (4.1.8)$$

Equations (4.1.1) through (4.1.8) are for Option 1. A summary of the symbols as used for calculating EAc1 Option 1 can be found in Table 4.1.2.

Options 2 and 3   Options 2 and 3 are not model-based and therefore do not have a method for applying on-site renewable energy sources to the point total for this credit. Their prescriptive requirements are well outlined in the LEED-NC 2.2 Reference Guide.

Special Circumstances and Exemplary Performance

EAc1 is also listed with the EB icon, primarily because the construction assembly of much of the exterior of the building impacts energy losses. In addition, the orientation of the building is a significant portion of Option 1.

In the summer of 2007, the USGBC opened a ballot to assess the possibility of requiring a minimum of two points to be earned in this credit subcategory for all projects. This would mean a minimum of two points if Option 1 were followed, and it would be fulfilled by the four-credit Option 2 for applicable projects. There was no guidance given on how the prescriptive Option 3 would count toward this proposed minimum. All new projects should first check if this was approved or other required energy minimum changes were approved prior to initiation. Reducing energy use as quickly as possible is a very dynamic goal for many organizations.

The Indoor Environmental Quality category promotes good indoor air quality, and one of the strategies used to do this is increased ventilation, which often increases the energy load. Ventilation strategies should be carefully evaluated in conjunction with energy efficiencies and impacts to maintain an optimum balance between these two goals.

There are no exemplary performance points associated with this credit as up to 10 points are already included.

Symbol	Description	Comments
BBP	Base building performance ($)	• Total includes both regulated and process loads • Average of four model orientations
BBPP	Base building performance process loads ($)	Default is BBPP = 25% of BBP
BBPR	Base building performance regulated loads ($)	
PDEM	Proposed design energy model ($)	Total includes both regulated and process loads
PDEMP	Proposed design energy model process loads ($)	Process is same for base and proposed unless specially addressed by the exceptional calculation method, but usually only regulated loads are reduced
PDEMR	Proposed design energy model regulated loads ($)	
REC	On-site renewable energy cost "savings" ($)	Equivalent "savings" of off-site utility costs for using on-site renewable energies
PBP	Proposed building performance ($)	Proposed design costs less rec savings: PBP = PDEM − REC
AECS	Annual energy cost savings (%)	$$\text{AECS} = 100\left(1 - \frac{\text{PBP}}{\text{BBP}}\right) \quad \text{or}$$ $$\text{AECS} = 100\left(1 - \frac{\text{PDEM}}{\text{BBP}} + \frac{\text{REC}}{\text{BBP}}\right)$$
ECM	Exceptional calculation method	Method to calculate changes to the proposed design for some process load changes or some energy cost savings for measures not readily included in the energy model used

TABLE 4.1.2   Summary of Symbols and Criteria Equations for EAc1 Option 1

4.2 EA Credit 2: *On-Site Renewable Energy*

The intention of EAc2 is to encourage the production of renewable energy sources on-site. This will serve to decrease the load on the off-site energy resources and also improve energy distribution efficiencies, as the potential energy losses that can be allocated to transport to the site are decreased. The energy offsets can earn one to three points.

USGBC Rating System

LEED-NC 2.2 lists the Intent, Requirements, and Potential Technologies and Strategies for EAc2 as follows:

Intent

Encourage and recognize increasing levels of on-site renewable energy self-supply in order to reduce environmental and economic impacts associated with fossil fuel energy use.

Requirements

Use on-site renewable energy systems to offset building energy cost. Calculate project performance by expressing the energy produced by the renewable systems as a percentage of the building annual energy cost and using the table below to determine the number of points achieved. Use the building annual energy cost calculated in EA Credit 1 or use the Department of Energy (DOE) Commercial Buildings Energy Consumption Survey (CBECS) database to determine the estimated electricity use. (Table of use for different building types is provided in the Reference Guide.)

% Renewable Energy	Points
2.5%	1
7.5%	2
12.5%	3

Potential Technologies and Strategies

Assess the project for non-polluting and renewable energy potential including solar, wind, geothermal, low-impact hydro, biomass and bio-gas strategies. When applying these strategies, take advantage of net metering with the local utility.

Calculations and Considerations

The project boundary is an important variable in the calculations since this credit requires these renewable energy sources to be provided on-site. There may be some exceptions to this if the building is part of a campus setting where on-campus renewable energies are shared, but usually large, facility wide renewable energy sources are considered to be a part of EAc6: *Green Power,* and are not counted as part of EAc2. There is no rule stating that the project area must always be contiguous, but justifications for the area being separated and other special circumstances should be appropriately documented and approved by the USGBC.

The LEED-NC 2.2 Reference Guide specifically lists the systems that are eligible for applying to EAc2 as on-site renewable energy (REC). In general, the system must produce electrical or thermal energy on-site. Other on-site systems that are not included can usually be used as energy offsets to EAc1, Option 1. If the renewable energy sources come from off-site, then they are not a part of EAc2, but may be a part of EAc 6: *Green Power.* The lists of eligible and noneligible on-site energy sources from the LEED-NC 2.2 Reference Guide are summarized in Table 4.2.1. Note that hydrogen fuel cells are not addressed. This is so because hydrogen fuel cells are not energy sources, but rather are energy storage devices. Their applicability would be dependent on whether the source for producing the hydrogen used in the fuel cell came from a renewable or nonrenewable source.

The credit gives one to three points based on the percent of an anticipated on-site *energy usage* that is provided by the eligible on-site renewable sources. As noted in EAc1, these energy cost savings are referred to as the renewable energy costs (RECs) and are given in dollars per year. They are virtual savings using the on-site energy generated times the applicable local or state average energy cost rate, as defined in EAc1. In earlier versions

Energy Source	Systems Eligible for EAc2	Systems Not Eligible for EAc2
Solar	Photovoltaic systems	Passive solar strategies
	Solar thermal systems	Daylighting
		Architectural features
Biofuel energy systems with the following fuels	Untreated wood wastes	Wood coated with paints, plastics, or formica or more than 1% treated with halogen or arsensic-based preservatives
	Wood mill residues	Other forestry biomass waste
	Agricultural crops or wastes	Municipal solid waste
	Animal wastes and other organic wastes	
	Landfill gas	
Ground source	Geothermal heating	Geoexchange heat pumps
	Geothermal electric	
Surface water	Low-impact hydroelectric	
	Wave and tidal power	
Wind	Wind power	

Source: LEED-NC 2.2 Reference Guide.

TABLE 4.2.1 Summary of On-Site Renewable Energy Systems

of LEED, this credit was given three parts (EAc2.1, EAc2.2, and EAc2.3), each worth one point. In LEED-NC 2.2, they are combined into EAc2, worth one to three points.

The denominator in the eligibility equations is more complex. If Option 1 of EAc1 is sought, then the denominator is the proposed building performance (PBP). If not, then the denominator is the *building annual energy cost* (BAEC) and is calculated using the local energy rates (CSAECSE$_i$ and CSAECSF$_j$), or if current local values are not available, the U.S. Department of Energy (DOE) Energy Information Administration (EIA) 2003 Commercial Sector Average Energy Costs by State as given in Table 4.1.1 in EAc1 along with default energy consumption of both electricity and nonelectrical fuel usage rates per square foot of the building, based on typical building usages. The applicable default usage rates have been compiled by the USGBC are the LEED-NC Reference Guide from median data in the DOE EIA 1999 Commercial Buildings Energy Consumption Survey (CBECS) and is also tabulated in Table 4.2.2.

For the calculations of the building annual energy cost (BAEC), the U.S. Department of Energy (DOE) Energy Information Administration (EIA) 2003 Commercial Sector Average Energy Costs by State (CSAECS) and the median data in the DOE EIA 1999 Commercial Buildings Energy Consumption Survey for electrical (CBECSE$_j$) and for fuel use (CBECSF$_j$) for usage j are usually used as default values, although it is preferred that actual local energy cost rates be used. Additional exceptions to these default values

Building Use	Median Electrical Use [kWh/(ft² . yr)] (CBECSE$_j$)	Median Fuel Use [kBtu/(ft² . yr)] (CBECSF$_j$)
Education	6.6	52.5
Food sales	58.9	143.3
Food service	28.7	137.8
Health care inpatient	21.5	50.2
Health care outpatient	9.7	56.5
Lodging	12.6	39.2
Retail (other than mall)	8.0	18.0
Enclosed and strip malls	14.5	50.8
Office	11.7	58.5
Public assembly	6.8	72.9
Public order and safety	4.1	23.7
Religious worship	2.5	103.6
Service	6.1	33.8
Warehouse and storage	3.0	96.9
Other	13.8	52.5

TABLE **4.2.2** Default Energy Consumption Intensities by Building Usage j [Based on DOE EIA 1999 Commercial Buildings Energy Consumption Survey -CBECS]

should be documented, and explanations should be given in the submittals with supporting data. So for all projects that do not seek Option 1 of EAc1, the design energy cost is this BAEC, whose default value is calculated by using Eq. (4.2.1). The GFA in Eq. (4.2.1) is the gross floor area of the building as defined in Chap. 2. (If the project is made up of several buildings or building portions with different uses as listed in Table 4.2.2, then the commercial building energy consumptions should be separately analyzed for each portion, multiplied by the gross floor area in that building portion, and summed to obtain the BAEC.)

$$\text{BAEC} = [(\text{CSAECSE}_i \times \text{CBECSE}_j) + (\text{CSAECSF}_i \times \text{CBECSF}_j)](\text{GFA})$$
$$\text{for state/district } i \text{ and use } j \quad \text{(if EAc1 Option 1 is not sought)} \quad (4.2.1)$$

Now the PBP for projects using Option 1 of EAc1 and the BAEC for all other projects are used to determine the percent renewable energy (PREPBP or PREBAEC) as a percent of the PBP or BAEC, respectively, to determine the credit points listed in the Requirements section for EAc2. For projects that seek points via Option 1 of EAc1:

$$\text{PREPBP} = 100 \times \frac{\text{REC}}{\text{PBP}} \quad \text{if EAc1 Option 1 is sought} \quad (4.2.2)$$

Project Type	Minimum REC Percent of PBP (PREPBP)	Minimum REC Percent of PDEM	Minimum REC Percent of BAEC (PREBAEC)	Available Points
Seeking EAc1 Option 1	2.5	2.4	n/a	1
	7.5	7.0	n/a	2
	12.5	11.1	n/a	3
Not seeking EAc1 Option 1	n/a	n/a	2.5	1
	n/a	n/a	7.5	2
	n/a	n/a	12.5	3

TABLE 4.2.3   EAc2 Point Values for Projects

or

$$PREPBP = 100 \times \frac{REC}{PDEM - REC} \qquad \text{if EAc1 Option 1 is sought} \qquad (4.2.3)$$

and for all other projects

$$PREBAEC = 100 \times \frac{REC}{BAEC} \qquad \text{if EAc1 Option 1 is not sought} \qquad (4.2.4)$$

Special Circumstances and Exemplary Performance (We Credits 3.1 and 3.2)

Note that increasing on-site renewable energy cost savings (REC) for a building that seeks EAc1 via Option 1 gives credit for the REC in two ways, both as an increase in the numerator and as a decrease in the denominator. The effective impact of the REC based on the proposed design energy model costs (PDEM) can be seen in Table 4.2.3.

More information on current renewable energy options can be found on the websites of the DOE National Renewable Energy Laboratory (NREL), the DOE Energy Information Administration (EIA), *Distributed Energy—The Journal for Onsite Power Solutions*, and the EPA Clean Energy program.

There are no exemplary performance points associated with this credit as up to three points are already included.

4.3 EA Credit 3: *Enhanced Commissioning*

The intent of EAc3 is to gain even more benefits for system efficiencies and reliabilities by enhancing the commissioning process.

USGBC Rating System

LEED-NC 2.2 lists the Intent, Requirements, and Potential Technologies and Strategies for EAc3 as follows:

Intent

Begin the commissioning process early during the design process and execute additional activities after systems performance verification is completed.

Requirements

Implement, or have a contract in place to implement, the following additional commissioning process activities in addition to the requirements of EA Prerequisite 1 and in accordance with the LEED-NC 2.2 Reference Guide:

1. Prior to the start of the construction documents phase, designate an independent Commissioning Authority (CxA) to lead, review, and oversee the completion of all commissioning process activities. The CxA shall, at a minimum, perform Tasks 2, 3 and 6. Other team members may perform Tasks 4 and 5.
   a. The CxA shall have documented commissioning authority experience in at least two building projects.
   b. The individual serving as the CxA shall be:
      i. independent of the work of design and construction;
      ii. not an employee of the design firm, though they may be contracted through them;
      iii. not an employee of, or contracted through, a contractor or construction manager holding construction contracts; and
      iv. (can be) a qualified employee or consultant of the Owner.
   c. The CxA shall report results, findings and recommendations directly to the Owner.
   d. This requirement has no deviation for project size.

2. The CxA shall conduct, at a minimum, one commissioning design review of the Owner's Project Requirements (OPR), Basis of Design (BOD), and design documents prior to mid-construction documents phase and back-check the review comments in the subsequent design submission.

3. The CxA shall review contractor submittals applicable to systems being commissioned for compliance with the OPR and BOD. This review shall be concurrent with A/E reviews and submitted to the design team and the Owner.

4. Develop a systems manual that provides future operating staff the information needed to understand and optimally operate the commissioned systems.

5. Verify that the requirements for training operating personnel and building occupants are completed.

6. Assure the involvement by the CxA in reviewing building operation within 10 months after substantial completion with O&M staff and occupants. Include a plan for resolution of outstanding commissioning-related issues.

Potential Technologies and Strategies

Although it is preferable that the CxA be contracted by the Owner, for the enhanced commissioning credit, the CxA may also be contracted through the design firms or construction management firms not holding construction contracts. The LEED-NC 2.2 Reference Guide provides detailed guidance on the rigor expected for following process activities:

- Commissioning design review
- Commissioning submittal review
- Systems manual

Calculations and Considerations

EAc3 requires a more comprehensive commissioning process in addition to the fundamental commissioning requirements as outlined in EAp1. It is important that the commissioning authority be familiar with the LEED requirements. The LEED-NC 2.2 Reference Guide gives an excellent outline of the six specific types of activities that are required to meet this credit. A summary table of the additional steps as compared to fundamental commissioning has been included in the LEED-NC 2.2 Reference Guide and was included as Table 4.0.3. More details can be found in the LEED-NC 2.2 Reference Guide. Definitions of CxA, BOD, and OPR as given in the LEED-NC 2.2 Reference Guide are listed in App. B.

Special Circumstances and Exemplary Performance

EAc3 has been listed as a construction phase submittal, primarily because many of the activities cannot be performed or verified until during or after substantial completion of the project. It is also listed with the EB icon. The potential for many benefits associated with a commissioning-type review process for system efficiencies that do not necessitate substantial investment of additional monies for future retrofits is lost if not initiated in the early phases of the initial project. There are no exemplary performance points associated with this credit.

4.4 EA Credit 4: *Enhanced Refrigerant Management*

As mentioned in EAp3, there is an effort internationally to reduce the use of certain refrigerants and fire suppression chemicals that have negatively impacted the amount of ozone in the stratosphere. The Montreal Protocol, as mentioned in the credit intent, is the international agreement that sets timetables for the elimination of production and reduction in use of many of these chemicals. Its full title is the Montreal Protocol on Substances that Deplete the Ozone Layer. There is an international panel related to the Montreal Protocol called the Technology and Assessment Panel (TEAP).

The intention of EA credit 4 is to reduce the use of refrigeration and fire suppression chemicals that have a large ozone depletion potential (ODP), including many as listed in the Montreal Protocol, while also considering the global warming potential (GWP) of the alternatives. The global warming potential of a substance is an index that was established by the International Panel on Climate Change (IPCC).

USGBC Rating System

LEED-NC 2.2 lists the Intent, Requirements, and Potential Technologies and Strategies for EAc4 as follows:

Intent
Reduce ozone depletion and support early compliance with the Montreal Protocol while minimizing direct contributions to global warming.

Requirements
OPTION 1

Do not use refrigerants.

OR

OPTION 2

Select refrigerants and HVAC&R that minimize or eliminate the emission of compounds that contribute to ozone depletion and global warming. The base building HVAC&R

equipment shall comply with the following formula, which sets a maximum threshold for the combined contributions to ozone depletion and global warming potential:

$$LCGWP = LCODP \times 105 \leq 100$$

Where:

LCODP = [ODPr × (Lr × Life + Mr) × Rc]/Life

LCGWP = [GWPr × (Lr × Life + Mr) × Rc]/Life

LCODP: Lifecycle Ozone Depletion Potential ($lbCFC_{11}$/Ton-Year)

LCGWP: Lifecycle Direct Global Warming Potential ($lbCO_2$/Ton-Year)

GWPr: Global Warming Potential of Refrigerant (0 to 12,000 $lbCO_2$/lbr)

ODPr: Ozone Depletion Potential of Refrigerant (0 to 0.2 $lbCFC_{11}$/lbr)

Lr: Refrigerant Leakage Rate (0.5% to 2.0%; default of 2% unless otherwise demonstrated)

Mr: End-of-life Refrigerant Loss (2% to 10%; default of 10% unless otherwise demonstrated)

Rc: Refrigerant Charge (0.5 to 5.0 lb of refrigerant per ton of cooling capacity)

Life: Equipment Life (10 years; default based on equipment type, unless otherwise demonstrated)

For multiple types of equipment, a weighted average of all base building level HVAC&R equipment shall be applied using the following formula:

$$[\Sigma \ (LCGWP + LCODP \times 105) \times Qunit \ ]/Qtotal \leq 100$$

Where:

Qunit = Cooling capacity of an individual HVAC or refrigeration unit (Tons)

Qtotal = Total cooling capacity of all HVAC or refrigeration

Small HVAC units (defined as containing less than 0.5 lbs of refrigerant), and other equipment such as standard refrigerators, small water coolers, and any other cooling equipment that contains less than 0.5 lbs of refrigerant, are not considered part of the "base building" system and are not subject to the requirements of this credit.

AND

Do not install fire suppression systems that contain ozone-depleting substances (CFCs, HCFCs or halons).

Potential Technologies and Strategies

Design and operate the facility without mechanical cooling and refrigeration equipment. Where mechanical cooling is used, utilize base building HVAC and refrigeration systems for the refrigeration cycle that minimize direct impact on ozone depletion and global warming. Select HVAC&R equipment with reduced refrigerant charge and increased equipment life. Maintain equipment to prevent leakage of refrigerant to the atmosphere. Utilize fire suppression systems that do not contain HCFCs or halons.

Calculations and Considerations

Two issues must be addressed to fulfill the requirements of this credit. The first deals with the refrigerant processes in heating, ventilation, air conditioning, and refrigeration (HVAC&R) systems, and the second deals with fire suppression. The two options listed in the credit are the two options for HVAC&R alternatives, and one of these must be met. The last item listed in the Requirements section is for fire suppression systems and must be abided by in all cases.

As mentioned in EA prerequisite 3, CFCs (chlorofluorocarbons) are being phased out due to their potential impact on the ozone levels in the stratosphere. (Many of these CFCs and other refrigerants commonly go by the DuPont trademark name Freon.)

However, some of the alternative compounds may also impact the stratospheric ozone, but not necessarily to the extent that the CFCs do. Two examples of alternatives are the groups of chemicals referred to as the hydrochlorofluorocarbons (HCFCs) and hydrofluorocarbons (HFCs). HCFCs are similar to CFCs but also contain some hydrogen atoms instead of chlorine atoms and therefore have a lower potential for ozone depletion than do the CFCs. HFCs are also similar to HCFCs and CFCs but with no chlorine atoms, and therefore the ozone depletion potential of HFCs is negligible.

In addition, other chemicals used for fire suppression may impact the ozone in the stratosphere. Of particular concern are halons, which are organic compounds that contain bromine atoms in addition to chlorine and fluorine. A bromine atom can have an even greater impact than a chlorine atom has on the concentration of ozone in the stratosphere.

To compare the impact of the various refrigerants and fire suppression chemicals on the ozone in the stratosphere, a scale has been developed called the *ozone depletion potential* (ODP). One of the refrigerants typically referred to as CFC-11 has been given the ranking of 1 on the ODP scale, and the other chemicals are compared to CFC-11 on a mass basis. Most compounds are ranked with a value less than 1, except for those with bromine atoms, which may be higher.

There is concern that some of the alternatives to CFCs and other chemicals used for refrigeration and fire suppression may have an impact on another global atmospheric phenomenon, specifically global warming. Global climate change is of major interest, as changes in the global climate may cause weather and ocean level variations that can adversely impact humans and the environment. Carbon dioxide (CO_2) is a common compound in the troposphere that most living creatures expire. Scientists have shown that its concentration in the troposphere has been increasing in the last few centuries, and there is concern that the additional CO_2 and other chemicals that may come from anthropogenic processes in the atmosphere may have a blanketing effect, causing average temperatures to rise. There is much debate over the importance of carbon dioxide and these other gases in overall climate change.

Carbon dioxide has been given a global warming potential (GWP) of 1. Several refrigerants and fire suppression chemicals have also been given a GWP ranking based on the GWP of CO_2 on a mass basis. Many different sources list many different values for ODPs and GWPs as their values are estimated from very complex models and scenarios. The GWPs and ODPs of many refrigerants are given in the LEED-NC 2.2 Reference Guide. These and several other compounds are listed in Table 4.4.1 and are based on Tables 1-5 and 1-6 of *The Scientific Assessment of Ozone Depletion, 2002*, a report of the World Meteorological Association's Global Ozone Research and Monitoring Project; those that are not updated in this 2002 report are from *The Scientific Assessment of Ozone Depletion, 1998*.

The ODPs and GWPs are based on the mass of the noted refrigerant or fire suppression chemical with respect to the mass of the associated base chemical, CFC-11 and CO_2, respectively. In the case of the GWP, note that there are significantly larger quantities of carbon dioxide emitted on a mass basis into the atmosphere than there are of most other refrigerants. In addition, there is much scientific debate on the relative impact that water vapor has on the global climate. Regardless, the other refrigerants and fire suppression chemicals have been ranked with a GWP as compared to that of carbon dioxide, so that designers and regulators can make informed decisions as to the use of these chemicals from both an ODP and a GWP perspective. More information on these compounds and other ozone-depleting compounds can be found on the U.S. Environmental Protection Agency Ozone Depletion website: http://www.epa.gov/ozone/strathome.html.

Refrigerant or Fire Suppression Compound	Also Referred to as	ODP$_r$*	GWP$_r$*	Common Building Application
Chlorofluorocarbons				
CFC-11		1	4,680	Centrifugal chillers
CFC-12		1	10,720	Refrigerators, chillers
CFC-114		0.94	9,800	Centrifugal chillers
CFC-500		0.605	7,900	Centrifugal chillers, humidifiers
CFC-502		0.221	4,600	Low-temperature refrigeration
Halons				
Halon 1211		6	1,860	Fire suppression
Halon 1301		12	7,030	Fire suppression
Halon 2402		<8.6	1,620	Fire suppression
Hydrochlorofluoro-carbons				
HCFC-22	R-22	0.04	1,780	Air conditioning, chillers
HCFC-123	R-123	0.02	76	CFC-11 replacement
Hydrofluorocarbons				
HFC-23		~0	12,240	Ultralow-temperature refrigeration
HFC-134a	R-134a	~0	1,320	CFC-12 or HCFC-22 replacement
HFC-245fa	R-245fa	~0	1,020	Insulation agent, centrifugal chillers
HFC-404a		~0	3,900	Low-temperature refrigeration
HFC-407c	R-407c	~0	1,700	HCFC-22 replacement
HFC-410a	R-410a	~0	1,890	Air conditioning
HFC-507a		~0	3,900	Low-temperature refrigeration
Natural refrigerants				
Carbon dioxide (CO_2)		~0	1	
Ammonia (NH_3)		~0	0	
Propane		~0	3	

*The units for ODP$_r$ are lb$_{CFC11}$/lb$r$, and the units for GWP$_r$ are lb$_{CO2}$/lb$_r$ for each refrigerant compound r, where lb$_r$ represents the mass (in pounds) of the respective refrigerant or fire suppression chemical, lb$_{CFC11}$ represents the mass (in pounds) of CFC-11, and lb$_{CO2}$ represents the mass (in pounds) of CO_2.
Source: Courtesy LEED-NC 2.2 Reference Guide.

TABLE 4.4.1   Some Relative Ozone Depletion Potentials (ODPs) and Global Warming Potentials (GWPs)

The ODPs and GWPs are both sets of comparative environmental scales of certain chemicals. If chemicals are going to be chosen based on both of these environmental concerns, then there also needs to be a comparative environmental ranking between the scales. The USGBC has developed a formula for use in LEED certification that gives a comparative ranking of 1 to 100,000 between the relative importance of the mass of CO_2 to the mass of CFC-11 for global warming and ozone depletion potential, respectively. The relative multiplier 100,000 lb_{CO2}/lb_{CFC11} will be used in the overall certification equation presented later in this section [Eq. (4.4.5)].

EAc4 is worth one point. To obtain this point, the project must comply with either Option 1 or Option 2 relating to refrigeration and HVAC systems, *and* the project must also comply with the requirement that the fire suppression systems do not contain ozone-depleting substances (CFCs, HCFCs, or halons).

Option 1 and Option 2 only apply to systems or units that contain more than 0.5 lb of refrigerant. Items such as standard household refrigerators and small water coolers are excluded.

Option 1   Option 1, which relates to refrigeration and HVAC systems, is very simple. If the facility uses no refrigerants except for those small units which are exempted, then this option is fulfilled. In addition, if there are only natural refrigerants used for the applicable systems (CO_2, NH_3, water, or propane), then this option is fulfilled. No calculations are needed.

Option 2   Option 2 is more complex. It requires a series of calculations on the base building HVAC systems using the following 10 variables:

GWP_r Global warming potential of refrigerant r (These are listed in Table 4.4.1 or can be obtained from the EPA website. They typically range from 0 to 13,000 lb_{CO2}/lb_r.)

$LCGWP_i$ The equivalent annual life-cycle-based global warming potential for system i in lb_{CO2}/ton of cooling capacity of system i per year $[lb_{CO2}/(ton_{ac} \cdot yr)]$

$LCODP_i$ The equivalent annual life-cycle-based ozone depleting potential for system i in lb_{CFC11}/ton of cooling capacity of system i per year $[lb_{CFC11}/(ton_{ac} \cdot yr)]$

$Life_i$ Equipment life of system i (These are usually based on the 2003 ASHRAE *Applications Handbook*. The values as listed in the LEED-NC 2.2 Reference Guide are listed in Table 4.4.2. Other values may be substituted if appropriately documented by manufacturers' data and warranties.)

Lr_i Annual refrigerant leakage rate of system i (The values range from 0.005 to 0.02 lb refrigerant leaked per pound of refrigerant in the unit per year with a default of 0.02 unless otherwise shown, such as with manufacturers' test data. This is based on an annual leakage of 0.5 to 2.0 percent of the total system refrigerant. Zero leakage may not be claimed.)

Mr_i End-of-life refrigerant loss of system i (The values range from 0.02 to 0.10 lb refrigerant lost per pound of refrigerant in the unit with a default of 0.10 unless otherwise shown, such as with manufacturers' test data. This is based on the assumption that 2 to 10 percent of the total system refrigerant will not be recovered when the system life is over, based on either installation or removal losses. Zero loss may not be claimed.)

Type of Equipment System i	Life$_i$ (Years)
Window air-conditioning units and heat pumps	10
Unitary, split, and packaged air-conditioning units and heat pumps	15
Reciprocating compressors and reciprocating chillers	20
Centrifugal and absorption chillers	23
All other HVAC&R equipment (unless otherwise noted)	15

TABLE 4.4.2   Equipment Life as Given in LEED-NC 2.2 Reference Guide

ODP_r — Ozone depletion potential of refrigerant r (These are listed in Table 4.4.1 or can be obtained from the EPA website. For refrigerants other than the CFCs, they typically range from 0 to 0.2 lb_{CFC11}/lb_r.)

$Q_{unit\,i}$ — The cooling capacity of an individual HVAC or refrigeration unit i in $ton_{ac} \cdot s$ (A ton_{ac} of cooling capacity is based on an older system of rating refrigeration when tons of ice were delivered for iceboxes. The tonnage given is actually in tons of ice melted in 24 h. One ton_{ac} of cooling can be directly converted to 12,000 Btu/h.)

Q_{total} — The total cooling capacity of all HVAC or refrigeration in $ton_{ac} \cdot s$ ($\Sigma Q_{unit\,i}$ over all systems i)

Rc_i — Refrigerant charge of system i (This usually ranges from 0.5 to 5.0 lb of refrigerant per ton_{ac} of cooling capacity.)

$LCODP_i$ and $LCGWP_i$, the equivalent annual life-cycle-based ozone depleting potential and the equivalent annual life-cycle-based global warming potential for system i are calculated by Eqs. (4.4.1) and (4.4.2). In each equation, the variable to the left side of the addition sign represents the contributions from annual leakage and the variable to the right side of the addition sign represents an annualized contribution from installation of the unit and disposing of the unit at the end of its useful life.

$$LCODP_i = [Lr_i + (Mr_i/Life_i)] \, (ODP_r \times Rc_i) \qquad (4.4.1)$$

$$LCGWP_i = [Lr_i + (Mr_i/Life_i)] \, (GWP_r \times Rc_i) \qquad (4.4.2)$$

For systems with the default values of leakage and loss and a 10-year life, these equations simplify to Eqs. (4.4.3) and (4.4.4). Similar simplified equations can be written for the various default values and equipment lives.

$$LCODP_i = 0.03(ODP_r \times Rc_i) \qquad \text{for default leakage and loss, 10-year life} \qquad (4.4.3)$$

$$LCGWP_i = 0.03(GWP_r \times Rc_i) \qquad \text{for default leakage and loss, 10-year life} \qquad (4.4.4)$$

Finally, a weighted average of all base building level HVAC&R equipment systems shall be used over all systems i such that

$$\{\Sigma_i [LCGWP_i + (LCODP_i \times 100{,}000 \; lb_{CO2}/lb_{CFC11})]Q_{unit\,i}\}/Q_{total} \leq 100 \; lb_{CO2}/(ton_{ac} \cdot yr)$$
$$(4.4.5)$$

For projects with multiple systems, each individual system does not need to meet the criteria based on the inequality in Eq. (4.4.5), but the overall project average must meet the criteria. The LEED-NC 2.2 Reference Guide has a table in the EAc4 section that gives default maximum allowable equipment refrigerant charges in pounds of refrigerant per ton of cooling for several refrigerants based on the default losses and leaks and applicable equipment life. These default maxima are for the individual pieces of equipment. These default maxima do not have to be met if the overall project complies with Eq. (4.4.5).

It is important that the refrigerant also be chosen based on compliance with the Montreal Protocol, as many refrigerants are being phased out over the next two decades. It is recommended that project designers refer to the EPA ozone depletion and global warming potential lists on its website for information on new or alternative refrigerants. In addition, in 2004 several major corporations formed an initiative called *Refrigerants, Naturally!* The intent of this initiative is to promote and develop refrigeration technologies which are HFC-free. Its website contains information on many alternative systems, such as those with carbon dioxide and solar cooling. Tri-State Generation and Transmission Association, Inc., also has a Web-based *Energy Library* which gives information on many alternatives.

Special Circumstances and Exemplary Performance

In summary, one point will be awarded in this category if no CFCs, HCFCs, or halons are used for fire suppression systems, *and* either the criteria in Eq. (4.4.5) are met or there are no refrigerants used other than natural refrigerants in the base building HVAC&R systems. There is no EP point available for this credit.

4.5 EA Credit 5: *Measurement and Verification*

Designing a system to be more efficient and the system actually performing in that manner in various situations with different occupancies are two separate issues. Therefore, credit is also given if it can be proved that the systems are operating as intended. A side benefit of this credit is that it gives the owners the knowledge of which systems may not be operating as efficiently as desired and, therefore, the opportunity for modification and improvement.

USGBC Rating System

LEED-NC 2.2 lists the Intent, Requirements, and Potential Technologies and Strategies for EAc5 as follows:

Intent
Provide for the ongoing accountability of building energy consumption over time.

Requirements

- Develop and implement a Measurement & Verification (M&V) Plan consistent with Option D: Calibrated Simulation (Savings Estimation Method 2), or Option B: Energy Conservation Measure Isolation, as specified in the *International Performance Measurement & Verification Protocol (IPMVP) Volume III: Concepts and Options for Determining Energy Savings in New Construction*, April, 2003.
- The M&V period shall cover a period of no less than one year of post-construction occupancy.

Potential Technologies and Strategies

Develop an M&V Plan to evaluate building and/or energy system performance. Characterize the building and/or energy systems through energy simulation or engineering analysis. Install the necessary metering equipment to measure energy use. Track performance by comparing predicted performance to actual performance, broken down by component or system as appropriate. Evaluate energy efficiency by comparing actual performance to baseline performance. While the IPMVP describes specific actions for verifying savings associated with energy conservation measures (ECMs) and strategies, this LEED credit expands upon typical IPMVP M&V objectives. M&V activities should not necessarily be confined to energy systems where ECMs or energy conservation strategies have been implemented. The IPMVP provides guidance on M&V strategies and their appropriate applications for various situations. These strategies should be used in conjunction with monitoring and trend logging of significant energy systems to provide for the ongoing accountability of building energy performance.

Calculations and Considerations

EAc5 gives an additional avenue for enhanced energy performance by implementing a program that measures and verifies (M&V) energy performance in the built environment. If this is done, then there is a potential for detecting and modifying energy inefficiencies over the lifetime of a building. The credit is given if the Measurement and Verification (M&V) Plan is developed and implemented through at least one year of postconstruction occupancy. The engineering details for obtaining this credit are varied, and it is not the intention of this book to go into these details. However, there are a few items useful for all participants in a green building project to understand.

This credit is worth one point and is based on documents developed by the Efficiency Valuation Organization (EVO), which is a nonprofit organization devoted to global energy efficiency (www.evo-world.org/ipmvp.php). The relevant standard is the *International Performance Measurement and Verification Protocol (IPMVP), Volume III: Concepts and Options for Determining Energy Savings in New Construction, April 2003*. There are two different approaches that can be taken. For smaller or less complex buildings where it is relatively easy to isolate the separate energy-saving and energy systems, the credit may be obtained by following the Energy Conservation Measure (ECM) Isolation method (Option B) of the IPMVP Volume III. For more complex energy systems, the Whole Building Calibration Simulation method (Option D) of the IPMVP Volume III may be chosen. In addition to the development of an M&V Plan, there may be a substantial amount of metering which may be needed on various building systems to measure the energy usages and savings. Table 4.5.1 summarizes these two approaches.

It is important that the submittals include the IPMVP option chosen, a copy of the M&V Plan as written according to the requirements in IPMVP Volume III, appropriate models and calculations based on the data collected, and a narrative if there are any

Building Energy System Types	IPMVP Vol. III Option	Method Name
Simple, easily isolated	Option B	Energy Conservation Measure (ECM) Isolation
Complex	Option D	Whole Building Calibration Simulation

TABLE 4.5.1 Summary of IPMVP Approaches for EAc5

special circumstances or explanations of calculation methods. This submittal must cover at least a full year of postconstruction occupancy, with this minimum one year time period beginning after the building has reached a reasonable occupancy and use level. Therefore, it is not usually reasonable to schedule the submittal exactly one year from the certificate of occupancy (CofO), which may be issued by the local building official or building authority. Instead, the project team should estimate an occupancy schedule, keep track of the building occupancies, and have the one-year period begin after most of the uses are established. In the case where one group plans to be the sole occupant of the new building and move the entire workforce within a month of issuance of the CofO, then the one-year period may start relatively quickly after construction ends. In the case where an office building is constructed that has multiple occupants, there may be some floors or areas that may take a year or so after construction to find tenants, and the verification of this credit may take longer.

Special Circumstance and Exemplary Performance

EAc5 has been given an EB notation by the USGBC. It is also designated as a credit that needs a construction phase submittal, as much of the activity is during or post construction. There is no EP point available for this credit.

4.6 EA Credit 6: *Green Power*

The intent of EAc6 is to encourage the use of power from a power grid that has been made from renewable energy sources.

USGBC Rating System

LEED-NC 2.2 lists the Intent, Requirements, and Potential Technologies and Strategies for EAc2 as follows:

Intent

Encourage the development and use of grid-source, renewable energy technologies on a net zero pollution basis.

Requirements

Provide at least 35% of the building's electricity from renewable sources by engaging in at least a two-year renewable energy contract. Renewable sources are as defined by the Center for Resource Solutions (CRS) Green-e products certification requirements.

DETERMINE THE BASELINE ELECTRICITY USE

Use the annual electricity consumption from the results of EA Credit 1.

OR

Use the Department of Energy (DOE) Commercial Buildings Energy Consumption Survey (CBECS) database to determine the estimated electricity use.

Potential Technologies and Strategies

Determine the energy needs of the building and investigate opportunities to engage in a green power contract. Green power is derived from solar, wind, geothermal, biomass or low impact hydro sources. Visit www.green-e.org for details about the Green-e program. The power product purchased to comply with credit requirements need not be Green-e certified. Other sources of green power are eligible if they satisfy the Green-e program's technical requirements. Renewable energy certificates (RECs), tradable renewable certificates (TRCs), green tags and other forms of green power that comply with Green-e's technical requirements can be used to document compliance with EA Credit 6 requirements.

Calculations and Considerations

EAc6 is based on either purchasing or trading electricity from renewable off-site power sources for at least 35 percent of the defined electricity needs for the project during two years. The definition of renewable power sources is per the Center for Resource Solutions' Green-e Product Certification Requirements, which can be found at www.green-e.org. The electricity can be purchased from the local power provider, if available. If the "green" electricity is not directly available, but the area provider has a green power program, then the contract can include enrollment into this program with its additional premium to facilitate green power usage elsewhere. If the area provider cannot offer either of these options, then there is also the opportunity to purchase Green-e accredited Tradable Renewable Certificates (TRCs). In this case, power is purchased locally, and the TRCs for the calculated amount are purchased. The TRCs represent the premium needed for purchasing renewable power elsewhere in the country and facilitate its widespread use.

The threshold value of green power purchased is based on the electrical power needs portion of the energy model if the EAc1 Option 1 is sought. Otherwise, it is based on the electrical power needs portion of the building annual energy cost (BAEC) calculation as outlined in EAc2. Either way, electricity cost rates do not have to be evaluated since only the electrical power needs are being analyzed, and there is no need to have a cost-based bridge for adding different power usage impacts. The sets of calculations based on the BAEC method and the EAc1 Option 1 energy model method are described in the following paragraphs.

Based on the BAEC method when the EAc1 Option 1 is not sought, the building annual energy usage of electricity (BAEE) can be found by using the median annual electrical intensities for the particular building type (CBECSE$_j$) as listed in Table 4.2.2 under EAc2. The calculation for BAEE is given in Eq. (4.6.1). [This is the electricity portion of Eq. (4.2.1) and does not include cost rates.] The GFA in Eq. (4.6.1) is the gross floor area of the building as defined in Chap. 2. If the project is made up of several buildings or building portions with different uses as listed in Table 4.2.2, then the commercial building energy consumptions (CBECSE$_j$) for electricity should be separately analyzed for each portion, multiplied by the gross floor area in that building portion, and summed to obtain the BAEE.

$$BAEE = (CBECSE_j)(GFA) \qquad \text{if EAc1 Option 1 is not sought} \qquad (4.6.1)$$

On the other hand, the basis for the electric usage when EAc1 Option 1 is sought is a portion of the Proposed Building Performance (PBP) as previously calculated in Eq. (4.1.1). In LEED-NC 2.1, a point was earned in EAc6 if the green power purchased was at least 50 percent of the difference between all regulated electricity loads and the electricity provided by on-site renewal energy sources. For this earlier version, the EAc1 calculations were all based on regulated loads. In LEED-NC 2.2, the calculations for EAc1 Option 1 are based on both regulated and nonregulated loads for which the nonregulated loads are assumed to be approximately 25 percent of the regulated loads. (Nonregulated loads are sometimes referred to as *process loads*.) Therefore, the threshold in LEED-NC 2.2 has been reduced to 35 percent to take into account the calculation changes. In LEED-NC 2.1, the modeled energy electrical use was further modified to account for on-site renewable energy production. Keeping this modification in LEED-NC 2.2, the calculation appropriate for the first compliance path (EAc1 Option 1) should again include this reduction in the denominator of the on-site renewable energy, so that the percent green power purchased is truly only the percent of the demand from the off-site grid.

Keeping this modification in mind, let PBE (proposed building electrical performance) be the electrical energy portion of the PBP as defined in EAc1 Option 1. If PDEE is the proposed design electrical energy portion (in electrical usage units) as given from the proposed design energy model (PDEM), then the PBE is the difference in the electrical usage portion of the PDEM in units of electrical power per year (PDEE) and the electrical portion of the annual on-site renewable energy (REC) in units of electrical power per year (REE).

$$PBE = PDEE - REE \qquad \text{if EAc1 Option 1 is sought} \qquad (4.6.2)$$

Using the annual amount of green electricity (GE) purchased or traded from off-site, and the definitions for proposed building energy needs, PBE and BAEE, based on the EAc1 Option 1 compliance pathway and other energy calculation pathways, respectively, the percent green electricity (PGE) can be calculated for each as

$$PGEPB = 100 \times \frac{GE}{PBE} \qquad \text{if EAc1 Option 1 is sought} \qquad (4.6.3)$$

or alternatively

$$PGEPB = 100 \times \frac{GE}{PDEE - REE} \qquad \text{if EAc1 Option 1 is sought} \qquad (4.6.4)$$

and

$$PGEBAE = 100 \times \frac{GE}{BAEE} \qquad \text{if EAc1 Option 1 is not sought} \qquad (4.6.5)$$

PGEPB or PGEBAE must be greater than or equal to 35 for a point to be awarded in EAc6. Table 4.6.1 gives a summary of the symbols and equations.

Special Circumstances and Exemplary Performance

EAc6 has been designated as a construction phase submittal, mainly because the utility contracts are usually not fully developed until construction is nearly complete. If there are special circumstances for these credits, a narrative must be submitted describing these circumstances.

Exemplary performance (EP) credit can be achieved by doubling the requirement, either by doubling the amount of green electricity purchased or by doubling the length of the contract.

4.7 Discussion and Overview

The Energy and Atmosphere category seeks to reduce energy usage and the impacts of energy usage on the environment. There are also additional comments which can be made about the multiple benefits or effects of some of the measures used to obtain these

Symbol	Description	Comments	Relevant Equations	Restrictions
GE	Annual amount of green electricity from off-site	Can be purchased or traded		
BAEE	Building annual energy usage of electricity	Electrical energy part of BAEC, the building annual energy cost method	$BAEE = (CBECSE)(GFA)$	Use if EAc1 Option 1 is not sought
PGEBAE	Percent green electricity based on the BAEE	Based on the BAEC method	$PGEBAE = 100 \times \dfrac{GE}{BAEE}$	Use if EAc1 Option 1 is not sought
REE	Annual on-site renewable electrical energy provided	Electrical energy portion of the annual on-site renewable energy (REC)		
PDEE	Annual proposed design electrical energy	Electrical energy part of the proposed design energy model (PDEM)		Use if EAc1 Option 1 is sought
PBE	Annual proposed building electrical performance	Electrical energy part of the proposed building performance (PBP)	$PBE = PDEE - REE$	Use if EAc1 Option 1 is sought
PGEPB	Percent green electricity based on the PBE	Based on the proposed design energy model (PDEM)	$PGEPB = 100 \times \dfrac{GE}{PBE}$ or $PGEPB = 100 \times \dfrac{GE}{PDEE - REE}$	Use if EAc1 Option 1 is sought

TABLE 4.6.1   Green Power Credit Summary of Symbols and Equations

goals. Two examples relate to the promotion of on-site renewable energy sources and to the location of the "breathing zone" within the interior spaces. Also, it might be useful to address the use of hydrogen fuel cells and how that may fit in with the energy goals. These are briefly discussed in the following sections.

On-Site Renewable Energy   As previously discussed, EAc1, EAc2, and EAc6 all seek to reduce the use of nonrenewable energy sources for both environmental and economic reasons. Combined, the three credit subcategories give extra credit if on-site renewable energy sources are used in this reduction. There are many reasons for this, including the reliability of the overall power options when there are multiple sources, the importance of distributing the use of energy sources so that they are not as concentrated and therefore do not have such large, concentrated impacts on the environment in only a few areas, and security issues in the energy network. Not only are there points available in EAc2 for on-site renewable energy sources, but also use of these on-site sources makes it possible to earn more points in EAc1 Option 1 and to reach the 35 percent threshold in EAc6 for the same building energy needs.

For example, let us look at new building project X where the project has equal cost values of electrical and natural gas needs. In the proposed design if both the electrical and natural gas loads are reduced 21 percent from the baseline design and there is no on-site renewable electricity production, then four points may be earned based on EAc1 Option 1 as shown in Eq. (4.7.1).

$$\text{AECS} = 100 \times \left(1 - \frac{\text{PBP}}{\text{BBP}}\right) = 21 \tag{4.7.1}$$

So PBP/BBP is 0.79, but since PBP equals PDEM less the REC, and REC is zero, then PDEM/BBP is 0.79.

Let us then hypothetically reevaluate the situation if 24 percent of the electrical needs (12 percent of the total cost value of the total project energy needs or REC = 0.12PDEM) can be generated by renewable sources on-site. Then the following two equations apply:

$$\text{AECS} = 100 \times \left(1 - \frac{\text{PDEM}}{\text{BBP}} + \frac{\text{REC}}{\text{BBP}}\right) = 30.5 \tag{4.7.2}$$

and

$$\text{PREPBP} = 100 \times \left(\frac{\text{REC}}{\text{PDEM} - \text{REC}}\right) = 13.3 \tag{4.7.3}$$

The total points available now from EAc1 and EAc2 are nine (six and three for each, respectively). So the point total has more than doubled.

In addition, without the on-site renewable energy sources, the threshold of the percent of purchased green power needed based on the design electrical use decreases from 35 percent to just under 27 percent.

Initially, without including on-site renewable energy sources, to obtain the point for EAc6, the following must be true:

$$35 \geq 100(\text{GE}/\text{PBE}) \tag{4.7.4}$$

but if REE = 0.24PBE, then

$$35 \geq 100[\text{GE}/(\text{PBE} - 0.24\text{PBE})] = 0.76[100(\text{GE}/\text{PBE})] \tag{4.7.5}$$

or

$$26.6 \geq 100 \, (\text{GE}/\text{PBE}) \tag{4.7.6}$$

In summary, the incorporation of on-site renewable energy into a project scope can facilitate obtaining many additional points in a project, several more than if just EAc2 is evaluated. This helps justify additional first costs that may be realized for installing these on-site systems.

Hydrogen Fuel Cells   Hydrogen fuel cells are currently under development to help meet our energy needs. They are not specifically addressed in the LEED-NC 2.2 categories, but may be considered for additional innovative points under certain circumstances. The hydrogen in the fuel cells by itself is not an energy source, but rather an intermediary energy storage fuel, and the hydrogen is made from other compounds such as H_2O or CH_4 with energy used in the formation process. Since there are many options for the energy used in the hydrogen production process and the source compounds for the hydrogen, the use of hydrogen fuel cells may or may not be considered renewable. However, there are many other potential environmental benefits of using hydrogen fuel cells on a project. One is the decrease in the on-site emissions of many criteria air pollutants, such as nitrogen oxides (NO_x), carbon monoxide (CO), and volatile organic compounds (VOCs), as compared to many alternative-fuel choices. If hydrogen fuel cells are used on the site, then several issues should be considered by the designers to maintain the effectiveness of the system. The hydrogen fuel cells usually intake ambient air as the oxygen (O_2) source to produce the main product, water. Similar to the air intakes for makeup air in a building, care must be taken in the location of the air intakes for the fuel cells so that the source air is not contaminated with pollutants. Air intakes located near fossil fuel motor vehicle loading areas or drives may direct exhausts into the cells that can foul the membranes. Figure 4.7.1 shows a fuel cell on a LEED-NC certified building in Columbia, S.C. In addition to heating water in the building, the waste energy in the fuel cell coolant system is used to preheat the water.

Figure 4.7.1   Fuel Cell at West Quad on the University of South Carolina Campus in Columbia, S.C.

Breathing Zone   The designers of the energy systems should be cognizant of many parameters that affect other credits. One is the breathing zone. This is used in the Indoor Environmental Quality category for EQc1 if any of the options for carbon dioxide (CO_2) monitoring are used, and for EQc2 if the air quality testing option is sought. The *breathing zone* is an area defined as being between two imaginary planes that are located 3 and 6 ft above the floor, and are bound vertically by planes that are around 2 ft off from any wall or fixed air unit. The air quality parameter measurements are taken within the breathing zone as specified. If the breathing zone represents the volumes that are typically important for good air quality, then designing the energy systems in the building such that improved air quality is focused in these volumes would be very important for a green building.

References

Andrepoint, J. (2006), "Developments in Thermal Energy Storage: Large Applications, Low Temps, High Efficiency, and Capital Savings," *Energy Engineering*, 103(4): 7–18.

ASHRAE (2004), *90.1-2004: Energy Standard for Buildings Except Low-Rise Residential Building*, American Society of Heating, Refrigerating and Air-Conditioning Engineers, Atlanta, Ga.

ASHRAE (2004), *Advanced Energy Design Guide for Small Office Buildings*, American Society of Heating, Refrigerating and Air-Conditioning Engineers, Atlanta, Ga.

ASHRAE (2007), www.ashrae.org, website accessed July 12, 2007.

Cooper, C. D., and F. C. Alley (2002), *Air Pollution Control, A Design Approach*, Waveland Press, Inc., Prospect Heights, Ill.

De Nevers, N. (2000), *Air Pollution Control Engineering*, 2d ed., McGraw-Hill, New York.

Distributed Energy (2007), www.distributedenergy.com, *Distributed Energy, The Journal for Onsite Power Solutions*, website accessed July 12, 2007.

EIA (2001), *Emissions of Greenhouse Gases in the United States 2001: Alternatives to Chlorofluorocarbons: Lowering Ozone Depletion Potentials vs. Raising Global Warming Potentials*, Energy Information Administration, U.S. Department of Energy, Washington, D.C., accessed webpage July 12, 2007, http://www.eia.doe.gov/oiaf/1605/gg02rpt/ozone.html

EIA (2007), *Energy Market and Economic Impacts of a Proposal to Reduce Greenhouse Gas Intensity with a Cap and Trade System*, January 2007, Energy Information Administration, U.S. Department of Energy, Washington, D.C.

EPA (2007), http://www.epa.gov/solar/, U.S. Environmental Protection Agency Clean Energy Program, website accessed July 12, 2007.

Fay, J. A., and D. S. Golomb (2002), *Energy and the Environment*, Oxford University Press, New York.

Gonzalez, M. J., and J. G. Navarro (2006), "Assessment of the Decrease of CO_2 Emissions in the Construction Field through the Selection of Materials: Practical Case Study of Three Houses of Low Environmental Impact," *Building and Environment*, 41: 902–909, Elsevier, Amsterdam.

Green-e (2007), www.green-e.org, The Center for Resource Solutions (CRS), San Francisco, Calif., website accessed July 12, 2007.

Heinsohn, R. J., and R. L. Kabel (1999), *Sources and Control of Air Pollution*, Prentice-Hall, Upper Saddle River, N.J.

NREL (2007), http://www.nrel.gov/, National Renewable Energy Laboratory, Denver, Colo., U.S. Department of Energy, webpage accessed July 12, 2007.

Refrigerants, Naturally! (2007), http://www.refrigerantsnaturally.com/, website accessed July 12, 2007.

Tri-State (2007), http://tristate.apogee.net/, Tri-State Generation and Transmission Association, Inc., Energy Library, Westminster, Colo., accessed July 12, 2007.

USGBC (2003), *LEED-NC for New Construction, Reference Guide*, Version 2.1, 2d ed., May, U.S. Green Building Council, Washington, D.C.

USGBC (2005–2007), *LEED-NC for New Construction, Reference Guide*, Version 2.2, 1st ed., U.S. Green Building Council, Washington, D.C., October 2005 with errata posted through Spring 2007.

Varodompun, J., and M. Navvab (2007), "HVAC Ventilation Strategies: The Contribution for Thermal Comfort, Energy Efficiency, and Indoor Air Quality," *Journal of Green Building*, Spring, 2(2): 131–150.

Wark, K., C. F. Warner, and W. T. Davis (1998), *Air Pollution, Its Origin and Control*, Addison Wesley Longman, Menlo Park, Calif.

WMO (1998), *The Scientific Assessment of Ozone Depletion, 1998*, World Meteorological Association's Global Ozone Research and Monitoring Project, Geneva, Switzerland.

WMO (2002), *The Scientific Assessment of Ozone Depletion, 2002*, World Meteorological Association's Global Ozone Research and Monitoring Project, Geneva, Switzerland.

Exercises

1.   You are designing a new building in South Carolina where electricity costs about $0.085 per kWh and natural gas costs about $5.32 per MMBtu. The average of the four orientations of the base design uses about 12 kW of electricity and 130 MMBtu of natural gas each month for regulated uses.

   A. Determine on an annual basis the BBP (baseline building performance).

   B. You decide to implement some energy-saving programs for the regulated uses that reduce these electrical uses by 30 percent and natural gas uses by 25 percent. You are not going to use any renewable energy sources on-site. Determine the PDEM (proposed design energy model) cost and the PBP (proposed building performance).

   C. Determine your AECS (annual energy cost savings) and the available points for EA credit 1.

2.   You are designing a new building in a state where electricity costs about $0.095 per kWh and natural gas costs about $6.32 per MMBtu. The average of the four orientations of the base design uses about 14 kW of electricity and 140 MMBtu of natural gas each month for regulated uses.

   A. Determine on an annual basis the BBP (baseline building performance).

   B. You decide to implement some energy-saving programs for the regulated uses that reduce these electrical uses by 30 percent and natural gas uses by 25 percent. In addition, you would like to use on-site renewable sources of electricity with photovoltaic cells for 2 kW of the electricity. Determine the PDEM cost and the PBP.

   C. Determine your AECS and the available points for EA credits 1 and 2.

3.   Determine how much green power you need to purchase to get a point for credit EAc6 for the new building in Exercise 1.

4.   Energy modelers have determined that the average BBP for your new project in St. Louis is about $1500 per month. Of this, 55 percent represents costs due to electrical usages, and the remainder of the costs is from natural gas usage. The PDEM gives a value of $1200 per month with 50 percent of the costs from electrical usages. All costs are based on the default energy costs of 2003.

Determine the points available for EAc1 and the amount of green power that must be purchased to obtain the point for EAc6.

5.   Energy modelers have determined that the average BBP for your new project in Portland is about $1500 per month. Of this, 60 percent represents costs due to electrical usages, and the remainder is from natural gas. The PDEM gives a value of $1200 per month with 50 percent of the costs from electrical usages. All costs are based on the default energy costs of 2003. You are planning to install photovoltaic cells at the site that would generate on average 2 kW of power for on-site usage. Determine the points available for EAc1 and the amount of green power that must be purchased to obtain the point for EAc6.

6.   There are many options for bulbs in light fixtures. For instance, emergency exit lighting could possibly use incandescent, fluorescent, or LED (light-emitting diode) bulbs. In this exercise, assume that the three different types of lamps use 40, 10, and 5 W and have typical lamp lives of 0.7, 1.7, and 80 years, respectively.
  A. Calculate the energy usage for a single emergency exit lighting fixture with incandescent, compact fluorescent, and LED emergency bulb over 10 years.
  B. In your state, at current electrical rates, what are the electricity costs for 10 years?
  C. Estimate the additional bulb replacement costs as appropriate (both material and labor), and give the estimated 10-year life-cycle cost for these three types.

7.   You are building a 10,000 ft² retail facility in Nebraska. You have opted to obtain some points for EAc1 by the prescriptive methods and will not do the whole building energy simulation. However, you would like to see what size solar power you would need to install on-site for EAc2. Determine the minimum average power generation required from on-site solar power to obtain one point for EAc2. Determine the amount of green power you would need to purchase from off-site sources to obtain the EAc6 credit, with and without installation of the minimum on-site solar cell.

8.   You are building a 4600 ft² pharmacy on Long Island. You will not do the whole building energy simulation due to the small size of the facility. However, there is a green power provider in the area that sells renewable energy certificates at 18.5 cents/kWh. Current electricity rates in the area are 17.3 cents/kWh. What is your annual cost premium to obtain EAc6?

9.   You are designing a commercial strip with 10 storefronts and offices above. Each store uses a 5-ton packaged HVAC unit, and each office above needs a 2-ton split HVAC unit. The small refrigerators and water coolers in the store back rooms all use less than 0.5 lb of refrigerant and are exempt from this calculation. You have found manufacturers' specifications for the package units which list them as using 1.8 lb of refrigerant per ton of air conditioning for each unit and HFC-410A for the refrigerant. The cut sheet for the split units lists 3.1 lb of HCFC-22 per ton of air conditioning.
  Calculate the equivalent annual life-cycle-based global warming potential (LCGWP$_i$) and the equivalent annual life-cycle-based ozone depleting potential (LCODP$_i$) for each unit type. Determine if EAc4 is met.

10.   You are designing a commercial office tower with 10 offices or stores on the first floor. Each of the first-floor spaces needs a 2-ton split HVAC unit listed at 3.1 lb of HCFC-22 per ton of air conditioning. The upper floors are all used for an engineering consulting firm, and they are cooled by a single 400-ton$_{ac}$ centrifugal chiller using HCFC-123, which lists 1.8 lb of refrigerant per ton of air conditioning.
  Calculate the equivalent annual life-cycle-based global warming potential (LCGWP$_i$) and the equivalent annual life-cycle-based ozone depleting potential (LCODP$_i$) for each unit type. Determine if EAc4 is met.

11. For a typical 10-year life HVAC&R unit, calculate the maximum allowable equipment refrigerant charge (Rc_i) per ton$_{ac}$ that a single unit may have based on the default values for leakages and losses in order to be compliant with EAc4 for the following refrigerants:

    A. HCFC-22
    B. HCFC-123
    C. HFC-134a
    D. HFC-245fa
    E. HFC-407c
    F. HFC-410a

12. For a typical 15-year life HVAC&R unit, calculate the maximum allowable equipment refrigerant charge (Rc_i) per ton$_{ac}$ that a single unit may have based on the default values for leakages and losses in order to be compliant with EAc4 for the following refrigerants:

    A. HCFC-22
    B. HCFC-123
    C. HFC-134a
    D. HFC-245fa
    E. HFC-407c
    F. HFC-410a

13. For a typical 20-year life HVAC&R unit, calculate the maximum allowable equipment refrigerant charge (Rc_i) per ton$_{ac}$ that a single unit may have based on the default values for leakages and losses in order to be compliant with EAc4 for the following refrigerants:

    A. HCFC-22
    B. HCFC-123
    C. HFC-134a
    D. HFC-245fa
    E. HFC-407c
    F. HFC-410a

14. For a typical 23-year life HVAC&R unit, calculate the maximum allowable equipment refrigerant charge (Rc_i) per ton$_{ac}$ that a single unit may have based on the default values for leakages and losses in order to be compliant with EA credit 4 for the following refrigerants:

    A. HCFC-22
    B. HCFC-123
    C. HFC-134a
    D. HFC-245fa
    E. HFC-407c
    F. HFC-410a

15 Simplify Eqs. (4.4.1) and (4.4.2) for a unit with a 15-year life using the default values of leakage and loss.

$$LCODP_i = [Lr_i + (Mr_i/Life_i)](ODP_r \times Rc_i) \qquad (4.4.1)$$

$$LCGWP_i = [Lr_i + (Mr_i/Life_i)](GWP_r \times Rc_i) \qquad (4.4.2)$$

16 Simplify Eqs. (4.4.1) and (4.4.2) for a unit with a 20-year life using the default values of leakage and loss.

17.   Simplify Eqs. (4.4.1) and (4.4.2) for a unit with a 23-year life using the default values of leakage and loss.

18.   Find out what the current status is of EA Credit 1: *Optimize Energy Performance*. Has the USGBC adopted a minimum number of required points? How many? If this number is more than the one point allotted for Option 3, how has Option 3 been modified to account for this?

19.   Derive Eqs. (4.7.2) and (4.7.3) from the information given in the text and verify the point total.

CHAPTER 5

LEED Materials and Resources

The Materials and Resources category of the U.S. Green Building Council (USGBC) rating system for new construction version 2.2 (LEED-NC 2.2) consists of one prerequisite subcategory and seven credit subcategories which together may earn a possible 13 points (credits) maximum, not including EP points. The notation format for the prerequisites and credits is, for example, MRp1 for Materials and Resources prerequisite 1 and MRc1.1 for Materials and Resources credit 1.1. Also in this chapter, to facilitate easier cross-referencing between this text and the USGBC rating system, the second digit in a section heading, equation, table, or figure number represents the credit subcategory number for sections that deal directly with a USGBC LEED-NC 2.2 credit subcategory.

The Materials and Resources (MR) portion deals with issues that reduce the use of new materials and resources, encourages the use of materials and resources that have a smaller impact on the environment, and promotes the reuse or recycling of materials so that more virgin materials and resources are not used on LEED certified projects. The life cycles of many products and materials are taken into account to reduce the impact on the environment of their use. This may include transportation impacts, harvesting impacts, manufacturing impacts, and the benefit of using recycled materials in the production of the product.

The one prerequisite subcategory is MR Prerequisite 1: *Storage and Collection of Recyclables* (MRp1). As is typical for prerequisites, there are no points associated with MRp1.

The points for the credits in the Materials and Resources category are individually related to a specific credit under a subcategory as was done in the Sustainable Sites and Water Efficiency categories. There are seven credit subcategories, five of which have multiple credits and the last two which each have just one credit and associated point. The seven credit subcategories are MR Credit Subcategory 1: *Building Reuse*, MR Credit Subcategory 2: *Construction Waste Management*, MR Credit Subcategory 3: *Materials Reuse*, MR Credit Subcategory 4: *Recycled Content*, MR Credit Subcategory 5: *Regional Materials*, MR Credit Subcategory 6: *Rapidly Renewable Materials*, and MR Credit Subcategory 7: *Certified Wood*. MR Credit Subcategory 1 has three credits; MR Credit Subcategories 2, 3, 4, and 5 each have two credits; and MR Credit Subcategories 6 and 7 are both single credits. Each credit is worth one point not including EP points, which will be discussed later.

The prerequisites and credits in the Materials and Resources category can be grouped into the following three overall sets of goals:

- MRp1 and the first four Materials and Resources credit subcategories (1 through 4) all encourage the continued use, reuse, or recycling of materials or products to decrease the use of more raw materials and decrease the impact to the waste disposal stream.

- The two credits in the Materials and Resources subcategory 5 encourage the use of materials or products that do not require long transportation distances to reach the site. This reduces the negative environmental impacts of transportation and increases sustainability in the area.

- The Materials and Resources subcategories 6 and 7 encourage stewardship in the use of products made from renewable materials, particularly promoting rapidly renewable materials in MRc6.

The one prerequisite and the 13 credits in the Materials and Resources category are as follows:

- **MR Prerequisite 1** (MRp1): *Storage and Collection of Recyclables*

- **MR Credit 1.1** (MRc1.1): *Building Reuse—Maintain 75% of Existing Walls, Floors and Roof*

- **MR Credit 1.2** (MRc1.2): *Building Reuse—Maintain 95% of Existing Walls, Floors and Roof* (previously referred to *as Building Reuse—Maintain 100% of Existing Walls, Floors and Roof* in LEED-NC 2.1)

- **MR Credit 1.3** (MRc1.3): *Building Reuse—Maintain 50% of Interior Non-Structural Elements* (previously referred to as *Building Reuse—Maintain 100% of Shell/ Structure and 50% of Non-Shell/Non-Structure* in LEED-NC 2.1)

- **MR Credit 2.1** (MRc2.1): *Construction Waste Management—Divert 50% from Disposal* (previously referred to as *Construction Waste Management—Divert 50% from Landfill* in LEED-NC 2.1)

- **MR Credit 2.2** (MRc2.2): *Construction Waste Management—Divert 75% from Disposal* (previously referred to as *Construction Waste Management—Divert 75% from Landfill* in LEED-NC 2.1)

- **MR Credit 3.1** (MRc3.1): *Materials Reuse—5%* (previously referred to as *Resource Reuse—5%* in LEED-NC 2.1)

- **MR Credit 3.2** (MRc3.2): *Materials Reuse—10%* (previously referred to as *Resource Reuse—10%* in LEED-NC 2.1)

- **MR Credit 4.1** (MRc4.1): *Recycled Content—10% (post-consumer + ½ pre-consumer)* [previously referred to as *Recycled Content—5% (post-consumer + ½ post-industrial)* in LEED-NC 2.1] (EB)

- **MR Credit 4.2** (MRc4.2): *Recycled Content—20% (post-consumer + ½ pre-consumer)* [previously referred to as *Recycled Content—10% (post-consumer + ½ post-industrial)* in LEED-NC 2.1] (EB)

- **MR Credit 5.1** (MRc5.1): *Regional Materials—10% Extracted, Processed and Manufactured Regionally* (previously referred to as *Regional Materials—20% Manufactured Regionally* in LEED-NC 2.1) (EB)

- **MR Credit 5.2** (MRc5.2): *Regional Materials—20% Extracted, Processed and Manufactured Regionally* (previously referred to as *Regional Materials—50% Extracted Regionally* in LEED-NC 2.1) (EB)

- **MR Credit 6** (MRc6): *Rapidly Renewable Materials* (EB)

- **MR Credit 7** (MRc7): *Certified Wood* (EB)

LEED-NC 2.2 provides a table which summarizes some of these prerequisite and credit characteristics, and the information from it notes the following three items:

- A significant change from LEED version 2.1

- The main phase of the project for the submittal document preparation (design or construction)

- The project team member(s) who have the most decision-making responsibility as to whether this credit may or may not be achieved (owner, design team, or contractor).

The submittals for each credit or prerequisite have been sectioned into either Design Submittals and/or Construction Submittals. Even though the entire LEED process encompasses all phases of a project, the prerequisites or credits which are noted as Construction Submittal credits may have major actions going on through the construction phase, which, unless documented through that time, may not be able to be verified from preconstruction documents or the built project. Sometimes these credits have activities associated with them that cannot or need not be performed until or subsequent to the construction phase. All the Materials and Resources credits have been included in the construction submittal column and are thus noted in Table 5.0.1. Table 5.0.1 also contains a column explaining the point count.

Four of the credit subcategories listed in Table 5.0.1 have the icon EB noted after their title (Recycled Content, Regional Materials, Rapidly Renewable Materials, and Certified Wood). This icon is a tool to help those who wish to proceed with the continuing LEED-EB certification (Existing Building) after certification of LEED-NC is obtained. Those credits noted with this icon are usually significantly more cost-effective to implement during the construction of the building than later during its operation. They are also shown in Table 5.0.1.

There are EP points under the Innovation and Design category available for exceeding each of the respective credit criteria to a minimum level for some of the credits. All the credit subcategories other than Building Reuse (MRc1.1, 1.2, and 1.3) have an EP point available as noted in Table 5.0.1.

One of the variables that must be determined for each project and that must be kept consistent throughout the credits in a LEED submittal is the size and location of the project site. The location of the site boundary is particularly important to the construction waste management credits (MRc2.1 and 2.2) since existing concrete, masonry, or asphalt from pavement on a site which is crushed and reused on-site or elsewhere should be included in the weight or volume of the materials diverted from disposal. There may be exceptions to this for special circumstances such as campus settings. Therefore, there might be a need for narratives describing any such variations and how the project still meets the intent of the credit. The importance of the LEED boundary to MRc2.1 and 2.2 is also noted in Table 5.0.1.

Prerequisite or Credit	Construction Submittal	EB Icon	Credit Point (Not Including Exemplary Performance or ID points)	Exemplary Performance and Other ID Point Availability	Exemplary Performance Point Criterion	Location of Site Boundary and Site Area Important in Calculations
MRp1: Storage and Collection of Recyclables		n.a.	n.a.	n.a.	n.a.	No
MRc1.1: Building Reuse—Maintain 75% of Existing Walls, Floors and Roof	*	No	1	No	n.a.	No
MRc1.2: Building Reuse—Maintain 95% of Existing Walls, Floors and Roof	*	No	1 in addition to 1 for MRc1.1	No	n.a.	No
MRc1.3: Building Reuse—Maintain 50% of Interior Non-Structural Elements	*	No	1	No	n.a.	No
MRc2.1: Construction Waste Management—Divert 50% from Disposal	*	No	1	Yes (1)	95%	Yes
MRc2.2: Construction Waste Management—Divert 75% from Disposal	*		1 in addition to 1 for MRc2.1			
MRc3.1: Materials Reuse—5%	*	No	1	Yes (1)	15%	No
MRc3.2: Materials Reuse—10%	*		1 in addition to 1 for MRc3.1			
MRc4.1: Recycled Content—10% (post-consumer + ½ pre-consumer)	*	Yes	1	Yes (1)	30%	No
MRc4.2: Recycled Content—20% (post-consumer + ½ pre-consumer)	*		1 in addition to 1 for MRc4.1			
MRc5.1: Regional Materials— 10% Extracted, Processed and Manufactured Regionally	*	Yes	1	Yes (1)	40%	No
MRc5.2: Regional Materials— 20% Extracted, Processed and Manufactured Regionally	*		1 in addition to 1 for MRc5.1			
MRc6: Rapidly Renewable Materials (>2.5%)	*	Yes	1	Yes (1)	5%	No
MRc7: Certified Wood (>50%)	*	Yes	1	Yes (1)	95%	No

TABLE 5.0.1   MR EB Icon, Exemplary Performance Point Availability, and Miscellaneous

Material and Resources Credit Calculation Summaries

There are so many different types and uses of materials that go into a building, and they have vastly different values based on weight, cost, or application. Therefore, to determine percentages of material usages, it is important to define which materials are included in the calculations and what units the calculations are based on. The LEED-NC 2.2 Reference Guide provides a great table of MR credit metrics which help in organizing and standardizing the calculations and credit criteria. The process for classification can be summarized in four steps:

1. Segregation of materials into material uses
2. Segregation of the materials for the MR credits into applicable material types or items
3. Determination of the units
4. Setting of the bases (denominator) for each calculation

For the first step, LEED uses the same materials use groupings as the Construction Specifications Institute (CSI), which divides typical construction categories into a sequential summary list of divisions each with many subsets. The division summary list (The CSI Master Format Division List) of the main 16 divisions is as noted in the following list:

Division 1	General Requirements
Division 2	Site Construction
Division 3	Concrete
Division 4	Masonry
Division 5	Metals
Division 6	Woods and Plastics
Division 7	Thermal and Moisture Protection
Division 8	Doors and Windows
Division 9	Finishes
Division 10	Specialties
Division 11	Equipment
Division 12	Furnishings
Division 13	Special Construction
Division 14	Conveying Systems
Division 15	Mechanical (Includes Plumbing)
Division 16	Electrical

A summary of the uses based on the CSI divisions included for the material calculations for each of the credits in the Materials and Resources category can be found in Table 5.0.2.

Note that items used for mechanical, electrical, and plumbing (MEP) purposes are not included in any of the calculations in the Materials and Resources category except for construction waste management. The calculations for the items in subcategories 3

Credits	CSI Divisions Used for Credit Calculations				
	2–10	12 (Furniture and Furnishings)	15 (Mechanical)	15 (Plumbing)	16 (Electrical)
MRc1.1, 1.2, and 1.3: Building Reuse	Yes	n.a.	n.a.	n.a.	n.a.
MRc2.1 and 2.2: Construction Waste Management	Yes	Yes	Yes	Yes	Yes
MRc3.1 and 3.2: Materials Reuse	Yes	Items may be included if consistent for credits 3–7	n.a.	n.a.	n.a.
MRc4.1 and 4.2: Recycled Content	Yes		n.a.	n.a.	n.a.
MRc5.1 and 5.2: Regional Materials	Yes		n.a.	n.a.	n.a.
MRc6; Rapidly Renewable Materials	Yes		n.a.	n.a.	n.a.
MRc7: Certified Wood	Yes		n.a.	n.a.	n.a.

TABLE 5.0.2   CSI Division Uses Applicable for Materials and Resources Calculations for Each MR Credit Subcategory

through 7 are based on value or cost. MEP items tend to be costly per unit weight compared to many of the other bulk building materials. The intent of these subcategories, to the greatest extent possible, is to be able to recycle or reuse or purchase local materials. Therefore, it is assumed that having the high-dollar MEP items in these calculations might bias the equations, without having the bulk materials included.

Note also that furniture and furnishings are optional, but only if fully included consistently in subcategories 3 through 7. The choice is up to the project team. There are many manufacturers of furnishings and furniture items which are providing product options that can meet recycle content, certified wood, and regional material needs among others. Figure 5.0.1a and b depicts many furniture options provided by some manufacturers in the United States.

The approximate classifications used in LEED-NC 2.2 for steps 2 through 4 are summarized in Table 5.0.3. These will also be further explained under each individual subcategory.

The last summary detail in the credit calculations for this category is to understand which items can be included in multiple credits and which cannot. This is summarized in Table 5.0.4. Note that some individual items may be made of several types of materials. Values and allocations for these are usually by weight percent of the item.

(a)

(b)

FIGURE 5.0.1   Examples of furniture that may optionally be included in various MR credits. The furniture is from (a) Steelcase, Inc., and (b) Haworth, Inc. The carpets are from Milliken & Company and are specially adapted for recycling. (*Photographs taken at the Strom Thurmond Building on Fort Jackson, Columbia, S.C., June 2007.*)

Credits	Items	Unit of Items	Basis	Unit of Basis	Default Unit of Basis
MRc1.1 and 1.2: Building Reuse: Existing Walls, Floors and Roof	Existing, to remain, structural walls and decks	Area (exclude doors, windows unsound or hazardous materials)	All existing structural walls and decks	Area (exclude doors, windows unsound or hazardous materials)	n.a.
MRc1.3: Building Reuse—Interior Non-Structural Elements	Existing, to be reused, non-structural interior walls, doors, windows, flooring, and ceiling materials	Visible area of all these surfaces, but only count one side of interior doors and windows	Final design non-structural interior walls, doors, windows, flooring and ceiling materials	Visible area of all these surfaces, but only count one side of interior doors and windows	n.a.
MRc2.1 and 2.2: Construction Waste Management	Diverted C&D debris*	Weight or volume	All C&D debris*	Weight or volume	n.a.
MRc3.1 and 3.2: Materials Reuse	Installed reused or salvaged items†	Greater $ value of new or salvage cost of these items	All installed materials in applicable CSI uses†	$ Value of all installed applicable CSI materials	45% of total construction costs ($)†
MRc4.1 and 4.2: Recycled Content	Installed items all or partially made from recycled materials†,‡	$ Value of recycled portions* of these materials (based on weight %)	All installed materials in applicable CSI uses†	$ Value of all installed applicable CSI materials	45% of total construction costs ($)†
MRc5.1 and 5.2: Regional Materials	Installed materials extracted, processed, and manufactured regionally†	$ Value of regional materials (based on weight %)	All installed materials in applicable CSI uses†	$ Value of all installed applicable CSI materials	45% of total construction costs ($)†

components. The first is to make sure that there are easily accessible and marked recycling areas in the building design so that all the occupants can recycle these items readily. Areas should be designated for easy user access in commercial and institutional facilities and, as appropriate, for central collection, especially for residential complexes. The second is to make sure that recycling is part of the operations and is actually implemented.

USGBC Rating System

LEED-NC 2.2 lists the Intent, Requirements, and Potential Technologies and Strategies for this prerequisite as follows:

Intent
Facilitate the reduction of waste generated by building occupants that is hauled to and disposed of in landfills.

Requirements
Provide an easily accessible area that serves the entire building and is dedicated to the collection and storage of hazardous materials for recycling, including (at a minimum) paper, corrugated cardboard, glass, plastics and metals.

Potential Technologies and Strategies
Coordinate the size and functionality of the recycling areas with the anticipated collection services for glass, plastic, office paper, newspaper, cardboard and organic wastes to maximize the effectiveness of the dedicated areas. Consider employing cardboard balers, aluminum can crushers, recycling chutes and collection bins at individual workstations to further enhance the recycling program.

Calculations and Considerations

There are no calculations required for the submittal, but floor plans and recycling schedules/contracts should be provided to verify compliance. The main consideration in the design of a facility is to provide adequate and appropriate space for the recycling activities. The LEED-NC 2.2 Reference Guide does provide a guideline for the amounts of gross floor area (GFA) that might typically service different commercial building sizes. These are reiterated and smoothed into a graph in Figure 5.0.2.

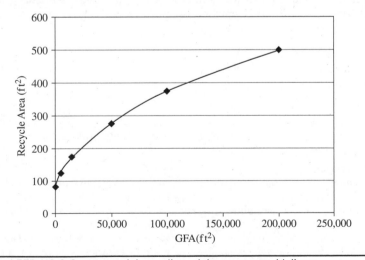

FIGURE 5.0.2   LEED-NC 2.2 commercial recycling minimum area guidelines.

5.1 MR Credit Subcategory 1: *Building Reuse*

MR Credit 1.1: *Building Reuse—Maintain 75% of Existing Walls, Floors and Roof*

MR Credit 1.2: *Building Reuse—Maintain 95% of Existing Walls, Floors and Roof*

MR credit 1.1 is worth one point, and MR credit 1.2 is worth one point in addition to MR credit 1.1. Both of these credits encourage the reuse of structural materials of existing structures if building is on a site with an existing facility.

USGBC Rating System MR Credit 1.1

LEED-NC 2.2 lists the Intent, Requirements, and Potential Technologies and Strategies for this credit as follows:

Intent

Extend the life cycle of existing building stock, conserve resources, retain cultural resources, reduce waste and reduce environmental impacts of new buildings as they relate to materials manufacturing and transport.

Requirements

Maintain at least 75% (based on surface area) of existing building structure (including structural floor and roof decking) and envelope (exterior skin and framing, excluding window assemblies and non-structural roofing material). Hazardous materials that are remediated as a part of the project scope shall be excluded from the calculation of the percentage maintained. If the project includes an addition to an existing building, this credit is not applicable if the square footage of the addition is more than 2 times the square footage of the existing building.

Potential Technologies and Strategies

Consider reuse of existing, previously occupied buildings, including structure, envelope and elements. Remove elements that pose contamination risk to building occupants and upgrade components that would improve energy and water efficiency such as windows, mechanical systems and plumbing fixtures.

USGBC Rating System MR Credit 1.2

LEED-NC 2.2 lists the Intent, Requirements, and Potential Technologies and Strategies for this credit as follows:

Intent

Extend the life cycle of existing building stock, conserve resources, retain cultural resources, reduce waste and reduce environmental impacts of new buildings as they relate to materials manufacturing and transport.

Requirements

Maintain an additional 20% (95% total, based on surface area) of existing building structure (including structural floor and roof decking) and envelope (exterior skin and framing, excluding window assemblies and non-structural roofing material). Hazardous materials that are remediated as a part of the project scope shall be excluded from the calculation of the percentage maintained. If the project includes an addition to an existing building, this credit is not applicable if the square footage of the addition is more than 2 times the square footage of the existing building.

Potential Technologies and Strategies
Consider reuse of existing buildings, including structure, envelope and elements. Remove elements
that pose contamination risk to building occupants and upgrade components that would improve
energy and water efficiency such as windows, mechanical systems and plumbing fixtures.

Calculations and Considerations MR Credits 1.1 and 1.2

Neither of these two credits (MRc1.1 or MRc1.2) is listed with the EB icon, presumably
because they are not really applicable for existing building reviews. This credit is not
applicable if there are substantial additions to the building, totaling more than double the
existing gross floor area of the project. If GFA_{PEX} is defined as being the gross floor area of the
existing building portion of the project which is to remain and be added on to, and GFA_p is
the proposed total gross floor area of the project, both existing and new, then neither MRc1.1
nor MRc1.2 can be obtained if

$$GFA_p > 3 \times GFA_{PEX} \tag{5.1.1}$$

It is important to note that the calculations are based on the area of the structural walls
or deckings (floor or roof) and the cladding less the window and door areas. This is
simplified from the LEED-NC 2.1 calculations which had some volume and some area
calculations based on the volume of many structural units and the area of the cladding
items. A simplified summary of how to itemize each of the components is as follows:

- *Deckings:* Itemize separately the square footage of structural roof decking,
  interior floor decking, and the foundation with decking or slab on grade.
  Exclude any areas that are considered to be structurally unsound or hazardous
  materials. Exclude non-structural roof surfaces such as asphalt shingle. The
  total summation of these can be referred to as $STRDECK_T$. The summation of
  the decking areas that are to be retained can be referred to as $STRDECK_{RE}$.

- *Exterior walls:* Itemize separately each exterior wall area excluding windows
  and doors. Exclude any areas that are considered to be structurally unsound or
  hazardous materials. The total summation of these can be referred to as
  $STREXWALL_T$. The summation of the exterior wall areas that are to be retained
  can be referred to as $STREXWALL_{RE}$.

- *Interior structural walls:* Itemize separately each interior *structural* wall area
  excluding openings, and count both sides of the wall elements. It is unclear if
  this would mean the inside of exterior walls too, if included. Exclude any areas
  that are considered to be structurally unsound or hazardous materials. Also, do
  not include any non-structural interior wall sections. The total summation of
  these can be referred to as $STRINWALL_T$. The summation of the interior structural
  wall areas that are to be retained can be referred to as $STRINWALL_{RE}$.

One point is earned for MRc1.1 if the following is true:

$$(STRDECK_{RE} + STREXWALL_{RE} + STRINWALL_{RE})$$
$$\geq 0.75(STRDECK_T + STREXWALL_T + STRINWALL_T) \tag{5.1.2}$$

An additional point is earned for MRc1.2 if the following is true:

$$(STRDECK_{RE} + STREXWALL_{RE} + STRINWALL_{RE})$$
$$\geq 0.95(STRDECK_T + STREXWALL_T + STRINWALL_T) \tag{5.1.3}$$

	Not Applicable If	Worth	Criterion
MRc1.1	$GFA_P > 3 \times GFA_{PEX}$	1 Point	$(STRDECK_{RE} + STREXWALL_{RE} + STRINWALL_{RE}) \geq 0.75(STRDECK_T + STREXWALL_T + STRINWALL_T)$
MRc1.2	$GFA_P > 3 \times GFA_{PEX}$	1 Point in addition to MRc1.1	$(STRDECK_{RE} + STREXWALL_{RE} + STRINWALL_{RE}) \geq 0.95(STRDECK_T + STREXWALL_T + STRINWALL_T)$
MRc1.3	$GFA_P > 3 \times GFA_{PEX}$	1 Point	$INTFIN_{RE} \geq 0.5 \times INTFIN_T$

TABLE 5.1.1   Building Reuse Equation Summary

Special Circumstances and Exemplary Performance MR Credits 1.1 and 1.2

The calculations for MR credits 1.1 and 1.2 are summarized in Table 5.1.1. They will be further explained in later sections as they relate to the other credits. Note that MRc2.1 or 2.2 (construction waste management) may include the weight or volume of existing structural items which are to remain if and only if MRc1.1 is not applied for. There is no exemplary point available for these credits as 95 percent is already a very high achievement.

MR Credit 1.3: *Building Reuse—Maintain 50% of Interior Non-Structural Elements*

MR credit 1.3 is worth one point and is not dependent on applying for MR credit 1.1. This credit encourages the reuse of interior non-structural materials of existing structures if building is on a site with an existing facility. ·

USGBC Rating System

LEED-NC 2.2 lists the Intent, Requirements, and Potential Technologies and Strategies for this credit as follows:

Intent
Extend the life cycle of existing building stock, conserve resources, retain cultural resources, reduce waste and reduce environmental impacts of new buildings as they relate to materials manufacturing and transport.

Requirements
Use existing interior non-structural elements (interior walls, doors, floor coverings and ceiling systems) in at least 50% (by area) of the completed building (including additions). If the project includes an addition to an existing building, this credit is not applicable if the square footage of the addition is more than 2 times the square footage of the existing building.

Potential Technologies and Strategies
Consider reuse of existing buildings, including structure, envelope and interior non-structural elements. Remove elements that pose contamination risk to building occupants and upgrade components that would improve energy and water efficiency, such as mechanical systems and plumbing fixtures. Quantify the extent of building reuse.

Calculations and Considerations

This credit is not listed with the EB icon, presumably because it is not really applicable for existing building reviews. Just as for MRc1.1 and MRc1.2, this credit is not applicable if there are substantial additions to the building, totaling more than double the existing GFA. If GFA_{PEX} is defined as being the gross floor area of the existing building portion of the project which is to be retained and added on to, and GFA_p is the proposed total gross floor area of the project including both the retained and the new areas, then MRc1.3 cannot be applied for if

$$GFA_p > 3 \times GFA_{PEX} \qquad (5.1.1)$$

This credit is intended to reuse as much as possible of the interior finished (non-structural) surfaces in the final structure. First, all interior non-structural surfaces in the *proposed design* (including the additions if applicable) shall be totaled. Since it is usually based on the proposed design and the finished product, there is no need to exclude hazardous materials, as these should not be included in any newly finished or renovated building in the first place. These interior structural surfaces include all finished interior ceilings, flooring surfaces, doors, windows, casework, demountable partitions (count both sides), walls (count both sides), and visible casework surfaces. It is easy to understand if one pictures the finished interior surfaces that can be seen in all the rooms, but with the interior doors and windows and sidelights counting only once. Do not include any exterior doors and windows, even their interior side. Let the total of all these areas be represented by $INTFIN_T$. Let $INTFIN_{RE}$ be the total of all these finished interior surfaces that have been reused from the existing building, even those which may have been moved. Then one point is earned for MRc1.3 if

$$INTFIN_{RE} \geq 0.5 \times INTFIN_T \qquad (5.1.4)$$

Special Circumstances and Exemplary Performance

The calculations are also summarized in Table 5.1.1 and will be further explained in later sections as they relate to the other credits. Note that MRc2.1 or 2.2 (construction waste management) may include the weight or volume of existing non-structural interior items which are to remain if and only if MRc1.3 is not applied for. Also, MRc1.3 is not dependent on obtaining MRc1.1.

There is no EP point related to MRc1.3. It is the author's opinion that this is so because it is not very reasonable to encourage more than a 50 percent reuse of interior surfaces as many are worn over time or may have been made from materials not considered to be as sustainable as products encouraged in the USGBC rating system.

5.2 MR Credit Subcategory 2: *Construction Waste Management*

The construction waste management subcategory covers what is commonly called *construction and demolition* (C&D) debris. One point is awarded for MRc2.1, and an additional one point can be awarded for MRc2.2 in addition to the one for MR credit 2.1.

MR Credit 2.1: *Construction Waste Management--Divert 50% From Disposal*

MR Credit 2.2: *Construction Waste Management--Divert 75% From Disposal*

USGBC Rating System MR Credit 2.1

LEED-NC 2.2 lists the Intent, Requirements, and Potential Technologies and Strategies for this credit as follows:

Intent

Divert construction and demolition debris from disposal in landfills and incinerators. Redirect recyclable recovered resources back to the manufacturing process. Redirect reusable materials to appropriate sites.

Requirements

Recycle and/or salvage at least 50% of non-hazardous construction and demolition debris. Develop and implement a construction waste management plan that, at a minimum, identifies the materials to be diverted from disposal and whether the materials will be sorted on-site or commingled.

Excavated soil and land-clearing debris do not contribute to this credit. Calculations can be done by weight or volume, but must be consistent throughout.

Potential Technologies and Strategies

Establish goals for diversion from disposal in landfills and incinerators and adopt a construction waste management plan to achieve these goals. Consider recycling cardboard, metal, brick, acoustical tile, concrete, plastic, clean wood, glass, gypsum wallboard, carpet and insulation. Designate a specific area(s) on the construction site for segregated or commingled collection of recyclable materials, and track recycling efforts throughout the construction process. Identify construction haulers and recyclers to handle the designated materials. Note that diversion may include donation of materials to charitable organizations and salvage of materials on-site.

USGBC Rating System MR Credit 2.2

LEED-NC 2.2 lists the Intent, Requirements, and Potential Technologies and Strategies for this credit as follows:

Intent

Divert construction and demolition debris from disposal in landfills and incinerators. Redirect recyclable recovered resources back to the manufacturing process. Redirect reusable materials to appropriate sites.

Requirements

Recycle and/or salvage an additional 25% beyond MR Credit 2.1 (75% total) of non-hazardous construction and demolition debris. Excavated soil and land-clearing debris do not contribute to this credit. Calculations can be done by weight or volume, but must be consistent throughout.

Potential Technologies and Strategies

Establish goals for diversion from disposal in landfills and incinerators and adopt a construction waste management plan to achieve these goals. Consider recycling cardboard, metal, brick, acoustical tile, concrete, plastic, clean wood, glass, gypsum wallboard, carpet and insulation. Designate a specific area(s) on the construction site for segregated or commingled collection of recyclable materials, and track recycling efforts throughout the construction process. Identify construction haulers and recyclers to handle the designated materials. Note that diversion may include donation of materials to charitable organizations and salvage of materials on-site.

Calculations and Considerations MR Credits 2.1 and 2.2

The intention of these two credits and the EP point for diversion of 50, 75, or 95 percent, respectively, of the construction and demolition (C&D) waste stream from landfills is to find reuses for as many materials as possible from either demolition activities or construction activities. Note that there are two different sources of these waste materials.

One source is from the existing manmade buildings or other structural items on the site which are to be either reused or demolished, and one source is from the construction process. If a project has no demolition on-site since the land was originally not built upon, then the source is solely from the construction phase.

It is very important to establish a construction waste management plan that involves the construction contractor to adequately verify that these goals have been met. Good waste management plans will include proper designations of areas on-site for the recycling activities, plans to train the workers in the recycling procedures, some feedback or reporting system to oversee the activities, and proper labeling of all recycling facilities, particularly in the languages of the workers. Care should be taken to avoid tight or not readily accessible areas where either disposal or pickup will be difficult. Care should also be taken to avoid problems with stormwater in and around the recycling facilities.

A significant amount of documentation is needed to verify the items throughout the demolition and construction phases. These might include tipping fee receipts from the recyclers or landfills and accounting of all the dumpster and waste removal procedures on-site. For instance, dumpsters come in different sizes and may be removed when full or partially full, depending on the pickup times. It is difficult to determine the volume of a waste in a container if it is not documented as to its fullness at the time of pickup. In some cases the weights will be given regardless from the tipping accounts, but some debris items are valuable (such as metals) and recyclers may contract to haul them free of charge, in which case the general contractor may not receive the necessary records unless they are specifically requested in the contracts.

Good waste management plans will have the contracts and oversight in place necessary to have dumpsters usually as full as possible when picked up, since this will typically reduce overall hauling costs and environmental impacts of the transport phase. Commonly used dumpsters in the United States range from a small size of 6 yards to a large size of 30 yards or more. A yard in this case actually refers to a cubic yard. Dumpster sizes are usually chosen based on the expected amount of different types of wastes generated. Some wastes may also never be put into dumpsters, but instead either reused on-site or loaded into dump trucks or other hauling vehicles and then transported off-site. Typical examples of this are recycled asphalt or concrete pavements which are usually first crushed and then piled on-site before they are reused.

The calculations are fairly simple if there is adequate control, recording, and documentation resulting from the waste management plan. What is important is to make sure that the items included in these calculations have been adequately documented.

The items used in the calculations for MRc2.1 and MRc2.2 are usually defined by the following criteria:

- In all cases, hazardous materials, land-clearing debris, and soils removed from the site are not included in the calculations.

- All the demolition wastes that go into the dumpsters and are hauled off-site are included.

- All the construction wastes that go into the dumpsters and are hauled off-site are included.

- Any pavements that are crushed and reused on-site are included.

- Existing building items which are to remain as addressed in MRc1.1 and 1.2 (building reuse—structural) can also be included but only if MRc1.1 is not applied for.

- Existing building items which are to remain as addressed in MRc1.3 (building reuse—non-structural) can also be included but only if MRc1.3 is not applied for.
- According to the errata posted by the USGBC in the spring of 2007, other building materials salvaged and reused on-site can contribute to this credit if they are not included in the MR credit 3 calculations.

Using these criteria, the following variables can be defined:

DeBLDGREUSE	The items reused in the building as defined in MRc1.1 and MRc1.2 if MRc.1.1 is *not* applied for, and the items reused in the building as defined in MRc1.3 if MRc.1.3 is *not* applied for and any other building materials salvaged and reused on-site if they are not included in MRc3.1 calculations
DeCARD	Cardboard recycled off-site
DeDIVERT	C&D debris diverted
DeGYPSUM	Gypsum-type wallboard recycled off-site
DeLANDFILL	All C&D debris excluding hazardous wastes, land-clearing debris, and soils that are disposed of off-site in a landfill or other waste facility without recycling or reuse
DeMETAL	Metals recycled off-site
DeMISC	Miscellaneous items recycled off-site
DePAVEREUSE	Existing pavement debris reused on-site
DeRUBBLE	Any rubble (bricks, concrete, asphalt, masonry, etc.) recycled off-site
DeWOOD	Any wood recycled off-site

The units on these variables can be either volume-based or weight-based, but must be consistent throughout. There has been very limited research on the conversion factors needed to bring some volume measurements into weight measurements (or vice versa). The LEED-NC 2.2 Reference Guide specifies that either the solid waste conversion factors listed in its Table 2 under MRc2.1and 2.2 can be used, or another justifiable set of conversion factors. The default LEED conversion factors are represented in Table 5.2.1. Typically, these are assumed to be waste management dumpster volumes. In addition, another set of some of the conversion factors was developed based on construction debris (no demolition debris was included in these factors) from a LEED certified project in Columbia, S.C., in 2005. These conversion factors were derived from the weights and volumes of the dumpsters and were referred to as "dumpster densities" in that report. They are also given in Table 5.2.1 as are other conversion factors found in a compilation of studies on recycling operations elsewhere, but which were usually from receiving facilities and not necessarily LEED-related. Table 5.2.1 is taken from L. Haselbach and S. Bruner, "Determining Construction Debris Recycling Dumpster Densities," *Journal of Green Building*, Summer 2006, 1(3).

Note that the density of the solid source material as given in Table 5.2.1 is always much greater than the conversion factor. This is so because there are many void spaces in the dumpsters and hauling trucks since the contents are not compacted to the density of the material. The source material densities may be better used in converting volumes to weight of reused building reuse items. It is recommended that the Columbia study value for steel and metal be used for cases where there is only construction debris. It

Solid Waste Material	AISC (1980) Source Material Density (lb/CY)	LEED-NC 2.2 Conversion Factors (lb/CY)	Cal-Recovery Bulk Density (lb/CY)	Columbia, S.C., Study Dumpster Avg. Density (lb/CY)	Columbia, S.C., Study Density Statistics (lb/CY)	No. Data Points in Study
General trash (mixed waste)	n.a.	350	n.a.	600	Min. 280 Max. 1540 SD 290	34
Masonry (rubble)	3000–4000	1400	1610 (brick) 1855 (conc.)	1680	Min. 1220 Max. 2620 SD 320	15
Wood	680–1100	300	330	280	Min. 140 Max. 480 SD 110	6
Steel (metals)	13,200	1000	440	320	Min. 260 Max. 380 SD 50	7
Sheetrock (mypsum wallboard)	1300–1950	500	390	440	Min. 280 Max. 620 SD 170	4
Cardboard	—	100	—	Not studied	Not studied	—

Source: Courtesy Journal of Green Building Vol. 1, No. 3, College Publishing.

TABLE 5.2.1 Calculated Average, Maximum, Minimum, and Standard Deviation (SD) of Solid Waste Conversion Factors from the Columbia, S.C., Study by Haselbach and Bruner as Compared to LEED-NC 2.2, Cal-Recovery Values and Density of the Solid Source Material

was noted in that study that recycled metals in the dumpsters for construction-only phases were rarely very large, dense pieces such as structural steel, but were rather usually items such as pieces of ductwork, piping, and flashing.

The equations needed to determine if the construction waste management credits in the construction waste management subcategory are earned are as follows:

$$\text{DeDIVERT} = \text{DeBLDGREUSE} + \text{DeCARD} + \text{DeGYPSUM} + \text{DeMETAL}$$
$$+ \text{DeMISC} + \text{DePAVEREUSE} + \text{DeRUBBLE} + \text{DeWOOD} \quad (5.2.1)$$

One point for MRc2.1 is earned if

$$\text{DeDIVERT} \geq 0.50(\text{DeLANDFILL} + \text{DeDIVERT}) \quad (5.2.2)$$

One additional point for MRc2.2 is earned if

$$\text{DeDIVERT} \geq 0.75(\text{DeLANDFILL} + \text{DeDIVERT}) \quad (5.2.3)$$

Figure 5.2.1 shows how one type of material can get recycled and save energy and transportation costs for many different groups. It is a pile of crushed concrete from the demolition of a dormitory at the University of South Carolina. It was hauled only a few miles away to an asphalt plant where it is stockpiled for reuse in many ways.

Figure 5.2.1 Stockpile of crushed concrete at the Sloan Construction asphalt plant in Columbia, S.C., ready for reuse. The concrete comes from the demolition of dormitories at the University of South Carolina, only a few miles away. (*Photograph taken August 2007.*)

Special Circumstances and Exemplary Performance MR Credits 2.1 and 2.2

An EP point can be awarded in the construction waste management subcategory if the diversion rate reaches at least 95 percent. One additional point for EP is earned if

$$DeDIVERT \geq 0.95(DeLANDFILL + DeDIVERT) \qquad (5.2.4)$$

5.3 MR Credit Subcategory 3: *Materials Reuse*

MR Credit 3.1: *Materials Reuse—5%*

MR Credit 3.2: *Materials Reuse—10%*

MRc3.1 is worth one point, and MRc3.2 is worth one point in addition to the point for MRc3.1. The intention of this subcategory is to promote the reuse of salvaged goods in the marketplace.

USGBC Rating System MR Credit 3.1

LEED-NC 2.2 lists the Intent, Requirements, and Potential Technologies and Strategies for this credit as follows:

Intent
Reuse building materials and products in order to reduce demand for virgin materials and to reduce waste, thereby reducing impacts associated with the extraction and processing of virgin resources.

Requirements
Use salvaged, refurbished or reused materials such that the sum of these materials constitutes at least 5%, based on cost, of the total value of materials on the project. Mechanical, electrical and plumbing components and specialty items such as elevators and equipment shall not be included in this calculation. Only include materials permanently installed in the project. Furniture may be included, providing it is included consistently in MR Credits 3–7.

Potential Technologies and Strategies
Identify opportunities to incorporate salvaged materials into building design and research potential material suppliers. Consider salvaged materials such as beams and posts, flooring, paneling, doors and frames, cabinetry and furniture, brick and decorative items.

USGBC Rating System MR Credit 3.2

LEED-NC 2.2 lists the Intent, Requirements, and Potential Technologies and Strategies for this credit as follows:

Intent
Reuse building materials and products in order to reduce demand for virgin materials and to reduce waste, thereby reducing impacts associated with the extraction and processing of virgin resources.

Requirements
Use salvaged, refurbished or reused materials for an additional 5% beyond MR Credit 3.1 (10% total, based on cost). Mechanical, electrical and plumbing components and specialty items such as elevators and equipment shall not be included in this calculation. Only include

materials permanently installed in the project. Furniture may be included, providing it is included consistently in MR Credits 3–7.

Potential Technologies and Strategies

Identify opportunities to incorporate salvaged materials into building design and research potential material suppliers. Consider salvaged materials such as beams and posts, flooring, paneling, doors and frames, cabinetry and furniture, brick and decorative items.

Calculations and Considerations MR Credits 3.1 and 3.2

The items that can be included in the Materials Reuse category can come from many sources. Usually, only permanent items in the project from the CSI categories 2 through 10 are included, but furniture and furnishings can be included if they are consistently used throughout credit subcategories 3 through 7 as noted in Table 5.0.4. If they are included, then they should also be added into the denominators in the calculations.

In general, there are two different classes of what may be referred to as salvaged items. First, there are the items that are reused or salvaged from an existing facility on-site, and then there are items salvaged from elsewhere.

Most of these items from the site are either already included in the first credit subcategory for Building Reuse, or if MRc1.1 is not applied for, then are included as being diverted per the second credit subcategory (construction waste management) and cannot be included in the Materials Reuse (salvage) category. However, some additional items can be included. For instance, exterior doors and windows and associated hardware are items which are not always appropriately included in building reuse. If they are appropriately refurbished or refinished and reused somewhere in the project, then they can be included, even if used for a different purpose.

According to the errata as posted in the spring of 2007, materials contributing toward achievement of MRc3 cannot be applied to MR credits 1, 2, 4, 6, or 7; but if MRc3 is not being attempted, applicable materials can be applied to another LEED credit if eligible.

The second group includes items that have been used in another project somewhere and have been salvaged, many times with refurbishing or refinishing, and reused in the project. Some areas of the country have a large market for salvaged items. Some items commonly salvaged include wood flooring and beams, ornate ironwork, and bricks. Figures 5.3.1 and 5.3.2 show some features made from salvaged wood at a LEED-NC certified building in Columbia, S.C.

It is impossible to compare different types of construction materials unless there is a basis by which they can be compared. In the energy category, when there are different energy sources and types, this common rating system is cost or value. Again, in the remaining Materials and Resources categories, the calculations are based on cost or value. This is summarized in Table 5.0.3.

To promote as many salvaging activities as possible, and at the same time promote as economical a project as possible, the cost values of the salvaged items used in the calculations represent either the cost of salvaging (either the costs to refurbish the on-site items or the cost to purchase the off-site items) or the market value of the equivalent new item, whichever is greater.

Let the following variables be defined:

$MATL\$_T$ Total cost of all project materials in CSI categories 2 through 10 plus any furniture or furnishing costs included consistently in MR subcategories 2 through 7 (Total default costs if not itemized can be calculated as 45 percent

FIGURE 5.3.1   Salvaged wood staircase at Cox and Dinkins, Engineers and Surveyors, Columbia, S.C. (*Photograph taken June 2007.*)

FIGURE 5.3.2   Salvaged wood alcove area at Cox and Dinkins, Engineers and Surveyors, Columbia, S.C. (*Photograph taken June 2007.*)

of total construction costs plus any furniture or furnishing costs included consistently in MR subcategories 2 through 7, and salvaged items are valued at the greater of market value or actual salvage cost.)

$MATL\$_{SAL}$ Total cost of salvaged materials plus any salvaged furniture or furnishing costs if included consistently in MR subcategories 2 through 7 (Salvaged items are valued at the greater of market value or actual cost.)

One point is earned for MRc3.1 if the following is true:

$$100(MATL\$_{SAL}/MATL\$_T) \geq 5 \qquad (5.3.1)$$

An additional point is earned for MRc3.2 if the following is true:

$$100(MATL\$_{SAL}/MATL\$_T) \geq 10 \qquad (5.3.2)$$

Special Circumstances and Exemplary Performance MR Credits 3.1 and 3.2

An EP point can be earned in the MR subcategory for increasing the performance by 50 percent more than required for MRc3.2, that is, for using salvaged or reused items which are valued at 15 percent or more of the applicable project material costs. Therefore, an EP point can be earned for material reuse if the following is true:

$$100(MATL\$_{SAL}/MATL\$_T) \geq 15 \qquad (5.3.3)$$

Since it is usually not certain whether salvaged materials can be appropriately refurbished or reused for many applications in the design phase, this subcategory is listed as a construction phase submittal. The contractor should keep records of salvaging activities and costs.

5.4 MR Credit Subcategory 4: *Recycled Content*

The intent of this credit subcategory is to encourage the use of items made from recycled materials to lessen the resource and environmental impacts of making many products from raw materials and to lessen the solid waste load in the country. MRc4.1 is worth one point, and MRc4.2 is worth one point in addition to the point for MRc4.1.

MR Credit 4.1: *Recycled Content—10% (post-consumer + ½ pre-consumer)*

MR Credit 4.2: *Recycled Conten —20% (post-consumer + ½ pre-consumer)*

USGBC Rating System MR Credit 4.1

LEED-NC 2.2 lists the Intent, Requirements, and Potential Technologies and Strategies for this credit as follows:

Intent
Increase demand for building products that incorporate recycled content materials, thereby reducing impacts resulting from extraction and processing of virgin materials.

Requirements
Use materials with recycled content such that the sum of post-consumer recycled content plus one-half of the pre-consumer content constitutes at least 10% (based on cost) of the

total value of the materials in the project. The recycled content value of a material assembly shall be determined by weight. The recycled fraction of the assembly is then multiplied by the cost of assembly to determine the recycled content value. Mechanical, electrical and plumbing components and specialty items such as elevators shall not be included in this calculation. Only include materials permanently installed in the project. Furniture may be included, providing it is included consistently in MR Credits 3–7. Recycled content shall be defined in accordance with the International Organization of Standards document, *ISO 14021—Environmental labels and declarations—Self-declared environmental claims (Type II environmental labeling).* Post-consumer material is defined as waste material generated by households or by commercial, industrial and institutional facilities in their role as end-users of the product, which can no longer be used for its intended purpose. Pre-consumer material is defined as material diverted from the waste stream during the manufacturing process. Excluded is reutilization of materials such as rework, regrind or scrap generated in a process and capable of being reclaimed within the same process that generated it.

Potential Technologies and Strategies
Establish a project goal for recycled content materials and identify material suppliers that can achieve this goal. During construction, ensure that the specified recycled content materials are installed. Consider a range of environmental, economic and performance attributes when selecting products and materials.

USGBC Rating System MR Credit 4.2
LEED-NC 2.2 lists the Intent, Requirements, and Potential Technologies and Strategies for this credit as follows:

Intent
Increase demand for building products that incorporate recycled content materials, thereby reducing the impacts resulting from extraction and processing of virgin materials.

Requirements
Use materials with recycled content such that the sum of post-consumer recycled content plus one-half of the pre-consumer content constitutes an additional 10% beyond MR Credit 4.1 (total of 20%, based on cost) of the total value of the materials in the project. The recycled content value of a material assembly shall be determined by weight. The recycled fraction of the assembly is then multiplied by the cost of assembly to determine the recycled content value. Mechanical, electrical and plumbing components and specialty items such as elevators shall not be included in this calculation. Only include materials permanently installed in the project. Furniture may be included, providing it is included consistently in MR Credits 3–7. Recycled content shall be defined in accordance with the International Organization of Standards document, *ISO 14021—Environmental labels and declarations— Self-declared environmental claims (Type II environmental labeling).* Post-consumer material is defined as waste material generated by households or by commercial, industrial and institutional facilities in their role as end-users of the product, which can no longer be used for its intended purpose. Pre-consumer material is defined as material diverted from the waste stream during the manufacturing process. Excluded is reutilization of materials such as rework, regrind or scrap generated in a process and capable of being reclaimed within the same process that generated it.

Potential Technologies and Strategies
Establish a project goal for recycled content materials and identify material suppliers that can achieve this goal. During construction, ensure that the specified recycled content materials are installed. Consider a range of environmental, economic and performance attributes when selecting products and materials.

Calculations and Considerations MR Credits 4.1 and 4.2

The intention of these credits is to use items that are made from materials which have been made from recycled materials to reduce the use and costs (economic, social, and environmental) of using virgin material sources. In this case recycled differs from salvaged in the sense that the salvaged items as addressed in the materials reuse credits (MRc3.1 and 3.2) are recycled in their final functional form and recycled items in the recycled content credits (MRc4.1 and 4.2) are made in part from materials recycled either after final use (postconsumer) or from by-products of industrial processes (preconsumer). Materials that are preconsumer but that could be put back into the industrial stream from whence they came are not eligible. There is greater value given for the items recycled postconsumer, since they have already gone through two types of material streams: manufacturing and final product use.

As in many of the MR credit subcategories, these credits are based on material cost and values. There are so many different types of materials and material assemblies (items) installed in the built environment, and cost (or value) is really the only common element that can be used for comparison in typical construction analyses. The credit criteria are based on the percentage of total material costs that come from preconsumer and postconsumer recycled materials. The values of the preconsumer and postconsumer recycled portions of the materials can be defined as follows:

$MATL\$_{RECPOC}$ Total value of postconsumer recycled portion of materials (CSI 2 through 10) plus any value of postconsumer recycled portions of furniture or furnishings if included consistently in MR subcategories 2 through 7. Value determinations are based on postconsumer recycled weight percent of the value of each of the individual items.

$MATL\$_{RECPRC}$ Total value of pre-consumer recycled portion of materials (CSI 2 through 10) plus any value of preconsumer recycled portions of furniture or furnishings if included consistently in MR subcategories 2 through 7. Value determinations are based on preconsumer recycled weight percent of the value of each of the individual items.

Given these, one point is earned for MRc4.1 if the following is true:

$$100[MATL\$_{RECPOC} + (0.5 \times MATL\$_{RECPRC})]/MATL\$_{T} \geq 10 \qquad (5.4.1)$$

An additional point is earned for MRc4.2 if the following is true:

$$100[MATL\$_{RECPOC} + (0.5 \times MATL\$_{RECPRC})]/MATL\$_{T} \geq 20 \qquad (5.4.2)$$

Most construction items are not 100 percent made from recycled materials. Therefore, the recycled value is only a portion of the value of the item, whether these items are one piece (such as a steel beam) or an assembly (such as a prebuilt window unit). The value determination is usually made on a mass percent basis. So for any item the following can be defined:

$MASS_{RECPOCi}$ Mass (weight) of the postconsumer recycled portion of unit i

$MASS_{RECPRCi}$ Mass (weight) of the preconsumer recycled portion of unit i

$MASS_{UNITi}$ Total mass (weight) of unit i

$\text{MATL\$}_{\text{UNIT}i}$ Total cost of unit i

$\text{MATL\$}_{\text{RECPOC}i}$ Value of postconsumer recycled portions of unit i

$\text{MATL\$}_{\text{RECPRC}i}$ Value of preconsumer recycled portions of unit i

The value of the preconsumer and postconsumer recycled portions for each individual unit i can be determined from the following equations:

$$\text{MATL\$}_{\text{RECPOC}i} = \text{MATL\$}_{\text{UNIT}i}\,(\text{MASS}_{\text{RECPOC}i})/\text{MASS}_{\text{UNIT}i} \qquad (5.4.3)$$

$$\text{MATL\$}_{\text{RECPRC}i} = \text{MATL\$}_{\text{UNIT}i}\,(\text{MASS}_{\text{RECPRC}i})/\text{MASS}_{\text{UNIT}i} \qquad (5.4.4)$$

And, in similar manner, the total values of the recycled portions needed in Eqs. (5.4.1) and (5.4.2) can be determined from the summation of the individual unit contributions as follows:

$$\text{MATL\$}_{\text{RECPOC}} = \Sigma\,\text{MATL\$}_{\text{RECPOC}i} \qquad \text{over all units } i \qquad (5.4.5)$$

$$\text{MATL\$}_{\text{RECPRC}} = \Sigma\,\text{MATL\$}_{\text{RECPRC}i} \qquad \text{over all units } i \qquad (5.4.6)$$

Special Circumstances and Exemplary Performance MR Credits 4.1 and 4.2

There are some cases in which the value of the recycled content in some materials is disproportionately high compared to the costs of the other material contents. In these cases, it may be appropriate to use an alternative to the listed weight-based calculation, but these should be justified. An already accepted alternative is in the case of Portland cement–based concrete. Many concretes are made with a combination of Portland cement and fly ash (usually up to 25 percent). Costs for cement and other cementitious materials are generally a large percent of the cost of concrete, and therefore it is appropriate to calculate the value of the recycled cementitious material content in concrete as the weight percent of only the cementitious materials and multiply by the cost of the cementitious materials.

As related to the recycled content credit subcategory, this use of fly ash is promoted as an environmental benefit as it recycles waste from another process which might otherwise be landfilled, and it also promotes a reduction in the use of energy and the production of carbon dioxide from the cement-making process.

There is also a default postconsumer recycle content value for steel of 25 percent which can be used regardless of documentation as steel is usually made from at least 25 percent postconsumer recycled steel. In cases where the steel recycled content is greater than 25 percent, documentation should be provided.

An additional EP point can be earned for recycled content if the following is true:

$$100[\text{MATL\$}_{\text{RECPOC}} + (0.5 \times \text{MATL\$}_{\text{RECPRC}})]/\text{MATL\$}_{T} \geq 30 \qquad (5.4.7)$$

Since it is usually not certain during the design phase of the exact amounts of recycled content that might be purchased for each item, this subcategory is listed as a construction phase submittal. The contractor should keep records of recycled content and costs.

	Postconsumer Recycled Content Value of Unit *i* ($MATL\$_{RECPOCi}$)	Total Postconsumer Recycled Content Value ($MATL\$_{RECPOC}$)	Preconsumer Recycled Content Value of Unit *i* ($MATL\$_{RECPRCi}$)	Total Preconsumer Recycled Content Value ($MATL\$_{RECPRC}$)
Equations	$MATL\$_{UNITi} \times (MASS_{RECPOCi})/ MASS_{UNITi}$	$\sum MATL\$_{RECPOCi}$	$MATL\$_{UNITi} \times (MASS_{RECPRCi})/ MASS_{UNITi}$	$\sum MATL\$_{RECPRCi}$
Some exceptions	Value of recycled steel content is at a minimum of 25% or greater with documentation		Recycled supplementary cementitious materials used in concrete are valued per weight percent of cementitious materials	

TABLE 5.4.1   Summary of Recycled Content Material Value Equations

The equations for this credit subcategory are summarized in Tables 5.4.1 and 5.4.2. Some examples of carpet systems made with materials which can be included in the recycle content subcategory can be seen in Figure 5.4.1.

5.5   MR Credit Subcategory 5: *Regional Materials*

MR Credit 5.1: *Regional Materials—10% Extracted, Processed and Manufactured Regionally*

MR Credit 5.2: *Regional Material—20% Extracted, Processed and Manufactured Regionally*

The intent of this credit subcategory is to encourage the use of regional materials to lessen the environmental impacts of transportation and to make regions more self-sufficient.

	Worth	Criterion
MRc4.1	1 point	$100[MATL\$_{RECPOC} + (0.5 \times MATL\$_{RECPRC})]/MATL\$_{T} \geq 10$
MRc4.2	1 point additional to MRc4.1	$100[MATL\$_{RECPOC} + (0.5 \times MATL\$_{RECPRC})]/MATL\$_{T} \geq 20$
Exemplary performance	1 point additional to MRc4.2	$100[MATL\$_{RECPOC} + (0.5 \times MATL\$_{RECPRC})]/MATL\$_{T} \geq 30$

TABLE 5.4.2   Recycled Content Equation Summary

Figure 5.4.1   Carpets which may count toward the recycle content credit as provided by Interface, Inc., and Lees Carpets. (*Photograph taken at the Sustainable Interiors ribbon cutting at the Strom Thurmond Building on Fort Jackson, Columbia, S.C., June 2007.*)

MRc5.1 is worth one point, and MRc5.2 is worth one point in addition to the point for MRc5.1.

USGBC Rating System MR Credit 5.1

LEED-NC 2.2 lists the Intent, Requirements, and Potential Technologies and Strategies for this credit as follows:

Intent

Increase demand for building materials and products that are extracted and manufactured within the region, thereby supporting the use of indigenous resources and reducing the environmental impacts resulting from transportation.

Requirements

Use building materials or products that have been extracted, harvested or recovered, as well as manufactured, within 500 miles of the project site for a minimum of 10% (based on cost) of the total materials value. If only a fraction of a product or material is extracted/harvested/recovered and manufactured locally, then only that percentage (by weight) shall contribute to the regional value. Mechanical, electrical and plumbing components and specialty items such as elevators and equipment shall not be included in this calculation. Only include materials permanently installed in the project. Furniture may be included, providing it is included consistently in MR Credits 3–7.

Potential Technologies and Strategies

Establish a project goal for locally sourced materials, and identify materials and material suppliers that can achieve this goal. During construction, ensure that the specified local materials are installed and quantify the total percentage of local materials installed. Consider a range of environmental, economic and performance attributes when selecting products and materials.

USGBC Rating System MR Credit 5.2

LEED-NC 2.2 lists the Intent, Requirements, and Potential Technologies and Strategies for this credit as follows:

Intent

Increase demand for building materials and products that are extracted and manufactured within the region, thereby supporting the use of indigenous resources and reducing the environmental impacts resulting from transportation.

Requirements

Use building materials or products that have been extracted, harvested or recovered, as well as manufactured, within 500 miles of the project site for an additional 10% beyond MR Credit 5.1 (total of 20%, based on cost) of the total materials value. If only a fraction of the material is extracted/harvested/recovered and manufactured locally, then only that percentage (by weight) shall contribute to the regional value.

Potential Technologies and Strategies

Establish a project goal for locally sourced materials and identify materials and material suppliers that can achieve this goal. During construction, ensure that the specified local materials are installed. Consider a range of environmental, economic and performance attributes when selecting products and materials.

Calculations and Considerations MR Credits 5.1 and 5.2

As in many of the other MR credit subcategories, these credits are based on material cost and values. There are so many different types of materials and material assemblies (items) installed in the built environment, and cost (or value) is really the only common element that can be used for comparison in typical construction analyses. The credit criteria are based on the percentage of the total material costs that represents items or portions of items which have been extracted, processed, *and* manufactured within a 500-mi radius of the project site. For salvaged or reused items, the point of extraction is considered to be the location that it is being salvaged from, and the manufacturing point is the location of the salvage dealer.

Given the following definition:

$MATL\$_{REG}$ Total value of regionally extracted, processed, and manufactured portions of materials (CSI 2 through 10) plus any similar portions of furniture or furnishings if included consistently in MR subcategories 2 through 7. Value determinations are based on the weight percent of the regional portion of each of the individual item values.

Then one point is earned for MRc5.1 if the following is true:

$$100(MATL\$_{REG})/MATL\$_T \geq 10 \qquad (5.5.1)$$

An additional point is earned for MRc5.2 if the following is true:

$$100(MATL\$_{REG})/MATL\$_T \geq 20 \qquad (5.5.2)$$

Most construction items are not 100 percent made from regional materials. Therefore, the regional value is only a portion of the value of the item, whether these items are one piece (such as a steel beam) or an assembly (such as a prebuilt window unit). The value determination is made on a mass percent basis. So for any item the following can be defined:

	Regional Content Value of Unit i (MATL$\$_{REGi}$)	Total Regional Content Value (MATL$\$_{REG}$)
Equations	MATL$\$_{UNITi}$(MASS$_{REGi}$)/MASS$_{UNITi}$	ΣMATL$\$_{REGi}$ over all units i

TABLE 5.5.1 Summary of Regional Materials Value Equations

MASS$_{REGi}$ Mass (weight) of the regional portion of unit i

MASS$_{UNITi}$ Total mass (weight) of unit i

MATL$\$_{UNITi}$ Total cost of unit i

MATL$\$_{REGi}$ Value of the regional portions of unit i

The value of the regional portions for each individual unit i can be determined from the following:

$$MATL\$_{REGi} = MATL\$_{UNITi} (MASS_{REGi})/MASS_{UNITi} \qquad (5.5.3)$$

And, in a similar manner, the total values of the regional portions needed in Eqs. (5.5.1) and (5.5.2) can be determined from the summation of the individual unit contributions as follows:

$$MATL\$_{REG} = \Sigma\ MATL\$_{REGi} \qquad \text{over all units } i \qquad (5.5.4)$$

Special Circumstances and Exemplary Performance MR Credits 5.1 and 5.2

Since it is usually not certain during the design phase of the exact amounts of regional items or content that might be purchased, this subcategory is listed as a construction phase submittal. The contractor should keep records of the locations of extraction, processing, and manufacturing.

Any materials used for obtaining the points in this category can also be used for MR subcategories 3 (Materials Reuse), 4 (Recycled Content), 6 (Rapidly Renewable Materials), and 7 (Certified Wood).

An additional EP point can be earned for regional content if the following is true:

$$100(MATL\$_{REG})/MATL\$_{T} \geq 40 \qquad (5.5.5)$$

The equations for this credit subcategory are summarized in Tables 5.5.1 and 5.5.2.

	Worth	Criterion
MRc5.1	1 point	$100(MATL\$_{REG})/MATL\$_{T} \geq 10$
MRc5.2	1 point in addition to MRc5.1	$100(MATL\$_{REG})/MATL\$_{T} \geq 20$
Exemplary performance	1 point in addition to MRc5.2	$100(MATL\$_{REG})/MATL\$_{T} \geq 40$

TABLE 5.5.2 Regional Materials Equation Summary

5.6 MR Credit 6: *Rapidly Renewable Materials*

The intent of MR credit 6 is to encourage the use of rapidly renewable materials, that is, materials that are made from living products that have a short growth span. One point is available for MRc6.

USGBC Rating System

LEED-NC 2.2 lists the Intent, Requirements, and Potential Technologies and Strategies for this credit as follows:

Intent
Reduce the use and depletion of finite raw materials and long-cycle renewable materials by replacing them with rapidly renewable materials.

Requirements
Use rapidly renewable building materials and products (made from plants that are typically harvested within a ten-year cycle or shorter) for 2.5% of the total value of all building materials and products used in the project, based on cost.

Potential Technologies and Strategies
Establish a project goal for rapidly renewable materials and identify products and suppliers that can support achievement of this goal. Consider materials such as bamboo, wool, cotton insulation, agrifiber, linoleum, wheatboard, strawboard and cork. During construction, ensure that the specified renewable materials are installed.

Calculations and Considerations

Rapidly renewable materials are considered to be sustainable since they tend to use less land and have a faster economic cycle than materials with a growing span longer than 10 years. They are not limited to plant products, but may also be animal-based. Some common examples of products made wholly or partially from rapidly renewable materials are linoleum, bamboo flooring, wheatboard, cotton and wool products, and cork.

The determinations are based on the following definitions and equations.

$MATL\$_{RRM}$ Total value of rapidly renewable materials (CSI 2 through 10) plus any similar portions of furniture or furnishings made from rapidly renewable materials if furniture and furnishings are included consistently in MR subcategories 2 through 7. Value determinations are based on the weight percent of the rapidly renewable portion of each of the individual item values.

Then one point is earned for MRc6.1 if the following is true:

$$100(MATL\$_{RRM})/MATL\$_T \geq 2.5 \qquad (5.6.1)$$

Most applicable construction items are not 100 percent made from rapidly renewable materials, but rather are composites such as bamboo flooring or linoleum. Therefore, the rapidly renewable value is only a portion of the value of the item, whether these items are one piece (such as linoleum flooring) or an assembly (such as wheatboard casework). The value determination is made on a mass percent basis. So for any item the following can be defined:

$MASS_{RRMi}$ Mass (weight) of the rapidly renewable portion of unit i

$MASS_{UNITi}$ Total mass (weight) of unit i

$\text{MATL\$}_{\text{UNIT}i}$ Total cost of unit i

$\text{MATL\$}_{\text{RRM}i}$ Value of the rapidly renewable portions of unit i

The value of the rapidly renewable portions for each individual unit i can be determined from the following:

$$\text{MATL\$}_{\text{RRM}i} = \text{MATL\$}_{\text{UNIT}i}\,(\text{MASS}_{\text{RRM}i})/\text{MASS}_{\text{UNIT}i} \qquad (5.6.2)$$

And in similar manner, the total value of the rapidly renewable portion needed in Eq. (5.6.1) can be determined from the summation of the individual unit contributions as follows:

$$\text{MATL\$}_{\text{RRM}} = \Sigma\text{MATL\$}_{\text{RRM}i} \qquad \text{over all units } i \qquad (5.6.3)$$

Special Circumstances and Exemplary Performance

There may be some concern that many of these rapidly renewable products are not available regionally or are exotic. For instance, bamboo is not commonly associated with forests in the United States. However, many of the rapidly renewable products can be grown and produced within various regions of the United States. Figure 5.6.1 shows a wild growing bamboo stand in an uncultivated area around Lake Murray in South Carolina.

FIGURE 5.6.1 Bamboo growing wild near Lake Murray, S.C. (*Photograph taken July 2007.*)

	Rapidly Renewable Content Value of Unit i (MATLS$_{RRMi}$)	Total Rapidly Renewable Content Value (MATLS$_{RRM}$)
Equations	MATL$\$_{UNITi}$(MASS$_{RRMi}$)/MASS$_{UNITi}$	ΣMATL$\$_{RRMi}$ over all units i

TABLE **5.6.1** Summary of Rapidly Renewable Materials Value Equations

An additional EP point can be earned for rapidly renewable content if the following is true:

$$100(\text{MATL}\$_{RRM})/\text{MATL}\$_T \geq 5 \tag{5.6.4}$$

The equations for this credit subcategory are summarized in Tables 5.6.1 and 5.6.2.

5.7 MR Credit 7: *Certified Wood*

The intention of this credit is to use wood products which are made from wood that has been grown and processed in a manner that is environmentally friendly. One point is available for MRc7.

USGBC Rating System

LEED-NC 2.2 lists the Intent, Requirements, and Potential Technologies and Strategies for this credit as follows:

Intent

Encourage environmentally responsible forest management.

Requirements

Use a minimum of 50% of wood-based materials and products, which are certified in accordance with the Forest Stewardship Council's (FSC) Principles and Criteria, for wood building components. These components include, but are not limited to, structural framing and general dimensional framing, flooring, sub-flooring, wood doors and finishes.

Include materials permanently installed in the project. Furniture may be included, providing it is included consistently in MR Credits 3–7. Wood products purchased for temporary use on the project (e.g. formwork, bracing, scaffolding, sidewalk protection, and guard rails) may be included in the calculation at the project team's discretion. If any such materials are included, all such materials must be included in the calculation. If such materials are purchased for use on multiple projects, the applicant may include these materials for only one project, at its discretion.

Potential Technologies and Strategies

Establish a project goal for FSC-certified wood products and identify suppliers that can achieve this goal. During construction, ensure that the FSC-certified wood products are installed and quantify the total percentage of FSC-certified wood products installed.

	Worth	Criterion
MRc6	1 point	$100(\text{MATL}\$_{RRM})/\text{MATL}\$_T \geq 2.5$
Exemplary performance	1 point in addition to MRc6	$100(\text{MATL}\$_{RRM})/\text{MATL}\$_T \geq 5$

TABLE **5.6.2** Rapidly Renewable Materials Equation Summary

Calculations and Considerations

This credit encourages the use of Forest Stewardship Council labeled wood products. Forest Stewardship Council (FSC) products will be labeled as either FSC Pure or FSC Mixed, and both types of labels ensure that the wood portions of the item are 100 percent FSC-certified. FSC-certified products will have a Chain-of-Custody (CoC) certification. CoC numbers for all items are required in the submittals.

Note that this credit is based only on new wood, in both the numerator and the denominator. Any reused or salvaged wood and wood products are included in other credits. Also, even though the equations developed in this chapter to determine the values of the FSC new wood and all the new wood on the project are based on mass (weight) percentages of the items, there is the option for any item to be instead based on volume percent or value percent if these are easier or more readily available ways to determine the percent value. For instance, a cost percent value may be easier to do with veneers. Some of the variables used in the equations are as follows:

$WOOD\$_T$	Total value of *new* wood portions of all items installed in the project
$WOOD\$_{FSC}$	Total value of FSC *new* wood portions of all items installed in the project
$MASS_{FSCi}$	Mass (weight) of the FSC *new* wood portion of unit i (Volume or value can be substituted for mass as appropriate for an item.)
$MASS_{WOODi}$	Total mass (weight) of all *new* wood in unit i (Volume or value can be substituted for mass as appropriate for an item.)
$MASS_{UNITi}$	Total mass (weight) of unit i (Volume or value can be substituted for mass as appropriate for an item.)
$MATL\$_{UNITi}$	Total cost of unit i
$MATL\$_{FSCi}$	Value of the FSC *new* wood portions of unit i
$MATL\$_{WOODi}$	Value of all the *new* wood portions of unit i

One point is earned for MRc7.1 if the following is true:

$$100(MATL\$_{FSC})/MATL\$_{WOOD} \geq 50 \qquad (5.7.1)$$

Most items are not made entirely of new wood, and therefore the values of the certified new wood and the wood portions are determined as a mass (weight), volume, or cost percent of the total items. Usually the mass percent is used, as in the following equations:

$$MATL\$_{FSCi} = MATL\$_{UNITi} (MASS_{FSCi})/MASS_{UNITi} \qquad (5.7.2)$$

$$MATL\$_{WOODi} = MATL\$_{UNITi} (MASS_{WOODi})/MASS_{UNITi} \qquad (5.7.3)$$

And in a similar manner, the total values of the FSC *new* wood and total *new* wood portions needed in Eq. (5.7.1) can be determined from the summation of the individual unit contributions as follows:

$$MATL\$_{FSC} = \Sigma MATL\$_{FSCi} \qquad \text{over all units } i \qquad (5.7.4)$$

$$MATL\$_{WOOD} = \Sigma MATL\$_{WOODi} \qquad \text{over all units } i \qquad (5.7.5)$$

	Rapidly Renewable Content Value of Unit *i* (MATLS$_{RRMi}$)	Total Rapidly Renewable Content Value (MATLS$_{RRM}$)
Equations	MATLS$_{UNITi}$ (MASS$_{RRMi}$)/MASS$_{UNITi}$	ΣMATLS$_{RRMi}$ over all units *i*

TABLE 5.7.1 Summary of Certified Wood Material Value Equations

Special Circumstances and Exemplary Performance

An additional EP point can be earned for certified wood if the following is true:

$$100(\text{MATL\$}_{FSC})/\text{MATL\$}_{WOOD} \geq 95 \qquad (5.7.6)$$

The equations for this credit subcategory are summarized in Tables 5.7.1 and 5.7.2.

5.8 Discussion and Overview

The LEED-NC 2.2 Reference Manual lists all the credits in the Materials and Resources category as requiring Construction Phase Submittals. This is so for a variety of reasons. Three of the main reasons are the following:

- Contractors have control over the waste materials and their disposal. It is important to keep track of tipping sheets, etc.

- Many of the material specifications and shop drawings are done or furnished by the contractor. These determine a lot of the material makeup.

- Most of the materials are purchased by the contractor so the contractor has decision-making capabilities for where the products come from with respect to both local materials and makeup of the materials for recycling, reuse, FSC certification, etc.

It is very important that an effective material tracking program be set up and maintained during the entire construction phase. These requirements should be explained well in the construction documents and periodically reviewed to make sure that the expectations are being met.

	Worth	Criterion
MRc5.1	1 point	100(MATLS$_{RRM}$)/MATLS$_T \geq 2.5$
Exemplary performance	1 point additional	100(MATLS$_{RRM}$)/MATLS$_T \geq 5$

TABLE 5.7.2 Certified Wood Equation Summary

References

AISC (1980), *Manual of Steel Construction,* 8th ed., American Institute of Steel Construction Inc., Chicago ©1980, first revised printing 11/84, Second Impression 2/86.

Albermani, F., G. Y. Goh, and S. L. Chan (2006), "Lightweight Bamboo Double Layer Grid System," Engineering Structures, http://www.sciencedirect.com/science/journal/01410296> Volume 29, Issue 7<http://www.sciencedirect.com/science?_ob=PublicationURL&_tockey=%23TOC%235715%232007%23999709992%23659670%23FLA%23&_cdi=5715&_pubType=J&_auth=y&_acct=C000050221&_version=1&_urlVersion=0&_userid=10&md5=de3b4f08c26c73b1658aee5e404d6a74>, July 2007, pp. 1499–1506.

Cal Recovery (1991), "Conversion Factors for Individual Material Types," Prepared for the California Integrated Waste Management Board by Cal Recover, Inc., December.

CSI (2004), *The Project Resource Manual—CSI Manual of Practice,* 5th ed., Construction Specifications Institute, Alexandria, Va.

FSC (2007), http://www.fsc.org/en/, Forest Stewardship Council website accessed July 18, 2007.

Haselbach, L., and S. Bruner (2006), "Determining Construction Debris Recycling Dumpster Densities," *Journal of Green Building,* Summer, 1(3).

Spiegel, R., and D. Meadows (2006), *Green Building Materials: A Guide to Product Selection and Specification,* 2d ed., CSI, The Construction Specifications Institute, Alexandria, Va.

USGBC (2003), *LEED-NC for New Construction, Reference Guide,* Version 2.1, 2d ed., May, U.S. Green Building Council, Washington, D.C.

USGBC (2005–2007), *LEED-NC for New Construction, Reference Guide,* Version 2.2, 1st ed., U.S. Green Building Council, Washington, D.C., October 2005 with errata posted through Spring 2007.

Exercises

1. Put together a CPM (critical path method) schedule for items for C&D debris. (See Chap. 8 for information on a CPM.)

2. Make a site plan with staging areas for C&D debris recycling for a project with demolition.
A. Calculate the number of dumpsters needed and sizes.
B. Determine the area needed for the dumpsters and access.
C. Sketch these items on the plan.

3. Make a site plan with staging areas for C&D debris recycling for a project with no demolition.
A. Calculate the number of dumpsters needed and sizes.
B. Determine the area needed for the dumpsters and access.
C. Sketch these items on the plan.

4. Total construction cost of your new building on a site (project A) with no demolition is $10,000,000, and Table 5.E.1 represents the materials tracked as applicable to MR credits 3 through 7. This table does not represent all the materials used, only those tracked for LEED purposes. All new wood and all the furniture items are listed. Include the listed furniture items (wood furniture and workstations) in the calculations.
A. Calculate the eligible points based on Table 5.E.1 for MR credits 3, 4, 5, 6, and 7.
B. Is this project eligible for any EP points as related to MR credits 3, 4, 5, 6, and 7? Why?

5. Total construction cost of your new building (project A) on a site with no demolition is $10,000,000, and Table 5.E.1 represents the materials tracked as applicable to MR credits 3 through 7.

Product	Company	Postconsumer Recycled %	Preconsumer Recycled %	Content Info Source	Manufacturing Location (mi) from Site	% Raw Materials within 500 mi	% Rapidly Renewable Materials	FSC Certified Wood (%)	Product Cost ($)
Used brick*	Ali and Allison Salvagers	n.a.	n.a.	n.a.	95	100	n.a.	n.a.	50,000
New brick	Bower's Brick	0	15	Factory letter	600	0	0	n.a.	30,000
Concrete fill*	Bramblett Concrete	n.a.	n.a.	Contractor	69	100	n.a.	n.a.	10,000
Steel	Dorn Metal	60	0	Manufacturer	355	0	0	n.a.	165,000
Ready mix concrete	Amanda Mix	5	0	Brochure	227	100	0	n.a.	32,000
Compost	Tab and Kim Mulch	100	0	Common knowledge	26	100	n.a.	n.a.	5,000
Wheatboard	Martin Walls	0	0	Cut sheet	59	75	75	n.a.	93,000
Metal siding	Neal and Quinn McSiding	25	0	Cut sheet	552	0	0	n.a.	45,000
Metal roofing	Minor Roof	85	0	Brochure	342	25	0	n.a.	47,000
Ceramic tile	Munson Tile	95	0	Brochure	343	45	0	n.a.	12,000
Acoustical tile	Randolph and Roney Ceilings	90	0	Cut sheet	34	0	0	n.a.	15,000
Toilet partitions	Rowland Supply	0	100	Brochure	78	0	n.a.	n.a.	7,000
Carpet	Ashley Carpet	40	0	Cut sheet	115	0	n.a.	n.a.	50,000
Carpet pad	Wilder Padding	0	100	Brochure	56	0	n.a.	n.a.	4,000
Linoleum	Wiseman Floor	0	0	Cut sheet	3,000	0	80	n.a.	18,000

Material	Supplier							
Special finish wood†	Amponsah and Deep Finishes	0	0	Brochure	250	100	0	100	15,000
Misc. finish carpentry†	Sanford Trees	0	0	Brochure	330	100	0	34	32,000
Pine boards†	Gaither Wood	0	0	Cut sheet	45	100	0	50	40,000
Salvaged wood flooring*	Grant Interiors	n.a.	n.a.	Contractor	340	100	n.a.	n.a.	62,000
Wood doors, frames†	Del Porto Accessways	0	0	Brochures	552	0	0	65	120,000
Wood furniture†	Floyd and Frick Furniture	0	0	Manufacturer	123	0	0	30	135,000
Wood roof structure†	Guidt Roof Salvaging	n.a.	n.a.	Brochure	145	100	n.a.	n.a.	115,000
Workstations	Lockard Stations	45	20	Brochure	222	0	0	0	35,000

*Salvaged materials.
†New wood.

TABLE 5.E.1 Materials and Resources Project A Tracking Summary

This table does not represent all the materials used, only those tracked for LEED purposes. All new wood and all the furniture items are listed. Do not include the listed furniture items (wood furniture and workstations) in the calculations.

    A. Calculate the eligible points based on Table 5.E.1 for MR credits 3, 4, 5, 6, and 7.

    B. Is this project eligible for any EP points as related to MR credits 3, 4, 5, 6, and 7? Why?

6. Total construction cost of your new building on a site (project B) with no demolition is $20,000,000, and Table 5.E.2 represents the materials tracked as applicable to MR credits 3 through 7. This table does not represent all the materials used, only those tracked for LEED purposes. All new wood and all the furniture items are listed. Include the listed furniture items (wood furniture and workstations) in the calculations.

    A. Calculate the eligible points based on Table 5.E.2 for MR credits 3, 4, 5, 6, and 7.

    B. Is this project eligible for any EP points as related to MR credits 3, 4, 5, 6, and 7? Why?

7. Total construction cost of your new building on a site (project B) with no demolition is $20,000,000, and Table 5.E.2 represents the materials tracked as applicable to MR credits 3 through 7. This table does not represent all the materials used, only those tracked for LEED purposes. All new wood and all the furniture items are listed. Do not include the listed furniture items (wood furniture and workstations) in the calculations.

    A. Calculate the eligible points based on Table 5.E.2 for MR credits 3, 4, 5, 6, and 7.

    B. Is this project eligible for any EP points as related to MR credits 3, 4, 5, 6, and 7? Why?

8. You have a small commercial building that is going to be renovated, and you are going for LEED-NC (instead of LEED-EB). The building footprint is 40 ft by 40 ft. It is one story with an exterior wall height of 12 ft and a flat roof. You can assume the following:

- Slab on grade of 12-ft depth
- Four exterior doors (7 ft by 3 ft)
- Six exterior windows (4 ft by 3 ft)
- No interior structural walls

You are interested in going for the building reuse credits for the structural items. You plan to keep all the exterior walls and the foundation and slab, but do not plan to keep the roof structure as your client would like a pitched roof, which would provide attic storage. Calculate the percent of building reuse and whether MR credits 1.1 and 1.2 are earned.

9. You have a small open-room warehouse that is going to be renovated into a commercial building, and you are going for LEED-NC (instead of LEED-EB). The building footprint is 60 ft by 40 ft. It is one story with an exterior wall height of 12 ft and a flat roof. All interior rooms are 10 ft high. The new interior will consist of the following:

- A 39-ft by 30-ft front retail customer area with a 7-ft by 3-ft exterior door and two 8-ft by 4-ft exterior windows
- A 27-ft by 20-ft office area in the back left with a 3-ft by 6-ft interior window to the customer area
- An 11-ft by 10-ft restroom-utility room on the far back right
- An 11-ft by 16-ft storage area on the middle back right
- A 6-ft by 27-ft hallway down the center of the back part to an existing exit door

These dimensions are interior wall-to-wall measurements. There are four interior doors (3 ft by 7 ft) off the hallway—one to the customer area, one to the office, one to the storage room, and the last one to the restroom-utility room.

    You want to see if you can get MR credit 1.3 by keeping the original hardwood flooring in all areas except the restroom-utility room, keeping the drop ceiling in all areas except the retail customer area, and keeping the gypsum along all the interior sides of the exterior walls.

    A. Sketch the design.

    B. Determine if you are eligible for MR credit 1.3.

Product	Company	Postconsumer Recycled %	Preconsumer Recycled %	Content Info Source	Manufacturing Location (Mi from Site)	% Raw Materials within 500 mi	% Rapidly Renewable Materials	FSC Certified Wood (%)	Product Cost ($)
Used brick*	Bagley Salvagers	n.a.	n.a.	n.a.	95	100	n.a.	n.a.	100,000
New brick	Koty's Brick	0	15	Factory letter	600	0	0	n.a.	65,000
Concrete fill*	Barwick Rubble	n.a.	n.a.	Contractor	69	100	n.a.	n.a.	20,000
Steel	Burns Metal	60	0	Manufacturer	355	0	0	n.a.	310,000
Ready mix concrete	Kendrick Concrete	5	0	Brochure	227	100	0	n.a.	65,000
Compost	Mogan Mulch	100	0	Common knowledge	26	100	n.a.	n.a.	10,000
Wheatboard	Wall and Wall Walls	0	0	Cut sheet	59	75	75	n.a.	190,000
Metal siding	Byrd Siding	25	0	Cut sheet	552	0	0	n.a.	95,000
Metal roofing	LeCroy Roof	85	0	Brochure	342	25	0	n.a.	94,000
Ceramic tile	Whetstone Tile	95	0	Brochure	343	45	0	n.a.	27,000
Acoustic tile	Powell Ceilings	90	0	Cut sheet	34	0	0	n.a.	32,000
WC partitions	Tristan Supply	0	100	Brochure	78	0	n.a.	n.a.	15,000
Carpet	Bethany Carpet	40	0	Cut sheet	115	0	n.a.	n.a.	110,000
Carpet pad	Sipes Padding	0	100	Brochure	56	0	n.a.	n.a.	24,000
Linoleum	Kimberly Floor	0	0	Cut sheet	3,000	0	80	n.a.	36,000
Special finish wood†	Tune Finishes	0	0	Brochure	250	100	0	100	30,000

Table 5.E.2 Materials and Resources Project B Tracking Summary

Product	Company	Postconsumer Recycled %	Preconsumer Recycled %	Content Info Source	Manufacturing Location (Mi from Site)	% Raw Materials within 500 mi	% Rapidly Renewable Materials	FSC Certified Wood (%)	Product Cost ($)
Misc. finish carpentry†	Wade Trees	0	0	Brochure	330	100	0	34	65,000
Pine boards†	Wagner Wood	0	0	Cut sheet	45	100	0	50	80,000
Reused wood flooring*	Young and Old Interiors	n.a.	n.a.	Contractor	340	100	n.a.	n.a.	130,000
Wood doors, frames†	Gorman Accessways	0	0	Brochures	552	0	0	65	220,000
Wood furniture†	Bridgette Furniture	0	0	Manufacturer	123	0	0	30	260,000
Wood roof structure*	Pearce Roof Salvaging	n.a.	n.a.	Brochure	145	100	n.a.	n.a.	230,000
Workstations	Worthy Stations	45	20	Brochure	222	0	0	0	75,000

*Salvaged materials.
†New wood.

TABLE 5.E.2 Materials and Resources Project B Tracking Summary (Continued)

Material	Dumpster Size (yards)	Number of Dumpsters Filled
Cardboard	30	3
Gypsum wallboard	30	1
Steel	6	2
Wood	30	3
Mixed waste to landfill	30	6
Rubble (concrete/brick)	20	3

TABLE 5.E.3 Project C&D Debris Table

10. You set up a C&D debris program at your site and recycle all the cardboard, gypsum wallboard, steel, concrete and brick, and wood for a new construction project. Your dumpster count at the end of the project is as listed in Table 5.E.3. Calculate the percent recycled C&D debris by both volume and weight using the LEED conversion factors. With the weight calculation which credits might you get? With the volume calculations which credits might you get?

11. You are putting up a new building at a currently vacant site. You set up a construction debris program at your site and recycle all the cardboard, gypsum wallboard, steel, concrete and brick, and wood for a new construction project. Your dumpster count at the end of the project is as listed in Table 5.E.3. Calculate the percent recycled C&D debris by both volume and weight using the Columbia, S.C., study dumpster densities. With the weight calculation which credits might you get? With the volume calculations which credits might you get?

12. You are planning to start designing a project, and the site engineer asks how and where the C&D debris area should be located. Using the six categories given by LEED for typical conversion factors, estimate dumpster sizes needed and the footprint on the ground needed for this C&D debris recycling area. Make a sketch of the footprint. Make sure that proper access is available to the location for both the recycling contractors and the workers, and sketch in these accessways. Be sure to include fencing as needed.

13. Total construction cost of your new building on a site (project C) with no demolition is $15,000,000, and Table 5.E.4 represents the materials tracked as applicable to MR credits 3 through 7. This table does not represent all the materials used, only those tracked for LEED purposes. All new wood and all the furniture items are listed. Include the listed furniture items (wood furniture and workstations) in the calculations.
   A. Calculate the eligible points based on Table 5.E.4 for MR credits 3, 4, 5, 6, and 7.
   B. Is this project eligible for any EP points as related to MR credits 3, 4, 5, 6, and 7. Why?

14. Total construction cost of your new building on a site (project C) with no demolition is $15,000,000, and Table 5.E.4 represents the materials tracked as applicable to MR credits 3 through 7. This table does not represent all the materials used, only those tracked for LEED purposes. All new wood and all the furniture items are listed. Do not include the listed furniture items (wood furniture and workstations) in the calculations.
   A. Calculate the eligible points based on Table 5.E.4 for MR credits 3, 4, 5, 6, and 7.
   B. Is this project eligible for any EP points as related to MR credits 3, 4, 5, 6, and 7. Why?

Product	Company	Postconsumer Recycled %	Preconsumer Recycled %	Content Info Source	Manufacturing Location (MI from Site)	% Raw Materials within 500 MI	% Rapidly Renewable Materials	FSC Certified Wood (%)	Product Cost ($)
Used brick*	Bagley Salvagers	n.a.	n.a.	n.a.	95	100	n.a.	n.a.	90,000
New brick	Koty's Brick	0	15	Factory letter	600	0	0	n.a.	65,000
Concrete fill*	Barwick Rubble	n.a.	n.a.	Contractor	69	100	n.a.	n.a.	20,000
Steel	Burns Metal	60	0	Manufacturer	355	50	0	n.a.	310,000
Ready mix concrete	Kendrick Concrete	5	0	Brochure	227	100	0	n.a.	65,000
Compost	Mogan Mulch	100	0	Common knowledge	26	100	n.a.	n.a.	10,000
Wheatboard	Wall and Wall Walls	0	0	Cut sheet	59	75	75	n.a.	190,000
Metal siding	Byrd Siding	55	0	Cut sheet	552	0	0	n.a.	95,000
Metal roofing	LeCroy Roof	85	0	Brochure	342	25	0	n.a.	94,000
Ceramic tile	Whetstone Tile	95	0	Brochure	343	45	0	n.a.	27,000
Acoustic tile	Powell Ceilings	90	0	Cut sheet	34	50	0	n.a.	32,000
WC partitions	Tristan Supply	0	100	Brochure	78	0	n.a.	n.a.	15,000
Carpet	Bethany Carpet	40	0	Cut sheet	115	50	n.a.	n.a.	110,000
Carpet pad	Sipes Padding	0	100	Brochure	56	0	n.a.	n.a.	24,000
Linoleum	Kimberly Floor	0	0	Cut sheet	3,000	0	80	n.a.	36,000
Special finish wood†	Tune Finishes	0	0	Brochure	250	100	0	100	30,000

Material	Supplier								
Misc. finish carpentry†	Wade Trees	0	0	Brochure	330	100	0	34	65,000
Pine boards†	Wagner Wood	0	0	Cut sheet	45	100	0	50	80,000
Reused wood flooring*	Young and Old Interiors	n.a.	n.a.	Contractor	340	100	n.a.	n.a.	130,000
Wood doors, frames†	Gorman Accessways	0	0	Brochures	552	0	0	65	220,000
Wood furniture†	Bridgette Furniture	0	0	Manufacturer	123	0	0	30	160,000
Wood roof Structure*	Pearce Roof Salvaging	n.a.	n.a.	Brochure	145	100	n.a.	n.a.	230,000
Workstations	Worthy Stations	45	20	Brochure	222	0	0	0	75,000

*Salvaged materials.
†New wood.

TABLE 5.E.4 Materials and Resources Project C Tracking Summary

CHAPTER 6

LEED Indoor Environmental Quality

The Indoor Environmental Quality (IEQ) category of the U.S. Green Building Council (USGBC) rating system for new construction version 2.2 (LEED-NC 2.2) consists of two prerequisite subcategories and eight credit subcategories that together may earn a possible 15 points (credits) maximum, not including exemplary performance (EP) points. The notation format for the prerequisites and credits is, for example, EQp1 for Indoor Environmental Quality prerequisite 1 and EQc1 for Indoor Environmental Quality credit 1. To facilitate cross-referencing between this text and the USGBC rating system, the second digit in a section heading, equation, table, or figure number represents the credit subcategory number for sections that deal directly with a USGBC LEED-NC 2.2 credit subcategory.

The Indoor Environmental Quality (EQ) portion deals with materials and systems inside the building that affect the health and comfort of the occupants and construction workers.

The two prerequisite subcategories are EQ Prerequisite 1: Minimum IAQ Performance (EQp1) and EQ Prerequisite 2: Environmental Tobacco Smoke (ETS) Control (EQp2). As is typical for prerequisites, there are no points associated with EQp1 and EQp2.

The points for the credits in the Indoor Environmental Quality category are individually related to a specific credit under a subcategory as was previously done in the Sustainable Sites and Water Efficiency categories. There are eight credit subcategories, five of which have multiple credits and three of which have just one credit and associated point. The eight credit subcategories are EQ Credit Subcategory 1: Outdoor Air Delivery Monitoring, EQ Credit Subcategory 2: Increased Ventilation, EQ Credit Subcategory 3: Construction IAQ Management Plan, EQ Credit Subcategory 4: Low-Emitting Materials, EQ Credit Subcategory 5: Indoor Chemical and Pollutant Source Control, EQ Credit Subcategory 6: Controllability of Systems, EQ Credit Subcategory 7: Thermal Comfort, and EQ Credit Subcategory 8: Daylighting & Views.

EQ credit subcategory 4 has four credits; EQ credit subcategories 3, 6, 7, and 8 each have two credits; and EQ credit subcategories 1, 2, and 5 are all single credits. Each credit is worth one point, not including EP points, which will be discussed later.

The prerequisites and credits in the Indoor Environmental Quality category can be grouped into the following two overall sets of goals:

- The two prerequisites and the first five credit subcategories promote good indoor air quality (IAQ). EQp1, EQc1, and EQc2 are all related to indoor air exchange with outside air, a corrective measure for potentially poor indoor quality. EQp1 sets a minimum air exchange rate. EQc1 gives two ways to monitor adequate air exchanges by monitoring air flow rates and by utilizing carbon dioxide concentrations as the marker compound. EQc2 promotes an air exchange rate that is higher than the minimum set established in EQp1. On the other hand, EQp2, EQc3.1, EQc3.2, EQc4.1, EQc4.2, EQc4.3, EQc4.4, and EQc5 all represent measures for preventing potential air pollutants from entering into the indoor airspaces.

- The last three credit subcategories promote the well-being and comfort of the occupants with respect to thermal comfort, lighting, and views. EQc7.1, EQc7.2, and EQc6.2 are all related to thermal comfort, with EQc7.1 related to design, EQc7.2 related to verification of performance, and EQc6.2 giving preferred thermal comfort control flexibility to individual or groups of occupants. EQc6.1 and EQc8.1 pertain to lighting. EQc6.1 gives preferred lighting control flexibility to individual or group occupants, and EQc8.1 addresses increasing natural lighting (daylighting) to supplement the other indoor lighting. EQc8.2 addresses a minimum amount of views for occupants.

The two prerequisites and the 15 credits in the Indoor Environmental Quality category are as follows:

- **EQ Prerequisite 1** (EQp1): *Minimum IAQ Performance*
- **EQ Prerequisite 2** (EQp2): *Environmental Tobacco Smoke Control*
- **EQ Credit 1** (EQc1): *Outdoor Air Delivery Monitoring* (previously referred to as *Carbon Dioxide (CO_2) Monitoring* in LEED-NC 2.1) (EB)
- **EQ Credit 2** (EQc2): *Increased Ventilation* (previously referred to as *Ventilation Effectiveness* in LEED-NC 2.1)
- **EQ Credit 3.1:**(EQc3.1): *Construction IAQ Management Plan—During Construction*
- **EQ Credit 3.2** (EQc3.2): *Construction IAQ Management Plan—Before Occupancy* (previously referred to as *Construction IAQ Management Plan—After Construction/ Before Occupancy* in LEED-NC 2.1)
- **EQ Credit 4.1** (EQc4.1): *Low-Emitting Materials—Adhesives and Sealants*
- **EQ Credit 4.2** (EQc4.2): *Low-Emitting Materials—Paints and Coatings*
- **EQ Credit 4.3** (EQc4.3): *Low-Emitting Materials—Carpet Systems* (previously referred to as *Low-Emitting Materials—Carpet* in LEED-NC 2.1)
- **EQ Credit 4.4** (EQc4.4): *Low-Emitting Materials—Composite Wood and Agrifiber* (previously referred to as *Low-Emitting Materials—Composite Wood* in LEED-NC 2.1)
- **EQ Credit 5** (EQc5): *Indoor Chemical and Pollutant Source Control*
- **EQ Credit 6.1** (EQc6.1): *Controllability of Systems—Lighting* (previously referred to as *Controllability of Systems—Perimeter Spaces* in LEED-NC 2.1)
- **EQ Credit 6.2** (EQc6.2): *Controllability of Systems—Thermal Comfort* (previously referred to as *Controllability of Systems—Non-Perimeter Spaces* in LEED-NC 2.1) (EB)
- **EQ Credit 7.1** (EQc7.1): *Thermal Comfort—Design* (previously referred to as *Thermal Comfort—Compliance with ASHARE 55-1992* in LEED-NC 2.1) (EB)

- **EQ Credit 7.2** (EQc7.2): *Thermal Comfort—Verification* (previously referred to as *Thermal Comfort—Permanent Monitoring System* in LEED-NC 2.1) (EB)
- **EQ Credit 8.1** (EQc8.1): *Daylighting and Views—Daylighting 75% of Spaces* (EB)
- **EQ Credit 8.2** (EQc8.2): *Daylighting and Views—Views for 90% of Spaces* (EB)

LEED-NC 2.2 provides a table that summarizes some of these prerequisite and credit characteristics, and the information from it can be grouped in the following three areas:

- A significant change from LEED version 2.1
- The main phase of the project for the submittal document preparation (Design or Construction)
- The project team member(s) who has (have) the most decision-making responsibility as to whether this credit may or may not be achieved (owner, design team, or contractor).

As previously mentioned in the other categories, the submittals for each credit or prerequisite have been sectioned into either Design Submittals and/or Construction Submittals. Even though the entire LEED process encompasses all phases of a project, the prerequisites or credits that are noted as Construction Submittal credits may have major actions going on through the construction phase which, unless documented through that time, may not be verifiable from preconstruction documents or the built project. Sometimes these credits have activities associated with them that cannot or need not be performed until or subsequent to the construction phase. All the credits in both the Construction IAQ Management Plan and the Low-Emitting Materials subcategories have been noted in the construction submittal column, because the credits of the former cover construction phase activities and the credits of the latter pertain to materials mentioned that are frequently chosen and used during the construction phase. This designation is noted in Table 6.0.1. There is also a column explaining the point count.

Six of the credit subcategories listed in Table 6.0.1 have the icon EB noted after their title (Outdoor Air Delivery Monitoring, Controllability of System—Thermal Comfort, Thermal Comfort—Design, Thermal Comfort—Verification, Daylighting and Views—Daylight 75% of the Spaces, and Daylighting and Views—Views for 90% of the Spaces). This icon is a tool to help those who wish to proceed with the continuing LEED-EB certification (Existing Building) after certification of LEED-NC (New Construction or Major Renovation) is obtained. Those credits noted with this icon are usually more cost-effective to implement during the construction of the building than later during its operation. They are also shown in Table 6.0.1.

There are EP points under the Innovation and Design category available that relate to several items in other categories. These are available for exceeding each of the respective credit criteria to a minimum level as noted in the respective credit descriptions. EP points are only available for the Daylighting and Views subcategory of the EQ credits. This is also noted in Table 6.0.1.

There are many important parameters used to attain and verify the credits. Some are important for multiple credits, such as the determination of the site boundary, which was previously discussed as being an important parameter in many of the credits in the other categories, but is not really directly applicable to any of the EQ credits. One

Prerequisite or Credit	Construction Submittal	EB Icon	Credit Point (Not Including EP or ID points)	EP and Other ID Point Availability	EP Point Criterion
EQp1: Minimum IAQ Performance			n.a.		
EQp2: Environmental Tobacco Smoke (ETS) Control			n.a.		
EQc1: Outdoor Air Delivery Monitoring		Yes	1	No	n.a.
EQc2: Increased Ventilation		No	1	No	n.a.
EQc3.1: Construction IAQ Management Plan, During Construction	*	No	1	No	n.a.
EQc3.2: Construction IAQ Management Plan, Before Occupancy	*	No	1	No	n.a.
EQc4.1: Low-Emitting Materials: Adhesives and Sealants	*	No	1	No	n.a.
EQc4.2: Low-Emitting Materials: Paints and Coatings	*	No	1	No	n.a.
EQc4.3: Low-Emitting Materials: Carpet Systems	*	No	1	No	n.a.
EQc4.4: Low-Emitting Materials: Composite Wood and Agrifiber	*	No	1	No	n.a.
EQc5: Indoor Chemical and Pollutant Source Control		No	1	No	n.a.
EQc6.1: Controllability of Systems: Lighting		No	1	No	n.a.
EQc6.2: Controllability of Systems: Thermal Comfort		Yes	1	No	n.a.
EQc7.1: Thermal Comfort: Design		Yes	1	No	n.a.
EQc7.2: Thermal Comfort: Verification		Yes	1	No	n.a.
EQc8.1: Daylighting and Views: Daylight 75% of Spaces		Yes	1	Yes	95%
EQc8.2: Daylighting and Views: Views for 90% of Spaces		Yes	1	Yes	None set

TABLE 6.0.1 EQ EB Icon, EP Point Availability, and Miscellaneous

important parameter in the Indoor Environmental Quality category is the location of the breathing zone. The measurements are taken within the breathing zone to test for the air quality (AQ) parameters as specified for EQc1 if any of the options for carbon dioxide (CO_2) monitoring are used, EQc2 for the air quality testing option, and EQc3.2 for the air quality testing option and for the calculations for both EQp1 and EQc2. It is important to design a building for improved AQ in this volume.

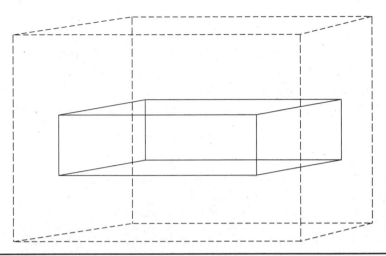

FIGURE 6.0.1 Typical rectangular breathing zone for air quality testing in a room.

The breathing zone is an area defined between two imaginary planes that are located 3 ft and 6 ft above the floor and bounded vertically by planes that are approximately 2 ft off from any wall or fixed air unit. (See Fig. 6.0.1 for a zone in a room approximately 10 ft high.) The measurements are taken within the breathing zone to test for the air quality parameters as specified. The breathing zone represents the volumes that are typically important for good air quality; therefore designing the building such that improved air quality is focused in these volumes is important for a green building. There is no special information in LEED-NC 2.2 for spaces with exceptionally high ceilings, but these should be carefully addressed by the designers.

In summary, most of the Indoor Environmental Quality credits address problems associated with health, comfort, and well-being in the indoor space. Although several of the credits in the Indoor Environmental Quality category might include a capital cost investment, many provide a means to potentially reduce the associated costs over the life of the building, which is done by increasing occupant productivity by maintaining an environment that is more healthful and more comfortable for the occupants. The two prerequisite subcategories *must* be adhered to and verified for each project.

Indoor Environmental Quality Prerequisites

EQ Prerequisite 1: *Minimum IAQ Performance*
This first prerequisite quantifies a minimum exchange of indoor air with outdoor air to prevent buildup of contaminants in the indoor air. This includes contamination due to moisture such as in residential bathrooms and kitchens and in shower or pool facilities. The minimum is set by the recommendations of the American Society of Heating, Refrigerating and Air-Conditioning Engineers in ASHRAE 62.1-2004, *Ventilation for Acceptable Indoor Air Quality*, Sections 4 through 7, with natural ventilation systems specifically adhering to the minimum requirements as set forth in paragraph 5.1. Mixed-mode ventilation, where there are both mechanical and natural systems, should still adhere to the mechanical ventilation requirements. ASHRAE 62 has been separated into

two parts: 62.1 for most buildings and then 62.2 which is for low-rise residential facilities. Therefore, in some cases for low-rise residential facilities, this second part, ASHRAE 62.2-2004, *Ventilation and Acceptable Indoor Air Quality in Low-Rise Residential Buildings*, should be used instead.

USGBC Rating System

LEED-NC 2.2 lists the Intent, Requirements, and Potential Technologies and Strategies for EQp1 as follows:

Intent

Establish minimum indoor air quality (IAQ) performance to enhance indoor air quality in buildings, thus contributing to the comfort and well-being of the occupants.

Requirements

Meet the minimum requirements of Sections 4 through 7 of ASHRAE 62.1-2004, Ventilation for Acceptable Indoor Air Quality. Mechanical ventilation systems shall be designed using the Ventilation Rate Procedure or the applicable local code, whichever is more stringent. Naturally ventilated buildings shall comply with ASHRAE 62.1-2004, paragraph 5.1.

Potential Technologies and Strategies

Design ventilation systems to meet or exceed the minimum outdoor air ventilation rates as described in the ASHRAE standard. Balance the impacts of ventilation rates on energy use and indoor air quality to optimize for energy efficiency and occupant health. Use the ASHRAE 62 Users Manual for detailed guidance on meeting the referenced requirements.

Calculations and Considerations

All ventilation should be designed to prevent the uptake of contaminants into the system. In addition, intakes should be located far from items such as cooling towers, sanitary vents, and vehicular exhaust sources (for example, parking garages, loading docks, and street traffic). Usually a minimum 25-ft setback is required with a 40-ft setback preferred.

Mechanical and Mixed-Mode Ventilation The ventilation rate procedure is used for the mechanical and mixed-mode ventilation areas. This procedure includes both people-related and space-related zone source pollutants. Examples of a person-related air pollutant are dander and cold germs, while a space-related air pollutant may be formaldehyde emitted from a carpet system. The zone is the area of the breathing zone in each interior space i. There are two correction factors included in the calculations: one is for the zone distribution effectiveness (EZ_i), and the other is the overall system ventilation effectiveness (EV). The calculations are usually done by experienced HVAC professionals, but the gist can be approximately summarized by the following definitions and sets of equations.

$OAIZ_i$ Outdoor air intake to zone i (cfm)

BZA_i Breathing zone horizontal area for zone i (ft^2)

$BZAR$ Breathing zone minimum area rate (cfm/ft^2). It is usually 0.06 cfm/ft^2 as given in the standard.

$BZOD_i$ Breathing zone occupant density for zone i (people/ft^2)

$BZPR_j$ Breathing zone minimum people rate for use j (cfm/person) as given in the standard (For example, this is 5 cfm/person for general office spaces, conference rooms, and meeting spaces, and it is 7.5 cfm/person for lecture rooms and classrooms.)

EV   Overall system ventilation effectiveness

EZ_i Zone distribution effectiveness. (This is typically 1.0 for overhead distribution systems and 1.2 for underfloor distribution systems.)

For all zones i with mechanical or mixed-mode ventilation systems, the standard requires that

$$OAIZ_i \geq BZA_i[(BZOD_i)(BZPR_j) + (BZAR)]/(EZ_i)(EV) \qquad (6.0.1)$$

Natural Ventilation   For naturally ventilated areas, there are two main requirements:

- There must be an operable window or opening (which can be opened) within 25 ft of any occupiable area in a room. Note that a roof opening is also acceptable.
- The total operable vertical areas of the openings in the room must be at least 4 percent of the net occupiable horizontal area in that space or room.

For the requisite submittals when natural ventilation alone is chosen, floor plans must be submitted showing the minimum 25-ft access to operable windows and both the operable vertical areas of each of the openings and the horizontal occupiable areas of the spaces. Summaries of these areas showing compliance should also be included. Figure 6.0.2 shows an example layout of operable windows that does not comply with the prerequisite.

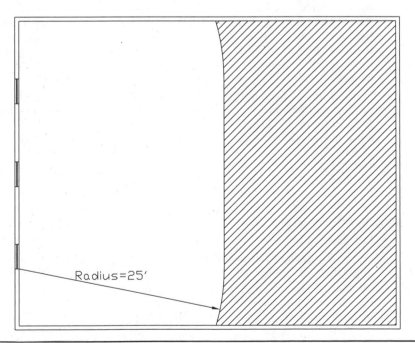

FIGURE 6.0.2 This does not meet EQ Prerequisite 1 for natural ventilation. Hatched area is not within 25 ft of operable windows.

EQ Prerequisite 2: *Environmental Tobacco Smoke (ETS) Control*

The purpose of this prerequisite is to prevent secondhand smoke from going into areas other than designated or desired smoking areas of buildings.

USGBC Rating System

LEED-NC 2.2 lists the Intent, Requirements, and Potential Technologies and Strategies for EQp2 as follows:

Intent

Minimize exposure of building occupants, indoor surfaces, and ventilation air distribution systems to Environmental Tobacco Smoke (ETS).

Requirements

OPTION 1

- Prohibit smoking in the building.
- Locate any exterior designated smoking areas at least 25 feet away from entries, outdoor air intakes and operable windows.

OR

OPTION 2

- Prohibit smoking in the building except in designated smoking areas.
- Locate any exterior designated smoking areas at least 25 feet away from entries, outdoor air intakes and operable windows.
- Locate designated smoking rooms to effectively contain, capture and remove ETS from the building. At a minimum, the smoking room must be directly exhausted to the outdoors with no re-circulation of ETS-containing air to the non-smoking area of the building, and enclosed with impermeable deck-to-deck partitions. With the doors to the smoking room closed, operate exhaust sufficient to create a negative pressure with respect to the adjacent spaces of at least an average of 5 Pa (0.02 inches of water gauge) and with a minimum of 1 Pa (0.004 inches of water gauge).
- Performance of the smoking room differential air pressures shall be verified by conducting 15 minutes of measurement, with a minimum of one measurement every 10 seconds, of the differential pressure in the smoking room with respect to each adjacent area and in each adjacent vertical chase with the doors to the smoking room closed. The testing will be conducted with each space configured for worst case conditions of transport of air from the smoking rooms to adjacent spaces with the smoking rooms' doors closed to the adjacent spaces.

OR

OPTION 3 (For residential buildings only)

- Prohibit smoking in all common areas of the building.
- Locate any exterior designated smoking areas at least 25 feet away from entries, outdoor air intakes and operable windows opening to common areas.
- Minimize uncontrolled pathways for ETS transfer between individual residential units by sealing penetrations in walls, ceilings and floors in the residential units, and by sealing vertical chases adjacent to the units.
- All doors in the residential units leading to common hallways shall be weather-stripped to minimize air leakage into the hallway. If the common hallways are pressurized with respect to the residential units then doors in the residential units leading to the common

hallways need not be weather-stripped provided that the positive differential pressure is demonstrated as in Option 2, considering the residential unit as the smoking room.

- Acceptable sealing of residential units shall be demonstrated by a blower door test conducted in accordance with ANSI/ASTM-E779-03, Standard Test Method for Determining Air Leakage Rate By Fan Pressurization, AND use the progressive sampling methodology defined in Chapter 4 (Compliance Through Quality Construction) of the Residential Manual for Compliance with California's 2001 Energy Efficiency Standards (http://www.energy.ca.gov/title24/2001standards/residential_manual/index.html). Residential units must demonstrate less than 1.25 square inches leakage area per 100 square feet of enclosure area (i.e. sum of all wall, ceiling and floor areas).

Potential Technologies and Strategies

Prohibit smoking in commercial buildings or effectively control the ventilation air in smoking rooms. For residential buildings, prohibit smoking in common areas, design building envelope and systems to minimize ETS transfer among dwelling units.

Calculations and Considerations

Either no smoking is allowed in a building, or if there are smoking areas, such as designated areas in a nonresidential facility, or residential apartments where the occupants may wish to smoke, then these areas are adequately segregated by a combination of physical barriers and negative pressure so that the contaminants from tobacco smoke do not enter the other areas. There is also the potential for secondhand smoke to enter a facility if smoking is allowed near openings to the outdoors and air intakes. Therefore, if outdoor smoking areas are allowed on a site, they must be located away from these potential avenues for indoor contamination. The requirements listed previously in the USGBC LEED-NC 2.2 rating system section give very good explanations of what is required. A useful summary of the various options and requirements can be found in Fig. 6.0.3. Note the following unit conversions:

$$1 \text{ atm} \sim 101{,}300 \text{ Pa}$$

$$1 \text{ in of water} \sim 249 \text{ Pa}$$

6.1 EQ Credit 1: *Outdoor Air Delivery Monitoring*

The intent of this credit is to verify the presence of adequate air exchanges with the outside. It is assumed that if there is sufficient air exchange with the outside, then potential contaminants will not build up in the indoor airspace. This credit has changed significantly from the requirements in LEED-NC 2.1.

USGBC Rating System

LEED-NC-2.2 lists the Intent, Requirements, and Potential Technologies and Strategies for EAQc1 as follows:

Intent

Provide capacity for ventilation system monitoring to help sustain occupant comfort and well-being.

Requirements

Install permanent monitoring systems that provide feedback on ventilation system performance to ensure that ventilation systems maintain design minimum ventilation

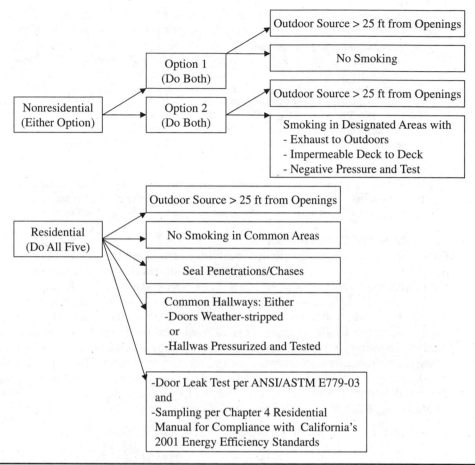

FIGURE 6.0.3 Summary of ETS requirements: all the listed items for each option must be followed.

requirements. Configure all monitoring equipment to generate an alarm when the conditions vary by 10% or more from setpoint, via either a building automation system alarm to the building operator or via a visual or audible alert to the building occupants.

FOR MECHANICALLY VENTILATED SPACES

- Monitor carbon dioxide concentrations within all densely occupied spaces (those with a design occupant density greater than or equal to 25 people per 1000 sq. ft.). CO_2 monitoring locations shall be between 3 feet and 6 feet above the floor.
- For each mechanical ventilation system serving non-densely occupied spaces, provide a direct outdoor airflow measurement device capable of measuring the minimum outdoor airflow rate with an accuracy of plus or minus 15% of the design minimum outdoor air rate, as defined by ASHRAE 62.1-2004.

FOR NATURALLY VENTILATED SPACES
Monitor CO_2 concentrations within all naturally ventilated spaces. CO_2 monitoring shall be located within the room between 3 feet and 6 feet above the floor. One CO_2 sensor may be used to represent multiple spaces if the natural ventilation design uses passive stack(s) or

other means to induce airflow through those spaces equally and simultaneously without intervention by building occupants.

Potential Technologies and Strategies
Install carbon dioxide and airflow measurement equipment and feed the information to the HVAC system and/or Building Automation System (BAS) to trigger corrective action, if applicable. If such automatic controls are not feasible with the building systems, use the measurement equipment to trigger alarms that inform building operators or occupants of a possible deficiency in outdoor air delivery.

Calculations and Considerations
The verification of adequate air exchange in spaces where mechanical ventilation systems are used is done in two ways, depending on whether the space will be densely or sparsely occupied. The reason for this segregation is that sparsely occupied areas do not usually need much variation in the airflow, and verification can be with a simple air rate check. Areas that are densely occupied at times will need a much higher air exchange rate during periods of high occupancy than during other times. Many of these occupied spaces are highly occupied for only a few hours during the day, and it would lead to energy inefficiencies to require a higher rate during the times of low occupation. An example might be a conference room or a classroom that may be packed at certain times and either vacant or sparsely used the rest of the time. The definition of *densely occupied* is a maximum occupancy of 25 people per 1000 ft^2 (40 ft^2 per person). Areas that are naturally ventilated have only one option for verification.

Mechanically Ventilated, Non-Densely Occupied Spaces   For areas of the building that are never densely occupied and are mechanically ventilated, the verification is actually a flow measurement of the rate of outdoor airflow in. The rate should not drop more than 10 percent lower than the design rate. For some multiple small spaces, the measuring device can be in the combined return air duct. The accuracy of the device should be no less than plus or minus 15 percent of this design rate. Note that the rate requirements are based on the air exchange designs of either EQp1 or EQc2 if that is sought, whichever has higher air exchange rates. Therefore, for projects not seeking EQc2, the alarm should be activated when the outdoor air rate into indoor zone i is less than 90 percent of the design; i.e., the alarm should be activated if

$$OAIZ_i \leq 0.9 \times BZA_i[(BZOD_i)(BZPR_j) + (BZAR)]/(EZ_i)(EV) \qquad \text{(not seeking EQc2)} \qquad (6.1.1)$$

or

$$OAIZ_i \leq 0.9 \times 1.3 \times BZA_i[(BZOD_i)(BZPR_j) + (BZAR)]/(EZ_i)(EV) \qquad \text{(seeking EQc2)} \qquad (6.1.2)$$

Mechanically Ventilated, Densely Occupied Spaces and Naturally Ventilated Spaces   For mechanically ventilated areas that may be crowded with human occupants at times and for all naturally ventilated areas, a special marker compound is used that can be indicative of the number of people in the room and their activity level. This marker compound is carbon dioxide (CO_2). To better understand the significance of the required air exchange control with respect to the levels of carbon dioxide, one should become familiar with the impact that humans have on the concentration of carbon dioxide in a space.

As we breathe, we intake air laden with oxygen, use a lot of the oxygen, and exhale air that has less oxygen and increased levels of CO_2. If there are more people in a room, there is more oxygen used. Likewise if the people in the room are exercising or performing other high-energy activities, they will need more oxygen for the increased energy levels, and more carbon dioxide is released.

The uptake of oxygen by a human is related to the metabolic rate of the person. Metabolism is the process of burning energy by living creatures. When an energy source (food) is burned in the human body, oxygen is used to convert the carbon-based compounds in the food to CO_2 and water with a subsequent release of energy. It is important to note that each person burns calories at different rates. The metabolic rate at which a person burns calories while at rest is referred to as 1 *Met*, and it is unique to each person. It may be unique to each person at different stages in his or her life, and as a person becomes more active, the Mets are increased. A metabolic rate of two to four times the resting rate is considered to be low to moderate activity. For very intense activity, the increase may be up to approximately 12 Mets, but usually vigorous exercise results in approximately 6 Mets. So how do we relate this to the carbon dioxide levels in a room?

When we breathe outside air, we inhale air with oxygen at about 21 percent (volumetric) and carbon dioxide (CO_2) at around 380 ppm (0.038 percent volumetric). Outdoor levels do vary regionally due to other anthropogenic and natural activities that produce carbon dioxide, and these can vary from 300 to 500 ppm, but on average the concentration was approximately 385 ppm in 2007. This average level has been steadily rising over the last century.

When we exhale air, there is a little less oxygen (and more water) and around 4.5 percent (volumetric) carbon dioxide, which is then rapidly diluted by the air around it. (Note that when referring to the concentration of a vapor-phase substance, *ppm* stands for molar parts per million or the number of molecules of the substance with respect to 1 million molecules of the air mixture. Usually ppm is not a weight-based measure in the gaseous phase. At typical temperatures and pressures that humans occupy, the gaseous phase is approximated to act as an ideal gas, and molar parts per million can be assumed to be equivalent to volumetric parts per million. Therefore, the concentration ppm as used in this text for vapor-phase concentrations is always molar-based.) When we are outdoors, the space is so large that the contribution from our breathing does not noticeably increase the ambient CO_2 concentration; but when we are indoors, the concentration of the CO_2 is elevated in the room since we are adding CO_2 to this set volume. Typically, you can estimate that sedentary activities in a typical space with proper ventilation will *raise* the CO_2 concentrations by about 300 to 600 ppm depending on the human density of the room. Better estimates can be made by doing an air concentration model of the indoor space, but the approximate amounts just stated are a good rule-of-thumb initial estimate for typical indoor carbon dioxide level increases.

CO_2 is used as a marker compound for several reasons. First, an increase in CO_2 levels usually means a decrease in oxygen levels due to consumption by the occupants. Lower oxygen levels tend to make us more lethargic. Second, if there are higher CO_2 levels, then higher levels of many other air pollutants from the sources of the carbon dioxide (humans) may be reasonably expected to be in the air. These other pollutants include bacteria and other pathogens that humans cough or breathe out. In addition, more activity tends to increase the potential for settled air pollutants to reentrain into the airspace and for other pollutants to get knocked off of the sources (people) carrying them and into the air. Having monitors that detect higher levels of CO_2 will allow for either automatic or manual increases in the air exchange rate, so that oxygen levels increase and other pollutant levels decrease if the air exchange is made with a less polluted air source. An automatic system is referred to as a *demand-controlled ventilation* (DCV) system. Usually the outdoor air is less polluted and contains more oxygen, but air exchanges through filters can also be designed if there are concerns with the outside air. Usually, at a minimum, mechanical ventilation systems have some sort of a particulate matter filter.

As mentioned previously, one important parameter in the Indoor Environmental Quality category is the location of the breathing zone. This is used for both EQc1 and EQc2 for the air quality testing options pertaining to the location of the marker compound monitors. A schematic of a breathing zone as designated for LEED air quality purposes can be seen in Fig. 6.0.1. Here the typical breathing zone is a 3-ft-high rectangle located 3 ft off the floor, and 2 ft from any wall or air handling unit.

Special Circumstances and Exemplary Performance
There is no EP point related to EQc1.

6.2 EQ Credit 2: *Increased Ventilation*
The intention of this credit is to improve air quality by providing air exchange rates that are higher than the minimums established in the prerequisite. However, increased air exchange rates with outdoor rates may have an impact on energy use, so additional strategies such as countercurrent flows for preheating or cooling may be needed to improve both air quality and energy efficiency. Ventilation strategies should be carefully evaluated in conjunction with energy efficiencies and impacts to maintain an optimum balance between these two goals.

USGBC Rating System
LEED-NC 2.2 lists the Intent, Requirements, and Potential Technologies and Strategies for EQc2 as follows:

Intent
Provide additional outdoor air ventilation to improve indoor air quality for improved occupant comfort, well-being and productivity.

Requirements
FOR MECHANICALLY VENTILATED SPACES

- Increase breathing zone outdoor air ventilation rates to all occupied spaces by at least 30% above the minimum rates required by ASHRAE Standard 62.1-2004 as determined by EQ Prerequisite 1.

FOR NATURALLY VENTILATED SPACES
Design natural ventilation systems for occupied spaces to meet the recommendations set forth in the Carbon Trust "Good Practice Guide 237" [1999]. Determine that natural ventilation is an effective strategy for the project by following the flow diagram process shown in Figure 1.18 of the Chartered Institution of Building Services Engineers (CIBSE) Applications Manual 10: 2005, Natural ventilation in non-domestic buildings.
AND

- Use diagrams and calculations to show that the design of the natural ventilation systems meets the recommendations set forth in the CIBSE Applications Manual 10: 2005, Natural ventilation in non-domestic buildings.

OR

- Use a macroscopic, multi-zone, analytic model to predict that room-by-room airflows will effectively naturally ventilate, defined as providing the minimum ventilation rates required by ASHRAE 62.1-2004 Chapter 6, for at least 90% of occupied spaces.

Potential Technologies and Strategies

For mechanically ventilated spaces: Use heat recovery, where appropriate, to minimize the additional energy consumption associated with higher ventilation rates. For naturally ventilated spaces: Follow the eight design steps described in the Carbon Trust Good Practice Guide 237 – 1) Develop design requirements, 2) Plan airflow paths, 3) Identify building uses and features that might require special attention, 4) Determine ventilation requirements, 5) Estimate external driving pressures, 6) Select types of ventilation devices, 7) Size ventilation devices, 8) Analyze the design. Use public domain software such as NIST's CONTAM, Multizone Modeling Software, along with LoopDA, Natural Ventilation Sizing Tool, to analytically predict room-by-room airflows.

Calculations and Considerations

Mechanical and Mixed-Mode Ventilation   The required ventilation rate increase of 30 percent for mechanically ventilated spaces (and mixed-mode areas) can be determined by either a modeling method called the *ventilation rate procedure* or a testing method called the *indoor air quality procedure*. These can both be found in the referenced ASHRAE Standard 62.1-2004, Section 6. However, the ventilation rate procedure is the most commonly used and is the prescribed method for EQp1. Alternatively, for low-rise residential buildings ASHRAE 62.2-2004 may be used, as previously mentioned in EQp1. Therefore, for all the mechanically ventilated zones i, the requirements can be summarized by the expanded version of Eq. (6.1.1) as given in Eq. (6.2.1).

$$\text{OAIZ}_i \geq 1.3 \times \text{BZA}_i[(\text{BZOD}_i)(\text{BZPR}_j) + (\text{BZAR})]/(\text{EZ}_i)(\text{EV}) \qquad (6.2.1)$$

An important parameter in the Indoor Environmental Quality category is the location of the breathing zone. This is used for EQc1 for the air quality testing option and was shown previously in Fig. 6.0.1. The option for mechanically ventilated spaces requires that the 30 percent increase in air exchanges be in the breathing zones.

Natural Ventilation   For all naturally ventilated spaces it is necessary to have *both* special designs and a model calculation verification as in the following:

- **Design (comply with all the following)**
  - Per *Carbon Trust Good Practice Guide 237* (1999) eight design steps:
    1. Develop design requirements.
    2. Plan airflow paths.
    3. Identify special attention items.
    4. Determine ventilation requirements.
    5. Estimate external driving pressures.
    6. Select types of ventilation devices.
    7. Size ventilation devices.
    8. Analyze the design.
  - Determine if effective per the flow diagram in Fig. 1.18 of the Chartered Institution of Building Services Engineers (CIBSE) Applications Manual 10: *Natural Ventilation in Non-Domestic Buildings*, 2005.

- **Calculations (use one of the following methods)**
  - CIBSE Applications Manual 10 (2005)
  - Macroscopic, multizone analytic model of room-by-room airspaces for >90 percent of occupied spaces per ASHRAE 62.1- 2004, Chapter 6.

The *Carbon Trust Good Practice Guide 237* (1999) can be accessed at http://www.carbontrust.co.uk/Publications/publicationdetail.htm?productid=GPG237&meta NoCache=1. The requirements for naturally ventilated spaces are summarized in Fig. 6.2.1.

Special Circumstances and Exemplary Performance

At this time, improved indoor air quality is primarily sought by exchanging indoor air with air of better quality from the outside. However, this may not be the best solution in all cases. There are two main concerns that frequently occur. The first is the possibility of poor ambient air quality in some areas during different seasons, and the second is the increased energy use required to condition the additional outside airflow. Therefore, there are also other methods for improved indoor air quality that are being researched and recommended. Many are specific to a certain class of indoor air pollutants such as particulate matter or VOCs and may be important for some applications. These might serve as a special circumstance or an innovative design credit at this time, although future versions of the LEED rating system might add options for special cases. One of the most commonly used indoor air quality mechanisms is filtration, which decreases particulate matter, a pollutant that may carry other contaminants such as microbes or allergens. Improved filtration is partially addressed in the EQ credit subcategories 3 and 5. Some other methods are mentioned in the Discussion and Overview section of this chapter.

There is no EP point associated with EQ credit 2.

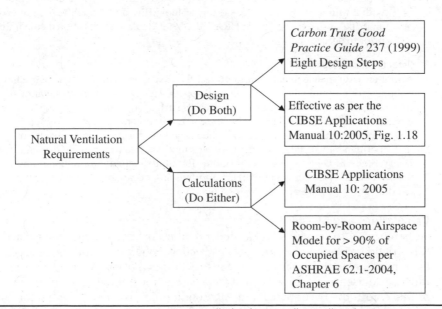

FIGURE 6.2.1 Requirements for increased ventilation in naturally ventilated spaces.

6.3 EQ Credit Subcategory 3: *Construction IAQ Management Plan*

EQ Credit 3.1: *Construction IAQ Management Plan—During Construction*

Many activities during construction can have an impact on the indoor air quality of portions of buildings that are occupied during the construction phase and on the indoor air quality of the entire building after construction. To reduce this, there are a series of recommended steps to be taken during construction that will keep these impacts to a minimum. The steps range from keeping construction areas physically segregated from built areas as much as possible, to ensuring that building materials do not get contaminated with moisture or chemicals prior to or after installation.

USGBC Rating System

LEED-NC 2.2 lists the Intent, Requirements, and Potential Technologies and Strategies for EQc3.1 as follows:

Intent

Reduce indoor air quality problems resulting from the construction/renovation process in order to help sustain the comfort and well-being of construction workers and building occupants.

Requirements

Develop and implement an Indoor Air Quality (IAQ) Management Plan for the construction and pre-occupancy phases of the building as follows:

- During construction meet or exceed the recommended Control Measures of the Sheet Metal and Air Conditioning National Contractors Association (SMACNA) IAQ Guidelines for Occupied Buildings under Construction, 1995, Chapter 3.
- Protect stored on-site or installed absorptive materials from moisture damage.
- If permanently installed air handlers are used during construction, filtration media with a Minimum Efficiency Reporting Value (MERV) of 8 shall be used at each return air grille, as determined by ASHRAE 52.2-1999. Replace all filtration media immediately prior to occupancy.

Potential Technologies and Strategies

Adopt an IAQ management plan to protect the HVAC system during construction, control pollutant sources and interrupt contamination pathways. Sequence the installation of materials to avoid contamination of absorptive materials such as insulation, carpeting, ceiling tile and gypsum wallboard. Coordinate with Indoor Environmental Quality Credits 3.2 and 5 to determine the appropriate specifications and schedules for filtration media. If possible, avoid using permanently installed air handlers for temporary heating/cooling during construction. Consult the LEED-NC v2.2 Reference Guide for more detailed information on how to ensure the well-being of construction workers and building occupants if permanently installed air handlers must be used during construction.

Calculations and Considerations

Adherence to this credit is highly dependent on the construction phase activities, and this should be planned for prior to construction. Scheduling and sequencing are critical, as is management of many daily construction activities. An IAQ management plan must be devised and adhered to that incorporates the following three main requirements of this credit: the Sheet Metal and Air Conditioning National Contractors Association (SMACNA) *IAQ Guidelines for Occupied Buildings under Construction*, 1995, Chap. 3, must be adhered to; absorptive materials must be protected from moisture; and permanent air handlers must have filtration media installed during construction and replaced prior to occupancy. The SMACNA guidelines are further subdivided into five main categories representing the five SMACNA principles. The SMACNA principles shall be met even if the facility is not occupied during construction as applicable.

SMACNA   The five SMACNA principles are

- Protection of permanent HVAC equipment
- Air pollutant source control
- Air pollutant pathway interruption
- Housekeeping to prevent air pollution
- Scheduling to prevent contamination

Protection of permanent HVAC equipment can best be achieved by *not* using permanent HVAC equipment during the construction phase, whenever possible. Temporary heating sources may be a good alternative. In addition, all openings in the system should be sealed, if possible, during construction; and if it is not possible, then filters should be installed and replaced prior to occupancy. Unducted plenums represent other potential sources for contamination into the HVAC system that cannot be easily controlled at point sources. Special consideration should be given to these areas and work scheduled so that the plenums are isolated as quickly as possible with ceilings. Another benefit of protecting HVAC equipment is for continued energy performance. Studies have revealed that fouled HVAC equipment can significantly impact the efficiency of the units.

One additional recommendation for protection of permanent HVAC equipment is not to store construction materials or other potential air pollutant sources in mechanical rooms. These preventive procedures will aid in minimizing particulate matter (PM) contamination of the equipment and ductways. This leads to suggestions for other air pollutant source controls.

The best way to ensure air pollutant source control is to minimize many sources of air pollutants such as toxic finish materials or high-VOC (volatile organic compound or carbon) materials used on the site. However, there are still some materials that probably will be used that might cause a problem if not handled properly. Housekeeping chemicals and used chemical containers should be stored and disposed of responsibly. Idling vehicles and other sources of emissions such as petroleum-fueled construction tools should be reduced if possible and located in areas remote from enclosed areas or accessways to enclosed areas such as HVAC intakes. This leads to suggestions for preventing on-site air pollutants from entering constructed areas.

Air pollutant pathway interruptions can be made by isolating the areas under construction from clean or occupied areas by physical barriers (permanent or temporary) or ventilation and depressurization schemes.

Housekeeping is arguably the most important way to prevent air pollutant contaminants from affecting the facility. Frequent cleaning, vacuuming, and dust prevention measures, such as wetting agents for exposed soils, can prevent much of the PM pollutants from entering the facility and affecting the construction phase workers. Housekeeping also includes properly storing, staging, and protecting absorbent materials so that they do not become moist or absorb unwanted contaminants. Proper housekeeping can be facilitated by scheduling these activities into the construction phase.

Scheduling is a constant theme in the SMACNA principles. Scheduling housekeeping activities into the daily operations of the construction phase is important. Sequencing installation of permanent or temporary barriers between finished and unfinished areas can help in pathway interruption. Scheduling the use of high-VOC or toxic materials or air pollutant emitting equipment prior to installation of absorbent materials or away from enclosed areas can keep the sources of air pollutants down, as can scheduling protection of the HVAC systems and installation of good filtering media.

Moisture Protection   In addition to the SMACNA principles, this credit emphasizes that absorptive material such as sheetrock shall be protected from moisture damage while on-site at all times, from storage through staging, installation, and after installation. Moisture damage can result in other air pollutant problems such as mold after occupancy.

Filtration Media   Finally, this credit requires that any air handlers used during construction have filtering media on each return grill, and this medium shall be replaced prior to occupancy. The filtering media used for construction phase IAQ management should meet minimum standards for filtering of particulate matter. In earlier versions of LEED, those with a *minimum efficiency rating value* (MERV) rating of 13 were recommended, but this credit is now based on a MERV rating of 8 and the higher rating of 13 is part of EQc5. If the project seeks to receive credit for EQc5, then all the HVAC equipment should be sized to accept filters with the MERV rating of 13.

These filters are used to remove particulate matter from the air. The MERV ratings are based on a combination of removal efficiencies for several different size ranges of particulate matter and a maximum pressure drop. Particulate matter is made up of many different materials and can be either liquid or solid. Most of the larger particles are solids such as dust and soot. Particulate matter comes in all shapes and sizes, and they are classified by listing the size of a particle by the diameter of a sphere of the same volume. The USEPA regulates ambient (outside) concentrations of PM in two different size ranges. The two ranges comprise all particulate matter less than or equal to 10 micrometers (10 µm) in equivalent diameter and all particulate matter less than or equal to 2.5 µm in equivalent diameter. These are commonly referred to as PM10 and PM2.5. Also PM2.5 is sometimes referred to as *fines*. Particulate matter in the air larger than 10 µm in diameter is usually entrained particles that will rapidly settle out and can easily be filtered by our nostrils. Since particulate matter is made up of many substances, both organic and inorganic, it may have a wide range of densities. Usually an average specific gravity is estimated as 2, which is in between the value for many liquids and organics which are closer to 1, and the value for minerals, which is closer to 3. Metals usually have a specific gravity greater than 3.

For the MERV ratings for HVAC equipment filtering relating to indoor air, particulate matter is subdivided into three size ranges: 3 to 10 µm, 1 to 3 µm, and 0.3 to 1 µm (large, medium, and small). For higher MERV values, the removal efficiency will be greater. The MERV ratings also address the allowed pressure drop across a filter. Some typical MERV ratings and removal efficiencies are listed in Table 6.3.1. These were taken from ASHRAE 52.2. Note that there is a minimum size given of 0.3 µm. Usually particles that are smaller than this have so little mass that they are unstable as a solid. [Typical molecules range in size from about 74 picometers (pm) for the smaller hydrogen molecule to 50 angstroms (Å) for large molecules. This converts to about 0.000074 to 0.005 µm. Many of the smaller molecules exist

Group Number	MERV Rating	Average Removal Efficiency (%) 0.3 to 1 μm	Average Removal Efficiency (%) 1 to 3 μm	Average Removal Efficiency (%) 3 to 10 μm	Minimum Final Resistance (in. water gage)
1	1–4	—	—	<20	0.3
2	5	—	—	20–34.9	0.6
	6	—	—	35–49.9	
	7	—	—	50–69.9	
	8	—	—	70–84.9	
3	9	—	<50	≥85	1.0
	10	—	50–64.9	≥85	
	11	—	65–79.7	≥85	
	12	—	80–89.9	≥90	
4	13	<75	≥90	≥90	1.4
	14	75–84.9	≥90	≥90	
	15	85–94.9	≥90	≥90	
	16	≥95	≥95	≥95	

Source: American Society of Heating, Refrigerating and Air-Conditioning Engineers, Inc., *www.ashrae.org*.

TABLE 6.3.1   MERV Ratings from ASHRAE 52.2-1999: *Method of Testing General Ventilation Air-Cleaning Devices for Removal Efficiency by Particle Size.*

as gases, while most of the larger molecules are liquids or solids at room temperature. There are some larger polymers and supermolecules, but they are not a common ambient air pollution concern. For the sake of a physical comparison, note that a typical textbook may have a page thickness of about 70 μm.]

Special Circumstances and Exemplary Performance
There is no EP point for this credit.

EQ Credit 3.2: *Construction IAQ Management Plan—Before Occupancy*
EQ credit 3.1 lists a series of measures that can be taken during the construction phase to reduce the potential for air pollution from construction affecting both the workers and the occupants after construction. Then EQc3.2 lists a series of additional measures which can be taken just before occupancy to further ensure good air quality.

USGBC Rating System
LEED-NC 2.2 lists the Intent, Requirements, and Potential Technologies and Strategies for EQc3.2 as follows:

Intent
Reduce indoor air quality problems resulting from the construction/renovation process in order to help sustain the comfort and well-being of construction workers and building occupants.

Requirements

Develop and implement an Indoor Air Quality (IAQ) Management Plan for the pre-occupancy phase as follows:

OPTION 1 — Flush-Out

- After construction ends, prior to occupancy and with all interior finishes installed, perform a building flush-out by supplying a total air volume of 14,000 cu. ft. of outdoor air per sq. ft. of floor area while maintaining an internal temperature of at least 60 degrees F and relative humidity no higher than 60%.

OR

- If occupancy is desired prior to completion of the flush-out, the space may be occupied following delivery of a minimum of 3,500 cu. ft. of outdoor air per sq. ft. of floor area to the space. Once a space is occupied, it shall be ventilated at a minimum rate of 0.30 cfm/sq. ft. of outside air or the design minimum outside air rate determined in EQ Prerequisite 1, whichever is greater. During each day of the flush-out period, ventilation shall begin a minimum of three hours prior to occupancy and continue during occupancy. These conditions shall be maintained until a total of 14,000 cu. ft./sq. ft. of outside air has been delivered to the space.

OR

OPTION 2 — Air Testing

- Conduct baseline IAQ testing, after construction ends and prior to occupancy, using testing protocols consistent with the United States Environmental Protection Agency Compendium of Methods for the Determination of Air Pollutants in Indoor Air and as additionally detailed in the Reference Guide.

- Demonstrate that the contaminant maximum concentrations listed 'below' are not exceeded.

CONTAMINANT	MAXIMUM CONCENTRATION
Formaldehyde	50 parts per billion
Particulates (PM10)	50 micrograms per cubic meter
Total Volatile Organic Compounds (TVOC)	500 micrograms per cubic meter
4-Phenylcyclohexene (4-PCH)*	6.5 micrograms per cubic meter
Carbon Monoxide (CO)	9 parts per million and no greater than 2 parts per million above outdoor levels

*This test is only required if carpets and fabrics with styrene butadiene rubber (SBR) latex backing material are installed as part of the base building systems.

- For each sampling point where the maximum concentration limits are exceeded conduct additional flush-out with outside air and retest the specific parameter(s) exceeded to indicate the requirements are achieved. Repeat procedure until all requirements have been met. When retesting non-complying building areas, take samples from the same locations as in the first test.

- The air sample testing shall be conducted as follows:
  1) All measurements shall be conducted prior to occupancy, but during normal occupied hours, and with the building ventilation system starting at the normal daily start time and operated at the minimum outside air flow rate for the occupied mode throughout the duration of the air testing.

2) The building shall have all interior finishes installed, including but not limited to millwork, doors, paint, carpet and acoustic tiles. Non-fixed furnishings such as workstations and partitions are encouraged, but not required, to be in place for the testing.

3) The number of sampling locations will vary depending upon the size of the building and number of ventilation systems. For each portion of the building served by a separate ventilation system, the number of sampling points shall not be less than one per 25,000 sq. ft., or for each contiguous floor area, whichever is larger, and include areas with the least ventilation and greatest presumed source strength.

4) Air samples shall be collected between 3 feet and 6 feet from the floor to represent the breathing zone of occupants, and over a minimum 4-hour period.

Potential Technologies and Strategies

Prior to occupancy, perform a building flush-out or test the air contaminant levels in the building. The flush-out is often used where occupancy is not required immediately upon substantial completion of construction. IAQ testing can minimize schedule impacts but may be more costly. Coordinate with Indoor Environmental Quality Credits 3.1 and 5 to determine the appropriate specifications and schedules for filtration media.

Calculations and Considerations

The intention of this credit is to ensure that a building has good air quality after construction. The end of construction is defined as the completion of the stage when all finishes are applied to the base building components. Furniture and furnishings are not a part of the base building. Also, this credit does not apply to the core and shell portion of LEED until the interior finishes are complete.

There are two main methods of compliance: either there is a substantial flush-out of the building and HVAC systems in either of two prescribed manners, or the team proves that several marker air pollutants are below a certain minimum level after some flush-out. Since it may be costly or unrealistic to delay occupancy after construction is complete for many projects, the second flush-out option allows for the building to be occupied earlier. Also, if there are separate HVAC systems in distinct areas of the building, areas which are completed can be occupied prior to completion of all the areas, if the compliant areas are also isolated and protected by the SMACNA principles as outlined in EQc3.1. Just as in EQc3.1, the filters should be replaced following any flush-out, and the filter media should meet the minimum requirements of the design or MERV 13 if EQc5 is also being sought. There are standards set for the temperature and relative humidity (RH) of the outside air used for flush-out to prevent excessive moisture in the building and also to promote evaporation of the applicable volatile compounds. Volatility increases rapidly with temperature for many organic compounds that are intended to be flushed out of the building. Therefore, to facilitate a rapid flush-out, the temperature of the air used in the flush-out should not be too low.

Prescribed Flush-Out Options   The first flush-out option requires a substantial amount of air and may take a few weeks. The second flush-out option requires that a portion (at least 25 percent) of the flush-out be completed prior to occupancy and then can continue at a minimum rate during occupancy until complete.

Air Quality Option   The air quality option allows for occupancy as soon as it can be proved that the air quality adheres to a minimum standard. Usually improved air quality is attained by flushing out the spaces. Attaining these minimum standards may require several more days of flush-out if areas are found not in compliance. There are five groups of compounds for this air quality option: formaldehyde, particulate matter of 10-μm diameter or less (PM10), total volatile organic compounds (TVOC), carbon monoxide (CO), and 4-phenylcyclohexene (4-PCH). (The 4-PCH needs only be measured if there are carpets or fabrics installed as part of the base building with styrene butadiene rubber latex backing material.) The maximum allowed concentrations as measured per the prescribed methods are listed in the Requirements section.

Formaldehyde has the chemical formula H_2CO and goes by many other common names, such as *methyl aldehyde*. It is one of the most commonly used chemicals for manufacturing and is commonly used as a preservative for many everyday goods. In buildings, it is emitted from many pressed wood building materials, such as particleboard, as an off-gas from the resins used. Formaldehyde is also one of the simplest organic compounds and is a by-product of many chemical reactions including combustion, such as in cigarette smoking or fuel burning or even from reactions within our own bodies. Therefore it is found throughout the spaces we occupy at low levels. Formaldehyde is a good compound to test, because its presence at higher levels may mean that either there is still substantial off-gassing of many construction materials and/or there are unacceptable levels of combustion emissions entering the facility. A level of 0.1 ppm is considered to be the level above which there may be eye, throat, or lung irritation. Common levels in homes are usually well below this threshold for human irritation. However, homes with certain insulations or a lot of pressed wood products, and poor ventilation, may reach levels of up to 0.3 ppm. The U.S. Department of Labor Occupational Safety and Health Administration (OSHA) has set a permissible exposure limit (PEL) for formaldehyde of 0.75 ppm [time-weighted average (TWA) in an 8-h workday] and a short-term exposure limit (STEL) of 2 ppm for 15 min as per 29 CFR 1910.1048. The U.S. Department of Housing and Urban Development (HUD) has established a maximum level of 0.4 ppm for mobile homes. The standard in this credit is for formaldehyde to be less than 50 parts per billion (ppb) or, equivalently, 0.05 ppm.

TVOC is also commonly referred to as VOC and stands for total volatile organic compounds (or carbons). It consists of many hundreds of organic compounds that are volatilized. Organic compounds are defined as compounds which contain carbon excluding the inorganic carbon compounds such as carbon monoxide, carbon dioxide, carbonic acid, metallic carbides, and carbonates. The term *VOCs* usually includes all the volatilized organic compounds except for a few exclusions due to low reactivity of these specific chemicals. Emissions are regulated for VOCs by the USEPA because many VOCs participate in atmospheric photochemical reactions, such as the smog reaction which causes ozone to develop in the troposphere during the day. However, there are no ambient air quality standards for VOCs as a total group. Some of the VOCs are considered to be carcinogenic and may have individual concentrations set, such as formaldehyde and benzene. Many VOCs are natural, such as limonene emitted by citrus fruits and many emitted by pine trees or cut grass, which give those distinctive smells to the air. VOCs can be by-products of combustion and represent much of the evaporative portion of paints, glues, and other surface-applied products. Since it is not reasonable to evaluate each and every VOC, they are regulated as a group. In addition, their additive impact may be greater than the individual compound concentrations, so minimizing the concentration of VOCs as a group is also helpful to health and property. Since the VOCs contain a range of many compounds of various molecular weights, the limits given are in the units of mass per volume of air instead of ppm (parts per million molar or volumetric).

PM10 has been described in detail earlier in EQc3.1. Particulate matter contains a range of many compounds, and these are in either solid or liquid form. Since particular matter is not separate gaseous compounds, the limits given are in the units of mass per volume of air instead of ppm. Unacceptable levels of PM10 may mean that there are still many leftover sources from the construction phase in the building or HVAC system and/or that there are pathways from other sources into the occupied areas.

4-Phenylcyclohexene (4-PCH) is applicable only if certain carpet and fabric backing binders are used. Its odor is commonly referred to as "new carpet smell." Each molecule of 4-PCH is made up of 12 carbon atoms and 14 hydrogen atoms.

Carbon monoxide is a commonly known air pollutant that, in high enough quantities, can cause death from asphyxiation. It is a product of incomplete combustion from any carbonaceous fuel source. There are two indoor limits given. The first is a maximum in any case, and the second is a limit for the value above the outdoor levels. Many buildings are located in urban areas where motor vehicle emissions and other emissions from combustion can give an elevated outdoor CO level. In all cases, the maximum value inside should not be more than 2 ppm above the outside. As per the EPA, the national ambient air quality standard (NAAQS) for CO is 9 ppm 8-h nonoverlapping average not to be exceeded more than once per year. (There are some areas in California, Montana, Nevada, Oregon, and Texas which are currently in nonattainment of the standard.)

As outlined in the LEED-NC 2.2 Reference Guide, the air quality testing should be performed in the breathing zone (see Fig. 6.0.1) during typical occupancy hours and with the minimum air rates that will occur during occupancy, in addition to several other detailed prescriptive requirements. In this way, the tests will measure what is expected to be the worst background conditions related to the facility but not directly related to human activity in the building. If tests fail, then after corrective action, new tests need to be performed in the same locations.

Special Circumstances and Exemplary Performance

It may be difficult to fully understand the varied units and multiple requirements in this credit, and for other indoor air quality requirements for green facilities. Therefore, some additional information about air pollution calculations and conversions is given at the end of this chapter, and a listing of the units used can be found in App. C.

Several criteria should be considered in the design and construction phase if this credit is sought:

- The flush-out flow rates should be considered, and if the permanent HVAC equipment is not expected to accommodate these flush-out rates in a timely fashion, then some additional temporary units may be needed and/or openings, such as windows, positioned to facilitate the desired level of air exchange.

- If a project is done in phases, then the flush-outs may be phased, but all the SNACNA principles must be adhered to between each phase.

- As mentioned previously, this credit is not directly applicable to core and shell projects until the interiors are complete.

- Chemicals are often used prior to occupancy to prepare and "clean" a facility. If high-VOC chemicals are used, then these may alter the results. They should be avoided.

- Areas designated for the air quality sampling should be representative of areas with the least effective ventilation (worst-case scenarios).

- It may seem difficult to meet the air quality requirements for CO and perhaps PM10 if the facility is built in an area where the outside ambient air has high concentrations, especially since the testing is usually done during occupancy hours that are usually the worst hours for ambient air concentrations. However, there are many ways in which this problem can be addressed. Some of the poor air pollutant concentrations are close to sources such as roadways in urban areas. If the intakes for the HVAC systems or natural ventilation openings are located away from the roadways and potential sources, then the air used for the facility may be adequate to meet good indoor air quality standards.

There is no EP point for EQ credit 3.2.

6.4 EQ Credit Subcategory 4: *Low-Emitting Materials*

The intention of this credit subcategory is to reduce surface coverings or coatings in the interior spaces that might release VOCs into the indoor airspace. In addition to overall VOC concentrations, a few specific organic compounds that are irritable to humans at certain concentrations and some threshold levels are considered. Figure 6.4.1 shows a reception area that has been built with many surfaces that have reduced VOC emissions as compared to other older products.

FIGURE 6.4.1 Various examples of low VOC-emitting interior surfaces. (*Photograph taken at the Sustainable Interiors ribbon cutting at the Strom Thurmond Building on Fort Jackson, Columbia, S.C., June 2007. The paints were provided by Sherwin-Williams Company and Duron Paints & Wallcoverings. Also shown are carpets and flooring provided by Shaw Industries Group, Lees Carpets and Interface, Inc. These flooring materials may also count toward several MR credits.*)

There are four credits in this subcategory. The four credits (EQc4.1, EQc4.2, EQc4.3, and EQc4.4) cover the following potential indoor sources in the base building of volatile organic compounds:

- Adhesives and sealants
- Paints and coatings
- Carpet systems
- Composite wood and agrifiber products

All four credits in the Low-Emitting Materials subcategory have the same intent as given in the LEED-NC 2.2 Reference Guide.

EQ Credit 4.1: *Low-Emitting Materials—Adhesives and Sealants*

USGBC Rating System
LEED-NC 2.2 lists the Intent, Requirements, and Potential Technologies and Strategies for EQc4.1 as follows:

Intent
Reduce the quantity of indoor air contaminants that are odorous, irritating and/or harmful to the comfort and well-being of installers and occupants.

Requirements
All adhesives and sealants used on the interior of the building (defined as inside of the weatherproofing system and applied on-site) shall comply with the requirements of the following reference standards:

- Adhesives, Sealants and Sealant Primers: South Coast Air Quality Management District (SCAQMD) Rule #1168. VOC limits are listed in Table 6.4.1 and correspond to an effective date of July 1, 2005 and rule amendment date of January 7, 2005.
- Aerosol Adhesives: Green Seal Standard for Commercial Adhesives GS-36 requirements in effect on October 19, 2000 (see Table 6.4.2).

Potential Technologies and Strategies
Specify low-VOC materials in construction documents. Ensure that VOC limits are clearly stated in each section of the specifications where adhesives and sealants are addressed. Common products to evaluate include: general construction adhesives, flooring adhesives, fire-stopping sealants, caulking, duct sealants, plumbing adhesives, and cove base adhesives.

EQ Credit 4.2: *Low-Emitting Materials—Paints and Coatings*

USGBC Rating System
LEED-NC 2.2 lists the Intent, Requirements, and Potential Technologies and Strategies for EQc4.2 as follows:

Intent
Reduce the quantity of indoor air contaminants that are odorous, irritating and/or harmful to the comfort and well-being of installers and occupants.

Requirements
Paints and coatings used on the interior of the building (defined as inside of the weatherproofing system and applied on-site) shall comply with the following criteria:

ADHESIVES					
Architectural Applications	**VOC Limit***	**Specialty Applications**	**VOC Limit***	**Substrate Specific Applications**	**VOC Limit***
Indoor carpet adhesives	50	PVC welding	510	Metal to metal	30
Carpet pad adhesives	50	CPVC welding	490	Plastic foams	50
Wood flooring adhesives	100	ABS welding	325	Porous material (except wood)	50
Rubber floor adhesives	60	Plastic cement welding	250	Wood	30
Subfloor adhesives	50	Adhesive primer for plastic	550	Fiberglass	80
Ceramic tile adhesives	65	Contact adhesive	80		
VCT and asphalt adhesives	50	Special purpose contact adhesive	250		
Drywall and panel adhesives	50	Structural wood member adhesive	140		
Cove base adhesives	50	Sheet applied rubber lining operations	850		
Multipurpose construction adhesives	70	Top and trim adhesive	250		
Structural glazing adhesives	100				

Sealants	**VOC Limit***	**Sealant Primers**	**VOC Limit***
Architectural	250	Architectural nonporous	250
Nonmembrane roof	300	Architectural porous	775
Roadway	250	Other	750
Single-ply roof membrane	450		
Other	420		

* VOC limits are given in grams of VOC to liter of the adhesive not including the water component of the adhesive or other exempt components ($g_{voc}/L_{adhesive\ less\ water}$).

TABLE 6.4.1 SCAQMD VOC Limits for Adhesives, Sealants, and Sealant Primers

Adhesive	% VOC by Weight (less water)
General purpose mist spray	65
General purpose web spray	55
Special purpose aerosol adhesives (all types)	70

*Aerosol adhesives: Green Seal Standard for Commercial Adhesives GS-36 requirements in effect on October 19, 2000.

TABLE 6.4.2   Green Seal VOC Limits for Aerosol Adhesives

- Architectural paints, coatings and primers applied to interior walls and ceilings: Do not exceed the VOC content limits established in Green Seal Standard GS-11, Paints, First Edition, May 20, 1993.
    - Flats: 50 g/L (grams of VOCs per liter of product minus water)
    - Non-Flats: 150 g/L (grams of VOCs per liter of product minus water)
    - Primers: Primers must meet the VOC limit for non-flat paints
- Anti-corrosive and anti-rust paints applied to interior ferrous metal substrates: Do not exceed the VOC content limit of 250 g/L (grams of VOCs per liter of product minus water) established in Green Seal Standard GC-03, Anti-Corrosive Paints, Second Edition, January 7, 1997.
- Clear wood finishes, floor coatings, stains, and shellacs applied to interior elements: Do not exceed the VOC content limits established in South Coast Air Quality Management District (SCAQMD) Rule 1113, Architectural Coatings, rules in effect on January 1, 2004. The units are in grams of VOCs per liter of product minus water. The following list of SCAQMD VOC limits are examples. Refer to the standards for complete details.
    - Clear wood finishes: varnish 350 g/L; lacquer 550 g/L
    - Floor coatings: 100 g/L
    - Sealers: waterproofing sealers 250 g/L; sanding sealers 275 g/L; all other sealers 200 g/L
    - Shellacs: Clear 730 g/L; pigmented 550 g/L
    - Stains: 250 g/L

Potential Technologies and Strategies

Specify low-VOC paints and coatings in construction documents. Ensure that VOC limits are clearly stated in each section of the specifications where paints and coatings are addressed. Track the VOC content of all interior paints and coatings during construction.

EQ Credit 4.3: *Low-Emitting Materials—Carpet Systems*

USGBC Rating System

LEED-NC 2.2 lists the Intent, Requirements, and Potential Technologies and Strategies for EQc4.3 as follows:

Intent

Reduce the quantity of indoor air contaminants that are odorous, irritating and/or harmful to the comfort and well-being of installers and occupants.

Requirements

All carpet installed in the building interior shall meet the testing and product requirements of the Carpet and Rug Institute's Green Label Plus program.

All carpet cushion installed in the building interior shall meet the requirements of the Carpet and Rug Institute Green Label program.

All carpet adhesive shall meet the requirements of EQ Credit 4.1: VOC limit of 50 g/L (the units are in grams of VOCs per liter of product minus water).

Potential Technologies and Strategies

Clearly specify requirements for product testing and/or certification in the construction documents. Select products either that are certified under the Green Label Plus program or for which testing has been done by qualified independent laboratories in accordance with the appropriate requirements. The Green Label Plus program for carpets and its associated VOC emission criteria in micrograms per square meter per hour, along with information on testing method and sample collection developed by the Carpet and Rug Institute (CRI) in coordination with California's Sustainable Building Task Force and the California Department of Health Services (DHS), are described in Section 9, Acceptable Emissions Testing for Carpet, DHS Standard Practice CA/DHS/EHLB/R-174, dated 07/15/04. This document is available at http://www.cal-iaq.org/VOC/Section01350_7_15_2004_FINAL_PLUS_ADDENDUM-2004-01.pdf . [It is also published as Section 01350 Section 9 (dated 2004)] by the Collaborative for High Performance Schools (www.chps.net).]

EQ Credit 4.4: *Low-Emitting Materials—Composite Wood and Agrifiber Products*

USGBC Rating System

LEED NC 2.2 lists the Intent, Requirements, and Potential Technologies and Strategies for EQc4.4 as follows:

Intent

Reduce the quantity of indoor air contaminants that are odorous, irritating and/or harmful to the comfort and well-being of installers and occupants.

Requirements

Composite wood and agrifiber products used on the interior of the building (defined as inside of the weatherproofing system) shall contain no added urea-formaldehyde resins. Laminating adhesives used to fabricate on-site and shop-applied composite wood and agrifiber assemblies shall contain no added urea-formaldehyde resins. Composite wood and agrifiber products are defined as: particleboard, medium density fiberboard (MDF), plywood, wheatboard, strawboard, panel substrates and door cores. Furniture and equipment are not considered base building elements and are not included.

Potential Technologies and Strategies

Specify wood and agrifiber products that contain no added urea-formaldehyde resins. Specify laminating adhesives for field and shop applied assemblies that contain no added urea-formaldehyde resins.

Calculations and Considerations EQ Credit Subcategory 4: *Low-Emitting Materials*

The various referenced rules and organization standards can be found at the following websites:

- The SCAQMD Rule # 1168 can be found at http://www.aqmd.gov/rules/reg/reg11/r1168.pdf. The Rule 102 which gives the definition of exempt compounds and other definitions in Rule #1168 can be downloaded from http://www.aqmd.gov/rules/reg/reg01/r102.pdf. The exempt compounds are a group of mainly hydrochlorofluorocarbons (HCFCs), hydrofluorocarbons (HFCs), and chlorofluorocarbons (CFCs). They are considered to be less reactive than most other VOCs with respect to the production of tropospheric ozone, although many are now banned due to their stratospheric ozone depletion potential. (Refer to EAp3 and EAc4 for more information on the ozone depletion potential.)

- *Green Seal Standard, GS36, Commercial Adhesives*, October 19, 2000, can be found at http://www.greenseal.org/certification/standards/commercialadhesives.cfm.

- *Green Seal Standard, GS-11, Paints*, 1st ed., May 20, 1993, can be found at http://www.greenseal.org/certification/standards/paints.cfm.

- *Green Seal Standard, GC-03, Anti-Corrosive Paints*, 2d ed., January 7, 1997, can be found at http://www.greenseal.org/certification/standards/anti-corrosivepaints.cfm.

- The SCAQMD *Rule 1113, Architectural Coatings*, as amended beyond the cited date to June 9, 2006, can be found at http://www.aqmd.gov/rules/reg/reg11/r1113.pdf.

- Information about the Carpet and Rug Institute's (CRI's) Green Label and Green Label Plus Programs can be found at http://www.carpet-rug.org/.

Note that the carpet in EQc4.3 must meet the CRI Green Label Plus requirements, and the carpet cushion must meet the CRI Green Label requirements. The cushion requirement is based on a test of manufacturers' products for emissions of certain VOCs, which are updated on an annual basis. The compounds and the maximum emission factors as given on the CRI Green Label website are listed in Table 6.4.3.

The Green Label Plus requirement in EQc4.3 is for carpets. This program is listed on the Green Seal website and is a certification that carpets go through a testing for the release of total VOCs (TVOCs) and 13 individual chemicals. These chemicals are

- Acetaldehyde
- Benzene
- Caprolactam
- 2-Ethylhexanoic acid
- Formaldehyde
- 1-Methyl-2-pyrrolidinone
- Naphthalene
- Nonanal

VOC	Maximum Emission Factor [mg/(m²·h)]	
	Carpet Cushion	**Carpet**
TVOC (total volatile organic compounds)	1.00	0.5
BHT (butylated hydroxytoluene)	0.30	—
Formaldehyde	0.05	0.05*
4-PC (4-phenylcyclohexene)	0.05	0.05
Styrene	—	0.4

*To prove that none is used.

TABLE 6.4.3 Maximum Emission Factors for VOCs from Carpet Cushions and from Carpets per CRI Green Label Website April 11, 2007

- Octanal
- 4-Phenylcyclohexene
- Styrene
- Toluene
- Vinyl acetate

The specific emissions for each are not given on the website, but the TVOCs and the three individual ones that were previously regulated under the Green Label program are given with their maximum emission levels for the Green Label program in Table 6.4.3.

EQc4.4 refers to a resin used in many products and formerly used in a foam insulation (UFFI) called urea-formaldehyde. Urea-formaldehyde is made in a chemical reaction involving urea [chemical formula CON_2H_4 or $(NH_2)_2CO$] and formaldehyde (H_2CO). Unfortunately, products made with it tend to emit formaldehyde, and their use is discouraged. The requirement states that urea-formaldehyde not be put in any material used for the base building, either made off-site or installed on-site. This requirement is not for furniture or equipment used by the occupants, which perhaps leaves room for another potential innovative credit for these types of materials.

Note again that many of the maximum concentrations given for the products used for surface treatment in this subcategory list these concentrations in milligrams of VOC per liter of the product less the water. In other words, if you purchased a gallon of the paint or adhesive or other listed type product and it contained 50 percent by volume water, then the VOC concentration used to compare to the standard would be double what the actual VOC concentration is in the can. This way, a product cannot be simply watered down to meet the criteria.

Special Circumstances and Exemplary Performance EQ Credit Subcategory 4:
Low-Emitting Materials

It is really important that all the requirements and material specifications needed for any of the EQc4 subcategory credits be included in all portions of the construction documents and specifications. It is also essential that information on all the adhesives, paints, and other VOC-containing products used on the site be carefully documented for the project. This may include cut sheets or manufacturers' specifications with the appropriate certification designations, MSDS (Material Safety Data Sheets), and test reports as applicable. Originally, this may have been a large change for construction projects. However, as materials adhering to these requirements and as certified by the organizations become more common, it may be easier for the contractors to maintain the appropriate records without additional effort. If these products are used instead of more polluting products as were previously the standard for a company, then they will become more cost-effective and will also have a positive impact on the environment on all the facilities where they are used, even those not going for green certification.

This entire subcategory has an alternative compliance path referred to as the VOC budget. There may be instances when there are no appropriate low-VOC options, and the team is given the opportunity to do a VOC budget on all the potential sources listed to prove that the overall emissions rate is lower than in a facility that otherwise complied. This budget approach is good for another reason. The first two credits require low-emitting materials, but they do not state a limit on the amounts of these used. A facility that uses very little of these materials, even if the emissions per liter are high, may have a lower total rate of VOC emissions than a facility that uses a lot of low-VOC products.

The VOC budget option allows for a better model of low-VOC product using facilities, but it is not as simple to model or comply with as the prescriptive requirements. This alternative also allows for compliance when there are some mistakes made during the construction phase, such as the inadvertent use of a paint product that is not on the approved list by a subcontractor. Actions that compensate for these mistakes will help attain good indoor air quality and are a way to attain these credits.

There are no EP points for the EQ Credit Subcategory 4: *Low-Emitting Materials.*

6.5 EQ Credit 5: *Indoor Chemical and Pollutant Source Control*

The intention of EQc5 is to keep potential air pollutants from entering occupied portions of the building after occupancy. Three strategies are employed: preventing dust and grime from entering via pedestrian entrances, keeping indoor chemical sources segregated from occupied areas by physical means, and keeping dust and grime out of the HVAC system by using very high-efficiency filter media.

USGBC Rating System

LEED-NC 2.2 lists the Intent, Requirements, and Potential Technologies and Strategies for EQc5 as follows:

Intent
Minimize exposure of building occupants to potentially hazardous particulates and chemical pollutants.

Requirements
Design to minimize and control pollutant entry into buildings and later cross-contamination of regularly occupied areas:

- Employ permanent entryway systems at least six feet long in the primary direction of travel to capture dirt and particulates from entering the building at entryways that are directly connected to the outdoors and that serve as regular entry points for building users. Acceptable entryway systems include permanently installed grates, grilles, or slotted systems that allow for cleaning underneath. Roll-out mats are only acceptable when maintained on a weekly basis by a contracted service organization.

- Where hazardous gases or chemicals may be present or used (including garages, housekeeping/laundry areas and copying/printing rooms), exhaust each space sufficiently to create negative pressure with respect to adjacent spaces with the doors to the room closed. For each of these spaces, provide self-closing doors and deck to deck partitions or a hard lid ceiling. The exhaust rate shall be at least 0.50 cfm/sq. ft., with no air recirculation. The pressure differential with the surrounding spaces shall be at least 5 Pa (0.02 inches of water gauge) on average and 1 Pa (0.004 inches of water) at a minimum when the doors to the rooms are closed.

- In mechanically ventilated buildings, provide regularly occupied areas of the building with air filtration media prior to occupancy that provides a Minimum Efficiency Reporting Value (MERV) of 13 or better. Filtration should be applied to process both return and outside air that is to be delivered as supply air.

Potential Technologies and Strategies
Design facility cleaning and maintenance areas with isolated exhaust systems for contaminants. Maintain physical isolation from the rest of the regularly occupied areas of the building. Install permanent architectural entryway systems such as grills or grates to prevent occupant-borne contaminants from entering the building. Install high-level

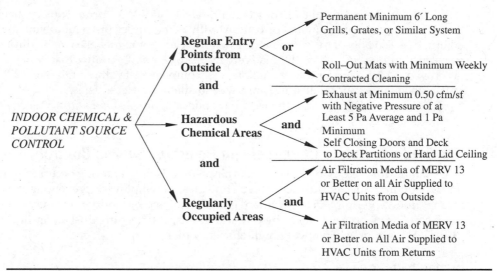

FIGURE 6.5.1 Summary of requirements for EQ Credit 5: *Indoor Chemical and Source Pollutant Control.*

filtration systems in air handling units processing both return air and outside supply air. Ensure that air handling units can accommodate required filter sizes and pressure drops.

Considerations and Exemplary Performance

The prescriptive items in EQc5 are well detailed in the requirements section already noted. They are summarized in Fig. 6.5.1. It should be added that in the design phase, care should be taken to specify HVAC equipment that filter media with a MERV rating of 13 or better can fit, and that these filters are installed and used on the equipment at all times, even prior to occupancy (see EQc3.1 and 3.2).

There is no EP point associated with this credit.

6.6 EQ Credit Subcategory 6: *Controllability of Systems*

The intention of this subcategory is to allow the building occupants greater control over both lighting and thermal comfort. Not only may it aid in indoor environmental quality and comfort for the occupants, but also if used wisely, it may save on energy costs.

EQ Credit 6.1: *Controllability of Systems—Lighting*

This credit allows for greater occupant control over the lighting where one works or performs regular tasks. LEED-NC 2.2 lists the Intent, Requirements, and Potential Technologies and Strategies for EQc6.1 as follows:

Intent

Provide a high level of lighting system control by individual occupants or by specific groups in multi-occupant spaces (i.e. classrooms or conference areas) to promote the productivity, comfort and well-being of building occupants.

Requirements

Provide individual lighting controls for 90% (minimum) of the building occupants to enable adjustments to suit individual task needs and preferences.

AND

Provide lighting system controllability for all shared multi-occupant spaces to enable lighting adjustment that meets group needs and preferences.

Potential Technologies and Strategies

Design the building with occupant controls for lighting. Strategies to consider include lighting controls and task lighting. Integrate lighting systems controllability into the overall lighting design, providing ambient and task lighting while managing the overall energy use of the building.

Calculations and Considerations

This requirement has two parts that must both be met. The first is for the regular occupants of the building at their daily workstations, and the second is for shared spaces that may be used periodically by regular occupants and also by transient occupants. The definitions from the LEED-NC 2.2 Reference Guide for *individual occupant spaces, nonoccupied spaces, nonregularly occupied spaces,* and *shared (group) multioccupant spaces* can be found in App. B.

Individual Control   The first requirement is that 90 percent of the occupants have control over their task lighting at their typical workstation or living space. Task lighting can be permanently wired or can be portable lighting, such as desk lamps. For commercial or institutional uses, typical workstations can be seen as individual offices, desks or tables in cubicles, booths, and other usual locations where an employee performs most of his or her work. To determine the occupancy for these nonresidential uses, it would be most appropriate to start with the full-time employee equivalents (FTEs) as defined in the Sustainable Sites category in Eqs. (2.4.2) through (2.4.3), for the following definitions:

FTE_j Full-time equivalent occupant during shift j

$FTE_{j,i}$ Full-time equivalent occupancy of employee i during shift j

$$FTE_j = \sum FTE_{j,i} \qquad \text{for all employees in shift } j \qquad (2.4.2)$$

$$FTE = \text{Maximum } FTE_j \qquad (2.4.3)$$

For occupancies where shift employees share the same workstations as other shift employees depending on the time, and for uses where there is really only one main "shift," providing individual controls at locations equal to 90 percent of FTE should adequately meet the intent of the requirement. If there are alternative regular uses, where many occupants in separate shifts have different workstation locations, then the definition of the required number of individual lighting controls will need to be analyzed and validated differently. However, for most commercial and institutional uses, the minimum number of individual lighting control locations (ILCLs) can be calculated as

$$ILCL \geq 0.9 \times FTE \qquad (6.6.1)$$

Group Control   The second part of this credit is to make sure that there are lighting controls in group spaces, such as conference rooms and classrooms, so that the groups gathered in these areas have greater control over the lighting as appropriate for their group needs. If some of the lighting comes from daylighting, then glare control and room-darkening shades or features should be available too.

Special Circumstances and Exemplary Performance

The individual lighting loads resulting from implementation of the requirements of this credit should be included in the energy calculations as performed in the Energy and Atmosphere category. In many cases there will be a resultant decrease in overall lighting load, as some areas might not need the maximum lighting during all times, but only when the individual prefers the higher lighting intensities. Since individual control may also result in a decrease in energy efficiency if the controls are left on during unoccupied times, education into energy conservation should be a part of this credit. An example might be posted notices to please turn off the individually controlled lights when exiting the area, or automatic sensors that detect when the room is unoccupied. There is no EP point for this credit.

EQ Credit 6.2: *Controllability of Systems—Thermal Comfort*

This credit allows for increased occupant control over the thermal environment where one works or performs regular tasks. Thermal comfort is based on many environmental factors and personal factors. Four common environmental factors are air temperature, radiant temperature, humidity, and airspeed, with air temperature being the predominant factor. The personal factors include clothing and activity levels. Clothing is dependent on not only the weather or season, but also the standard attire for the occupants. Thermal comfort is addressed in this credit and in both EQc71 and EQc7.2. This credit (6.2) gives flexibility in the thermal control for many individuals or groups so that they can readily change one or more of the environmental factors to suit their preferences. Thermal comfort is considered to be very important, particularly for economic reasons, as comfort is a main factor in productivity and job satisfaction.

USGBC Rating System

LEED-NC 2.2 lists the Intent, Requirements, and Potential Technologies and Strategies for EQc6.2 as follows:

Intent

Provide a high level of thermal comfort system control by individual occupants or by specific groups in multi-occupant spaces (i.e. classrooms or conference areas) to promote the productivity, comfort and well-being of building occupants.

Requirements

Provide individual comfort controls for 50% (minimum) of the building occupants to enable adjustments to suit individual task needs and preferences. Operable windows can be used in lieu of comfort controls for occupants of areas that are 20 feet inside of and 10 feet to either side of the operable part of the window. The areas of operable window must meet the requirements of ASHRAE 62.1-2004 paragraph 5.1 Natural Ventilation.

AND

Provide comfort system controls for all shared multi-occupant spaces to enable adjustments to suit group needs and preferences. Conditions for thermal comfort are described in ASHRAE Standard 55-2004 to include the primary factors of air temperature, radiant temperature, air speed and humidity. Comfort system control for the purposes of this credit is defined as the provision of control over at least one of these primary factors in the occupant's local environment.

Potential Technologies and Strategies

Design the building and systems with comfort controls to allow adjustments to suit individual needs or those of groups in shared spaces. ASHRAE Standard 55-2004 identifies the factors of thermal comfort and a process for developing comfort criteria for building spaces that suit the needs of the occupants involved in their daily activities. Control strategies can be developed to expand on the comfort criteria to allow adjustments to suit individual needs and preferences. These may involve system designs incorporating operable windows, hybrid systems integrating operable windows and mechanical systems, or mechanical systems alone. Individual adjustments may involve individual thermostat controls, local diffusers at floor, desk or overhead levels, or control of individual radiant panels, or other means integrated into the overall building, thermal comfort systems, and energy systems design. In addition, designers should evaluate the closely tied interactions between thermal comfort (as required by ASHRAE Standard 55-2004) and acceptable indoor air quality (as required by ASHRAE Standard 62.1-2004, whether natural or mechanical ventilation).

Calculations and Considerations

Just as in EQc6.1, this requirement has two parts that must both be met. The first is for the regular occupants of the building at their typical workstations, and the second is for shared spaces that may be used periodically by regular occupants and also by transient occupants. The definitions from the LEED-NC 2.2 Reference Guide for *individual occupant spaces, nonoccupied spaces, nonregularly occupied spaces,* and *shared (group) multioccupant spaces* can be found in App. B.

Individual Control   The first requirement is that 50 percent of the occupants have control over their thermal comfort at their typical workstation or living space. Thermal comfort controls usually refer to some form of *conditioning* and are usually for both heating and cooling. This conditioning can be active (mechanical HVAC systems) or passive (natural ventilation). A typical type of control may be individual diffusers, or access to an operable window. (The occupant location should be no more than 10 ft sideways from the edge of an operable window and no more than 20 ft away in front of the window.) As in EQc6.1, for commercial or institutional uses, typical workstations can be seen as individual offices, desks, or tables in cubicles, booths, and other usual locations where an employee performs most of his or her work. To determine the occupancy for these nonresidential uses, it would be most appropriate to again start with the full-time employee equivalents as defined in the Sustainable Sites category in Eqs. (2.4.1) through (2.4.3), as were used in EQ credit 6.1:

For occupancies where shift employees share the same workstations as other shift employees at their different work times, and for uses where there is really only one main shift, providing individual controls at locations equal to 50 percent of FTE should adequately meet the intent of the requirement. If there are alternative regular uses, where many occupants in different shifts work in alternative locations, then the definition of the required number of individual thermal controls will need to be analyzed and validated differently. However, for most commercial and institutional uses, the minimum number of individual comfort control locations (ICCLs) can be calculated as

$$ICCL \geq 0.5 \times FTE \qquad (6.6.2)$$

Group Control   The second part of this credit is to make sure that there are thermal comfort controls in group spaces, such as conference rooms and classrooms, so that the groups gathered in these areas have greater control as appropriate for their group needs.

Special Circumstances and Exemplary Performance   Thermal comfort control as described in ASHRAE Standard 55-2004 should be balanced with the minimum indoor air quality requirements, as required per ASHRAE Standard 62.1-2004. Also, since many of the features for thermal comfort relate to windows and the HVAC systems, this credit has been labeled with an EB from the USGBC. It is more difficult to change in an existing building. There is no exemplary performance point for this credit.

6.7 EQ Credit Subcategory 7: *Thermal Comfort*

EQ Credit 7.1: *Thermal Comfort—Design*

As stated in EQc6.2, thermal comfort is based on many environmental and personal factors, including air temperature, radiant temperature, humidity, airspeed, clothing, and activity levels. Thermal comfort is also addressed in EQc6.2 and EQc7.2. This credit, EQc7.1, deals with designing the HVAC and other energy-related systems with added emphasis on thermal comfort. Figure 6.7.1 depicts a dessicant wheel used in HVAC systems to help reduce outside humidity levels prior to the cooling coils. Humidity is frequently the factor used for thermal comfort control.

USGBC Rating System

LEED-NC 2.2 lists the Intent, Requirements, and Potential Technologies and Strategies for EQc7.1 as follows:

> #### Intent
> Provide a comfortable thermal environment that supports the productivity and well-being of building occupants.

Figure 6.7.1   Trane CDQ (Cool, Dry, Quiet) is a dessicant wheel which is placed in series with the cooling coil and can help provide dry air to a space for better humidity control. (*Photograph Courtesy Trane.*)

Requirements

Design HVAC systems and the building envelope to meet the requirements of ASHRAE Standard 55-2004, Thermal Comfort Conditions for Human Occupancy. Demonstrate design compliance in accordance with the Section 6.1.1 Documentation.

Potential Technologies and Strategies

Establish comfort criteria per ASHRAE Standard 55-2004 that support the desired quality and occupant satisfaction with building performance. Design building envelope and systems with the capability to deliver performance to the comfort criteria under expected environmental and use conditions. Evaluate air temperature, radiant temperature, air speed, and relative humidity in an integrated fashion and coordinate these criteria with EQ Prerequisite 1, EQ Credit 1, and EQ Credit 2.

Calculations and Considerations

Thermal comfort controls usually refer to some form of conditioning and are for both heating and cooling. This conditioning can be active (mechanical HVAC systems) or passive (natural ventilation) or a combination of both (mixed-mode conditioning). Levels of comfort are different for active and passive systems, and ASHRAE Standard 55-2004 separately addresses these two. For the mechanical systems, there is a *predicted mean vote* (PMV) comfort model which deals with both a thermal balance and a thermal comfort. PMV is based on a thermal sensation scale that ranges from cold (−3) to hot (+3) for low airspeeds (< 40 ft/min).

Special Circumstances and Exemplary Performance   As mentioned in EQc6.2, thermal comfort control as described in ASHRAE Standard 55-2004 should be balanced with the minimum indoor air quality requirements, as required per ASHRAE Standard 62.1-2004. In addition, lighting can affect thermal comfort, and its impacts should be considered in the designs. Also, since many of the features for thermal comfort relate to windows and the HVAC systems, this credit has been labeled with an EB from the USGBC. It is more difficult to change in an existing building. There is no exemplary performance point for this credit.

EQ Credit 7.2: *Thermal Comfort—Verification*

Again, thermal comfort is based on many environmental and personal factors, and it is also addressed in EQc6.2 and EQc7.1. This credit, EQc7.2, deals with verifying that the occupants of the building feel that they have an adequate level of thermal comfort.

USGBC Rating System

LEED-NC 2.2 lists the Intent, Requirements, and Potential Technologies and Strategies for EQc7.2 as follows:

Intent

Provide for the assessment of building thermal comfort over time.

Requirements

Agree to implement a thermal comfort survey of building occupants within a period of six to 18 months after occupancy. This survey should collect anonymous responses about thermal comfort in the building including an assessment of overall satisfaction with thermal performance and identification of thermal comfort-related problems. Agree to develop a plan for corrective action if the survey results indicate that more than 20% of occupants are dissatisfied with thermal comfort in the building. This plan should include measurement of relevant environmental variables in problem areas in accordance with ASHRAE Standard 55-2004.

	Very Dissatisfied −3	Dissatisfied −2	Somewhat Dissatisfied −1	Neutral 0	Somewhat Satisfied 1	Satisfied 2	Very Satisfied 3
Are you satisfied with the thermal comfort of your workspace?							

TABLE **6.7.1** Thermal Comfort Questionnaire

Potential Technologies and Strategies
ASHRAE Standard 55-2004 provides guidance for establishing thermal comfort criteria and the documentation and validation of building performance to the criteria. While the standard is not intended for purposes of continuous monitoring and maintenance of the thermal environment, the principles expressed in the standard provide a basis for design of monitoring and corrective action systems.

Calculations and Considerations
This credit is well described in the LEED write-up. The survey should be based on a variation of a seven-point Likert-type scale where the lowest is −3, representing very dissatisfied, and the highest is +3, representing very satisfied. A typical question may look like the one in Table 6.7.1.

Anyone responding with a number less than zero (neutral) should be included in the dissatisfied category. Additional questions should allow the respondents to explain how they are dissatisfied (or satisfied) so that ideas for corrective action can be developed. Since any answer less than neutral is included in the rating, why even have seven distinct rating levels? This is to help in the determination of corrective action. For instance, a person who is only mildly dissatisfied, perhaps on hot summer days, may become satisfied with a simple policy change, such as a more casual clothing requirement during heat waves or the use of a personal fan. Similarly, corrective action for a very dissatisfied individual may require installing additional controls or items in the main HVAC system. Additional questions may be asked about factors such as humidity ratings and lighting levels, etc., to aid in designing means for improved satisfaction with the workspace.

Special Circumstances and Exemplary Performance
As mentioned in EQc6.2 and EQc7.1, many of the features for thermal comfort relate to windows and the HVAC systems, so this credit has been labeled with an EB from the USGBC. It may be more difficult to take corrective action in an existing building. There is no exemplary performance point for this credit.

6.8 EQ Credit Subcategory 8: *Daylighting and Views*
The intentions of these two credits are to allow for more natural lighting during the day, known as daylighting, and for more views of the outside for the occupants in the regularly used areas in a building. Wall surfaces which are part of the inside of an

exterior wall that are less than 2.5 ft from the floor surface are not usually included in view or daylighting calculations. These same wall surfaces from 2.5 ft above the floor to 7.5 ft above the floor are considered to be "view" surfaces and can also count toward daylighting. Any glazed wall or ceiling surface (sky or toplights, light tubes, etc.) above 7.5 ft from the floor are usually designed for daylighting purposes (but not for views).

EQ Credit 8.1: *Daylighting and Views—Daylight 75% of Spaces*

USGBC Rating System

LEED-NC 2.2 lists the Intent, Requirements, and Potential Technologies and Strategies for EQc8.1 as follows:

Intent

Provide for the building occupants a connection between indoor spaces and the outdoors through the introduction of daylight and views into the regularly occupied areas of the building.

Requirements

OPTION 1 — CALCULATION

Achieve a minimum glazing factor of 2% in a minimum of 75% of all regularly occupied areas.

The glazing factor is calculated as follows:

Glazing Factor = Window Geometry Factor × (Window Area [SF]/Floor Area [SF]) × (Actual T_{vis}/ Minimum T_{vis}) × Window Height Factor

OR

OPTION 2 — SIMULATION

Demonstrate, through computer simulation, that a minimum daylight illumination level of 25 footcandles has been achieved in a minimum of 75% of all regularly occupied areas. Modeling must demonstrate 25 horizontal footcandles under clear sky conditions, at noon, on the equinox, at 30 inches above the floor.

OR

OPTION 3 — MEASUREMENT

Demonstrate, through records of indoor light measurements, that a minimum daylight illumination level of 25 footcandles has been achieved in at least 75% of all regularly occupied areas. Measurements must be taken on a 10-foot grid for all occupied spaces and must be recorded on building floor plans. In all cases, only the square footage associated with the portions of rooms or spaces meeting the minimum illumination requirements can be applied towards the 75% of total area calculation required to qualify for this credit. In all cases, provide daylight redirection and/or glare control devices to avoid high-contrast situations that could impede visual tasks. Exceptions for areas where tasks would be hindered by the use of daylight will be considered on their merits.

Potential Technologies and Strategies

Design the building to maximize interior daylighting. Strategies to consider include building orientation, shallow floor plates, increased building perimeter, exterior and interior permanent shading devices, high performance glazing and automatic photocell-based controls. Predict daylight factors via manual calculations or model daylighting strategies with a physical or computer model to assess footcandle levels and daylight factors achieved.

Calculations and Considerations

The definitions for *glazing, glazing factor, daylighting, incident light, nonoccupied spaces, nonregularly occupied spaces, regularly occupied spaces,* and *visible light transmittance* T_{vis} are listed in App. B. The most important definition is that of *regularly occupied areas.* Regularly occupied areas are where most occupants actually work on a regular basis. Maintenance work is not included. In residential facilities, they are the living or family areas. In these areas, having daylighting from more "indirect" sources is promoted.

Figure 6.8.1 depicts some of the typical types of daylighting glazed surfaces and their locations. To obtain this credit, there are restrictions on these various locations for daylighting and views so that glare is controlled. Typical glare control practices are listed in the LEED-NC 2.2 Reference Guide.

There are three options for credit compliance. The first two are by some form of lighting model, and the third is based on actual measurements. The first is based on a glazing factor, which is dependent on the window types and orientations. The second and third options are based on an estimate (model) or measurement, respectively, of horizontal footcandles (illumination) in the regularly occupied areas.

The three options all require having 75 percent of the regularly occupied floor areas in the building meet a minimum daylight illumination level. However, there are some differences in how the 75 percent is determined. The applicable regularly occupied floor areas can be segregated into different modeling units. They can be in modeling units of "rooms" or "areas." For the modeling options, most small rooms are taken as one unit, and large rooms can be segregated into applicable areas. The square footage of all the modeling units that meet the minimum daylight illumination level can be used for the 75 percent determination. For the third option, the one based on measuring illumination levels, the applicable regularly occupied floor areas are segregated into 10-ft by 10-ft grids. All the 100-ft² squares that meet the minimum daylighting levels count toward the 75 percent determination. Only the first model will be explained in greater detail

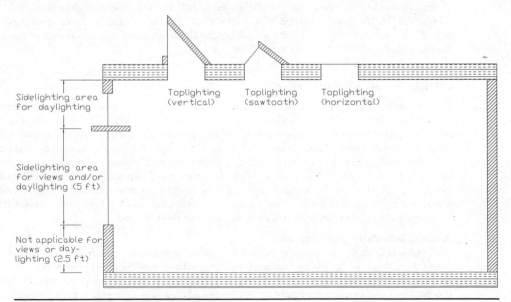

FIGURE 6.8.1 Sidelighting and toplighting.

here, as the second model requires more complex simulations and the third option requires actual measurements after construction.

Option 1   The first option gives a simple method that can be used in the design phase to quickly evaluate options that may or may not meet the credit criteria. As can be seen in the LEED-NC 2.2 Requirements, it is based on determining if the glazing factor for at least 75 percent of the regularly occupied areas meets a minimum value of 0.02 (2 percent). The equation given in the Requirements section is based on the window's location (height), angle (geometry factor), and a minimum visible transmittance T_{vis} based on the window location. Together, these parameters can be combined to give a *combined daylight factor* for each window type i, denoted by CDF_i. Typical minimum values for each combined daylight factor as specified by the parameters set in LEED-NC 2.2 are listed in Table 6.8.1. Note that each window type may vary, and the manufacturers' information should be used for the final calculations.

By using these factors, the area of each window type, and the areas of the regularly occupied spaces, compliance can be calculated using the following definitions:

DC   Daylighting compliance for the project (in percent)

GF_j Glazing factor of regularly occupied room area j

RA_j The floor area of regularly occupied room area j

RAT   The total floor area of all regularly occupied room areas

WA_{ij} The sum of the areas of window type i in room area j

The glazing factor of regularly occupied room area j can be determined as

$$GF_j = \frac{\sum_i (CDF_i)(WA_{ij})}{RA_j} \quad \text{for all window types } i \text{ in regularly occupied area } j \quad (6.8.1)$$

The total floor area of all regularly occupied room areas can be determined as

$$RAT = \sum_j RA_j \quad \text{for all regularly occupied room areas } j \quad (6.8.2)$$

Glazing Type	Combined Daylighting Factor CDF_i
Sidelighting (2.5–7.5 ft above the floor)	0.20
Sidelighting (>7.5 ft above the floor)	0.20
Toplighting, vertical	0.50
Toplighting, sawtooth	0.83
Toplighting, horizontal	1.25

TABLE 6.8.1   Typical Minimum Combined Daylighting Factors (CDF_i)

Daylighting compliance for the project can be calculated as

$$DC = 100 \times \frac{\sum_{j} RA_j}{RAT} \qquad \text{for all regularly occupied areas j where } GF_j \geq 0.02 \qquad (6.8.3)$$

And the project complies with EQc8.1 if

$$DC \geq 75 \qquad (6.8.4)$$

Options 2 and 3 These options are based on a horizontal illumination level. The requisite amount is at least 25 horizontal footcandles by natural lighting in a minimum of 75 percent of the regularly occupied areas. Illumination engineers are experienced in providing *isoluxes* of areas. Isoluxes are plan views giving horizontal illumination contours. They are usually given at the floor level, but for Option 2 the isoluxes would be on the horizontal plane at 30 in above the floor. Isolux stands for "same light," and the term *lux* (abbreviated lx) is also the SI unit for illuminance. One footcandle equals approximately 10.764 lx. Option 3 does not have a height listed, but it is assumed that the measurements are also taken at 30 in above the floor. Both of these are based upon natural light that would be expected or is measured on a clear day at noon on the equinox (either vernal or autumnal).

Special Circumstances and Exemplary Performance
Since the location of windows and other daylighting features is an intricate part of the architecture of a building, this credit has been labeled with an EB from the USGBC. It is more difficult to change in an existing building. Figure 6.8.2 shows an area with natural daylighting at a LEED-NC certified building in Columbia, S.C. Figure 6.8.3a and b show the outdoor and indoor views of skylights at a mall in South Carolina.

FIGURE 6.8.2 Natural lighting at Cox and Dinkins, Engineers and Surveyors, Columbia, S.C. (*Photograph taken June 2007.*)

(a)

(b)

FIGURE 6.8.3   Skylights and celestories at Columbiana Mall in Irmo, S.C. (a) Exterior view; (b) interior view. (*Photograph taken July 2007.*)

EQc8.1 has also been listed as available for an EP credit in LEED-NC 2.2. This additional point can be achieved if at least 95 percent of the regularly occupied spaces are daylighted by the minimum glazing factor as in Option 1, or the minimum 25 fc as simulated or measured for Options 2 and 3, respectively. For Option 1 this corresponds to the following requirement:

$$DC \geq 95 \qquad \text{additional EP point criterion} \qquad (6.8.5)$$

Since Option 3 represents a measurement that can only be taken on two days of the year, and under restrictive weather conditions, it may cause a scheduling conflict for certification if the weather is not favorable. In that case, a description of special circumstances regarding the measurements taken on alternate dates may need to be included in the submittals.

EQ Credit 8.2: *Daylight and Views—Views for 90% of Spaces*

USGBC Rating System
LEED-NC 2.2 lists the Intent, Requirements, and Potential Technologies and Strategies for EQc8.2 as follows:

Intent
Provide for the building occupants a connection between indoor spaces and the outdoors through the introduction of daylight and views into the regularly occupied areas of the building.

Requirements
Achieve direct line of sight to the outdoor environment via vision glazing between 2 feet 6 inches and 7 feet 6 inches above finish floor for building occupants in 90% of all regularly occupied areas. Determine the area with direct line of sight by totaling the regularly occupied square footage that meets the following criteria:

- In plan view, the area is within sight lines drawn from perimeter vision glazing.
- In section view, a direct sight line can be drawn from the area to perimeter vision glazing. Line of sight may be drawn through interior glazing. For private offices, the entire square footage of the office can be counted if 75% or more of the area has direct line of sight to perimeter vision glazing. For multi-occupant spaces, the actual square footage with direct line of sight to perimeter vision glazing is counted.

Potential Technologies and Strategies
Design the space to maximize daylighting and view opportunities. Strategies to consider include lower partition heights, interior shading devices, interior glazing, and automatic photocell-based controls.

Calculations and Considerations
Compliance with EQc8.2 requires a minimum number of regularly occupied areas in the building to have direct lines of sight to exterior views from a specified height above the floor level. This height is usually given as approximately 3.5 ft, which is the average seated eye height of most occupants, but it can vary if there is reasonable justification for a lower or higher height. Since this is a three-dimensional requirement, lines must be drawn on a plan view showing that there are direct sight lines to a view from the horizontal planar area, and these areas must also be drawn in elevation to show that there are sight lines from the eye height that

reach the vision glazing. (The definition of vision glazing as given by LEED-NC 2.2 can be found in the definitions section in App. B and is the glazed exterior wall space between 2.5 and 7.5 ft from the floor as depicted in Fig. 6.8.1.) The sight lines can go through interior glazing.

The 90 percent compliance rate is for all regularly occupied areas, with one main exception. Private offices need only have 75 percent of their floor area within the sight line criterion for the entire office space to count toward the credit. Why? Private offices are usually more spacious (in square feet per occupant) and are occupied by a sole occupant who usually has the opportunity to move his or her desk or work area to a place of choice. Therefore, there need not be views from all corners of the room. Unfortunately, employees in multioccupied office areas, conference rooms, or cafeterias usually do not have a choice where they sit or work. If private offices have a minimum of 75 percent of the floor area compliant, then the entire floor area counts. If not, then only the fraction of the floor area that is compliant counts. This encourages designers and owners to put private offices in the interior spaces with glazing through interior walls to exterior walls. In this way, privacy in private offices can be maintained with shades and blinds, while views are still available and not obstructed from other occupants.

A direct consequence of this requirement is the challenge to provide cubicle areas with views. In response, many companies are designing alternative cubicle designs that have some attributes that allow for views from the work areas. Figure 6.8.4a through d shows some newer cubicle designs.

View compliancy (VC) with EQc8.2 can be determined in the following way using the following variables:

VA_j — Area compliant with both view criteria in regularly occupied room area j (including private offices)

$VAPO_k$ — Area compliant with both view criteria in private office k (the $VAPO_k$ values represent a subset of the VA_j values)

VC — View compliance for the project (in percent)

RA_j — Floor area of regularly occupied room area j (including private offices)

$RAPO_k$ — Floor area of private office k ($RAPO_k$ values represent a subset of RA_j values)

RAT — Total floor area of all regularly occupied room areas

The total floor area of all regularly occupied room areas, as previously listed in EQc8.1, is

$$RAT = \sum_j RA_j \qquad \text{for all regularly occupied room areas } j \qquad (6.8.2)$$

And therefore view compliance (VC) for the project can be calculated as

$$VC = 100 \times \left[\left(\frac{\sum_j VA_j}{RAT} \right) + \left(\frac{\sum_k (RAPO_k - VAPO_k)}{RAT} \right) \right] \qquad \begin{array}{l} \text{for all } k \text{ where } (VAPO_k)/ \\ (RAPO_k) \geq 0.75 \end{array} \qquad (6.8.6)$$

(a) (b)

(c) (d)

FIGURE 6.8.4 Examples of innovative cubicle designs that may aid in addressing views. The manufacturer is (a) Steelcase, Inc., (b) Herman Miller, Inc., (c) Kimball International, Inc., and (d) Knoll, Inc. The chairs in front were provided by Haworth, Inc., and may optionally count toward other MR credits if furniture is included in the calculations. (*Photographs taken at the Strom Thurmond Building on Fort Jackson, Columbia, S.C., in June 2007.*)

The term to the left of the addition sign in Eq. (6.8.6) represents the total floor areas in regularly occupied spaces that comply with both view criteria, and the terms to the right of the addition sign represent the additional amount that may be included for private offices where at least 75 percent of the private office area meets both criteria. The project then complies with EQc8.2 if

$$VC \geq 90 \tag{6.8.7}$$

Special Circumstances and Exemplary Performance

Since the location of windows and large work areas is an intricate part of the architecture of a building, this credit has been labeled with an EB from the USGBC. It is more difficult to change in an existing building.

There is an EP point available for EQc8.2, although no specific criterion is set for it. Each project would be evaluated separately for special view features. The obvious implication is that the subjects of the views may aid in additional benefits.

6.9 Discussion and Overview

This category covers indoor air quality, indoor thermal comfort, indoor lighting, and views. In the future, these and other important indoor environmental concerns will be expanded upon. For example, in the USGBC LEED for Schools rating system released in 2007, both a prerequisite and a credit have been added that relate to acoustics or sound pollution. Currently, however, the major focus is on indoor air quality as the impacts on the human occupants of the buildings have been shown to be important.

There are other alternatives for promoting good indoor air quality. There may be the potential for many other types of monitoring other than CO_2, and other ways to "clean" indoor air than through outdoor air exchanges, which is very important in dense urban areas or other areas where outside air quality is not always up to the National Ambient Air Quality Standards (NAAQS). These technological alternatives may eventually become other supplemental options in the indoor air quality credit subcategories or earn innovation credits. Two new technologies that are being developed to improve air quality with respect to microbes and pathogens in indoor environments are UV lighting and copper components such as heat exchanger fins, cooling coils, and condensate drip pans in the HVAC systems. More information about the possible advantages of copper can be found on the website of the Copper Development Association (CDA) (*http://www.copper.org/*). Figure 6.9.1 depicts a factory installation of UV lights in a Trane air handler.

ASHRAE was awarded a grant in 2006 from the EPA to develop a special publication entitled *Advanced Indoor Air Quality (IAQ) Design Guide for Non-Residential Buildings*. The document is expected to be complete in late 2008, and the work is being done in conjunction with many other organizations including the USGCB, AIA, and SMACNA. This document should be useful in comparing air quality alternatives and designing innovative control methods.

Also, ASTM International is developing a standard that addresses methods for the evaluation of indoor air quality. It is Standard D7297 and is under the jurisdiction of Subcommittee D22.05 on Indoor Air, which is part of ASTM Committee D22, Air Quality.

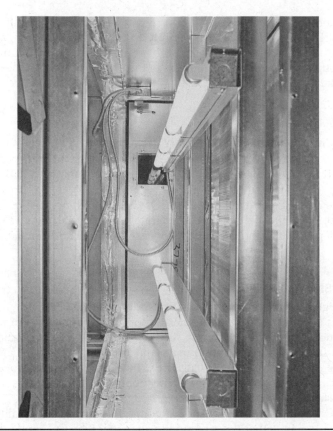

Figure 6.9.1   UV lights in a Trane air handler which help keep the surface of the coil and drain pan clean. (*Photograph Courtesy Trane.*)

Air Pollution Calculation Conversions and Concentrations

Air pollutant concentrations and emission rates are typically reported in many units. Conversion of these units is not always straightforward and is also dependent on the temperature and pressure of the readings. Typically the values given from standard testing are converted to standard temperature and pressure, which is at 1 atmosphere (atm) (14.7 psia) and between 20 and 25°C. The following items can aid in understanding these concepts and conversions to estimate air pollution levels.

Temperature

The SI temperature scale is in degrees Celsius (°C), and the absolute SI scale is in kelvins (K) such that

$$K = °C + 273 \tag{6.9.1}$$

The temperature scale in the United States, degrees Fahrenheit (°F) can be converted to its absolute scale, degrees Rankine (°R), by the following:

$$°R = °F + 460 \tag{6.9.2}$$

To convert between Celsius and Fahrenheit, the following equation is used:

$$°C = 1.8(°F - 32)$$ (6.9.3)

Ideal Gas Law

The ideal gas law is a model that quantifies how the volumes and concentrations of gases change with temperature and pressure. It can be used to model most gases and mixtures of gases at typical ambient temperatures and pressures. The ideal gas law is Eq. (6.9.4) and is based on the following variables:

P absolute pressure

V volume of gas

n moles of gas

R_{Ideal} ideal gas law constant; one value is $0.0821 \ L \cdot atm \cdot (mol \cdot K)$.

T absolute temperature

$$PV = nR_{Ideal}T$$ (6.9.4)

If the concentration of a gas or a mixture of gases is given in moles per volume (n/V), then the changed concentration due to temperature variations can readily be estimated by multiplying the initial concentration by the ratio of the two temperatures. Note that the volume increases with an increase in temperature, so the concentration decreases with an increase in temperature, and the temperatures used must be based on one of the absolute scales (Kelvin or Rankine).

If the concentration of a gas or a mixture of gases is given in moles per volume (n/V), then the changed concentration due to pressure variations can readily be estimated by multiplying the initial concentration by the ratio of the two pressures. Note that the volume increases with a decrease in pressure, so the concentration decreases with a decrease in pressure, and the pressures used must be absolute pressures, not gauge pressures.

Some Typical Properties of Ambient Air

The following properties of ambient air are useful in performing indoor air pollutant concentration calculations:

- At room temperature and 1 atm, there is approximately 24 liters per mole (L/mol) of air.

- At the earth's surface, dry air is predominantly composed of 78 percent nitrogen, 21 percent oxygen, and 1 percent argon. This is a mole or volume percent.

- Dry air composed of approximately 78 percent nitrogen, 21 percent oxygen, and 1 percent argon has an average molecular weight of approximately 29 g/mol.

Concentrations

Air pollutant concentrations are typically given either by mass of the pollutant per volume of air or by a molar ratio. The first unit type (mass/volume) can be used for gaseous, gaseous mixtures, liquid, or solid air pollutants. One common unit is micrograms per meter cubed ($\mu g/m^3$).

The molar ratios can only be used for the gaseous pollutants. They are usually given in parts per million (ppm) or parts per billion (ppb). When these units are used for gaseous phase concentrations, they are in parts per million or billion mole. They are not based on mass. Since most of the air pollution concentrations that are of interest inside buildings and in ambient air can be modeled by the ideal gas law, the ppm or ppb on a molar basis is equal to the ppm or ppb on a volumetric basis.

If masses of air pollutants are known, the molecular weights must be used to convert the units to ppm or ppb. This cannot be done for the air pollutant categories that combine several types of pollutants, such as total VOCs, even though the VOCs are gases. Therefore, these concentrations are always given in mass per volume.

Box Models

Simple air pollutant box models based on the conservation of mass can be used to estimate indoor air pollutant concentrations (see Chap. 10 for a description of simple box models). Although the examples given in Chap. 10 are based on water and water pollutant concentrations, similar box models can also be used for both air mass balances and air pollution mass balances. However, care must be taken when balancing the air masses and air pollutant masses so that variations in flows, volumes, and concentrations due to temperature and pressure changes are addressed.

References

ASHRAE (1999), ASHRAE 52.2-1999: *Method of Testing General Ventilation Air-Cleaning Devices for Removal Efficiency by Particle Size*, American Society of Heating, Refrigerating and Air-Conditioning Engineers, Atlanta, Ga.

ASHRAE (2004), ASHRAE 62.1-2004: *Ventilation for Acceptable Indoor Air Quality*, American Society of Heating, Refrigerating and Air-Conditioning Engineers, Atlanta, Ga.

ASHRAE (2004), ASHRAE 62.2-2004: *Ventilation and Acceptable Indoor Air Quality in Low-Rise Residential Buildings*, American Society of Heating, Refrigerating and Air-Conditioning Engineers, Atlanta, Ga.

ASTM (2003), ANSI/ASTM-E779-03: *Standard Test Method for Determining Air Leakage Rate By Fan Pressurization*, ASTM, West Conshohocken, Pa.

ASTM (2007), "Indoor Air Quality," Global Notebook Section of *ASTM Standardization News*, July, 35(7): 4.

ASTM (2007), "New Practice Will Consistently Evaluate Indoor Air Quality Problems," Technical News Section of *ASTM Standardization News*, July, 35(7): 10.

Carbon Trust (1999), "Natural Ventilation in Non-Domestic Buildings—A Guide for Designers; Developers and Owners," *Carbon Trust Good Practice Guide 237*, The Carbon Trust, London, United Kingdom.

CDA (2007), http://www.copper.org/health/, Copper Development Association website accessed July 19, 2007.

CEC (2001), *Residential Manual For Compliance with California's 2001 Energy Efficiency Standards*, California Energy Commission Pub. No. P400-01-022, June 1, 2001, Sacramento, Calif.

CIBSE (2005), *Natural Ventilation in Non-Domestic Buildings*, Chartered Institution of Building Services Engineers, London, United Kingdom.

Cooper, C. D., and F. C. Alley (2002), *Air Pollution Control, A Design Approach*, Waveland Press, Inc., Prospect Heights, Ill.

CRI (2006), http://www.carpet-rug.com/, Carpet and Rug Institute website accessed July 24, 2007,

De Nevers, N. (2000), *Air Pollution Control Engineering*, 2d ed., McGraw-Hill, New York.

EPA OAR (1991), *Building Air Quality: A Guide for Building Owners and Facility Managers*, EPA Document 402-F-91-102, December, U.S. Environmental Protection Agency, Office of Air and Radiation, Washington, D.C.

EPA OAR (1998), *Building Air Quality Action Plan*, EPA Document 402-K-98-001, June, U.S. Environmental Protection Agency, Office of Air and Radiation, Washington, D.C.

EPA OAR (2006), http://www.epa.gov/iaq/formalde.html, U.S. Environmental Protection Agency, Office of Air and Radiation, Washington, D.C., website accessed October 19, 2006.

EPA OAR (2006), http://www.epa.gov/iaq/voc.html, U.S. Environmental Protection Agency, Office of Air and Radiation, Washington, D.C., website accessed October 19, 2006.

EPA OAR (2006), http://www.epa.gov/oar/oaqps/greenbk/o3co.html, U.S. Environmental Protection Agency, Office of Air and Radiation, Washington, D.C., website accessed October 19, 2006.

EPA OAR (2006), http://www.epa.gov/pmdesignations/basicinfo.htm, U.S. Environmental Protection Agency, Office of Air and Radiation, Washington, D.C., website accessed October 19, 2006.

Green Seal (1993), *The Green Seal Standard GS-11, Paints*, 1st ed., May 20, 1993, http://www.greenseal.org/certification/standards/paints.cfm, Green Seal, Washington, D.C.

Green Seal (1997), *The Green Seal Standard GC-03, Anti-Corrosive Paints*, 2d ed., January 7, 1997, http://www.greenseal.org/certification/standards/anti-corrosivepaints.cfm, Green Seal, Washington, D.C.

Green Seal (2000), *Green Seal Standard, GS36, Commercial Adhesives*, October 19, 2000, http://www.greenseal.org/certification/standards/commercialadhesives.cfm, Green Seal, Washington, D.C.

Heinsohn, R. J., and R. L. Kabel (1999), *Sources and Control of Air Pollution*, Prentice-Hall, Upper Saddle River, N.J.

Likert, R. (1932), "A Technique for the Measurement of Attitudes," *Archives of Psychology* 140: 55.

Michels, H. T. (2006), "Anti-Microbial Characteristics of Copper," *ASTM Standardization News*, October, 34(10): 3–6.

SCAQMD (2004), *Rule 1113, Architectural Coatings*, January 1, 2004, South Coast Air Quality Management District, Diamond Bar, Calif.

SCAQMD (2005), *Rule 1168, Adhesive and Sealant Applications*, as Amended January 7, 2005, South Coast Air Quality Management District, Diamond Bar, Calif.

Siegel, J., I. Walker, and M. Sherman (2002), "Dirty Air Conditioners: Energy Implications of Coil Fouling," Lawrence Berkeley National Laboratory Paper LBNL-49757, March 1.

SMACNA (1995), *IAQ Guidelines for Occupied Buildings under Construction*, Chap. 3, Sheet Metal and Air Conditioning National Contractors Association, Chantilly, Va.

USGBC (2003), *LEED-NC for New Construction, Reference Guide*, Version 2.1, 2d ed., May, U.S. Green Building Council, Washington, D.C.

USGBC (2005–2007), *LEED-NC for New Construction, Reference Guide*, Version 2.2, 1st ed., U.S. Green Building Council, Washington, D.C., October 2005 with errata posted through Spring 2007.

USGBC (2007), *LEED for Schools for New Construction and Major Renovations*, Approved 2007 Version, April, U.S. Green Building Council, Washington, D.C.

Varodompun, J., and M. Navvab (2007), "HVAC Ventilation Strategies: The Contribution for Thermal Comfort, Energy Efficiency, and Indoor Air Quality," *Journal of Green Building*, Spring, 2(2): 131–150.

Wark, K., C. F. Warner, and W. T. Davis (1998), *Air Pollution, Its Origin and Control*, Addison Wesley Longman, Menlo Park, Calif.

Wikipedia (2006), http://en.wikipedia.org/wiki/Metabolic_equivalent, as accessed October 16, 2006.

Exercises

1. Put together a CPM schedule for items for SMACNA and the other requirements for EQc3.1 and EQc3.2. (See Chap. 8 for information on a CPM.)

A. With procurement staggered

B. With on-site protected storage

C. For sequencing installation of absorbent materials

2. Make a site plan with staging areas for material protection.

A. Calculate the size of the covered areas needed.

B. Determine the area needed for the staging and access.

C. Put these items on the plan.

3. When we breathe outside air, we inhale air with oxygen at about 21 percent (volumetric) and carbon dioxide (CO_2) at around 380 ppm (0.038 percent volumetric) and exhale air that has a little less oxygen (a bunch more water) and around 4.5 percent carbon dioxide, which is then rapidly diluted by the air around it. Typically, you can estimate that sedentary activities in a typical space with proper ventilation will raise the CO_2 concentrations by about 350 ppm. Sedentary activities are said to be 1 Met (metabolic rate), which is a measure of the rate at which we expend energy, which can also represent the rate at which we exhale CO_2. Vigorous exercise might be about 6 Met.

A. In a room with computer programmers and typical ventilation, estimate what you might expect the concentration of CO_2 to be (in ppm and also percent volumetric).

B. Estimate what this concentration might be in a jazzercise room if you do not increase the ventilation from the level in the computer room (in ppm and percent volumetric). (Many times the CO_2 alarm is set at 1000 ppm. CO_2 is considered toxic at about 5 percent volumetric.)

4. For EQc3.1 and EQc5 there are requirements to have particulate filtering capabilities on the air handlers based on the ASHRAE MERV (minimum efficiency reporting value) ratings. For these ratings, typical airborne particulates are categorized into the following size ranges: 0.3 to 1 μm, 1 to 3 μm, and 3 to 10 μm. The micrometer (abbreviated μm) is given as an average diameter measurement and is 10^{-6} m. If you assume that these particles are round and have a specific gravity of 2, what are their volume ranges and mass ranges?

5. Credit 4.3 limits total VOC emissions and also several specific compounds emissions, including formaldehyde from carpets and carpet cushioning. It also limits the VOC level (concentration) in the carpet adhesive as per EQc4.1. You are carpeting a room that is 4.5 m by 6 m wide and 3.2 m tall, and it is totally sealed so that there are no air exchanges. If you install carpet cushioning and carpet on the floor that all are at the emissions limits as established by the Carpet and Rug Institute (CRI) for formaldehyde and as listed in Table 6.4.3., how long (in hours) will it take until the concentration in the room of formaldehyde reaches the threshold for human irritation (around 0.1 ppm)? Remember, we are assuming a perfectly sealed room. You can assume that both the formaldehyde and the air are ideal gases.

6. You are carpeting a room that is 4.5 m by 6 m wide and 3.2 m tall, and the room changes out its full volume of air every 4 h. If you install carpet cushioning and carpet on the floor that all are at the emissions limits as established by the Carpet and Rug Institute (CRI) for formaldehyde as listed in Table 6.4.3, what is the steady-state concentration of formaldehyde in the room? Please note the following:

- *Steady state* means that the mass rate of formaldehyde leaving the room in the air which is leaving equals the mass rate of the formaldehyde which is being emitted into the room from the carpet system as noted. (See Chap. 10 for mass balance box models.)
- You can assume that both the formaldehyde and the air are ideal gases.

7. You are carpeting a room that is 4.5 m by 4.5 m wide and 3.0 m tall, and the room changes out its full volume of air every 2 h. If you install carpet cushioning and carpet on the floor that all are at the emissions limits as established by the Carpet and Rug Institute for formaldehyde as listed in Table 6.4.3, what is the steady-state concentration of formaldehyde in the room? Please note the following:

- Steady state means that the mass rate of formaldehyde leaving the room in the air which is leaving equals the mass rate of the formaldehyde which is being emitted into the room from the carpet system as noted. (See Chap. 10 for mass balance box models.)
- You can assume that both the formaldehyde and the air are ideal gases.

8. Your room has equal mass concentrations of particulate matter in the 1- to 3-μm range and in the 3- to 10-μm range and an insignificant amount of the smaller particulate matter on a mass basis. What is the overall efficiency of a MERV 8 filter and a MERV 13 filter for this room based on Table 6.3.1?

9. Your room has approximately 25 percent on a mass basis of the particulate matter in the 1- to 3-μm range, approximately 20 percent in the 3- to 10-μm range, and the remainder in the 0.3- to 1.0-μm range. What is the overall efficiency of a MERV 8 filter and a MERV 13 filter for this room based on Table 6.3.1?

10. Ambient air PM2.5 1997 maximum standards are 15 μg/m³ on an average annual basis and 65 μg/m³ on an average 24-h basis in 1997. The 24-h standard was tightened to 35 μg/m³ in 2006. You have a room that is 4 m by 4 m by 3 m. If you flush out the volume of this room and it is at the maximum annual ambient air standard for PM2.5, what mass of PM2.5 would you expect to collect on a MERV 13 filter?

11. Estimate the time it would take to flush out a building after occupancy for the minimum rates set in EQ credit 3.2 Option 1. The standard gives a total of 14,000 ft³ of outside air per square foot. This alternate flush-out option allows for occupancy after 3500 ft³ is complete. At the minimum rate of 0.30 ft³/(min · ft²) for a typical 9-h day occupancy (8 h plus lunch), how many extra days will this take? (Note that the system must start at least 3 h prior to occupancy.)

12. The maximum allowed concentration for formaldehyde for EQ credit 3.2 based on air quality testing is 50 ppb (parts per billion). What is the concentration in micrograms per cubic meter? Formaldehyde is CH_2O. You can assume that both the formaldehyde and the air are ideal gases.

13. The maximum allowed concentration for 4-phenylcyclohexene for EQ credit 3.2 based on air quality testing is 6.5 μg/m³. What is the concentration in ppm? You can assume that both the 4-PCH and the air are ideal gases.

14. The maximum allowed concentration for carbon monoxide for EQ credit 3.2 based on air quality testing is 9 ppm. What is the concentration is micrograms per cubic meter? You can assume that both the CO and air are ideal gases.

15. A typical diameter of a mechanical pencil lead is 0.5 mm. How many PM particles of 10-μm size would need to be lined up side by side to form this diameter?

16. Look at your classroom or lecture hall. Make a sketch of the room and determine if it meets the natural ventilation requirement for operable windows as per EQ prerequisite 1. If it does not, then make a plan sketch and elevation of the room which would meet the requirements if this is feasible. If it is not, then explain why.

17. Look at your classroom or lecture hall. Make a sketch of the room and determine if it meets the view requirements of EQ credit 8.2. If it does not, then make a plan and an elevation sketch of the room which would meet the requirements. If this is not feasible, explain why.

18. Look at your classroom or lecture hall. Make a sketch of the room and determine if it meets the daylighting requirements of EQ credit 8.1, using the typical minimum values of the combined daylighting factors in Table 6.8.1. If it does not, then make a plan and an elevation sketch of the room with the additional features which would meet the requirements. If this is not feasible, explain why.

19. You are designing the window arrangement in a new office space in a single-story flat-roof building. The room is 30 ft wide by 80 ft long with an exterior wall along one length. Design windows and openings so that both the natural ventilation requirements in EQ prerequisite 1 and the view requirements of EQ credit 8.2 are met. Show this with both a plan and an elevation sketch of the room.

20. You are designing the window arrangement in a new office space in a single-story flat-roof building. The space is 40 ft wide by 90 ft long with an exterior wall along one length. There need to be six private 10-ft by 10-ft offices in this office space, with the remainder open for cubicles and meeting areas. Design windows and openings so that both the daylighting requirements of EQ credit 8.1 and the view requirements of EQ credit 8.2 are met. Show this with plan and elevation sketches of the spaces. (Use the typical minimum values of the combined daylighting factors in Table 6.8.1.)

21. You have performed a survey of all your employees and have collected the number of responses as listed in Table 6.E.1. Do you qualify for a point for EQc7.2?

22. Perform a comfort survey on the students in your class or another group as specified by your lecturer. Summarize your findings.

	Very Dissatisfied −3	Dissatisfied −2	Somewhat Dissatisfied −1	Neutral 0	Somewhat Satisfied 1	Satisfied 2	Very Satisfied 3
Are you satisfied with the thermal comfort of your workspace?	0	18	32	94	85	8	2

TABLE 6.E.1 Thermal Comfort Survey Results

CHAPTER 7

LEED Innovation and Design Process

T he Innovation and Design Process portion deals with issues otherwise not included in the other categories, or which exceed to a specified degree some of the intents from the other credit categories. This category of the U.S. Green Building Council (USGBC) rating system for new construction version 2.2 (LEED-NC 2.2) consists of two credit subcategories which together may earn a possible five points (credits) maximum. Up to four of this maximum of five can come from EP (EP) points earned in the other categories. The notation format for credits is, for example, IDc1.1 for Innovation and Design Process credit 1.1. Also, in this chapter, to facilitate easier cross-referencing between this text and the USGBC rating system, the second digit in a section heading, equation, table, or figure number represents the credit subcategory number for sections that deal directly with a USGBC LEED-NC 2.2 credit subcategory.

The points for the credits in the Innovation and Design Process category are individually related to a specific credit under a subcategory, as was done in the Sustainable Sites and Water Efficiency categories. There are two credit subcategories. The two credit subcategories are ID Credit Subcategory 1: Innovation in Design and ID Credit Subcategory 2: LEED Accredited Professional. The first credit subcategory has a possible four credits, each with an associated point, and the last has just one credit and associated point.

Basically, the credits in the Innovation and Design Process category can be grouped into the following three overall sets of goals:

- Any of the four credits in the Innovation in Design subcategory can be used for projects which substantially exceed the intent of certain credits in the other categories. These are the EP options as listed in each section.

- Any of the four credits in the Innovation in Design subcategory can alternatively be used for other innovative strategies that are not included in the previous categories which can be shown to meet the intent of LEED.

- The one credit in the LEED Accredited Professional subcategory can be used to ensure that one of the team members is familiar with the overall LEED process. The intent is to have someone who can understand the systematic approach needed in sustainable design.

The five credits in the Innovation and Design Process category are as follows:

- **ID Credit 1.1** (IDc1.1): *Innovation in Design*
- **ID Credit 1.2** (IDc1.2): *Innovation in Design*
- **ID Credit 1.3** (IDc1.3): *Innovation in Design*
- **ID Credit 1.4** (IDc1.4): *Innovation in Design*
- **ID Credit 2** (IDc2): *LEED Accredited Professional*

None of these credits have been significantly changed from LEED version 2.1. LEED-NC 2.2 provides a table which summarizes whether the credit is considered a design submittal or a construction submittal. IDc2 is a construction submittal as the professional is expected to participate throughout the project phases. The four possible Innovation in Design credits can be either design or construction submittals depending on their intent and requirements. Likewise, any of or all the project team members may be the main decision makers for these credits depending again on their content. In the same manner it can only be determined by the particular intent and requirements of an innovation credit as to whether it should have the EB icon associated with it. IDc2 does not have an EB icon, as a LEED Accredited Professional can be used in any subsequent phases of a project life.

7.1 ID Credit Subcategory 1: *Innovation in Design*

ID Credits 1.1–1.4: *Innovation in Design*

USGBC Rating System

LEED-NC 2.2 lists the Intent, Requirements, and Potential Technologies and Strategies for IDc1.1 through IDc1.4 as follows:

Intent

To provide design teams and projects the opportunity to be awarded points for exceptional performance above the requirements set by the LEED-NC Green Building Rating System and/or innovative performance in Green Building categories not specifically addressed by the LEED-NC Green Building Rating System.

Requirements

Credit 1.1 (1 point): In writing, identify the intent of the proposed innovation credit, the proposed requirement for compliance, the proposed submittals to demonstrate compliance, and the design approach (strategies) that might be used to meet the requirements.

Credit 1.2 (1 point): Same as Credit 1.1

Credit 1.3 (1 point): Same as Credit 1.1

Credit 1.4 (1 point): Same as Credit 1.1

Potential Technologies and Strategies

Substantially exceed a LEED-NC performance credit such as energy performance or water efficiency. Apply strategies or measures that demonstrate a comprehensive approach and quantifiable environment and/or health benefits.

Calculations and Considerations

As mentioned previously, there are two main avenues for obtaining any one of the four available innovation credits. Either exceed one of the LEED credit criteria from the previous categories or provide an innovative strategy not specifically addressed by LEED-NC 2.2.

In the first case, the exceedance is usually double the standard or the next incremental step. Any previous credit subcategory that may be available for one of these ID credits is listed with the availability of EP noted in previous chapters.

In the second case, many different options can be pursued. The LEED-NC Reference Guide gives three specific criteria. The environmental performance associated with the innovation credit must be quantifiable, this performance must be comprehensive throughout the project, and it must be such that it could be applicable to other projects. If an innovation credit is awarded to one project, it does not mean that it will automatically be awarded to other projects; and conversely, if an innovation credit is not awarded for a specific project, it does not mean that it cannot be awarded for another project with other circumstances. A good source for information about some possible innovation credits for a particular project can be found in the other LEED rating systems and the LEED application guidances which were listed in Tables 1.1.1 and 1.2.1.

Since education is considered to be an extremely important avenue for the development and implementation of sustainable construction, there are many projects which choose an educational outlet for one of the innovation in design credits. This type of activity must be unique and sustainable, not simply a one-time lecture or event. The importance of education has been further emphasized in the LEED for Schools rating system for K-12 schools as adopted in 2007. In this rating system a sixth credit has been added to the Innovation and Design category which specifically addresses continuing education opportunities.

The four innovation credits on a specific project must be substantially different from one another, and none can be technically similar to other credits, except as noted for the EP option. They can be new types of credits not related to any other subcategory in LEED-NC 2.2, or they can be any of the EP credit options. Table 7.1.1 gives a summary of the subcategories for which exemplary performance IDc1 points may be given. Although there are 17 points listed in Table 7.1.1, only 4 can be applied to any one project.

Special Circumstances and Exemplary Performance

One of the variables which must be determined for each project and which must be kept consistent throughout the credits in a LEED submittal is the size and location of the project site. It has been noted that there may be exceptions to this for special circumstances such as campus settings. Since campus and multiple-building projects may have varying boundaries depending on the phases of construction, additional LEED guidances have been developed to aid in the project interpretations for several credits. Likewise, many projects have other unique criteria that may not fit well into the credit intents and requirements as listed in LEED-NC 2.2.

There is no EP point available for this credit subcategory as this credit subcategory is the repository of the EP points from other credit subcategories.

Credit	Exemplary Performance and Other ID Point Availability	Exemplary Performance Point Criterion
SSc2: Development Density and Community Connectivity	Yes, 1 point for Option 1 only	Site density is double the area density or Area density is doubled for double the radius
SSc4.1: Alternative Transportation: Public Transportation Access	Yes, 1 point for either option	Double and also a minimum 200 transit rides daily Option 1: 2 stations and minimum 200 transit rides/day Option 2: 4 stops and minimum 200 transit rides/day
SSc4.1: Alternative Transportation: Public Transportation Access	Yes, 1 point for entire SSc4 subcategory	Comprehensive plan
SSc4.2: Alternative Transportation: Bicycle Storage and Changing Rooms		
SSc4.3: Alternative Transportation: Low-Emitting and Fuel-Efficient Vehicles		
SSc4.4: Alternative Transportation: Parking Capacity		
SSc5.1: Site Development: Protect or Restore Habitat (For previously developed sites, criterion is 50%)	Yes, 1 point for previously developed sites	75%
SSc5.2: Site Development: Maximize Open Space: Option 1: Exceed Open Space Requirement by 25% Option 2: Match the Building Footprint with Open Space if No Zoning Option 3: 20% Open Space When None Required	Yes, 1 point	Double Option 1: Exceed by 50% Option 2: Double Building Footprint with Open Space Option 3: 20% Open Space When None Required
SSc7.1: Heat Island Effect: Non-Roof (50%)	Yes, 1 point	100%
SSc7.2: Heat Island Effect: Roof Option 2: 50% Green Roof	Yes, 1 point for Option 2 only	100% Green Roof
WEc2: Innovative Wastewater Technologies (50%)	Yes, 1 point for either option	1 point for 100%
WEc3.1: Water Use Reduction: 20% EPAct WEc3.2: Water Use Reduction: 30% EPAct	Yes, 1 or 2 points	1 point for 40% EPAct and/or 1 point for 10% Non-EPAct

TABLE 7.1.1   Summary of Exemplary Performance Point Availability

Credit	Exemplary Performance and Other ID Point Availability	Exemplary Performance Point Criterion
EAc6: Green Power	Yes, 1 point	Doubling the minimum amount of electricity to 70% or Doubling the minimum contract length to 4 years
MRc2.1: Construction Waste Management—Divert 50% from Disposal	Yes, 1 point	95%
MRc2.2: Construction Waste Management—Divert 75% from Disposal		
MRc3.1: Materials Reuse—5%	Yes, 1 point	15%
MRc3.2: Materials Reuse—10%		
MRc4.1: Recycled Content—10% (post-consumer + ½ pre-consumer)	Yes, 1 point	30%
MRc4.2: Recycled Content—20% (post-consumer + ½ pre-consumer)		
MRc5.1: Regional Materials—10% Extracted, Processed and Manufactured Regionally	Yes, 1 point	40%
MRc5.2: Regional Materials—20% Extracted, Processed and Manufactured Regionally		
MRc6: Rapidly Renewable Materials (>2.5%)	Yes, 1 point	5%
MRc7: Certified Wood (>50%)	Yes, 1 point	95%
EQc8.1: Daylighting and Views: Daylight 75% of Spaces	Yes, 1 point	95%
EQc8.2: Daylighting and Views: Views for 90% of Spaces	Yes, 1 point	None set

TABLE 7.1.1   *(Continued)*

7.2 ID Credit 2: LEED Accredited Professional

USGBC Rating System
LEED-NC 2.2 lists the Intent, Requirements, and Potential Technologies and Strategies for IDc2 as follows:

Intent
To support and encourage the design integration required by a LEED-NC green building project and to streamline the application and certification process.

Requirements
At least one principal participant of the project team shall be a LEED Accredited Professional (AP).

Potential Technologies and Strategies

Educate the project team members about green building design and construction and application of the LEED Rating System early in the life of the project. Consider assigning the LEED AP as a facilitator of an integrated design and construction process.

Considerations and Exemplary Performance

IDc2 has been designated as a construction phase submittal. Mainly, this is done because the LEED Accredited Professional should participate in all aspects of the project including construction. There is no EP point available for this credit.

7.3 Discussion and Overview

As previously mentioned in Chap. 1, the USGBC is developing guidances specific to many special subsets of project application types, to facilitate LEED certification and sustainable practices for these special types of projects. They are referred to as application guides and are listed in Table 1.2.1. There are many special options for Innovation in Design credits as outlined in the various guidances being developed by the USGBC and its partners for specific project type categories. In addition, sometimes EP points are handled differently in these guidances and in the other LEED rating sytems. There are options in some where a credit point that is listed as an EP in LEED-NC 2.2 may be an additional listed credit in the guidance or system for the particular project type. If what was once an EP point is counted as a credit, then this sometimes leaves open the opportunity for more innovation and design credits to be incorporated into the project. (Note that there may also be an increase in the number of points required for some rating systems which may counterbalance this.)

An example of the renotation of an EP point to a credit is in the guidance on schools (LEED for Schools adopted in 2007). This guidance mainly deals with K-12 schools. Building green schools is considered to be a great opportunity for improved efficiency, reduced cost, and improved student productivity in the United States. It has been found that saving water in these types of facilities to the required 40 percent reduction needed for exemplary performance for WE credit 3.2 is often readily achievable. Therefore, in this guidance, this level of potable water savings can be used to obtain a second point for WE credit 3.2. The Water Efficiency credit category in this guidance also has an example of a case where an EP point is modified and then allowed to count as a credit point. It establishes the new WE credit 4, which is partially based on the WE credits 3.1 and 3.2 EP option related to process water. Here, in addition to a requisite decrease in process water use (the number of systems and percentage reduction are specifically outlined in the guidance), the project must also fulfill two additional requirements: no garbage disposals and no refrigeration equipment with once-through cooling water.

Certification Level	LEED-NC 2.2 Required Points	LEED for Schools Required Points
Certified	26–32	29–36
Silver	33–38	37–43
Gold	39–51	44–57
Platinum	52–69	58–79

TABLE 7.3.1 Certification Points for LEED-NC 2.2 and LEED for Schools (2007)

	LEED-NC 2.2 Available Points	LEED for Schools Available Points	LEED-NC 2.2 Prerequisites	LEED for Schools Prerequisites
SS	14	16	1	2
WE	5	7	0	0
EA	17	17*	3	3
MR	13	13	1	1
EQ	15	20	2	3
ID	5	6	0	0

*A minimum of two points are mandatory as of April 2007 for EAc1 in LEED for Schools.

TABLE 7.3.2 Category Points for LEED-NC 2.2 and LEED for Schools (2007)

Again, having this become a credit point instead of an EP point may afford the project an opportunity for even more innovation points in sustainable design.

However, in the LEED for Schools rating system, not only did the total number of possible credits increase, but so did the number of credits necessary to reach each certification level and so did the number of prerequisites. The total points required for LEED for Schools credit rankings are compared to the LEED-NC 2.2 credit rankings in Table 7.3.1. The number of prerequisites and points available in each category for these two rating systems are summarized in Table 7.3.2.

References

Katz, G. (2006), *Greening America's Schools Costs and Benefits,* October, A Capital E Report, Capital E, Washington, D.C.

USGBC (2003), *LEED-NC for New Construction, Reference Guide,* Version 2.1, 2d ed., May, U.S. Green Building Council, Washington, D.C.

USGBC (2004–2005), *LEED-CI Green Building Rating System for Commercial Interiors,* Version 2.0, November 2004, updated December 2005, U.S. Green Building Council, Washington, D.C.

USGBC (2004–2005), *LEED-EB Green Building Rating System for Existing Buildings, Upgrades, Operations and Maintenance,* Version 2, October 2004, updated July 2005, U.S. Green Building Council, Washington, D.C.

USGBC (2005), *LEED-NC Application Guide for Multiple Buildings and On-Campus Building Projects, for Use with the LEED-NC Green Building Rating System Versions 2.1 and 2.2,* October, U.S. Green Building Council, Washington, D.C.

USGBC (2005–2007), *LEED-NC for New Construction, Reference Guide,* Version 2.2, 1st ed., U.S. Green Building Council, Washington, D.C., October 2005 with errata posted through Spring 2007.

USGBC (2006), *LEED Green Building Rating System for Core and Shell Development,* Version 2.0, July, U.S. Green Building Council, Washington, D.C.

USGBC (2007), *LEED for Homes Program Pilot Rating System,* Version 2.11a, January, U.S. Green Building Council, Washington, D.C.

USGBC (2007), *LEED for Neighborhood Development Rating System*, Pilot, Congress for the New Urbanism, Natural Resources Defense Council, U.S. Green Building Council, Washington, D.C.

USGBC (2007), *LEED for Retail—New Construction and Major Renovations*, Pilot Version 2, April, U.S. Green Building Council, Washington, D.C.

USGBC (2007), *LEED for Schools for New Construction and Major Renovations*, Approved 2007 Version, April, U.S. Green Building Council, Washington, D.C.

Exercises

1. You are redeveloping an acre of land that was originally an asphalt parking lot. Along with the building and parking, you plan to turn 10,000 ft^2 of the lot into landscaped areas with natural and adapted plantings. What credits might this aid you in earning? What EP points might you consider?

2. True or false? The following action may be eligible for EP credit: A stormwater management plan is made for the site that reduces the rate and quantity of the existing runoff (1- and 2-year, 24-h storm) by 20 percent. The site currently has an imperviousness of approximately 35 percent. Why or why not?

3. True or false? The following action may be eligible for EP credit: The fire suppression systems in the new building will not use any CFCs, any halons, or any HCFCs. Why or why not?

4. True or false? The following action may be eligible for EP credit: Of the materials for the construction of the new building, except for plumbing, mechanical, furniture, and electrical materials, 50 percent (based on their cost value) are extracted or harvested and manufactured within 500 mi of the site. Why or why not?

5. True or false? The following action may be eligible for EP credit: Individual lighting controls will be available in every office and cubicle and at every desk in the new office building. Why or why not?

6. True or false? The following action may be eligible for EP credit: A covered staging area will be provided at the construction site to store materials away from the rain. Why or why not?

7. Research a LEED certified project in your area and find out which ID credits were obtained.

8. There are several projects that have used education as an ID credit. Find a specific example and summarize it. Do you think this will be applicable to all projects, or is it only specific to certain projects? Explain your opinion.

9. Look at a nearby construction project, and think of a possible innovative credit, other than EP, that you might like to consider for that project. Describe it and explain why you think it is innovative.

CHAPTER **8**

A Systematic View of Green

8.1 Green Building from the Project Viewpoint

How to Get Started

The trick to getting started is to meet, then meet again, and then meet again. Integrated building design is the cornerstone for creating sustainable projects. If the systems and components that make up the facility are only looked at independently, then the result usually will not be optimal for performance, sustainability, and cost. It is important that all aspects of the project be viewed as interrelated.

Initial meetings with all the project team members who have already been identified are crucial to the success of the project. At these initial meetings, basic project scopes and requirements are reviewed. With respect to the incorporation of green concepts in the project, it is beneficial to start reviewing the prerequisites and credit opportunities of the LEED rating system. Project members should be asked to start identifying the necessary steps and design features needed to meet all the prerequisites. They should also be asked to review the list of possible credits and start identifying credits that may or may not be feasible to incorporate into the design while maintaining the project scope and objectives.

Many large projects hold early formal meetings commonly referred to as *charrettes*. Invited participants are the owner, stakeholders, identified project team member representatives, and other potential participants in the design and construction community, in addition to public and regulatory participation as appropriate. Charrettes are intensely focused, collaborative efforts. They can help create momentum to meet project goals in addition to fostering trust and buy-in from the regulatory agencies and community. Collaborative agreement early in a project can help avoid costly design changes and also identify potential partners and opportunities for increased sustainability and performance from the project. Education is a vital component.

Another set of obstacles can involve the local and state regulatory agencies if they are not familiar with some of the practices used. Likewise local statutes may prohibit, discourage, or not address some of the techniques suggested for more sustainable construction. Therefore it is important to work closely with local and state officials and see which practices may be incorporated and what steps may be taken to overcome some of these obstacles. As the sustainable practices become more common, some of these obstacles will disappear.

Several states have started to adopt green building policies. This serves to facilitate the process in two ways. First, adoption of green policies by states helps direct state and local agencies in the manner that green projects will be permitted and evaluated. Second, requiring green practices for many state projects helps improve the learning curve and familiarity in the region with respect to sustainable construction. Understanding and knowledge are always some of the best tools for implementation of new practices. By 2005 two states, Washington and Nevada, had passed legislation relating to sustainable construction; and nine other states—California, Arizona, Colorado, Michigan, Pennsylvania, New York, New Jersey, Maryland, and Maine—had Executive Orders in place relating to either sustainable construction or energy efficiency. Most also referred to the USGBC LEED rating systems in their policy directives.

How to Keep Track

At the initial meetings and charrettes, project objectives are outlined. It is important that these objectives be referred to repeatedly throughout the project life and reevaluated and revised as needed. One portion of the objectives is to establish the green design criteria. LEED issues should be incorporated into a project as early as possible, to facilitate an effective and efficient project completion and to minimize costs and changes. Just as in other project and construction management philosophies which over the years have been successfully incorporated into design and construction, such as Quality Assurance and Quality Control (QA/QC), the incorporation of green building practices may at first seem awkward. However, they should soon become routine and part of an established procedure with its benefits accepted and appreciated. In similar manner to QA/QC, green building issues are continually reevaluated and improved upon during the life of a project.

So where should it all start? It starts first with the owner and project conception. In some cases the owner may voluntarily opt to seek LEED certification; in other cases it may be required by local or state mandate. Either way, the intention to seek LEED certification is usually first identified in an early draft of the specifications set by the owner for a project. The LEED-NC process refers to the specifications set by the owner as the "Owner's Project Requirements" or OPR. As the OPR evolve, the specific LEED credits preferred or recommended by the owner eventually get listed.

These green criteria can be easily kept track of in checklists that are based on the LEED credit lists. The checklists are dynamic documents that change and improve with the project. They can be readily downloaded from the USGBC website. A blank checklist is shown as Fig. 1.2.2. For each credit the question is asked whether the project should attempt to achieve it. The answers can be yes, maybe (?), or no. Each credit is checked off in the Yes, the Maybe (?), or the No column. "Yes" means that at this point in the project, it is the intention that this credit will be sought. The question mark is for those credits that are still being evaluated for cost and project effectiveness, and "no" means that the associated credit will probably not be considered for certification of the subject project. Throughout the life of a project, any of the notations may be moved into different columns for each credit based on more detailed design, information, and project team decisions.

These estimated credit checklists are very good tools for directing all the project team members during the project. However, it may be difficult to fill out the early versions. This may be particularly difficult if the project is the first for an owner, or the first of its kind in a region, since it may be difficult to determine which credits are most applicable and effective for the particular project type and for the regional resources. Various members of the USGBC and researchers have started to develop ranked lists with columns expressing the "ease" or possibility of obtaining each credit based on local experience. Table 8.1.1 gives examples of three types of rankings that have been used to start the initial checklist. One ranking is a subjective ranking of ease of obtaining, the second ranking is the percent of regional projects that have received the credit, and the third ranking is a ranking that gives an initial estimate of viability ranked by the certification levels in the LEED-NC rating system. The items in Table 8.1.1 are for illustrative purposes

Credit	Ease of Earning			Local Project Experience (% Earned)	Probable Certification Level			
	Easy	Neutral	Hard		Certified	Silver	Gold	Platinum
SSc1	✓			80	✓			
SSc2*		✓		60		✓		
SSc3*		✓		40			✓	
SSc4.1*		✓		50			✓	
SSc4.2	✓			95	✓			
SSc4.3	✓			80	✓			
SSc4.4	✓			85	✓			
SSc5.1			✓	30				✓
SSc5.2		✓		65		✓		
SSc6.1	✓			95	✓			
SSc6.2		✓		65		✓		
SSc7.1	✓			80	✓			
SSc7.2	✓			80	✓			
SSc8			✓	45				✓
WEc1.1	✓			85	✓			
WEc1.2		✓		45			✓	
WEc2	✓			95	✓			
WEc3.1	✓			100	✓			
WEc3.2	✓			90	✓			
EAc1†	✓ (3)	✓ (5)	✓ (10)	50 (5)	✓ (3)	✓ (4)	✓ (6)	✓ (10)
EAc2†		✓ (1)	✓ (3)	50 (1)			✓ (1)	✓ (3)
EAc3		✓		60			✓	
EAc4		✓		65		✓		
EAc5			✓	15				✓
EAc6			✓	25				✓
MRc1.1*		✓		60		✓		
MRc1.2*		✓		40			✓	
MRc1.3*		✓		50		✓		
MRc2.1	✓			100	✓			
MRc2.2		✓		60		✓		
MRc3.1		✓		55			✓	
MRc3.2			✓	25				✓

TABLE 8.1.1 Example of Three Types of Ranked List of Credits for Initial Estimates. This list is an example only and should be appropriately altered based on local experience

Credit	Ease of Earning			Local Project Experience (% Earned)	Probable Certification Level			
	Easy	Neutral	Hard		Certified	Silver	Gold	Platinum
MRc4.1	✓			100	✓			
MRc4.2		✓		55			✓	
MRc5.1	✓			90	✓			
MRc5.2		✓		60		✓		
MRc6			✓	5				✓
MRc7*		✓		50			✓	
EQc1			✓	30				✓
EQc2	✓			85	✓			
EQc3.1		✓		55			✓	
EQc3.2			✓	20				✓
EQc4.1	✓			90	✓			
EQc4.2	✓			95	✓			
EQc4.3	✓			95	✓			
EQc4.4	✓			85	✓			
EQc5		✓		60		✓		
EQc6.1			✓	20				✓
EQc6.2			✓	10				✓
EQc7.1		✓		40			✓	
EQc7.2		✓		60		✓		
EQc8.1	✓			90	✓			
EQc8.2		✓		60			✓	
IDc1.1– 1.4	✓ (2)	✓ (3)	✓ (4)	50 (3)	✓ (2)	✓ (3)		✓ (4)
IDc2	✓			100	✓			
Total	26	25	18	n.a.	26	12	14	17

*Very site-specific. Each has been given a median rating for illustrative purposes only.
†May earn more than one point. The total cumulative number of points earned as ranked is in the parenthesis.

TABLE 8.1.1 Example of Three Types of Ranked List of Credits for Initial Estimates. This list is an example only and should be appropriately altered based on local experience (*Continued*)

only and do not represent the actual ease based on any scientific or region-specific analyses. Prerequisites have been excluded from the list as they are required.

The ranked list used is subject to change readily as the demand for green products changes and the knowledge or experience with green construction evolves in an area. This ranked list is also partially based on projects adhering to earlier versions of LEED, and subsequent versions may make certain credits more or less applicable for similar projects in the future. Regardless, past experiences can still give a reasonable starting point for initial estimates of LEED-NC certification feasibility.

The University of South Carolina (USC) has worked on many LEED registered projects. One of the projects was unique in that it represented a proposed baseball stadium, which is significantly different from the types of projects that the LEED-NC 2.2 rating system is usually based on. With this project, it was not easy to readily identify the sustainable opportunities as outlined in LEED-NC, but USC wanted to explore the possibility of LEED certification. To deal with this, the project team first developed a checklist from its current ranked list based on prior experience with other projects. One month later, after some preliminary design was complete, the list was revised based on further information and reviews of CIRs. Table 8.1.2 depicts the changes in the credit checklist for these two periods in the early design stage for the baseball stadium project. As the project proceeded beyond this early design phase, there were fewer and fewer changes. Table 8.1.2 shows how a checklist can help initiate the process even with a project for which there is much uncertainty as to the best sustainable route, and for which there are few previous similar projects to use for initial estimates.

Project Methods and Scheduling

Project Methods and Phases

Project schedules and interactions differ depending upon the type of project and the type of construction process used. For example, some projects follow the more traditional design-bid-build process, whereas others might have one design-build contract. Since continuity and a systems approach are important for effective incorporation of sustainable construction into projects, the different project delivery methods do impact the manner in which these are handled. The method of interaction between the project participants and their responsibilities with respect to sustainable construction will be different for the different methods. It may be easier to have a single LEED lead for the design-build method, whereas, with the design-bid-build process, it may be necessary to transfer the lead from perhaps an architectural firm designee to a construction firm designee as the project proceeds. Depending on the prerequisite and credit, the importance of the impact of the delivery method chosen will differ.

Regardless of the project delivery method, the stages of a project can usually be summarized into a few main categories:

- Project conceptualization
- Schematic design
- Planning and zoning (P&Z) and other municipal planning organization (MPO) reviews
- Design development (detailed design)
- Permits
- Construction document development
- Bid and procurement
- Construction
- Closeout
- Operations

Credit	September 2006			October 2006		
	Yes	?	No	Yes	?	No
SSc1	✓				✓*	
SSc2		✓			✓	
SSc3		✓		✓*		
SSc4.1	✓			✓		
SSc4.2	✓				✓*	
SSc4.3		✓				✓*
SSc4.4		✓			✓	
SSc5.1		✓			✓	
SSc5.2	✓			✓		
SSc6.1		✓			✓	
SSc6.2	✓			✓		
SSc7.1	✓				✓*	
SSc7.2		✓			✓	
SSc8		✓				✓*
WEc1.1	✓				✓*	
WEc1.2		✓			✓	
WEc2			✓			✓
WEc3.1	✓			✓		
WEc3.2	✓			✓		
EAc1†	✓ (2–4)			✓* (2–4)	✓* (2–4)	
EAc2†			✓			✓
EAc3	✓			✓		
EAc4	✓			✓		
EAc5	✓			✓		
EAc6			✓			✓
MRc1.1			✓			✓
MRc1.2			✓			✓
MRc1.3			✓			✓
MRc2.1	✓			✓		
MRc2.2		✓		✓*		
MRc3.1			✓			✓
MRc3.2			✓			✓

TABLE 8.1.2   LEED-NC Version 2.2 Registered Project Checklist with Credit Estimates for a USC Baseball Stadium in Columbia, S.C., September 2006 and October 2006

Credit	September 2006			October 2006		
	Yes	**?**	**No**	**Yes**	**?**	**No**
MRc4.1	✓			✓		
MRc4.2		✓		✓*		
MRc5.1	✓			✓		
MRc5.2		✓		✓*		
MRc6			✓			✓
MRc7			✓			✓
EQc1	✓			✓		
EQc2		✓			✓	
EQc3.1	✓			✓		
EQc3.2		✓			✓	
EQc4.1	✓			✓		
EQc4.2	✓			✓		
EQc4.3	✓			✓		
EQc4.4		✓		✓*		
EQc5	✓				✓*	
EQc6.1		✓			✓	
EQc6.2		✓			✓	
EQc7.1	✓			✓		
EQc7.2	✓			✓		
EQc8.1	✓			✓		
EQc8.2		✓			✓	
IDc1.1–1.4[†]	✓ (1–2)	✓ (3–4)		✓ (1–2)	✓ (3–4)	
IDc2	✓			✓		
Total	29–31	20	18–20	27–31	18–22	20

*Represents a change.
[†]May earn more than one point. The range of points earned as ranked is in the parenthesis.

TABLE 8.1.2   (Continued)

As previously mentioned, many of the credits must be considered in the early stages of project conception to be effective for the project, whereas others may slowly evolve as the design phase progresses into greater detail. It is important that there be interaction regarding the LEED process early on and frequently throughout each step.

CPM Schedules

The term *schedule* actually refers to two different project design and management tools. One type of schedule is a tally of specific items. An example is a door schedule on an architectural plan. A door schedule is the list of the number of types of doors needed for the project. This type of schedule usually includes information on sizes, materials, other characteristics, and manufacturers. The other tool called a schedule refers to a listing or diagram of activities with relationship to time. It is this type of schedule, a time-activity schedule, that is being further evaluated in this section. Note that both types of schedules, item-related and time-activity-related, are important to projects. For instance, in Chap. 10 where erosion and sedimentation control measures are further explored, the design package might require both an itemized schedule of materials or plants needed for erosion control and a time-activity schedule to sequence the erosion control activities so that the soils are adequately protected from erosion during all the construction stages.

Activity-time scheduling is a critical tool in any design-construction process. There are many types of scheduling techniques used, but most can be summarized in the following two formats:

- Bar charts (Gantt chart)
- Critical path method (CPM)

Bar or Gantt charts are very common and represent activities in a vertical list with time on the horizontal and bars along the horizontal which give some idea when these activities will take place. Table 8.1.3 depicts a simple bar chart of some typical project phases.

CPMs are similar listings of activities except that a CPM schedule also shows a dependency of activities on other activities. For instance, a building should be designed before it is constructed, not constructed before it is designed. CPMs are usually depicted as diagrams with interconnected arrows. If the activity is listed on the arrow, then it is referred to as an "activity on the arrow" CPM, and if the activity is depicted in a box between arrows, then it is referred to as an "activity on the node" CPM. Time can be included in the CPM schedule by either listing time requirements with the activity or placing the activity on a figure similar to the bar chart where time is depicted on the horizontal axis. The term *critical path method* comes from the act of adding time along any path of arrows and determining the path which would take the longest time. The activities along this path are considered critical since any delay in them will potentially delay the completion of the entire project scheduled.

CPMs can be evaluated manually or with many established project scheduling programs. Some programs are simple, and others are quite complex and include additional opportunities to relate many features to each activity such as the workforce needs or material needs. Adding these is usually referred to as *resource loading*. Other versions add opportunities for delayed actions or only partial completion of activities before the next starts. CPMs are frequently used and are considered to be valuable tools in project management. Figure 8.1.1 depicts a very simplified CPM of the activities listed in Table 8.1.3 in the activity on the node format with some delayed or early starts and finishes on some of the activities. This is done because some of the activities within a node may start before a predecessor is totally complete. Determination of the critical path would need to include these delayed or early start activity estimates.

Scheduling is even more important for several reasons in sustainable construction. First, the additional activities needed to design, collect data, and verify systems, etc., which are a part of the LEED process must be integrated into the design and construction schedules. There may be cost or schedule conflicts if many of the LEED

Project Phase	Year 1									Year 2								
Project concept	x																	
Schematic design	x	x	x															
MPO reviews		x	x	x	x													
Design development			x	x	x	x												
Permits					x	x	x											
Construction document development					x	x	x											
Bid and procure							x	x										
Construction								x	x	x	x	x	x	x	x	x		
Closeout															x	x	x	
Operations																		x

TABLE 8.1.3   Simple Bar Chart of Typical Project Phases

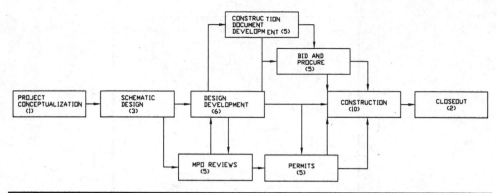

Figure 8.1.1 Simplified CPM schedule for some major construction project activities with some overlapping starts and finishes. (*Overall activity time in months is given.*)

related activities are not performed in a timely fashion. The second reason pertains to a more detailed scheduling level. As noted in Chap. 6 on indoor environmental quality, if construction activities are not scheduled in special ways, there are opportunities for impacts to indoor air quality such as with mold or particulate matter. Scheduling is one of the five principles of SMACNA for good indoor air quality. This is the same for many other activities such as erosion and sedimentation control. If the sediment fence is not installed on a graded site prior to a storm event, then major erosion may result. Usually the best scheduling technique for these types of activities is the CPM.

Because of the need for detailed schedules with sequenced activities related to several of the LEED credits, there is also a need for staging areas for these activities to be noted on the plans. Some examples are as follows:

- Staging areas need to be noted for areas where materials may need to be covered from the elements.

- Areas need to be designated for C&D debris separation and storage prior to removal.

- Erosion and sediment control measures need to be designated on the site plans.

Lead People and LEED

Throughout the project it is important to make sure that there is some consistency in the incorporation of LEED into the project. All summary information relating to the LEED aspects should be filtered through one entity established for this, and frequent contact should be made between the central LEED clearinghouse person and all the project team members and interested parties. This entity may change through the course of a project and the LEED information be passed to another, but there should always be a central clearinghouse established at all times.

The inclusion of parties in LEED decisions outside of the core design team may be crucial even at the onset of the project. For instance, modifications may impact earlier zoning approvals, and many zoning-related changes may take months for approval, severely impacting a project schedule. During the zoning stage, the lead for the project may be, for instance, an attorney or planner, and later during detailed design, the lead may pass to an architectural or other design firm. Nonetheless, there should be a central contact for the green design aspects in contact with the project lead person who can see if any changes have major impacts on the green process. In addition, many projects have specialty designers and occupants or owner subsets, who should be involved. The baseball stadium project at the University of South Carolina represents a case where there are specialty designers for the baseball field as well as special needs during operations of the team and managers which should be considered in decisions, including many LEED related decisions.

So who should this green contact person be? This is a very good question, and the answer may be that there should be several people following the LEED process, but with only one designated at a time as the main clearinghouse. If the owner of a project is an entity which does a lot of construction and has a construction management department, then for consistency and continual improvement, the main LEED contact might be an owner representative. If the owner of the project is not experienced in LEED or if the owner does not expect to do multiple projects, then perhaps the LEED contact should come from the design and construction community. Which field should the individual come from? This is a function of experience, project scope, and project phase. For many projects, the most influential LEED contact in the conceptual phase may be a civil and environmental engineer or a landscape architect, as many of the initial decisions are site-related. Then, through detailed design, a contact with architectural or mechanical engineering design experience might be the most knowledgeable. Similarly, in the construction phase it may be easiest for a commissioning or contractor representative to oversee the main part of the LEED process. Regardless, as long as there is consistency and a smooth transition from the main contacts throughout the project life, the process should work well.

So what needs to be done to make sure that the LEED clearinghouse contact is adequately involved in all stages of the project? Communication, communication, communication! It is not necessary to call special meetings with the green contact all the time that might add to the project schedule and costs. It is necessary, however, that this person be included on all meeting and important decision correspondence. Then on a periodic basis it is helpful to have a portion of a meeting or a special charrette dedicated to the LEED process. This can be done simply by adding the green contact person to the project team meetings and including a section on LEED in construction and other meetings, such as is done with QA/QC. The project checklist with credit estimates as portrayed in Fig. 1.2.2 and Table 8.1.2 is a good example of a document that should be updated by the green contact person and distributed to all project team members frequently throughout a project.

The really important part is that the entire project be looked upon as a system and the green credits be interpreted as a package. It has been found that if individual entities each evaluate specific credits, there is a tendency to add or delete credits based on limited economic and/or environmental perspectives. This may lead to going green becoming an added cost to a project. However, if credits are evaluated in a systemwide manner with many project features clustered to obtain multiple credits and environmental goals, then this reduces the costs of construction and the project. The only way that this can be done effectively is to always have one green lead, or "director of sustainability" as others may wish to call the person, for a project at any time. Any transfers of personnel for this position must be done well and comprehensively.

Miscellaneous

There will be many additional rules of thumb developed to aid in the green process as the green rating systems become more prevalent. However, already several suggestions can be made based on past experiences. The best way to learn about many of these is to stay informed about green building and, with respect to the LEED rating system, become familiar with the CIRs and other documents available. Some additional suggestions are listed in the following sections.

Preliminary Site Assessment

Many existing site features can impact the green decisions that are made. Having a standard questionnaire or template for preliminary site assessment that includes questions about these features can be a useful tool. Including a summary of these features on the documents in all phases of the project, including project conceptualization, may facilitate a more efficient and economical incorporation of sustainability into the project. Some of the items include

- Prevailing wind direction
- Building rotation
- Neighboring shading features such as foliage and buildings
- Floodplain locations
- Wetlands, streams, rock outcroppings, and other natural features
- Soil types
- Existing on-site vegetation
- Grade considerations
- Existing area and on-site utilities
- Prior site uses and any existing facilities such as pavement, pads, and buildings

Predesign Site Selection

Predesign site selection can have a large impact on the ease of obtaining many of the credits. Some of the credits may not seem attainable at first glance, or may be obtained if there is some forethought about how they may be accomplished. For instance,

- The development density option in SSc2 may be more applicable than expected for campuses and other projects where large areas of land may be used for recreational or community purposes.
- The community connectivity option in SSc2 may be more applicable if there are community programs that the proposed project can support, which will enhance this credit, such as contributing or supporting "rails to trails" projects or donating land or funds for the municipality to improve pedestrian access, etc.
- The brownfields credit may be obtained for projects that have some contamination in only small areas, such as an old UST.
- The project team should check for possible planned SSc4.1 public transportation access or perhaps develop a unique plan for project-provided access to these systems.
- All three of the remaining Alternative Transportation credits (SSc4.2, 4.3, and 4.4) may be fairly easy to accomplish if parking is looked at from a holistic view, especially in campus situations, where a comprehensive parking plan may be applicable to multiple projects.

Project Area and Project Boundary

As mentioned in earlier chapters, both the project area and the project boundary are important variables in many of the credit determinations for the USGBC rating system. For many projects these are easily set as the lot in question. However, for campus projects and projects on land areas which also contain other buildings or facilities, there is some flexibility in determining the project area and the project boundary, or extent of disturbance. Careful consideration of the project area and extent of disturbance can aid in obtaining LEED credits and also have a positive influence on the impact of the project. There is no one good solution. Sometimes it is useful to have greater area included if the additional space contains beneficial features such as open space or natural habitat. Sometimes it is useful to have a smaller project area if a goal is to maintain dense urban communities to better use existing infrastructure. It is important that the project area be viewed as a variable that can have a positive impact on sustainability and that the decisions include the various project participants.

There are some cases in which the project area may be made up of land portions that are not contiguous. This may be common on campuses or other land areas where there are existing facilities that are to remain or there are utility sources that are shared by many buildings. It is more difficult to differentiate whether renewable energy sources or other resources are part of the project when they may be shared by many facilities, both existing and proposed. The University of South Carolina was recently working on a library addition project where the project area was bisected by some of the existing facilities. Due to the existing layout, it was determined that the addition should consist of two new wings, one on the west side of the existing library and one on the east side. Figure 8.1.2 is a rough sketch of what a design for the additions might look like. As can

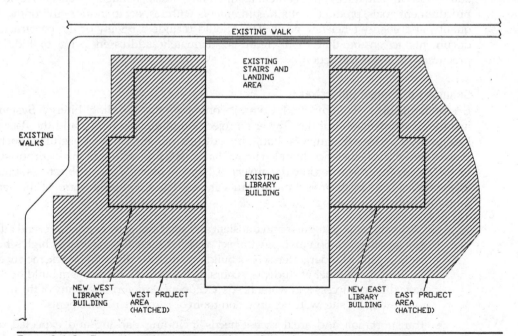

FIGURE 8.1.2   Example project where portions of the site area are not contiguous.

be seen, the project area consists of two noncontiguous parts. This makes the determination of some of the LEED criteria variables such as the density radius and the density boundary more difficult. However, the goal in sustainable construction is to meet the intent of a credit, and extenuating circumstances can always be explained in the submittals.

Designating the project area does not relieve the project from responsibility for certain actions which may be outside of the boundary. One example includes the ESC measures that are part of the SS requirements. These must be implemented anywhere that the project might impact, which may include protecting existing catch basins off-site. For most projects, there will be off-site work relating to the project such as drives and utilities in the state DOT or local right-of-way. Appropriate ESC measures must be taken there too.

Designation of Responsible Parties

As a project develops, it is also useful to keep a table listing the responsible party for many of the specific data collection, reporting, and oversight needs for sustainable construction. The simplest way to start is to designate responsible parties for each of the credits listed on the project checklist. This can then be further subdivided into actual activities as needed for accurate project control. The preliminary designation of responsible parties should also be made a part of the bid package and construction documents so that the contractors are well aware of their responsibilities from the onset. This helps prevent costly change orders and also lets the green criteria become an initial part of the project quality control and quality assurance program. To facilitate this transfer of responsibility to certain individuals for specific LEED items, there is also a need for special material and construction requirements to be included in the detailed design package and the construction documents and given to both the contractors and the procurement agents. Some example items are recycle content requirements, waste management needs, regional procurement goals, product content requirements with respect to indoor environmental quality, and energy efficiency goals. This should include a review of the construction documents to ensure that each credit is adequately addressed prior to bidding, procurement, and construction.

Summary of Project Objectives

EA prerequisite 1, Fundamental Commissioning of the Building Energy Systems, includes documentation of the Owner's Project Requirements (OPR) and the Basis of Design (BOD). These documents should include not only information useful for energy system design and efficiency, but also items that can be used to direct the incorporation of other sustainable features into the project. A listing of overall project objectives should be made available to all project participants and updated regularly. Some of the items might include the following:

- Occupancies, both regular and transient and subdivided by gender, age, and other special categorization that may impact project design. For instance, a high school building and a kindergarten school building will have different bicycle rack needs for the same number of students. A nursing home and an apartment building will have different access and water needs. Occupancies determine many of the needs for sustainable site, water, energy, and room or furniture requirements.

- Transportation and areal access needs including parking and expected trip generation.

- Interior space needs.
- Exterior space objectives.
- Hours and frequencies of operation.
- Maintenance needs.
- Special needs including
  - Security
  - Special chemical use or storage needs
  - Special facilities such as laboratories or examination rooms

This list is in no way complete, but can aid in asking initial questions of the owner or decision makers so that design and construction can proceed in a more efficient and sustainable manner.

8.2 Summary of Relevant Codes and Credits

Summary of Relevant Codes

The LEED-NC 2.2 rating system and Reference Guide list many codes, standards, and other guidances that are used in the various prerequisite and credit determinations. Table 8.2.1 lists many of these by prerequisite or credit as a summary tool for obtaining any supplemental information needed to aid in the project execution.

Summary of the Credits

Sustainable construction is composed of many diverse objectives, criteria, impacts, and design features. It is a systematic way to build that includes multiple disciplines and various requirements that sometimes have possibly conflicting outcomes. Due to the complexity of sustainability, it is difficult to keep track of all the objectives. The USGBC LEED rating system and other sustainability guides do an excellent job of organizing and ranking many of these complex system objectives. However, it is still difficult to remember all the various parts, particularly since most people cannot be experts in everything. To help overcome the enormity of it all, it is recommended to have as many project participants as possible familiar with the prerequisites and credits. One way to accomplish this is to have many project decision makers accredited in LEED or by other environmental certifications. To facilitate both studying for the LEED accreditation and using it as a tool for any interested party to become familiar with the prerequisites and credits, the author has developed a study table. Two examples of the template for study tables are presented in Tables 8.2.2 and 8.2.3 for Sustainable Sites and Water Efficiency, respectively.

The first thing one might notice is that most of the cells in Tables 8.2.2 and 8.2.3 are empty. This is so that individuals can complete the table by reviewing the Reference Guide or the chapters in this text and complete them in the manner most useful for that individual. A civil engineer will need different information in various cells to understand the items than a person trained in architecture or planning or mechanical engineering. Thus, these tables can be used both for reviewing the prerequisites and credits and for customizing the information to each individual. One of the most import columns to complete is the variable/concepts column. What fits here will vary tremendously from credit to credit, especially since many credit criteria are based on quantitative variables and other criteria on qualitative concepts. For instance, for SSc2, there are specific quantitative

Prerequisite or Credit	Referenced Standards
SSp1: Construction Activity Pollution Prevention	- USEPA Document No. 832R92005, Chap. 3 - 2005 EPA CGP (ESC)
SSc1: Site Selection	- 7 CFR 657.5 (Prime Farmland) - FEMA 100-year flood definition - U.S. Fish and Wildlife Service Threatened or Endangered Species List - 40 CFR Parts 230–233 and Part 22 (Wetlands)
SSc2: Development Density and Community Connectivity	None
SSc3: Brownfield Redevelopment	- ASTM E1903-97 Phase II Environmental Site Assessment - EPA brownfields definition
SSc4.1: Alternative Transportation: Public Transportation Access	None
SSc4.2: Alternative Transportation: Bicycle Storage and Changing Rooms	None
SSc4.3: Alternative Transportation: Low-Emitting and Fuel-Efficient Vehicles	- CARB definition of zero emissions vehicles or - ACEEE Annual Vehicle Rating Guide
SSc4.4: Alternative Transportation: Parking Capacity	Maximum parking allowed by - Local zoning code Or if not available - Portland, Oregon, Zoning Code: Title 33, Chapter 33.266 (Parking and Loading) Or if not applicable - Institute of Transportation Engineers, *Parking Generation*, 3d ed., Washington D.C., 2004
SSc5.1: Site Development: Protect or Restore Habitat	None
SSc5.2: Site Development: Maximize Open Space	None
SSc6.1: Stormwater Management: Quantity Control	None
SSc6.2: Stormwater Management: Quality Control	- USEPA Document No. EPA 840B92002 or - TARP (Washington State) BMP monitoring
SSc7.1: Heat Island Effect: Non-Roof	- SRI: ASTM E1980-01 - Emittance: ASTM E408-71(1996)e1 or C1371-04 - Reflectance: ASTM E903-96, E1918-97, or C1549-04

TABLE 8.2.1   Listing of Referenced Standards

Prerequisite or Credit	Referenced Standards
SSc7.2: Heat Island Effect: Roof	- SRI: ASTM E1980-01 - Emittance: ASTM E408-71(1996)e1 or C1371-04 - Reflectance: ASTM E903-96, E1918-97, or C1549-04
SSc8: Light Pollution Reduction	- IESNA RP-33 - ASHRAE/IESNA Standard 90.1-2004 (w/o amendments), Section 9 (Exterior)
WEc1.1: Water Efficient Landscaping: Reduce by 50%	- Irrigation Association reference evapotranspiration rates
WEc1.2: Water Efficient Landscaping: No Potable Water Use or No Irrigation	- Irrigation Association reference evapotranspiration rates
WEc2: Innovative Wastewater Technologies	- EPAct: Energy Policy Act of 1992
WEc3.1: Water Use Reduction: 20%	- EPAct: Energy Policy Act of 1992
WEc3.2: Water Use Reduction: 30%	- EPAct: Energy Policy Act of 1992
EAp1: Fundamental Commissioning of the Building Energy Systems	None
EAp2: Minimum Energy Performance	- ASHRAE/IESNA Standard 90.1-2004 (w/o amendments) (Mandatory of Sections 5.4, 6.4, 7.4, 9.4, 10.4) And (Prescriptive of Sections 5.5, 6.5, 7.5, 9.5, or performance of Section 11)
EAp3: Fundamental Refrigerant Management	None
EAc1: Optimize Energy Performance	- ASHRAE/IESNA Standard 90.1-2004 (w/o amendments), Appendix G Or - ASHRAE Advanced Energy Design Guide for Small Buildings, 2004 Or - Advanced Buildings Benchmark, Version 1.1 (Except sections 1.7, 1.11, and 1.14)
EAc2: On-Site Renewable Energy	- ASHRAE/IESNA Standard 90.1-2004 (w/o amendments), Appendix G (if EAc1 Option 1)
EAc3: Enhanced Commissioning	None
EAc4: Enhanced Refrigerant Management	None
EAc5: Measurement and Verification	International Performance Measurement and Verification Protocol (IPMVP), Vol. III: April 2003; Option D (complex buildings) or Option B (small buildings)

TABLE 8.2.1   (Continued)

Prerequisite or Credit	Referenced Standards
EAc6: Green Power	- ASHRAE/IESNA Standard 90.1-2004 (w/o amendments), Appendix G (if EAc1 Option 1) - Center for Resource Solutions Green-e program technical requirements
Materials and Resources Category Overall	CSI MasterFormat Division List
MRp1: Storage and Collection of Recyclables	None
MRc1.1–1.3: Building Reuse	None
MRc2.1–2.2: Construction Waste Management	None
MRc3.1–3.2: Materials Reuse	None
MRc4.1–4.2: Recycled Content	ISO 14021—Environmental labels and declarations—Self-declared environmental claims (Type II environmental labeling)
MRc5.1–5.2: Regional Materials	None
MRc6: Rapidly Renewable Materials (>2.5%)	None
MRc7: Certified Wood (>50%)	Forest Stewardship Council's Certification
EQp1: Minimum IAQ Performance	For mechanically ventilated areas: - ASHRAE 62.1-2004, Ventilation for Acceptable Indoor Air Quality, Sections 4–7 For naturally ventilated areas: - ASHRAE 62.1-2004, paragraph 5.1 For some low-rise residential facilities: - ASHRAE 62.2-2004 Ventilation and Acceptable Indoor Air Quality in Low-Rise Residential Buildings
EQp2: Environmental Tobacco Smoke (ETS) Control	Residential facilities if not weather-stripped: - ANSI/ASTM-E779-03, Standard Test Method for Determining Air Leakage Rate by Fan Pressurization and - Residential Manual for Compliance with California's 2001 Energy Efficiency Standards, Chapter 4
EQc1: Outdoor Air Delivery Monitoring	None
EQc2: Increased Ventilation	For mechanically ventilated areas: - ASHRAE Standard 62.1-2004 (30% greater) or for low-rise residential facilities - ASHRAE 62.2-2004

TABLE 8.2.1   Listing of Referenced Standards (*Continued*)

Prerequisite or Credit	Referenced Standards
	For naturally ventilated areas: - *Carbon Trust Good Practice Guide 237* (1999) and - Chartered Institution of Building Services Engineers (CIBSE) Applications Manual 10: 2005, natural ventilation in nondomestic buildings, effective as per Fig. 1.18 And model by either - Chartered Institution of Building Services Engineers (CIBSE) Applications Manual 10: 2005, natural ventilation in nondomestic buildings Or - ASHRAE 62.1-2004, Chapter 6, for at least 90% of occupied spaces
EQc3.1: Construction IAQ Management Plan, During Construction	- Sheet Metal and Air Conditioning National Contractors Association (SMACNA) IAQ Guidelines for Occupied Buildings under Construction, 1995, Chapter 3 - ASHRAE 52.2-1999 for MERV ratings
EQc3.2: Construction IAQ Management Plan, Before Occupancy	Option 2: - EPA Compendium of Methods for the Determination of Air Pollutants in Indoor Air
EQc4.1: Low-Emitting Materials: Adhesives and Sealants	Adhesives, sealants, and Sealant Primers: - South Coast Air Quality Management District (SCAQMD) Rule #1168, through 01/07/2005 Aerosol adhesives: - *Green Seal Standard for Commercial Adhesives GS-36* requirements 10/19/
EQc4.2: Low-Emitting Materials: Paints and Coatings	Architectural paints, coatings, and primers applied to interior walls and ceilings: - VOC content limits in *Green Seal Standard GS-11, Paints*, 1st ed., May 20, 1993. Anti-corrosive and antirust paints applied to interior ferrous metal substrates: - VOC content limit of 250 g/L established in *Green Seal Standard GC-03, Anti-Corrosive Paints*, 2d ed., January 7, 1997. Clear wood finishes, floor coatings, stains, and shellacs applied to interior elements: - VOC content limits established in South Coast Air Quality Management District (SCAQMD) Rule 1113, Architectural Coatings 10/1/2004

TABLE 8.2.1   *(Continued)*

Prerequisite or Credit	Referenced Standards
EQc4.3: Low-Emitting Materials: Carpet Systems	Carpet and cushion: - *CRI's Green Label Plus program* Adhesives: See EQc4.1
EQc4.4: Low-Emitting Materials: Composite Wood and Agrifiber	None
EQc5: Indoor Chemical and Pollutant Source Control	- ASHRAE 52.2-1999 for MERV ratings
EQc6.1: Controllability of Systems: Lighting	None
EQc6.2: Controllability of Systems: Thermal Comfort	- ASHRAE Standard 55-2004, Thermal Comfort Conditions for Human Occupancy - Balance with ASHRAE Standard 62.1-2004 for Acceptable Indoor Air Quality
EQc7.1: Thermal Comfort: Design	- ASHRAE Standard 55-2004, Thermal Comfort Conditions for Human Occupancy
EQc7.2: Thermal Comfort: Verification	- ASHRAE Standard 55-2004, Thermal Comfort Conditions for Human Occupancy
EQc8.1: Daylighting and Views: Daylight 75% of Spaces	None
EQc8.2: Daylighting and Views: Views for 90% of Spaces	None
IDc1.1–1.4: Innovation in Design	None or as listed previously for applicable EP point subcategories
IDc2: LEED Accredited Professional	None

TABLE 8.2.1   Listing of Referenced Standards (*Continued*)

variables such as density radius and development density that are used, whereas for EQc3.1 the concepts include the five principles of SMACNA.

It is intended that the individual make similar tables for the other LEED-NC categories and complete them for a comprehensive review of the rating system and as a tool for future reference.

8.3   Additional Tools and Education

Energy Star, BEES, and More

Two national rating tools have been developed in recent years that can aid in many of the sustainable design decisions. One has been developed by the EPA and rates individual building products based on energy consumption. It is the Energy Star program. The Energy Star label is a common feature on many consumer products also. The second tool has been developed by the National Institute of Standards Technology (NIST). It is called Building

Credit	Exemplary Performance Point Availability	Exemplary Performance Point Criterion	Options	Standards/Codes	Criterion	Submittals	Comments	Variables/Concepts
SSp1: Construction Activity Pollution Prevention	n.a.							
SSc1: Site Selection	No							
SSc2: Development Density and Community Connectivity	Yes	Double						
SSc3: Brownfield Redevelopment	No							
SSc4.1: Alternative Transportation—Public Transportation Access	Yes	Double plus Minimum Transit Rides/Day						
SSc4.1: Alternative Transportation: Public Transportation Access	Yes	Plan						
SSc4.2: Alternative Transportation: Bicycle Storage and Changing Rooms								
SSc4.3: Alternative Transportation: Low-Emitting and Fuel-Efficient Vehicles								
SSc4.4: Alternative Transportation: Parking Capacity								
SSc5.1: Site Development: Protect or Restore Habitat	Yes	75%						
SSc5.2: Site Development: Maximize Open Space	Yes	Double						
SSc6.1: Stormwater Management: Quantity Control	No							
SSc6.2: Stormwater Management: Quality Control	No							
SSc7.1: Heat Island Effect: Non-Roof	Yes	100%						
SSc7.2: Heat Island Effect: Roof	Yes	100% Green						
SSc8: Light Pollution Reduction	No							

TABLE 8.2.2   SS Prerequisite and Credit Summary Table Study Guide

Credit	Exemplary Performance Point Availability	Exemplary Performance Point Criterion	Options	Standards/Codes	Criterion	Submittals	Comments	Variables/Concepts
WEc1.1: Water Efficient Landscaping: Reduce by 50%	No							
WEc1.2: Water Efficient Landscaping: No Potable Water Use or No Irrigation	No							
WEc2: Innovative Wastewater Technologies	Yes (1)	1 point for 100%						
WEc3.1: Water Use Reduction: 20%		1 point for 40% EPAct and/or 1 point for 10% Non-EPAct						
WEc3.2: Water Use Reduction: 30%	Yes (2)							

TABLE 8.2.3 WE Credit Summary Table Study Guide

for Environmental and Economic Sustainability (BEES) and can be used to aid in choosing building materials based on both environmental and economic considerations. The current version in 2007 was *BEES 4.0*. It can be downloaded for free from the NIST website.

The BEES model usually assesses products based on a 50-year model, and it looks into various stages of the product life. The main stages are

- Raw materials acquisition
- Product manufacture
- Transportation
- Installation
- Operation and maintenance
- Recycling and waste management

Various impacts are evaluated in the BEES model, and initially the main environmental items included were GWP, acid-rain impact, nitrification potential, natural resource depletion, solid waste impact, and indoor air quality considerations. As the model developed, additional factors were added. The various subcategories evaluated in Version 4 are listed in Table 8.3.1.

Environmental Categories	Economic Categories
Global warming	First Costs
Acidification	Future Costs
Eutrophication	
Fossil fuel depletion	
Indoor air quality	
Habitat alteration	
Water intake	
Criteria air pollutants	
Human health	
Smog	
Ozone depletion	
Ecological toxicity	

TABLE 8.3.1   Various Modules Assessed in BEES 4.0

Other decision tools for choosing sustainable materials and products are being developed by many organizations, both public and private. As the movement evolves, these will become more readily accessible and inclusive. Even trade organizations are developing sustainable design manuals and guides for their product and services customers. An example is the National Ready Mixed Concrete Association (NRMCA) which has published a guide to LEED 2.2 with respect to concrete applications. It is a good idea to ask the various trade organizations representing the building materials that one uses if they can supply you with their current updated sustainability guide.

Incorporation into CSI Formats

There has been much interest into how best to incorporate sustainable construction into the established building process methods in the United States. Even though there may be many methods tried and recommended, one of the most important things is to establish some consistency so that the lessons learned and experiences can be evaluated and handed down to the next generation as effectively as possible. With this is mind, it is important to establish some consistent reporting mechanisms. Since the Construction Specification Institute (CSI) has developed the CSI MasterFormat Division List, which is used extensively in the United States for construction management and data reporting, particularly with respect to materials and products, it makes sense to also follow this format with the management of data with respect to the LEED rating system. In response to this, the CSI has recently started developing the GreenFormat. GreenFormat was under test in 2007. This format requests data on products and is organized into 14 categories of questions:

- MasterFormat 2004 section number and name of material/product
- Manufacturer's information

- Product description
- Regulatory agency sustainability approvals
- Sustainable standards and certifications
- Sustainable performance criteria
- Sustainable compositions of product
- Material extraction and transportation
- Manufacturing phase
- Construction phase
- Facility operations phase
- Deconstruction and recycling phase
- Additional information
- Certification

Additional information about the status of GreenFormat can be found at www.greenformat.com.

Education

Recent research has shown that increased familiarity with the LEED system is one of the major factors in improving performance and the cost-effectiveness of green building. The two major components of this are experience with the system and education about the system. Several universities across the country have set the standard with green building, making commitments to using the LEED rating system for all new campus construction. At the University of South Carolina in Columbia, it was shown that this commitment had a significant positive impact on the familiarity of the various stakeholders in the area. A study also showed that regions of the state where universities and colleges had forged ahead with green building initiatives also became pockets of additional sustainable construction. Both experience and continual education are important.

The USGBC offers many workshops on the rating system and the various guidances. In addition, many facilities have been built that are becoming showcases for various sustainable alternatives. Some examples in South Carolina are the West Quad, Learning Center, at the University of South Carolina (see Fig. 2.7.3), the Sustainable Interiors Showcase at Fort Jackson in Columbia, and the Edisto Interpretive Center at Edisto Beach. The Southface Energy Institute in Atlanta, Georgia, and the Rocky Mountain Institute in Old Snowmass, Colorado, are both national showcases and offer a range of educational opportunities. On a federal level, in addition to the resources previously mentioned from the Department of Defense (DoD) there is the Office of the Federal Environmental Executive (OFFE) where additional resources have been collated and also documented in a report entitled *The Federal Commitment to Green Building: Experiences and Expectations* (http://www.ofee.gov/sb/fgb_report.html) and of course the EPA websites for the various environmental concerns, including a website devoted to green buildings (http://www.epa.gov/greenbuilding/).

As mentioned in Chap. 1, there is also a movement among higher education presidents and faculty to incorporate sustainability education into the curriculum. It is particularly important to add sustainability concepts to the tools that engineers use for future designs and innovations.

References

Bilec, M., and R. Ries (2007), "Preliminary Study of Green Design and Project Delivery Methods in the Public Sector," *Journal of Green Building*, Spring, 2(2): 151–160.

CSI (2004), *The Project Resource Manual—CSI Manual of Practice*, 5th ed., Construction Specifications Institute, Alexandria, Va.

CSI (2006), "Supporting Sustainability with GreenFormat," *The Construction Specifier*, 59 (12), accessed at http://www.csinet.org/s_csi/docs/13700/13644.pdf, July 26, 2007.

Davidson, C. I., et al. (2007), "Adding Sustainability to the Engineer's Toolbox: A Challenge for Engineering Educators," *Environmental Science and Technology*, 41 (July 15): 4847–4850.

DuBose, J. R., S. J. Bosch, and A. R. Pearce (2007), "Analysis of State-Wide Green Building Policies," *Journal of Green Building*, Spring, 2(2): 161–177.

EPA (2007), http://www.epa.gov/greenbuilding/, U.S. Environmental Protection Agency webpage, accessed July 10, 2007.

Hooper, P.A. (2007), "The Construction Specification Institute's Sustainable Reporting Data Format: GreenFormat," *Journal of Green Building*, Spring, 2(2): 3–13.

Horman, M. J., et al. (2006), "Delivering Green Buildings: Process Improvements for Sustainable Construction," *Journal of Green Building*, Winter, 1(1).

Huff, W. (2007), "Obstacles and Rewards of Enterprise Sustainability and the Need for a Director of Sustainability," *Journal of Green Building*, Spring, 2(2): 62–69.

Kibert, C. J. (2005), *Sustainable Construction; Green Building Design and Delivery*, John Wiley & Sons, Hoboken, N.J.

NIST (2007), *BEES 4.0*, National Institute of Standards Technology, Gaithersburg, Md., http://www.bfrl.nist.gov/oae/software/bees/ website accessed August 2, 2007.

NRMCA (2005), *Ready Mixed Concrete Industry LEED Reference Guide—Updated with LEED Version 2.2 Information*, National Ready Mixed Concrete Association, Silver Spring, Md.

OFEE (2003), *The Federal Commitment to Green Building: Experiences and Expectations*, Office of the Federal Environmental Executive, http://www.ofee.gov/sb/fgb_report.html accessed July 10, 2007.

Portney, K. E. (2002), "Taking Sustainable Cities Seriously: A Comparative Analysis of Twenty-Four US Cities," *Local Environment*, 7(4): 363–380.

RMI (2007), http://www.rmi.org/, Rocky Mountain Institute webpage, accessed July 10, 2007.

Schexnayder, C. J., and R. E. Mayo (2004), *Construction Management Fundamentals*, McGraw-Hill, New York.

Southface (2007), http://www.southface.org/, Southface Energy Institute webpage accessed July 10, 2007.

Thomas, D., and O. J. Furuseth (1997), "The Realities of Incorporating Sustainable Development into Local-Level Planning: A Case Study of Davidson, North Carolina," *Cities*, 14(4): 219–226.

USGBC (2003), *LEED-NC for New Construction, Reference Guide*, Version 2.1, 2d ed., May, U.S. Green Building Council, Washington, D.C.

USGBC (2005–2007), *LEED-NC for New Construction, Reference Guide*, Version 2.2, 1st ed., U.S. Green Building Council, Washington, D.C., October 2005 with errata posted through Spring 2007.

Watson, C. H. (2006), *The University of South Carolina: A Catalyst for Sustainable Development?*, MS thesis, University of South Carolina, Columbia.

Exercises

1. Put together a CPM schedule for sequencing the installation of absorbent materials related to SMACNA for EQc3.1 and EQc3.2 for materials stored outside. Two options that might be used are staggered procurement and on-site protected storage.

2. Put together a CPM schedule for sequencing the installation of absorbent materials related to SMACNA for EQc3.1 and EQc3.2 with respect to interior painting and finishing and VOCs.

3. Look up a paint product on the GreenFormat system and list the information.

4. Look up a wood furniture product on the GreenFormat system and list the information.

5. Look up a plumbing product on the GreenFormat system and list the information.

6. Download the BEES program and evaluate the performance of a particular product.

7. Go to the appliance store and compare the Energy Star labels on four different refrigeration units.

8. Compare the Energy Star labels on four different television sets.

9. Compare the Energy Star labels on four different driers.

10. Identify three different materials or products used to construct your classroom. Find out if green information is available on the product websites.

11. Go to the paint department in a store and compare the labels on similar paint products from three different manufacturers. Compile and compare any environmental data. Go to the manufacturers' websites and see if green standards are listed for these products. Summarize them.

12. Go to the carpet department in a store and find the names of three different manufacturers. Go to the manufacturers' websites and see if green standards are listed for these products. Summarize them.

13. Go to a furniture department and find three manufacturers of a specific type of furniture product (for example, wood chair, mattress, couch, dining table). Go to the manufacturers' websites and see if green standards are listed for these products. Summarize them.

14. Visit a vacant lot in your area and make a list of existing site features that you can see from outside the property line and other areal information such as direction, access, and nearby utilities.

15. Go online and see what additional information you can collect on the site in Exercise 14. This may include mapping programs, land record databases, MPO GIS websites, and state websites. There are also websites that list areal environmental items such as designated brownfields and floodplain information.

16. Develop an initial site assessment questionnaire that can be used for preliminary site assessments from a site visit.

17. Develop an expanded site assessment questionnaire that can be used to collect information from local land record offices, departments of public works, and other public record agencies that will be useful in initial site assessments.

18. Look up one of the Code of Federal Regulation (CFR) codes referenced in Table 8.2.1. Summarize your findings. These can usually be found online.

19. Look up one of the EPA referenced items in Table 8.2.1. Summarize your findings. These can usually be found online.

20. Go to a library (or other resource) and read one of the ASTM standards referenced in Table 8.2.1. Summarize your findings.

21. Go to a library (or other resource) and read one of the ASHRAE standards referenced in Table 8.2.1. Summarize your findings.

22. Look up EPAct 1992 as referenced in Table 8.2.1. Summarize your findings.

23. Look up one of the Green Seal standards as referenced in Table 8.2.1. Summarize your findings.

24. Update Table 8.2.1 based on any errata to LEED-NC 2.2 or for the current version of LEED-NC.

25. Complete Table 8.2.2.

26. Complete Table 8.2.3.

27. Make and complete a summary study guide similar to Table 8.2.2 for the EA category.

28. Make and complete a summary study guide similar to Table 8.2.2 for the MR category.

29. Make and complete a summary study guide similar to Table 8.2.2 for the EQ category.

CHAPTER 9

Department of Defense (DoD) Sustainable Construction and Indoor Air Quality (IAQ)

onstruction spending in the United States as of December 2006 totaled approximately $1200 billion annually. The federal government alone pumped over $19 billion a year into the economy for the construction of new facilities, and the U.S. government had an inventory of approximately 445,000 buildings occupying nearly 3.1 billion ft^2 of floor space. These facilities consumed approximately 0.4 percent of the nation's total energy usage and spewed out about 2 percent of all building-related greenhouse gases. The federal government is a major consumer of construction materials and a significant user of energy and natural resources. To minimize the impact that federal facilities have on the environment, the U.S. government has become a leader in implementing policy and objectives for energy reduction and green building.

9.1 Government Mandates

The President of the United States exercises authority over all the federal agencies within the executive branch of the U.S. government. One of the ways the President provides guidance is through written directives called Executive Orders (EOs). In the late 1990s President Clinton began issuing a series of EOs aimed at "Greening the Government." Among other objectives, these have served to set the baseline for energy reduction and implementation of green building within all federal agencies. EO 13101, "Greening the Government Through Waste Prevention, Recycling, and Federal Acquisition," was the first of this series of EOs and was signed in September 1998. Some of the requirements of this order included promotion of waste prevention and recycling and the use of environmentally preferable products. Federal agencies are required to establish procurement programs and develop specifications to ensure all products meet or exceed Environmental Protection Agency (EPA) guidelines. This requires maximizing the use of materials that are reused, recovered, or recycled and have reduced or eliminated toxicity. Additionally, this order required each agency to establish goals for solid waste prevention and recycling. In June 1999, EO 13123, "Greening the Government Through Efficient Energy Management," was signed, setting aggressive goals for reducing the government's energy consumption. This order established the following targets:

- Reduce building-related greenhouse gases by 30 percent by 2010 compared to 1990 levels.

- Reduce energy consumption for facilities by 30 percent by 2005 and 35 percent by 2010 relative to 1985 energy usage.

- Install 2000 solar energy systems for federal facilities by the year 2000 and 20,000 by the year 2010.

- Reduce petroleum use by switching to natural gas or renewable fuels or by eliminating unnecessary use.

- Reduce water consumption to meet Department of Energy water conservation goals.

This order also established objectives for federal agencies to maximize use of Energy Star products, pursue Energy Star building criteria, and implement sustainable building design. Energy Star is a program of the U.S. Environmental Protection Agency and the U.S. Department of Energy which has the purpose of helping the United States save money and protect the environment through the use of energy-efficient products and practices. The sustainable design principles developed by the Department of Defense and General Services Administration were also specifically cited as benchmarks for other agencies.

Another federal publication, the Office of Management and Budget's Circular A-11, provided more details and updates to the goals behind EO 13123 and specifically encouraged agencies to evaluate the use of Energy Star or LEED standards and guidelines for incorporation in designs for new building construction or renovation in Section 55 in its 2002 version. A summary of how the different federal agencies have responded to this order as reported by the Office of the Environmental Executive can be found in the 2003 report *The Federal Commitment to Green Building: Experiences and Expectations.*

Pollution prevention (P2) and environmentally preferable purchasing (EPP) were programs set forth by EO 13148, "Greening the Government Through Leadership in Environmental Management," signed in April 2000. This order furthered the use of life-cycle considerations and implemented additional acquisition and procurement practices to select products and develop facilities that have reduced toxic chemicals, hazardous substances, and pollutants. This order also required the planning and development of environmentally and economically beneficial landscaping.

Then in January of 2007 President Bush signed Executive Order 13423, "Strengthening Federal Environmental, Energy, and Transportation Management." This order further identified sustainable strategies and goals for the agencies of the federal government. Each agency was ordered to meet certain energy, transportation, and environmental goals and to maintain an environmental management system (EMS) structure. The specific manners in which these were met are to be developed by the agencies. EO 13423 also revoked the three previously mentioned orders signed by President Clinton. A summary of these orders is presented in Table 9.1.1.

Executive Order	Title	Federal Register Date	Status as of July 2007
EO 13101	Greening the Government through Waste Prevention, Recycling, and Federal Acquisition	September 16, 1998	Revoked Jan. 26, 2007
EO 13123	Greening the Government through Efficient Energy Management	June 8, 1999	Revoked Jan. 26, 2007
EO 13148	Greening the Government through Leadership in Environmental Management	April 26, 2000	Revoked Jan. 26, 2007
EO 13423	Strengthening Federal Environmental, Energy, and Transportation Management	January 26, 2007	Current

TABLE 9.1.1   Recent U.S. Sustainable Construction Related Executive Orders

9.2 Department of Defense Facilities

In 2006, the Department of Defense (DoD) owned a total of over 330,000 buildings occupying more than 2.0 billion ft^2 of floor space. The DoD accounted for approximately two-thirds of the federal government's buildings by number and square footage and used 244 trillion Btu of energy annually at a cost of $2.6 billion. The DoD continues to grow and modernize through new facility construction. Its military construction (MILCON) budget has averaged $6 billion per year during the 5-year period from 2001 through 2005 and grew to $8 billion for fiscal year 2006 (FY06). Based on the volume of projects for the development of facilities for new and emerging requirements, renovation and modernization of aging facilities, and support for moving commands and service members due to base realignment and closure (BRAC), the DoD budget showed a MILCON volume of nearly $8 billion for FY07 and almost doubled to $15 billion per year for FY08 through FY11. The DoD has become a leader among government agencies in implementing sustainable principles, such as energy reduction, recycling and waste minimization, and lessening the impact on the environment, for planning, design, and construction though experience and by necessity. Its implementation involves not only buildings (vertical construction), but also sustainability of the ranges and other horizontal construction. The DoD has also adopted *low-impact development* (LID) practices at many facilities.

U.S. Army and the U.S. Army Corps of Engineers

Within the Department of the Army, installation management services are provided by the Office of the Assistant Chief of Staff for Installation Management (ACSIM). This includes establishing policy and program management for MILCON. However, the actual coordination and management of MILCON projects are the responsibility of the U.S. Army Corps of Engineers (USACE). The Army is the largest federal facility owner of all agencies with over 160,000 buildings occupying 1.1 billion ft^2 of floor space. This is nearly one-third of all federal buildings and one-half of those owned by the DoD. The USACE also manages an equivalent percentage of the MILCON budget each year. Following the signing of EO 13123, the Office of the Deputy Assistant Secretary of the

Army for Installations and Housing (DASA-I&H) issued a memo in April of 2000 addressing sustainable design and development. DASA-I&H directed ACSIM to implement policy and USACE to develop technical guidance for sustainable construction. In May 2000, ACSIM published the *sustainable design and development* (SDD) policy, defining *SDD* as "the systematic consideration of current and future impacts of an activity, product or decision on the environment, energy use, natural resources, the economy and quality of life" and mandating SDD considerations for Army installations. One year later, USACE developed and released the Sustainable Project Rating Tool (SPiRiT) and provided technical guidance for implementation in Technical Letter ETL 1110-3-491, Sustainable Design for Military Facilities. SPiRiT is a tool similar to and based on LEED, although it incorporates operations and maintenance issues, allows for flexibility in design for building modifications based on need changes, and is a self-rating tool that does not require third-party certification as with LEED. ACSIM issued a policy statement in May 2001 for the implementation of SPiRiT with the initial requirement of a bronze rating for all new construction projects. Over the next several years, the sustainable rating requirements were raised to silver and then to gold, and several version changes were made to further refine SPiRiT. In late 2005, the USACE's Construction Engineering Research Laboratory (CERL) released the results of a study comparing SPiRiT rated projects to LEED and recommended that the Army adopt LEED as the primary rating tool. On January 5, 2006, the Office of the Assistant Secretary of the Army for Installations and the Environment (OASA-I&E) released a memorandum mandating the change from SPiRiT to LEED effective with the FY08 MILCON program and establishing LEED silver as the standard. The U.S. Army has aggressively pursued sustainable construction in the MILCON program and seems poised to continue raising the standard as LEED grows as the industry benchmark.

The Army has been progressively establishing 25-year sustainability goals at many of its facilities. This effort is led out of the Sustainability Division ODEP (Office of Director of Environmental Programs) in the Pentagon. In response to this initiative, in 2007, a group at Fort Jackson in Columbia, S.C., developed a showcase for "Sustainable Interiors" as both a way to incorporate sustainability into the purchasing procedure on post and an educational outlet for the Army and the community. Figure 9.2.1 shows an educational cutaway in the showcase.

Naval Facilities Engineering Command

For the Department of the Navy, the Commander, Naval Installations Command (CNIC), owns and manages all the Navy's installations and facilities thereon. CNIC relies on Naval Facilities Engineering Command (NAVFAC) for all engineering support, including the establishment of technical guidance and policy, related to facility sustainability and execution of the MILCON program. The Department of the Navy, including the Marine Corps, owns about 30 percent of the DoD's buildings and has recently had a robust MILCON program, managing above normal construction volume mainly due to the naval installations on the Gulf Coast that sustained significant damage due to hurricanes Ivan in 2004 and Katrina, Rita, and Wilma in 2005. The Navy was the first federal agency to pursue implementation of sustainable design features into MILCON projects. The *Whole Building Design Guide* (www.wbdg.org) was launched by the Navy in 1997 to begin incorporating sustainable requirements into construction standards, specifications, and guidelines. The Construction Criteria Base (CCB) was

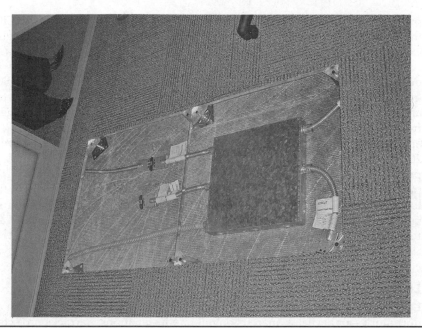

FIGURE 9.2.1   Army Sustainable Interiors demonstration and education site. The cutaway shows a TecCrete variable-height floor system by Haworth, Inc. Raised floor systems can help provide space to facilitate energy-efficient controls and air distribution in buildings as related to several EA credits. (*Photograph taken at the Sustainable Interiors ribbon cutting at the Strom Thurmond Building on Fort Jackson, Columbia, S.C., June 2007.*)

built within the WBDG website as a data store for standardizing construction specifications for all the military services. The WBDG is now maintained by the National Institute for Building Sciences and is supported by eight federal agency partners.

NAVFAC completed DoD's first LEED certified facility with the construction of a 365,000 ft², $55 million Bachelor Enlisted Quarters at Great Lakes Naval Training Center in FY99. Since then several major renovation projects at Washington Navy Yard and the Pentagon have been showcases for applying LEED for existing buildings and the use of LID. In July 2002, NAVFAC's Chief of Engineering issued a memorandum adopting the LEED rating system as a tool and metric for MILCON. In June 2003, NAVFAC Instruction 9830.1, Sustainable Development Policy, was issued requiring all construction, renovation, and repair projects over $750,000 to attain the LEED certified level and Navy Family Housing to implement the EPA Energy Star Label Homes Program. The Department of the Navy was a pioneer in making sustainable design and development a standard for military construction and continues these efforts through utilization of LEED.

Air Force Center for Environmental Excellence

The U.S. Air Force programs and budgets for MILCON requirements go through the Office of the Civil Engineer. Field agencies within this office include the Air Force Civil Engineer Support Agency (AFCESA), which establishes construction standards and design criteria, and the Air Force Center for Environmental Excellence (AFCEE), which provides technical support and guidance. Although it is the smallest facility owner of the military services, the Air Force still maintains a sizable MILCON budget; however, execution of Air Force MILCON projects is divided between USACE and NAVFAC (the Air Force does not staff its own construction agent). The Air Force was the first department within the DoD to adopt the U.S. Green Building Council's (USGBC) LEED green building rating system as the standard for measuring sustainable development and construction. The Sustainable Development Policy was issued by a memorandum from the U.S. Air Force (USAF) Civil Engineer Deputy Chief of Staff for Installations and Logistics in December 2001. The goals set forth in this memo were to achieve 20 percent of Air Force MILCON projects selected as LEED pilots by FY04 and all projects to become LEED certified by FY09. The Air Force, through AFCEE, has also published a number of guides to assist project teams with developing specific action plans to incorporate LEED guidance into their design and construction planning. From the beginning, the Air Force has supported LEED and continues to develop and publish tools to aid in implementation of sustainable development.

9.3 LEED Application Guide for Lodging

Lodging facilities for soldiers, sailors, airmen, and marines are referred to as barracks, dormitories, or bachelor quarters on military installations. Barracks projects make up a large percentage of military construction projects, as the military forces grow and troops are relocated from one installation to another, and in recent years are part of major recapitalization programs to replace older facilities dating back to the World War II era or older. The quality and design of barracks have a vital role in the health, welfare, and morale of the service members who call them home; therefore, sustainable design contributes benefits not only to the global environment but also to the local living environment for men and women in uniform.

The most common design for barracks facilities is a three- or four-story building or grouping of these buildings in a "campus" area. Since LEED-NC 2.2 was primarily developed for new and renovated commercial buildings and general office facilities, different challenges are encountered, particularly in achieving energy efficiency and indoor environmental quality, for low-rise residential buildings (less than four stories) which are designed for continuous occupancy. To provide guidance and promote sustainable construction for barracks facilities, AFCEE worked with the USGBC to develop and publish *Application Guide for Lodging Using the LEED Green Building Rating System*. This publication was prepared by Paladino and Company, Inc., in 2001 and is more commonly referred to as the LEED Application Guide for Lodging. The intent of this guide is to provide interpretation of LEED credits in application to low-rise, lodging building projects and specific to Air Force dormitory construction, although it also allows for generic application on other similar facilities.

The LEED Application Guide for Lodging addresses each of the six LEED categories: Sustainable Sites (SS), Water Efficiency (WE), Energy and Atmosphere (EA), Materials and Resources (MR), Indoor Environmental Quality (EQ), and Innovation and Design Process (ID). The guide is laid out in a format that details the official intent and requirements of LEED for each credit; then it provides special considerations for lodging facilities and unique applications and requirements for USAF dormitories. Also, for each credit, a list of steps is provided to evaluate or achieve the credit, including a summary of the official LEED submittals required and applicable Air Force references.

The intent of this format is to ensure the project planning and design teams have consolidated information for review and consideration as a tool to interpret LEED for lodging facilities.

The LEED Application Guide for Lodging is a useful tool for design and project teams involved in construction of low-rise lodging facilities and specifically MILCON for U.S. Air Force dormitories. This guide expands on the USGBC's LEED green building rating system by giving suggestions for interpreting the intent of many of the credits and how they can be applied to the unique considerations for facilities of this type and those located on Air Force installations. In addition to the information provided relative to each credit and the steps to evaluate and achieve the credit, the references listed in the guide provide a direct link to additional resources available to project teams engaged in USAF lodging designs. The following summary is provided as examples of the information included in the LEED Application Guide for Lodging for each of the LEED categories and specifically for USAF dormitory projects.

Sustainable Sites

There are 15 credits in the sustainable sites category of LEED. The write-up in the USAF application guide has a significant amount of guidance for interpretation in this category, perhaps due to the unique considerations for development on federal government property and the planning and programming methodology used by the Air Force for construction projects.

The LEED Application Guide for Lodging addresses credits such as site selection, redevelopment, and alternative transportation, stating that achieving these credits may not be possible due to the existing installation location, relationship and layout of existing facilities, and installation general plan (which is the basis for site selection and style of construction). Also, since much of the project planning is done in advance to get funding approval for MILCON, the design and construction architect-engineer (AE) may not be able to have any impact on these issues once a project is approved. Credits for reducing site disturbance and providing for stormwater management many times rely on local zoning regulations and defined (subdivided) parcels of land. However, zoning and subdividing are usually not applicable on government property, although information is provided to interpret the intent as it would apply for an Air Force project.

In some cases, design features that would meet the intent of the credit cannot be achieved due to restrictions in military design. One example is construction of parking underneath a building to reduce the heat island effect. This is prohibited for government facilities due to Anti-Terrorism/Force Protection (AT/FP) concerns.

However, other useful sustainable features are promoted throughout the guide, such as use of pervious concrete and design of constructed wetlands. The primary references for this area include the USAF Landscape Design Guide, the Land Use Planning Bulletin, and general plan for the specific USAF installation.

Water Efficiency

There are only three credit subcategories in the area for Water Efficiency, each of which is well addressed in current Air Force guidances. Most of these guidances are available at the the Air Force Center for Engineering and the Environment which provides Air Force leaders with the expertise on facilities management and construction. The Air Force Water Conservation Program under the Air Force Civil Engineer Support Agency addresses wastewater technologies and water use reduction.

Energy and Atmosphere

The prerequisites in the Energy and Atmosphere category are covered by federal, DoD, and Air Force policy or MILCON specifications; however, some of the six credit categories may be difficult or costly to achieve. Energy reduction and the related LEED credit points are a high-priority goal for DoD, and targets are set by Air Force policy. Guides such as the USAF Passive Solar Handbook (available from the Air Force Center for Engineering and the Environment) encourage designs that promote use of natural light to help reduce energy use. Certain sustainable strategies, such as the use of many types of renewable energy and green power, are completely dependent on availability and are currently not as likely to be attained on military installations. However, many installations are looking into opportunities for future energy self-sufficiency.

Credits for additional commissioning, measurement and verification, and ozone depletion may be achievable, but first cost is an issue that must also be considered in the budget cycles. The Air Force has implemented a process of life-cycle assessment (LCA) for MILCON projects to evaluate issues such as these not only for first cost, but also with consideration of what the return would be over the life of the facility.

Materials and Resources

The LEED section for Materials and Resources contains one prerequisite and eight credit subcategories. In this area, some sustainable elements are standard practice in the military, but others that would otherwise seem achievable may be hampered by procurement regulations. The prerequisite for storage and collection of recyclables is a given, as it has become common practice on military installations. The building reuse credit depends on the project but is not common since many older facilities have exceeded their design life and do not meet current military design requirements; therefore, they would be too expensive to upgrade rather than replace. The credit for construction waste management is covered by the Construction and Demolition (C&D) Waste Management Guide (available from the Air Force Center for Engineering and the Environment), and the Defense Reutilization and Marketing Service (DRMS) is a practical avenue to incorporate resource reuse.

What is most interesting in this section is that credits for recycled content, local and regional materials, rapidly renewable materials, and certified wood all should be achievable with proper planning and product selection, although the LEED Application Guide for Lodging includes a disclaimer stating, "Government procurement regulations may prohibit achievement of this credit because there may only be a single supplier for the qualifying material." Recent efforts at many military installations have helped ease the procurement obstacles. An example is the Sustainable Interiors Showcase that opened in 2007 at Fort Jackson in South Carolina, where procurement and design teams collaborated to develop a grouping of interior finishes and furnishings which could be easily purchased and used to attain many credits.

Indoor Environmental Quality

This section includes two prerequisites and 16 credits. The prerequisites are easily met since IAQ performance requirements are included in construction specifications, and environmental tobacco smoke is typically achieved by smoking bans and the establishment of outdoor designated smoking areas. Some credits are fairly easy, such as daylight views and operable windows (for increased ventilation and controllability of systems), which are inherent in lodging design. Some credits are not as feasible, such as CO_2 monitoring, since it may be difficult and expensive due to the decentralized HVAC design common to lodging facilities. Other credits could be accomplished by simply making them a requirement in the design standards and construction specifications. Although the LEED Application Guide for Lodging cites the importance of IAQ due to the long lifetime of lodging facilities and the extended duration of exposure to contaminants by occupants, it also states that there are no standards or product selection criteria for construction IAQ management and low-emitting materials which could easily aid in mandating achievement of these credits. However, as noted in the Materials and Resources section, recent initiatives at many military facilities to form sustainable initiatives between procurement and design teams have resulted in these materials becoming more readily accessible for installation use. Indoor air quality (IAQ) is one of the main components of this category and is one of the most important sustainable features for lodging facilities and Air Force housing projects since it is of major importance for the health of occupants and has a direct impact on training and mission if service members become ill. Items related to IAQ in Air Force facilities are further detailed in Sec. 9.4.

Innovation and Design Process

No specific guidance is provided for these credits as they involve individually substantiated evidence that additional innovation beyond the other categories and credits was applied to the specific project. There is very little that could be added to help interpret these broadly defined credits to a lodging faculty or Air Force project.

Many of the items in the LEED Application Guide for Lodging can be applied in so many other industries in the United States. Many states are promoting "Green Lodging" programs in the hotel and tourism industry too. Considering that lodging is the fourth-largest energy consumer in the U.S. commercial sector, it does make both environmental and economic sense for this industry to also go green.

9.4 Importance of Indoor Air Quality

IAQ is one of the most vital design considerations for military lodging facilities since indoor pollution-related illnesses can create losses of personnel for critical training or mission requirements. IAQ also has a higher relative significance in residential or lodging facilities due to the longer sustained time period residents occupy these facilities. The health and comfort of occupants of facilities and the impact this has on productivity are a recognized value of a sustainable facility. Poor indoor air quality may result in temporarily degrading an individual's physical well-being and can lead to serious long-term illnesses. Types of symptoms commonly linked to poor indoor air quality include headaches, fatigue, shortness of breath, sinus congestion, coughing, sneezing, irritation of the eye, nose, and throat, dizziness, and nausea. Immediate reactions may occur after a single exposure and can be further exacerbated by repeated exposures. These health effects could be a minor inconvenience or significantly debilitating. LEED-NC 2.2 contains two prerequisites and nine credits pertaining specifically to IAQ.

Indoor air pollution results from a variety of contaminants from sources in the outside air, equipment used in the facility, building components and furnishings, and human activities conducted within the facility. Outside air contaminants include pollen, dust, industrial pollutants, vehicle exhaust, and soil contamination such as radon or even moisture, which may breed mold or other microbial growth. Facility equipment pollutants result from dust or microbes in HVAC components, use of office equipment and supplies, and operations of elevator motors or other mechanical systems. Building components and furnishings contribute to poor IAQ from dust collected within carpeting and hard-to-clean areas, microbiological growth from water damage or moisture, and volatile organic compounds (VOCs) released from furnishings and building products. Human activities produce indoor pollution from cleaning solvents, cooking, cosmetics, pesticides, and VOCs from maintenance products (e.g., paint, caulk, adhesives). Additionally, inadequate ventilation may increase the indoor pollutant levels by not diluting the indoor sources and taking the pollutants out.

Although the primary function of HVAC systems is to control the temperature and humidity of supply air, they are increasingly being designed for maintaining IAQ as well. According to the EPA IAQ Guide, "a properly designed and functioning HVAC system provides thermal comfort, distributes adequate amounts of outdoor air to meet ventilation needs of all building occupants, and isolates and removes odors and contaminants through pressure control, filtration, and exhaust fans." Ultimately, there are three primary strategies for controlling IAQ: source control, ventilation, and air cleaning.

Source Control

Source control can have an immediate and ongoing impact on IAQ by limiting contaminant-producing materials used in the construction and furnishing of the facility and implementing controls to reduce cleaning and maintenance products which produce indoor pollutants. According to the EPA IAQ Guide, "source control is generally the most cost effective approach to mitigating IAQ problems in which point sources of contaminants can be identified." Source control is directly addressed by one of the prerequisites and three of the subcategories in the Indoor Environmental Quality category of LEED-NC 2.2. They are as follows:

- EQ Prerequisite 2: Environmental Tobacco Smoke (ETS) Control prohibits or restricts smoking within or near the facility.

- EQ credits 3.1 and 3.2 require construction IAQ management plans to minimize the potential of contaminants from the construction phase having an impact after occupancy.

- EQ credits 4.1 through 4.4 require selection of low-emitting materials for construction products.

- EQ credit 5 addresses entryway and other source area designs to minimize the intrusion of outdoor contaminants.

There are also other design issues not specifically credited through LEED that help prevent poor IAQ. The siting of the facility and the location of ventilation intakes are related to source control in that these design considerations can help maximize separation from outdoor sources of contaminants. Additionally, it is important that safeguards, policies, or practices be put in place with sufficient related enforcement to ensure product selection for cleaning and maintenance activities conducted throughout the life of the facility that utilizes low-emitting solutions and compounds. It is evident from the number of credits available within LEED that source control is an important part of IAQ management, although there are additional measures that need to be considered for both the design and the operation of facilities to maintain a healthy IAQ.

Ventilation

The common catch phrase "dilution is the solution to pollution" has actual application as a control mechanism for IAQ. Increasing the amount of outdoor air brought in through the HVAC system is another approach to lowering the concentrations of indoor air contaminants. According to the EPA IAQ Guide, some of the main design considerations for HVAC systems to dilute indoor air contaminants are to "increase the total quantity of supply air (including outdoor air), increase the proportion of outdoor air to total air, and improve air distribution." To measure this effectiveness, "the term 'ventilation efficiency' is used to describe the ability of the ventilation system to distribute supply air and remove internally generated pollutants." LEED-NC 2.2 stresses proper ventilation though EQ Prerequisite 1: Minimum IAQ Performance which establishes ASHRAE 62.1-2004, Ventilation for Acceptable Indoor Air Quality, as the standard ventilation system design. Additionally, EQ credit 1 provides for monitoring CO_2 concentrations as a means to ensure ventilation is maintained as designed ,and EQ credit 2 recognizes efforts to increase ventilation above the ASHRAE 62.1-2004 standard. Special ventilation system considerations for establishing negative pressure in specific spaces are addressed within EQ prerequisite 1 for ETS control and EQ credit 5 for indoor chemical and pollutant source control. The EPA IAQ Guide also specifies that although introduction of sufficient outdoor air for proper dilution of contaminants is key, "inadequate distribution of ventilation air can also produce IAQ problems." This comment emphasizes the importance of commissioning and proper testing and balancing of HVAC systems for enhancing ventilation efficiency. EA prerequisite 1 and EA credit 3 relate to the commissioning requirements, and EA credit 5 addresses proper measurement and verification. Although these credits are within the Energy and Atmosphere category of LEED-NC 2.2, they have direct contributions to IAQ as well.

Lodging facilities have other unique considerations when it comes to proper ventilation for good IAQ. For example, the use of decentralized HVAC systems is a fairly common practice; however, design solutions need to be evaluated for other cost-effective alternatives that still allow some degree of occupant control while maintaining minimum ventilation requirements. Another concern for lodging is proper ventilation of cooking equipment and fireplaces, if installed. Since not all sources of indoor air pollution can be eliminated, proper ventilation is critical to maintaining good IAQ within a facility. LEED contains several direct and related prerequisites and credits that emphasize the importance of ventilation for IAQ management.

Air Cleaning

Air cleaning, or filtration, should not be considered a primary control mechanism for IAQ management since it is most effective when used in conjunction with either source control or ventilation. Filtration has two main functions—removing contaminants from the airstream that are circulated within the facility and preventing buildup of contaminants within the HVAC system which can reduce equipment efficiency, lead to reentrainment of contaminants in the airstream, and serve as a breeding ground for other microbes. Some technologies that remove contaminants from air are particulate filtration, electrostatic precipitation, negative ion generation, and gas sorption. The first three technologies are designed to remove particulates, and the fourth is designed to remove gaseous contaminants.

Particulate filtration involves the use of mechanical devices placed in the airstream often using tightly woven fibers to trap the particulates as air flows through the filter medium. Filters are classified by their minimum efficiency reporting value (MERV) which relates to the efficiency the filter has for removing particles from 0.3 to 10 μm. High-efficiency particulate air (HEPA) filters and ultra-low penetration air (ULPA) filters are also available for the removal of even smaller particulates. The disadvantage of particulate filtration systems is that the higher the efficiency of the filter, the more it will increase the pressure drop within the air distribution system and reduce total airflow. This in turn can overwork fan motors causing increase in energy usage and can reduce the air exchange rate critical to dilution of indoor air pollutants. Also, particulate filtration systems may require frequent maintenance for replacement or cleaning of filters.

Electrostatic precipitation and negative ion generation are similar in principle because they apply a charge to particles to enable removal from the airstream. With electrostatic precipitators (ESPs), the charged particles become attracted to an oppositely charged surface on which they are collected. These systems provide relatively high-efficiency filtration of small respirable particles at low pressure losses, although ESPs must be serviced regularly and produce ozone. Negative ion generators (ionizers) apply a static charge to remove particles from the indoor air. The charged particles then become attracted to surfaces such as walls, floors, draperies, or even occupants. This is not always an ideal method since it requires deliberate cleaning of the space to actually remove the particles, although more effective designs include collectors to attract particles back into the unit. Like ESPs, ionizers require frequent maintenance and may produce ozone.

Gas sorption is a process for the control of gaseous contaminants, such as formaldehyde, sulfur dioxide, ozone, or VOCs. It involves the use of a sorption material (e.g., activated carbon, chemically treated clays) and processing of the gaseous contaminant, such as chemical reaction with the sorbent, binding of the contaminant with the sorbent, or diffusion of the contaminant from areas of higher concentration to areas of lower concentration. There are currently few standards for rating the performance of these systems, which makes design and evaluation problematic. Also, operating expenses of gas sorption systems can be quite high, and sorbent filters can reach breakthrough conditions at higher temperatures and relative humidity which causes them to desorb VOCs back into the airstream. Recent system development and testing for gaseous contaminant removal have been conducted with photocatalytic oxidation (PCO). PCO is a process involving a photocatalytic filter and UV light for the elimination of VOCs through an oxidation reaction producing carbon dioxide and water.

Air cleaning through filtration systems can be a critical link to long-term maintenance of good IAQ. It is not likely that sources of contaminants can be completely avoided through product selection, and dilution through increased ventilation does not fully eliminate contaminants; therefore, filtration can drastically improve the effectiveness of removing indoor pollutants as the third leg of the IAQ strategic triad. Currently, LEED-NC 2.2 provides no specific credits for application of air cleaning or filtration systems, although if a unique system design is selected with appropriate calculations demonstrating removal efficiency, credits could be applied for under the Innovation and Design Process (ID) category.

Future Combined Technologies

Several other methods are being developed for improvement of IAQ. Many combine both source control and air cleaning in the HVAC system. One mentioned previously related to the UV disinfection and VOC removal methods. Another related to microbes and mold production is the introduction of antimicrobial surfaces such as copper and other antimicrobial products into the HVAC systems. Due to the importance of improved IAQ within military facilities, particularly in housing, medical facilities, and other 24-h occupied spaces such as submarines and ships, the DoD is interested in research into many of these areas. Current research on the opportunities for copper as an antimicrobial in HVAC systems is being led by the Copper Development Association (CDA) and funded through a Congressional Special Interest (CSI) Medical Program overseen by TATRC, the Telemedicine and Advanced Technology Research Center of the U.S. Army Medical Research & Materiel Command (USAMRMC).

The occupants of a facility are perhaps the most significant portion of the "environment" that should be considered when designing for sustainability. Reduction or elimination of indoor air contaminants is vital to enabling healthy working or living conditions. Ultimately, the most effective means of maintaining good IAQ involves effective facility design and proper application of ventilation systems considering all strategies for IAQ management.

9.5 Summary

As the largest facility manager within the federal government, the DoD bears a great responsibility for leadership in adopting strategies and developing guidance for reducing energy consumption and minimizing environmental impact. Although all the military services started from different approaches, they have merged in recent years by incorporating sustainable requirements into standardized construction specifications and uniformly establishing the LEED green building rating system as the primary metric for sustainable military construction. Since the construction of military facilities, particularly low-rise lodging facilities such as dormitories or barracks, often entails unique requirements, it has been recognized that additional guidance and interpretation of the intent of certain USGBC LEED green building rating system credits are necessary. The U.S. Air Force, through AFCEE, has developed one such guide, the LEED Application Guide for Lodging. One of the greatest impacts LEED has on the performance of military personnel comes from the Environmental Quality (EQ) category, specifically related to the prerequisites and credits for IAQ. Source control, ventilation, and air cleaning are the three primary strategies for controlling IAQ. LEED-NC 2.2 provides specific credits for the first two, though not currently for air cleaning. Regardless, it is important for designers and engineers to consider all possible strategies and technologies available to maximize the facility's IAQ and provide a healthy working and living environment for occupants.

References

AFCESA (2007), http://www.afcesa.af.mil/ces/cesc/water/cesc_watercons.asp, Air Force Water Conservation Program of the Air Force Civil Engineer Support Agency website, accessed July 23, 2007.

Brzozowski, C. (2007), "Sleeping Green," *Distributed Energy, The Journal for Onsite Power Solutions*, July/August, 5(4): 24–33.

Bush, G. W. (2007), "Executive Order 13423—Strengthening Federal Environmental, Energy, and Transportation Management," *Federal Register*, 72 FR 3919, January 26.

CDA (2007), http://www.copper.org/health/, Copper Development Association webpage, accessed July 10, 2007.

Clinton, W. J. (1998), "Executive Order 13101 Greening the Government Through Waste Prevention, Recycling, and Federal Acquisition," *Federal Register,* 63 FR 49643, September 16.

Clinton, W. J. (1999), "Executive Order 13123 Greening the Government Through Efficient Energy Management," *Federal Register,* 64 FR 30851, June 8.

Clinton, W. J. (1999), "Executive Order 13148 Greening the Government Through Leadership in Environmental Management, *Federal Register* 65 FR 24595, April 26.

DoD (2004), *Unified Facilities Criteria (UFC) Design: Low Impact Development Manual,* Department of Defense, October 24, 2004, accessed July 10, 2007, http://www.lowimpactdevelopment.org/lid%20articles/ufc_3_210_10.pdf.

DRMS (2007), http://www.drms.dla.mil/, Defense Reutilization and Marketing Services website, accessed July 23, 2007.

EPA (2007), http://www.energystar.gov/, Energy Star website, U.S. Environmental Protection Agency, U.S. Department of Energy, Washington, D.C., accessed July 21, 2007.

EPA OAR (1991), *Building Air Quality: A Guide for Building Owners and Facility Managers,* EPA Document 402-F-91-102, December, U.S. Environmental Protection Agency, Office of Air and Radiation, Washington, D.C.

EPA OAR (1998), *Building Air Quality Action Plan,* EPA Document 402-K-98-001, June, U.S. Environmental Protection Agency, Office of Air and Radiation, Washington, D.C.

EPA and U.S. Consumer Product Safety Commission Office of Radiation and Indoor Air (1995), *The Inside Story: A Guide to Indoor Air Quality,* EPA Document # 402-K-93-007, April 1995, U.S. Environmental Protection Agency, Washington, D.C.

Ingersoll, W., and L. M. Haselbach (2007), "RS-VSP: A New Approach to Support Military Range Sustainability," *Environmental Engineering Science,* May, 24(4): 525–534.

Michels, H. T. (2006), "Anti-Microbial Characteristics of Copper," *ASTM Standardization News,* October, 34(10): 3–6.

OFEE (2003), *The Federal Commitment to Green Building: Experiences and Expectations,* Office of the Federal Environmental Executive, http://www.ofee.gov/sb/fgb_report.html accessed July 10, 2007.

OMB (2002), Circular A-11, Section 55, Office of Management and Budget, Washington, D.C.

OMB (2007), *Budget of the United States Government Fiscal Year 2008,* Office of Management and Budget, http://www.whitehouse.gov/omb/budget/fy2008/, accessed April 3, 2007.

USAF (2006), *The Air Force Center for Environmental Excellence Environmental Quality and Compliance Resources, November 20, 2006,* Air Force Center for Environmental Excellence, http://www.afcee.brooks.af.mil/eq/sustain/ExecOrderCrossReference.doc, accessed April 3, 2007.

USAF, USGBC, and Paladino and Company, Inc. (2001), *Application Guide for Lodging Using the LEED Green Building Rating System,* U.S. Air Force Center for Environmental Excellence, U.S. Green Building Council, Washington, D.C.

USAMRMC (2007), https://mrmc-www.army.mil/crpindex.asp, website on the Congressional Research Programs managed by the U.S. Army Medical Research and Materiel Command accessed July 18, 2007.

U.S. Census Bureau (2007), *Construction Spending,* U.S. Census Bureau Manufacturing, Mining, and Construction Statistics, http://www.census.gov/const/www/c30index.html, accessed April 2007.

USGBC (2004–2005), *LEED-EB Green Building Rating System for Existing Buildings, Upgrades, Operations and Maintenance*, Version 2, October 2004, updated July 2005, U.S. Green Building Council, Washington, D.C.

USGBC (2005), *LEED-NC Application Guide for Multiple Buildings and On-Campus Building Projects, for Use with the LEED-NC Green Building Rating System Versions 2.1 and 2.2*, October, U.S. Green Building Council, Washington, D.C.

USGBC (2005–2007), *LEED-NC for New Construction, Reference Guide*, Version 2.2, 1st ed., U.S. Green Building Council, Washington, D.C., October 2005 with errata posted through Spring 2007.

USGBC (2007), *LEED for Schools for New Construction and Major Renovations*, approved 2007 version, April, U.S. Green Building Council, Washington, D.C.

U.S. GPO (2007), *Access Budget of the United States Government, January 30, 2007*, U.S. Government Printing Office, http://www.gpoaccess.gov/usbudget/browse.html, accessed April 3, 2007.

Yu, K.-P.,et al. (2006), "Effectiveness of Photocatalytic Filter for Removing Volatile Organic Compounds in the Heating, Ventilation, and Air Conditioning System," *Journal of the Air & Waste Management Association*, May, 56: 666–674.

Exercises

1. What is the current status of EO 13423? Has it been superseded or enhanced by any other Executive Orders, and if so, what are they?

2. What is the current legislation in your state with respect to green building of state facilities?

3. Visit the Army sustainability website. What are some current projects and programs in sustainability in the Army?

4. Many Army bases have developed 25-year sustainability plans. Which ones have adopted these plans?

5. Many Army bases are developing 25-year sustainability plans. What is the status of the plans for Army facilities in your region of the country?

6. The website http://www.FedCenter.gov is a website for Federal Facilities Environmental Stewardship and Compliance Assistance. What current regulations does it list with respect to green building?

7. The Office of Management and Budget's Circular A-11, which specifically encouraged agencies to evaluate the use of Energy Star or LEED standards and guidelines for incorporation in designs for new building construction or renovation in Section 55 in its 2002 version, was still valid when President Bush issued EO 13423. What is its status now?

8. What are some current practices with respect to the sustainability of live firing ranges at DoD facilities?

CHAPTER 10

Low-Impact Development and Stormwater Issues

Stormwater management is an important design and construction practice. Without proper management, the resulting runoff can cause many problems including flooding and poor water quality. Traditionally, the design of stormwater facilities is taught in hydrology and watershed management classes. These types of courses are important to stormwater management, but focus more on the big picture, not individual construction projects. The purpose of this chapter is to look at stormwater management from the construction and individual site perspective so that it can be better integrated into the LEED or green building process. However, the actual design of any stormwater management system should be done by practitioners experienced in civil engineering systems, hydrology, and watershed management in addition to these newer concepts.

The concepts that are of most interest for stormwater management are the rate of stormwater runoff from a site or area, the amount of stormwater runoff from an area, the pollutant loadings from the runoff, and the impact on erosion and subsequent pollution that runoff has on the downstream areas. Over the decades many methods have been developed to estimate runoff. They are used for calculating rates, volumes, and a concept referred to as the time of concentration. A simplified definition of the *time of concentration* is the time it takes for various runoff contributions from a defined site or area to come together to a point producing a peak rate at that point. Akan and Houghtalen (2003) list three methods for determining the time of concentration: the SCS (Soil Conservation Service) time-of-concentration method, kinematic time-of-concentration formulas, and the Kirpich formula. They also list the following runoff rate calculation methods: unit hydrograph method, TR-55 graphical peak discharge method and tabular hydrograph method, the Santa Barbara urban hydrograph method, U.S. Geological Survey (USGS) regressions equations, the rational method, and the kinematic rational methods.

The concept of percent imperviousness as used by the USGBC, and as discussed in Chap. 2, is based on the rational method. The rational method is a very old (nearly a century in use) and fairly simplified method. It is applicable to smaller watersheds, usually much smaller than 100 acres. Most project areas associated with the LEED certification process are relatively small (10 acres or less) and therefore fall into this category. It is mainly a method to estimate peak runoff based on a 2- to 5-year storm. The calculations for both peak runoff rates and volumes usually require some stormwater path information, and therefore the other hydrological models are frequently used, expecially the TR-55 method for larger watersheds. Regardless, the following discussions concern methods to control runoff from a project, and proper hydrological modeling should be used to determine the rates and volumes. This text is not intended to teach hydrology; there are many excellent references that already do that. Instead, it is intended to guide the designers to approach stormwater management in a new evolving and inclusive way for more sustainable results.

The field of stormwater management is rapidly growing for a number of reasons, of which two important ones come to mind. The first is the ever-increasing development of the land in both the United States and the rest of the world, as populations grow and gain higher standards of living. This development usually translates to an increase in impervious surfaces and changes to site hydrologics and thus usually increases stormwater runoff. The second reason is based on the concern about global climate change, which may result in alterations to precipitation and weather patterns in many areas of the world and then could impact the hydrology and subsequent runoff, even from existing developments.

10.1 Nonpoint Source Pollution, BMPs, and LID

Most of the focus in the last century has been on collecting stormwater runoff and discharging it somewhere else to prevent flooding in the immediate area caused by development. However, in the last few decades it has been recognized that this general design may not be the best alternative in all cases. Collecting stormwater can cause powerful flows off-site that may result in property and ecological damage in addition to poor water quality. This pollution from stormwater runoff is commonly referred to as *nonpoint source* (NPS) pollution, as it has areal sources.

In response to the concerns about stormwater runoff and NPS pollution, the United States has adopted a permitting program, with the intent that it will help decrease off-site impacts of stormwater runoff. It is called the National Pollutant Discharge Elimination System (NPDES). Phases of NPDES have slowly been implemented in the 1990s and early 2000s, and the practices and policies resulting from their implementation are considered to be important for stormwater management sustainability, both in the construction phase and during the operational life of a facility.

There is no standard set of solutions to manage stormwater and NPS pollution. Each site and each watershed has its own needs and special conditions. Therefore, stormwater management design in the 1990s developed into a concept called Best Management Practices (BMPs). BMPs are a compilation of various methods and policies to choose from that are intended to serve as stormwater quantity controls and stormwater quality control measures. Several BMPs together may "best" serve to reduce stormwater impacts of construction and development.

BMPs are frequently subdivided into several broad categories. They can be grouped by project phase, by the form they take, and by function. Differentiation by project phase is whether they serve to reduce NPS pollution during the demolition and construction phase, or whether they are used to manage stormwater and reduce water quality impacts during the operational phase of a facility, as depicted in Fig. 10.1.1.

BMPs can also be differentiated by form, that is, whether they are structural or nonstructural, as noted in Fig. 10.1.2. A structural BMP is one that is physically put in place, such as a detention pond, swale, or a catch basin. A nonstructural BMP is a policy or practice. (Note that the LEED-NC 2.2 Reference Guide tends to put the more natural BMPs such as swales also in the nonstructural category.) An example of a BMP that is a policy might be a regulation that all facilities which change motor oil also have facilities

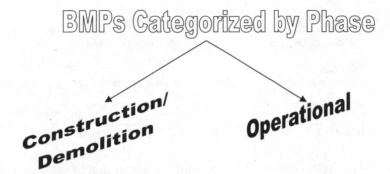

FIGURE 10.1.1   BMPs can be categorized by the life-cycle phase of the facility.

for accepting used motor oil for recycle at no charge. This helps reduce the potential for people to put used motor oil on the ground or into stormwater devices and lessens the chance for oil pollution downstream of a site. Another example of a practice is a change in fertilizer schedules based on weather patterns so that excess fertilizer does not get carried off-site during a storm. A nonstructural practice common in the Bay Area of California is to label stormwater inlets with pictures of fish to warn people that the inlet collects runoff that discharges into the Bay and anything put into the inlet might harm life in the Bay. Table 10.1.1 lists some categorizations of BMPs by form.

Finally, BMPs can be broadly categorized by function. However, there are many different ways that BMPs are organized by function. Sometimes BMPs are segregated as quality or quantity control measures. Sometimes they are separated into groupings based on whether they control erosion or control sedimentation. When they are categorized by the latter, they are typically referred to as *erosion and sedimentation control* (ESC) measures. Figure 10.1.3 gives some different categorizations by function.

Specific structural BMPs are listed under various names and functions that have been adopted regionally or locally or by various types of practitioners. Since they are too numerous to list or organize, an attempt will be made in the following sections to give a few examples of BMP terminology and categorization. These examples come from entities such as a watershed-related agency in the Chesapeake Bay area, a southeastern state department of transportation (DOT) as related to ESC, and research into BMPs by the Transportation Research Board (TRB) National Cooperative Highway Research

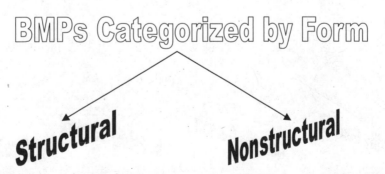

FIGURE 10.1.2   BMPs can be categorized by form.

Structural BMPs	Nonstructural BMPs
Detention ponds	Reused oil collection programs
Swales	Weather-based fertilizer application schedules
Catch basins	Stormwater inlet warning painting practices
Oil/water separators	Seasonal grading schedules

TABLE 10.1.1   Some Examples of BMPs Segregated by Form

Program (NCHRP). They are not intended to be comprehensive or inclusive for the United States, but together give a good introduction to some of the terminology used.

The use of BMPs is starting to achieve wide acceptance in the United States. It is a method to increase the sustainability of stormwater management systems. However, like the older, more traditional stormwater management systems, it again focuses on what is discharged from a site. So, to also start addressing what remains on a site or in a development, an even more focused concept is being introduced. It is commonly referred to as *low-impact development* (LID), although some refer to a similar concept as IMPs (Integrated Management Practices). These LID practices use many of the concepts from traditional hydrologic models and many BMPs, but they also focus on how the stormwater impacts the site and other "preventive" land development techniques to keep the site hydrology as close to the natural state as possible. Traditionally, hydrology management deals with the larger storms. LID focuses more on the frequently occurring storms, which are of a short duration and high frequency. These smaller, more frequent storms commonly carry an abundance of pollutants from a site, and even though they may not be individually as destructive as a larger storm, since they are more frequent, if not controlled, collectively they may cause substantial erosion and other deleterious impacts.

The following sections of this chapter will

- Outline some of the reasons why BMPs were developed
- Outline many structural BMPs as categorized by one agency in the Chesapeake area
- Outline many BMPs as categorized by a state DOT for ESC
- Outline many structural BMPs as evaluated by the NCHRP
- Introduce LID and IMPs concepts

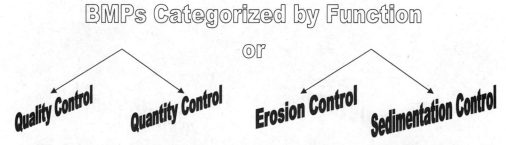

FIGURE 10.1.3   BMPs can be broadly categorized by function.

Percent Impervious Cover	Effect on Downstream Receiving Waters
<10	Sensitive
10–25	Impacted
>25	Nonsupporting

TABLE 10.1.2   Downstream Effect Rankings on Receiving Waters from Nonmitigated Impervious Cover for Some Example Watersheds (CWP, 2000)

Why BMPs Were Developed: Stormwater Impacts of Urbanization

The Center for Watershed Protection (CWP) in Ellicott City, Md., has compiled, reviewed, and synthesized extensive research and data on the impacts of urbanization on stormwater runoff and watershed protection methods. It has been shown that one of the main parameters that increases stormwater runoff and downstream impacts is an increase in impervious areas caused by development. These increases can be both on the ground (pavements) and from roofs. Each watershed is different and responds uniquely to the amount and location of impervious cover, so there is no single simple model based only on impervious cover. Table 10.1.2 shows a range of impacts that was found to occur for a number of watersheds studied.

In these, the impacts of urbanization are usually categorized into four areas—hydrological, geomorphological, water quality, and habitat-related. Some of the effects and impacts in each of these categories are listed in Table 10.1.3.

The Center for Watershed Protection also summarized some of the methods or tools which can be used to protect the watershed and puts them into eight approximate groupings. These are summarized in Table 10.1.4.

Hydrological	Geomorphological	Water Quality	Habitat/Wildlife
Increased frequency of flooding	Widened streams	Increased temperature	Decreased habitat value
Increased volumes of flooding	Modified access for fish passage	Increased beach closures	Decreased large woody debris
Higher peak flows	Loss of riffle pools	Increased pollutants - SS (suspended solids)	Increased fish barriers
Increased variation in storage needs	Fragmentation of tree cover	- Nutrients (N and P) - Metals (Cu, Zn, Pb, Cd)	Shift in food chain
Modified channelization	Silt embedded in banks and bottom	- Oil and greases - Bacteria	Decreased speciation
Lower dry weather flows	Degraded bottom substrate quality	- Various toxics (pesticides/ herbicides)	Decreased weather impact buffers
	Eroded stream banks		Increased algal problems

TABLE 10.1.3   Some Impacts of Urbanization and Increased Impervious Cover (CWP, 2000)

Methodologies	Examples
Land use planning	Stormwater "hot spots" away from water bodies
Increased land conservation	Conservation easements, cluster housing
Increase aquatic buffers	Setbacks from banks for development
Better site design	Less impervious cover
Erosion and sediment control	Geofabrics, planting schedules
BMPs	Sedimentation basins, infiltration swales
Control of nonstormwater discharges	Point source pollutant control measures
Watershed stewardship	Cleanup days, educational programs

TABLE 10.1.4 Various Methodologies for Watershed Protection (CWP, 2000)

The remainder of this chapter focuses on many of the methodologies in the ESC and BMP grouping, which are typically part of the engineering design of a project.

BMP Terminology

There are many different names for similar BMPs in the United States and many variations of BMPs that are regionally or locally used. There are even variations on the term *BMP*, such as calling some of them STPs (Stormwater Treatment Practices). Sometimes the variations are more random, but at other times the variations are based on adaptations to local conditions such as soil types or topography. Very frequently the variations are due to the focus of the group or profession that has categorized them. For instance Tables 10.1.5 and 10.1.6 list some typical categorizations of BMPs as given by a watershed group and a highway-related research organization. In the ESC section, there will be a table referring to some categorization of these methods typical of many state DOTs. Table 10.1.5 gives a listing of many BMPs that have been categorized by the CWP.

Pond-Type BMPs	Wetlands	Infiltration	Filtering	Open Channels
Micropool ED* pond	Shallow marsh	Infiltration trench	Surfaces and filter	Dry swale
Wet pond	ED shallow wetlands	Infiltration basin	Underground sand filter	Wet swale
Multiple pond system	Pond/wetland system	Porous pavement	Perimeter sand filter	Grass channel
Pocket ponds	Pocket marsh		Organic filter	
	Submerged gravel wetland		Pocket sand filter	
			Bioretention	

*Extended detention.

TABLE 10.1.5 Some Typical Categorizations of Structural BMPs with a Watershed Focus (CWP, 2000)

Runoff Flow and Volume Control	Runoff Volume Control	Physical, Chemical, or Biological Sorption, Ion Exchange	Physical Screening	Force-Related Physical Screening	Aerating, Flocculating, Volatilizing, Coagulating	Disinfection	Microbial Mediated Processes
ED* basins	Infiltration basins	Engineered media	Screens, racks	ED basins	Sprinklers	Chlorine	Wetlands
Retention ponds	Infiltration trenches	Granular AC,† compost or peat filters	Biofilters	Retention basins	Aerators	Ultraviolet systems	Bioretention
Detention ponds	Permeable pavements	Sand/gravel filters	Permeable pavements	Wetlands	Coagulant systems	Ozone	Biofilters
Tanks, vaults	Dry swales	Subsurface wetlands	Infiltration basins	Settling basins	Flocculent systems	Advanced systems	Retention ponds
Wetlands	Dry wells	Zeolites or complexation media	Infiltration trenches	Tanks, vaults			Engineered, sand, compost filters
Equalization basins	ED basins	Bioretention	Bioretention systems	Check dams			
		Infiltration trenches/ basins	Engineered media	Oil/grit separators			
		Wetlands	Sand/gravel filters	Hydrodynamic catch basins			
		Biofilters	Catch basin inserts				

*Extended detention.
†Activated carbon.

TABLE 10.1.6 Functional Unit Operations of Many BMPs (NCHRP, 2006)

There are also components to a BMP that provide additional pretreatment or water quality protection. Some of these are referred to as water quality (WQ) inlets, and they may take many forms and names. Some common pretreatment units are forebays or sedimentation basins before a pond, or specialty premanufactured items such as oil/ grit separators or cyclonic catch basin inserts. There are also structural additions to the BMP systems whose primary purpose is to decrease the flow rate of runoff, which is important for erosion and damage protection in the receiving structure or water body. Many of these are referred to as level spreaders or energy dissipators, although sometimes they are just called by the name of the material they are made of such as riprap. Figure 10.1.4 depicts a detention pond where additional riprap has been used for erosion control.

Table 10.1.6 lists many of the structural BMPs that have been segregated by functional unit operations by the NCHRP (2006). Many of the BMPs serve multifunctional purposes. Figure 10.1.5 depicts a detention pond with added retention which aids in stormwater quality control.

In addition to the BMPs already listed in the tables, there are many new or combined variations on all these themes. For instance, there is a BMP called an *ecology ditch*. It can best be described as a combination of a swale and a bioretention cell.

There are also classifications of BMPs with respect to adaptation for climate change. In these cases, other BMPs and functions are added to the mix. Some of the additional functions considered are

- Urban heat island reduction
- Urban forestry support

Figure 10.1.4   Detention pond with riprap erosion control around the inlets and banks in Irmo, S.C. (*Photograph taken July 2007.*)

FIGURE 10.1.5   Detention pond with some retention for additional quality control in Irmo, S.C. (*Photograph taken July 2007.*)

- Carbon sequestration
- Micro hydropower potential
- Combined sewer overflow (CSO) reduction

ESC

Erosion and sediment control measures are important for both the construction phase and the operational (built) phase, but are particularly important during the construction phase of a project. Construction sites are sometimes referred to as nonpoint source pollution "hot spots" as the pollution in the runoff, if not controlled and treated, is frequently several times higher than from a developed site.

ESCs are varied in function, form, and applicability just as other BMPs are. The best practices are usually site-specific and sometimes season-specific. Therefore, there are many options to choose from and many manufacturers. Since there are such a broad range of products, there are various testing facilities which are analyzing the options throughout the country. Some of the testing facilities include

- The Hydraulics, Sedimentation, and Erosion Control Laboratory operated by the Texas Transportation Institute and the Texas Department of Transportation
- The San Diego State University Soil Erosion Research Laboratory
- The St. Anthony Falls Laboratory operated by the University of Minnesota
- Alden Research Laboratory in Holden, Mass.

Information from these and other testing facilities, from other research reports, and from the various manufacturers can be used to determine which ESC measures may be applicable for a site, but there is an immense amount of information and with the different options, a need developed to establish some form of organized inspection and organization of the methods. From this has sprung the programs of the Certified Professional in Erosion Prevention and Sediment Control (CPESC) and the Certified Erosion Prevention and Sedimentation Control Inspector (CEPSCI) which are being recognized by many state agencies as a means to facilitate erosion and sediment control on project sites, particularly in departments of transportation, where there are always many horizontal maintenance and construction projects underway. Table 10.1.7 lists the various categories of erosion and sedimentation control practices that are taught in the CEPSCI certification classes in South Carolina.

The most common ESC measures used during construction are the silt fence around the perimeter of the disturbed areas so that sediments do not wash out from the site, construction entrances at the openings to the disturbed areas so that vehicles can come in and out, and some form of inlet protection on any stormwater device installed prior to or during construction so that it does not become a conduit for sediment flow off the site during the construction phase. Based on weather and soil conditions, some type of dust control may also be needed to prevent soil from escaping the site by wind mechanisms.

When the construction is nearing completion, additional measures are permanently installed on the site so that erosion does not occur after construction. These range from vegetative measures to using riprap or constructed erosion prevention devices.

Erosion Prevention Practices	Sediment Control Practices
Surface roughening	Sediment basin
Mulching	Multipurpose basin
Erosion control blankets (ECBs)	Sediment dam (trap)
Turf reinforcement mats (TRMs)	Silt fence
Dust control	Ditch check • Rock • Sediment tube
Polyacrylamide (PAM)	Stabilized construction entrance
Stabilization • Temporary • Riprap for channel • Final	Inlet protection: • Filter fabric • Block and gravel • Hardware fabric and stone • Block and gravel curb • Prefabricated
Outlet protection	Rock sediment dikes

TABLE 10.1.7 Typical Erosion Prevention and Sediment Control Practices Taught in ESC Certification Classes (CEPSCI, 2005)

LID

The main concepts of low-impact development (LID) are to minimize the impacts of development, mainly the stormwater-related impacts. This is done usually by mimicking the natural hydrologic cycle at the site or project as much as possible and mimicking the natural hydrologic runoff cycle. So how does one implement the design goal of mimicking the former site hydrology with LID features? There are essentially four main parameters of site hydrology which can be used to do this:

- Storage
- Infiltration and the recharge of groundwater
- Evaporation and evapotranspiration
- Detention

And there are a series of many concepts that can help do this. The main ones as listed in the Prince George's County (1999) LID manual and as amended are as follows:

- Reduce added imperviousness.
- Conserve natural resources.
- Maintain natural drainage courses.
- Reduce use of pipes, use open-channel sections.
- Minimize clearing and grading.
- Provide dispersed areas for stormwater storage uniformly through the site.
- Disconnect flow paths.
- Keep predevelopment time of concentration.
- Employ public education and operational procedures.
- Maintain sustainability of LID practices and BMPs.

These can be summarized as follows: reduce new development, conserve existing and natural systems, and disperse and disconnect the site stormwater.

New development introduces many conveyance mechanisms for stormwater which need to be disconnected, reduced, or staged and planned. If these are not reduced or mediated, then they can facilitate a rapid transport of runoff off-site and the resulting environmental and flooding problems. Keeping water on-site and disconnecting flow paths can also reduce pollutant transport off-site.

Typical built items that can become flow paths include

- Impervious roadways
- Roofs
- Gutters
- Downspouts
- Impervious drives
- Curbs
- Pipes
- Swales

- Impervious parking areas
- Grading

Reducing these items or designing them in alternative fashions can aid in lowering the impacts of development and can also mean a reduction in construction costs. However, development means we are building or engineering something, so for the site to be accessible and functional, many of the surface features that may also become flow paths will be installed. Some measures will need to be taken to minimize the impacts these have on the site hydrology. The main focus of LID is to think micromanagement and integrated management practices.

These practices can be multifunctional. In fact, many of the LID and IMP concepts can be implemented in otherwise "nonfunctional" spaces. Dr. Allen Davis gives the example of a grass swale or bioretention landscape strip being located in the area needed for the overhang of fronts or backs of cars in parking spaces.

In response to the economic benefit of multifunctional best management practices as promoted in the LID concepts, many agencies have expanded their stormwater guides to include in detail some of the IMPs. On July 2, 2007, the North Carolina Department of the Environment and Natural Resources released its new BMP manual. There are chapters in this manual devoted to bioretention, permeable pavements, and other BMPs that aid in attaining the LID concepts.

Then, in 2007, the Environmental and Water Resources Institute (EWRI) of the American Society of Civil Engineers (ASCE) formed an LID committee in its Urban Water Resources Research Council (UWRRC). It has been recognized that there is a great need for research and development of manuals, best practices, and standards with respect to the LID concepts. The first three initiatives started by this committee are to look into the technology associated with green roofs, porous and permeable pavements, and rainwater harvesting.

Another recent action that is helping to foster the LID trend in design and development is a new credit added to the USGBC LEED for Schools rating system in 2007. In the Sustainable Sites category an additional credit was added called *site master plan*. Basically, this credit requires a master plan to be prepared for a school location covering current and planned development, and in this master plan the future plans should include adhering to at least four out of six specified SS credits. The specified credits are SSc1: Site Selection; SSc5.1: Site Development: Protect or Restore Habitat; SSc5.2: Site Development: Maximize Open Space; SSc6.1: Stormwater Management: Quantity Control; SSc6.2: Stormwater Management: Quality Control; and SSc7.1: Heat Island Effect: Non-Roof. If one thinks about these credits, they together promote many of the LID concepts.

Combining a number of credits is very similar to the concept of integrating various management practices. For instance, in most cases one needs landscape areas for aesthetics, but why not also use these for stormwater management? Shading is needed for urban heat island effect reduction, but if trees are used, the tree islands can also be used for stormwater management.

There are a number of issues that must be addressed when adopting LID practices for good design. Some of these include the following:

- Make sure that flood control is adequately considered.
- Be careful that water is not infiltrated where it might cause structural damages such as under some impervious pavements (hydrostatic forces) and under buildings (water damage).

- Some of the quasi-structural BMPs may not be sustainable since they may not appear to have readily noticeable site functionality. They need some way of fostering permanence. Some methods to do this may be structural such as adding rock walls; others may be nonstructural such as establishing conservation easements for infiltration areas.

- The decrease in paved areas must take into consideration safety and access concerns.

LID and Land Development Design

This chapter in no way attempts to be a comprehensive resource for LID. It would take an entire book or manual to even start any detailed design. However, a few of the main concepts have been reviewed, and this chapter gives some site design examples and an example method to do some calculations on one of the newer stormwater management technologies that may be used for LID designs. Three other commonly referenced resources for information on how to incorporate LID into land development design are

- Prince George's County, Department of Environmental Resources (1999), *Low-Impact Development Design Strategies: An Integrated Design Approach*, Prince George's County, Md.

- CWP (1998), *Better Site Design*, Center for Watershed Protection, Ellicott City, Md.

- DoD (2004), *Unified Facilities Criteria (UFC) Design: Low Impact Development Manual*, Department of Defense, October 24.

When one is developing a site, many local land development and zoning regulations may also impact LID designs. Some common ones are listed in Table 10.1.8.

In addition to considering the items listed in Table 10.1.8, it may be helpful to keep a list of other concepts in the engineering toolbox for site design to aid in attaining LID goals. Many of these are listed or reiterated in Table 10.1.9.

There are several important variables used in LID calculations which are many times also used for local zoning calculations. Table 10.1.10 summarizes several of the common basic ones.

The variables C, DF, %imp, and %impness are explained in Chap. 2 and will not be repeated here. However, there is a need for a closer look at CN, FAR, and GFA.

CN is similar to the simpler rational runoff coefficient, but also includes other site factors such as soil types. It is used in very well-established hydrology modeling, and more information about its use and application can be found in most hydrology texts and from the Natural Resources Conservation Service. How to modify the curve number to adequately take into account LID design modifications is under development. Some mechanisms for doing this have been introduced by Prince George's County Department of Environmental Resources (1999) and the DoD (2004).

The gross floor area of a building (GFA_i) can have many definitions depending on the local ordinances used. Usually it includes all the floors of a building, taken from the outer dimensions, except for unfinished attic and basement spaces. Chapter 2 discusses some modifications which can be used to also include parking garages for the purposes of the LEED calculations. For every case, the actual areas used as such should be summarized.

Zoning Requirement	Meaning	Applicability to LID
Minimum landscape areas	A minimum percentage of the lot must be landscaped	These areas will reduce impervious areas and may also be used for IMPs for stormwater management such as bioretention cells
Required landscape buffers	A minimum amount of setback or buffer must be landscaped between the development on the site and neighbors	These areas will reduce impervious areas and may also be used for IMPs for stormwater management such as bioretention cells
Minimum trees in parking areas	A minimum amount of tree shading is required in parking areas	These areas will reduce impervious areas and may also be used for IMPs for stormwater management such as bioretention cells
Maximum slopes on drives/paved areas	For vehicular safety and access reasons there are maximum allowed slopes in paved areas	Minimizing slopes on paved surfaces decreases the flow rate of runoff
Maximum ADA (Americans with Disabilities Act) slopes	For ADA access reasons there are maximum allowed slopes along the accessibility route	Minimizing slopes on surfaces decreases the flow rate of runoff
FAR	Minimum or maximum floor area ratios are sometimes established to limit either the use or the land disturbance impact on the site, respectively	Balancing FARs with lowered impervious cover can help maintain more areas on the lot which mimic the natural hydrology
Minimum and maximum parking	Minimum parking requirements establish adequate parking for the use and maximum limits site impacts	Minimizing parking areas will decrease impervious surfaces
Road/drive width requirements	There are usually minimum and maximum road and drive widths for safety and access reasons	Minimizing paved areas will decrease impervious surfaces

Table 10.1.8   Example Zoning or Land Development Requirements That Impact LID Design Features

The *floor area ratio* (FAR) is usually defined as the gross floor area of all the buildings on a site divided by the total site area, as in Eq. (10.1.1).

$$\text{FAR} = \frac{\sum \text{GFA}_i}{\text{A}_T} \qquad \text{for all buildings } i \text{ on the site} \qquad (10.1.1)$$

Why is it necessary to have so many variables? Well, each is useful in aiding in the determination of different impacts of construction. Several examples are also listed in Table 10.1.10. One simple example to explain the importance of determining many of these variables for sustainable construction purposes can be seen if the floor area ratio is examined. A metamodel called Sustainable Futures 2100 has been recently developed

Tool	Meaning	Applicability to LID
Limit development envelope and minimize grading	Only disturb during and after construction the minimum amount of land necessary	Limited disturbance will reduce impervious areas and maintain natural areas for stormwater management
Remember "out of sight" may be "out of mind"	Do not install LID features that will not be known to users	Making LID features a prominent part of the site will aid in proper use for stormwater management
Choose alternatives to lawns	Consider alternative landscapes to lawns	Alternative landscapes may reduce runoff and aid in evapotranspiration
Spread concentrated flows	Use level spreaders or energy dissipators to decrease flow energy	Concentrated flows from gutters and leaders can cause increased erosion and runoff
Modify curb types	Use no curbing or modified curbing to not concentrate flow and allow infiltration on-site	Concentrated flows from curb gutters can cause increased erosion and runoff
Reduce imperviousness	Limit impervious surfaces and seek alternative hardscapes that are more pervious	Reduced imperviousness helps limit runoff and aids in on-site infiltration
Conserve natural resources	Conserve natural areas that may aid in LID such as wetlands and good infiltrating soil areas	This will help maintain on-site infiltration and limit runoff
Conserve natural drainage	Maintain natural drainage courses	Preexisting drainage courses usually have already stabilized
Reduce pipe use	Use open channels where possible	This increases awareness and may also allow for more infiltration and evaporation
Disperse	Provide smaller stormwater storage and infiltration features throughout the site	This decreases flow rates at individual areas and helps in infiltrating and watering throughout the site
Disconnect	Disconnect flow paths such as from leaders to drives	This limits the increase of spot runoff rates by superposition, and disruptive flows have more time to infiltrate
Maintain time of concentration	Maintain predevelopment time of concentration t_c where possible	Altering the t_c may mean that runoff will combine with runoff from other areas in a different way, which could increase flows
Educate	Consider options for educating designers and users from the activities on the site	Site users should understand the importance of the LID features and be familiar with maintenance requirements. Also consider using the site for education of others, whether with signage, publications, or visits

TABLE 10.1.9   Other Miscellaneous Tools for LID Designs

Tool	Meaning	Applicability to LID
Sustainability of LID and BMPs	Consider if the practice will remain effective over time	BMPs/LID practices may require maintenance for continued performance. This must be incorporated into the site operations. Also, some practices are removed because of the lack of knowledge with respect to their importance by future users or designers. Making the features more structurally evident or using educational tools will improve their sustainability

TABLE 10.1.9   Other Miscellaneous Tools for LID Designs (*Continued*)

Variable	Title	Chapter Introduced	Use
C	Rational runoff coefficient	2	To estimate the runoff from a site simply based on site surface types
CN	Curve number	10	To estimate the runoff from a site using more advanced models
DF	Development footprint	2	To determine total building and hardscape areas. Useful to estimate the impact on the land with respect to hydrology, vegetation, and heat island issues
FAR	Floor area ratio (also refer to building density in App. B)	10	To estimate the intensity of use of the site based on built floor area. Useful to estimate the impact on areal utilities, transportation, and services
GFA	Gross floor area	2, 10	Usually the built floor area used for the FAR and development density
%imp	Percent impervious	2	To estimate the percent of the site with impervious hardscape. Useful to estimate the impact on the land with respect to hydrology, vegetation, and heat island issues
%impness	Percent imperviousness	2	To estimate overall imperviousness of the site. Useful to estimate the impact on the land with respect to hydrology. Can correlate with the rational runoff coefficient

TABLE 10.1.10   Some Common Variables Used for LID Determinations

to examine many of the complex environmental and natural resource impacts of construction, both on site and regional. It was used to compare a single-story house and a two-story house of similar gross floor area, which in turn means similar FARs. It was determined that the single-story home had less of an energy impact during construction, but the two-story home had a higher operational energy demand. Hydrologic impacts are different too and were not compared in that analysis, but obviously the single-story home will use more of the land and therefore leave less for keeping the natural hydrologic cycle of the land without additional LID measures. So, knowledge of only the FAR is not sufficient for determining the complex impacts of construction, although it is a good indicator of many items such as utility and regional transportation usage related to the site.

10.2 Modeling BMPs

There are many commercial stormwater models that are commonly used for design and assessment. Many of these have opportunities for including some of the LID scenarios for stormwater management into the assessment. In the future, more individual LID practices and technologies will be included in them. However, given the many options for LID and the site-specific needs and variations for each, it is helpful to have some simple methods to perform preliminary design and performance estimates of the practices. The following section gives some simple mass balance methods for preliminary design and assessment of many LID technologies.

Overall Mass Balances

In stormwater management there are two types of mass balances that are of interest: a mass balance on the water and a mass balance on the pollutant in question. Regardless of the type of substance for which its mass is being balanced, the balances can be summarized with Eqs. (10.2.1) and (10.2.2). Given the "system" or unit operation that is being evaluated, Eq. (10.2.1) represents a dynamic situation where the mass of the substance in question within the system changes with time. Equation (10.2.2) represents a steady-state situation where the mass of the substance in question within the system is constant.

$$\text{Rate of accumulation} = \Sigma\,\text{Rates in} - \Sigma\,\text{Rates out} \pm \Sigma\,\text{Internal reactions/processes}$$
$$\text{dynamic mass balance} \qquad (10.2.1)$$

$$0 = \Sigma\,\text{Rates in} - \Sigma\,\text{Rates out} \pm \Sigma\,\text{Internal reactions/processes}$$
$$\text{steady-state mass balance} \qquad (10.2.2)$$

Stormwater Mass Balances

The first mass balance explained here is the total stormwater mass balance around a system such as a BMP or other LID practice. Let the BMP be represented by a box, and let the typical flows in and out be as shown in Fig. 10.2.1.

In this simple box model of a stormwater mass balance around a BMP, the following variables are used:

A_{BMP} Surface area of BMP (length squared, usually acres or ft^2)

E Evaporation rate (length/time, usually in/h)

F Infiltration rate (length/time, usually in/h)

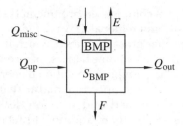

FIGURE 10.2.1   Simple box model of a stormwater mass balance on a BMP.

I	Rainfall rate (length/time, usually in/h)
Q_{misc}	Miscellaneous flows into a BMP (length³/time, usually ft³/s or cfs)
Q_{out}	Total runoff out of a BMP (length³/time, usually ft³/s)
Q_{up}	Runoff from upslope areas into a BMP (length³/time, usually ft³/s)
S_{BMP}	Storage volume in a BMP (length³, usually ft³)
t	Time

Assuming that there are no internal reactions within the BMP that remove or add water, these variables can be substituted into Eqs. (10.2.1) and (10.2.2) and the following water mass balance equations result:

$$\text{Accumulation rate} = Q_{misc} + Q_{up} + I(A_{BMP}) - E(A_{BMP}) - F(A_{BMP}) - Q_{out}$$
$$\text{stormwater dynamic mass balance} \quad (10.2.3)$$

$$0 = Q_{misc} + Q_{up} + I(A_{BMP}) - E(A_{BMP}) - F(A_{BMP}) - Q_{out}$$
$$\text{stormwater steady-state mass balance} \quad (10.2.4)$$

In addition to these equations, the initial state of the BMP with respect to the mass of water within it must be known. For the initial condition where the storage volume within the BMP is not full, Eq. (10.2.3) represents the condition when the storage volume within the BMP is filling with water if the flows in are greater than the flows out; or if there is any water within the BMP, then losing water when the flows out are greater than the flows in. Equation (10.2.4) represents the condition when the amount of stormwater that comes in equals what goes out. This usually means the storage volume is full, or that the storage volume is not "filling" since the potential for flow out is greater than what comes in.

What is needed now is some way to estimate the flows. The simplest model to use for many of the flows in the stormwater mass balance is based on the rational runoff method, and this is presented here. (However, flows from other models can be substituted into the overall mass balances if they are used.) In the *rational runoff method*, the governing generic equation relates the total runoff Q to the average rational runoff coefficient C, the rainfall rate I, and the land area A_T in the following manner:

$$Q = CIA_T \quad \text{generic rational method equation} \quad (10.2.5)$$

[If the units used are runoff in cubic feet per second, rainfall rate in inches per hour, and area in acres, then it just so happens that the conversion factor from acre-inches per hour (acre·in/h) to cubic feet per second is approximately 1, and no unit conversions are needed.]

In the case of most BMPs, we may have upslope overland flow into the system, rainfall directly on the top surface of the BMP, some additional water sources (such as pipe flows from elsewhere), evaporation from the top of the BMP, infiltration into the ground/subsurface underneath, and then the flow out of the BMP, which is the main flow rate we are interested in determining. The direct rainfall flow into the BMP has been estimated in Eqs. (10.2.3) and (10.2.4) as the rainfall rate times the surface area of the BMP, and both the soil infiltration flow out (exfiltration) and the evaporation loss have been estimated by some sort of areal soil/subsurface infiltration rate and some sort of areal evaporation rate (see also Chap. 3 for evapotranspiration rates). The flow from the upslope areas can be estimated by Eq. (10.2.5), using the following variables:

A_{up} Land area upslope of BMP that drains into it (length squared, usually acres)

C_{up} Average stormwater runoff coefficient of land surface upslope of BMP

This results in Eq. (10.2.6).

$$Q_{up} = C_{up}IA_{up} \qquad \text{upslope rational method runoff estimate contribution} \qquad (10.2.6)$$

By using these equations, many different situations can be modeled. For the first scenario, assume that it is raining at a constant rainfall rate I, that the only flows in are from upslope runoff and the rain, that there is negligible evaporation, and that there was no water initially stored in the BMP. The equations can be used to estimate the time it might take to start runoff out of the BMP (which is essentially the time to fill the storage volume during which the runoff out of the BMP is zero) if the flows in are greater than the infiltration rate.

$$t = S_{BMP}/[C_{up}IA_{up} + I(A_{BMP}) - F(A_{BMP})] \qquad \text{time to fill in first scenario} \qquad (10.2.7)$$

Now, assume that the storage volume S_{BMP} has filled and the rain continues. The equations can also be used to estimate the steady-state runoff out of this BMP. In this case, the outlet for the flow out is above the storage volume height, a common condition in many cases. If there are multiple outlets, then the equations will need to be solved for each stage of storage within the BMP.

$$Q_{out} = C_{up}IA_{up} + I(A_{BMP}) - F(A_{BMP}) \qquad \text{steady-state flow out in second scenario} \qquad (10.2.8)$$

The main way to model the stormwater mass balance around any BMP is to first make a simplified sketch of the system, labeling potential flows in and out and internal items such as storage as in Fig. 10.2.1. Then the initial conditions are used along with Eqs. (10.2.3) and (10.2.4) to model the system. For solutions, the equations are further simplified by allowing negligible flows to be neglected for the model and by using accepted models and parameters such as the rational method or soil infiltration rates to substitute for many of the variables.

Pollutant Mass Balances

The mass balance of pollutants around a BMP can be modeled by using the box model in the stormwater mass balance model and including the concentration of the pollutant in each flow, or within the BMP, using the following definitions:

C_{PBMP} Average concentration of a pollutant in a BMP

C_{PF} Average concentration of a pollutant infiltrating into ground

C_{PI} Average concentration of a pollutant in rainfall

C_{Pmisc} Average concentration of a pollutant coming from miscellaneous flows into a BMP

C_{Pout} Average concentration of a pollutant flowing out of BMP

C_{Pup} Average concentration of a pollutant in upslope runoff

Note that the concentration of a pollutant has been ignored in the evaporation portion, but this may need to be included as a pollutant evaporation rate if the contaminant in question is volatile. Also, many pollutants may have reactions internal to the BMP, such as microbial degradation of organics to inorganic compounds, and the various internal reaction rates would need to be added. However, for cases where neither internal reactions nor pollutant evaporation is important, the stormwater equations can be rewritten as pollutant mass balances as in Eqs. (10.2.9) and (10.2.10).

Accumulation rate of pollutants at time $t = Q_{misc}C_{Pmisc} + Q_{up}C_{Pup} + IA_{BMP}C_{PI} - FA_{BMP}C_{PF} - Q_{out}C_{Pout}$
    pollutant dynamic mass balance, no internal reactions or evaporation   (10.2.9)

$0 = Q_{misc}C_{Pmisc} + Q_{up}C_{Pup} + IA_{BMP}C_{PI} - FA_{BMP}C_{PF} - Q_{out}C_{Pout}$
    pollutant steady-state mass balance, no internal reactions or evaporation   (10.2.10)

The storage volume is then used along with the time and the equations to calculate the concentration within the BMP (C_{PBMP}).

There are additional models that are needed to aid in this estimation, and these are models of how the pollutant is distributed within the BMP so that it can be determined what the pollutant concentrations might be of the flows out of the system. The two extremes are usually the completely mixed assumption (the concentration is equal throughout the BMP) and the plug flow assumption. Plug flow means that there is no mixing in the direction of flow within the BMP so that there is a pollutant concentration gradient in this direction. If it is assumed that the pollutant is evenly distributed within the BMP (completely mixed model), then it can be assumed that C_{PBMP} is equal to both C_{PF} and C_{Pout}. Plug flow and other models are beyond the scope of this text and are better presented in environmental modeling or chemical operation texts, but may be important in some BMP modeling, particularly in systems such as bioretention cells where there may be nutrient uptake within some layers of the cell.

If the completely mixed model is assumed for the pollutant mass balances, then Eqs. (10.2.9) and (10.2.10) simplify to

Accumulation rate of pollutants at time $t = Q_{misc}C_{Pmisc} + Q_{up}C_{Pup} + IA_{BMP}C_{PI} - (FA_{BMP} + Q_{out})C_{PBMP}$
    pollutant dynamic completely mixed model, no internal reactions/evaporation

(10.2.11)

$0 = Q_{misc}C_{Pmisc} + Q_{up}C_{Pup} + IA_{BMP}C_{PI} - (FA_{BMP} + Q_{out})C_{PBMP}$
    pollutant steady-state completely mixed, no internal reactions/evaporation   (10.2.12)

Pervious Concrete

Pervious concrete is an alternative paving surface that allows infiltration into the subbase and ground below. It is a novel paving material that is being developed to aid in maintaining preexisting hydrologic conditions while still giving a structural base for vehicular and other uses. There is a need to be able to model the various pervious concrete systems and their impacts on hydrology and pollutant transport.

Many studies show that the infiltration rate through the pervious concrete itself is usually much greater than that through any natural soil type. The infiltration rate may therefore be limited by the subbase and subgrade below. Typically, pervious concrete can be placed directly over well-drained soils such as sands, but should be placed over additional storage subbases such as gravel, when used over poorly drained soils such as clays. This allows for additional time for the stormwater to infiltrate into the ground instead of running off. Usually overflow pipes are also included for stormwater management purposes for large storms.

Based on the LEED-NC 2.2 Reference Guide, pervious concrete areas can be considered to be pervious. However, there are also local or state interpretations of how to count the highly pervious concrete and asphalt pavements with respect to imperviousness, and these must be taken into consideration for local requirements and interpretations. LEED usually defers to the local code if it is more stringent than the LEED credit criteria. For example, in North Carolina, pervious concrete and pervious asphalt may count toward 40 to 60 percent as managed grass with respect to percent imperviousness for a site.

A pervious concrete application is a good method by which to use the simple box model to estimate runoff from a site. Consider a system consisting of pervious concrete with a gravel subbase and then soil below and with the pavement system essentially flat. This system receives upslope runoff and direct rainfall. The condition being modeled is a storm with no antecedent precipitation (storage volume initially empty) and during the time when evaporation can be assumed negligible (during the precipitation event). There are two questions being asked:

- How long until the onset of runoff?
- After the storage volume is full, what is the steady-state rate of runoff?

The system is depicted as a two-layer simple box model as shown in Fig. 10.2.2. Since there are two different layers to the pavement system, the total storage can be estimated as the sum of the pervious concrete and the gravel layer storage volumes. Given:

$A_{\text{BMP-PC}}$ Surface area of pervious concrete BMP

S_{Gravel} Storage volume in gravel layer (length^3, usually ft^3)

S_{Pervious} Storage volume in pervious concrete layer (length^3, usually ft^3)

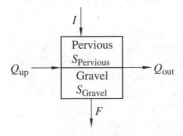

Figure 10.2.2 Simple box model of pervious concrete over gravel system.

Equations (10.2.7) and (10.2.8) can thus be modified for the pervious concrete system model as follows:

$$t = (S_{\text{Gravel}} + S_{\text{Pervious}})/[C_{\text{up}}IA_{\text{up}} + (I - F)(A_{\text{BMP-PC}})] \quad \text{time to fill} \quad (10.2.13)$$

$$Q_{\text{out}} = C_{\text{up}}IA_{\text{up}} + (I - F)(A_{\text{BMP-PC}}) \quad \text{steady-state flow out} \quad (10.2.14)$$

Many other initial conditions and configurations can be modeled for pervious concrete systems. In like manner, other BMP systems can be modeled using the basic concepts of a box model and mass balances.

References

Akan, A. O., and R. J. Houghtalen (2003), *Urban Hydrology, Hydraulics and Stormwater Quality, Engineering Applications and Computer Modeling,* John Wiley & Sons, Hoboken, N.J.

Atlanta Regional Commission and Georgia Department of Natural Resources (2001), *Georgia Stormwater Management Manual,* vol. 2: *Technical Handbook.*

Barber, M. E., et al. (2003), "Ecology Ditch: A Best Management Practice for Storm Water Runoff Mitigation," *ASCE Journal of Hydrologic Engineering,* May/June, American Society of Civil Engineers, Reston, Va.

Bean, E. Z. (2005), *A Field Study to Evaluate Permeable Pavement Surface Infiltration Rates Runoff Quantity, Runoff Quality and Exfiltrate Quality.* MS thesis, North Carolina State University, Raleigh.

Berke, P. R., et al. (2003), "Greening Development to Protect Watersheds, Does New Urbanism Make a Difference?" *Journal of American Planning Association,* 69(4).

Brady, N. C., and R. Weil (2007), *The Nature and Property of Soils,* 14th ed. Prentice-Hall, Upper Saddle River, N.J.

CPESC (2007), http://www.cpesc.net/, Certified Professional in Erosion and Sediment Control website accessed May 14, 2007.

CEPSCI (2005), *Erosion Prevention and Sediment Control Inspector Training Handout,* CEPSCI South Carolina, Clemson University.

CEPSCI (2007), https://www.clemson.edu/t3s/cepsci/index.htm, CEPSCI South Carolina, Clemson University website accessed July 27, 2007.

CWP (1998), *Better Site Design,* CD, Center for Watershed Protection, Ellicott City, Md.

CWP (2000), *8 Tools of Watershed Protection,* CD, Center for Watershed Protection, Ellicott City, Md.

CWP (2000), *Impacts of Urbanization,* CD, Center for Watershed Protection, Ellicott City, Md.

CWP (2000), *The Practice of Watershed Protection,* Center for Watershed Protection, Ellicott City, Md.

CWP (2000), *A Review of Stormwater Treatment Practices,* CD, Center for Watershed Protection, Ellicott City, Md.

Davis, A. P. (2005), "Green Engineering Principles, Promote Low Impact Development," *ES&T,* August 15, pp. 338A–344A.

Davis, A. P., and R. H. McCuen (2005), *Stormwater Management for Smart Growth,* Springer, New York.

Debo, T., and A. J. Reese (2003), *Municipal Stormwater Management,* 2d ed., Forester Press, Santa Barbara, Calif.

Deletic, A., and T. D. Fletcher (2006), "Performance of Grass Filters Used for Stormwater Treatment—A Field and Modelling Study," *Journal of Hydrology*, 317: 261–275.

DoD (2004), *Unified Facilities Criteria (UFC) Design: Low Impact Development Manual*, Department of Defense, October 24, 2004, accessed July 10, 2007, http://www.lowimpactdevelopment.org/lid%20articles/ufc_3_210_10.pdf.

Driscoll, E. (1986), "Methodology for Analysis of Detention Basins for Control of Urban Runoff Quality," EPA440/5-87-001, U.S. Environmental Protection Agency, Office of Water, Nonpoint Source Branch, Washington, D.C.

Elliot, A. H., and S. A. Trowsdale (2007), "A Review of Models for Low Impact Urban Stormwater Drainage," *Environmental Modelling & Software*, 22(3): 394–405.

EPA (1986), *Preliminary Data Summary of Urban Storm Water Best Management Practices*, EPA-821-R-99-012, U.S. Environmental Protection Agency, Washington, D.C.

EPA (2000), *Introduction to Phytoremediation*, EPA-600-R-99-107, U.S. Environmental Protection Agency, Cincinnati, OH.

EPA (2007), http://cfpub.epa.gov/npdes/, Environmental Protection Agency NPDES publication website accessed May 24, 2007.

EPA OW (2007), http://www.epa.gov/watersense/, U.S. Environmental Protection Agency Office of Water, Water Sense; Efficiency Made Easy website accessed July 10, 2007.

Ferguson, B. (2005), *Porous Pavements*, Taylor & Francis CRC Press, Boca Raton, Fla.

FHWA (1971), *Hydraulic Engineering*, Circular No. 9 (HEC-9), U.S. Department of Transportation, Federal Highway Administration.

FHWA (1989), *Hydraulic Engineering*, Circular No. 11 (HEC-11), U.S. Department of Transportation, Federal Highway Administration.

Fifield, J. S. (2002), *Field Manual on Sediment and Erosion Control Best Management Practices for Contractors and Inspectors*, Forester Press, Santa Barbara, Calif.

Funkhouser, L. (2007), "Stormwater Management as Adaptation to Climate Change," *Stormwater, The Journal for Surface Water Quality Professionals*, July/August, 8(5): 50–71, Forester Communications, Santa Barbara, Calif.

Haselbach, L. M., and R. M. Freeman (2006), "Vertical Porosity Distributions in Pervious Concrete Pavement," American Concrete Institute *ACI Materials Journal*, November–December, 103(6): 452–458.

Haselbach, L. M., S. R. Loew, and M. E. Meadows (2005), "Compliance Rates for Storm Water Detention Facility Installation," *ASCE Journal of Infrastructure Systems*, March, 11(1): 61–63.

Haselbach, L. M., S. Valavala, and F. Montes (2006), "Permeability Predictions for Sand Clogged Portland Cement Pervious Concrete Pavement Systems," *Journal of Environmental Management*, October, 81(1): 42–49.

Hinman, C. (2005), *Low Impact Development: Technical Guidance Manual for Puget Sound*, Puget Sound Action Team, Washington State University, Pierce County Extension, Olympia, Wash.

Holman-Dodds, J., A. Bradley, and K. Potter (2003), "Evaluation of Hydrologic Benefits of Infiltration Based Urban Storm Water Management," *Journal of American Water Resources Association*, February, 39(1): 205–215.

Hunt, W. F., and B. A. Doll (2000), "Designing Stormwater Wetlands for Small Watersheds," *Urban Waterways Series*, AG-588-2, North Carolina State University, Raleigh.

Hunt, W. F., and W. G. Lord (2006), "Bioretention Performance, Design, Construction and Maintenance," *Urban Waterways Series*, AGW-588-05, North Carolina State University, Raleigh.

Hunt, W. F., and N. M. White (2001), "Designing Rain Gardens/Bioretention Areas," *North Carolina Cooperative Extension Service Bulletin, Urban Waterfronts Series*, AG-588-3, North Carolina State University, Raleigh.

Kadlec, R. H., and R. L. Knight (1996), *Treatment Wetlands*, Lewis Publishers, Boca Raton, Fla.

Kim, H., E. A. Seagren, and A. P. Davis (2003), "Engineered Bioretention for Removal of Nitrate from Stormwater Runoff," *Water Environment Research*, 75: 355–367.

Li, K., et al. (2007), "Development of a Framework for Quantifying the Environmental Impacts of Urban Development and Construction Practices," *Environmental Science and Technology*, 41(14): 5130–5136.

Loew, S. R., L. M. Haselbach, and M. E. Meadows (2004), "Life Cycle Analysis Factors for Construction Phase BMPs in Residential Subdivisions," *Proceedings of the ASCE-EWRI 2004 World Water and Environmental Resources Congress*, 6/27–7/1/04, Salt Lake City, Utah, Best Management Practices (BMP) Technology Symposium: Current and Future Directions.

Malcom, H. R. (1989), *Elements of Urban Stormwater Design*, North Carolina State University, Raleigh.

Millen, J. A., A. R. Jarrett, and J. W. Faircloth (1996), "Reducing Sediment Discharge from Sediment Basins with Barriers and a Skimmer," ASAE Microfiche 96-2056.

Montes, F., and L. M. Haselbach (2006), "Measuring Hydraulic Conductivity in Pervious Concrete," *Environmental Engineering Science*, November, 23(6): 956–965.

Montes, F., S. Valavala, and L. M. Haselbach (2005), "A New Test Method for Porosity Measurements of Portland Cement Pervious Concrete," *Journal of ASTM International*, January, 2(1).

Moran, A. C., W. F. Hunt, and G. D. Jennings (2004), "A North Carolina Field Study to Evaluate Green Roof Quantity, Runoff Quality, and Plant Growth," in *Proceedings of Green Roofs for Healthy Cities Conference*, Portland, Ore.

NCDENR, DLR (2006), *Erosion and Sediment Control Planning and Design Manual*, North Carolina Department of Environment and Natural Resources, Division of Land Resources, Raleigh.

NCDENR, DWQ (1999), *Neuse River Basin: Model Stormwater Program for Nitrogen Control*, July, North Carolina Department of Environment and Natural Resources, Division of Water Quality, Raleigh, http://h2o.enr.state.nc.us/su/Neuse_SWProgram_Documents.htm.

NCDENR, DWQ (2003), *Tar-Pamlico River Basin: Model Stormwater Program for Nutrient Control*, North Carolina Department of Environment and Natural Resources, Division of Water Quality, Raleigh, http://h2o.enr.state.nc.us/nps/Model-FINAL9-12-03.doc.

NCDENR, DWQ (2004), *Memorandum:Updates to Stormwater BMP Efficiencies*. September 8, North Carolina Department of Environment and Natural Resources, Division of Water Quality, Raleigh.

NCDENR, DWQ (2007), *Stormwater Best Management Practices Manual*, July 2, 2007, North Carolina Department of Environment and Natural Resources, Division of Water Quality, Raleigh, http://h2o.enr.state.nc.us/su/bmp_manual.htm, website accessed July 21, 2007.

NCDOT (2003), *Best Management Practices for Construction and Maintenance Activities*, August, North Carolina Department of Transportation, Raleigh.

NCHRP (2006), *Evaluation of Best Management Practices for Highway Runoff Control*; National Cooperative Highway Research Program (NCHRP) Report 565, Transportation Research Board (TRB), Washington, D.C.

NIPC (1993), *Urban Stormwater Best Management Practices for Northeastern Illinois*, Northeastern Illinois Planning Commission.

NRCS (2007), http://www.nrcs.usda.gov/, Natural Resources Conservation Service website accessed July 27, 2007.

Pitt, R., S. Clark, and D. Lake (2006), *Construction Site Erosion and Sediment Controls: Planning, Design and Performance*, Forester Press, Santa Barbara, Calif.

Prince George's County, Department of Environmental Resources (1999), *Low-Impact Development Design Strategies: An Integrated Design Approach*, Prince George's County, Md.

Prince George's County, Department of Environmental Resource (2002), *Bioremediation Manual*, Prince George's County, Md., accessed May 10, 2007, http://www.goprincegeorgescounty.com/Government/AgencyIndex/DER/ESD/Bioretention/bioretention.asp.

Rafter, D. (2007), "The Art of Testing: Putting Stormwater, Erosion, and Sediment Control BMPs through Their Paces," *Stormwater, The Journal for Surface Water Quality Professionals*, July/August, 8(5): 24–32, Forester Communications, Santa Barbara, Calif.

SCDHEC (2003), *South Carolina Stormwater Management and Sediment Control Handbook for Land Disturbance Activities*, August, accessed July 6, 2007, South Carolina Department of Health and Environmental Control; Bureau of Water, Office of Ocean and Coastal Resource Management http://www.scdhec.gov/environment/ocrm/pubs/docs/swmanual.pdf .

Schnoor, J. L. (1996), *Environmental Modeling: Fate and Transport of Pollutants in Water, Air, and Soil*, John Wiley & Sons, New York.

Schueler, T. R. (1987), *Controlling Urban Runoff: A Practical Manual for Planning and Designing Urban BMPs*, Department of Environmental Programs, Metropolitan Washington Council of Governments.

Schueler, T. R. (1992), *Design of Stormwater Wetland Systems*, Department of Environmental Programs, Metropolitan Washington Council of Governments.

Schueler, T., and R. Claytor (1996), *Design of Stormwater Filtering Systems*, Chesapeake Research Consortium, Center for Watershed Protection, Ellicott City, Md.

U.S. Soil Conservation Service (1986), *Urban Hydrology for Small Watersheds*, Technical Release 55, Washington, D.C.

Valavala, S., F. Montes, and L. M. Haselbach (2006), "Area Rated Rational Runoff Coefficient Values for Portland Cement Pervious Concrete Pavement," *ASCE Journal of Hydrologic Engineering*, May–June, 11(3): 257–260.

Waschbusch, R. J., R. Selbig, and R. T. Bannerman (1999), *Sources of Phosphorus in Stormwater and Street Dirt from Two Urban Residential Basins in Madison, Wisconsin, 1994–95*, Water Resources Investigations Report 99-4021, U.S. Geological Survey.

Wossink, A., and W. Hunt (2003), *An Evaluation of Cost and Benefits of Structural Stormwater Best Management Practices in North Carolina*, North Carolina Cooperative Extension Service, Raleigh.

Zhang, Y.-K., and K. E. Schilling (2006), "Effects of Land Cover on Water Table, Soil Moisture, Evapotranspiration, and Groundwater Recharge: A Field Observation and Analysis," *Journal of Hydrology*, 319: 328–338.

Exercises

1. You are going to pave a parking lot on 0.5 acre of land. You are required to maintain a minimum of 10 percent of the area as landscaped or green space. You decide to incorporate bioretention cells into three landscape areas in the parking lot.

    A. Sketch an example parking lot and show the stormwater mass balances around the parking lot.

    B. Sketch one of the bioretention cells and make a box model around it showing the overall water mass balance. Define each variable and incorporate them into Eqs. (10.2.3) and (10.2.4). Assume that there is no overflow pipe in the bioretention cell.

2. Most bioretention cells have overflow pipes that are in the top of the gravel layer or somewhere else in the interior. This changes the amount of the storage volume filled until there is flow out from the cell. Using the site in Exercise 1, modify the equations to incorporate this additional outflow option and show a sketch of the new box model.

3. You are going to develop a 1-acre site with a building, parking areas, and a drive. You opt to use a detention pond for the site.

    A. Sketch the site and depict the stormwater flows on the plan.

    B. Sketch the detention pond and make a box model around it showing the overall water mass balance. Define each variable and incorporate them into Eqs. (10.2.3) and (10.2.4).

4. You are going to develop a 1-acre site with a building, parking areas, and a drive. You opt to use a retention pond for the site.

    A. Sketch the site and depict the stormwater flows on the plan.

    B. Sketch the retention pond and make a box model around it showing the overall water mass balance. Define each variable and incorporate them into Eqs. (10.2.3) and (10.2.4).

5. You are going to develop a 1-acre site with a building, parking areas, and a drive. You opt to use swales which will connect to an existing stormwater pipe in the street for drainage.

    A. Sketch the site and depict the stormwater flows on the plan.

    B. Sketch one of the swales and make a box model around it showing the overall water mass balance. Define each variable and incorporate them into Eqs. (10.2.3) and (10.2.4).

6. You are going to develop a 1-acre site with a building, parking areas, and a drive. You opt to use a small manmade wetland for the site.

    A. Sketch the site and depict the stormwater flows on the plan.

    B. Sketch the wetland and make a box model around it showing the overall water mass balance. Define each variable and incorporate them into Eqs. (10.2.3) and (10.2.4).

7. You are asked about phosphorus removal in the wetland from Exercise 6. The reeds in the wetland are known to uptake phosphorus. Sketch the wetland and make a box model around it showing the overall phosphorus mass balance. Assume that the concentration of phosphorus in the water in the wetland is completely mixed. Define each variable and incorporate them into Eqs. (10.2.11) and (10.2.12).

8. You are asked about phosphorus removal in the bioretention cell from Exercise 2. The organic layer in the cell has been shown to uptake phosphorus. Sketch the bioretention cell with the various layers and make a box model around and within it showing the overall phosphorus mass balance. Assume that the concentration of phosphorus in the water in each layer is completely mixed. Define each variable and incorporate them into Eqs. (10.2.11) and (10.2.12). (You will probably need a series of mass balance equations.)

9. You are asked about phosphorus removal in the swale from Exercise 5. The grass along the swale has been shown to uptake phosphorus at a specified rate based linearly on concentration. In this case, a completely mixed model is not appropriate, since phosphorus is being removed along the length of flow through the swale. Sketch the swale and make a box model around and within it showing the overall phosphorus mass balance. Define each variable and incorporate them into Eqs. (10.2.9) and (10.2.10).

10. You are building a home on a lot in Columbia, S.C., that is 160 ft^2 and currently undeveloped (wooded). Your local municipality has initiated a stormwater utility, and you will be taxed for any impervious surfaces that are developed. Please compare building a split-level (two-story) home to a traditional single-story home as to impervious area, percent impervious on the lot, and floor area ratio (FAR). You should make a rough sketch of your site options with all pertinent dimensions shown and labeled. The design should meet the following criteria:

- You should have 2400 ft^2 of gross floor area in the single-story home, which includes an attached two-car garage. To compensate for the stairwell, the split-level home would have 2500 ft^2 of gross floor area, and the garage is a side-access drive-under to the right of the house. The split-level home is the same dimension on both floors including the garage.
- To protect the exterior siding, you plan to put a 1-ft overhang/gutter setup around the entire roof (both types of homes).
- The front wall of the house must be at least 20 ft set back from the front property line (both types of homes).
- Minimum backout area from drive-under side-access garages is 30 ft (split-level home).
- The driveway will be paved and so will a 3-ft-wide walkway from the front door to the driveway (don't forget a landing). Your spouse insists that the front door be centered on the house (both types of homes).

11. The South Carolina DOT recommends using the rational method ($Q = CIA$) for runoff volume calculations for sites under 100 acres. For the site in Exercise 10 calculate the estimated runoff for each option (single-story versus split-level). You may assume that the overhangs are included in the impervious area. You may assume that the runoff coefficient is 0.9 for all paved and roofed surfaces and 0.30 for all other surfaces. Because of the small size of the lot, the time of concentration t_c can be assumed to be 5 min. Rainfall intensities are available from the SCDOT rainfall intensity charts (available on the SCDOT website). The municipality wants to know the value for both a 2-year and a 10-year storm.

12. Many MPOs have local requirements regulating the minimum volume in special BMPs for pollutant removal to handle what is believed to be the most frequent volumes of the most polluted runoff. According to many researchers, this is the first 0.5 in of rain to fall on impervious areas during a storm. You are building on a site with approximately 19,000 ft^2 of impervious surface area. You are installing a grass swale IMP, which is expected to have a removal efficiency of 50 to 60 percent for total phosphorus and is designed to handle this water quality volume. Typically, runoff from impervious surfaces in this area tends to have P concentrations of about 60 mg/L. Calculate the estimated uptake of phosphorus by the grass in pounds for a storm event.

13. We are interested in rainwater harvesting, and our uncle at the pickle factory is donating several 55-gal rain barrels for our new ranch house. The house has a roofed surface area of 2800 ft^2. If we have 4 in of rain, how many rain barrels would we need to capture the full rainfall without having to drain them? Please note that there is some storage, even on a roof. The roof is a typical 4V:12H asphalt shingle roof. Please use and reference a typical runoff coefficient.

14. You are placing a pervious concrete parking area over a natural sand subbase. The pervious has an average porosity of 25 percent, and the sand has an infiltration rate of 0.022 cm/s. The parking area will be about 0.27 acre and is essentially flat. It will also accept the drainage from an upslope area of about 5.38 acres. This upslope area is mixed use and has about one-half with a runoff coefficient of 0.80 and the other half around 0.30 (rational method). If you get a constant rainfall of 2.9 in/h, what would you expect the steady-state (after the storage in the pervious concrete is filled) runoff rate to be (volumetric flow) from the parking area from the entire drainage area? Draw a rough site sketch. Note that the rate of infiltration through the pervious concrete is much greater than the rate through the sand.

15.   The parking lot in Exercise 10.14 is experiencing a lot of sand drifting and may occasionally become fully covered with blowing sand, causing a reduction in the effective permeability of the pervious system. It has been found by researchers that this reduces the system infiltration rate to a fraction of the sand infiltration rate, and this fraction is the same as the fraction of the porosity in the pervious, since this more accurately represents the flow paths to the sand. What is the worst volumetric steady-state runoff rate that might be expected at steady state for the same conditions in Exercise 14 if the pervious is fully clogged (covered) with this sand?

16.   In Exercise 14, the engineer decides to alter the design and have this top 6-in layer of pervious concrete of 25 percent porosity over a 6-in layer of gravel with a porosity of 45 percent. This will be placed over the same native sand of 0.022 cm/s permeability. Assume the same upslope active runoff conditions and the same steady rainfall of 2.9 in/h. What is the expected time to the start of runoff from the pervious parking lot if the full upslope active runoff and the direct rainfall start at the same time? Assume the lot to be fairly flat. (Note that there is a time delay to maximum runoff from other areas based on the time of concentrations, but we are ignoring that here.)

17.   Many engineers are worried that if placed on slopes, water will just pour out of pervious pavement, especially if it is placed over poorly draining soils. Typically soils have the following infiltration rate ranges:

Soil	Infiltration Rate Order of Magnitudes
Gravel	>1 cm/s
Sand	10^{-2} cm/s
Silt	10^{-6} cm/s
Clay	$<10^{-8}$ cm/s

If the rainfall rate is greater than the soil percolation rate, then when the "available" storage in the pervious and subbase layers is full, there will be runoff. In previous exercises we assumed that the available storage was all the voids in the pavement and gravel layers, but in this case the available storage can be reduced when built on a slope. You are proposing a 5-ft-wide sidewalk, 100 ft long with the length sloping with the grade (width is approximately horizontal). The sidewalk is made with a 4-in-thick top layer of pervious concrete of 25 percent porosity and a 4-in-thick subbase of gravel of 40 percent porosity. Calculate, for a rainfall rate of 1 in/h over silt, an estimate of the time until runoff starts if the sidewalk is on a 2 percent grade. Make the following assumptions to simplify the analysis. Assume that friction through the pervious and gravel is negligible, so the water fills the storage volumes right away. (Typically the flow rate through unclogged pervious concrete is much greater than 0.3 cm/s.) Draw a rough sketch to explain your estimates. Is it realistic to assume that the rainfall falls at this rate for the time estimated?

18.   Repeat Exercise 17 if the sidewalk is on a 10 percent grade.

19.   Repeat Exercise 17 if the sidewalk is built over sand.

20.   Repeat Exercise 18 if the sidewalk is built over sand.

Notation

Acronyms and Chemical Symbols

The following is a list of commonly used acronyms and notations for chemical compounds as found throughout this text.

ACEEE	American Council for an Energy Efficient Economy
ACSIM	Office of the Assistant Chief of Staff for Installation Management, U.S. Army
ADA	Americans with Disabilities Act
AE	Architect-engineer
AFCEE	Air Force Center for Environmental Excellence
AFCESA	Air Force Civil Engineer Support Agency
AIA	The American Institute of Architects
AISC	American Institute of Steel Construction
ANSI	American National Standards Institute
ASAE	American Society of Agricultural Engineers
ASCE	American Society of Civil Engineers
ASLA	American Society of Landscape Architects
ASTM	American Society for Testing and Materials
ASHRAE	American Society of Heating, Refrigerating, and Air-Conditioning Engineers
ASTM	ASTM International, a.k.a. the American Society for Testing and Materials
AT/FP	Antiterrorism/Force Protection
ATM	Automated teller machine
B20	Mixture of 20 percent biodiesel and 80 percent petroleum diesel
B100	100 percent biodiesel
BEES	Building for Environmental and Economic Sustainability
BHT	Butylated hydroxytoluene

BMPs	Best Management Practices for stormwater control
BOD	Basis of design
BOD_5	5-Day Biological oxygen demand
BRAC	Base realignment and closure
BREEAM	Building Research Establishment's Environmental Assessment Method
CARB	California Air Resources Board
CBECS	Commercial Buildings Energy Consumption Survey (U.S. DoE)
CCB	Construction criteria base
Cd	Cadmium
CDA	Copper Development Association
CEPSCI	Certified Erosion Prevention and Sedimentation Control Inspector
CEC	California Energy Commission
CERCLA	Comprehensive Environmental Response, Compensation, and Liability Act
CERL	USACE's Construction Engineering Research Laboratory
CFC	Chlorofluorocarbon
CFC11	Trichlorofluoromethane (a.k.a. R-11 or CCl3F)
CF_2BrCl	Bromochlorodifluoromethane (halon 1211)
CF_3Br	Bromotrifluoromethane (halon 1301)
CFR	U.S. Code of Federal Regulations
CGP	Construction general permit
CHPS	Collaborative for High Performance Schools
CIBSE	Chartered Institution of Building Services Engineers
CIR	Credit Interpretation Request (Credit Interpretation Ruling)
CN	Curve number
CNIC	Department of the Navy, Commander, Naval Installations Command
CO	Carbon monoxide
CO_2	Carbon dioxide
CON_2H_4	Urea [see also $(NH_2)_2CO$]
CoC	Forest Stewardship Council (FSC) Chain-of-Custody certification
CofO	Certificate of Occupancy
CPESC	Certified Professional in Erosion and Sediment Control
CPM	Critical path method (a scheduling method)
CRS	Center for Resource Solutions
CSAECS	Commercial Sector Average Energy Costs by State (U.S. DOE)
CSI	Congressional Special Interest medical program
CSO	Combined sewer overflow

Cu	Copper
Cx	Commissioning
CxA	Commissioning authority
C&D	Construction and demolition (usually refers to debris)
DASA-I&H	Office of the Deputy Assistant Secretary of the Army for Installations and Housing
DCV	Demand-controlled ventilation
DoD	U.S. Department of Defense
DOE	U.S. Department of Energy
DRMS	Defense Reutilization and Marketing Service
EA	Energy and Atmosphere
EB	Existing building
ECB	Energy cost budget (method)
or	
ECB	Erosion control blanket
ECM	Energy conservation measure (EAc5)
or	
ECM	Exceptional calculation method (EAc1)
EGB	Emerging Green Builders
EIA	Energy Information Administration, U.S. Department of Energy
EMS	Emergency medical services
EO	Executive Order
EPA	U.S. Environmental Protection Agency
EPAct	Energy Policy Act of 1992
EPDM	Ethylene propylene diene monomer, rubber roofing material
EPP	Environmentally preferable purchasing
EQ	Indoor Environmental Quality
ESC	Erosion and sedimentation control
ESP	Electrostatic precipitator
ETS	Environmental tobacco smoke
EVO	Efficiency valuation organization
EWRI	Environmental and Water Resources Institute of ASCE
FAR	Floor area ratio
FEMA	Federal Emergency Management Agency
FHWA	Federal Highway Administration
FIRM	Flood insurance rate map

FPT	Functional performance testing
FYxx	Fiscal Year 20xx or 19xx
GBI	Green Building Initiative
GFA	Gross floor area
GPO	Government Printing Office
GWP	Global warming potential
HCFC	Hydrochlorofluorocarbon
HEPA	High-efficiency particulate air (filters)
HFC	Hydrofluorocarbon
HOV	High-occupancy vehicle
HOV2	High-occupancy vehicle with two or more people
HOV3	High-occupancy vehicle with three or more people
HUD	Department of Housing and Urban Development
HVAC	Heating, ventilating, and air conditioning
HVAC&R	Heating, ventilating, air conditioning, and refrigeration
H_2CO	Formaldehyde
H_2O	Water
IA	Irrigation Association
IAQ	Indoor Air Quality
ID	Innovation and Design Process
IDA	International Dark-Sky Association
IESNA	Illuminating Engineering Society of North America
IMP	Integrated Management Practice
ImpEE	Improving Engineering Education project, University of Cambridge
IPC	International Plumbing Code
IPCC	International Panel on Climate Change
IPMVP	International Performance Measurement and Verification Protocol
ISO	International Organization for Standardization
ITE	Institute of Transportation Engineers
LBNL	Lawrence Berkeley National Laboratory
LCA	Life-cycle assessment or analysis
LCC	Life-cycle cost
LCI	Life-cycle inventory
LEED	Leadership in Energy and Environmental Design
LID	Low-impact development
LPD	Lighting power density (W/ft^2 or W/lin ft, or W/item)
LZ1	Light zone 1: "dark" ambient illumination

LZ2	Light zone 2: "low" ambient illumination
LZ3	Light zone 3: "medium" ambient illumination
LZ4	Light zone 4: "high" ambient illumination
MDF	Medium-density fiberboard
MEP	Mechanical, electrical, plumbing
MERV	Minimum efficiency reporting value, for particulate removal by filters
MET	Metabolic rate
MILCON	Military construction
MPO	Municipal planning organization
MR	Materials and Resources
MSDS	Material Safety Data Sheet
MTBE	Methyl-tertiary-butyl ether
M&V	Measurement and Verification
N	Nitrogen
NAAQS	National Ambient Air Quality Standard
NAHB	National Association of Home Builders
NAVFAC	Naval Facilities Engineering Command
NC	New construction or major renovation
NCHRP	National Cooperative Highway Research Program
NH_3	Ammonia
$(NH_2)_2CO$	Urea (see also CON_2H_4)
NIST	National Institute of Standards and Technology
NO_x	Nitrogen oxides
NPDES	National Pollutant Discharge Elimination System
NPS	Nonpoint source (pollution)
NRCS	Natural Resources Conservation Service [formerly the Soil Conservation Service (SCS)]
NREL	National Renewable Energy Laboratory, DOE
NRMCA	National Ready Mixed Concrete Association
O_2	Oxygen molecule
O_3	Ozone
OASA-I&E	Office of the Assistant Secretary of the Army for Installations and the Environment
ODEP	Office of Director of Environmental Programs (Army)
ODP	Ozone-depleting potential
OFEE	Office of the Federal Environmental Executive
OMB	Office of Management and Budget

OPR	Owner's project requirements
OSB	Oriented strand board
OSHA	Occupational Safety and Health Administration
O&M	Operations and Maintenance
P	Phosphorus
PAM	Polyacrylamide
Pb	Lead
PCO	Photocatalytic oxidation
PEL	Permissible exposure limit
PM	Particulate matter
PM10	Particulate matter of 10-μm diameter or less
PM2.5	Particulate matter of 2.5-mm diameter or less
ppb	Parts per billion
ppm	Parts per million
PVC	Polyvinyl chloride
P2	Pollution prevention
P&Z	Planning & zoning
QA/QC	Quality Assurance and Quality Control
RCRA	Resource Conservation and Recovery Act
REC	Renewable energy certificate (EAc6)
REC	Renewable energy cost (EAc1 and EAc2)
RH	Relative humidity
RMI	Rocky Mountain Institute
SBR	Styrene butadiene rubber
SCS	Soil Conservation Service [now the Natural Resources Conservation Service (NRCS)
SDD	Sustainable design and development
SEC	Soil and erosion control
SMACNA	Sheet Metal and Air Conditioning National Contractors Association
SOV	Single-occupancy (motor) vehicle
SPiRiT	Sustainable project rating tool
SS	Suspended solids; Sustainable Sites
SSI	Sustainable Sites Initiative
STEL	Short-term exposure limit
STP	Stormwater Treatment Practice
SWAT	Smart Water Applications Technologies
TARP	Technology Acceptance Reciprocity Partnership, Washington State Department of Ecology

TATRC	Telemedicine and Advanced Technology Research Center of the U.S. Army Medical Research & Materiel Command
TEAP	Technology and Assessment Panel (Montreal Protocol)
TES	Thermal energy storage
TRB	Transportation Research Board
TRC	Tradable renewable certificate (EAc6)
TRM	Turf reinforcement mat
TSS	Total suspended solids
TVOC	Total volatile organic compounds/carbons (see also VOC)
TWA	Time-weighted average
ULPA	Ultra-low-penetration air (filters)
UPC	Uniform Plumbing Code
USACE	U.S. Army Corps of Engineers
USAF	U.S. Air Force
USAMRMC	U.S. Army Medical Research & Materiel Command
USGBC	U.S. Green Building Council
USGS	U.S. Geological Survey
USBG	U.S. Botanic Garden
UST	Underground storage tank (usually petroleum)
UV	Ultraviolet
UWRRC	Urban Water Resources Research Council (ASCE-EWRI)
VOC	Volatile organic carbon or compound (see also TVOC)
WBDG	Whole Building Design Guide
WE	Water Efficiency
WMO	World Meteorological Association
ZEV	Zero emission vehicle
Zn	Zinc
4-PCH	4-Phenylcyclohexene

Variables and Parameters

The following is a list of symbols for variables and parameters developed for use in this text. In some cases units are given. These are ones typically used in this text, but they can vary depending on the preference of the user.

Symbol	Chapter	Definition (typical units as applicable)
A_{BMP}	10	Surface area of BMP
A_{BMP-PC}	10	Surface area of pervious concrete BMP

Symbol	Chapter	Definition (typical units as applicable)
A_i	2, 3	Land area of land or landscape surface i at a site
A_{imp}	2	Total land area of the site covered with impervious surfaces
A_T	2, 10	Total land area of the project lot or site
A_{Ti}	2	Total land area of a lot i
A_{up}	10	Land area upslope of BMP that drains into it
AECS	4	Annual energy cost savings (percent)
AFV	2	Minimum number of alternative-fuel vehicles that can be adequately serviced with fuel
BAEE	4	Building annual electrical energy (kWh/yr) (Option 2)
BAEC	4	Building annual energy cost (energy $/yr) (Option 2)
BBP	4	Baseline building performance (energy $/yr)
BBPP	4	Baseline building performance process load (energy $/yr)
BBPR	4	Baseline building performance regulated load (energy $/yr)
BF	2	Building footprint—planar projection of built structures onto land
BR	2	Minimum number of bicycle rack spaces (commercial/institutional uses)
BSF	2	Minimum number of bicycles for which secure and covered bicycle storage facilities are provided for residential uses
BWB	3	Blackwater baseline generation rate (gal/yr)
BWD	3	Blackwater design generation rate (gal/yr)
BZA_i	6	Breathing zone horizontal area for zone i (ft^2)
BZAR	6	Breathing zone minimum area rate [ft^3/(min·ft^2) or (cfm/sf)]; usually 0.06 cfm/sf as given in the standard
$BZOD_i$	6	Breathing zone occupant density for zone i (people/ft^2)
$BZPR_j$	6	Breathing zone minimum people rate for use j [ft^3/min per person (cfm per person)] as given in the standard (For example, this is 5 ft^3/min per person for general office spaces and conference or meeting rooms, and it is 7.5 ft^3/min per person for lecture rooms and classrooms)
C	2, 10	Overall stormwater runoff coefficient of the total site area (see %Impness)
C_i	2	Stormwater runoff coefficient of land surface i
C_{up}	10	Average stormwater runoff coefficient of land surface upslope of BMP

Symbol	Chapter	Definition (typical units as applicable)
C_{PBMP}	10	Average concentration of a pollutant in a BMP
C_{PF}	10	Average concentration of a pollutant infiltrating into ground
C_{PI}	10	Average concentration of a pollutant in rainfall
C_{Pmisc}	10	Average concentration of a pollutant coming from miscellaneous flows into a BMP
C_{Pout}	10	Average concentration of a pollutant flowing out of BMP
C_{Pup}	10	Average concentration of a pollutant in upslope runoff
$CBECSE_j$	4	Commercial buildings energy consumption survey median electrical use for building use j [kWh/(ft²·yr)]
$CBECSF_j$	4	Commercial buildings energy consumption survey median fuel use for building use j [kBtu/(ft²·yr)]
CDF_i	6	Combined daylight factor for each window type i
CE	3	Controller efficiency factor
CE_i	3	Controller efficiency factor for area i
CP	2	Number of covered parking (CP) spaces for which top roofing material above them has an SRI of at least 29. Under cover includes underground, under the building, and under shade structures such as canopies
$CSAECSE_i$	4	Local electrical energy cost rate or the commercial sector average "electrical" energy costs for state/district i ($/kWh)
$CSAECSF_i$	4	Local fuel cost rate or the commercial sector average "fuel" energy costs for state/district i ($/kBtu)
CVPP1	2	Minimum number of preferred parking spaces or passes for carpools or vanpools for SSc4.4 Option 1
CVPP2	2	Minimum number of preferred parking spaces or passes for carpools or vanpools for SSc4.4 Option 2
DB	2	Density boundary
DC	6	Daylighting compliance (percent)
DD	2	Development density (ft²/acre)
DD_{site}	2	Development density of site (ft²/acre)
DD_{area}	2	Development density of the area within the density boundary as determined by the density radius for the density boundary
DDD_{area}	2	Development density of double the surrounding area as determined by the density radius for double (DRD) the density boundary area

Symbol	Chapter	Definition (typical units as applicable)
DeBLDG-REUSE	5	Items reused in building as defined in MRc1.1 if MRc.1.1 is not applied for, and the items reused in the building as defined in MRc1.3 if MRc1.3 is not applied, and any other building materials salvaged and reused on-site if they are not included in credit 3 calculations
DeCARD	5	Cardboard recycled off-site
DeDIVERT	5	C&D debris diverted
DeGYPSUM	5	Gypsum-type wallboard recycled off-site
DeLAND-FILL	5	All C&D debris excluding hazardous wastes, land-clearing debris, and soils which are disposed of off-site in a landfill or other waste facility without recycling or reuse
DeMETAL	5	Metals recycled off-site
DeMISC	5	Miscellaneous items recycled off-site
DePAVE-REUSE	5	Existing pavement debris reused on-site
DeRUBBLE	5	Any rubble (bricks, concrete, asphalt, masonry, etc.) recycled off-site
DeWOOD	5	Any wood recycled off-site
DF	2	Development footprint: total building and hardscape areas
DO	2, 3	Design occupancy, residential
DR	2	Density radius for density boundary area (ft)
DRD	2	Density radius for double the density boundary area (ft)
E	10	Evaporation rate (length/time, usually in/h)
EQR	2	Total roof areas covered by equipment and/or solar appurtenances and other appurtenances
ET	3	Evapotranspiration rate (in/time)
ET_0	3	The reference evapotranspiration rate based on local climate and a reference plant (grass or alfalfa). It is usually also based on conditions in that area for the month of July (in/time)
ETB_{Li}	3	Baseline evapotranspiration rate for landscape area i
ETD_{Li}	3	Design evapotranspiration rate for landscape area i
EV	6	Overall system ventilation effectiveness
EZ_i	6	Zone distribution effectiveness. This is typically 1.0 for overhead distribution systems and 1.2 for under-floor distribution systems.
F	10	Infiltration rate (length/time, usually in/h)
FAR	10	Floor area ratio

Symbol	Chapter	Definition (typical units as applicable)
FTE	2	Full-time equivalent occupant during the busiest time of day
FTE_j	2,3	Full-time equivalent building occupant during shift j (people per shift)
$FTE_{j,i}$	2,3	Full-time equivalent building occupancy of employee i during shift j (persons per shift)
$FTEF_j$	3	Female full-time employee equivalent for shift j (people per shift)
$FTEM_j$	3	Male full-time employee equivalent for shift j (people per shift)
GE	4	Green electricity purchased or "traded" from off-site (kWh/yr)
GFA_i	2	Gross floor area of other building i (ft²)
GFA_p	2, 5	Gross floor area of the subject project (ft²)
GFA_{PEX}	2, 5	Gross floor area of existing building portion of the subject project to be added onto (ft²)
GF_j	6	Glazing factor of regularly occupied room area j
GW	3	Annual graywater or other alternative nonpotable water source usage in wastewater fixtures
GWP_r	4	Global warming potential of refrigerant r (lb_{CO_2}/lb_r)
HS	2	Hardscape: all nonbuilding manmade hard surface on the lot such as parking areas, walks, drives, patios. Hardpacked areas such as gravel parking areas are included. Pools, fountains, and similar water-covered areas are included if they are on impervious surfaces which can be exposed when the water is emptied
I	10	Rainfall intensity or rate (length/time, usually in/h)
ICCL	6	Minimum number of individual comfort control locations
IE	3	Irrigation efficiency factor
IEB_i	3	Baseline irrigation efficiency for area i
IED_i	3	Design irrigation efficiency for area i
ILCL	6	Minimum number of individual lighting control locations
$INTFIN_{RE}$	5	The total of all the surfaces included in $INTFIN_T$ which have been reused from the existing building, even those which may have been moved
$INTFIN_T$	5	Total of all interior nonstructural surfaces in the *proposed design* (including the additions if applicable). This includes all finished interior ceilings, flooring surfaces, doors, windows, casework, walls (count

Symbol	Chapter	Definition (typical units as applicable)
		both sides), and visible casework surfaces. It is easy to understand if one pictures the finished interior surfaces that can be seen in all the rooms, but with doors and windows only counting once
IWUB	3	Baseline indoor water usage (gal/yr)
$IWUB_{auto}$	3	Baseline indoor water usage (with autocontrols on lavatory faucets) (gal/yr)
$IWUB_{no\text{-}auto}$	3	Baseline indoor water usage (without autocontrols on lavatory faucets) (gal/yr)
IWUD	3	Design indoor water usage (gal/yr)
$IWUD_{auto}$	3	Design indoor water usage (with autocontrols on lavatory faucets) (gal/yr)
$IWUD_{no\text{-}auto}$	3	Design indoor water usage (without autocontrols on lavatory faucets) (gal/yr)
k_d	3	Irrigation vegetation density factor
k_{dB}	3	Irrigation vegetation density factor–Baseline
k_{dD}	3	Irrigation vegetation density factor–Design
k_{mc}	3	Irrigation site-specific microclimate factor
k_s	3	Irrigation vegetation species factor
k_{sB}	3	Irrigation vegetation species factor–Baseline
k_{sD}	3	Irrigation vegetation species factor–Design
KSR_i	3	Kitchen sink water use rate for fixture type i (usually gal/min or gpm)
KSU_i	3	Number of daily kitchen sink uses for occupant type i
$LCGWP_i$	4	Life-cycle-based global warming potential for system i in pounds of CO_2 per ton of cooling capacity of system i per year $[lb_{CO2}/(ton_{ac} \cdot yr)]$
$LCODP_i$	4	Life-cycle-based ozone-depleting potential for system i in pounds of CFC11 per ton of cooling capacity of system i per year $[lb_{CFC11}/(ton_{ac} \cdot yr)]$
LEFEVP1	2	Minimum number of low-emitting and fuel-efficient vehicles *and* preferred parking spaces or passes for SSc4.3 Option 1
LEFEVP2	2	Minimum number of low-emitting and fuel-efficient vehicle preferred parking spaces or passes for SSc4.3 Option 2
$Life_i$	4	Equipment life of system i (years)
LFR_i	3	Lavatory faucet water use rate for fixture type i (usually gal/min or gpm)

Symbol	Chapter	Definition (typical units as applicable)
LFU_i	3	Number of daily lavatory uses for occupant type i
Lr_i	4	Annual refrigerant leakage rate of system i (lb refrigerant leaked per lb of refrigerant in the unit per year with a default of 0.02)
LSMAN	2	Manmade or graded landscaped area totals
LSNAP	2	All previously disturbed areas on ground that are replanted with native and adapted plants
LSNAT	2	Natural landscaped area totals
lx	6	Abbreviation of unit of measure lux
$MASS_{FSCi}$	5	Mass (weight) of the FSC *new* wood portion of unit i (volume or value can be substituted for mass as appropriate for an item)
$MASS_{WOODi}$	5	Total mass (weight) of all *new* wood in unit i (volume or value can be substituted for mass as appropriate for an item)
$MASS_{RECPOCi}$	5	Mass (weight) of postconsumer recycled portion of unit i
$MASS_{RECPRCi}$	5	Mass (weight) of preconsumer recycled portion of unit i
$MASS_{REGi}$	5	Mass (weight) of regional portion of unit i
$MASS_{RRMi}$	5	Mass (weight) of rapidly renewable portion of unit i
$MASS_{UNITi}$	5	Total mass (weight) of unit i (volume or value can be substituted for mass as appropriate for an item in certified wood calculations)
$MATL\$_{FSCi}$	5	Value of FSC *new* wood portions of unit i
$MATL\$_{RECPOCi}$	5	Value of postconsumer recycled portions of unit i
$MATL\$_{RECPRCi}$	5	Value of preconsumer recycled portions of unit i
$MATL\$_{RECPOC}$	5	Total value of postconsumer recycled portion of materials (CSI 2-10) plus any value of postconsumer recycled portions of furniture or furnishing if included consistently in MR subcategories 2 to 7. Value determinations are based on postconsumer recycled weight percent of value of each of individual items
$MATL\$_{RECPRC}$	5	Total value of preconsumer recycled portion of materials (CSI 2-10) plus any value of preconsumer recycled portions of furniture or furnishing if included consistently in MR subcategories 2 to 7. Value deter-minations are based on preconsumer recycled weight percent of value of each of the individual items.

Symbol	Chapter	Definition (typical units as applicable)
MATL\$$_{REG}$	5	Total value of regionally extracted, processed, and manufactured portions of materials (CSI 2-10) plus any similar portions of furniture or furnishing if included consistently in MR subcategories 2 to 7. Value determinations are based on the weight percent of the regional portion of each of the individual item values
MATL\$$_{REGi}$	5	Value of the regional portions of unit i
MATL\$$_{RRM}$	5	Total value of rapidly renewable materials (CSI 2-10) plus any similar portions of furniture or furnishing if included consistently in MR subcategories 2 to 7. Value determinations are based on the weight percent of the rapidly renewable portion of each of the individual item values
MATL\$$_{RRMi}$	5	Value of the rapidly renewable portions of unit i
MATL\$$_{SAL}$	5	Total cost of salvaged materials plus any salvaged furniture/furnishing costs if included consistently in MR subcategories 2 to 7 (salvaged items are valued at greater of market value or actual cost)
MATL\$$_T$	5	Total cost of all project materials in CSI categories 2 to 10 plus any furniture/furnishing costs included consistently in MR subcategories 2 to 7 (salvaged items are valued at greater of market value or actual salvage cost; default is 45% of total construction costs plus any furniture or furnishing costs included consistently in MR subcategories 2 through 7)
MATL\$$_{UNITi}$	5	Total cost of unit i
MATL\$$_{WOODi}$	5	Value of all the *new* wood portions of unit i
MIDAL	2	Maximum allowed percent of total initial designed fixture lumens emitted at an angle of 90 degrees or higher from nadir
Mr$_i$	4	End-of-life refrigerant loss of system i (in pounds refrigerant lost per pound of refrigerant in the unit with a default of 0.10)
n	6	Number of moles of a gas
ND$_{FTE}$	3	Number of days in a year that FTE-type occupants use the building
ND$_{TOW}$	3	Number of days in a year that transient-type occupants use the building
NonEPActB	3	Annual baseline water use for applicable non-EPAct fixtures
NonEPActD	3	Annual design water use for applicable non-EPAct fixtures

Symbol	Chapter	Definition (typical units as applicable)
O	2	O is defined as the area of the hardscape which is covered in open-grid pavement systems which are at least 50% pervious with vegetation in the open cells
O_{mod}	2	Sum of the areas with open grid pavement systems that are not shaded or have high SRI values
$OAIZ_i$	6	Outdoor air intake to zone i (ft^3/min or cfm)
ODP_r	4	Ozone depletion potential of refrigerant r (lb$_{CFC11}$/lb$_r$)
OS	2	Open space: all natural areas and landscaped areas, and wet areas with vegetated and low slopes
$OSZONE$	2	The percent of the total area required to be open space by local code, if such a requirement exists
P	6	Absolute pressure
PBE	4	Proposed building electrical performance (kWh/yr) (the PDEE less on-site renewable electrical energy credits-REE)
PBP	4	Proposed building performance (energy \$/yr) (the PDEM less on-site renewable energy cost credits-REC)
PBU_j	2	Peak building users during shift j (people per shift)
PD	2	Previously developed: all areas that have been built upon, landscaped, or graded (this does not include areas that may have been farmed in centuries past and have returned to natural vegetation)
$PDEE$	4	Proposed design electrical energy from the PDEM model (kWh/yr)
$PDEM$	4	Proposed design energy model (energy \$/yr)
$PDEMP$	4	Proposed design energy model process load (energy \$/yr)
$PDEMR$	4	Proposed design energy model regulated load (energy \$/yr)
$PED25LS$	2	The special pedestrian areas with a minimum of 25% landscaping
$PGEPB$	4	Percent green electricity (from off-site) based on the proposed building performance (PBP) model if EAc1 Option 1 is sought
$PGEBAE$	4	Percent green electricity (from off-site) based on the building annual energy (BAE) model if EAc1 Option 1 is not sought
$PREPBP$	4	Percent renewable energy based on the proposed building performance (Option 1)
$PREBAEC$	4	Percent renewable energy based on the building annual energy cost (Option 2)

Symbol	Chapter	Definition (typical units as applicable)
Q	10	Runoff rate (length^3/time, usually ft^3/s or cfs)
Q_{misc}	10	Miscellaneous flows into a BMP (length^3/time, usually ft^3/s or cfs)
Q_{out}	10	Total runoff out of a BMP (length^3/time, usually ft^3/s or cfs)
Q_{up}	10	Runoff from upslope areas into a BMP (length^3/time, usually ft^3/s or cfs)
$Q_{unit i}$	4	The cooling capacity of an individual HVAC or refrigeration unit i in ton$_{ac}$. (A ton$_{ac}$ of cooling capacity is based on an older system of rating refrigeration when tons of ice were delivered for iceboxes. The tonnage given is actually in tons of ice melted in 24 h. One ton$_{ac}$ of cooling can be directly converted to 12,000 Btu/h)
Q_{total}	4	The total cooling capacity of all HVAC or refrigeration (ton$_{ac}$)
R_{Ideal}	6	Ideal gas law constant. One value is 0.0821 L·atm/(mol · K)
R	2	Defined as the area of the hardscape that has an SRI of at least 29
R_{mod}	2	Sum of the areas with high SRI excluding those areas that are shaded
RA_j	6	Floor area of regularly occupied room area j
RAT	6	Total floor area of all regularly occupied room areas
Rc_i	4	Refrigerant charge of system i (lb of refrigerant/ton$_{ac}$ of cooling capacity)
REC	4	Renewable energy cost credit for on-site renewable energy (energy $/yr)
REE	4	Renewable electrical energy credit for on-site renewable energy (kWh/yr)
RH	3	Relative humidity (%)
RRL	2	Sum of the low-slope roofed areas with an *areal weighted average* SRI greater than 78
RRS	2	Sum of the steep-slope roofed areas with an areal weighted average SRI greater than 29 (RRS)
RW	3	Reuse water used for irrigation
S	2	Effective area of the hardscape that is shaded
S_{BMP}	10	Storage volume in a BMP (length^3, usually ft^3)
S_{Gravel}	10	Storage volume in the gravel layer (length^3, usually ft^3)
$S_{Pervious}$	10	Storage volume in the pervious concrete layer (length^3, usually ft^3)

Symbol	Chapter	Definition (typical units as applicable)
SD	5	Standard deviation
SHOWERS	2	Number of showers required
SR_i	3	Shower water use rate for fixture type i (usually gal/min or gpm)
SRI	2	Solar reflectance index
$STRDECK_{RE}$	5	Total square footage of structural roof decking, interior floor decking, and the foundation with decking or slab on grade which are intended to be reused from the existing structure, excluding any areas that are considered to be structurally unsound or hazardous materials
$STRDECK_T$	5	Total square footage of structural roof decking, interior floor decking, and the foundation with decking or slab on grade, excluding any areas that are considered to be structurally unsound or hazardous materials
$STREXWALL_{RE}$	5	Total square footage of exterior wall areas excluding windows and doors which are intended to be reused from the existing structure. Exclude any areas that are considered to be structurally unsound or hazardous materials
$STREXWALL_T$	5	Total square footage of exterior wall areas excluding windows and doors. Exclude any areas that are considered to be structurally unsound or hazardous materials
$STRINWALL_{RE}$	5	Total square footage of interior *structural* wall area excluding openings and counting both sides of the wall elements which are intended to be reused from the existing structure, excluding any areas that are considered to be structurally unsound or hazardous materials
$STRINWALL_T$	5	Total square footage of interior *structural* wall area excluding openings and counting both sides of the wall elements, excluding any areas that are considered to be structurally unsound or hazardous materials
SU_i	3	Number of daily shower uses for occupant type i
T	6	Absolute temperature
T_{vis}	6	Visible transmittance
t	10	Time
t_c	10	Time of concentration
TBW	3	Annual volume of blackwater treated and infiltrated and/or reused on-site
TIDAL	2	Total initial designed angled lumens for all exterior lighting that breaks the horizontal plane
TIDL	2	Sum of the total initial designed lumens for all exterior lighting

Symbol	Chapter	Definition (typical units as applicable)
TKSU	3	Total number of annual kitchen sink uses in a building
TLFU	3	Total number of annual lavatory faucet uses in a building
TO_j	2	Transient occupancy during shift j (people per shift)
TOW	3	Total daily transient occupancy for water usage calculations
TOWF	3	Total daily female transient occupancy for water usage calculations
TOWM	3	Total daily male transient occupancy for water usage calculations
TP	2	Total number of parking spaces
TPWA	3	Total potable water applied for irrigation for the project (design)
TR	2	Total roof area which is defined as the sum of the horizontal projections of the roofs on the site plan less the sum of the roofed areas that hold equipment, solar energy panels, or other appurtenances (BF-EQR)
TSU	3	Total number of annual shower uses in a building
TUU	3	Total number of annual urinal uses in the facility
TWCU	3	Total number of annual water closet uses in the facility
TWA	3	Total water applied (for irrigation)
TWAB	3	baseline total water applied for irrigation for the project
TWAD	3	Design total water applied for irrigation for the project
$TWAB_i$	3	Baseline total water applied for landscape area i
$TWAD_i$	3	Design total water applied for landscape area i
UR_i	3	Urinal water use per flush for fixture type i
UUF_i	3	Number of daily urinal uses by a female for occupant type i
UUM_i	3	Number of daily urinal uses by a male for occupant type i
V	6	Volume of a gas
VNAPR	2	All the roofed areas that are planted with native and adapted plants
VR	2	Sum of the vegetated roof areas
WA_{ij}	6	The sum of the areas of window type i in room area j
WCR_i	3	Water closet water use per flush for fixture type i
$WCUF_i$	3	Number of daily water closet uses by a female for occupant type i

WCUM$_i$	3	Number of daily water closet uses by a male for occupant type i
WETMAN	2	Total manmade wetlands, ponds, or pool areas
WETMANLS	2	Total manmade wetlands, ponds, or pools with vegetated low slopes (<25%) leading to its edges. Does not include pools and fountains with impervious bottoms as these are part of the hardscape
WETMANSS	2	Total manmade wetland, ponds, or pools with any steep ($\geq$25%) or unvegetated slopes leading to its edges. Does not include pools and fountains with impervious bottoms as these are part of the hardscape
WETNAT	2	Total natural wetlands, ponds, stream, or pool areas
WOOD\$$_T$	5	Total value of *new* wood portions of all items installed in the project
WOOD\$$_{FSC}$	5	Total value of FSC *new* wood portions of all items installed in the project
WQ$_{VA}$	2	Water quality volume for arid climates which must be infiltrated on-site, used on-site, or treated prior to discharge off-site
WQ$_{VH}$	2	Water quality volume for humid climates which must be infiltrated on-site, used on-site, or treated prior to discharge off-site
WQ$_{VS}$	2	Water quality volume for semiarid climates which must be infiltrated on-site, used on-site, or treated prior to discharge off-site
%Imp	2	Percent of the land area that is covered with "impervious" surfaces
%Impness	2	Relative percent imperviousness of all the surfaces at a site (see runoff coefficient)
2V:12H	2	Slope such that for every 2 units raised in the vertical direction, 12 units are moved in the horizontal direction

APPENDIX B
Definitions

The following definitions as quoted come with permission from the LEED-NC 2.2 Reference Guide with additional modifications or explanations as noted. [USGBC (2005–2007), *LEED-NC for New Construction, Reference Guide*, Version 2.2, 1st ed., U.S. Green Building Council, Washington, D.C., October 2005 with errata posted through Spring 2007.]

Acid Rain 'The precipitation of dilute solutions of strong mineral acids, formed by the mixing in the atmosphere of various industrial pollutants (primarily sulfur dioxide and nitrogen oxides) with naturally occurring oxygen and water vapor.'

Adapted (or Introduced) Plants 'Plants that reliably grow well in a given habitat with minimal attention from humans in the form of winter protection, pest protection, water irrigation, or fertilization once root systems are established in the soil. Adapted plants are considered to be low-maintenance but not invasive.'

Adaptive Reuse 'The renovation of a building or site to include elements that allow a particular use or uses to occupy a space that originally was intended for a different use.'

Adhesive 'Any substance that is used to bond one surface to another surface by attachment. Adhesives include adhesive bonding primers, adhesive primers, adhesive primers for plastics, and any other primer.'

Aerosol Adhesive 'An adhesive packaged as an aerosol product in which the spray mechanism is permanently housed in a nonrefillable can designed for handheld application without the need for ancillary hoses or spray equipment. Aerosol adhesives include special-purpose spray adhesives, mist spray adhesives, and web spray adhesives.'

Agrifiber Board 'A composite panel product derived from recovered agricultural waste fiber from sources including, but not limited to, cereal straw, sugarcane bagasse, sunflower husk, walnut shells, coconut husks, and agricultural prunings. The raw fibers are processed and mixed with resins to produce panel products with characteristics similar to those derived from wood fiber. The products must comply with the following requirements:

1. The product is inside of the building's waterproofing system.
2. Composite components used in assemblies are to be included (e.g., door cores, panel substrates).
3. The product is part of the base building systems.'

Air Changes per Hour (ACH)   'The number of times per hour a volume of air, equivalent to the volume of space, enters that space.'

Air Conditioning   Refer to ASHRAE 62.1-2004. [ASHRAE (2004), ASHRAE 62.1-2004: *Ventilation for Acceptable Indoor Air Quality*, American Society of Heating, Refrigerating and Air-Conditioning Engineers, Atlanta, Ga.]

Albedo   'Synonymous with solar reflectance.'

Alternative-Fuel Vehicles   'Vehicles that use low-polluting, nongasoline fuels such as electricity, hydrogen, propane or compressed natural gas, liquid natural gas, methanol, and ethanol. Efficient gas-electric hybrid vehicles are included in this group for LEED purposes.' Please note the additional comments about this in SS credit 4.3.

Angle of Maximum Candela   'The direction in which the luminaire emits the greatest luminous intensity'

Anti-corrosive Paints   'Coatings formulated and recommended for use in preventing the corrosion of ferrous metal substrates.'

Aquatic Systems   'Ecologically designed treatment systems that utilize a diverse community of biological organisms (e.g., bacteria, plants, and fish) to treat wastewater to advanced levels.'

Aquifer   'An underground water-bearing rock formation or group of formations, which supplies groundwater, wells, or springs.'

Assembled Recycled Content   This 'includes the percentages of postconsumer and preconsumer content. The determination is made by dividing the weight of the recycled content by the overall weight of the assembly.'

Automatic Fixture Sensors   'Motion sensors that automatically turn on and off lavatories, sinks, water closets, and urinals. Sensors may be hardwired or battery-operated.'

Baseline Building Performance   'The annual energy cost for a building design intended for use as a baseline for rating above standard design, as defined in ASHRAE 90.1-2204 Informative App. G.'

Basis of Design (BOD)   'Design information necessary to accomplish the owner's project requirements, including system descriptions, indoor environmental quality criteria, other pertinent design assumptions (such as weather data), and references to applicable codes, standards, regulations, and guidelines.'

Biodiversity   'The variety of life in all forms, levels, and combinations, including ecosystem diversity, species diversity, and genetic diversity.'

Biomass   'Plant material such as trees, grasses and crops that can be converted to heat energy to produce energy.'

Bioremediation   'Involves the use of microorganisms and vegetation to remove contaminants from water and soils. Bioremediation is generally a form of in situ remediation, and it can be a viable alternative to landfilling or incineration.'

Blackwater   'Blackwater does not have a single definition that is accepted nationwide. Wastewater from toilets and urinals is, however, always considered blackwater. Wastewater from kitchen sinks (perhaps differentiated by the use of a garbage disposal), showers, or bathtubs may be considered blackwater by state or local codes. Project teams should comply with the blackwater definitions as established by the authority having jurisdiction in their areas.'

Breathing Zone 'The region within an occupied space between planes 3 and 6 ft above the floor and more than 2 ft from the walls or fixed air-conditioning equipment.'

Building Density 'The floor area of the building divided by the total area of the site (square feet per acre).' See FAR in Chap. 10.

Building Envelope 'The exterior surface of a building's construction—the walls, windows, roof, and floor. Also referred to as the "building shell."'

Building Footprint 'The area on a project site that is used by the building structure and is defined by the perimeter of the building plan. Parking lots, landscapes, and other nonbuilding facilities are not included in the building footprint.'

Carpool 'An arrangement in which two or more people share a vehicle for transportation.' Many state transportation agencies also provide special commuting lanes for carpools or vanpools. These are referred to as HOV (high-occupancy vehicle) lanes. Sometimes the definition of an HOV is for two or more people, at other times it is for three or more. Many times these are therefore further designated HOV2 or HOV3, respectively.

Car Sharing 'A system under which multiple households share a pool of automobiles, either through cooperative ownership or through some other mechanism.'

CERCLA 'Comprehensive Environmental Response, Compensation, and Liability Act. (CERCLA), commonly known as Superfund. CERCLA addresses abandoned or historical waste sites and contamination. It was enacted in 1980 to create a tax on the chemical and petroleum industries and provided federal authority to respond to releases of hazardous substances.'

Chain-of-Custody 'A document that tracks the movement of a wood product from the forest to a vendor and is used to verify compliance with FSC guidelines. A *vendor* is defined as the company that supplies wood products to project contractors or subcontractors for on-site installation.'

Chlorofluorocarbons (CFCs): Chlorofluorocarbons are *haloalkanes* that are thought to aid in the depletion of the stratospheric ozone layer. They are carbon-based organic compounds (alkanes) which, in addition to carbon, include chlorine and fluorine atoms. Some are commonly referred to as *Freons*.

Cogeneration: 'The simultaneous production of electrical or mechanical energy (power) and useful thermal energy from the same fuel/energy source such as oil, coal, gas, biomass, or solar.'

Commissioning (Cx) 'The process of ensuring that systems are designed, installed, functionally tested, and capable of being operated and maintained to perform in conformity with the owner's project requirements.'

Commissioning Plan 'A document defining the commissioning process, which is developed in increasing detail as the project progresses through its various phases.'

Commissioning Report 'The document that records the results of the commissioning process, including the as-built performance of the HVAC system and unresolved issues.'

Commissioning Specification 'The contract document that details the objective, scope and implementation of the construction and acceptance phases of the commissioning process as developed in the design-phase commissioning plan.'

Commissioning Team 'Those people responsible for working together to carry out the commissioning process.'

Community 'An interacting population of individuals living in a specific area.'

Completed Design Area 'The total area of finished ceilings, finished floors, full-height walls and demountable partitions, interior doors, and built-in case goods in the space when the project is completed: exterior windows and exterior doors are not considered.'

Composite Wood 'A product consisting of wood or plant particles or fibers bonded together by a synthetic resin or binder (i.e., plywood, particleboard, OSB, MDF, composite door cores). For the purposes of LEED-NC requirements, products must comply with the following conditions:

1. The product is inside of the building's waterproofing system.
2. Composite wood components used in assemblies are included (e.g., door cores, panel substrates, plywood sections of I-beams).
3. The product is part of the base building system.'

Composting Toilet Systems 'Dry plumbing fixtures that contain and treat human waste via microbiological processes.'

Construction and Demolition (C&D) Debris 'Waste and recyclables generated from construction, renovation, and demolition or deconstruction of preexisting structures. Land-clearing debris including soil, vegetation, rocks, etc., is not to be included.'

Conventional Irrigation 'Most common irrigation system used in the region where the building is located. A common conventional irrigation system uses pressure to deliver water and distributes it through sprinkler heads above the ground.'

Curfew Hours 'Locally determined times when greater lighting restrictions are imposed. When no local or regional restrictions are in place, 10:00 pm is regarded as a default curfew time.'

Cutoff Angle Angle between the vertical axis of a luminaire and the first line of sight (of a luminaire) at which the light source is no longer visible (as per LEED-NC 2.1).

Daylighting 'The controlled admission of natural light into a space through glazing with the intent of reducing or eliminating electric lighting. By utilizing solar light, daylighting creates a stimulating and productive environment for building occupants.'

Development Footprint 'The area on the project site that has been impacted by any development activity. Hardscape, access roads, parking lots, nonbuilding facilities, and building structure are all included in the development footprint.'

Direct Line of Sight to Perimeter Vision Glazing 'The approach used to determine the calculated area of regularly occupied areas with direct line of sight to perimeter vision glazing. The area determination includes full-height partitions and other fixed construction prior to installation of furniture.'

Drip Irrigation 'A high-efficiency irrigation method in which water is delivered at low pressure through buried mains and submains. From the submains, water is distributed to the soil from a network of perforated tubes or emitters. Drip irrigation is a type of microirrigation.'

Ecosystem   'A basic unit of nature that includes a community of organisms and their nonliving environment linked by biological, chemical, and physical processes.'

Embodied Energy   'Energy that is used during the entire life cycle of the commodity for manufacturing, transporting and disposing of the commodity as well as the inherent energy captured within the product itself.'

Emissivity   'The ratio of the radiation emitted by a surface to the radiation emitted by a black body at the same temperature'

Endangered Species   'An animal or plant species that is in danger of becoming extinct throughout all or significant portion of its range due to harmful human activities or environmental factors.'

Energy Conservation Measures (ECMs)   'Installations of equipment or systems, or modifications of equipment or systems, for the purpose of reducing energy use and/or costs.'

Energy Star   'The rating a building earns using the ENERGYSTAR Portfolio Manager to compare building energy performance to that of similar buildings in similar climates. A score of 50 represents average building performance.'

Environmental Attributes of Green Power   'Emission reduction benefits that result from green power being used instead of conventional power sources.'

Environmental Tobacco Smoke (ETS)   'Also known as secondhand smoke, consists of airborne particles emitted from the burning end of cigarettes, pipes and cigars, and exhaled by smokers. These particles contain about 4000 different compounds, up to 40 of which are known to cause cancer.'

Environmentally Preferable Products   'Products identified as having a lesser or reduced effect on health and the environment when compared with competing products that serve the same purpose.'

Environmentally Preferable Purchasing   'A U.S. federalwide program (Executive Order 13101) that encourages and assists executive agencies in the purchasing of environmentally preferable products and services.'

Erosion   'A combination of processes in which materials of the earth's surface are loosened, dissolved, or worn away and transported from one place to another by natural agents (such as water, wind, or gravity).'

Exhaust Air   'The air removed from a space and discharged to outside the building by means of mechanical or natural ventilation systems.'

Ex Situ Remediation   'Removal of contaminated soil and groundwater. Treatment of the contaminated media occurs in another location, typically a treatment facility. A traditional method of ex situ remediation is pump-and-treat technology that uses carbon filters and incineration. More advanced methods of ex situ remediation include chemical treatment or biological reactors.'

Flat Coatings   'Coatings that register a gloss of less than 15 on an 85-degree meter or less than 5 on a 60-degree meter.'

Fly Ash   'The solid residue derived from incineration processes. Fly ash can be used as substitute for portland cement in concrete.' Concrete is frequently made with around 25 percent by weight of its cementitious material as fly ash with the remainder portland cement.

Footcandle (fc) (1) A unit of illuminance equal to one lumen of light falling on a one-square-foot area from a one candela light source at a distance of one foot (as per LEED-NC 2.2) or (2) a measure of light falling on a given surface. One footcandle (1 fc) is equal to the quantity of light falling on a one-square-foot (1-ft^2) area from a one candela (1-cd) light source at a distance of one foot (1 ft). Footcandles can be measured both horizontally and vertically by a footcandle or lightmeter (as per LEED-NC 2.1). The second definition is more comprehensive.

Formaldehyde 'A naturally occurring VOC found in small amounts in animals and plants, but that is carcinogenic and an irritant to most people when present in high concentrations, causing headaches, dizziness, mental impairment, and other symptoms. When present in the air at levels above 0.1 ppm (parts per million), it can cause watery eyes; burning sensations in the eyes, nose, and throat; nausea; coughing; chest tightness; wheezing; skin rashes; and asthmatic and allergic reactions.'

Full Cutoff Luminaire A full cutoff luminaire has zero candela (0-cd) intensity at an angle of 90 degrees above the vertical axis (nadir) and at all angles greater than 90 degrees from nadir. Additionally, the candela per 1000 lamp lumens does not numerically exceed 100 (10 percent) at an angle of 80 degrees above nadir. This applies to all lateral angles around the luminaire (as per LEED-NC 2.1).

Functional Performance Testing (FPT) 'The process of determining the ability of the commissioned system to perform in accordance with the owner's project requirements, basis of design, and construction documents.'

GFA Gross floor area. See Square Footage of a Building.

Glare Sensation produced by luminance within the visual field that is significantly greater than the luminance to which the eyes are adapted, which causes annoyance, discomfort, or loss in visual performance and visibility (as per LEED-NC 2.1).

Glazing Transparent part of a wall and usually made of glass or plastic. Wikipedia, 10/10/06.

Glazing Factor 'The ratio of interior illuminance at a given point on a given plane (usually the work plane) to the exterior illuminance under known overcast sky conditions. LEED uses a simplified approach for its credit compliance calculations. The variables used to determine the daylight factor include the floor area, window area, window geometry, visible transmittance T_{vis}, and window height.'

Graywater or Greywater 'Term defined by the Uniform Plumbing Code (UPC) in its App. G, titled "Graywater Systems for Single-Family Dwellings," as "untreated household waste-water which has not come into contact with toilet waste. Graywater includes used water from bathtubs, showers, bathroom wash basins, and water from clothes-washer and laundry tubs. It shall not include wastewater from kitchen sinks or dishwashers." The International Plumbing Code (IPC) defines graywater in its App. C, titled "Graywater Recycling Systems," as "wastewater discharged from lavatories, bathtubs, showers, clothes washers, and laundry sinks." Some states and local authorities allow kitchen sink wastewater to be included in graywater. Other differences with the UPC and IPC definitions can probably be found in state and local codes. Project teams should comply with the graywater definitions as established by the authority having jurisdiction in their areas.'

Greenfield Sites 'Those that are not developed or graded and remain in a natural state.'

Greenfields 'Sites that have not been previously developed or graded and remain in a natural state.'

Greenhouse Gases 'Gases such as carbon dioxide, methane and CFCs that are relatively transparent to the higher-energy sunlight, but trap lower-energy infrared radiation.'

Halons 'Substances used in fire suppression systems and fire extinguishers in buildings. These substances deplete the stratospheric ozone layer.' Halons are haloalkanes as CFCs are, but with bromine as well as chlorine or fluorine groups. Two common ones are bromochlorodifluoromethane (halon 1211, CF_2BrCl) and bromotrifluoromethane (halon 1301, CF_3Br).

Heat Island Effect Effect that 'occurs when warmer temperatures are experienced in urban landscapes compared to adjacent rural areas as a result of solar energy retention on constructed surfaces. Principal surfaces that contribute to the heat island effect include streets, sidewalks, parking lots, and buildings.'

Horizontal View at 42 in 'The approach used to confirm that the direct line of sight to perimeter vision glazing remains available from seated position. It uses section drawings that include the installed furniture to make the determination.'

HVAC Systems 'Heating, ventilating, and air-conditioning systems used to provide thermal comfort and ventilation for building interiors.'

Hybrid Vehicles 'Vehicles that use a gasoline engine to drive an electric generator and use the electric generator and/or storage batteries to power electric motors that drive the vehicle's wheels.'

Hydrochlorofluorocarbons (HCFCs) 'Refrigerants used in building equipment that deplete the stratospheric ozone layer, but to a lesser extent than CFCs.' They are carbon-based organic compounds (haloalkanes) that in addition to carbon include chlorine and fluorine atoms and definitely hydrogen atoms. The hydrogen atoms are located in bonds that might have had chlorine atoms if they were CFCs and therefore have less of a chance of releasing chlorine atoms into the stratosphere. The chlorine atoms can cause chain reactions which destroy the ozone molecule.

Hydrofluorocarbons (HFCs) 'Refrigerants that do not deplete the stratospheric ozone layer. However, some HFCs have high global warming potential and thus are not environmentally benign.' HFCs are haloalkanes as CFCs are, but with only hydrogen and fluorine groups attached and no chlorine.

Illuminance Amount of light falling on a surface, measured in units of footcandles (fc) or lux (lx) (as per LEED-NC 2.1).

Impervious Surfaces 'Surfaces that promote runoff of precipitation volumes instead of infiltration into the subsurface. The imperviousness or degree of runoff potential can be estimated for different surface materials.'

Incident Light Light that strikes a surface. The angle between a ray of light and the perpendicular to the surface is the angle of incidence. Wikipedia, accessed 10/10/06, http://en.wikipedia.org/wiki/Main_Page.

Individual Occupant Spaces 'Typically private offices and open office plans with workstations.'

Indoor Adhesive, Sealant, and/or Sealant Primer Product 'An adhesive or sealant product applied on-site, inside of the building's weatherproofing system.'

Indoor Air Quality 'The nature of air inside the space that affects the health and well-being of building occupants.'

Indoor Carpet System   'Carpet, carpet adhesive, or carpet cushion product installed on-site, inside of the building's weatherproofing system.'

Indoor Composite Wood or Agrifiber Product   'Composite wood or agrifiber product installed on-site, inside of the building's weatherproofing system.'

Indoor Paint or Coating Product   'A paint or coating product applied on-site, inside of the building's weatherproofing system.'

Infrared Emittance   (As defined in SSc7.1) 'A parameter between 0 and 1 that indicates the ability of a material to shed infrared radiation. The wavelength range for this radiant energy is roughly 5 to 40 μm. Most building materials (including glass) are opaque in this part of the spectru, and have an emittance of roughly 0.9. Materials such as clean, bare metals are the most important exceptions to the 0.9 rule. Thus clean, untarnished galvanized steel has low emittance, and aluminum roof coatings have intermediate emittance levels.'

Infrared or Thermal Emittance   (As defined in SSc7.2) 'A parameter between 0 and 1 (or 0 and 100 percent) that indicates the ability of a material to shed infrared radiation (heat). The wavelength range for this radiant energy is roughly 3 to 40 μm. Most building materials (including glass) are opaque in this part of the spectrum and have an emittance of roughly 0.9. Materials such as clean, bare metals are the most important exceptions to the 0.9 rule. Thus clean, untarnished galvanized steel has low emittance, and aluminum roof coatings have intermediate emittance levels.'

In Situ Remediation   'Treatment of contaminants in place using technologies such as injection wells or reactive trenches. These methods utilize the natural hydraulic gradient of groundwater and usually require only minimal disturbance of the site.'

Installation Inspection   'The process of inspecting components of the commissioned systems to determine if they are installed properly and ready for systems performance testing.'

Interior Lighting Power Allowance   'The maximum light power in watts allowed for the interior of a building.'

Interior Nonstructural Components Reuse   'Factor determined by dividing the total area of retained interior, nonstructural components by the total area of the interior, nonstructural components in the completed design' (as per errata Spring 2007).

Invasive Plants   'Both indigenous and nonindigenous species or strains that are characteristically adaptable, aggressive, have a high reproductive capacity, and tend to overrun the ecosystems which they inhabit. Collectively, they are one of the great threats to biodiversity and ecosystem stability.'

Laminate Adhesive   'An adhesive used in wood and agrifiber products (veneered panels, composite wood products contained in engineered lumber, door assemblies, etc.).'

Landfill   'A waste disposal site for the deposit of solid waste from human activities.'

Landscape Area   'Area of the site that is equal to the total site area less the building footprint, paved surfaces, water bodies, patios, etc.'

Life-Cycle Analysis (LCA)   'An evaluation of the environmental effects of a product or activity holistically, by analyzing the entire life cycle of a particular material, process, product, technology, service, or activity.'

Life-Cycle Cost (LCC) Method   'A technique of economic evaluation that sums over a given study period the costs of initial investment (less resale value), replacement, operations (including energy use), and maintenance and repair of an investment decision (expressed in present or annual value terms).'

Life-Cycle Inventory (LCI)   'An accounting of the energy and waste associated with the creation of a new product through use and disposal.'

Light Pollution   'Waste light from building sites that produces glare is directed upward to the sky or is directed off the site.'

Lighting Power Density (LPD)   'The installed lighting power, per unit area.'

Local Zoning Requirements   'Local government regulations imposed to promote orderly development of private lands and to prevent land use conflicts.'

Low-Emitting and Fuel-Efficient Vehicles   'For the purpose of this credit (SS credit 4.3), vehicles that either are classified as zero emission vehicles (ZEVs) by the California Air Resources Board (CARB) or have achieved a minimum green score of 40 on the American Council for an Energy Efficient Economy (ACEEE) annual vehicle rating guide.'

Luminance   Commonly called brightness or the light coming from a surface or light source. Luminance is composed of the intensity of light striking an object or surface and the amount of that light reflected back toward the eye. Luminance is measured in foot-lamberts (fL) or candelas per square meter (cd/m²) (as per LEED-NC 2.1).

Mass Transit   'Transportation facilities designed to transport large groups of persons in a single vehicle such as buses or trains.'

Mass Transit Vehicles   'Vehicles typically capable of serving 10 or more occupants, such as buses, trolleys, light rail, etc.'

Metering Controls   'Generally manual on/automatic off controls which are used to limit the flow of water. These types of controls are most commonly installed on lavatory faucets and on showers.'

Microirrigation   'Irrigation systems with small sprinklers and microjets or drippers designed to apply small volumes of water. The sprinklers and microjets are installed within a few centimeters of the ground, while drippers are laid on or below grade.'

Mixed-Mode Ventilation   'A ventilation strategy that combines natural ventilation with mechanical ventilation, allowing the building to be ventilated either mechanically or naturally, and at times both mechanically and naturally simultaneously.'

Native (or Indigenous) Plants   'Plants that have adapted to a given area during a defined time period and are not invasive. In America, the term often refers to plants growing in a region prior to the time of settlement by people of European descent.'

Net Metering   'A metering and billing arrangement that allows on-site generators to send excess electricity flows to the regional power grid. These electricity flows offset a portion of the electricity flows drawn from the grid. For more information on net metering in individual states, visit the DOE's Green Power Network website at www.eere.energy.gov/greenpower/netmetering.shtml.'

Nonflat Coatings   'Coatings that register a gloss of 5 or greater on a 60-degree meter and a gloss of 15 or greater on an 85-degree meter.'

Nonoccupied Spaces   'All rooms used by maintenance personnel that are not open for use by occupants. Included in this category are janitorial, storage and equipment rooms, and closets.'

Nonporous Sealant 'A substance used as a sealant on nonporous materials. Nonporous materials do not have openings in which fluids may be absorbed or discharged. Such materials include, but are not limited to, plastic and metal.'

Nonpotable Water 'Water that is not suitable for human consumption without treatment that meets or exceeds EPA drinking water standards.'

Non-Regularly Occupied Spaces 'Corridors, hallways, lobbies, break rooms, copy rooms, storage rooms, kitchens, restrooms, stairwells, etc.'

Nonroof Impervious Surfaces 'All surfaces on the site with a perviousness of less than 50 percent, not including the roof of the building. Examples of typically impervious surfaces include parking lots, roads, sidewalks, and plazas.' This definition is inconsistent with the typical definitions of impervious surfaces and the calculations in Chaps. 2 and 10. Typically a surface with a perviousness of less than about 10 percent is considered to be an impervious surface, and that is the assumption made in this text.

Non-Water-Using Urinal 'A urinal that uses no water, but instead replaces the water flush with a specially designed trap that contains a layer of buoyant liquid which floats above the urine layer, blocking sewer gas and urine odors from the room.' See Urinal, Non-Water-Using.

Off-gassing 'The emission of volatile organic compounds from synthetic and natural products.'

On-Site Wastewater Treatment 'Use of localized treatment systems to transport, store, treat, and dispose of wastewater volumes generated on the project site.'

Open-Grid Pavement 'For LEED purposes, pavement that is less than 50 percent impervious and contains vegetation in the open cells.'

Open-Space Area 'Open Space Area is as defined by local zoning requirements. If local zoning requirements do not clearly define open space, it is defined for the purposes of LEED calculations as the property area minus the development footprint; and it must be vegetated and pervious, with exceptions only as noted in the credit requirements section. For projects located in urban areas that earn SS credit 2, open space also includes nonvehicular, pedestrian-oriented hardscape spaces.'

Open-Grid Pavement 'For LEED purposes, pavement that is less than 50 percent impervious and contains vegetation in the open cells.'

Outdoor Lighting Zone Definitions 'Definitions developed by IDA for the Model Lighting Ordinance that provide a general description of the site environment/context and basic lighting criteria.' The IDA is the International Dark-Sky Association and can be found at www.darksky.org.

Owner's Project Requirements (OPRs) 'Written document that details the functional requirements of a project and the expectations of how it will be used and operated; or an explanation of the ideas, concepts, and criteria that are determined by the owner to be important to the success of the project (previously called the design intent).'

Paints 'Liquid, liquefiable or mastic compositions that are converted to a solid protective, decorative, or functional adherent film after application as a thin layer. These coatings are intended for on-site application to interior or exterior surfaces of residential, commercial, institutional or industrial buildings.'

Pedestrian Access 'Implies that pedestrians can walk to the services without being blocked by walls, freeways or other barriers.'

Percentage Improvement 'Percentage Improvement is the percent energy cost savings for the proposed building performance versus the baseline building performance.'

Perviousness 'The percent of the surface area of a paving material that is open and allows moisture to pass through the material and soak into the earth below the paving system.'

Phenol Formaldehyde Chemical that 'off-gases only at high temperature and is used for exterior products; although many of those products are suitable for interior applications.'

Porous Sealant 'A substance used as a sealant on porous materials. Porous materials have tiny openings, often microscopic, in which fluids may be absorbed or discharged. Such materials include, but are not limited to, wood, fabric, paper, corrugated paperboard and plastic foam.'

Postconsumer Descriptive of 'waste material generated by households or by commercial, industrial, and institutional facilities in their role as end-users of the product, which can no longer be used for its intended purpose. This includes returns of materials from distribution chains (source: ISO 14021). Examples of this category include construction and demolition debris, material collected through curbside and dropoff recycling programs, broken pallets (if from a pallet refurbishing company, not a pallet making company), discarded products (e.g., furniture, cabinetry, and decking), and urban maintenance waste (e.g., leaves, grass clippings, tree trimmings).'

Potable Water 'Water suitable for drinking and supplied from wells or municipal water systems.'

Preconsumer content 'Material diverted from the waste stream during the manufacturing processes. Excluded is reutilization of materials such as rework, regrind, or scrap generated in a process and capable of being reclaimed within the same process that generated it (source: ISO 14021). Examples in this category include planer shavings, ply trim, sawdust, chips, bagasse, sunflower seed hulls, walnut shells, culls, trimmed materials, print overruns, overissue publications, and obsolete inventories. (Previously referred to as postindustrial content.)'

Preferred Parking 'Parking spots that are closest to the main entrance of the project exclusive of spaces designated for handicapped, or to parking passes provided at a discounted price.'

Previously Developed Sites 'Sites that previously contained buildings, roadways, or parking lots or were graded or altered by direct human activities.'

Primer 'A material applied to substrate to improve adhesion of a subsequently applied adhesive.'

Prior Condition Area 'The total area of finished ceilings, finished floors, full-height walls, and demountable partitions, interior doors, and built-in case goods that existed when the project area was selected; exterior windows and exterior doors are not considered.'

Process Water 'Water used for industrial processes and building systems such as towers, boilers and chillers.'

Property Area 'The total area within the legal property boundaries of a site that encompasses all areas of the site including constructed areas and nonconstructed areas.' However, for instances where the subject site area is part of a campus or other larger unit, the site area of the project can be defined by the project team, but must be consistent throughout the LEED certification process.

Proposed Building Performance   'The annual energy cost calculated for a proposed design, as defined in ASHRAE 90.1-2004 Informative App. G.'

Public Transportation   'Bus, rail or other transportation service for the general public, operating on a regular, continual basis that is publicly or privately owned.'

Rapidly Renewable Materials   'Materials considered to be an agricultural product, both fiber and animal, that takes 10 years or less to grow or raise, and to harvest in an ongoing and sustainable fashion.'

Rated Power   'The nameplate power on a piece of equipment. It represents the capacity of the unit and is the maximum a unit will draw.'

RCRA   'Resource Conservation and Recovery Act. RCRA focuses on active and future facilities. It was enacted in 1976 to give the EPA authority to control hazardous wastes from cradle to grave, including generation, transportation, treatment, storage, and disposal. Some nonhazardous wastes are also covered under RCRA.'

Receptacle Load   'All equipment that is plugged into the electrical system, from office equipment to refrigerators.'

Recycling   'The collection, reprocessing, marketing and use of materials that were diverted or recovered from the solid waste stream.'

Refrigerants   'The working fluids of refrigeration cycles. They absorb heat from a reservoir at low temperatures and reject heat at higher temperatures.'

Regionally Extracted Materials   'For LEED-NC purposes, materials whose source is within a 500-mi radius of the project site.'

Regionally Manufactured Materials   'For LEED-NC purposes, materials that must be assembled as a finished product within a 500-mi radius of the project site. Assembly, as used for this LEED definition, does not include on-site assembly, erection, or installation of finished components, as in structural steel, miscellaneous iron, or systems furniture.'

Regularly Occupied Spaces   'Areas where workers are seated or standing as they work inside a building. (They include office spaces, conference rooms, and cafeterias.) In residential applications it refers to living and family rooms.' Opposite of Nonregularly Occupied Spaces.

Remediation   'The process of cleaning up a contaminated site by physical, chemical or biological means. Remediation processes are typically applied to contaminated soil and groundwater.'

Renewable Energy Certificate (REC)   'Representation of the environmental attributes of green power. RECs are sold separately from the electrons that make up the electricity. RECs allow the purchase of green power even when the electrons are not purchased.'

Retained Components   'The portions of the finished ceilings, finished floors, full-length walls and demountable partitions, interior doors, and built-in case goods that existed in the prior condition and remained in the completed design.'

Reuse   'A strategy to return materials to active use in the same or related capacity.'

Risk Assessment 'A methodology used to analyze for potential health effects caused by contaminants in the environment. Information from the risk assessment is used to determine cleanup levels.'

Salvaged Materials 'Construction materials recovered from existing buildings or construction sites and reused in other buildings. Common salvaged materials include structural beams and posts, flooring, doors, cabinetry, brick and decorative items.'

Sealant 'Any material with adhesive properties that is formulated primarily to fill, seal or waterproof gaps or joints between two surfaces. Sealants include sealant primers and caulks.'

Sedimentation 'The addition of soils to water bodies by natural and human-related activities. Sedimentation decreases water quality and accelerates the aging process of lakes, rivers and streams.'

Shared (Group) Multioccupant Spaces 'Conference rooms, classrooms, and other indoor spaces used as a place of congregation for presentations, trainings, etc. Individuals using these spaces share the lighting and temperature controls, and they should have, at a minimum, a separate zone with accessible thermostat and an airflow control.'

Shielding A nontechnical term that describes devices or techniques that are used as part of a luminaire or lamp to limit glare, light trespass, and light pollution (as per LEED-NC 2.1).

Site Area Same as Property Area.

Site Assessment 'An evaluation of aboveground (including facilities) and subsurface characteristics, including the geology and hydrology of the site, to determine if a release has occurred, as well as the extent and concentration of the release. Information generated during a site assessment is used to support remedial action decisions.'

Solar Reflectance (Albedo) (As defined in SSc7.2) 'Ratio of the reflected solar energy to the incoming solar energy over wavelengths of approximately 0.3 to 2.5 μm. A reflectance of 100 percent means that all the energy striking a reflecting surface is reflected into the atmosphere and none of the energy is absorbed by the surface. The best standard technique for its determination uses spectrophotometric measurements with an integrating sphere to determine the reflectance at each different wavelength. An averaging process using a standard solar spectrum then determines the average reflectance (see ASTM Standard E903).'

Solar Reflectance Index (SRI) 'A measure of the constructed surface's ability to reflect solar heat, as shown by a small temperature rise. It is defined so that a standard black (reflectance 0.05, emittance 0.90) is 0 and a standard white (reflectance 0.80, emittance 0.90) is 100. To calculate the SRI for a given material, obtain the reflectance value and emittance value for the material. SRI is calculated according to ASTM E 1980-01. Reflectance is measured according to ASTM E 903, ASTM E 1918, or ASTM C 1549. Emittance is measured according to ASTM E 408 or ASTM C 1371. Default values for some materials will be available in the LEED-NC 2.2 Reference Guide.'

For example, a standard black surface has a temperature rise of 90°F (50°C) in full sun, and a standard white surface has a temperature rise of 14.6°F (8.1°C). Once the maximum rise of a given material has been computed, the SRI can be computed by interpolating between the values for white and black. Materials with the highest SRI values are the coolest choices for paving. Due to the way SRI is defined, particularly hot materials can even take slightly negative values, and particularly cool materials can even exceed 100 (Lawrence Berkeley National Laboratory Cool Roofing Materials Database).

Square Footage of a Building   'Total area in square feet of all rooms including corridors, elevators, stairwells, and shaft spaces.' This is also referred to as the gross floor area (GFA). Typically GFAs are given from outer wall to outer wall. The inclusion or exclusion of parking garages has been defined in the LEED errata posted in the spring of 2007 as follows "Only 2 stories of a parking structure may be counted as part of building square footage. Surface parking (only 1 story of parking) cannot count as part of building square footage; this is to ensure efficient use of land adjacent to the building footprint. Both structured and stacked parking are allowable in square footage calculations."

Stormwater Runoff   'Water volumes that are created during precipitation events and flow over surfaces into sewer systems or receiving waters. All precipitation waters that leave project site boundaries on the surface are considered to be stormwater runoff volumes.' Note that this also means from the site through storm sewers.

Sustainable Forestry   'The practice of managing forest resources to meet the long-term forest product needs of humans while maintaining the biodiversity of forested landscapes. The primary goal is to restore, enhance, and sustain a full range of forest values—economic, social, and ecological.'

Systems Performance Testing   'The process of determining the ability of the commissioned systems to perform in accordance with the Owner's Project Requirements, Basis of Design, and construction documents.'

Tertiary Treatment   'The highest form of wastewater treatment that includes the removal of nutrients, organic and solid material, along with biological or chemical polishing (generally to effluent limits of 10 mg/L BOD_5 and 10 mg/L TSS).'

Thermal Comfort   'A condition of mind experienced by building occupants expressing satisfaction with the thermal environment.'

Threatened Species   'An animal or plant species that is likely to become endangered within the foreseeable future.'

Tipping Fees   'Fees charged by a landfill for disposal of waste volumes. The fee is typically quoted for 1 ton of waste.'

Total Suspended Solids (TSS)   'Particles or flocs that are too small or light to be removed from stormwater via gravity settling. Suspended solid concentrations are typically removed via filtration.'

Underground Parking   'A "tuck-under" or stacked parking structure that reduces the exposed parking surface area.'

Urea Formaldehyde   'A combination of urea and formaldehyde that is used in some glues and may emit formaldehyde at room temperature.'

Urinal, Non-Water-Using   'A urinal that does not use water, but instead replaces the water flush with a specially designed trap that contains a layer of buoyant liquid which floats above the urine layer, blocking sewer gas and urine odors from the room.'

Verification   'The full range of checks and tests carried out to determine if all components, subsystems, systems, and interfaces between systems operate in accordance with the contract documents. In this context, *operate* includes all modes and sequences of control operation, interlocks and conditional control responses, and specified responses to abnormal or emergency conditions.'

Visible Light Transmittance T_{vis}   'The ratio of total transmitted light to total incident light. In other words, it is the amount of visible spectrum (380 to 780 nm) light passing through a glazing surface divided by the amount of light striking the glazing surface. A higher T_{vis} value indicates that a greater amount of visible spectrum incident light is passing through the glazing.'

Vision Glazing   'The portion of exterior windows above 2 ft 6 in and below 7 ft 6 in that permits a view to the outside of the project space.'

VOCs (Volatile Organic Compounds)   'Carbon compounds that participate in atmospheric photochemical reactions (excluding carbon monoxide, carbon dioxide, carbonic acid, metallic carbides and carbonates, and ammonium carbonate). The compounds vaporize (become a gas) at normal room temperatures.'

Wetland Vegetation   'Plants that require saturated soils to survive as well as certain tree and other plant species that can tolerate prolonged wet soil conditions.'

APPENDIX C
Units

The following is a list of abbreviations or notations for units as found throughout this text.

Å	Angstrom (1×10^{-10} m)
acre	acre
ACH	air changes per hour (see App. B)
atm	atmosphere
Btu	British thermal units
cf	cubic feet
cu. ft.	cubic feet
cfm	cubic feet per minute
cfs	cubic feet per second
cm	centimeter
fc	footcandles
fpm	feet per minute
ft	feet
ft/min	feet per minute
ft^2	square feet
ft^3	cubic feet
ft^3/min	cubic feet per minute
ft^3/s	cubic feet per second
g	gram
gal	gallon
gal/min	gallons per minute
gpf	gallons per flush
gpm	gallons per minute
h	hour
in	inch
K	kelvin
kBtu	thousand British thermal units

kW	kilowatt
kWh	kilowatthour
L	liter
lb	pound
lf	lineal foot
m	meter
Met	metabolic rate
mg	milligram
mils	millimeters
mm	millimeter
MMBtu	million British thermal units
mol	mole
Pa	Pascal (1 Pa is approximately 0.004 in of water)
pm	picometer (1×10^{-12} m)
ppb	parts per billion (molar-based for gaseous phase; may be volumetric- based for ideal gases. Usually mass-based for aqueous phases, but should be specified)
ppm	parts per million (Molar-based for gaseous phase. May be volumetric-based for ideal gases. Usually mass-based for aqueous phases, but should be specified)
psia	pounds per square inch absolute
s	second
sf	square feet
sq. ft.	square feet
square	100 ft^2 when referring to roofing
ton$_{ac}$	ton of cooling (an energy unit: 1 ton$_{ac}$ of cooling is 12,000 Btu/h)
W	watt
yd	cubic yard if referring to quantities of building materials or dumpster size
yr	year
μg	microgram (1×10^{-6} g)
μm	micrometer (1×10^{-6} m)
°C	degrees Celsius
°F	degrees Fahrenheit
°R	degrees Rankine
$	U.S. dollar
%	percent

Index

━ L ━

QUEEN VICTORIA

in her Letters and Journals

QUEEN VICTORIA
IN HER
Letters and Journals

A SELECTION BY

Christopher Hibbert

VIKING

VIKING
Viking Penguin Inc.,
40 West 23rd Street,
New York, New York 10010, U.S.A.

First American edition
Published in 1985 by Viking Penguin Inc.
© Selection and editorial matter Christopher Hibbert 1984
All rights reserved
ISBN 0-670-80430-4
Library of Congress Catalog Card Number: 84-40398

Set in Ehrhardt by Keyspools Ltd
Printed in Great Britain

CONTENTS

ILLUSTRATIONS

INTRODUCTION

Surely no monarch in the history of the world wrote as much, and few wrote so well, as Queen Victoria. It has been calculated by Giles St. Aubyn that she wrote on average about 2,500 words every day of her adult life, achieving a total of some sixty million in the course of her reign, and that if she had been a novelist her complete works would have run into seven hundred volumes, published at the rate of one a month. When Henry Ponsonby was appointed to succeed General Grey as her Private Secretary in 1870 he was soon made well aware of the Queen's extraordinary industry as a correspondent, a diarist, and as a writer of notes, minutes, annotations and memoranda on all manner of subjects. A seemingly endless stream of paper passed in locked despatch boxes between her room and his, from lengthy state documents to terse notes and orders scribbled on the bottom of the separate sheets upon which each particular point for her consideration had to be submitted, or upon mourning paper into whose broad, black edges many of the words, heavily underlined and often abbreviated, trailed indecipherably. Her hand, even on the white part of the paper, was often by then exceptionally difficult to read; and Ponsonby would spend minute after minute poring over some peculiarly intractable word. One short missive, in which the Queen complained that the 'atrocious and disgraceful writing' of a young nobleman at the Colonial Office was '*too* dreadful', took him a quarter of an hour to make out, while the letter that gave rise to Her Majesty's complaint was, if childishly written, perfectly legible.

Yet, difficult, demanding and capricious as his employer was, Ponsonby had cause to be grateful for her concise and definite decisions. He was rarely left waiting for a judgement upon any matter and never left in doubt by ambiguity. In his biography of his father, Arthur Ponsonby gave several examples of the Queen's forthright, trenchant and assertive style. A painter asks leave to engrave one of the pictures he painted for the Queen: 'Certainly not. They are not good and he is very pushing.' A lady seeks permission for her daughter to write an article on the Royal Mews: 'This is a dreadful and dangerous woman. She better take the facts from the other papers.' A clergyman on retirement is credited with having given satisfaction as chaplain at Hampton Court: 'This is a mistake. He never gave satisfaction and was most interfering and disagreeable.' An opinion is sought on the Duke of Wellington's statue which had been erected on the arch at Hyde Park Corner: 'The D. of W.'s statue is a perfect disgrace. Pray say it ought to be covered over and hidden from sight.' Oscar Wilde asks if he may copy some of the poetry which Her Majesty wrote as a child: 'Really what will people not say and invent. Never c^d the Queen in her whole life write one line of poetry serious or comic or make a Rhyme even. This is therefore all invention and a myth.' May an author present Her

Majesty with a copy of a book against vivisection? 'The Queen will readily accept it. The subject causes her whole nature to boil over agᵗ these "Butchers" (Doctors and Surgeons).'

As with her instructions to her Household, so with her correspondence, both with her Ministers and her family, the Queen's views are clearly and honestly expressed with all the outspoken forthrightness of her nature. Indeed, if we had no other evidence upon which to base our estimate of her character, her letters and journals would provide the essential clues. No one reading them could doubt her simplicity and practicality, her sound common sense, her deep capacity for affection, the undeviating and sometimes highly uncomfortable regard for truth, the stubborn imperiousness protecting an inner insecurity and awareness of her own limitations. Her writings also reflect the development of her character from the shy young Queen whom her first Prime Minister, Lord Melbourne, loved as he might protectively have loved his own daughter, to the formidable old lady of whom even Bismarck stood in awe. At first she appears so vulnerable, apprehensive of appearing stupid and therefore disinclined to venture opinions on matters in which she felt out of her depth, being conscious when she did so, as she confessed herself, of sometimes 'saying stupid things in conversation'. Towards the end, when she had at last overcome the scarcely endurable grief of early widowhood, she appears completely confident in her opinions, masterful in her manner, so uncompromising in her truthfulness that the occasional revelations of diffidence, susceptibility and loneliness are all the more touching and poignant.

As well as revealing her character, the Queen's letters and journals are an indispensable guide to her interests and opinions. It is clear from her earliest correspondence how deeply concerned she was with foreign affairs and from her later letters, though she so often protested her dislike of politics and business, how conscientious she was in her attention to them. In Melbourne's time she was a committed Whig, declaring that Tories were amongst the things, like insects and turtle soup, that she detested most in all the world; and by Gladstone's day she had become an emphatic anti-Radical, praising the wisdom of the Tory Disraeli almost as often as she railed against the 'ridiculous' ideas of the 'deluded old fanatic' Gladstone. Yet, although a convinced and unyielding imperialist, she was never a die-hard Conservative and her views on many topics, so forcefully and lucidly expressed, would certainly have dismayed a large proportion of the more conventional members of the upper class.

With both Melbourne and Gladstone, as indeed with all her prime ministers, she maintained a voluminous and regular correspondence, often in times of crisis writing every day or even twice a day and never allowing any minister to forget that she followed the course of government policies with minute and critical watchfulness. Her relationships with her ministers are clearly charted in these letters. With Melbourne, charming, amusing and relaxed, she was completely at her ease, allowing him to guide and influence her to such an extent that the thought of losing him drove her first to unquenchable tears and then, in her early dealings with his political opponent, Sir Robert Peel, to that kind of pert obstinacy which the Duke of Wellington, the grand old man of the Tory Party, described as 'missy' after an interview during which she had worked herself up into a state of 'high passion and excitement'. Eventually, however, she had to accept Peel as prime minister; and she found him just as nervous and awkward as she had expected, 'cold, unfeeling and disagreeable', infecting her with his own embarrassment which she, on her part, attempted to conceal behind a cold and haughty reserve. But it was not long before she began to recognize the fine qualities beneath Peel's manner; and for this her beloved husband, the astute Prince Albert, was largely responsible.

At the beginning of her marriage Albert was permitted to take little part in political business, his role being limited to what his wife ingenuously called a little 'help with the blotting paper'; and it is quite distressing to read the bossy letters in which she put him in his place as to the constitution of his Household and the duration of his honeymoon. It was not long, however, before the Prince, patient, forbearing and persuasive as well as shrewd, had assumed that benign yet overwhelming influence over his wife that was to last beyond his death. Thereafter ministers were judged by their attitude towards him and by his feelings towards them. Peel held the Prince's qualities in high regard, as the Prince did Peel's; therefore the Queen, too, regarded Peel in a new and favourable light, and his enforced resignation in 1845 reduced her to floods of tears almost as prolonged as those induced by the resignation of Lord Melbourne. As it was with Peel, so it was with Lord John Russell, her next prime minister, whom the Prince distrusted and the Queen described as a 'dreadful' little man; and so it was with Russell's successor, Lord Derby, whom both the Prince and the Queen found dutiful and accommodating. They were equally in agreement about Lord Palmerston. As her journals show, the Queen had enjoyed the company of Palmerston before her marriage, finding him 'always so gay and amusing'. She confessed, to Lord Melbourne's amusement, that she liked him less after he had married Lord Melbourne's widowed sister at the age of fifty-five; and she had to agree that he had once behaved unforgivably at Windsor where, searching for the room of a friend with whom it was his occasional pleasure to spend the night, he had stumbled into the room of a less indulgent lady who had screamed for protection. But it was not until he became Foreign Secretary in 1846 that she and the Prince had occasion to find fault with those policies and high-handed methods which are castigated in letter after letter and entry after entry in her journal. Even so, after he had become prime minister during the Crimean War, Palmerston's conduct was so proper that the Queen agreed that of all the prime ministers they had had he was the one who 'gave the least trouble, and [was] the most amenable to reason and most ready to adopt suggestions'.

The monarch has every constitutional right to make suggestions, and it was a right which Prince Albert frequently exercised in the Queen's name. But occasionally the Prince – while justly given credit for guiding the Queen towards the creation of a new English monarchical tradition which placed the throne above party – overstepped the limits of constitutional propriety. And after his death the Queen, much taken with the mistaken doctrine of their adviser, Baron Stockmar, that the monarch was the 'permanent Premier' and the prime minister merely the 'temporary head of the Cabinet', followed her late husband's example. Often in the pages of this book the reader will detect the Queen behaving in a highly improper manner, writing to her royal relations on political matters without the knowledge of the foreign secretary, corresponding with generals without reference to the War Office and with former prime ministers behind the backs of their successors, threatening to abdicate if the Government pursued policies of which she disapproved, actively supporting and encouraging the opposition, and reacting angrily at any hint of criticism. She would not 'stand dictation', she countered furiously; she would not be a machine; it was *'very officious'* and *'impertinent'* to hint that she had been indiscreet. It was 'a miserable thing to be a constitutional Queen', she told her eldest daughter; and when Lord Salisbury's Conservative Government was defeated at the polls and she was obliged to send once more for Gladstone, by then a 'really wicked', 'wild incomprehensible man of eighty-two and a half', it seemed to her a bad defect in the country's 'much famed constitution' to have to do so 'merely on account of the number of votes'.

Her letters and diary entries thereafter continued to be replete with attacks upon Gladstone, the'*half-mad* firebrand'; but their meetings were not as strained as such epithets and their mutually antipathetical characters might lead the reader to expect. Always disliking interviews in which her opinions might be challenged, she preferred to leave criticism to the written rather than the spoken word; and so when she and Gladstone had to meet she contrived to restrict their talk to what he described as 'various nothings'. He respected her, so he wrote, 'for the scrupulous avoidance of anything which could have seemed to indicate a desire on her part to claim anything in common with me', and he was grateful to her, as his diaries make clear, for the graciousness with which she clothed her dislike of him, her 'altogether pleasant' manner, her 'perfect humour'. He could not, however, fail to feel 'rather melancholy' that he could never hope to achieve with her the kind of rapport she had enjoyed with 'dear Mr. Disraeli'. And it naturally distressed him that at his last audience with her, at which, of course, she was far too honest to express regret at his departure, she did not even discuss with him the choice of his successor whom he hoped would be Lord Spencer. She sent for Lord Rosebery.

Her relationship with Rosebery, as reflected in her letters to him, was less strained than it had been with Gladstone. But she treated him as though he were an inexperienced child in constant need of her advice and admonition, and she wrote to him frequently to reproach him for making an objectionable speech, for employing a tone not 'befitting a Prime Minister', for advancing controversial opinions without first 'obtaining her sanction'. He might have been her grandson. With her own grandson, the Kaiser, she could be equally severe.

Her letters to her family outnumber those to her ministers, and were written not so much because she had to write them but because she wanted to, out of love not duty. The two principal recipients were her uncle, King Leopold of the Belgians, and her eldest daughter, the Princess Royal, who was to become the Empress Frederick of Germany. Her correspondence with her 'dearest uncle' began in 1828, and was continued regularly until his death in 1865. Between these years he gave her the benefit of his generally sensible advice, answered her questions with the utmost care, and, at least in the early years, instructed her as best he could in the intricacies of European politics. She, in turn, kept him regularly informed of all her doings, hardly ever disappointing him, even if it meant writing under the most trying circumstances: 'I beg you to excuse this being so badly written, but my feet are being rubbed, and as I have got the box on which I am writing on my knee, it is not easy to write quite straight—but you must *not* think my hand trembles.'

The Queen's letters to her daughter are quite as artless and expressive as those to her uncle, more intimate, more spontaneous, more outspoken—particularly those carried by messenger which were not in danger of being opened in the ordinary post—and even more frequent. They were sent at the rate of about two a week and sometimes two a day from the year of Princess Victoria's marriage in 1858 until her mother died in 1901. The originals are now kept, bound in some sixty blue volumes, at Friedrichshof, the house near Frankfurt which the Empress built and named in honour of her husband. They are the property of the Kurhessische Hausstiftung. The copyright in them, as in all Queen Victoria's letters, belongs to Her Majesty Queen Elizabeth II. We are grateful for their permission to publish. Selections from these letters, about a third of them, have been skilfully edited by Sir Roger Fulford and published in five volumes by Evans Brothers (now Bell & Hyman) between 1964 and 1981. A sixth volume is yet to come.

Most of the Queen's letters to her daughter which have been printed here have been taken from Sir Roger's edition with his kind permission and that of his publishers. The Queen's letters to King Leopold are to be found in the earlier of the nine volumes of *The Letters of Queen Victoria: A Selection from Her Majesty's Correspondence* published in three series of three volumes each by John Murray between 1907 and 1932. The first series of these letters was edited by Arthur Christopher Benson and Viscount Esher and the second and third series by George Earle Buckle. They contain some two million words, both from the Queen's letters and her journals, extending over five thousand pages; and even so they represent but a small proportion of the papers which the Queen methodically collected and had bound for preservation in the Royal Archives at Windsor.

Regrettably not all of the Queen's papers survived intact. For, in fulfilment of a charge imposed upon her by her mother, the Queen's youngest child, Princess Beatrice, transcribed passages from the journals and burned the originals when she had finished with them. She often, in fact, went further than this, destroying whole entries which she thought unsuitable for transcription and substantially altering numerous passages which she did transcribe. But fortunately, unknown to her and with the connivance of the King and Queen, Lord Esher made a copy of the earlier journals so that from the time they were begun in 1832 to the death of the Prince Consort in 1861 a complete typed version of them does exist.

Princess Beatrice was not alone responsible for the mutilation of Queen Victoria's papers. On the instructions of her brother, King Edward VII, letters from his mother to Lord Granville and papers concerning the Lady Flora Hastings affair were also destroyed, as were letters written by the Queen to Disraeli about various members of her family. The papers which survived, however, far outnumbered those that were burned and, in addition to the material contained in the fourteen volumes mentioned above, this selection has been able to draw upon letters from the Queen to the Empress Augusta in *Further Letters of Queen Victoria* from the archives of the House of Brandenburg-Prussia, edited by Hector Bolitho (Thornton Butterworth, Ltd., 1938); upon the Queen's correspondence with Lady Canning in Virginia Surtees's *Charlotte Canning* (John Murray, 1972); upon extracts from her correspondence published in Monypenny and Buckle's six-volume *Life of Benjamin Disraeli* (John Murray, 1910–1920), in John Morley's *Life of William Ewart Gladstone* (three volumes, Macmillan, 1903), and in Arthur Ponsonby's *Henry Ponsonby, Queen Victoria's Private Secretary* (Macmillan, 1942); and upon those entries from her journals published in the two volumes of Lord Esher's *The Girlhood of Queen Victoria* (John Murray, 1912), and in the five volumes of Theodore Martin's *Life of the Prince Consort* (1875–1880).

As well as at Windsor, there are large numbers of the Queen's papers at Broadlands. Brian Connell used a selection of these for his *Regina v. Palmerston: The Correspondence between Queen Victoria and Her Foreign and Prime Minister, 1837–1865* (Evans Brothers Ltd., 1962); and Richard Hough made use of the Queen's letters to a much loved grandchild in *Advice to a Grand-daughter: Letters from Queen Victoria to Princess Victoria of Hesse* (William Heinemann, Ltd., 1975). To both of these editors and their publishers we are most grateful for permission to reprint extracts here.

There remain at Windsor many letters from Queen Victoria and much material from her journals which have never before been printed. Mr. John Murray and I have been kindly allowed to consult these papers and to reproduce those parts of them which are here accordingly published for the first time. We have to acknowledge the gracious permission of Her Majesty the Queen for their publication, as we do for the publication of all the other material in which she holds the copyright. We are deeply indebted for their help to Sir Robin Mackworth-Young, the Queen's Librarian, and to Miss Jane Langton, Registrar of the Royal Archives.

As is well known, the Queen was much given to heavy underlinings in her correspondence, usually by one but sometimes by two or even three or four strokes of her pen. Raymond Mortimer, in commending her style as making 'everything personal and expressive', said that these underscorings added a vehement emphasis to her writing which made it all the more vivid. And so, indeed, they do. But in print much of their characteristic force is lost; and since it would be so expensive to indicate throughout the text the Queen's underlinings in such typical passages as this, 'I never NEVER spent such an evening!! My dearest DEAREST DEAR Albert sat on a footstool by my side ... How can I ever be thankful enough to have such a Husband!', it has reluctantly been decided to abandon the attempt altogether and to print everything in the same type. It has also been decided, for the sake of easier reading, to modify the Queen's punctuation which is on occasion highly idiosyncratic; to write out in full words which she habitually abbreviated; and to translate the German words and phrases that came so readily to her pen, apart from those such as *Gemütlich* (comfortable) which she used often and whose meaning is more generally known. For economy of space the beginning and conclusion of letters have been omitted. The King of the Belgians was addressed as 'My dearest uncle' by his 'ever truly devoted niece'; the Princess Royal as 'my Beloved Child', 'my Dearest Child' and later as 'my Darling Child' (a form of endearment which reverted to 'Dearest' if her mother happened to be cross with her) by her 'ever devoted Mama V.R.'. Her favourite granddaughter was almost invariably 'Darling Victoria' to her 'ever devoted Grandmama'. The Prince of Wales (not an active correspondent) was usually just 'Dear Bertie', though he became 'Dearest Bertie' in later years.

The Queen's spelling has not been corrected. Nor, except in the rarest instances, does it stand in need of correction. 'Extacies' and 'schocking' are examples of a most uncommon failing which she was quick to notice in the writing of others and to reprove. 'I must tell you that you have mis-spelt some words several times, which you must attend to,' she told her daughter, 'for if others saw it, it might make them think you did not attend to orthography and had not been taught well. You wrote in two letters—appeal and appreciate each with one p.' The Princess Royal was also scolded for using unnecessary capital letters, for not numbering her pages correctly, for writing on 'enormous' paper which would 'not go into any box or book', for forgetting birthdays and other anniversaries, for not answering questions, for not telling 'enough about the people' she met, and 'above all,' so the Queen wrote in one letter, 'for not telling me what you do. My good dear child never liked matter of fact things—but Mama does, and when Lady Churchill leaves you I shall know nothing of what is going on which makes me sad. I tell you what is going on, so that you may follow everything, daily . . .'

This was quite true and it is the supreme virtue of the Queen's letters that they shed a clear and revealing light not only upon her personality but also upon the details of her everyday life, vigorously setting forth her opinions, disclosing her fondnesses and bugbears, her tastes and sympathies, her prejudices and erratic discernment.

So as to provide as much of the Queen's own writing as possible in the space available, linking passages have been kept to a minimum. Obscure allusions and references are briefly explained in square brackets, or in footnotes where an interpolation in the text might seem obtrusive; while additional information about persons mentioned in the text will be found in the index where the nicknames by which so many of the Royal Family and their relations were known to each other are also included. Readers looking for further information about the events touched upon in the following pages could not do better than refer to Elizabeth Longford's admirable biography, *Victoria R.I.*

When the first of the letters and the earliest diary entries printed here were written, Princess Victoria was living at Kensington Palace where she had been born on 24 May 1819. Her father was the Duke of Kent, a younger brother of King George IV and of King William IV; her mother was the fourth daughter of the Hereditary Prince (later Duke) of Saxe-Coburg-Saalfeld. A family tree elaborating her relationships is provided at the end of the book. The Queen's father died suddenly in January 1820; and thereafter her susceptible mother fell increasingly under the influence of the Comptroller of her Household, Sir John Conroy, an overbearing and scheming Irishman with a kind of coarse, self-satisfied charm. The

Princess hated Conroy of whom Lord Melbourne once said, 'My God! I don't like this man. There seems to be something very odd about him.' His intimate relationship with her mother helped to render the Princess's childhood more clouded than it appears in her own account of it with which we will open this book.

CHILDHOOD

1819—1837

One day in 1872 Queen Victoria wrote down some reminiscences of her early childhood:

My earliest recollections are connected with Kensington Palace, where I can remember crawling on a yellow carpet spread out for that purpose—and being told that if I cried and was naughty my 'Uncle Sussex' [Augustus Frederick, Duke of Sussex] would hear me and punish me, for which reason I always screamed when I saw him! I had a great horror of bishops on account of their wigs and aprons, but recollect this being partially got over in the case of the then Bishop of Salisbury [John Fisher] by his kneeling down and letting me play with his badge of Chancellor of the Order of the Garter. With another bishop, however, the persuasion of showing him my 'pretty shoes' was of no use. Claremont* remains as the brightest epoch of my otherwise rather melancholy childhood ... I used to ride a donkey given me by my Uncle, the Duke of York, who was very kind to me. I remember him well—tall, rather large, very kind but extremely shy. He always gave me beautiful presents.

To Ramsgate we used to go frequently in the summer ...

To Tunbridge Wells we also went, living at a house called Mt. Pleasant, now an hotel [the Calverley Hotel]. Many pleasant days were spent here, and the return to Kensington in October or November was generally a day of tears.

I was brought up very simply—never had a room to myself till I was nearly grown up—always slept in my Mother's room till I came to the Throne. At Claremont, and in the small houses at the bathing-places, I sat and took my lessons in my Governess's bedroom. I was not fond of learning as a little child—and baffled every attempt to teach me my letters up to 5 years old—when I consented to learn them by their being written down before me.

I remember going to Carlton House, when George IV lived there, as quite a little child before a dinner the King gave. In the year '26 (I think) George IV asked my Mother, my Sister [Victoria's half-sister, Princess Feodora] and me down to Windsor for the first time; he had been on bad terms with my poor father when he died,—and took hardly any notice of the poor widow and little fatherless girl, who

---

*Claremont Park, which had been built for Lord Clive, had been bought for £69,000 for the Queen's uncle, Prince Leopold of Saxe-Coburg-Saalfeld, later King of the Belgians, upon his marriage to George IV's daughter, Princess Charlotte.

were so poor at the time of his (the Duke of Kent's) death, that they could not have travelled back to Kensington Palace had it not been for the kind assistance of my dear Uncle, Prince Leopold. We went to Cumberland Lodge, the King living at the Royal Lodge. Aunt Gloucester [Princess Mary, the King's sister] was there at the same time. When we arrived at the Royal Lodge the King took me by the hand, saying: 'Give me your little paw.' He was large and gouty but with a wonderful dignity and charm of manner. He wore the wig which was so much worn in those days. Then he said he would give me something for me to wear, and that was his picture set in diamonds, which was worn by the Princesses as an order to a blue ribbon on the left shoulder. I was very proud of this,—and Lady Conyngham [the King's mistress] pinned it on my shoulder.

Lady Maria Conyngham [Lady Conyngham's younger daughter], then quite young, and Lord Graves [Comptroller of the Duke of Cumberland's Household] were desired to take me a drive to amuse me. I went with them, and Baroness (then Miss) Lehzen (my governess) in a pony carriage and 4, with 4 grey ponies (like my own), and was driven about the Park and taken to Sandpit Gate where the King had a Menagerie—with wapitis, gazelles, chamois, etc., etc. Then we went (I think the next day) to Virginia Water, and met the King in his phaeton in which he was driving the Duchess of Gloucester,—and he said 'Pop her in,' and I was lifted in and placed between him and Aunt Gloucester, who held me round the waist. (Mamma was much frightened.) I was greatly pleased, and remember that I looked with great respect at the scarlet liveries, etc. (the Royal Family had crimson and green liveries and only the King scarlet and blue in those days). We drove round the nicest part of Virginia Water and stopped at the Fishing Temple. Here there was a large barge and every one went on board and fished, while a band played in another! There were numbers of great people there, amongst whom was the last Duke of Dorset, then Master of the Horse. The King paid great attention to my Sister, and some people fancied he might marry her!! [she married in 1828 Prince Ernest Christian Charles of Hohenlohe Langenberg]. She was very lovely then—about 18—and had charming manners, about which the King was extremely particular. I afterwards went with Baroness Lehzen and Lady Maria C. to the Page Whiting's cottage ... Here I had some fruit and amused myself by cramming one of Whiting's children, a little girl, with peaches. I came after dinner to hear the band play in the Conservatory, which is still standing, and which was lit up by coloured lamps—the King, Royal Family, etc., sitting in a corner of the large saloon.

We lived in a very simple, plain manner; breakfast was at half-past eight, luncheon at half-past one, dinner at seven—to which I came generally (when it was no regular large dinner party)—eating my bread and milk out of a small silver basin. Tea was only allowed as a great treat in later years.

In 1826 (I think) my dear Grandmother, the Dowager Duchess of Saxe-Coburg-Saalfeld, came to Claremont, in the summer. I recollect the excitement and anxiety I was in, at this event,—going down the great flight of steps to meet her when she got out of the carriage, and hearing her say, when she sat down in her room, and fixed her fine clear blue eyes on her little grand-daughter whom she called in her letters 'the flower of May,' 'Ein schönes Kind'—'a fine child.' She was very clever and adored by her children but especially by her sons. She was a good deal bent and walked with a stick, and frequently with her hands on her back. She took long drives in an open carriage and I was frequently sent out with her, which I am sorry to confess I did not like, as, like most children of that age, I preferred running about. She was excessively kind to children, but could not bear naughty ones—and I shall never forget her coming into the room when I had been crying and naughty at my lessons and scolding me severely, which had a very salutary effect . . . I was very ill at that time, of dysentery, which illness increased to an alarming degree; many children died of it in the village of Esher. The Doctor lost his head, having lost his own child from it, and almost every doctor in London was away . . . I remember well being very cross and screaming dreadfully at having to wear, for a time, flannel next my skin. Up to my 5th year I had been very much indulged by every one, and set pretty well all at defiance.

At 5 years old, Miss Lehzen was placed about me, and though she was most kind, she was very firm and I had a proper respect for her. I was naturally very passionate, but always most contrite afterwards. I was taught from the first to beg my maid's pardon for any naughtiness or rudeness towards her; a feeling I have ever retained, and think every one should own their fault in a kind way to any one, be he or she the lowest.

At the end of July 1832, the Princess began to keep a journal. The earliest entries, written in pencil and inked over by an adult hand, describe a journey she made to Wales with her mother, her governess, Sir John Conroy and Conroy's daughter, Victoire. She was distressed by the sight of a coalmining area in the Midlands:

2 August 1832

We have just changed horses at Birmingham where I was two years ago and we visited the manufactories which are very curious. It rains very hard. We just passed through a town where all coal mines are and you see the fire glimmer at a distance in the engines in many places. The men, women, children, country and houses are all black. But I can not by any description give an idea of its strange and extraordinary appearance. The country is very desolate every where; there are coals about, and the grass is quite blasted and black. I just now see an extraordinary building flaming with fire. The country continues black, engines flaming, coals, in abundance, every where, smoking and burning coal heaps, intermingled with wretched huts and carts and little ragged children.

On 24 October the Princess and her party arrived at Alton Towers, the seat of the Earl of Shrewsbury:

This is an extraordinary house. On arriving one goes into a sort of gallery filled with armour, guns, swords, pistols, models, flags, etc., etc., then into a gallery filled with beautiful pictures and then into a conservatory with birds. We lunched there and the luncheon was served on splendid gold plate. I awoke at 7 and got up at ½ past 7. At ½ past 9 we breakfasted in the drawing-room, for the gentlemen who were going to hunt breakfasted in the other room, all the ladies and Sir John [Conroy] breakfasting with us. After breakfast at about ½ past 10 we went into the room where they were, and they gave us a toast with many cheers. After that we walked out to see the hunt. We saw them set off. It was an immense field of horsemen, who in their red jackets and black hats looked lively and gave an animating appearance to the whole. They had a large pack of hounds and three huntsmen or whippers-in. They drew a covert near here in hopes of finding a fox, but as they did not they returned and we got into the carriage ... When we came to a field, they drew another covert and succeeded; we saw the fox dash past and all the people and hounds after him, the hounds in full cry. The hounds killed him in a wood quite close by. The huntsman then brought him out and cutting off the brush Sir Edward Smith (to whom the hounds belong) brought it to me. Then the huntsmen cut off for themselves the ears and 4 paws, and lastly they threw it to the dogs, who tore it from side to side till there was nothing left. At ½ past 3 we came home. At ½ past 6 we dined, and I received my brush which had been fixed on a stick by the huntsman; it is a beautiful one.

On 8 November the party set out for Oxford in a closed carriage:

We got out first at the Divinity College, and walked from thence to the theatre [the Sheldonian], which was built by Sir Christopher Wren. The ceiling is painted with allegorical figures. The galleries are ornamented with carving enriched with gold. It was filled to excess. We were most warmly and enthusiastically received. They hurrayed and applauded us immensely for there were all the students there; all in their gowns and caps.

The royal party returned to Kensington Palace in time for Christmas.

23 April 1833

We rode a little way in the park, but the fog was so thick that we turned round and rode down by Gloucester Road, and turned up by Phillimore Place, where it was very fine and not at all foggy. Dear Rosa went beautifully. We came home at ½ past 1. At ½ past 1 we lunched ... At a ¼ to 7 we dined. Sir John dined here, and I dressed dear sweet little Dash [a King Charles spaniel which Conroy had given the Duchess] for the second time after dinner in a scarlet jacket and blue trousers. At 20 minutes past 8 Mamma went with Sir John to the Opera. I stayed up till ½ past 8.

24 May 1833

To-day is my birthday. I am to-day fourteen years old! How very old!! I awoke at ½ past 5 and got up at ½ past 7. I received from Mamma a lovely hyacinth brooch and a china pen tray. From Uncle Leopold a very kind letter, also one from sister Feodora. I gave Mamma a little ring. From Lehzen I got a pretty little china figure, and a lovely little china basket. I gave her a golden chain ...

At ½ past 7 we went to a Juvenile Ball that was given in honour of my birthday at St. James's by the King and Queen. We went into the Closet. Soon after, the doors were opened, and the King leading me went into the ball-room ... I danced [with several young gentlemen] ... We then went to supper. It was ½ past 11; the King leading me again. I sat between the King and Queen. We left supper soon. My health was drunk ... I danced in all 8 quadrilles. We came home at ½ past 12. I was very much amused.

The next month the Princess's cousins, Princes Alexander and Ernst of Württemberg came to stay at Kensington Palace, and proved a delightful contrast to her usual companion, Victoire Conroy, whom she disliked.

16 June 1833

They are both extremely tall, Alexander is very handsome and Ernst has a very kind expression. They are both extremely amiable.

12 July 1833

Alas! It is the last time that my dear cousins dine here, for they go tomorrow. I am very sorry that they are going, they were so agreeable and so amiable ... After dinner I played spillikins with Ernst, and put together dissected prints.

13 July 1833

I awoke at 6 and got up at ½ past 6. At 7 we breakfasted. It was a sad breakfast, for us indeed, as my dear cousins were going so soon. At about a ¼ to 8 we walked down our pier with them and there took leave of them, which made us both very unhappy. We saw them get into the barge, and watched them sailing away for some time on the beach. They were ... always satisfied, always good humoured; Alexander took such care of me in getting out of the boat, and rode next to me; so did Ernst. They talked about such interesting things, about their Turkish Campaign, about Russia, etc., etc. We shall miss them at breakfast, at luncheon, at dinner, riding, sailing, driving, walking, in fact everywhere.

5 June 1834

At 11 arrived my dearest sister Feodora whom I had not seen for 6 years. She is accompanied by Ernest, her husband, and her two eldest children Charles and Eliza. Dear Feodora looks very well but is grown much stouter since I saw her.

28 July 1834

The separation [from Feodora] was indeed dreadful. I clasped her in my arms, and kissed her and cried as if my heart would break, so did she dearest Sister. We then tore ourselves from each other in the deepest grief ... When I came home I was in such a state of grief that I knew not what to do with myself. I sobbed and cried most violently the whole morning . . . My dearest best sister was friend, sister, companion all to me, we agreed so well together in all our feelings and amusements . . . I love no one better than her.

From the age of nine the Princess maintained a regular, and at the beginning quite clearly supervised, correspondence with her Uncle, King Leopold of the Belgians.

TO KING LEOPOLD   22 October 1834

My dearest Uncle,—You cannot conceive how happy you have made me, by your very kind letter, which, instead of tiring, delights me beyond everything. I must likewise say how very grateful I feel for the kind and excellent advice you gave me in it.

For the autographs I beg to return my best thanks. They are most valuable and interesting, and will be great additions to my collections. As I have not got Sully's Memoirs, I shall be delighted if you will be so good as to give them to me. Reading history is one of my greatest delights . . . I am very fond of making tables of the Kings and Queens, as I go on, and I have lately finished one of the English Sovereigns and their consorts, as, of course, the history of my own country is one of my first duties.

19 November 1834

I am much obliged to you, dear Uncle, for the extract about Queen Anne, but must beg you, as you have sent me to show what a Queen ought not to be, that you will send me what a Queen ought to be.

2 February 1835

I know not how to thank you sufficiently for the most valuable autographs you were kind enough to send me. I am particularly delighted with that of Louis Quatorze, 'le grand Roi,' and my great admiration. You will not, I hope, think me very troublesome if I venture to ask for two more autographs which I should very particularly like to have; they are Mme. de Sévigné's and Racine's; as I am reading the letters of the former, and the tragedies of the latter, I should prize them highly.

JOURNAL   30 July 1835

I awoke at 7 and got up at 8. I gave Mamma a little pin and drawing done by me in recollection of today [the day of her confirmation], I gave Lehzen a ring, also in recollection of today . . . I went [to St. James's Palace] with the firm determination to become a true Christian, to try and comfort my dear Mamma in all her griefs, trials and anxieties, and to become a dutiful and affectionate daughter to her. Also to be obedient to dear Lehzen who has done so much for me. I was dressed in a white lace dress, with a white crape bonnet with a wreath of white roses round it.

9 September 1835

I must say with the exception of a few Choruses and one or two songs it is very heavy and tiresome [Handel's *Messiah* which the Princess heard in York Minster during a visit to the north] . . . I am not at all fond of Handel's music, I like the present Italian school such as Rossini, Bellini, Donizetti etc., much better.

29 September 1835

There was an immense concourse of people on the pier to see them [the King and Queen of the Belgians] arrive. After about half an hour's time, the steamer entered the Harbour, amidst loud cheering and the salute of guns from the pier, with the Belgian flag on its mast. My dearest Uncle Leopold and dearest Aunt Louisa [she was the daughter of Louis Philippe, King of the French] were very warmly received. It was but the people's duty to do so, as dear Uncle has lived for so long in England and was so much beloved. After another $\frac{1}{4}$ of an hour of anxious suspense, the waiter told us that "Their Majesties were coming." We hastened downstairs to receive them. There was an immense crowd before the door. At length Uncle appeared, having Aunt Louisa at his arm. What a happiness was it for me to throw myself in the arms of that dearest of Uncles, who has always been to me like a father, and whom I love so very dearly! I had not seen him for 4 years and 2 months. I was also delighted to make the acquaintance of that dear Aunt who is such a perfection and who has been always so kind to me, without knowing me. We hastened upstairs, where Uncle Leopold and Aunt Louisa showed themselves at the window and were loudly cheered; as they ought to be. I do not find dear Uncle at all changed. On the contrary I think he looks better than he did when I last saw him. Aunt Louisa is not quite so tall as Mamma, and has a very pretty slight figure. Her hair is of a lovely fair colour; her nose is aquiline, her eyes are quite lovely; they are light blue and have such a charming expression. She has such a sweet mouth and smile too. She is delightful, and was so affectionate to me.

5 November 1835

Dear good Lehzen takes such care of me [at Ramsgate where the Princess had been very ill, possibly with tonsillitis] and is so unceasing in her attentions to me that I shall never be able to repay her sufficiently for it but by my love and gratitude. I never can sufficiently repay her for all she has borne and done for me. She is the most affectionate, devoted, attached and disinterested friend I have and I love her most dearly . . . I feel that I gain strength every day.

29 February 1836

At $\frac{1}{2}$ past 7 we went to the play to Mme.Vestris's Olympic [Olympic Theatre, Wych Street, Strand]. I had never been there before; it is a very small but pretty, clean little theatre. It was the burletta of *One Hour* or *The Carnival Ball* in one act. The principal characters are: Mr. Charles Swiftly, Mr. Charles Mathews [Mme. Vestris's second husband], a most delightful and charming actor; he is son to the celebrated old Mathews who died last year. He is quite a young man, I should say not more than five or six and twenty. His face is not good-looking, but very clever and pleasing; he has a very slight, pretty figure, with very small feet and is very graceful and immensely active; he skips and runs about the stage in a most agile manner. He is so natural and amusing, and never vulgar but always very gentlemanlike. He is a most charming actor.

2 March 1836

Lady Burghersh [the Duke of Wellington's niece] told me that she knew Charles Mathews very well when she was in Florence, where he was come for the purpose of studying architecture; she said she had often acted with him in their private theatricals and that he always showed a great talent for acting, and that he then performed as a gentleman; he now acts quite like a gentleman, and looks so too; he is a charming performer I think. Lady Burghersh also said that he looks younger than he is, for that he must be 3 or 4 and thirty [he was 33]. He told her when at Florence that he had a great passion for the stage, but, as his father was greatly averse to his son becoming an actor, he refrained from doing it during his father's lifetime.

17 March 1836

We reached Windsor Castle at 6. We went to the Queen's room where Ferdinand [the Princess's cousin, Prince Ferdinand of Saxe-Coburg] and Augustus [Prince Ferdinand's brother, Prince Augustus of Saxe-Coburg] were presented to the King. We then went to our rooms. At ½ past 7 we dined in St. George's Hall with an immense number of people. Ferdinand looked very well. He wore the 3 Portuguese Orders in one ribbon, which he has the right of doing as husband to the Queen of Portugal. Ferdinand led the Queen in to dinner and the King led Mamma and I. I sat between the King and George Cambridge and opposite dear Ferdinand. After dinner we went into a beautiful new drawing-room [later known as the State Drawing-Room] where we remained till the gentlemen came from dinner. We then all went into the Waterloo Gallery where the ball was. The King went in first, then the Queen and Mamma, and then dear Ferdinand with me at his arm. I danced 3 quadrilles; 1st with dear Ferdinand, then with George Cambridge, and lastly with dear Augustus. During the evening dear Ferdinand came and sat near me and talked so dearly and so sensibly. I do so love him. Dear Augustus also sat near me and talked with me and he is also a dear good young man, and is very handsome. He is extremely quiet and silent, but there is a great deal in him. I am so fond too of my Uncle Ferdinand [the Princes' father].

18 March 1836

Dear good Augustus came up at ½ past 10 and stayed till 11. These visits please me very much; he is so quiet, and goes about looking at the things in the room, sits down and reads the newspapers, and never is in the way. He is a dear boy, and is so extremely good, kind and gentle; he has such a sweet expression and kind smile. I think Ferdinand handsomer than Augustus, his eyes are so beautiful, and he has such a lively, clever expression; both have such a sweet expression; Ferdinand has something quite beautiful in his expression when he speaks and smiles and he is so good. They are both very handsome and very dear!

To King Leopold   26 April 1836

You ask me about Sully's Memoirs, and if I have finished them. I have not finished them, but am reading them with great interest, and find there is a great deal in them which applies to the present times, and a great deal of good advice and reasoning in them. As you say, very truly, it is extremely necessary for me to follow the 'events of the day,' and to do so impartially. I am always both grateful and happy when you give me any advice, and hope you will continue to do so as long as I live.

Journal   3 May 1836

I like Lablache [Luigi Lablache, the opera singer, who gave the Princess singing lessons] very much, he is such a nice, good-natured, good-humoured man, and a very patient and excellent master; he is so merry too. *En profile* he has a very fine countenance, I think, an aquiline nose, dark arched eye-brows, and fine long eyelashes, and a very clever expression. He has a profusion of hair, which is very grey, and strangely mixed with some few black locks here and there. I liked my lesson extremely; I only wish I had one every day instead of one every week.

18 May 1836

At a ¼ to 2 we went down into the Hall, to receive my Uncle Ernest, Duke of Saxe-Coburg-Gotha*, and my Cousins, Ernest and Albert, his sons. My Uncle was here, now 5 years ago, and is looking extremely well. Ernest is as tall as Ferdinand and Augustus; he has dark hair, and fine dark eyes and eyebrows, but the nose and mouth are not good; he has a most kind, honest and intelligent expression in his countenance, and has a very good figure. Albert, who is just as tall as Ernest but stouter, is extremely handsome; his hair is about the same colour as mine; his eyes are large and blue, and he has a beautiful nose and a very sweet mouth with fine teeth; but the charm of his countenance is his expression, which is most delightful; *c'est à la fois* full of goodness and sweetness, and very clever and intelligent . . .

Both my Cousins are so kind and good; they are much more *formés* and men of the world than Augustus; they speak English very well, and I speak it with them. Ernest will be 18 years old on the 21st of June and Albert 17 on the 26th of August. Dear Uncle Ernest made me the present of a most delightful Lory, which is so tame that it remains on your hand, and you may put your finger into its beak, or do anything with it, without its ever attempting to bite.

To King Leopold   23 May 1836

Uncle Ernest and my cousins arrived here on Wednesday, *sains et saufs*. Uncle is looking remarkably well, and my cousins are most delightful young people. I will give you no detailed description of them, as you will so soon see them yourself. But I must say, that they are both very amiable, very kind and good, and extremely merry, just as young people should be; with all that, they are extremely sensible, and very fond of occupation.

---

*Ernest, Duke of Saxe-Coburg and Gotha, elder brother of King Leopold who was anxious that Princess Victoria should marry the Duke's younger son, Prince Albert.

JOURNAL   21 May 1836

I sat between my dear Cousins on the sofa and we looked at drawings. They both draw very well, particularly Albert, and are both exceedingly fond of music; they play very nicely on the piano. The more I see them the more I am delighted with them, and the more I love them. They are so natural, so kind, so very good and so well instructed and informed; they are so well bred, so truly merry and quite like children and yet very grown up in their manners and conversation. It is delightful to be with them; they are so fond of being occupied too; they are quite an example for any young person.

24 May 1836

Poor dear Albert, who had not been well the day before, looked very pale and felt very poorly. After being but a short while in the ball room and having only danced twice, turned as pale as ashes [at a state ball at St. James's Palace]; and we all feared he might faint; he therefore went home.

TO KING LEOPOLD   26 May 1836

I am sorry to say we have an invalid in the house in the person of Albert, who, though much better today, has had a smart bilious attack. He was not allowed to leave his room all day yesterday, but by dint of starvation, he is again restored to society, but looks pale and delicate.

31 May 1836

After I had gone down [the set in a country dance at Kensington Palace] I and my dear partner [Prince Albert who still felt ill] stopped, but the others kept it up for some time ... we all stayed up until $\frac{1}{2}$ past 3 and it was broad daylight when we left the room ... You are very kind dear Uncle to think of my health and of my not fatiguing myself; I can assure you all this dissipation does me a great deal of good.

7 June 1836

I must thank you, my beloved Uncle, for the prospect of great happiness you have contributed to give me, in the person of dear Albert. Allow me, then, my dearest Uncle, to tell you how delighted I am with him, and how much I like him in every way. He possesses every quality that could be desired to render me perfectly happy. He is so sensible, so kind, and so good, and so amiable too. He has, besides, the most pleasing and delightful exterior and appearance you can possibly see.

JOURNAL   10 June 1836

At 9 we all breakfasted for the last time together! It was our last happy happy breakfast with this dear Uncle and these dearest beloved Cousins, whom I do love so very, very dearly, much more dearly than any other Cousins in the world. Dearly as I love Ferdinand and also good Augustus, I love Ernest and Albert more than them, oh yes, much more. Augustus was like a good affectionate child, quite unacquainted with the world, phlegmatic, and talking but very little; but dearest Ernest and dearest Albert are so grown up in their manners ... and are very clever, naturally clever, particularly Albert who is the most reflecting of the two, and they like very much talking about serious and instructive things and yet are so very very merry and gay and happy, like young people ought to be; Albert used always to have some fun and some clever witty answer at breakfast and everywhere; he used to play and fondle Dash so funnily too. Both he and Ernest are extremely attentive to whatever they hear and see, and take interest in everything they see ... Though I wrote more when Uncle Ferdinand and Augustus went, in my Journal ... I feel this separation more deeply, though I do not lament as much as I did then, which came from my nerves not being strong then. I can bear more now.

3 August 1836

At a ¼ to 4 we went with Lehzen and Lady Flora [Lady Flora Hastings, one of the Ladies of the Duchess of Kent's Household] to Chiswick, to the Victoria Asylum or Children's Friend Society. It is a most interesting and delightful establishment, and has been founded almost entirely by Lady George and Miss Murray. It is for poor vagrant girls, who are received under the age of 15; and Miss Murray says that they have never had a girl 6 months who did not become a perfectly good child. Miss Murray told us many curious stories of the depraved and wretched state in which many arrive, and how soon they become reformed and good. There is one little girl in particular, a very pretty black-eyed girl, 11 years old, who was received two months ago from Newgate, and who boasted she could steal and tell lies better than anybody. She had been but two or three days in the school, and she got over 3 high walls, and stole a sheet; she was caught and brought back again. Miss Murray spoke to her, and found that the poor girl had no idea whatever of a God, and had a drunken father, a low Irishman; this man had lost his 1st wife and married again, and this step-mother taught the girl nothing but stealing and lying. Miss Murray told her of God, and spoke to her very seriously; the girl was put in solitary confinement for that night and was taken out the next morning; and ever since she has been a perfectly good girl.

16/18 September 1836

I love him so very very much [King Leopold who was on a visit to England]; oh! my love for him approaches to a sort of adoration ... He is indeed 'il mio secondo padre', or rather 'solo padre'! for he is indeed like my real father, as I have none! He is so clever, so mild, and so prudent; he alone can give me good advice on every thing. She [Queen Louise] is an Angel and I do so love her! How very much do I regret that she cannot come over and be with us here ... I can see dearest Aunt Louise sitting opposite to me ... with that angelic look and expression so peculiar to her; I can see her at dinner, walking, everywhere.

21 September 1836

Dear Uncle came up and fetched us down to breakfast, as he has done already once before, and twice for dinner. He always accompanied us upstairs when we went to bed. It was our last breakfast with him; I sat, as usual, near him. To hear dear Uncle speak on any subject is like reading a highly instructive book; his conversation is so enlightened, so clear. He is universally admitted to be one of the first politicians now extant. He speaks so mildly, yet firmly and impartially, about Politics ... It is dreadful how quickly this long looked for stay of dearest Uncle has come and is passed. Oh! it is dreadful ... that one is almost always separated from those one loves dearly and is encumbered with those one dislikes.

TO KING LEOPOLD   21 September 1836

As I hear that Mamma is going to send a letter to you which will reach you at Dover, and though it is only an hour and a half since we parted, I must write you one line to tell you how very, very sad I am that you have left us, and to repeat, what I think you know pretty well, how much I love you. When I think that but two hours ago we were happily together, and that now you are travelling every instant farther and farther away from us, and that I shall with all probability not see you for a year, it makes me cry.

JOURNAL   1 November 1836

Oh! Walter Scott is my *beau idéal* of a Poet; I do so admire him both in Poetry and Prose!

                                                             29 November 1836

We reached Canterbury in safety in spite of the rain and some wind, but not very long after we left it, it began to blow so dreadfully, accompanied by floods of rain at intervals, that our carriage swung and the post-boys could scarcely keep on their horses. As we approached Sittingbourne, the hurricane, for I cannot call it by any other name, became quite frightful and even alarming; corn stacks were flying about, trees torn up by their roots, and chimneys blown to atoms. We got out, or rather were blown out, at Sittingbourne. After staying there for a short while we got into the carriage ... which being larger and heavier than our post-chaise, would not shake so much. For the first 4 or 5 miles all went on more smoothly and I began to hope our difficulties were at an end. Alas! far from it. The wind blew worse than before and in going down the hill just before Chatham, the hurricane was so tremendous that the horses stopped for a minute, and I thought that we were undone, but by dint of whipping and very good management of the post-boys we reached Rochester in safety. Here we got out, and here it was determined that we must pass the night. Here we are therefore, and here we must remain, greatly to my annoyance, for I am totally unprepared, Lehzen's and my wardrobe maid are gone on to Claremont, and I hate sleeping at an inn.

                                                             29 December 1836

How I do wish I could do something for their [gipsies in an encampment near Claremont] spiritual and mental benefit and for the education of their children and in particular for the poor little baby who I have known since its birth, in the admirable manner Mr. Crabbe in his *Gipsies' Advocate* so strongly urges; he beseeches and urges those who have kind hearts and Christian feelings to think of these poor wanderers, who have many good qualities and who have many good people amongst them. He says, and alas! I too well know its truth, from experience, that whenever any poor Gipsies are encamped anywhere and crimes and robberies etc. occur, it is invariably laid to their account, which is shocking; and if they are always looked upon as vagabonds, how can they become good people? I trust in Heaven that the day may come when I may do something for these poor people, and for this particular family! I am sure that the little kindness which they have experienced from us will have a good and lasting effect on them!

To King Leopold   14 March 1837

We had a dinner on Saturday which amused me, as I am very fond of pleasant society, and we have been for these last three weeks immured within our old palace [Kensington], and I longed sadly for some gaiety. After being so very long in the country I was preparing to go out in right earnest, whereas I have only been twice to the play since our return, which is marvellous! However, we are to have another dinner to-morrow, and are going to the play and Opera. After Easter I trust I shall make ample amends for all this solitariness.

Journal   24 May 1837

Today is my 18th birthday! How old! and yet how far am I from being what I should be. I shall from this day take the firm resolution to study with renewed assiduity, to keep my attention always well fixed on whatever I am about, and to strive to become every day less trifling and more fit for what, if Heaven wills it, I'm some day to be! . . . At ½ p. 3 we drove out with dear Lehzen and came home at 5. The demonstrations of loyalty and affection from all the people were highly gratifying. The parks and streets were thronged and everything looked like a Gala day. Numbers of people put down their names and amongst others good old Lablache inscribed his . . . At ½ p. 10 we went to the ball at St. James's . . . I wished to dance with Count Waldstein who is such an amiable man, but he replied that he could not dance quadrilles, and as in my station I unfortunately cannot valse and gallop, I could not dance with him. The beauties there were (in my opinion) the Duchess of Sutherland, Lady Frances (or Fanny) Cowper, who is very pleasing, natural and clever-looking . . . The Courtyard and the streets were crammed when we went to the Ball, and the anxiety of the people to see poor stupid me was very great, and I must say I am quite touched by it, and feel proud which I always have done of my country and of the English Nation.

15 June 1837

The news of the King [William IV] are so very bad that all my lessons save the Dean's [her Principal Master, George Davys, Dean of Chester] are put off, including Lablache's . . . and we see nobody. I regret rather my singing-lesson, though it is only for a short period, but duty and proper feeling go before all pleasures.—10 minutes to 1,—I just hear that the Doctors think my poor Uncle the King cannot last more than 48 hours! Poor man! he was always kind to me, and he meant it well I know; I am grateful for it, and shall ever remember his kindness with gratitude. He was odd, very odd and singular, but his intentions were often ill interpreted! At about a ¼ p. 2 came Lord Liverpool [a friend of the family] and I had a highly important conversation with him—alone.

To King Leopold   19 June 1837

The King's state, I may fairly say, is hopeless; he may perhaps linger a few days, but he cannot recover ultimately. Yesterday the physicians declared he could not live till the morning, but to-day he is a little better; the great fear is his excessive weakness and no pulse at all. Poor old man! I feel sorry for him; he was always personally kind to me, and I should be ungrateful and devoid of feeling if I did not remember this.

I look forward to the event which it seems is likely to occur soon, with calmness and quietness; I am not alarmed at it, and yet I do not suppose myself quite equal to all; I trust, however, that with good-will, honesty, and courage I shall not, at all events, fail. Your advice is most excellent, and you may depend upon it I shall make use of it, and follow it.

THE YOUNG QUEEN

1837–1839

JOURNAL   20 June 1837

I was awoke at 6 o'clock by Mamma, who told me that the Archbishop of Canterbury [William Howley] and Lord Conyngham [the Lord Chamberlain] were here, and wished to see me. I got out of bed and went into my sitting-room (only in my dressing-gown), and alone, and saw them. Lord Conyngham then acquainted me that my poor Uncle, the King, was no more, and had expired at 12 minutes p. 2 this morning, and consequently that I am Queen. Lord Conyngham knelt down and kissed my hand, at the same time delivering to me the official announcement of the poor King's demise. The Archbishop then told me that the Queen [Adelaide] was desirous that he should come and tell me the details of the last moments of my poor, good Uncle; he said that he had directed his mind to religion and had died in a perfectly happy, quiet state of mind, and was quite prepared for his death. He added that the King's sufferings at the last were not very great but that there was a good deal of uneasiness. Lord Conyngham, who I charged to express my feelings of condolence and sorrow to the poor Queen, returned directly to Windsor. I then went to my room and dressed.

Since it has pleased Providence to place me in this station, I shall do my utmost to fulfil my duty towards my country; I am very young and perhaps in many, though not in all things, inexperienced, but I am sure, that very few have more real good will and more real desire to do what is fit and right than I have.

Breakfasted, during which time good faithful Stockmar [Baron Stockmar who had been sent by King Leopold to England as the Princess's adviser] came and talked to me. Wrote a letter to dear Uncle Leopold and a few words to dear good Feodore. Received a letter from Lord Melbourne [the Prime Minister] in which he said he would wait upon me at a little before 9. At 9 came Lord Melbourne, whom I saw in my room, and of course quite alone as I shall always do all my Ministers. He kissed my hand and I then acquainted him that it had long been my intention to retain him and the rest of the present Ministry at the head of affairs, and that it could not be in better hands than his. He then again kissed my hand. He then read to me the Declaration which I was to read to the Council, which he wrote himself and which is a very fine one. I then talked with him some little longer time after which he left me. He was in full dress. I like him very much and feel confidence in him. He is a very straightforward, honest, clever and good man. I then wrote a letter to the

Queen ... At about ½ p. 11 I went downstairs and held a Council in the red saloon. I went in of course quite alone, and remained seated the whole time. My two Uncles, the Dukes of Cumberland and Sussex, and Lord Melbourne conducted me ... I was not at all nervous and had the satisfaction of hearing that people were satisfied with what I had done and how I had done it. Receiving after this, Audiences of Lord Melbourne, Lord John Russell [the Home Secretary], Lord Albemarle (Master of the Horse), and the Archbishop of Canterbury, all in my room and alone. Saw Stockmar. Saw Clark, whom I named my Physician.

Wrote my journal. Took my dinner upstairs alone ... At about 20 minutes to 9 came Lord Melbourne and remained till near 10. I had a very important and a very comfortable conversation with him. Each time I see him I feel more confidence in him; I find him very kind in his manner too. Saw Stockmar. Went down and said good-night to Mamma etc. My dear Lehzen will always remain with me as my friend but will take no situation about me, and I think she is right.

24 June 1837

At 11 came Lord Melbourne and stayed till 12. He is a very honest, good and kind-hearted, as well as very clever man. He told me that Lady Tavistock had accepted the situation [of Lady of the Bedchamber] ... I really have immensely to do; I receive so many communications from my Ministers but I like it very much.

TO KING LEOPOLD   25 June 1837

Though I have an immense deal of business to do, I shall write you a few lines to thank you for your kind and useful letter of the 23rd, which I have just received. Your advice is always of the greatest importance to me ...

I am very well, sleep well, and drive every evening in the country; it is so hot that walking is out of the question. Before I go further let me pause to tell you how fortunate I am to have at the head of the Government a man like Lord Melbourne. I have seen him now every day, with the exception of Friday, and the more I see him, the more confidence I have in him; he is not only a clever statesman and an honest man, but a good and a kind-hearted man, whose aim is to do his duty for his country and not for a party. He is of the greatest use to me both politically and privately.

I have seen almost all my other Ministers, and do regular, hard, but to me delightful, work with them. It is to me the greatest pleasure to do my duty for my country and my people, and no fatigue, however great, will be burdensome to me if it is for the welfare of the nation. Stockmar will tell you all these things. I have reason to be highly pleased with all my Ministers.

JOURNAL   1 July 1837

I have so many communications from the Ministers, and from me to them, and I get so many papers to sign every day, that I have always a very great deal to do; but for want of time and space I do not write these things down. I delight in this work. At 10 minutes p. 2 came Lord Palmerston [the Foreign Secretary] and stayed till 6 minutes p. 3. We talked about Russia and Turkey a good deal etc. He is very agreeable, and clear in what he says.

To King Leopold   11 July 1837

I really and truly go into Buckingham Palace the day after to-morrow,* but I must say, though I am very glad to do so, I feel sorry to leave for ever my poor old birthplace.

                                                            25 July 1837

I shall not go out of town, I think, before the 20th or thereabouts of next month. Windsor requires thorough cleaning, and I must say I could not think of going in sooner after the poor King's death. Windsor always appears very melancholy to me, and there are so many sad associations with it. These will vanish, I daresay, if I see you there soon after my arrival there.

I have very pleasant large dinners every day. I invite my Premier generally once a week to dinner as I think it right to show publicly that I esteem him and have confidence in him, as he has behaved so well. Stockmar is of this opinion and is his great admirer.

In July the Queen drove to the House of Lords for the prorogation of Parliament.

Journal   17 July 1837

I went first to the Robing-room [in the House of Lords], but as there were so many people there I went to a Dressing-room where I put on the Robe which is enormously heavy. After this I entered the House of Lords preceded by all the Officers of State and Lord Melbourne bearing the Sword of State walking just before me. He stood quite close to me on the left-hand of the Throne, and I feel always a satisfaction to have him near me on such occasions, as he is such an honest, good, kind-hearted man and is my friend, I know it. The Lord Chancellor [Lord Cottenham] stood on my left. The house was very full and I felt somewhat (but very little) nervous before I read my speech, but it did very well, and I was happy to hear people were satisfied.

To King Leopold   19 July 1837

I have been so busy, I can say but two words more, which are that I prorogued Parliament yesterday in person, was very well received, and am not at all tired to-day, but quite frisky. There is to be no review this year, as I was determined to have it only if I could ride, and as I have not ridden for two years [because of her serious illness at Ramsgate], it was better not.

Journal   19 July 1837

I gave audiences to various foreign Ambassadors, amongst which were Count Orloff, sent by the Emperor of Russia to compliment me. He presented me with a

*Buckingham Palace had been designed by John Nash for George IV but neither he nor William IV had lived there. It was still not ready for occupation on the Queen's accession and it was not until alterations had been carried out under the supervision of Edward Blore that she was able to move in.

letter from the Empress of Russia accompanied by the Order of St. Catherine all set in diamonds. (I, of course, as I generally do every evening, wore the Garter.) The Levee began immediately after this and lasted till ½ p. 4 without one minute's interruption. I had my hand kissed nearly 3000 times! I then held a Council, at which were present almost all the Ministers. After this I saw Lord Melbourne for a little while, and then Lord Palmerston.

TO KING LEOPOLD   1 August 1837

I have resumed my singing lessons with Lablache twice a week, which form an agreeable recreation in the midst of all the business I have to do. He is such a good old soul, and greatly pleased that I go on with him. I admire the music of the Huguenots very much, but do not sing it, as I prefer Italian to French for singing greatly. I have been learning in the beginning of the season many of your old favourites, which I hope to sing with you when we meet.

JOURNAL   15 August 1837

Put on my habit and went ... to the Mews, which are in the garden. The Riding-house is very large ... I first rode a bay horse, a delightful one called Ottoman, and cantered about a good while. I then tried for a minute another horse which I did not like so well. I then remounted Ottoman. After him I mounted a beautiful and very powerful but delightful grey horse, a Hanoverian, called Fearon.

                                                                 29 August 1837

At 7 o'clock arrived my dearest most beloved Uncle Leopold and my dearest most beloved Aunt Louise. They are both, and look both, very well; dearest Aunt Louise is looking so well and is grown quite fat. I and Mamma as well as my whole court were all at the door to receive them. It is an inexpressible happiness and joy to me, to have these dearest beloved relations with me and in my own house. I took them to their rooms, and then hastened to dress for dinner. At 8 we dined ... I sat between dear Uncle and my good Lord Melbourne; two delightful neighbours. Dear Aunt Louise sat opposite. After dinner I sat on the sofa with dearest Aunt Louise, who is really an angel, and Lord Melbourne sat near me. Uncle talked with Lord Palmerston. It was a most delightful evening.

                                                              12 September 1837

I sat on the sofa with dearest Aunt Louise, who played a game at chess with me, to teach me, and Lord Melbourne sat near me ... Lord Melbourne, Lord Palmerston, Sir J. Hobhouse [President of the Board of Control], and later too Lord Conyngham, all gave me advice, and all different advice, about my playing at chess, and all got so eager that it was very amusing; in particular Lord Palmerston and Sir J. Hobhouse, who differed totally and got quite excited and serious about it. Between them all, I got quite beat, and Aunt Louise triumphed over my Council of Ministers!

After dinner I sat on the sofa part of the evening with Lady Tavistock, Lord Melbourne sitting near me, and the rest with my dearest Aunt Louise, with whom I played a game at chess, and beat her; Lord Palmerston, Lord Melbourne, and Lord Conyngham sat near me advising me. At 11, our last happy evening broke up, and Aunt Louise took leave in the kindest way imaginable of the whole party except my gentlemen; and good Lord Melbourne was touched to tears by this leave-taking. I cannot say how I shall miss my dearest Aunt Louise; she combines with great cleverness and learning, so much merriment, and has all the liveliness and fun of a girl of 16, with all the sense and deep thought of one of 30 and much older even. And I think she is so lovely, so graceful, she has such an angelic expression in her clear eyes; and she dresses so well, morning and evening. And then my beloved Uncle whom I look up to and love as a father, how I shall miss his protection out riding, and his conversation!

28 September 1837

I saluted them [a parade of Guardsmen and Lancers in Windsor Great Park] by putting my hand to my cap like the officers do, and was much admired for my manner of doing so. I then cantered up to the Lines with all the gentlemen and rode along them. Leopold behaved most beautifully, so quietly, the Bands really playing in his face. I then cantered back to my first position and there remained while the Troops marched by in slow and quick time, and when they manoeuvred, which they did beautifully ... The whole went off beautifully; and I felt for the first time like a man, as if I could fight myself at the head of my Troops.

I am very sorry indeed to go [from Windsor]! I passed such a very pleasant time here; the pleasantest summer I ever passed in my life, and I shall never forget this first summer of my Reign. I have had the great happiness of having my beloved Uncle and Aunt here with me, I have had very pleasant people and kind friends staying with me, and I have had delicious rides which have done me a world of good. Lord Melbourne rode near me the whole time. The more I see of him and the more I know of him, the more I like and appreciate his fine and honest character. I have seen a great deal of him, every day, these last 5 weeks, and I have always found him in good humour, kind, good, and most agreeable; I have seen him in my Closet for Political Affairs, I have ridden out with him (every day), I have sat near him constantly at and after dinner, and talked about all sorts of things, and have always found him a kind and most excellent and very agreeable man. I am very fond of him.

9 November 1837

Dressed for the Lord Mayor's dinner, in all my finery. At 2 I went in the state carriage and 8 horses with the Duchess of Sutherland [Mistress of the Robes] and Lord Albemarle; all my suite, the Royal Family, etc., went before me. I reached the Guildhall at a little before 4. Throughout my progress to the city, I met with the most gratifying, affectionate, hearty and brilliant reception from the greatest concourse of people I ever witnessed; the streets being immensely crowded as were also the windows, houses, churches, balconies, every where. I was then conducted by the Lord Chamberlain, the Lord Mayor [John Cowan] and Lady Mayoress preceding me, and my whole suite following me,—to a private drawing-room, where I found Mamma, the Duchess of Gloucester, the Duchess of Cambridge, and Augusta, and all their Ladies. All my Ladies came in there. After waiting some little time I sent for Lord Melbourne and Lord John Russell, to ask them some questions.

When the Lord Mayor was presented I said to Lord John Russell (what I had previously been told to do), 'I desire you to take proper measures for conferring the dignity of Baronet on the Lord Mayor.' I then knighted the Sheriffs, one of whom was Mr. Montefiore, a Jew, an excellent man [Sir Moses Montefiore who lived to be over 100 years old]; and I was very glad that I was the first to do what I think quite right, as it should be. The Lady Mayoress and all the Aldermen's wives were then presented. After this we returned, as before, to the Private Drawing room and remained there till a ¼ p. 5 when we went to dinner ... I drank a glass of wine with the Lord Mayor ... a quiet little old man of 70 (they say). When my health was given out, there was great cheering and applause.

The crowd [on our return] was, if possible, greater than it had been when I went in the day; and they cheered me excessively as I came along. The streets were beautifully illuminated on all sides, and looked very brilliant and gay. I got home by 20 m. to 10, and quite safely; I trust there have been no accidents. I cannot say how gratified, and how touched I am by the very brilliant, affectionate, cordial, enthusiastic and unanimous reception I met with in this the greatest Metropolis in the World; there was not a discontented look, not a sign of displeasure—all loyalty, affection and loud greeting from the immense multitude I passed through; and no disorder whatever. I feel deeply grateful for this display of affection and unfeigned loyalty and attachment from my good people. It is much more than I deserve, and I shall do my utmost to render myself worthy of all this love and affection

17 November 1837

After dinner I went to the play to Covent Garden, the Duchess of Sutherland and Lord Albemarle going with me in the carriage. I met with the same brilliant reception, the house being so full that there was a great piece of work for want of room, and many people had to be pulled out of the Pit by their wrists and arms into the Dress Circle. I never saw such an exhibition; My Ladies took it by turns, (their standing behind me, I mean).

7 December 1837

The figures and animals [in the two pictures which were shown to her by Edwin Landseer] were all most beautifully painted and grouped; and most exquisitely finished, so that I looked at them through a magnifying glass; I never saw anything so exquisite in every way. He also showed me a sketch in oils (small) of Lord Melbourne which is like, but too fat, and though flattering is not in my opinion half pleasing enough ... He had only had two sittings of Lord Melbourne. He certainly is the cleverest artist there is.

8 December 1837

There is no end to the amusing anecdotes and stories Lord Melbourne tells, and he tells them all in such an amusing funny way.

9 December 1837

Lord Melbourne spoke to me about several of the speakers in the House of Commons; spoke of Sir E. Sugden [Edward Sugden, afterwards Lord St. Leonards and Lord Chancellor, was elected Tory Member for Weymouth and Melcombe Regis in 1828], whom he says is a very clever lawyer, and said, 'His father was a haircutter; he cut my hair very often.' This is a singular thing. Told me of an affront which the "Demagogue Hunt" [Henry Hunt, the great orator, who had been Member for Preston] offered William Peel one day, in the House of Commons, on the latter's attacking him. William Peel said something derogatory about Hunt's extraction, upon which Hunt replied: 'If my father was the first gentleman of his family, your father was the last gentleman of his family.'

22 December 1837

I was delighted to see Lord Melbourne in excellent spirits, and looking much better. He was very clever and funny about education, at dinner; his ideas are excellent about it, I think. He said that he thought almost every body's character was formed by their Mother, and that if the children did not turn out well, the mothers should be punished for it.

27 December 1837

Lord Melbourne is very absent when in company, often, and talks to himself every now and then, loud enough to be heard but never loud enough to be understood. I am now, from habit, quite accustomed to it, but at first I turned round sometimes, thinking he was talking to me.

19 January 1838

Talked for some time with him [Lord Melbourne] and Lord Palmerston, about education, punishments, etc., Lord Melbourne was amazingly funny and amusing about this. I said I thought solitary confinement a good punishment: Lord Melbourne replied, 'I think it's a very stupefying punishment.' I mentioned the system of silence as a very good one and quoted myself as a proof of its having answered, which made them laugh very much. Lord Melbourne said, 'It may do very well with a lively child; but with one of a sulky grumpy disposition it would not answer.' . . . I said I thought it cruel to punish children by depriving them of their meals and saying they should go without their supper, etc. Lord Melbourne replied, 'Why, when I was a child, they had contrived to annoy me so, and had made me cry so much, that I had lost all appetite.'

21 January 1838

[Lord Melbourne] was very funny about a word which Lady Mary gave me to find out; she gave me the ivory letters and I was to find out the word; she gave me 'thermometer,' and she spelt it with an 'a' instead of an 'e,' and laughed very much at her bad spelling; upon which Lord Melbourne said, 'It is a very good way to spell it, but not the way,' which made us laugh. I said to him I was reading the first novel I had ever read—*The Bride of Lammermoor*; he said it was a very melancholy—a terrible story—but admires it; he mentioned *Old Mortality*, *Quentin Durward*, *The Fair Maid of Perth*, and *Kenilworth*, as Scott's best novels; he said there was 'a great deal of good' and 'a great deal of bad' in his novels.

26 January 1838

At 7 I went to Drury Lane ... It was Shakespear's tragedy of *Hamlet*, and we came in at the beginning of it. Mr. Charles Kean (son of old Kean) acted the part of Hamlet and I must say beautifully. His conception of this very difficult and I may almost say incomprehensible character, is admirable; his delivery of all the fine long speeches quite beautiful; he is excessively graceful and all his actions and attitudes are good, though not at all good-looking in face ... He fights uncommonly well too. All the other characters were very badly acted. I came away just as *Hamlet* was over. They would recognise me between the 2nd and 3rd acts,—I was compelled to come forward, curtsey, and hear 'God save the Queen' sung. The house was amazingly crowded and they received me admirably.

27 January 1838

Told Lord M. I had been much pleased with *Hamlet* last night; observed it was a very hard play to understand, which he agreed in; he said he thought the end of it 'awkward' and horrid; said he thought Hamlet was supposed to be mad, of a philosophical mind, and urged to do something which he did not like to do. He added that Mr. Fox always said that Hamlet possessed more of Shakespear's faults than almost any other play of Shakespear.

After dinner, talked (before I sat down) with all the gentlemen, etc. Spoke about Kean with Lord Melbourne; about Landseer and the sketches which Lord Melbourne saw and none of which he 'thought like,' he said, though very clever ... Lord Melbourne said that *Richard III.* by Shakespear was a very fine play; I observed that Richard was a very bad man; Lord Melbourne also thinks he was a horrid man; he believes him to have been deformed (which some people deny), and thinks 'there is no doubt that he murdered those two young Princes.'

1 February 1838

The curious old form of pricking the Sheriffs* was gone through; and I had to prick them all, with a huge pin. This was the first Council that I have yet held at which

---

*The names of the High Sheriffs nominated for each county were then and are still presented to the monarch by the Clerk of the Privy Council on a long vellum roll wound round a wooden roller and tied with green ribbons. The Clerk unwinds the roll and the monarch pierces the appropriate names with the spike of a brass bodkin, shaped like a doorknob, a custom believed to date from the time of Queen Elizabeth I who was presented with the Sheriff's Roll for her approval one day when she was sewing in the garden and, having no pen with which to make the usual dots against the chosen names, pricked them with her needle.

Lord Melbourne was not present, and I must say I felt sad not to see him in his place as I feel a peculiar satisfaction, nay I must own almost security, at seeing him present at these formal proceedings, as I know and feel that I have a friend near me, when I am as it were alone among so many strangers. This may sound almost childish, but it is not so.

4 February 1838

Lord Melbourne asked if I had seen *King Lear* (which I had half intended to do last week); I said I had not. He said (alluding to the manner in which it is being pèrformed at Covent Garden), 'It is *King Lear* as Shakespear wrote it; and which has not been performed so, since the time of Queen Anne.' As it is generally acted, Lord Melbourne told me, it is altered by Cibber, who 'put in a deal of stuff' of his own; that it was a much finer play as Shakespear wrote it, but 'most dreadfully tragic.' That Dr. Johnson had seen it performed in that way, and that 'it made such an impression on him that he never forgot it.' I observed to him that I feared that, and did not like all that madness on the stage. Lord Melbourne said, 'I can't bear that, but still it is a very fine play.'

5 February 1838

We came in [to Drury Lane theatre] before the performance had commenced. It was Shakespear's tragedy of *Richard III.*, and Charles Kean's first appearance (in London) as Richard. The house was crammed to the ceiling; and the applause was tremendous when Kean came on; he was unable to make himself heard for at least five minutes I should say ... It would be impossible for me to attempt to describe the admirable manner in which Kean delineated the ferocious and fiend-like Richard. It was quite a triumph and the latter part particularly so; he was applauded throughout in the most enthusiastic manner. He acted with such spirit too! ... He was uncommonly well disguised, and looked very deformed and wicked. All the other parts were very badly acted, and the three women were quite detestable.

6 February 1838

[Lord Melbourne] Spoke of the Duke of Wellington; he said 'The Duke of Wellington is amazingly sensible to attention; nothing pleases him so much as if one asks him his opinion about anything.' He added that many people were offended with the Duke's abrupt manner of speaking; I observed that I thought that was only a manner, and that he did not mean it so.

9 February 1838

We all agreed that it was very bad that no French was taught at the Public Schools, for that boys never learnt it afterwards. Lady Durham said that Lord Durham had had a great mind that their boy should learn no Latin at all, which however Lord Melbourne said he thought was a bad thing, for that he thought a man could not get on well in the world without Latin in the present state of society. I told Lord Melbourne that though Lehzen had often said that she had never seen such a passionate and naughty child as I was, still that I had never told a falsehood, though

I knew I would be punished; Lord Melbourne said: 'That is a fine character'; and I added that Lehzen entrusted me with things which I knew she would not like me to tell again, and that when I was ever so naughty, I never threatened to tell, or ever did tell them. Lord Melbourne observed: 'That is a fine trait.' I felt quite ashamed, on hearing this praise, that I had said so much about myself.

13 March 1838

I asked Lord Melbourne what was to take place concerning Slavery to-night.* Lord Melbourne then pulled out of his pocket the Bill or Act which is to be read to-night; he read to me the principal Heads of it explaining to me each part in the most clear and agreeable manner possible. I shall not have time or space to explain or name each head here, but before I do any, I must just observe that the necessity of this Act shows how shockingly cruel and cheating the Masters of the Slaves are, attempting to evade in every possible way what they are told to do, and what, as the Laws cannot be enforced on the spot, must be done by an Act of Parliament here.

14 March 1838

I asked Lord Melbourne how he liked my dress. He said he thought it 'very pretty' and that 'it did very well.' He is so natural and funny and nice about *toilette* and has a very good taste I think.

17 March 1838

Spoke of the Cabinet, which was just over; he said that they had been speaking about the Coronation in the Cabinet; and they all thought that it would be best to have it about the 25th or 26th of June, as it would end the Parliament well and make a good break.

4 April 1838

I said I feared there were so many rail-roads that they could not all answer; Lord Melbourne said he feared they would not, but that he was sorry for it, as he was engaged in one. 'I was fool enough to engage in one and to take 50 shares; I have already paid £1,000, and have lately had a call for £500 more,' he added. This rail-road is in Nottingham and he engaged in it about 4 years ago. I asked him if he liked rail-roads in general; he replied, 'I don't care about them,' which made me laugh; and he added that they were bad for the country as they brought such a shocking set of people 'who commit every horror.' 'They are picked men, who mind neither Lord nor laws, and commit every species of violence; nothing is safe,' he added; and 'it's more like a country in time of war' than peace.

6 April 1838

Spoke of my ride; rail-roads; that the Steam-Carriage could not be stopped under 150 yards' distance of an object; I observed that these Steam-Carriages are very

---

*The Government was to introduce a Bill ameliorating the conditions in which negro apprentices were employed in the West Indies.

dangerous; Lord Melbourne said, 'Oh! none of these modern inventions consider human life.'

Spoke of Byron, who Lord Melbourne said would not be 50 if he were alive; he said he was extremely handsome; had dark hair, was very lame and limped very much; I asked if the expression of his countenance was agreeable; he said not; 'he had a sarcastic, sardonic expression; a contemptuous expression.' I asked if he was not agreeable; he said 'He could be excessively so'; 'he had a pretty smile'; 'treacherous beyond conception; I believe he was fond of treachery.' Lord Melbourne added, 'he dazzled everybody,' and deceived them; 'for he could tell his story very well.' . . . Lord Melbourne said, 'The old King (George III.) had that hurried manner; but he was a shrewd, acute man, and most scrupulously civil.' He added that the King was rather tall, red in the face, large though not a corpulent man; prejudiced and obstinate beyond conception.

19 May 1838

'Very nice party' (my Concert), Lord M. said, 'and everybody very much pleased.' I smiled and said I feared I had done it very ill; that I was quite angry with myself and thought I had done it so ill; and was not civil enough. He said most kindly, 'Oh! no, quite the contrary, for I should have told you if it had been otherwise.' I then said I had felt so nervous and shy. 'That wasn't at all observed,' he said. I said that I often stood before a person not knowing what to say; and Lord Melbourne said that the longer one stood thinking the worse it was; and he really thought the best thing to do was to say anything commonplace and foolish, better than to say nothing.

To KING LEOPOLD   25 May 1838

I have been dancing till past four o'clock this morning; we have had a charming ball, and I have spent the happiest birthday that I have had for many years; oh, how different to last year! Everybody was so kind and so friendly to me.

JOURNAL   22 June 1838

At a ¼ p. 2 came Marshal Soult, Duc de Dalmatie [who had commanded Napoleon's forces in Spain during the Peninsular War and was now French Ambassador in London] . . . I was very curious to see him; he is not tall, but very broad, and one leg quite crooked from having been severely wounded; his complexion is dark, and he has the appearance of great age; his features are hard, and he speaks slowly and indistinctly. His eyes are piercing; he seemed much embarrassed.

27 June 1838

To Westminster Abbey to see all the Preparations for to-morrow [the day of the Queen's coronation]. The streets are full of people, and preparations of all kinds.

28 June 1838

I was awoke at four o'clock by the guns in the Park, and could not get much sleep afterwards on account of the noise of the people, bands, etc., etc. Got up at seven, feeling strong and well; the Park presented a curious spectacle, crowds of people up to Constitution Hill, soldiers, bands, etc. I dressed, having taken a little breakfast before I dressed, and a little after. At half-past 9 I went into the next room, dressed exactly in my House of Lords costume.

At 10 I got into the State Coach with the Duchess of Sutherland and Lord Albemarle and we began our Progress ... It was a fine day, and the crowds of people exceeded what I have ever seen; many as there were the day I went to the City, it was nothing, nothing to the multitudes, the millions of my loyal subjects, who were assembled in every spot to witness the Procession. Their good humour and excessive loyalty was beyond everything, and I really cannot say how proud I feel to be the Queen of such a Nation. I was alarmed at times for fear that the people would be crushed and squeezed on account of the tremendous rush and pressure.

I reached the Abbey amid deafening cheers at a little after half-past eleven; I first went into a robing-room quite close to the entrance where I found my eight train-bearers ... all dressed alike and beautifully in white satin and silver tissue with wreaths of silver corn-ears in front, and a small one of pink roses round the plait behind, and pink roses in the trimming of the dresses.

After putting on my mantle, and the young ladies having properly got hold of it and Lord Conyngham holding the end of it, I left the robing-room and the Procession began ... The sight was splendid; the bank of Peeresses quite beautiful all in their robes, and the Peers on the other side. My young train-bearers were always near me, and helped me whenever I wanted anything. The Bishop of Durham [Edward Maltby] stood on the side near me, but he was, as Lord Melbourne told me, remarkably *maladroit*, and never could tell me what was to take place. At the beginning of the Anthem ... I retired to St Edward's Chapel, a dark small place immediately behind the Altar, with my ladies and train-bearers—took off my crimson robe and kirtle, and put on the supertunica of cloth of gold, also in the shape of a kirtle, which was put over a singular sort of little gown of linen trimmed with lace; I also took off my circlet of diamonds and then proceeded bare-headed into the Abbey ... The Crown being placed on my head was, I must own, a most beautiful impressive moment; all the Peers and Peeresses put on their coronets at the same instant.

My excellent Lord Melbourne, who stood very close to me throughout the whole ceremony, was completely overcome at this moment, and very much affected; he gave me such a kind, and I may say fatherly look. The shouts, which were very great, the drums, the trumpets, the firing of the guns, all at the same instant, rendered the spectacle most imposing.

Poor old Lord Rolle, who is 82, and dreadfully infirm, in attempting to ascend the steps fell and rolled quite down, but was not the least hurt; when he attempted to re-ascend them I got up and advanced to the end of the steps, in order to prevent another fall.* When Lord Melbourne's turn to do Homage came, there was loud cheering; they also cheered Lord Grey and the Duke of Wellington; it's a pretty

*'Lord Rolle [in fact, 87] fell down as he was getting up the steps of the throne,' Charles Greville recorded in his journal. 'Her first impulse was to rise and when afterwards he came again to do homage she said, "May I not get up and meet him?" and then rose from the throne and advanced down one or two steps to prevent his coming up, an act of graciousness and kindness which caused a great sensation.'

ceremony; they first all touch the Crown, and then kiss my hand. When my good Lord Melbourne knelt down and kissed my hand, he pressed my hand and I grasped his with all my heart, at which he looked up with his eyes filled with tears and seemed much touched, as he was, I observed, throughout the whole ceremony. After the Homage was concluded I left the Throne, took off my Crown and received the Sacrament; I then put on my Crown again, and re-ascended the Throne, leaning on Lord Melbourne's arm. At the commencement of the Anthem I descended from the Throne, and went into St Edward's Chapel.

There was another most dear Being present at this ceremony, in the box immediately above the royal box, and who witnessed all; it was my dearly beloved angelic Lehzen, whose eyes I caught when on the Throne, and we exchanged smiles ... I then again descended from the Throne, and repaired with all the Peers bearing the Regalia, my Ladies and Train-bearers, to St Edward's Chapel, as it is called; but which, as Lord Melbourne said, was more unlike a Chapel than anything he had ever seen; for what was called an Altar was covered with sandwiches, bottles of wine, etc., etc. The Archbishop came in and ought to have delivered the Orb to me, but I had already got it, and he (as usual) was so confused and puzzled and knew nothing, and—went away. Here we waited some minutes. Lord Melbourne took a glass of wine, for he seemed completely tired. The Procession being formed, I replaced my Crown (which I had taken off for a few minutes), took the Orb in my left hand and the Sceptre in my right, and thus loaded, proceeded through the Abbey—which resounded with cheers, to the first robing-room ...

And here we waited for at least an hour, with all my ladies and train-bearers ... The Archbishop had (most awkwardly) put the ring on the wrong finger, and the consequence was that I had the greatest difficulty to take it off again, which I at last did with great pain ... At about half-past four I re-entered my carriage, the Crown on my head, and the Sceptre and Orb in my hands, and we proceeded the same way as we came—the crowds if possible having increased. The enthusiasm, affection, and loyalty were really touching, and I shall ever remember this day as the Proudest of my life! I came home at a little after six, really not feeling tired.

At eight we dined ... Lord Melbourne came up to me and said: 'I must congratulate you on this most brilliant day,' and that all had gone off so well. He said he was not tired, and was in high spirits ... He asked kindly if I was tired; said the Sword he carried (the first, the Sword of State) was excessively heavy. I said that the Crown hurt me a good deal ... We agreed that the whole thing was a very fine sight. He thought the robes, 'looked remarkably well.' 'And you did it all so well—excellent!' said he, with tears in his eyes. He said he thought I looked rather pale and 'moved by all the people' when I arrived; 'and that's natural; and that's better.' The Archbishop's and Dean's copes, which were remarkably handsome, were from James the Second's time; the very same that were worn at his Coronation, Lord Melbourne told me. Spoke of Soult having been very much struck by the ceremony of the Coronation; of the English being far too generous not to be kind to Soult.*

*Marshal Soult was applauded as he entered the Abbey, walking alone and followed by a numerous suite.

After dinner ... Lord Melbourne and I spoke of the numbers of Peers at the Coronation, which, Lord Melbourne said, with the tears in his eyes, was unprecedented. I observed that there were very few Viscounts; he said: 'There are very few Viscounts,' that they were an odd sort of title and not really English; that they came from Vice-Comités; that Dukes and Barons were the only real English titles; that Marquises were likewise not English; and that they made people Marquises when they did not wish to make them Dukes.

I said to Lord Melbourne when I first sat down that I felt a little tired on my feet; 'You must be very tired,' he said. Spoke of the weight of the Robes, etc., etc., the Coronets; and he turned round to me with the tears in his eyes, and said so kindly: 'And you did it beautifully—every part of it, with so much taste; it's a thing that you can't give a person advice upon; it must be left to a person.' To hear this, from this kind impartial friend, gave me great and real pleasure. Mamma and Feodore came back just after he said this. Spoke of the Bishops' Copes, about which he was very funny; of the Pages who were such a nice set of boys, and who were so handy ... Spoke again of the young ladies' dresses, about which he was very amusing ...He set off for the Abbey from his house at half-past eight, and was there long before anybody else; he only got home at half-past six and had to go round by Kensington. He said there was a large breakfast in the Jerusalem Chamber where they met before all began; he said, laughing, that whenever the Clergy, or a Dean and Chapter, had anything to do with anything, there's sure to be plenty to eat.

5 August 1838

I asked Lord Melbourne if he didn't think Johnson's Poetry very hard; he said he did, and that Garrick said, 'Hang it, it's a hard as Greek.' His Prose he admires, though he said pedantry was to be observed throughout it; and Lord Melbourne thinks what he said superior to what he wrote.

12 August 1838

Lord Melbourne was very funny about the Statue of the Duke of Wellington which is put up (in wood) only as a Trial, on the Archway on Constitution Hill,* and which we think looks dreadful and much too large; but Lord Melbourne said he thought a statue would look well there, and that it should be as large. We then observed what a pity Wyatt should do the statue, as we thought he did them so ill; and we mentioned George III.'s; but Lord Melbourne does not dislike that, and says it's exactly like George III, and like his way of bowing. He continued, 'I never will have anything to do with Artists; I wished to keep out of it all; for they're a waspish set of people.'

16 August 1838

'You were rather nervous,' [when reading her speech in the House of Lords] said Lord Melbourne; to which I replied, dreadfully so; 'More so than any time,' he continued. I asked if it was observed; he said, 'I don't think anyone else would have observed it, but I could see you were.' Spoke of my fear of reading it too low, or too

---

*This unsightly statue was removed to Aldershot and eventually replaced by Adrian Jones's *Quadriga* which stands on the arch today.

loud, or too quick; 'I thought you read it very well,' he said kindly. I spoke of my great nervousness, which I said I feared I never would get over. 'I won't flatter Your Majesty that you ever will; for I think people scarcely ever get over it; it belongs to a peculiar temperament, sensitive and susceptible; that shyness generally accompanies high and right feelings,' said Lord Melbourne most kindly; he was so kind and paternal to me.

29 August 1838

Spoke of George III's hand-writing; of mine, which Lord M. thinks very legible and generally very good; of my inclination to imitate hand-writings, and people,—which Lord M. said, showed quickness, and was in the Family; of George IV's mimickry. I said I kept a journal, which, as Lord Melbourne said, is very laborious, but a very good thing; for that it was astonishing in transacting business, how much one forgot, and how one forgot why one did the things.

2 September 1838

Spoke of Queen Caroline,* and of the feeling for her; 'I never saw anything like it in my life,' said Lord M.; 'it was very alarming; it even spread to the Troops.' 'George IV. never was popular,' Lord M. said. And whatever Queen Caroline did, had no weight with the people, for, they said, it was all his fault at first. Lord M. continued, that it was quite madness his (George IV.) conduct to her; for if he had only separated from her, and let her alone, that wouldn't have signified; but he persecuted her, and 'he cared as much about what she did, as if he had been very much in love with her,' which certainly was very odd. 'He (George IV.) was a clever man,' Lord M. said, but he thinks that he never was honestly advised about Queen Caroline; though, he continued, he very often disliked advice that was contrary to his wishes, and resented it; yet Lord M. thinks one can do anything with clever people, and if he had been properly talked to he might have listened.

3 September 1838

Mr. Pitt, Lord M. said, was a tall, thin man, with a red face; drank amazingly; so did Mr. Fox, Lord M. continued, and that neither had the slightest restraint over himself; they died the same year.

We then looked at 2 vols. of portraits of the Characters concerned in the French Revolution, which are very fine and very interesting. It was quite a delight and treat to look at them with Lord Melbourne, for there was hardly one character whom he did not know everything about, what they did, who they were; and he has such a charming, agreeable way of telling it all. When we came to Cambacérès [second consul in the constitution of 1799], Lord M. said 'he was a great gourmand; and Napoleon used to keep him at the Council while his dinner was spoiling.' And one day, Napoleon saw him writing a note, 'and he insisted on seeing it; it was to his Cook: "*Sauvez le rôtis; les entremets sont perdus.*"'

*Queen Caroline, the highly unconventional wife of George IV, was suspected of adultery. Proceedings were taken against her in the House of Lords. Public opinion at first strongly supported her against the King.

4 September 1838

I said that I couldn't understand the German books; Lord M. mentioned Schiller's *Thirty Years' War* (which he has read the Translation of) as a very good book. 'They are apt to be misty and obscure, the Germans,—and cloudy,' he said laughing. Spoke of my disliking Ancient History; of my having read many dull books; of my having disliked learning formerly, and particularly Latin, and being naughty at that, and at my Bible-lessons; Lord M. said it was a good thing to know a little Latin, on account of the construction of English; Greek he thinks unnecessary for a woman.

19 September 1838

Lord M. spoke of the people's great civility to me now, whereas they were rude to the King and put on their hats when they saw him, particularly in Buckinghamshire, 'which used to annoy the King very much'; 'but there was nothing to be done for it.'

21 September 1838

Spoke of eating and drinking, and dessert being unwholesome; as one eats it without real appetite; 'and I am of that opinion,' said Lord Melbourne; that he believed what one ate after dinner, and the few glasses of wine one drank, hurt one, and that if one was to get up before that it would be much better. 'That's why I should like to get up with the ladies,' said Lord Melbourne; 'but it never has been the practice here.' I said it was a very bad habit that of the gentlemen sitting in that way after dinner. He remembers, in the country, in the houses of foxhunters, sitting till 11 or 12, and 'coming in and finding all the women yawning.' 'I can't bear it,' said Lord M., 'though I did like it too, formerly. I believe the ladies like it; they like to have a little time to arrange their hair, and to talk.' I said I didn't. He continued, 'Of course the men were very much elevated by wine; but it tended to increase the gaiety of society, it produced diversity.' 'In every party,' said Lord M., 'there were generally 10 or 12 in that state.' Lord M. said he never saw any body eat and drink so much as George IV.; in 1798 it was beyond everything; and his spirits and love of fun beyond everything, too.

22 September 1838

Spoke for some time of church-going; and Lord Melbourne said he never used to go, after he left Eton; 'My Father and Mother never went,' he said. 'People didn't use to go so much formerly; it wasn't the fashion; but it is a right thing to do.' He said Uncle, last year, wanted to go twice, but Lord M. assured him (as it is) that that was unnecessary.

2 October 1838

Spoke of Lady Lyttelton [a Lady of the Bedchamber who later became Lady Superintendent to the royal children], who Lord M. said wasn't very young when she married, about 23. I said I thought 23 quite young enough to marry; 'So do I,' said Lord M., 'but girls begin to be nervous when they are past 19,' and think they'll never marry if 'they are turned 20.'

5 October 1838

Lady Lyttelton asked leave to put on spectacles for working; and Lord M. said, her asking leave showed she understood etiquette, for he said formerly nobody was allowed to come to Court in spectacles, or use glasses; that Mr. Burke, when he was first presented at Court, was told he must take off his spectacles; and that Lord M. said he remembered as long as anything, that no one (man) was allowed to wear gloves at Court. I praised Lady Lyttelton and said she was such a nice person.

11 December 1838

After dinner told Lord Melbourne how Mama teazed me about my drinking wine and told people I drank so much which schocked him much, and which he said was very wrong etc . . . Talked of my wishing but not daring to take tea.

23 December 1838

Read in *Eugene Aram* for some time while my hair was doing, and finished it; beautifully written and fearfully interesting as it is, I am glad I have finished it, for I never feel quite at ease or at home when I am reading a Novel, and therefore was really glad to go on to Guizot's *Révolution de l' Angleterre*.

26 December 1838

Talked of *Pickwick* and *Oliver Twist* by Dickens which is in very low life, 'I don't like those murders,' said Lord M. 'I wish to keep them away.'

28 December 1838

Read the first and part of the second part of *Nicholas Nickleby* which I think excessively amusing and clever.

30 December 1838

Lord M. said, 'I'm afraid to go to church for fear of hearing something very extraordinary.' I laughed and said he never went, and that he always managed to be very conveniently either unable to come down for a Sunday, or to be ill, which made him laugh very much. Talked of when my boxes arrived in London, and Lord M. said he always tried to prevent their bringing boxes to him when he was at dinner at Lady Holland's, for that she was always wanting to know what was in it, and would say, 'What's that? Let me see what that is.' That he always made as good a fight as he could, but that it was often very difficult to prevent her. Lord M. then said, 'No woman ever wrote a really good book; no sterling book.' Hannah More and Mme. de Sévigné were mentioned, and he admitted that those were both exceptions.

1 January 1839

Talked of my getting on in *Oliver Twist*; of the descriptions of 'squalid vice' in it; of the accounts of starvation in the Workhouses and Schools, Mr. Dickens gives in his books. Lord M. says, in many schools they give children the worst things to eat, and bad beer, to save expense; told him Mamma admonished me for reading light books.

2 January 1839

Read in *Oliver Twist* which is *too* interesting.

3 January 1839

Talked of my feeling low and ill, which as I had felt it both times I was here, at different seasons, was a proof, I thought, that the place [Windsor] disagreed, which he wouldn't allow. He said very funnily, 'You have got some fixed fancies; Your Majesty has settled in your mind certain things.' Lord M. asked if I had got on far with *Oliver Twist*; I said into the 2nd volume, and liked it so much and wished he would read it, which he said he would one day; talked of it, and of the story; of the *Beggars' Opera* by Gay, which Lord M. has seen very often, and which is coarse; but he says they have refined it down so much and scratched out so much as the times got more polished, that there was hardly anything good left.

7 April 1839

Lord M. was talking of some dish or other, and alluded to something in *Oliver Twist*; he read half of the 1st vol. at Panshanger [in Hertfordshire, seat of Earl Cowper]. 'It's all among Workhouses, and Coffin Makers, and Pickpockets,' he said; 'I don't like that low debasing style; it's all slang; it's just like *The Beggar's Opera*; I shouldn't think it would tend to raise morals; I don't like that low debasing view of mankind.' We defended Oliver very much, but in vain. 'I don't like those things; I wish to avoid them; I don't like them in reality, and therefore I don't wish to see them represented,' he continued; that everything one read should be pure and elevating.

2 June 1839

We then had great fun about the pronunciation of words; Rome – Room; gold – goold; *revenue* – re*ve*nue. Lord M. pronounces all in the latter manner. And then also about grammar; and saying *her* instead of – when it ought to be *she*; and Lord M. was so funny about, 'It is *she*!' He said someone, I don't remember who, said: 'Those who fight custom against grammar are fools.' . . . A delightful evening.

30 June 1839

I asked Lord M. if he was well, as he looked so pale. 'I've got a complaint in my bowels,' he said, 'which rather weakens me.' He ate too much, these last two days, he says . . . He was sickish before dinner; but felt no gout; he ate soup and fish but nothing else except dessert strawberries, which I told him were the very worst thing he could take.

4 October 1839

Talked of my dreaming so, and Lord M. said, 'That's not sleeping well. I had visions last night, and I attribute it all to those walnuts.' He repeated to himself 'several devils' he had seen, which made us laugh so, and I told him I often heard him damning people to himself, and wished to know who they could be, which made him laugh excessively. The other day I told Lord M. that he talked so much to himself out riding.

The Queen's contentment in the early years of her reign was overcast by her relationship with her mother who had moved to Buckingham Palace to act as an unwanted duenna and who, once installed there in a suite well separated from her daughter's rooms, did 'nothing but complain' and 'plague' her. She complained about the smallness of her rooms, about not being given her proper precedence, about her daughter's eating too much and drinking too much beer, about her treatment of Sir John Conroy, with whom the Queen was determined to have no further dealings, above all about her inadequate income.

TO THE DUCHESS OF KENT   17 August 1837

I thought you would not expect me to invite Sir John Conroy after his conduct towards me for some years past, and still more so after the unaccountable manner in which he behaved towards me, a short while before I came to the Throne ... I imagined you would have been amply satisfied with what I had done for Sir John Conroy, by giving him a pension of £3,000 a year, which only Ministers receive, and by making him a Baronet ... I thought you would have expected no more.

JOURNAL   15 January 1838

Got such a letter from Mama, oh! oh! such a letter.

                                                                18 February 1838

I showed him [Lord Melbourne] a letter I had got from Mama yesterday, relative to her debts, at which he was much schocked and grieved, I showed him also a list of the number of things I had paid and which money was owing to tradesmen; about my dress etc., etc. which had never been paid ... They (Mama and J.C.) ought to remember what incalculable falsehoods they have told about these debts. During the King's life [William IV's] they said there were no debts and that it was all calumny of the King's—which is really infamous.

There was soon to be further, serious trouble with the Duchess over one of her Ladies, Lady Flora Hastings, a religious, sharp-witted, intelligent young woman of attractive appearance but rather severe and unforthcoming manner. She was not a popular figure at court; and, since she was on friendly terms with Sir John Conroy, the Queen intensely disliked and distrusted her. When she complained of pain and swelling of the stomach, the Queen had no doubt what the matter with the 'detestable person' was.

JOURNAL   2 February 1839

Lady Flora had not been above two days in the house before Lehzen and I discovered how exceedingly suspicious her figure looked—more have since observed this and we have no doubt that she is—to use the plain words—with child!! Clark cannot deny the suspicion; the horrid cause of all this is the Monster and Demon Incarnate whose name I forbear to mention but which is the first word of the 2nd line of this page [Conroy]. Lady Tavistock accordingly with Lehzen's concurrence told Lord Melbourne of it, as it was a matter of serious importance. He accordingly replied to me this evening without—very properly—mentioning names, that 'The only way is to be quiet and watch it.' That it was a very ticklish thing for a physician to declare what might not be true, as so many deceptions had been practised upon physicians... Here ended this disgraceful subject which makes one loathe one's own sex. When they are bad how disgracefully and disgustingly servile and low women are!! I don't wonder at men considering the sex despicable.

Lady Flora was persuaded to undergo a medical examination and was pronounced a virgin. But Melbourne remained sceptical, and so did the Queen:

TO THE DUCHESS OF KENT   undated

Sir C[harles Clarke, the physician who examined her] had said that though she is a virgin still that it might be possible and one could not tell if such things could not happen. There was an enlargement in the womb like a child.

Even so the Queen decided that she must go to see Lady Flora to express her regret at what had happened.

JOURNAL   23 February 1839

She was dreadfully agitated and looked very ill, but on my embracing her, taking her by the hand, and expressing great concern and my wish that all should be forgotten,—she expressed herself exceedingly grateful to me, and said, that for Mama's sake she would suppress every wounded feeling and would forget it.

The Queen's relationship with her mother, who took Lady Flora's side, grew worse. It was, the Queen told Lord Melbourne, like 'having an enemy in the house'.

JOURNAL   17 April 1839

Said to [Lord Melbourne] how dreadful it was to have the prospect of torment for many years by Mama's living here, and he said it was dreadful, but what could be done? She had declared (some time ago) I said she would never leave me as long as I was unmarried. 'Well then, there's *that* way of settling it,' he said. That was a schocking alternative, I said.

7 June 1839

Talked of Lady Flora's being ill; (she hasn't appeared since Tuesday) and so sick. '*Sick?*' said Lord M. with a significant laugh.

12 June 1839

Talked of Lady Flora's being still ill, which he [Lord Melbourne] thought very odd, and wondered didn't excite more curiosity

16 June 1839

I told [Lord Melbourne] that she [the Duchess of Kent] said Lady Flora was very ill, could keep nothing upon her stomach, and had continued fever. 'Then she'll die,' said Lord M. . . . I said I never would travell with Mama again; he asked why; I told him what a trouble she was.

17 June 1839

Talked of Mama and her humour being so variable . . . of her being touchy and jealous. Talked of Lady Flora, and that some thought she would go abroad with the Conroys, which Lord M. thought she could never do . . . I sent for Lord Melbourne and told him that there had been such a piece of work before I went to dinner . . . and that Mama had been crying and saying that Lady Flora was dying; in short made a dreadful piece of work; and that I didn't believe she was so very ill as they said she was. 'As you say, Ma'am,' said Lord M., 'it would be very awkward if that woman was to die.'

22 June 1839

Lord M. asked how Lady Flora was; not so well, I said; that her own family were very fearful she would die, as they knew it would be said that she had died lying-in. Lord M. said, 'Exactly so.' Talked of the dreadful fuss they had made about it—and Lord M. said Lord Hastings made such a piece of work about it—as if there was no other family in the world but the Hastings's. 'I don't see,' said Lord M., 'if she was to die, that it's of such great importance if they say she was so [pregnant] or not; they make such a fuss about it; it is a very schocking thing and a very wrong thing, but such things *have* happened in families; it isn't unprecedented'; which is true enough.

Lady Flora was by now dangerously ill; and, on 27 June the Queen was advised that if she wished to see her before she died she must go at once.

JOURNAL   27 June 1839

I said I would be up again in a minute; and he [Melbourne] said 'Don't be in a hurry' . . . I went in alone; I found poor Lady Flora stretched on a couch looking as thin as anybody can be who is still alive; literally a skeleton, but the body very much swollen like a person who is with child; a searching look in her eyes, a look rather like a person who is dying; her voice like usual, and a good deal of strength in her hands; she was friendly, said she was very comfortable, and was very grateful for all I had done for her, and that she was glad to see me looking well. I said to her, I hoped to see her again when she was better, upon which she grasped my hand as if to say 'I shall not see you again'.

5 July 1839

The poor thing died without a struggle and only just raised her hands and gave one gasp.

A post-mortem revealed that the swelling of Lady Flora's stomach had been due to a growth on her liver. Newspapers expressed the outrage of the people. For a long time now the Queen's early popularity had fast been slipping away, and she had been feeling intermittently ill and generally downcast and unsure of herself.

JOURNAL   11 March 1839

I asked Lord Melbourne how he was and he said quite well, and I told him I thought I was going to have the Influenza, as I had pains all over; he replied most funnily, 'It's the best time to have it,—no Levée; you can't go through the year without being ill.'

13 March 1839

I said to Lord M. I knew I had been very disagreeable and cross in the morning, which he didn't allow. I said I had been exceedingly angry with John Russell for not letting me go to Drury Lane; Lord M. laughed and said, 'But it can't be.' I couldn't get my gloves on, and Lord M. said, 'It's those consumed rings; I never could bear them.' I said I was fond of them, and that it improved an ugly hand. 'Makes it worse,' he replied; I said I didn't wear them of a morning. 'Much better,' he said, 'and if you didn't wear them, nobody else would.' Ear-rings he thinks barbarous. I said I thought I was not getting stronger. 'Why, you have every appearance of getting stronger,' he said, and 'You should take the greatest care of your health; there's nothing like health; particularly in your situation; it makes you so independent; bad health puts you into the power of people.'

15 March 1839

I said I hoped he [Lord Melbourne] would always tell me whatever he heard; he said, 'I always do.' Not lately, I said; 'I haven't heard anything lately.' 'For,' I added, 'I was sure I made a great many mistakes'; 'No, I don't know that at all.' People said, he continued, that I was 'lofty, high, stern, and decided, but that's much better than that you should be thought familiar.' 'I said to [Lord] Stanley,' he continued, 'it's far better that the Queen should be thought high and decided, than that she should be thought weak. "By God!" he said, "they don't think that of her; you needn't be afraid of that."' Lord M. seemed to say this with pleasure. 'The natural thing,' he continued, 'would be to suppose that a girl would be weak and undecided; but they don't think that.' I said that I was often very childish, he must perceive; 'No, not at all, I don't see that in any respect,' he said.

17 March 1839

Talked of going to bed so much earlier formerly; of my going to sleep quickly; of Louis XIV. never being hungry till he came to dinner, and then after the 2 first spoonfuls eating quantities; I said it was quite the contrary with me, and that when I had had a little, all appetite went. 'That's not so well,' he said.

The Queen's unhappiness was increased by her fears that Melbourne's unsteady Government would fall.

22 March 1839

Got up at $\frac{1}{2}$ p. 9. Very anxious and nervous. Saw by the papers we were beat by 5; and they had sat till 4! I am in a sad state of suspense; it is now $\frac{3}{4}$ p. 12, and I have not yet heard from Lord Melbourne; I hear he was still asleep when my box arrived, and I desired they shouldn't wake him.

We were seated much as usual, my truly valuable and excellent Lord Melbourne being seated near me. I said to Lord M. that I was sure I never could bear up against difficulties; Lord M. turned round close to me, and said very earnestly and affectionately, 'Oh! you will; you must; it's in the lot of your Station, you must prepare yourself for it.' I said I never could, and he continued, 'Oh! you will; you always behaved very well.' I said to Lord M. I was sure he hadn't a doubt we should carry it [a vote of confidence in the House of Commons]. 'Upon my word I don't know,' he said.

27 March 1839

Lord Melbourne was rather silent during dinner; but I never saw him so much so as he was after dinner,—so completely absorbed, so totally disinclined to enter into any conversation whatever,—and merely just answering a question in as short a manner as possible, and then relapsing into the same silence; yet he did not look nor was he ill; I was quite grieved and distressed to see him so.

Melbourne survived, but not for long. At the beginning of May the Queen was distraught to learn that his Government, facing defeat on a colonial issue, would certainly have to resign.

7 May 1839

The state of agony, grief and despair into which this placed me may be easier imagined than described! All all my happiness gone! That happy peaceful life destroyed, that dearest kind Lord Melbourne no more my minister . . . I sobbed and cried much: could only put on my dressing gown. At 10 m past 12 came Lord Melbourne . . . It was some minutes before I could muster up courage to go in—and when I did, I really thought my heart would break; he was standing near the window; I took that kind, dear hand of his, and sobbed and grasped his hand in both mine and looked at him and sobbed out, 'You will not forsake me'; I held his hand for a little while, unable to leave go; and he gave me such a look of kindness, pity, and affection, and could hardly utter for tears, 'Oh! no,' in such a touching voice.

We then sat down as usual, and I strove to calm myself. He said, 'I was afraid this would happen.' There was a Warrant appointing an Inquiry into the Duchy of Cornwall which he begged me to sign; which I did. 'I'm afraid we can do nothing else,' he said (but resign). I said I feared he was right. 'But we shall see what they say at the Cabinet; I'll put down on paper the course I think you ought to pursue,' which I begged he would. He told me when he would come after the Cabinet. Wrote my journal. At 3 came Lord John [Russell], who said they had been discussing the whole in the Cabinet very much, but that they could come to no other determination but to resign; and he then thanked me for my kindness—which quite set me off crying, and I said it was a terrible thing for me.

At a ¼ p. 3 Lord Melbourne came to me and . . . said, pulling a paper out of his pocket, 'I have written down what I think you should do.' He then read to me what he had written down for me. The conclusion of the paper was, 'Your Majesty had better express your hope that none of Your Majesty's Household, except those who are engaged in Politics, may be removed.' Lord Melbourne said, 'I think you might ask him [the new Prime Minister] for that.' I quite agreed in this.

'I don't know who they'll put about you,' he said. I said it was so hard to have people forced upon you whom you disliked; Lord M. said, 'It is very hard, but it can't be helped.' I said I thought Lord John was very low. 'He was melancholy at seeing you melancholy,' said Lord M. . . . I said I was not going out, and I wished Lord M. would come to me. 'Yes, Ma'am, I will,' he said; and then after a pause he added, 'I don't think it would be right'; he said it would be observed; I pressed him and said it would not be, and if he would come after dinner; he said it wouldn't do; and 'I'm going to dine at Lady Holland's.' But I said he must come and see me. 'Oh! yes,' he replied, 'only not while these negotiations are going on.' I said, 'For I shall feel quite forsaken,' at which he gave me such a look of grief and feeling, and was much affected. he said, 'God bless you, Ma'am,' and kissed my hand . . . I was in a dreadful state of grief.

8 May 1839

I sobbed much, again held [Lord Melbourne's] hand in both mine, and kept holding his hand for some time fast in one of mine, as if I felt in doing so he could not leave me . . . He then got up, and we shook hands again and he kissed my hand, I crying dreadfully . . . I wrote once more to him. 'The Queen ventures to maintain one thing, which she thinks is possible; which is, that if she rode out tomorrow afternoon, she might just get a glimpse of Lord Melbourne in the Park; if he knew where she rode, she would meet him and it would be such a comfort; there surely could be no earthly harm in this; for, I may meet anyone; Lord Melbourne may think this childish but the Queen really is so anxious it might be; and she would bear thro' all her trials so much better if she could just see a friends face sometimes.' . . . I could eat nothing. Wrote one line to the Duke of Wellington to request him to come.

The Queen clung to the hope that the Duke of Wellington—who she felt sure would be more sympathetic to her wishes than the stiff and awkward Sir Robert Peel—could be persuaded to form a Government and thus render her having to deal with Tory ministers less intolerable.

To Melbourne   8 May 1839

The Queen thinks Lord Melbourne may possibly wish to know how she is this morning; the Queen is somewhat calmer; she was in a wretched state till nine o'clock last night, when she tried to occupy herself and try to think less gloomily of this dreadful change, and she succeeded in calming herself till she went to bed at twelve, and she slept well; but on waking this morning, all—all that had happened in one short eventful day came most forcibly to her mind, and brought back her grief; the Queen, however, feels better now; but she couldn't touch a morsel of food last night, nor can she this morning. The Queen trusts Lord Melbourne slept well, and is well this morning; and that he will come precisely at eleven o'clock. The Queen has received no answer from the Duke, which is very odd, for she knows he got her letter.

8 May 1839

The Queen told Lord Melbourne she would give him an account of what passed, which she is very anxious to do. She saw the Duke for about twenty minutes; the Queen said she supposed he knew why she sent for him, upon which the Duke said, No, he had no idea. The Queen then said that she had had the greatest confidence in her late Ministry, and had parted with them with the greatest reluctance; upon which the Duke observed that he could assure me no one felt more pain in hearing the announcement of their resignation than he did, and that he was deeply grieved at it. The Queen then continued, that as his party had been instrumental in removing them, that she must look to him to form a new Government. The Duke answered that he had no power whatever in the House of Commons, 'that if he was to say black was white [the Queen obviously meant to write 'black was black'], they would say it was not,' and that he advised me to send for Sir Robert Peel, in whom I could place confidence, and who was a gentleman and a man of honour and integrity. The Queen then said she hoped he would at all events have a place in the new Cabinet. The Duke at first rather refused, and said he was so deaf, and so old and unfit for any discussion, that if he were to consult his own feelings he would rather not do it, and remain quite aloof; but that as he was very anxious to do anything that would tend to the Queen's comfort, and would do everything and at all times that could be of use to the Queen, and therefore if she and her Prime Minister urged his accepting office, he would. The Queen said she had more confidence in him than in any of the others of his party. The Queen then mentioned the subject of the Household ... The Duke did not give any decisive answer about it, but advised the Queen not to begin with conditions of this sort, and wait till the matter was proposed. The Queen then said that she felt certain he would understand the great friendship she had for Lord Melbourne, who had been to her quite a parent, and the Duke said no one felt and knew that better than he did, and that no one could still be of greater use to the Queen than Lord Melbourne. The Duke spoke of his personal friendship for Lord Melbourne ... The Queen then mentioned her intention to prove her great fairness to her new Government in telling them, that they might know there was no unfair dealing, that I meant to see you often as a friend, as I owed so much to you. The Duke said he quite understood it, and knew I would not exercise this to weaken the Government, and that he would take my part about it, and felt for me. He was very kind, and said he called it 'a misfortune' that you had all left me.

The Queen wrote to Peel, who came after two, embarrassed and put out. The Queen repeated what she had said to the Duke about her former Government, and asked Sir Robert to form a new Ministry. He does not seem sanguine; says entering the Government in a minority is very difficult; he felt ... it arduous, and that he would require me to demonstrate confidence in the Government, and that my Household would be one of the marks of that.

JOURNAL   9 May 1839

Soon after this Sir Robert said, 'Now, about the Ladies', upon which I said I could not give up any of my Ladies, and never had imagined such a thing. He asked if I meant to retain all. 'All,' I said. 'The Mistress of the Robes and the Ladies of the Bedchamber?' I replied, 'All.' ... He said they were the wives of the opponents of the Government ... I said that [they] would not interfere; that I never talked politics with them, and that they were related, many of them, to Tories, and I enumerated those of my Bedchamber women and Maids of Honour; upon which he said he did not mean all the Bedchamber women and all the Maids of Honour, he meant the Mistress of the Robes and the Ladies of the Bedchamber; to which I replied they were of more consequence than the others, and that I could not consent, and that it had never been done before. He said I was a Queen Regnant, and that made the difference. 'Not here,' I said—and I maintained my right.

TO MELBOURNE   8 May 1839

The Queen then talked of her great friendship for, and gratitude to Lord Melbourne, and repeated what she had said to the Duke, in which Peel agreed; but he is such a cold, odd man she can't make out what he means. He said he couldn't expect me to have the confidence in him I had in you (and which he never can have) as he has not deserved it. My impression is, he is not happy and sanguine. He comes to me to-morrow at one to report progress in his formation of the new Government. The Queen don't like his manner after—oh! how different, how dreadfully different, to that frank, open, natural and most kind, warm manner of Lord Melbourne. The Duke I like by far better to Peel ... The Queen was very much collected, and betrayed no agitation during these two trying Audiences. But afterwards again all gave way. What is worst of all is the being deprived of seeing Lord Melbourne as she used to do.

9 May 1839

The Queen writes one line to prepare Lord Melbourne for what may happen in a very few hours. Sir Robert Peel has behaved very ill, and has insisted on my giving up my Ladies, to which I replied that I never would consent, and I never saw a man so frightened. He said he must go to the Duke of Wellington and consult with him, when both would return, and he said this must suspend all further proceedings ... He was quite perturbed—but this is infamous. I said, besides many other things, that if he or the Duke of Wellington had been at the head of the Government when I came to the Throne, perhaps there might have been a few more Tory Ladies, but that then if you had come into Office you would never have dreamt of changing them. I was calm but very decided, and I think you would have been pleased to see my composure and great firmness; the Queen of England will not submit to such trickery. Keep yourself in readiness, for you may soon be wanted.

9 May 1839

The Queen has received Lord Melbourne's letter. Lord Melbourne will since have heard what has taken place. Lord Melbourne must not think the Queen rash in her conduct; she saw both the Duke and Sir Robert again, and declared to them she could not change her opinion ... The Queen felt this was an attempt to see whether she could be led and managed like a child; if it should lead to Sir Robert Peel's refusing to undertake the formation of the Government, which would be absurd, the Queen will feel satisfied that she has only been defending her own rights, on a point which so nearly concerned her person, and which, if they had succeeded in, would have led to every sort of unfair attempt at power; the Queen maintains all her ladies,—and thinks her Prime Minister will cut a sorry figure indeed if he resigns on this. Sir Robert is gone to consult with his friends, and will return in two or three hours with his decision.

We shall see what will be done. The Queen would not have stood so firmly on the Grooms and Equerries, but her Ladies are entirely her own affair, and not the Ministers'.

On 10 May the Queen wrote to Peel in words suggested to her by Melbourne:

The Queen having considered the proposal made to her yesterday by Sir Robert Peel, to remove the Ladies of her Bedchamber, cannot consent to adopt a course which she conceives to be contrary to usage, and which is repugnant to her feelings.

She then wrote to Melbourne:

The Queen wrote the letter before she went to bed, and sent it at nine this morning; she has received no answer, and concludes she will receive none, as Sir Robert told the Queen if the Ladies were not removed, his party would fall directly, and could not go on, and that he only awaited the Queen's decision. The Queen therefore wishes to see Lord Melbourne about half-past twelve or one, if that would do.

The Queen fears Lord Melbourne has much trouble in consequence of all this; but the Queen was fully prepared, and fully intended to give these people a fair trial, though she always told Lord Melbourne she knew they couldn't stand; and she must rejoice at having got out of the hands of people who would have sacrificed every personal feeling and instinct of the Queen's to their bad party purposes.

How is Lord Melbourne this morning?

Sixty years later the Queen admitted in a conversation with her Private Secretary, Sir Arthur Bigge, that she had doubtless been too stubborn in successfully insisting upon retaining her Ladies. 'I was very young then,' she said, 'and perhaps I should act differently if it was all to be done again.' 'Yes, it was a mistake,' she conceded ... 'I was very hot about it. And so were my Ladies.' 'The simple truth,' commented Greville, 'is that the Queen could not endure the thought of parting with Melbourne who is everything to her. Her feelings ... are of a strength sufficient to tear down all prudential considerations.'

JOURNAL   12 May 1839

After dinner, when Lord Melbourne came into the room, he remained talking with me some time before we sat down, near the chimney. Talked of Sir Robert Peel, and my feeling so happy. 'You mustn't be sure that you have escaped yet,' he said.

Lord M. said, 'You must remember that he (Peel) is a man who is not accustomed to talk to Kings; a man of quite a different calibre; it's not like me; I've been brought up with Kings and Princes. I know the whole Family, and know exactly what to say to them; now he has not that ease, and probably you were not at your ease.' These are nearly his words I think ... We were seated much as usual, Lord Melbourne sitting near me. He was very much excited the whole evening, talking to himself and pulling his hair about, which always makes him look so much handsomer.

I caught his eye when he was frowning very much, and he smiled and rubbed his forehead and said, 'Never mind, I was only knitting my brows; I know it looks tremendous.' but that one shouldn't judge from expression, that very susceptible people constantly changed expression. I said he was very absent sometimes; 'Notoriously so,' he said, 'particularly when I've a great deal to do.'

18 May 1839

I said [to Lord Melbourne] that I had such an excessive dislike for Peel ... I was sure we should have quarrelled anyhow—very soon ... 'That's what he says,' continued Lord M. 'that you have such a dislike for him that you could never get on together.' I replied that was quite true. 'But that should not be,' said Lord M. ...

Lord M. said Peel had no great desire for office. I said I didn't believe that a bit ... I said the great difficulty was the concealing my satisfaction at getting rid of him ... Stayed up till 20m to 12. I told Lord M. he had been snoring, which made him smile.

19 May 1839

Lord M. said he was quite well, and when I said I thought him not well, the night before, he said, 'Only sleepy; that's not a sign of being ill; it's right to sleep after dinner; we ought all to lie down all round the room, and sleep,' which made me laugh very much.

Talked of the Duke of Wellington's and Peel's not having been at the Levée, which I thought very rude. Lord M. said, 'I don't think they mean that'; I replied that was all very well for Lord M., who was so kind and good, not to think people could mean things, but that there were very few like him (Ld. M.). 'I don't like you to have those feelings,' he said kindly. I said it was so foolish of Peel to act in this way, as by doing so he has made me dislike him. 'That's what his Party feels,' said Lord M. ... He went fast asleep for nearly half an hour which made me almost cross, for it always provokes me so, when he goes to sleep in that way of an evening, though it would be very wrong to wake him as he has much fatigue. When he woke I said to him I thought he was very tired. He said, 'No, only sleepy. I'm quite fresh now' ... Said I was getting more lazy every day. 'Mustn't do that,' said Lord M.

24 May 1839

This day I go out of my TEENS and become 20! It sounds so strange to me! I have much to be thankful for; and I feel I owe more to two people than I can ever repay! my dear Lehzen, and my dear excellent Lord Melbourne!

27 May 1839

I saw the Grand-Duke [Alexander of Russia, afterwards the Czar Alexander II] arrive [at Windsor] at 20m. to 7; he bowed up to my window ... We dined in St. George's Hall, which looked beautiful. The Grand-Duke led me in ... I really am quite in love with the Grand-Duke; he is a dear, delightful young man. At a little after 12 we went into the dining-room for supper; after supper they danced a Mazurka for $\frac{1}{2}$ an hour, I should think nearly; the Grand-Duke asked me to take a turn, which I did (never having done it before) and which is very pleasant; the Grand-Duke is so very strong, that in running round, you must follow quickly, and after that you are whisked round like in a Valse, which is very pleasant. After this we danced (what I had never even seen before) the 'Grossvater' or 'Rerraut,' and which is excessively amusing; I danced with the Grand-Duke, and we had such fun and laughter ... I never enjoyed myself more. We were all so merry; I got to bed by a $\frac{1}{4}$ to 3, but could not sleep till 5.

28 May 1839

The Grand-Duke talked of his very fine reception here, and said he would never forget it. 'Ce ne sont pas seulement des paroles, je vous assure, Madame, ' he said, but that it was what he felt, and that he never would forget these days here, which I'm sure I shall never also, for I really love this amiable and dear young man, who has such a sweet smile.

29 May 1839

I said all this excitement did me good. 'But you may suffer afterwards,' he [Lord Melbourne] said; 'you must take care of your health,—not to fall into bad health; you complain of that languor increasing, and dislike for exertion; now, it would be a dreadful thing for you if you were to take a dislike for business,' which I assured him I never should. 'You lead rather an unnatural life for a young person,' he continued; 'it's the life of a man.' I did feel it sometimes, I said . . .

I then went to the little blue room next my Dressing-room, where Lord Palmerston brought in the Grand-Duke to take leave. The Grand-Duke took my hand and pressed it warmly; he looked pale and his voice faltered, as he said, 'Les paroles me manquent pour exprimer tout ce que je sens'; and he mentioned how deeply grateful he felt for all the kindness he met with, that he hoped to return again, and that he trusted that all this would only tend to strengthen the ties of friendship between England and Russia. He then pressed and kissed my hand, and I kissed his cheek; upon which he kissed mine (cheek) in a very warm affectionate manner . . . I felt so sad to take leave of this dear amiable young man, whom I really think (talking jokingly) I was a little in love with.

30 May 1839

They played the Grand-Duke's and my favourite quadrilles, called 'Le Gay Loisir,' which made me quite melancholy, as it put me so in mind of all, and I felt sadly the change. Talked to Lord M. of my feeling the change, and of its being so seldom that I had young people of my own rank with me; of my having so disliked the idea of the Grand-Duke's coming, and that now I was so very very sorry at his going. 'Very often the case,' said Lord M. . . . Talked of the strange feeling when all the excitement was over, and that I feared I would feel the difference of not being able to have these sorts of dances; Lord M. said I could have them sometimes in London, though we agreed with difficulty; but certainly here; but I observed there must be many young people then; 'And you had a great posse of them,' said Lord M., and so nice, I observed. I said a young person like me must sometimes have young people to laugh with. 'Nothing so natural,' replied Lord M. with tears in his eyes.

3

THE BRIDE

1839–1840

The Queen had discussed with Lord Melbourne the possibility of marrying Prince Albert at the time of the Lady Flora Hastings affair when marriage had seemed 'a schocking alternative' to continuing to live with her mother.

JOURNAL   18 April 1839

Lord M. then said, 'Now, Ma'am, for this other matter.' I felt terrified (foolishly) when it came to the point; too silly of me to be frightened in talking to him. Well, I mustered up courage, and said that my Uncle's great wish—was—that I should marry my Cousin Albert—who was with Stockmar—and that I thought Stockmar might have told him (Ld. M.) so; Lord M. said, No—Stockmar had never mentioned a word; but, that I had said to my Uncle, I could decide nothing until I saw him again. 'That's the only way,' said Lord M. 'How would that be with the Duchess?' he asked. I assured him he need have no fear whatever on that score; then he said, 'Cousins are not very good things,' and 'Those Coburgs are not popular abroad; the Russians hate them.' I then said, who was there else? We enumerated the various Princes, of whom not one, I said, would do. For myself, I said, at present my feeling was quite against ever marrying. 'It's a great change in the situation,' he said. 'It's a very serious thing, both as it concerns the Political effect and your own happiness.' I praised Albert very much; said he was younger than me. I said Uncle Ernest pressed me much about it; Lord M. said if one was to make a man for it, one would hardly know what to make; he mustn't be stupid—nor cunning. I said, by all that I heard, Albert would just be the person. 'I think it would be wished for; still I don't think a foreigner would be popular,' said Lord M. I observed that marrying a subject was making yourself so much their equal, and brought you so in contact with the whole family. Lord M. quite agreed in this and said, 'I don't think it would be liked; there would be such jealousy.' I said, why need I marry at all for 3 or 4 years? did he see the necessity? I said I dreaded the thought of marrying; that I was so accustomed to have my own way, that I thought it was 10 to 1 that I shouldn't agree with any body. Lord M. said, 'Oh! but you would have it still' (my own way).

11 July 1839

I said to Lord M. I felt my temper was getting worse and worse. 'Oh no,' he said. 'That's worry.' I observed he had a great deal of worry, but a mild kind temper. 'I'm very passionate,' he said, 'very irritable', which I would hardly believe . . . Talked of my being so silent, which I thought wrong and uncivil, as I hated it in others. 'Silence is a good thing,' said Lord M. 'if you have nothing to say.'

12 July 1839

Talked of my Cousins Ernest and Albert coming over,—my having no great wish to see Albert, as the whole subject was an odious one, and one which I hated to decide about; there was no engagement between us, I said, but that the young man was aware that there was the possibility of such a union; I said it wasn't right to keep him on, and not right to decide before they came; and Lord M. said I should make them distinctly understand anyhow that I couldn't do anything for a year; I said it was disagreeable for me to see him though, and a disagreeable thing. "It's very disagreeable,' Lord M. said. I begged him to say nothing about it to anybody, or to answer questions about it, as it would be very disagreeable to me if other people knew it. Lord M. I didn't mind, as I told him everything. Talked of Albert's being younger. 'I don't know that that signifies,' said Lord M. . . . I said I wished if possible never to marry. 'I don't know about that,' he replied.

To King Leopold   15 July 1839

I shall send this letter by a courier, as I am anxious to put several questions to you, and to mention some feelings of mine upon the subject of my cousins' visit, which I am desirous should not transpire. First of all, I wish to know if Albert is aware of the wish of his Father and you relative to me? Secondly, if he knows that there is no engagement between us? I am anxious that you should acquaint Uncle Ernest, that if I should like Albert, that I can make no final promise this year, for, at the very earliest, any such event could not take place till two or three years hence.

Though all the reports of Albert are most favourable, and though I have little doubt I shall like him, still one can never answer beforehand for feelings, and I may not have the feeling for him which is requisite to ensure happiness. I may like him as a friend, and as a cousin, and as a brother, but not more; and should this be the case (which is not likely), I am very anxious that it should be understood that I am not guilty of any breach of promise, for I never gave any. I am sure you will understand my anxiety, for I should otherwise, were this not completely understood, be in a very painful position. As it is, I am rather nervous about the visit, for the subject I allude to is not an agreeable one to me.

Journal   28 August 1839

I went to look at Leslie's picture of the Coronation which is finished and so prettily; Lord Melbourne so like, but too old and not handsome enough. I went to look at Hayter's picture of the Coronation, which is a fine thing. I then went and sat for some little time to Landseer who has made of me in three days the likest little sketch in oils of me, that ever was done.

13 September 1839

Lord M. sat near me the whole evening and was very sleepy; he said it was 'right to be so' after dinner. We did not meet for that I said, and that I wondered he could sleep before so many ... I told him, in his sleep he often moved to be more comfortable ... Told him he had snored and ought to be in bed.

14 September 1839

I said I did not like people because they were handsome, but that I was not insensible to beauty. 'Exactly,' he said. He took port wine to keep himself awake which I feared might make him ill.

15 September 1839

Told Lord M. he had behaved so ill at church, was so fidgetty, and slept during the sermon, which he said, was right.

23 September 1839

I was sadly cross to Lord Melbourne when he came in, which was shameful: I fear he felt it, for he did not sit down of himself, as he usually does, but waited until I told him to do so. I can't think what possessed me,—for I love this dear excellent man who is kindness and forbearance itself, most dearly.

30 September 1839

Talked to Lord M. of my growing disinclination to business: 'You must conquer that,' he said, 'it isn't unnatural when the first novelty is over, but you mustn't let that feeling get the better, you must fight against it.' I had said it was unnatural for a young woman to like business.

TO KING LEOPOLD   1 October 1839

I had a letter from Albert yesterday saying they could not set off, he thought, before the 6th. I think they don't exhibit much *empressement* to come here, which rather shocks me.

I shall take care and send a gentleman and carriages to meet [them] either at Woolwich or the Tower, at whichever place you inform me they land at. The sooner they come the better. I have got the house full of Ministers. On Monday the Queen Dowager is coming to sleep here [at Windsor] for two nights; it is the first time, and will be a severe trial.

JOURNAL   10 October 1839

As we were returning along the new walk, one of my pages came running with a letter from Uncle Leopold, saying my cousins would be here very soon; they sent on the letter announcing their arrival. I said to Lord M. I was sure they would come this day, but he would never believe it. At ½ p. 7 I went to the top of the staircase and received my 2 dear cousins Ernest and Albert,—whom I found grown and changed, and embellished. It was with some emotion that I beheld Albert—who is beautiful. I embraced them both and took them to Mamma; having no clothes they couldn't appear at dinner ... Talked [to Lord Melbourne] of my cousins' bad passage; their not appearing on account of their *négligé*, which Lord M. thought they ought to have done, at dinner and certainly after. 'I don't know what's the dress I would appear in, if I was allowed,' said Lord M., which made us laugh.

After dinner my Cousins came in, in spite of their *négligé*, and I presented them to Lord Melbourne. I sat on the sofa with Lord Melbourne sitting near me, and Ernest near us and Albert opposite—(he is so handsome and pleasing), and several of the ladies and gentlemen round the sofa. I asked Lord M. if he thought Albert like me, which he is thought (and which is an immense compliment to me). 'Oh! yes, he is,' said Lord M., 'it struck me at once.'

11 October 1839

Albert really is quite charming, and so excessively handsome, such beautiful blue eyes, an exquisite nose, and such a pretty mouth with delicate moustachios and slight but very slight whiskers; a beautiful figure, broad in the shoulders and a fine waist; my heart is quite going ... It is quite a pleasure to look at Albert when he gallops and valses, he does it so beautifully, holds himself so well with that beautiful figure of his.

TO KING LEOPOLD   12 October 1839

The dear cousins arrived at half-past seven on Thursday, after a very bad and almost dangerous passage, but looking both very well, and much improved. Having no clothes, they could not appear at dinner, but nevertheless *débuté d* after dinner in their *négligé*. Ernest is grown quite handsome; Albert's beauty is most striking, and he so amiable and unaffected—in short, very fascinating; he is excessively admired here.

We rode out yesterday and danced after dinner. The young men are very amiable, delightful companions, and I am very happy to have them here; they are playing some Symphonies of Haydn under me at this very moment; they are passionately fond of music.

JOURNAL   13 October 1839

Talked of my cousins having gone to Frogmore; the length of their stay being left to me; and I said seeing them had a good deal changed my opinion (as to marrying), and that I must decide soon, which was a difficult thing. 'You would take another week,' said Lord M.; 'certainly a very fine young man, very good-looking,' in which I most readily agreed, and said he was so amiable and good tempered, and that I had such a bad temper; of my being the 1st now to own the advantage of beauty, which Lord M. said smiling he had told me was not to be despised, in spite of what I had said to him about it. Talked of my cousins being religious. 'That strong Protestant feeling is a good thing in this country,' he said, 'if it isn't intolerant,'—which I assured him it was not. I had great fun with my dear cousins after dinner. I sat on the sofa with dearest Albert; Lord Melbourne sitting near me, Ernest playing at chess, and many being seated round the table ... Eos [Prince Albert's greyhound] came in again and yawned. I played 2 games at Tactics with dear Albert, and 2 at Fox and Geese. Stayed up till 20 m. p. 11. A delightful evening.

14 October 1839

After a little pause I said to Lord M., that I had made up my mind (about marrying dearest Albert).—'You have?' he said; 'well then, about the time?' Not for a year, I thought; which he said was too long; that Parliament must be assembled in order to make a provision for him, and that if it was settled 'it shouldn't be talked about,' said Lord M.; 'it prevents any objection, though I don't think there'll be much; on the contrary,' he continued with tears in his eyes, 'I think it'll be very well received; for I hear there is an anxiety now that it should be; and I'm very glad of it; I think it is a very good thing, and you'll be much more comfortable; for a woman cannot stand alone for long, in whatever situation she is.' . . . Then I asked, if I hadn't better tell Albert of my decision soon, in which Lord M. agreed. How? I asked, for that in general such things were done the other way,—which made Lord M. laugh. When we got up, I took Lord M.'s hand, and said he was always so kind to me,—which he has always been; he was so kind, so fatherly about all this. I felt very happy . . . Talked to Lord Melbourne after dinner of my hearing Albert couldn't sleep these last few days; nor I either, I added; that he asked a good deal about England, about which I tried to give him the most agreeable idea.

                                                                    15 October 1839

At about $\frac{1}{2}$ p. 12 I sent for Albert; he came to the Closet where I was alone, and after a few minutes I said to him, that I thought he must be aware why I wished [him] to come here, and that it would make me too happy if he would consent to what I wished (to marry me); we embraced each other over and over again, and he was so kind, so affectionate; Oh! to feel I was, and am, loved by such an Angel as Albert was too great delight to describe! he is perfection; perfection in every way—in beauty—in everything! I told him I was quite unworthy of him and kissed his dear hand—he said he would be very happy [to share his life with her] and was so kind and seemed so happy, that I really felt it was the happiest brightest moment in my life, which made up for all I had suffered and endured. Oh! how I adore and love him, I cannot say!! how I will strive to make him feel as little as possible the great sacrifice he has made; I told him it was a great sacrifice,—which he wouldn't allow . . . I feel the happiest of human beings.

To King Leopold   15 October 1839

This letter will, I am sure, give you pleasure, for you have always shown and taken so warm an interest in all that concerns me. My mind is quite made up—and I told Albert this morning of it; the warm affection he showed me on learning this gave me great pleasure. He seems perfection, and I think that I have the prospect of very great happiness before me. I love him more than I can say, and I shall do everything in my power to render the sacrifice he has made (for a sacrifice in my opinion it is) as small as I can. He seems to have a very great tact—a very necessary thing in his position. These last few days have passed like a dream to me, and I am so much bewildered by it all that I know hardly how to write; but I do feel very, very happy . . . My feelings are a little changed, I must say, since last Spring, when I said I couldn't think of marrying for three or four years; but seeing Albert has changed all this . . . Before I proceed further, I wish just to mention one or two alterations in the plan of announcing the event.

As Parliament has nothing whatever to say respecting the marriage, can neither approve nor disapprove it (I mean in a manner which might affect it), it is now proposed that, as soon as the cousins are gone (which they now intend to do on the 12th or 14th of November, as time presses), I should assemble all the Privy Councillors and announce to them my intention ... Oh! dear Uncle, I do feel so happy! I do so adore Albert!

JOURNAL   17 October 1839

Talked of Mama's going away when I married; which we agreed she wouldn't like, and for which reason it was necessary she should hear of the thing in time..

19 October 1839

My dearest Albert came to me at 10 m. to 12 and stayed with me till 20 m. p. 1. Such a pleasant happy time. He looked over my shoulder and watched me writing to the Duchess of Northumberland, and to the Duchess of Sutherland; and he scraped out some mistakes I had made. I told him I felt so grateful to him and would do everything to make him happy. I gave him a ring with the date of the ever dear to me 15th engraved in it. I also gave him a little seal I used to wear. I asked if he would let me have a little of his dear hair.

22 October 1839

Talked of Albert's not greatly caring for beauty, and hating those beauties who are so fêted, and wishing to spite them always. Lord M. said, 'Ought to pay attention to the ladies.'

27 October 1839

I signed some papers and warrants etc. and he [Prince Albert] was so kind as to dry them with blotting paper for me. We talked a good deal together, and he clasped me so tenderly in his arms, and kissed me again and again ... and was so affectionate, so full of love! Oh! what happiness is this! How I do adore him!! I kissed his dear hand. He embraced me again so tenderly.

1 November 1839

At 7 m. p. 6 came my most beloved Albert and stayed with me till 10 m. p 7 ... He was so affectionate, so kind, so dear, we kissed each other again and again ... Oh! what too sweet delightful moments are these!! Oh! how blessed, how happy I am to think he is *really* mine; I can scarcely believe myself so blessed. I kissed his dear hand, and do feel so grateful to him; he is such an angel, such a very great angel!—We sit so nicely side by side on that little blue sofa; no two Lovers could ever be happier than we are! ... He took my hands in his, and said my hands were so little he could hardly believe they were hands, as he had hitherto only been accustomed to handle hands like Ernest's.

10 November 1839

I sat on the sofa with Albert and we played at that game of letters, out of which you are to make words, and we had great fun about them. Albert gave 'Pleasure,' and when I said to the people who were puzzling it out, it was a very common word Albert said, But not a very common thing, upon which Lord M. said, 'Is it truth, or honesty?' which made us all laugh.

14 November 1839

We kissed each other so often, and I leant on that dear soft cheek, fresh and pink like a rose ... It was ten o'clock and it was the time for his going ... I gave Albert a last kiss, and saw him get into the carriage and—drive off. I cried much, felt wretched, yet happy to think that we should meet again so soon! Oh! how I love him, how intensely, how devotedly, how ardently! I cried and felt so sad. Wrote my journal. Walked. Cried.

To Prince Albert   14 November 1839

It is desired here that the matter should be declared at Coburg as soon as possible, and immediately after that I shall send you the Order [of the Garter].

Your rank will be settled just before you come over, as also your rank in the Army. Everything will be very easily arranged. Lord Melbourne showed me yesterday the Declaration, which is very simple and nice. I will send it you as soon as possible ...

Lord Melbourne told me yesterday, that the whole Cabinet are strongly of opinion that you should not be made a Peer. I will write that to Uncle.

Journal   16 November 1839

Talked [to Lord Melbourne] of Mama and the necessity of speaking to her about ... leaving the house; Lord M. said he feared I should have great difficulty in getting her out of the house. 'There must be no harshness,' said Lord M., 'yet firm.'

23 November 1839

The room was full, but I hardly knew who was there. Lord M. I saw looking at me with tears in his eyes, but he was not near me. I then read my short Declaration. I felt my hands shook but I did not make one mistake. I felt more happy and thankful when it was over.

To Prince Albert   28 November 1839

This morning I received your dear, dear letter of the 21st. How happy do you make me with your love! Oh! my Angel Albert, I am quite enchanted with it! I do not deserve such love! Never, never did I think I could be loved so much.

Journal   5 December 1839

Talked of where she [her mother] was to go to, and her living in a hired house, Lord M. thought it would not be well looked upon etc.; and I got eager and dreaded she would make all this a pretext to stay in the house etc. ... Talked of my horror at being married in the Chapel Royal ... I complained that everything was always made so uncomfortable for Kings and Queens, and it was making this odious etc.—He looked ill and said he was unwell.

To Prince Albert   29 November 1839

I had a talk with Lord Melbourne about your arrangements for our marriage, and also about your official attendants, and he has told me that young [George] Anson (his Private Secretary), who is with him, greatly wishes to be with you. I am very much in favour of it, because he is an excellent young man, and very modest, very honest, very steady, very well-informed, and will be of much use to you. He is not a member of the House of Commons, which is also convenient; so long as Lord Melbourne is in office he remains his Secretary.

Prince Albert was much distressed by this letter. He had assumed that he would be allowed to choose for himself the gentlemen of his Household and, believing that the Crown should not display a preference for any political party, as King William IV had done towards the Tories and Queen Victoria towards the Whigs, he had hoped that the composition of his own Household would demonstrate his impartiality. His hopes were not to be realised.

To Prince Albert   8 December 1839

As to your wish about your gentlemen, my dear Albert, I must tell you quite honestly that it will not do. You may entirely rely upon me that the people who will be about you will be absolutely pleasant people, of high standing and good character. These gentlemen will not be in continual attendance on you; only on great occasions, and to accompany you when you go anywhere, and to dinners, etc.

You may rely upon my care that you shall have proper people, and not idle and not too young, and Lord Melbourne has already mentioned several to me who would be very suitable...

I have received to-day an ungracious letter from Uncle Leopold. He appears to me to be nettled because I no longer ask for his advice, but dear Uncle is given to believe that he must rule the roast [roost] everywhere. However, that is not a necessity.

Prince Albert protested in vain against his being obliged to take the Prime Minister's Secretary as his own. Would not this make him 'a partisan in the eyes of many?' 'Think of my position, dear Victoria,' he pleaded. 'I am leaving my home with all its old associations, all my bosom friends and going to a country in which everything is new and strange to me ... Is it not to be conceded that the two or three persons who are to have the charge of my private affairs should be persons who already command my confidence?' The Queen answered his question on 23 December, underlining almost every word:

It is, as you rightly suppose, my greatest, my most anxious wish to do everything most agreeable to you, but I must differ with you respecting Mr Anson ... What I said about Anson giving you advice, means, that if you like to ask him, he can and will be of the greatest use to you, as he is a very well-informed person. He will leave Lord Melbourne as soon as he is appointed about you. Though I am very anxious you should not appear to belong to a Party, still it is necessary that your Household should not form a too strong contrast to mine, else they will say, 'Oh, we know the Prince says he belongs to no party, but we are sure he is a Tory!' Therefore it is also necessary that it should appear that you went with me in having some of your people who are staunch Whigs; but Anson is not in Parliament, and never was, and therefore he is not a violent politician. Do not think because I urge this, Lord M. prefers it; on the contrary, he never urged it, and I only do it as I know it is for your own good. You will pardon this long story. It will also not do to wait till you come to appoint all your people. I am distressed to tell you what I fear you do not like, but it is necessary, my dearest, most excellent Albert. Once more I tell you that you can perfectly rely on me in these matters.

26 December 1839

I always think it safer to write in English, as I can explain myself better, and I hope you can read my English, as I try to be very legible. I am much grieved that you feel disappointed about my wish respecting your gentlemen, but very glad that you consent to it, and that you feel confidence in my choice.

To King Leopold   27 December 1839

Just two words (though you don't deserve half a one, as your silence is unpardonable) to say I have just heard from Albert, who, I am glad to say, consents to my choosing his people; so one essential point is gained.

Journal   31 December 1839

Talked of Albert's having such a fear of our not putting people of good character about him. Lord M. said, 'Lady William' [Russell] 'said the Prince's character is such as is highly approved at a German university, but which would be subject to some ridicule at ours'; as Lord M. said formerly, any attention to morality in universities was ridiculed, which I said was too shocking. I said funnily I thought Lord M. didn't like Albert so much as he would if he wasn't so strict. 'Oh! no, I highly respect it,' said Lord M. I then talked of A.'s saying I ought to be severe about people. 'Then you'll be liable to make every sort of mistake. In this country all should go by law and precedent,' said Lord M., 'and not by what you hear.'

10 January 1840

Talked of Albert's not quite understanding about his Household, which however, I said, I should make him easily understand. 'Don't let any difficulty stand in the way about George Anson,' said Lord M. kindly, but I said G. Anson was the fit person and that I should easily make him understand it.

19 January 1840

Talked of Albert's indifference about Ladies, and Lord M. said, 'A little dangerous, all that is,—it's very well if that holds, but it doesn't always,' Lord M. said. I said this was very wrong of him, and scolded him for it.

On 24 January Lord John Russell proposed an allowance of £50,000 a year for Prince Albert. A motion put forward by the Radical, Joseph Hume, to reduce this annuity to £21,000 was defeated. But on 27 January a Tory member, supported by Peel, proposed a reduction to £30,000. This amendment was carried, to the Queen's outraged disgust.

JOURNAL   27 January 1840

At near 11 Lord M. received a note saying that he had been beat by 104; that it had been made quite a party question (vile, confounded, infernal Tories), that Peel had spoken and voted against the 50,000 (nasty wretch) ... As long as I live, I'll never forgive these infernal scoundrels, with Peel at their head, as long as I live for this act of personal spite!!

Overruled by the Queen on the composition of his Household, the Prince was also overruled when he suggested that their honeymoon might perhaps be longer than the two or three days at Windsor which she had planned:

TO PRINCE ALBERT   31 January 1840

You have written to me in one of your letters about our stay at Windsor, but, dear Albert, you have not at all understood the matter. You forget, my dearest Love, that I am the Sovereign, and that business can stop and wait for nothing. Parliament is sitting, and something occurs almost every day, for which I may be required, and it is quite impossible for me to be absent from London; therefore two or three days is already a long time to be absent. I am never easy a moment, if I am not on the spot, and see and hear what is going on, and everybody, including all my Aunts (who are very knowing in all these things), says I must come out after the second day, for, as I must be surrounded by my Court, I cannot keep alone. This is also my own wish in every way.

JOURNAL   7 February 1840

We were seated as usual, Lord Melbourne sitting near me. Talked ... of the Marriage Ceremony; my being a little agitated and nervous; 'Most natural,' Lord M. replied warmly; 'how could it be otherwise?' Lord M. was so warm, so kind, and so affectionate, the whole evening, and so much touched in speaking of me and my affairs. Talked of my former resolution of never marrying. 'Depend upon it, it's right to marry,' he said earnestly; 'if ever there was a situation that formed an exception, it was yours; it's in human nature, it's natural to marry; the other is a very unnatural state of things; it's a great change—it has its inconveniences: everybody does their best, and depend upon it you've done well; difficulties may arise from it,' as they do of course from everything.

8 February 1840

At ½ p. 4 the Carriage and Escort appeared, drove through the centre gate, and up to the door; I stood at the very door; 1st stepped out Ernest, then Uncle Ernest, and then Albert, looking beautiful and so well; I embraced him and took him by the hand and led him up to my room; Mamma, Uncle Ernest, and Ernest following. I sat on the sofa with my beloved Albert, Lord Melbourne sitting near me ... Lord M. admired the diamond Garter which Albert had on, and said 'Very handsome.' I told him it was my gift; I also gave him (all before dinner) a diamond star I had worn, and badge. Lord M. made us laugh excessively about his new Coat, which he said, 'I expect it to be the thing most observed.'

On the morning of their wedding, which took place in the Chapel Royal, St. James's, the Queen sent a note by hand to the Prince:

10 February 1840

How are you to-day, and have you slept well? I have rested very well, and feel very comfortable to-day. What weather! I believe, however, the rain will cease.

Send one word when you, my most dearly loved bridegroom, will be ready. Thy ever-faithful, Victoria R.

JOURNAL   10 February 1840

Got up at a ¼ to 9—well, and having slept well; and breakfasted at ½ p. 9. Mamma came before and brought me a Nosegay of orange flowers. My dearest kindest Lehzen gave me a dear little ring ... Had my hair dressed and the wreath of orange flowers put on. Saw Albert for the last time alone, as my Bridegroom.

Saw Uncle, and Ernest whom dearest Albert brought up. At ½ p. 12 I set off, dearest Albert having gone before. I wore a white satin gown with a very deep flounce of Honiton lace, imitation of old. I wore my Turkish diamond necklace and earrings, and Albert's beautiful sapphire brooch. Mamma and the Duchess of Sutherland went in the carriage with me. I never saw such crowds of people as there were in the Park, and they cheered most enthusiastically. When I arrived at St. James's, I went into the dressing-room where my 12 young Train-bearers were, dressed all in white with white roses, which had a beautiful effect. Here I waited a little till dearest Albert's Procession had moved into the Chapel. I then went with my Train-bearers and ladies into the Throne-room, where the Procession formed; Lord Melbourne in his fine new dress-coat, bearing the Sword of State, and Lord Uxbridge and Lord Belfast [Captain of the Yeomen of the Guard] on either side of him walked immediately before me. Queen Anne's room was full of people, ranged on seats one higher than the other, as also in the Guard room, and by the Staircase,—all very friendly; the Procession looked beautiful going downstairs. Part of the Colour Court was also covered in and full of people who were very civil. The Flourish of Trumpets ceased as I entered the Chapel, and the organ began to play, which had a beautiful effect. At the Altar, to my right, stood Albert.

The Ceremony was very imposing, and fine and simple, and I think ought to make an everlasting impression on every one who promises at the Altar to keep what he or she promises. Dearest Albert repeated everything very distinctly. I felt so happy when the ring was put on, and by Albert. As soon as the Service was over, the

Procession returned as it came, with the exception that my beloved Albert led me out. The applause was very great, in the Colour Court as we came through; Lord Melbourne, good man, was very much affected during the Ceremony and at the applause ... I then returned to Buckingham Palace alone with Albert; they cheered us really most warmly and heartily; the crowd was immense; and the Hall at Buckingham Palace was full of people; they cheered us again and again ... I went and sat on the sofa in my dressing-room with Albert; and we talked together there from 10 m. to 2 till 20 m. p. 2. Then we went downstairs where all the Company was assembled and went into the dining-room—dearest Albert leading me in ... Talked to all after the breakfast, and to Lord Melbourne, whose fine coat I praised.

I went upstairs and undressed and put on a white silk gown trimmed with swansdown, and a bonnet with orange flowers. Albert went downstairs and undressed. At 20 m. to 4 Lord Melbourne came to me and stayed with me till 10 m. to 4. I shook hands with him and he kissed my hand. Talked of how well everything went off. 'Nothing could have gone off better,' he said, and of the people being in such good humour ... I begged him not to go to the party; he was a little tired; I would let him know when we arrived; I pressed his hand once more, and he said, 'God bless you, Ma'am,' most kindly, and with such a kind look. Dearest Albert came up and fetched me downstairs, where we took leave of Mamma and drove off at near 4; I and Albert alone.

As soon as we arrived [at Windsor] we went to our rooms; my large dressing room is our sitting room; the 3 little blue rooms are his ... After looking about our rooms for a little while, I went and changed my gown, and then came back to his small sitting room where dearest Albert was sitting and playing; he had put on his windsor coat; he took me on his knee, and kissed me and was so dear and kind. We had our dinner in our sitting room; but I had such a sick headache that I could eat nothing, and was obliged to lie down in the middle blue room for the remainder of the evening, on the sofa, but, ill or not, I never, never spent such an evening ... He called me names of tenderness, I have never yet heard used to me before—was bliss beyond belief! Oh! this was the happiest day of my life!—May God help me to do my duty as I ought and be worthy of such blessings.

11 February 1840

When day dawned (for we did not sleep much) and I beheld that beautiful angelic face by my side, it was more than I can express! He does look so beautiful in his shirt only, with his beautiful throat seen. We got up at $\frac{1}{4}$ p. 8. When I had laced I went to dearest Albert's room, and we breakfasted together. He had a black velvet jacket on, without any neckcloth on, and looked more beautiful than it is possible for me to say ... At 12 I walked out with my precious Angel, all alone—so delightful, on the Terrace and new Walk, arm in arm! Eos our only companion. We talked a great deal together. We came home at one, and had luncheon soon after. Poor dear Albert felt sick and uncomfortable, and lay down in my room ... He looked so dear, lying there and dozing.

To King Leopold   11 February 1840

I write to you from here [Windsor] the happiest, happiest Being that ever existed.

Really, I do not think it possible for any one in the world to be happier, or as happy as I am. He is an Angel, and his kindness and affection for me is really touching. To look in those dear eyes, and that dear sunny face, is enough to make me adore him. What I can do to make him happy will be my greatest delight. Independent of my great personal happiness, the reception we both met with yesterday was the most gratifying and enthusiastic I ever experienced; there was no end of the crowds in London, and all along the road. I was a good deal tired last night.

JOURNAL   12 February 1840

Already the 2nd day since our marriage; his love and gentleness is beyond everything, and to kiss that dear soft cheek, to press my lips to his, is heavenly bliss. I feel a purer more unearthly feel than I ever did. Oh! was ever woman so blessed as I am.

13 February 1840

My dearest Albert put on my stockings for me. I went in and saw him shave; a great delight for me.

THE QUEEN REGNANT
1841–1852

*The Queen returned to London with her husband after three 'very, very happy' days.
'His love and gentleness is beyond everything,' she wrote, 'and to kiss that dear soft
cheek, to press my lips to his, is heavenly bliss ... Oh! was ever woman so blessed as I
am.' She felt all the more blessed when, faced with the probable fall of Lord Melbourne's
Government in May 1841, she had the Prince to rely upon for comfort, support and
advice. On 11 May she wrote to King Leopold:*

I am sure you will forgive my writing a very short letter to-day, but I am so harassed
and occupied with business that I cannot find time to write letters. You will, I am
sure, feel for me; the probability of parting from so kind and excellent a being as
Lord Melbourne as a Minister (for a friend he will always remain) is very, very
painful, even if one feels it will not probably be for long; to take it philosophically is
my great wish, and quietly I certainly shall, but one cannot help feelings of affection
and gratitude. Albert is the greatest possible comfort to me in every way, and my
position is much more independent than it was before.

18 May 1841

I was sure you would feel for me. Since last Monday we have lived in the daily
expectation of a final event taking place, and the debate still continues, and it is not
certain whether it will even finish to-night, this being the eighth night ... Our plans
are so unsettled that I can tell you nothing, only that you may depend upon it
nothing will be done without having been duly, properly, and maturely weighed.
Lord Melbourne's conduct is as usual perfect; fair, calm, and totally disinterested,
and I am certain that in whatever position he is you will treat him just as you have
always done.

My dearest Angel is indeed a great comfort to me. He takes the greatest interest in
what goes on, feeling with and for me, and yet abstaining as he ought from biassing
me either way, though we talk much on the subject, and his judgment is, as you say,
good and mild.

31 May 1841

I beg you not to be alarmed about what is to be done; it is not for a Party triumph
that Parliament (the longest that has sat for many years) is to be dissolved; it is the

fairest and most constitutional mode of proceeding; and you may trust to the moderation and prudence of my whole Government that nothing will be done without due consideration; if the present Government get a majority by the elections they will go on prosperously; if not, the Tories will come in for a short time. The country is quiet and the people very well disposed.

15 June 1841

Affairs go on, and all will take some shape or other, but it keeps one in hot water all the time. In the meantime, however, the people are in the best possible humour, and I never was better received at Ascot, which is a great test, and also along the roads yesterday. This [Nuneham Courtenay, the family home of Edward Vernon Harcourt, Archbishop of York] is a most lovely place; pleasure grounds in the style of Claremont, only much larger, and with the river Thames winding along beneath them, and Oxford in the distance; a beautiful flower and kitchen garden, and all kept up in perfect order. I followed Albert here, faithful to my word, and he is gone to Oxford* for the whole day, to my great grief. And here I am all alone in a strange house, with not even Lehzen as a companion, in Albert's absence.

Parliament was dissolved by the Queen in person on 29 June, the Government having survived a vote of 'no confidence' by only one vote some weeks before. In the general election which followed the Tories, under Sir Robert Peel, were returned with a large majority. The Queen wrote to King Leopold on 3 August:

What is to come hangs over me like a baneful dream, as you will easily understand, and when I am often happy and merry, comes and damps it all. But God's will be done! and it is for our best we must feel . . .

Our little tour was most successful, and we enjoyed it of all things; nothing could be more enthusiastic or affectionate than our reception everywhere, and I am happy to hear that our presence has left a favourable impression, which I think will be of great use. The loyalty in this country is certainly very striking. We enjoyed Panshanger [Earl Cowper's house] still more than Woburn [The Duke of Bedford's]; the country is quite beautiful, and the house so pretty and *wohnlich*; the picture-gallery and pictures very splendid. The Cowpers are such good people too. The visit to Brocket [Lord Melbourne's] naturally interested us very much for our excellent Lord Melbourne's sake. The park and grounds are beautiful.

24 August 1841

You don't say that you sympathise with me in my present heavy trial, the heaviest I have ever had to endure, and which will be a sad heart-breaking to me—but I know

*The Prince had gone to Oxford to receive an address at Commemoration. He was 'enthusiastically received,' the Queen told King Leopold on 11 June, 'but the students . . . had the bad taste to show their party feeling in groans and hisses when the name of a Whig was mentioned, which they ought not to have done in my husband's presence.'

you do feel for me. I am quiet and prepared, but still I fell very sad, and God knows! very wretched at times, for myself and my country, that such a change must take place. But God in His mercy will support and guide me through all. Yet I feel that my constant headaches are caused by annoyance and vexation!

Thanks to her husband's help and influence the Queen did not find dealing with her new Prime Minister as much of an ordeal as she had expected. On 30 August she wrote to Melbourne:

The first interview with Sir Robert Peel has gone off well, and only lasted twenty minutes; and he sends the Queen to-morrow, in writing, the proposed arrangements, and will only come down on Wednesday morning ... He made many protestations of his sorrow, at what must give pain to the Queen (as she said to him it did), but of course said he accepted the task. The Duke of Wellington's health too uncertain, and himself too prone to sleep coming over him—as Peel expressed it—to admit of his taking an office in which he would have much to do, but to be in the Cabinet, which the Queen expressed her wish he should ... What the Queen felt when she parted from her dear, kind friend, Lord Melbourne, is better imagined than described; she was dreadfully affected for some time after, but is calm now. It is very, very sad; and she cannot quite believe it yet. The Prince felt it very, very much too, and really the Queen cannot say how kind and affectionate he is to her, and how anxious to do everything to lighten this heavy trial; he was quite affected at this sad parting. We do, and shall, miss you so dreadfully; Lord Melbourne will easily understand what a change it is, after these four years when she had the happiness of having Lord Melbourne always about her. But it will not be so long till we meet again. Happier and brighter times will come again.

To King Leopold   8 September 1841

Your kind letter gave me great pleasure, and I must own your silence on all that was going on distressed me very much! It has been indeed a sad time for me, and I am still bewildered, and can't believe that my excellent Lord Melbourne is no longer my Minister, but-he will be, as you say, and has already proved himself, very useful and valuable as my friend out of office. He writes to me often, and I write to him, and he gives really the fairest and most impartial advice possible. But after seeing him for four years, with very few exceptions—daily—you may imagine that I must feel the change; and the longer the time gets since we parted, the more I feel it. Eleven days was the longest I ever was without seeing him, and this time will be elapsed on Saturday, so you may imagine what the change must be. I cannot say what a comfort and support my beloved Angel is to me, and how well and how kindly and properly he behaves.

24 September 1841

I feel thankful for your praise of my conduct; all is going on well, but it would be needless to attempt to deny that I feel the change, and I own I am much happier when I need not see the Ministers; luckily they do not want to see me often.

Although she confided in George Anson that she felt she could 'not get over ... Peel's awkward manner', and although she had occasion to disapprove of some of his appointments and to deliver through him a sharp rebuke to the Foreign Secretary for allowing an ambassador to leave the country without an Audience, the Queen's relations with her new Prime Minister continued to improve. And, as the months passed, she wrote to her Uncle on other topics.

TO KING LEOPOLD   19 April 1842

I am quite bewildered with all the arrangements for our *bal costumé*, which I wish you could see; we are to be Edward III. and Queen Philippa, and a great number of our Court to be dressed like the people in those times ... but there is such asking, and so many silks and drawings and crowns, and God knows what, to look at, that I, who hate being troubled about dress, am quite *confuse*.

To get a little rest we mean to run down to Claremont from Friday to Monday. My last ball was very splendid, and I have a concert on Monday next.

31 May 1842

I wish to be the first to inform you of what happened yesterday evening, and to tell you that we are *saines et sauves*. On returning from the chapel on Sunday, Albert was observing how civil the people were, and then suddenly turned to me and said it appeared to him as though a man had held out a pistol to the carriage, and that it had hung fire. No one, however, who was with us, such as footmen, etc., had seen anything at all. Albert began to doubt what he believed he had seen. Well, yesterday morning (Monday) a lad came to Murray [Hon. Charles Augustus Murray, Master of the Household] (who of course knew nothing) and said that he saw a man in the crowd as we came home from church, present a pistol to the carriage, which, however, did not go off, and heard the man say, 'Fool that I was not to fire!' The man then vanished, and this boy followed another man (an old man) up St James's Street who repeated twice, 'How very extraordinary!' but instead of saying anything to the police, asked the boy for his direction and disappeared. The boy accordingly was sent to Sir Robert Peel, and (doubtful as it all still was) every precaution was taken, still keeping the thing completely secret, not a soul in the house knowing a word, and accordingly after some consultation, as nothing could be done, we drove out—many police then in plain clothes being distributed in and about the parks, and the two Equerries riding so close on each side that they must have been hit, if anybody had; still the feeling of looking out for such a man was not *des plus agréables;* however, we drove through the parks, up to Hampstead, and back again. All was so quiet that we almost thought of nothing,—when, as we drove down Constitution Hill, very fast, we heard the report of a pistol, but not at all loud, so that had we not been on the alert we should hardly have taken notice of it. We saw the man seized by a policeman next to whom he was standing when he fired, but we did not stop ... Others saw him take aim, but we only heard the report (looking both the other way). We felt both very glad that our drive had had the effect of having the man seized. Whether it was loaded or not we cannot yet tell, but we are again full of gratitude to Providence for invariably protecting us! The feeling of horror is very great in the public, and great affection is shown us. The man was yesterday examined at the Home Office, is called John Francis, is a cabinet-maker, and son of a machine-maker of Covent Garden Theatre, is good-looking (they say) ... Only twenty or twenty-one years old, and not the least mad—but very cunning. The boy identified him this morning, amongst many others. Everything is to be kept secret this time, which is very right, and altogether I think it is being well done. Every further particular you shall hear. I was really not at all frightened ...

Thank God, my Angel is also well!

6 June 1842

There seems no doubt whatever that Francis is totally without accomplices, and a *mauvais sujet*. We shall be able probably to tell you more when we see you.

Francis was condemned to death, a punishment which the Queen considered necessary. He was, however, reprieved on 1 July; and three days later, as the Queen had feared it might be, another attempt on her life was made, this time in the Mall, by a deformed youth no more than four feet tall. His pistol contained more tobacco than gunpowder. The Queen was unhurt and made no alteration in her travelling arrangements. In September she and the Prince went to Scotland.

To King Leopold   8 September 1842

I make no excuses for not having written, as I know that you will understand that when one is travelling about and seeing so much that is totally new, it is very difficult to find time to write ...

Albert has told you already how successfully everything had gone off hitherto, and how much pleased we were with Edinburgh, which is an unique town in its way. We left Dalkeith on Monday, and lunched at Dupplin, Lord Kinnoul's, a pretty place with quite a new house, and which poor Lord Kinnoul displayed so well as to fall head over heels down a steep bank, and was proceeding down another, if Albert had not caught him; I did not see it, but Albert and I have nearly died with laughing at the relation of it.

The next year the Queen paid a visit to King Louis Philippe, the first visit by an English to a French monarch since Henry VIII's to François I on the Field of the Cloth of Gold.

To King Leopold   4 September 1843

I write to you from this dear place [the Château d'Eu] where we are in the midst of this admirable and truly amiable family, and where we feel quite at home, and as if we were one of them. Our reception by the dear King and Queen has been most kind, and by the people really gratifying. I am highly interested and amused.

4 December 1843

We arrived at Chatsworth [from Belvoir Castle] on Friday, and left it at nine this morning, quite charmed and delighted with everything there. Splendour and comfort are so admirably combined, and the Duke [of Devonshire] does everything so well. I found many improvements since I was there eleven years ago. The conservatory [designed by Joseph Paxton] is out and out the finest thing imaginable of its kind. It is one mass of glass, 64 feet high, 300 long, and 134 wide. The grounds, with all the woods and cascades and fountains, are so beautiful too. The first evening there was a ball, and the next the cascades and fountains were illuminated, which had a beautiful effect. There was a large party there, including many of the Duke's family, the Bedfords, Buccleuchs, the Duke of Wellington, the Normanbys, Lord Melbourne (who is much better), and the Beauvales. We arrived here at half-past two, we perform our journey so delightfully on the railroad, so quickly and easily . . . Albert is going out hunting to-morrow, which I wish was over, but I am assured that the country is much better than the Windsor country.

The Prince continued to play a large and heroic part in her letters:

TO KING LEOPOLD   18 January 1842

Albert's great *fonction* [He had laid the foundation stone of the new Royal Exchange] yesterday went off beautifully, and he was so much admired in all ways; he always fascinates the people wherever he goes, by his very modest and unostentatious yet dignified ways. He only came back at twelve last night; it was very kind of him to come.

14 February 1843

I am only a little wee bit distressed at your writing on the 10th, and not taking any notice of the dearest, happiest day in my life [her wedding day], to which I owe the present great domestic happiness I now enjoy, and which is much greater than I deserve, though certainly my Kensington life for the last six or seven years had been one of great misery and oppression, and I may expect some little retribution, and, indeed, after my accession, there was a great deal of worry. Indeed I am grateful for possessing (really without vanity or flattery or blindness) the most perfect being as a husband in existence, or who ever did exist; and I doubt whether anybody every did love or respect another as I do my dear Angel! And indeed Providence has ever mercifully protected us, through manifold dangers and trials, and I feel confident will continue to do so, and then let outward storms and trials and sorrows be sent us, and we can bear all.

12 December 1843

Louise [Queen of the Belgians] will be able to tell you how well the remainder of our journey went off, and how well Albert's hunting answered [He had been out with the Belvoir hounds on 5 December]. One can hardly credit the absurdity of people here, but Albert's riding so boldly and hard has made such a sensation that it has been written all over the country, and they make much more of it than if he had done some great act!

It rather disgusts one, but still it had done, and does, good, for it has put an end to all impertinent sneering for the future about Albert's riding. This journey has done great good, and my beloved Angel in particular has had the greatest success; for instance, at Birmingham the good his visit has done has been immense, for Albert spoke to all these manufacturers in their own language, which they did not expect, and these poor people have only been accustomed to hear demagogues and Chartists.

The Prince had, indeed, been well received in Birmingham where the Mayor was a Chartist. But the disturbed state of the country was a cause of much concern to the Queen at this time. She voiced this concern to the Home Secretary, Sir James Graham:

23 June 1843

The Queen returns these communications to Sir James Graham, which are of a very unpleasant nature. The Queen trusts that measures of the greatest severity will be taken, as well to suppress the revolutionary spirit as to bring the culprits* to immediate trial and punishment. The Queen thinks this of the greatest importance with respect to the effect it may have in Ireland, likewise as proving that the Government is willing to show great forbearance, and to trust to the good sense of the people; but that if outrages are committed and it is called upon to act, it is not to be trifled with, but will visit wrong-doers with the utmost severity.

22 September 1843

The Queen has received Sir James Graham's letter of the 22nd. She has long seen with deep concern the lamentable state of turbulence in South Wales, and has repeatedly urged the necessity of its being put an end to, by vigorous efforts on the part of the Government. The Queen, therefore, willingly gives her sanction to the issuing of a special Commission for the trial of the offenders and to the issuing of a proclamation. Monday, the 2nd, being the earliest day at which, Sir James says, the necessary Council could be held, will suit the Queen very well; she begs, therefore, that Sir James will cause the Council to meet here on that day at three o'clock.

---

*The Rebecca rioters who, protesting against the Poor Law Amendment Act amongst other grievances, had destroyed turnpike toll houses and gates. Many of them were dressed as women. They took their name from the biblical prophecy that Rebecca's seed should 'possess the gate of those that hate them'.

The Queen was also worried that she might lose the service of Peel whom she now fully trusted. On 18 June 1844 she wrote to tell King Leopold that she was in 'the greatest possible danger' of being forced to accept a resignation of the Government 'without knowing to whom to turn'.

I am sure you will agree with me that Peel's resignation would not only be for us (for we cannot have a better and a safer Minister), but for the whole country, and for the peace of Europe—a great calamity. Our present people are all safe, and not led away by impulses and reckless passions. We must, however, take care and not get into another crisis; for I assure you we have been quite miserable and quite alarmed ever since Saturday . . . I should be equally sorry to lose Lord Aberdeen [the Foreign Secretary] as he is so very fair, and has served us personally, so kindly and truly.

At least she could reassure herself with the thought that she and Prince Albert were well received wherever they went :

To KING LEOPOLD   29 October 1844

I had the happiness of receiving your kind letter of the 26th while I was dressing to go to the City for the opening of the Royal Exchange. Nothing ever went off better, and the procession there, as well as all the proceedings at the Royal Exchange, were splendid and royal in the extreme. It was a fine and gratifying sight to see the myriads of people assembled—more than at the Coronation even, and all in such good humour, and so loyal; the articles in the papers, too, are most kind and gratifying; they say no Sovereign was more loved than I am (I am bold enough to say), and that, from our happy domestic home—which gives such a good example.

28 January 1845

The feeling of loyalty in this country is happily very strong, and wherever we show ourselves we are most heartily and warmly received, and the civilities and respect shown to us by those we visit is most satisfactory. I mention merely a trifling instance to show how respectful they are—the Duke of Buckingham, who is immensely proud, bringing the cup of coffee after dinner on a waiter to Albert himself. And everywhere my dearest Angel receives the respect and honours I receive.

But then, in December, the resignation of Peel could be delayed no longer. Both Lord Stanley and the Duke of Wellington declined to support him on Corn Law reform, and, as she reported to Lord Melbourne on 7 December, she was obliged to send for Lord John Russell :

Sir Robert Peel has informed the Queen that in consequence of differences prevailing in the Cabinet, he is very reluctantly compelled to solicit from the Queen the acceptance of his resignation, which she has as reluctantly accepted.

From the Queen's unabated confidence in Lord Melbourne, her first impulse was to request his immediate attendance here that she might have the benefit of his assistance and advice, but on reflection the Queen does not think herself justified, in the present state of Lord Melbourne's health, to ask him to make the sacrifice which the return to his former position of Prime Minister would, she fears, impose upon him.

It is this consideration, and this alone, that has induced the Queen to [send for] Lord John Russell ... The Queen hopes, however, that Lord Melbourne will not withhold from her new Government his advice, which would be so valuable to her.

It is of the utmost importance that the whole of this communication should be kept a most profound secret until the Queen has seen Lord John Russell.

On 12 December the Queen wrote to the Duke of Wellington:

The Queen has to inform the Duke of Wellington that, in consequence of Sir Robert Peel's having declared to her his inability to carry on any longer the Government, she has sent for Lord John Russell, who is not able at present to state whether he can form an Administration, and is gone to Town in order to consult his friends. Whatever the result of his enquiries may be, the Queen has a strong desire to see the Duke of Wellington remain at the head of her Army. The Queen appeals to the Duke's so often proved loyalty and attachment to her person, in asking him to give her this assurance. The Duke will thereby render the greatest service to the country and to her own person.

Lord John Russell failed in his efforts to form a Government, however; and, as Disraeli said, handed back the poisoned chalice to Peel. On 20 December the Queen wrote to Russell:

Sir Robert Peel has just been here. He expressed great regret that Lord John Russell had felt it necessary to decline the formation of a Government.

He said he should have acted towards Lord John Russell with the most scrupulous good faith, and that he should have done everything in his power to give Lord John support.

He thinks many would have been induced to follow his example.

Sir Robert Peel did not hesitate a moment in withdrawing his offer of resignation. He said he felt it his duty at once to resume his office, though he is deeply sensible of the difficulties with which he has to contend.

To King Leopold   23 December 1845

I have little to add to Albert's letter of yesterday, except my extreme admiration of our worthy Peel, who shows himself a man of unbounded loyalty, courage, patriotism, and high-mindedness, and his conduct towards me has been chivalrous almost, I might say. I never have seen him so excited or so determined, and such a good cause must succeed. We have indeed had an escape, for though Lord John's own notions were very good and moderate, he let himself be entirely twisted and twirled about by his violent friends, and all the moderate ones were crushed.

                                                                   30 December 1845

Many thanks for your kind letter of the 27th, by which I see how glad you are at our good Peel being again—and I sincerely and confidently hope for many years—my Minister. I have heard many instances of the confidence the country and all parties have in Peel; for instance, he was immensely cheered at Birmingham—a most Radical place; and Joseph Hume expressed great distress when Peel resigned, and the greatest contempt for Lord John Russell. The Members of the Government have behaved extremely well and with much disinterestedness. The Government has secured the services of Mr Gladstone [as Secretary of State for the Colonies] and Lord Ellenborough [as First Lord of the Admiralty], who will be of great use. Lord E. is become very quiet, and is a very good speaker.

We had a very happy Christmas. This weather is extremely unwholesome.

The Queen's hopes that Peel would remain her Prime Minister for many years were not to be realised. For in the summer of 1846 the Government, having achieved the repeal of the Corn Laws, were defeated on another issue. On 1 July the Queen wrote to Peel:

It does seem strange that at the moment of triumph the Government should have to resign. The Queen read Sir Robert Peel's speech with great admiration. The Queen seizes this opportunity (though she will see Sir Robert again) of expressing her deep concern at losing his services, which she regrets as much for the Country as for herself and the Prince. In whatever position Sir Robert Peel may be, we shall ever look on him as a kind and true friend, and ever have the greatest esteem and regard for him as a Minister and as a private individual.

To King Leopold   7 July 1846

Yesterday was a very hard day for me. I had to part with Sir R. Peel and Lord Aberdeen, who are irreparable losses to us and the Country; they were both so much overcome that it quite overset me, and we have in them two devoted friends. We felt so safe with them. Never, during the five years that they were with me, did they ever recommend a person or a thing which was not for my or the Country's best, and never for the Party's advantage only; and the contrast now is very striking; there is much less respect and much less high and pure feeling.

The Government which the Queen compared so unfavourably with Peel's was that of Lord John Russell which contained Lord Palmerston as Foreign Secretary. She soon had occasion to write sharply to both of them.

TO PALMERSTON   17 April 1847

The Queen has several times asked Lord Palmerston, through Lord John Russell and personally, to see that the drafts to our Foreign Ministers are not despatched previous to their being submitted to the Queen. Notwithstanding, this is still done, as for instance to-day with regard to the drafts for Lisbon. The Queen, therefore, once more repeats her desire that Lord Palmerston should prevent the recurrence of this practice.*

TO RUSSELL   14 October 1847

As to Mr Cobden's appointment to the Poor Law Board, the Queen thinks that he will be well qualified for the place in many respects, and that it will be advantageous to the Government and the Country that his talents should be secured to the service of the State, but the elevation to the Cabinet directly from Covent Garden [where Free Trade meetings had been held] strikes her as a very sudden step, calculated to cause much dissatisfaction in many quarters, and setting a dangerous example to agitators in general (for his main reputation Mr Cobden gained as a successful agitator). The Queen therefore thinks it best that Mr Cobden should first enter the service of the Crown, serve as a public functionary in Parliament, and be promoted subsequently to the Cabinet, which step will then become a very natural one.

TO RUSSELL   Undated

The Queen has seen with surprise in the *Gazette* the appointment of Mr Corigan [as Physician-in-Ordinary to Her Majesty in Ireland], about which she must complain to Lord John Russell. Not only had her pleasure not been taken upon it, but she had actually mentioned ... that she had her doubts about the true propriety of the appointment. Lord John will always have found the Queen desirous to meet his views with regard to all appointments and ready to listen to any reasons which he might adduce in favour of his recommendations, but she must insist upon appointments in her Household not being made without her previous sanction, and least of all such as that of a Physician to her person.

'It seems as if the whole face of Europe were changing,' the Queen wrote of 1848, the year of revolutions. 'I maintain that Revolutions are always bad for the country, and the cause of untold misery to the people.' She deplored the revolution in Paris which brought about the abdication of Louis Philippe; she lamented also the unrest in Austria and the spread of revolutionary fervour to Italy. She was as strong in her support of Austria as

---

*By now many of the Queen's letters were being drafted for her by Prince Albert, no doubt after being discussed between them. But it is clear that the Queen very rarely allowed him a free hand in their final composition.

Palmerston and Russell were in their support of the Italians now struggling for their independence under the uncertain leadership of Charles Albert, King of Sardinia. The Queen's letters to her Ministers became increasingly acerbic.

To RUSSELL   25 July 1848

The Queen must tell Lord John what she has repeatedly told Lord Palmerston, but without apparent effect, that the establishment of an *entente cordiale* with the French Republic, for the purpose of driving the Austrians out of their dominions in Italy, would be a disgrace to this country.

11 August 1848

The Queen is highly indignant at Lord Palmerston's behaviour now again with respect to Lord Normanby's appointment;* he knew perfectly well that Lord Normanby could not accept the post of Minister, and had written to the Queen before that such an offer could not be made, and has now made it after all, knowing that, by wasting time and getting the matter entangled at Paris, he would carry his point. If the French are so anxious to keep Lord Normanby as to make any sacrifice for that object, it ought to make us cautious, as it can only be on account of the ease with which they can make him serve their purposes. They, of course, like an *entente cordiale* with us at the expense of Austria ... but this can be no consideration for us ... The Queen has read the leading articles of the *Times* of yesterday and to-day on this subject with the greatest satisfaction as they express almost entirely the same views and feelings which she entertains. The Queen hopes that Lord John Russell will read them; indeed, the whole of the Press seem to be unanimous on this subject, and she can hardly understand how there can be two opinions upon it ... The Queen must say she is afraid that she will have no peace of mind and there will be no end of troubles so long as Lord Palmerston is at the head of the Foreign Office.

To PALMERSTON   20 August 1848

The Queen has received an autograph letter from the Archduke John which has been cut open at the Foreign Office. The Queen wishes Lord Palmerston to take care that this does not happen again. The opening of official letters even, addressed to the Queen, which she has of late observed, is really not becoming, and ought to be discontinued, as it used never to be the case formerly.

2 September 1848

The Queen has read in the papers the news that Austria and Sardinia have nearly settled their differences, and also 'that it was confidently stated that a French and British squadron, with troops on board, are to make a demonstration in the Adriatic.'

*Lord Normanby had been Ambassador in Paris. The Queen did not want to receive an Ambassador of the French Republic at her Court where he would be at the head of London Society and possibly prove to be a 'very awkward character'. She believed that Ministers only should be exchanged at first and that Normanby, as a former British Ambassador, should be withdrawn. But Palmerston wanted Normanby to remain in Paris where, in fact, he did remain.

Though the Queen cannot believe this, she thinks it right to inform Lord Palmerston without delay that, should such a thing be thought of, it is a step which the Queen could not give her consent to.

4 September 1848

The Queen since her arrival in Town has heard that the answer from Austria declining our mediation has some days ago been communicated to Lord Palmerston. The Queen is surprised that Lord Palmerston should have left her uninformed of so important an event.

7 September 1848

The Queen received the night before she left London (too late to write upon it) Lord Palmerston's letter and the long draft to Lord Normanby. As the draft is gone, the Queen will only remark that the passage expressing Lord Palmerston's agreement in the general argument of M. de Beaumont [the French Ambassador in London], 'as to the advantages which would arise from a previous concert between England and France as to any military operations which France might be compelled to undertake in Italy',—is most dangerous for the future peace of Europe . . . This is a line of policy to which the Queen cannot give her consent. It is quite immaterial, whether French troops alone are employed for such an unjust purpose, if this is done on previous concert with England.

The whole tone of the draft the Queen cannot approve.

TO RUSSELL   *7 September 1848*

The Queen must send the enclosed draft to Lord John Russell, with a copy of her letter to Lord Palmerston upon it. Lord Palmerston has as usual pretended not to have had time to submit the draft to the Queen before he had sent it off. What the Queen has long suspected and often warned against is on the point of happening, viz. Lord Palmerston's using the new entente cordiale for the purpose of wresting from Austria her Italian provinces by French arms. This would be a most iniquitous proceeding.

On 19 September the Queen wrote a memorandum describing a conversation she had had with Russell about Palmerston:

I said to Lord John Russell, that I must mention to him a subject, which was a serious one, one which I had delayed mentioning for some time, but which I felt I must speak quite openly to him upon now, namely about Lord Palmerston; that I felt really I could hardly go on with him, that I had no confidence in him, and that it made me seriously anxious and uneasy for the welfare of the country and for the peace of Europe in general, and that I felt very uneasy from one day to another as to what might happen.

To Russell   7 October 1848

The partiality of Lord Palmerston in this Italian question really surpasses all conception, and makes the Queen very uneasy on account of the character and honour of England, and on account of the danger to which the peace of Europe will be exposed.

To King Leopold   10 October 1848

The state of Germany is dreadful, and one does feel quite ashamed about that once really so peaceful and happy people. That there are still good people there I am sure, but they allow themselves to be worked upon in a frightful and shameful way ... In France a crisis seems at hand. What a very bad figure we cut in this mediation! Really it is quite immoral, with Ireland quivering in our grasp, and ready to throw off her allegiance at any moment, for us to force Austria to give up her lawful possessions. What shall we say if Canada, Malta, etc., begin to trouble us? It hurts me terribly.

21 November 1848

You will grieve to hear that our good, dear, old friend Melbourne is dying ... One cannot forget how good and kind and amiable he was, and it brings back so many recollections to my mind, though, God knows! I never wish that time back again.

27 November

Our poor old friend Melbourne died on the 24th. I sincerely regret him, for he was truly attached to me, and though not a firm Minister he was a noble, kind-hearted, generous being. Poor Lord Beauvale [Frederick Lamb, Lord Melbourne's younger brother] and Lady Palmerston [Lord Melbourne's sister] feel it very much. I wish it might soften the *caro sposo* of the latter-named person.

6 August 1849

Though this letter will only go tomorrow, I will begin it to-day and tell you that everything has gone off beautifully since we arrived in Ireland, and that our entrance into Dublin was really a magnificent thing. By my letter to Louise you will have heard of our arrival in the Cove of Cork. Our visit to Cork was very successful; the Mayor was knighted on deck (on board the *Fairy*), like in times of old. Cork is about seventeen miles up the River Lee, which is beautifully wooded and reminds us of Devonshire scenery. We had previously stepped on shore at Cove, a small place, to enable them to call it Queen's Town; the enthusiasm is immense, and at Cork there was more firing than I remember since the Rhine.

We left Cork with fair weather, but a head sea and contrary wind which made it rough and me very sick.

7th.—I was unable to continue till now ... We went into Waterford Harbour on Saturday afternoon, which is likewise a fine, large, safe harbour ... The next morning we received much the same report of the weather which we had done at Cork ... However we went out, as it could not be helped, and we might have remained there some days for no use. The first three hours were very nasty, but afterwards it cleared and the evening was beautiful. The entrance at seven o'clock into Kingston Harbour was splendid; we came in with ten steamers, and the whole harbour, wharf, and every surrounding place was covered with thousands and thousands of people, who received us with the greatest enthusiasm. We disembarked yesterday morning at ten o'clock, and took two hours to come here. The most perfect order was maintained in spite of the immense mass of people assembled, and a more good-humoured crowd I never saw, but noisy and excitable beyond belief, talking, jumping, and shrieking instead of cheering. There were numbers of troops out, and it really was a wonderful scene. This [The Lodge, Phoenix Park] is a very pretty place, and the house reminds me of dear Claremont. The view of the Wicklow Mountains from the windows is very beautiful, and the whole park is very extensive and full of very fine trees.

We drove out yesterday afternoon and were followed by jaunting-cars and riders and people running and screaming, which would have amused you. You see more ragged and wretched people here than I ever saw anywhere else. *En revanche*, the women are really handsome—quite in the lowest class—as well at Cork as here; such beautiful black eyes and hair and such fine colours and teeth.

To Russell   9 June 1850

The Queen has received Lord John Russell's two letters. If the Cabinet think it impossible to do otherwise, of course the Queen consents—though most reluctantly—to a compliance with the vote respecting the Post Office.* The Queen thinks it a very false notion of obeying God's will, to do what will be the cause of much annoyance and possibly of great distress to private families. At any rate, she thinks decidedly that great caution should be used with respect to any alteration in the transmission of the mails, so that at least some means of communication may still be possible.

To King Leopold   2 July 1850

By my letter to Louise you will have learnt all the details of this certainly very disgraceful and very inconceivable attack.† I have not suffered except from my head, which is still very tender, the blow having been extremely violent, and the brass end of the stick fell on my head so as to make a considerable noise. I own it makes me nervous out driving, and I stare at any person coming near the carriage, which I am afraid is natural.

Journal   2 July 1850

Certainly it is very hard and very horrid, that I, a woman—a defenceless young woman should be exposed to insults of this kind, and be unable to go out quietly for

---

*Lord Ashley, later 7th Earl of Shaftesbury, the philanthropist, had carried a resolution forbidding the delivery of letters on Sundays. A Committee of Inquiry was appointed to consider the proposed change which was, to the Queen's satisfaction, eventually abandoned.

†The Queen had been struck on the head with a cane by Robert Pate, an ex-lieutenant of the 10th Hussars, as she was leaving Cambridge House after a visit to the Duke of Cambridge. An attempt to prove Pate insane failed and he was sentenced to seven years' transportation.

a drive ... for a man to strike any woman is most brutal, and I, as well as everyone else, think this far worse than any attempt to shoot, which, wicked as it is, is at least more comprehensible and more courageous.*

To King Leopold   2 July 1850

We have, alas! now another cause of much greater anxiety in the person of our excellent Sir Robert Peel, who, as you will see, has had a most serious fall [while riding up Constitution Hill], and though going on well at first, was very ill last night; thank God! he is better again this morning, but I fear still in great danger. I cannot bear even to think of losing him; it would be the greatest loss for the whole country, and irreparable for us, for he is so trustworthy, and so entirely to be depended on. All parties are in great anxiety about him. I will leave my letter open to give you the latest news ...

I am happy to say that Sir Robert, though still very ill, is freer from pain, his pulse is less high, and he feels himself better; the Doctors think there is no vital injury, and nothing from which he cannot recover, but that he must be for some days in a precarious state.

9 July 1850

Poor dear Peel is to be buried to-day. The sorrow and grief at his death are most touching, and the country mourns over him as over a father. Every one seems to have lost a personal friend.

As I have much to write, you will forgive me ending here ... My poor dear Albert, who had been so fresh and well when we came back, looks so pale and fagged again. He has felt, and feels, Sir Robert's loss dreadfully. He feels he has lost a second father.

Memories of Peel's good conduct made Lord Palmerston's high-handed behaviour all the more intolerable to the Queen. His defence of Don David Pacifico, a Gibraltarian living in Athens who had had his house destroyed by a Greek mob, had been 'a most disagreeable business'. When the Greek Government had rejected Don Pacifico's excessive demands, Palmerston had sent a fleet to blockade the Piraeus and enforce them, almost precipitating a European war. Having triumphantly vindicated his conduct in a famous speech, in which he compared a British subject's rights anywhere in the world with the proud claim of an ancient Roman, 'Civis Romanus sum', Palmerston had become more unmanageable and self-assertive than ever. Encouraged by the Prince, the Queen renewed her efforts to have him replaced.

To Russell   28 July 1850

Lord John may be sure that she fully admits the great difficulties in the way of the projected alteration, but she on the other hand feels the duty she owes to the country

*Since John Francis's attempt to shoot the Queen, she had been fired upon on Constitution Hill by one William Hamilton whose pistol was fortunately charged only with powder. He was sentenced to seven years' transportation.

and to herself not to allow a man in whom she can have no confidence, who she knows has conducted himself in anything but a straightforward and proper manner to herself, to remain in the Foreign Office and thereby to expose herself to insults from other nations, and this country and to the constant risk of serious and alarming complications. The Queen considers these reasons as much graver than the other difficulties ...

Each time that we were in difficulty, the Government seemed to be determined to move Lord Palmerston and as soon as these difficulties were got over, those which present themselves in the carrying out of this removal appeared of so great a magnitude as to cause its relinquishment. There is no chance of Lord Palmerston reforming himself in his sixty-seventh year ... There is no question of delicacy and danger in which Lord Palmerston will not arbitrarily and without reference to his colleagues or sovereign engage this country.

12 August 1850

With reference to the conversation about Lord Palmerston which the Queen had with Lord John Russell the other day, and Lord Palmerston's disavowal that he ever intended any disrespect to her by the various neglects of which she has had so long and so often to complain, she thinks it right, in order to prevent any mistake for the future, shortly to explain what it is she expects from her Foreign Secretary. She requires: (1) That he will distinctly state what he proposes in a given case, in order that the Queen may know as distinctly to what she has given her Royal sanction; (2) Having once given her sanction to a measure, that it be not arbitrarily altered or modified by the Minister; such an act she must consider as failing in sincerity towards the Crown, and justly to be visited by the exercise of her Constitutional right of dismissing that Minister. She expects to be kept informed of what passes between him and the Foreign Ministers before important decisions are taken, based upon that intercourse; to receive the Foreign Despatches in good time, and to have the drafts for her approval sent to her in sufficient time to make herself acquainted with their contents before they must be sent off. The Queen thinks it best that Lord John Russell should show this letter to Lord Palmerston.*

TO PALMERSTON   12 October 1850

The Queen has received Lord Palmerston's letter respecting the draft to Baron Koller.† She cannot suppose that Baron Koller addressed his note to Lord Palmerston in order to receive in answer an expression of his own personal opinion;

---

*This letter was based upon a memorandum drawn up by Baron Stockmar. Russell showed it to Palmerston who, apparently contrite, came to apologise to Prince Albert. When Palmerston later overreached himself Russell read it out in the House.

†Baron Koller was the Austrian ambassador. Lord Palmerston had had to write to him about an attack made in London upon General Haynau, an Austrian officer who had become notorious as a brutal flogger of women in the Hungarian war. When visiting Barclay and Perkins brewery, the draymen had recognised him by his flamboyantly long mustachios (which Koller had advised him to cut off) and had assaulted him. In his reply to Koller's note Palmerston had expressed the opinion that Haynau had shown a 'want of propriety' in visiting England 'at the present moment'.

and if Lord Palmerston could not reconcile it to his own feelings to express the regret of the Queen's Government at the brutal attack and wanton outrage committed by a ferocious mob on a distinguished foreigner of past seventy years of age, who was quietly visiting a private establishment in this metropolis, without adding his censure of the want of propriety evinced by General Haynau in coming to England—he might have done so in a private letter, where his personal feelings could not be mistaken for the opinion of the Queen and her Government. She must repeat her request that Lord Palmerston will rectify this.

In the autumn of 1850 the Pope—having issued a brief dividing England into twelve bishoprics and having appointed Nicholas Wiseman as Archbishop of Westminster—announced that the English people, so long severed from Rome, were about to rejoin the Holy Church. Russell, appointing himself the Protestants' champion, violently attacked the Pope's aggression and thus lost the support not only of the Irish members but also of the Peelites, several of whom were High Churchmen and most of whom deplored Russell's intolerance.

To King Leopold   21 February 1851

I have only time just to write a few hasty lines to you from Stockmar's room, where I came up to speak to Albert and him, to tell you that we have got a Ministerial crisis; the Ministers were in a great minority last night, and though it was not a question vital to the Government, Lord John feels the support he has received so meagre, and the opposition of so many parties so great, that he must resign! This is very bad, because there is no chance of any other good Government, poor Peel being no longer alive, and not one man of talent except Lord Stanley in the Party . . . but Lord John is right not to go on when he is so ill supported, and it will raise him as a political man, and will strengthen his position for the future.

Whether Lord Stanley (to whom I must send to-morrow after the Government have resigned) will be able to form a Government or not, I cannot tell. Altogether, it is very vexatious, and will give us trouble. It is the more provoking, as this country is so very prosperous.

                                                                1 March 1851

I did not write to you yesterday, thinking I could perhaps give you some more positive news today, but I cannot. I am still without a Government [Lord Stanley having failed to form one], and I am still trying to hear and pause before I actually call to Lord John to undertake to form, or rather more to continue, the Government. We have passed an anxious, exciting week, and the difficulties are very peculiar; there are so many conflicting circumstances . . . but the 'Papal Question' is the real and almost insuperable difficulty.

No solution to the problem could be found, so Lord John and Palmerston remained in office. Meanwhile preparations for the Great Exhibition in Hyde Park continued.

JOURNAL   30 April 1851

Everyone is occupied with the great day and afternoon and my poor Albert is
terribly fagged. All day some question or other, or some difficulty, all of which my
beloved one takes with the greatest quiet and good temper ... The noise and bustle,
even greater than yesterday, as so many preparations are being made for the seating
of the spectators, and there is certainly still much to be done.

                                                            1 May 1851

This day is one of the greatest and most glorious days of our lives, with which, to my
pride and joy the name of my dearly beloved Albert is forever associated! It is a day
which makes my heart swell with thankfulness ... The Park presented a wonderful
spectacle, crowds streaming through it—carriages and troops passing, quite like the
Coronation Day, and for me, the same anxiety. The day was bright, and all bustle
and excitement. At ½ p. 11, the whole procession in 9 state carriages was set in
motion. Vicky and Bertie [the two eldest children] were in our carriage. Vicky was
dressed in lace over white satin, with a small wreath of pink wild roses, in her hair,
and looked very nice. Bertie was in full Highland dress. The Green Park and Hyde
Park were one mass of densely crowded human beings, in the highest good humour
and most enthusiastic. I never saw Hyde Park look as it did, being filled with crowds
as far as the eye could reach. A little rain fell, just as we started; but before we neared
the Crystal Palace, the sun shone and gleamed upon the gigantic edifice, upon which
the flags of every nation were flying. We drove up Rotten Row and got out of our
carriages at the entrance on that side. The glimpse through the iron gates of the
Transept, the moving palms and flowers, the myriads of people filling the galleries
and seats around, together with the flourish of trumpets, as we entered the building,
gave a sensation I shall never forget, and I felt much moved. We went for a moment
into a little room where we left our cloaks and found Mama and Mary. Outside all
the Princes were standing. In a few seconds we proceeded, Albert leading me having
Vicky at his hand, and Bertie holding mine. The sight as we came to the centre
where the steps and chair (on which I did not sit) was placed, facing the beautiful
crystal fountain was magic and impressive. The tremendous cheering, the joy
expressed in every face, the vastness of the building, with all its decorations and
exhibits, the sound of the organ (with 200 instruments and 600 voices, which
seemed nothing), and my beloved Husband the creator of this great 'Peace Festival',
uniting the industry and art of all nations of the earth, all this, was indeed moving,
and a day to live forever. God bless my dearest Albert, and my dear Country which
has shown itself so great today ... The Nave was full of people, which had not been
intended and deafening cheers and waving of handkerchiefs, continued the whole
time of our long walk from one end of the building, to the other. Every face was
bright, and smiling, and many even had tears in their eyes. Many Frenchmen called
out 'Vive la Reine'. One could of course see nothing, but what was high up in the
Nave, and nothing in the Courts. The organs were but little heard, but the Military
Band, at one end, had a very fine effect, playing the March from *Athalie*, as we
passed along. The old Duke of Wellington and Lord Anglesey walked arm in arm,
which was a touching sight. I saw many acquaintances, amongst those present. We

returned to our place and Albert told Lord Breadalbane to declare the Exhibition to be opened, which he did in a loud voice saying 'Her Majesty commands me to declare the Exhibition opened', when there was a flourish of trumpets, followed by immense cheering. Everyone was astounded and delighted. The return was equally satisfactory—the crowd most enthusiastic and perfect order kept. We reached the Palace at 20 m. past 1 and went out on the balcony, being loudly cheered. The Prince and Princess [of Prussia] were quite delighted and impressed. That we felt happy and thankful,—I need not say—proud of all that had passed and of my beloved one's success. Dearest Albert's name is for ever immortalised and the absurd reports of dangers of every kind and sort, set about by a set of people,—the 'soi-disant' fashionables and the most violent protectionists—are silenced. It is therefore doubly satisfactory that all should have gone off so well, and without the slightest accident or mishap.

Once the excitement was over, however, the Queen had once more to face the 'dreadful' Lord Palmerston who now gave offence by announcing that he intended to receive the Hungarian patriot, Lajos Kossuth, whose forthcoming visit to England was to arouse great fervour in Liberal and Radical circles.

To PALMERSTON   31 October 1851

The Queen mentioned to Lord Palmerston when he was last here at Windsor Castle that she thought it would not be advisable that he should receive M. Kossuth upon his arrival in England, as being wholly unnecessary, and likely to be misconstrued abroad. Since M. Kossuth's arrival in this country, and his violent denunciations of two Sovereigns with whom we are at peace, the Queen thinks that she owes it as a mark of respect to her Allies, and generally to all States at peace with this country, not to allow that a person endeavouring to excite a political agitation in this country against her Allies should be received by her Secretary of State for Foreign Affairs. Whether such a reception should take place at his official or private residence can make no difference as to the public nature of the act. The Queen must therefore demand that the reception of M. Kossuth by Lord Palmerston should not take place.

To RUSSELL   31 October 1851

The Queen has just received Lord John Russell's letter. She thinks it natural that Lord John should wish to bring a matter which may cause a rupture in the Government before the Cabinet, but thinks his having summoned the Cabinet only for Monday will leave Lord Palmerston at liberty in the intermediate time to have his reception of Kossuth, and then rest on his *fait accompli*. Unless, therefore, Lord John Russell can bind him over to good conduct, all the mischief which is apprehended from this step of his will result; and he will have, moreover, the triumph of having carried his point, and having set the Prime Minister at defiance.

1 November 1851

The Queen has to acknowledge Lord John Russell's letter of this day, and returns the copy of his to Lord Palmerston. She feels that she has the right and the duty to demand that one of her Ministers should not by his private acts, compromise her and the country, and therefore omitted in her letter to Lord Palmerston all reference to Lord John Russell's opinion; but she of course much prefers that she should be protected from the wilful indiscretions of Lord Palmerston by the attention of the Cabinet being drawn to his proceedings without her personal intervention.

20 November 1851

The Queen must write to-day to Lord John Russell on a subject which causes her much anxiety. Her feelings have again been deeply wounded by the official conduct of her Secretary of State for Foreign Affairs since the arrival of M. Kossuth in this country. The Queen feels the best interests of her people, the honour and dignity of her Crown, her public and personal obligations towards those Sovereigns with whom she professes to be on terms of peace and amity, most unjustifiably exposed ... These remarks seem to be especially called for after the report of the official interview between Lord Palmerston and the deputation from Finsbury,* and the Queen requests Lord John Russell to bring them under the notice of the Cabinet.

21 November 1851

The Queen cannot suppose that Lord John considers the official reception by the Secretary of State for Foreign Affairs of addresses, in which allied Sovereigns are called Despots and Assassins, as within that 'latitude' which he claims for every Minister.

Although Palmerston survived this crisis, he was not to survive the next. On 2 December Prince Louis Napoleon, nephew of Napoleon I, who had been elected President of the French Republic, engineered a coup d'état and proclaimed himself the Emperor Napoleon III. Two days later the Queen, who had harboured a hope that one of her Orléans relations might return to Paris on the Republic's demise, wrote to Russell:

The Queen has learnt with surprise and concern the events which have taken place at Paris. She thinks it is of great importance that Lord Normanby should be instructed to remain entirely passive, and to take no part whatever in what is passing. Any word from him might be misconstrued at such a moment.

13 December 1851

The Queen sends the enclosed despatch from Lord Normanby to Lord John

---

*Although he agreed not to receive Kossuth in his house, Palmerston did accept at the Foreign Office addresses from deputations of Radicals in which the Emperor of Austria and the Tsar were referred to as 'odious and detestable assassins' and 'merciless tyrants and despots'.

Russell, from which it appears that the French Government pretend to have received the entire approval of the late *coup d'état* by the British Government, as conveyed by Lord Palmerston to Count Walewski [the French Ambassador]. The Queen cannot believe in the truth of the assertion, as such an approval given by Lord Palmerston would have been in complete contradiction to the line of strict neutrality and passiveness which the Queen had expressed her desire to see followed with regard to the late convulsion at Paris, and which was approved by the Cabinet, as stated in Lord John Russell's letter of the 6th inst. Does Lord John know anything about the alleged approval, which, if true, would again expose the honesty and dignity of the Queen's Government in the eyes of the world?

19 December 1851

Lord John will readily conceive what must be her feelings in seeing matters go from bad to worse with respect to Lord Palmerston's conduct!

20 December 1851

The Queen has now to express to Lord John Russell her readiness to follow his advice, and her acceptance of the resignation of Lord Palmerston. She will be prepared to see Lord John after the Cabinet on Monday, as he proposes.

JOURNAL   *20 December 1851*

After luncheon we saw the correspondence, beginning with a long letter from Lord Palmerston, in which he hardly touches upon his improper conduct in telling Walewski he entirely approved of the coup d'état, trying to make out that each person 'coloured highly' what the other said, and that he had merely given his opinion. He entered into a long dissertation to prove the reason why Louis Napoleon was justified in doing what he did. To this, Lord John wrote: that this explanation was quite unsatisfactory,—that he discussed questions which had no bearing on the matter Lord John had called upon him to explain; that he felt the time had come when it was best he should no longer hold the Seals of the Foreign Office, for he perceived that his indiscretion led to endless misunderstandings and breaches of decorum, which endangered our relations with other countries. He offered Lord Palmerston the Lord-Lieutenancy of Ireland, and ended by (I consider rather unnecessary and uncalled for) praises. Lord Palmerston has answered very stiffly that he will be prepared to give up the Seals as soon as his successor is appointed.

Our relief was great and we felt quite excited by the news, for our anxiety and worry during the last five years and a half, which was indescribable, was mainly, if not entirely, caused by him! It is a great and unexpected mercy, for I really was on the point of declaring on my part that I could no longer retain Lord Palmerston, which would have been a most disagreeable task, and not unattended with a small amount of danger, inasmuch as it would have put me too prominently forward.

TO RUSSELL   *20 December 1851*

With respect to a successor to Lord Palmerston, the Queen must state, that after the sad experience which she has just had of the difficulties, annoyances, and dangers to which the Sovereign may be exposed by the personal character and qualities of the Secretary for Foreign Affairs, she must reserve to herself the unfettered right to approve or disapprove the choice of a Minister for this Office.

Lord Granville, whom Lord John Russell designates as the person best calculated for that post, would meet with her entire approval. The possible opinion of the Cabinet that more experience was required does not weigh much with the Queen. From her knowledge of Lord Granville's character, she is inclined to see no such disadvantage in the circumstance that he has not yet had practice in managing Foreign Affairs, as he will be the more ready to lean upon the advice and judgment of the Prime Minister where he may have diffidence in his own, and thereby will add strength to the Cabinet by maintaining unity in thought and action.

To King Leopold   23 December 1851

I have the greatest pleasure in announcing to you a piece of news which I know will give you as much satisfaction and relief as it does to us, and will do to the whole of the world. Lord Palmerston is no longer Foreign Secretary—and Lord Granville is already named his successor!! He had become of late really quite reckless.

Journal   26 December 1851

Half past 3 and no sign of Lord Palmerston. Lord John came in to say he thought Lord P. must not be intending to come. On my asking whether the Seals could be delivered up by another person, Lord John replied, they could, and that George III and William IV had sent for the Seals in anger, not allowing the minister to deliver them up; but it certainly always had been an understood thing that the minister ought to deliver up the Seals himself. Lord John however, at our suggestion, went to enquire about the trains, to see if any other was due ... By this time it being nearly 4, Lord John returned and advised me to receive the Seals from him, which I did, and then held the Council at which Lord Granville was sworn in, and I delivered the Seals to him. After the Council we saw him and when I expressed my satisfaction at seeing him in his new office, he was so overcome as to be hardly able to speak. We talked to him generally of what had passed ... Lord Granville seemed quite to understand everything and all he said was most discreet, giving us the impression that he was not alarmed at the task. He said he thought Lord Palmerston's faults were more easily to be avoided, then his great merits were to be copied ... The relief of having such a good, amiable, honest man, in the place of one whom I grieve to think, I must with truth, consider an unprincipled man, is not to be described, and I can hardly yet realise it.

To King Leopold   30 December 1851

All that you say about Lord Palmerston is but too true ... He *brouilléd* us and the country with every one ... It is too grievous to think how much misery and mischief might have been avoided. However, now he has done with the Foreign Office for ever, and 'the veteran statesman,' as the newspapers, to our great amusement and I am sure to his infinite annoyance, call him, must rest upon his laurels.

TO RUSSELL   31 December 1851

The Queen sees in the papers that there is to be a *Te Deum* at Paris on the 2nd for the success of the *coup d'état*, and that the Corps Diplomatique is to be present. She hopes that Lord Normanby will be told not to attend. Besides the impropriety of his taking part in such a ceremony, his doing so would entirely destroy the position of Lord John Russell opposite Lord Palmerston, who might with justice say that he merely expressed his personal approval of the *coup d'état* before, but since, the Queen's Ambassador had been ordered publicly to thank God for its success.

TO KING LEOPOLD   3 February 1852

Matters are very critical and all Van de Weyer [the Belgian Foreign Minister] has told us *n'est pas rassurant*. With such an extraordinary man as Louis Napoleon, one can never be for one instant safe. It makes me very melancholy; I love peace and quiet—in fact, I hate politics and turmoil, and I grieve to think that a spark may plunge us into the midst of war. Still I think that may be avoided. Any attempt on Belgium would be *casus belli* for us; that you may rely upon. Invasion I am not afraid of, but the spirit of the people here is very great—they are full of defending themselves—and the spirit of the olden times is in no way quenched.

Albert grows daily fonder and fonder of politics and business, and is so wonderfully fit for both—such perspicacity and such courage—and I grow daily to dislike them both more and more. We women are not made for governing—and if we are good women, we must dislike these masculine occupations; but there are times which force one to take interest in them *mal gré bon gré*, and I do, of course, intensely.

Lord Granville's first tenure of the Foreign Office proved to be short-lived, for towards the end of the month Lord John Russell's proposal to meet the possible threat from France by strengthening the local militia was deemed inadequate by Parliament and his Government fell. The Queen sent for Lord Derby.

TO KING LEOPOLD   24 February 1852

Great and not very pleasant events have happened since I wrote last to you. I know that Van de Weyer has informed you of everything, of the really (till the last day) unexpected defeat, and of Lord Derby's assumption of office, with a very sorry Cabinet. I believe, however, that it is quite necessary they should have a trial, and then have done with it. Provided the country remains quiet, and they are prudent in their Foreign Policy, I shall take the trial as patiently as I can . . .

Alas! your confidence in our excellent Lord Granville is no longer of any avail, though I hope ere long he will be at the Foreign Office again, [as he was to be in 1870–4 and 1880–5] and I cannot say that his successor [Lord Malmesbury], who has never been in office (as indeed is the case with almost all the new Ministers), inspires me with confidence.

9 March 1852

We have a most talented, capable, and courageous Prime Minister, but all his people have no experience—have never been in any sort of office before!

23 March 1852

Our acquaintance is confined almost entirely to Lord Derby, but then he is the Government. They do nothing without him. He has all the Departments to look after, and on being asked by somebody if he was not much tired, he said: 'I am quite well with my babies!'

30 March 1852

Mr Disraeli (alias Dizzy) [the Chancellor of the Exchequer] writes very curious reports to me of the House of Commons proceedings—much in the style of his books.

JOURNAL   16 March 1852

Disraeli's reports are just like his novels, highly coloured.

1 April 1852

She [Mrs Disraeli] is very vulgar, not so much in her appearance, as in her way of speaking, he is most singular,—thoroughly Jewish looking, a livid complexion, dark eyes and eyebrows and black ringlets. The expression is disagreeable, but I did not find him so to talk to. He has a very bland manner, and his language is very flowery.

TO KING LEOPOLD   17 September 1852

I am sure you will mourn with us over the loss we and this whole nation have experienced in the death of the dear and great old Duke of Wellington. It was a stroke, which was succeeded rapidly by others, and carried him off without any return of consciousness. For him it is a blessing that he should have been taken away in the possession of his great and powerful mind and without a lingering illness. But for this country, and for us, his loss—though it could not have been long delayed—is irreparable! He was the pride and the *bon génie*, as it were, of this country! He was the greatest man this country ever produced, and the most devoted and loyal subject, and the staunchest supporter the Crown ever had ... We shall soon stand sadly alone; Aberdeen is almost the only personal friend of that kind we have left. Melbourne, Peel, Liverpool—and now the Duke—all gone!

23 November 1852

Disraeli has been imprudent and blundering, and has done himself harm by a Speech he made about the Duke of Wellington, which was borrowed from an *éloge* by Thiers on a French Marshal!!! [Marshal Gouvion de St Cyr]

You will have heard ... how very touching the ceremony both in and out of doors was on the 18th [the day of the state funeral]. The behaviour of the millions assembled has been the topic of general admiration, and the foreigners have all assured me that they never could have believed such a number of people could have shown such feeling, such respect, for not a sound was heard! I cannot say what a deep and *wehmütige* impression it made on me! It was a beautiful sight. In the Cathedral it was much more touching still! The dear old Duke! he is an irreparable loss!

FAMILY LIFE
1841–1861

Happy as she was in her marriage, the Queen did not take kindly to the duties of child bearing which it imposed. It was not that she disliked children, but she did not at all care for babies when they were still incapable of more than what she called 'that terrible frog-like action'; and she was disgusted by the physical aspects of childbirth. She had been deeply upset to discover herself to be pregnant for the first time and, after the birth of the child, had been much annoyed when King Leopold expressed the hope that Princess Victoria would be the first of many children.

To KING LEOPOLD   5 January 1841

I think, dearest Uncle, you cannot really wish me to be the 'Mamma d'une *nombreuse* famille,' for I think you will see with me the great inconvenience a large family would be to us all, and particularly to the country, independent of the hardship and inconvenience to myself; men never think, at least seldom think, what a hard task it is for us women to go through this very often. God's will be done, and if He decrees that we are to have a great number of children why we must try to bring them up as useful and exemplary members of society ... I think you would be amused to see Albert dancing her in his arms; he makes a capital nurse (which I do not, and she is much too heavy for me to carry), and she already seems so happy to go to him.

The christening will be at Buckingham Palace on the 10th of February, our dear marriage-day.

Reluctant though she was to have a large family, the Queen gave birth with what she considered tiresome regularity. Her second child, the future King Edward VII, was born on 9 November 1841.

To KING LEOPOLD   29 November 1841

I would have written sooner, had I not been a little bilious, which made me very low, and not in spirits to write ... They think that I shall not get my appetite and spirits back till I can get out of town; we are therefore going in a week at latest. I am going for a drive this morning, and am certain it will do me good. In all essentials, I am better, if possible, than last year. Our little boy is a wonderfully strong and large child, with very large dark blue eyes, a finely formed but somewhat large nose, and a pretty little mouth; I hope and pray he may be like his dearest Papa. He is to be called Albert, and Edward is to be his second name. Pussy [Princess Victoria], dear child, is still the great pet amongst us all, and is getting so fat and strong again. She is not at all pleased with her brother . . .

I have been suffering so from lowness that it made me quite miserable, and I know how difficult it is to fight against it. I wonder very much who our little boy will be like. You will understand how fervent my prayers and I am [sure] everybody's must be, to see him resemble his angelic dearest Father in every, every respect, both in body and mind. Oh! my dearest Uncle, I am sure if you knew how happy, how blessed I feel, and how proud I feel in possessing such a perfect being as my husband, as he is, and if you think that you have been instrumental in bringing about this union, it must gladden your heart! How happy should I be to see our child grow up just like him! Dear Pussy travelled with us and behaved like a grown-up person, so quiet and looking about and coquetting with the Hussars on either side of the carriage. Now adieu!

18 January 1842

Our Claremont trip was very enjoyable, only we missed Pussy so much; another time we shall take her with us; the dear child was so pleased to see us again, particularly dear Albert, whom she is so fond of . . . We think of going to Brighton early in February, as the physicians think it will do the children great good, and perhaps it may me; for I am very strong as to fatigue and exertion, but not quite right otherwise; I am growing thinner, and there is a want of tone, which the sea may correct.

Although she did not mention it in her correspondence with her uncle, the Queen and Prince Albert were in the midst of their first serious quarrel. The source of the trouble was Baroness Lehzen who, so Anson said, let no opportunity of creating mischief escape her, and in her passionate jealousy was determined to retain her power over her former charge's children. She was 'a crazy, stupid intriguer . . . who regards herself as a demi-God,' wrote the Prince, 'and anyone who refuses to recognise her as such is a criminal.' He was convinced that her interference in the nursery was as much responsible for his daughter's weakness as was the incompetence of Dr. Clark and the nurses. He spoke to the Queen on the subject on 18 January and she flared up in anger, shouting that he could murder the child if he wanted to. Controlling his own anger, the Prince murmured, 'I must have patience.' He went downstairs to his own room where he gave vent to his fury in a note to his wife: 'Dr. Clark has mismanaged the child and poisoned her with calomel and you have starved her. I shall have nothing more to do with it; take the child away and do as you like and if she dies you will have it on your conscience.'

The Prince sent this note to Baron Stockmar with a covering letter in which he wrote, 'Victoria is too hasty and passionate for me to be able often to speak of my difficulties. She will not hear me out but flies into a rage and overwhelms me with reproaches of suspiciousness, want of trust, ambition, envy, etc. etc . . . All the disagreeableness I suffer comes from one and the same person [Baroness Lehzen] and that is precisely the person whom Victoria chooses for her friend and confidante.'

To Stockmar   19 January 1842

If A's note is full of hard words and other things that might make me angry and unhappy (as I know, he is unjust) don't show it me but tell me what he wants, for I don't wish to be angry with him and really my feelings of justice would be too violent to keep in, did I read what was too severe. If you think it is not calculated to do this, let me see it ... Albert must tell me what he dislikes and I will set about to remedy it, but he must also promise to listen to and believe me; when (on the contrary) I am in a passion which I trust I am not very often in now, he must not believe the stupid things I say like being miserable I ever married and so forth which come when I am unwell ... I have often heard Albert own that everybody recognised Lehzen's former services to me and my only wish is that she should have a quiet home in my house and see me sometimes. A. cannot object to my having her to talk to some times ... and I assure you upon my honour that I see her very seldom now and only for a few minutes, often to ask questions about papers and *toilette* for which she is of the greatest use to me. A. often and often thinks I see her, when I don't ... I tell you this as it is true, as you know me to be ... Dearest Angel Albert, God only knows how I love him. His position is difficult, heaven knows and we must do everything to make it easier.

20 January 1842

There is often an irritability in me which (like Sunday last which began the whole misery) makes me say cross and odious things which I don't myself believe and which I fear hurt A., but which he should not believe ... but I trust I shall be able to conquer it. Our position is tho' very different to any other married couples. A. is in my house and not I in his.—But I am ready to submit to his wishes as I love him so dearly.

Confronted by Prince Albert's forceful stand, the Queen was ready, in fact, to agree to the departure of Baroness Lehzen who was granted a generous pension and went to live with her sister in Buckeburg.

To King Leopold   20 September 1842

We found our dear little Victoria so grown and so improved, and speaking so plain, and become so independent; I think really few children are as forward as she is. She is quite a dear little companion. The Baby is sadly backward, but also grown, and very strong.

10 January 1843

Victoria plays with my old bricks, etc., and I think you would be pleased to see this and to see her running and jumping in the flower garden, as old—though I fear still little—Victoria of former days used to do. She is very well, and such an amusement to us, that I can't bear to move without her; she is so funny and speaks so well, and in French also, she knows almost everything.

4 April 1843

[Stockmar] will, I hope, tell you how prosperous he found us all; and how surprised and pleased he was with the children; he also is struck with Albert junior's likeness to his dearest papa, which everybody is struck with. Indeed, dearest Uncle, I will venture to say that not only no Royal *Ménage* is to be found equal to ours, but no other *ménage* is to be compared to ours, nor is any one to be compared, take him altogether, to my dearest Angel!

9 May 1843

Thank God I am stronger and better this time [after the birth of her third child, Princess Alice, on 25 April] than either time before, my nerves are so well which I am most thankful for ... My adored Angel has, as usual, been all kindness and goodness, and dear Pussy a very delightful companion. She is very tender with her little sister, who is a pretty and large baby and we think will be the beauty of the family.

16 May 1843

I am happy to give you still better accounts of myself. I have been out every day since Saturday, and have resumed all my usual habits almost (of course resting often on the sofa, and not having appeared in Society yet), and feel so strong and well; much better (independent of the nerves) than I have been either time.

Our little baby, who I really am proud of, for she is so very forward for her age, is to be called Alice, an old English name, and the other names are to be Maud (another old English name and the same as Matilda) and Mary, as she was born on Aunt Gloucester's birthday ... The King of Hanover ... Ernestus, [her Uncle Ernest who had been asked to be a sponsor] has never said when he will come, even now, but always threatens that he will.

6 June 1843

Our christening went off very brilliantly, and I wish you could have witnessed it ... The King of Hanover arrived just in time to be too late. He is grown very old and excessively thin, and bends a good deal. He is very gracious, for him. Pussy and Bertie (as we call the boy) were not at all afraid of him, fortunately; they appeared after the *déjeuner* on Friday, and I wish you could have seen them; they behaved so beautifully before that great number of people, and I must say looked very dear, all in white, and very *distingués;* they were much admired.

16 January 1844

We leave dear Claremont, as usual, with the greatest regret; we are so peaceable here; Windsor is beautiful and comfortable, but it is a palace, and God knows how willingly I would always live with my beloved Albert and our children in the quiet and retirement of private life, and not be the constant object of observation, and of newspaper articles. The children (Pussette and Bertie) have been most remarkably well, and so have we.

Prince Albert's father, the disreputable Duke of Saxe-Coburg and Gotha, died on 29 January 1844; and the Queen gave way to expressions of grief which, even for her generation and even though on this occasion she was principally concerned for her husband, must be considered extravagant.

To King Leopold   6 February 1844

You must now be the father of us two poor bereaved heart broken children. To describe to you all that we have suffered, all that we do suffer, would be difficult. God has heavily afflicted us. We feel crushed, overwhelmed, bowed down by the loss of one who was so deservedly loved, I may say adored, by his children and family. I loved him and looked on him as my own father.

25 March 1845

You will, I am sure, be pleased to hear that we have succeeded in purchasing Osborne in the Isle of Wight,* and if we can manage it, we shall probably run down there before we return to Town, for three nights. It sounds so snug and nice to have a place of one's own, quiet and retired.

To Melbourne   23 April 1845

The Queen had intended to have written to Lord Melbourne from Osborne to thank him for his last note, but we were so occupied, and so delighted with our new and really delightful home, that she hardly had time for anything; besides which the weather was so beautiful, that we were out almost all day ... She thinks it is impossible to imagine a prettier spot—valleys and woods which would be beautiful anywhere; but all this near the sea (the woods grow into the sea) is quite perfection; we have a charming beach quite to ourselves. The sea was so blue and calm that the Prince said it was like Naples. And then we can walk about anywhere by ourselves without being followed and mobbed, which Lord Melbourne will easily understand is delightful. And last, not least, we have Portsmouth and Spithead so close at hand, that we shall be able to watch what is going on, which will please the Navy, and be hereafter very useful for our boys [the Queen's second son, Prince Alfred, had been born in August the previous year].

To Peel   22 June 1845

We are more and more delighted with this lovely spot, the air is so pure and fresh,

---

*Osborne House was bought with an estate of about 1,000 acres. The house was too small; and the foundation stone of a new mansion, designed by Prince Albert with help from Thomas Cubitt, was laid on 23 June 1845. Anson estimated that the cost, which was to be borne by the Queen's income, would be £200,000.

and in spite of the hottest sun which oppresses one so dreadfully in London and even at Windsor ... really the combination of sea, trees, woods, flowers of all kinds, the purest air ... make it—to us—a perfect little Paradise.

To MELBOURNE   31 July 1845

We are comfortably and peacefully established here since the 19th, and derive the greatest benefit, pleasure, and satisfaction from our little possession here. The dear Prince is constantly occupied in directing the many necessary improvements which are to be made, and in watching our new house, which is a constant interest and amusement.

To KING LEOPOLD   3 March 1846

I wish you could be here, and hope you will come here for a few days during your stay, to see the innumerable alterations and improvements which have taken place. My dearest Albert is so happy here, out all day planting, directing, etc., and it is so good for him. It is a relief to be away from all the bitterness which people create for themselves in London.

                                                                                                          16 May 1848

The poor Duchess of Gloster is again in one of her nervous states, and gave us a dreadful fright at the Christening [of Princess Louise, the Queen's sixth child] by quite forgetting where she was, and coming and kneeling at my feet in the midst of the service. Imagine our horror!

                                                                                             30 December 1848

I write to you once more in this old and most dreadful year [of revolutions] ... but I must not include myself or my country in the misfortunes of this past year:—on the contrary I have nothing but thanks and most grateful thanks to offer up for all that has happened here.

The birth of Princess Louise in 1848 was followed by that of her seventh child, Prince Arthur, in 1850, and by that of her eighth, Prince Leopold, on 7 April 1853. She was given what she called 'that blessed chloroform' for the first time. The effect, she wrote in her journal on 22 April, was 'soothing, quieting and delightful beyond measure'.

To KING LEOPOLD   8 April 1853

Stockmar will have told you that Leopold is to be the name of our fourth young gentleman. It is a mark of love and affection which I hope you will not disapprove. It is a name which is the dearest to me after Albert, and one which recalls the almost only happy days of my sad childhood; to hear 'Prince Leopold' again, will make me think of all those days!

22 September 1855

I profit by your own messenger to confide to you, and to you alone, begging you not to mention it to your children, that our wishes on the subject of a future marriage for Vicky have been realised in the most gratifying and satisfactory manner.

On Thursday (20th) after breakfast, Fritz Wilhelm [Prince Frederick William of Prussia] said he was anxious to speak of a subject which he knew his parents had never broached to us—which was to belong to our Family; that this had long been his wish, that he had the entire concurrence and approval not only of his parents but of the King—and that finding Vicky so *allerliebst*, he could delay no longer in making this proposal ... He is a dear, excellent, charming young man, whom we shall give our dear child to with perfect confidence. What pleases us greatly is to see that he is really delighted with Vicky.

MEMORANDUM BY THE QUEEN　　　　　　　　　　29 September 1855

I must write down at once what has happened—what I feel and how grateful I am to God for one of the happiest days of my life! When we got off our ponies this afternoon Fritz gave me a look which implied that his little proposal to Vicky, which he had begged us to let him make—had succeeded ... He said in answer to my question whether anything had occurred, yes—that while riding with her—just at the very beginning, he began to speak of Germany, his hope that she would come there and stay there. They were interrupted in fact 3 times, upon one occasion by the picking up of some white heather, which he said was good luck—which he wished her—and she him. At last towards the end of the ride, he repeated again his observation about Prussia; she answered she would be happy to stay there for a year. He added he hoped always always—on which she became very red. He continued, he hoped he had said nothing which annoyed her—to which she replied 'Oh! no.' He added might he tell her parents? which she then expressed a wish to do herself. He then shook hands with her—said this was one of the happiest days of his life. I tell this all in a hurry. We approved all this ... Vicky came into my room, where we both were ... seemed very much agitated ... Her Papa asked her if she had nothing more to say 'Oh! yes a great deal'. We urged her to speak and she said: 'Oh! it is that I am very fond of the Prince'. We kissed and pressed the poor dear child in our arms and Albert then told her how the Prince ... on the 20th had spoken to us ... [how he] wished to see more and more of her. I asked did she wish the same? 'Oh, yes, everyday,' looking up joyously and happily in my face—she was kneeling. Had she always loved him? 'Oh always!' ... Albert came in to say that Fritz was there, and I took her in. She was nervous but did not hesitate or falter in giving her very decided answer ... He kissed her hand twice, I kissed him and when he kissed her hand again ... she threw herself into his arms, and kissed him with a warmth which was responded to by Fritz again and again and I would not for the world have missed so touching and beautiful a sight ... It is his first love! Vicky's great youth makes it even more striking, but she behaved as a girl of 18 would, so naturally, so quietly and modestly and yet showing how very strong her feelings are.

To PRINCESS AUGUSTA OF PRUSSIA   22 October 1855

I have hardly ever discussed Vicky with you, partly because it seemed somewhat immodest to mention her gradual development or to praise her unduly, and also because she was so largely the object of our secret hopes and desires. However, now that God has graciously granted our wishes I will keep silent no longer and will tell you all you wish to know. Already Fritz must have told you so much about her that I can have very little more to add. She has developed amazingly of late and her visit to France proved beneficial in every way. She is now slightly taller than I am and grows visibly. I find her very good company and this important event in her life has now brought us even closer together. I experience everything she feels, and since I myself still feel so young our relationship is more like that of two sisters . . . But she is still half a child and has to develop herself both physically and morally before their marriage takes place in two years' time.

To PALMERSTON   26 March 1856

The Chancellor speaks of people being inclined to make remarks as to its being wrong that the Princess Royal should be at so young an age bound to contract a marriage a year and a half hence, which would tend to fetter her future direction while she ought to be left a free agent.

He is however probably unaware that the Princess' choice, although made with the sanction and approval of her parents, has been one entirely of her own heart, and that she is as solemnly engaged by her own free will and wish to Prince Frederick William of Prussia, as anyone can be, and that before God she has pledged her word. Therefore, whether it will be publicly announced or not, she could not break this solemn engagement. The Princess is now confirmed and old enough to know her own feelings and wishes, though she may not be old enough to consummate the marriage and leave her parents' roof.

To PRINCESS AUGUSTA OF PRUSSIA   9 June 1856

The young couple are as happy as can be here! Fritz is unbelievably in love and shows such a touching faith in our child, which from some one so young was altogether unexpected and is, indeed, very flattering and delightful for Vicky. Her love and trust in him grows daily and yet she is very placid and sensible in herself.

                                                    6 October 1856

I had already realised with much sorrow that this separation [from her daughter upon her marriage] would affect you terribly and that the feeling of emptiness afterwards would be horribly painful and acute. But I do hope you will gradually accustom yourself to it, particularly as you were often separated from Luise for weeks on end.

With me the circumstances are quite different. I see the children much less and even here, where Albert is often away all day long, I find no especial pleasure or compensation in the company of the elder children. You will remember that I told you this at Osborne. Usually they go out with me in the afternoon (Vicky mostly, and the others also sometimes), or occasionally in the mornings when I drive or walk or ride, accompanied by my lady-in-waiting. And only very exceptionally do I find the rather intimate intercourse with them either agreeable or easy. You will not understand this, but it is caused by various factors. Firstly, I only feel properly *à mon aise* and quite happy when Albert is with me; secondly, I am used to carrying on my many affairs quite alone; and then I have grown up all alone, accustomed to the society of adult (and never with younger) people—lastly, I still cannot get used to the fact that Vicky is almost grown up. To me she still seems the same child, who had to be kept in order and therefore must not become too intimate. Here are my sincere feelings in contrast to yours. And this is why the separation, although in many ways very difficult and painful for me, will not be as acute and terrible as it is in your case, which is really lucky.

To Clarendon   25 October 1857

It would be well if Lord Clarendon would tell Lord Bloomfield [Minister at Berlin] not to entertain the possibility of such a question as the Princess Royal's marriage taking place at Berlin. The Queen never could consent to it, both for public and private reasons, and the assumption of its being too much for a Prince Royal of Prussia to come over to marry the Princess Royal of Great Britain in England is too absurd, to say the least. The Queen must say that there never was even the shadow of a doubt on Prince Frederick William's part as to where the marriage should take place, and she suspects this to be the mere gossip of the Berliners. Whatever may be the usual practice of Prussian Princes, it is not every day that one marries the eldest daughter of the Queen of England. The question therefore must be considered as settled and closed.*

To King Leopold   12 January 1858

It is a time of immense bustle and agitation; I feel it is terrible to give up one's poor child, and feel very nervous for the coming time, and for the departure. But I am glad to see Vicky is quite well again and *unberufen* has got over her cold and is very well.

To Princess Frederick William   2 February 1858

My first occupation on this sad, sad day [of the Princess' departure for Germany]—is to write to you. An hour is already past since you left—and I trust that you are recovering a little, but then will come that awful separation from dearest Papa! How I wish that was over for you, my beloved child! I struggle purposely against my feelings not to be too much overcome by them, as it is our duty to do, but I feel very sick when I think all, all is past, all that seemed so distant, all the excitement, every thing—and nothing here but a sad, sad blank! Yes it is cruel, very cruel—very trying for parents to give up their beloved children, and to see them go away from the happy peaceful home—where you used all to be around us!

---

*In accordance with the Queen's wishes the marriage took place at the Chapel Royal, St. James's on 25 January 1858. The Queen felt 'almost as if it were I that was being married over again, only much more nervous.'

... It is snowing away and everything is white and dreary! I could not go out—and shall see if I can this afternoon.

Poor dear Alice, whose sobs must have gone to your heart—is sitting near me writing to you ... Dearest, dearest child, may every blessing attend you both.

4 February 1858

I am better today, but my first thoughts on waking were very sad—and the tears are ever coming to my eyes and ready to flow again but I am feeling much better today. But the idea of not seeing you for so long seems unbearable. Every thing I do—or see makes me think of you, makes me long to tell you all about it ... Today we have been to the British Museum (this afternoon) and have seen the splendid new Reading Room and as we looked at the antiques—I kept thinking how our dear Vicky would have admired their beauties! Everything recalls you to our mind, and in every room we shall have your picture!

5 February 1858

You wrote dearest Papa such a beautiful letter, it made me cry so much, as indeed every thing does. I don't find I get any better. Even looking at your fine large photograph which I have mounted standing on a small easel before me—upsets me!

God bless you for your dear warm affectionate heart and for your love to your adored father. That will bring blessings on you both! How he deserves your worship—your confidence ... What a pride to be his child as it is for me to be his wife!

7 February 1858

Do only let Lady Churchill [the Queen's Lady of the Bedchamber who accompanied the Princess to Germany] describe all your rooms at the palace at Berlin and you must tell me exactly how your hours are—what you do—when you dress and undress and breakfast, etc., for you know all what we do to a minute but unfortunately we know nothing and that makes the separation so much more trying.

Now that you are established at your new home, you must try and answer my questions and enter into some of the subjects I mention else we can never replace conversation. You remember how vexed you always were when you did not get answers to your letters ... Get Jane C[hurchill]. to tell me all about your rooms—the railway carriages etc. Has the railway carriage got a small room to it? And (you will think me as bad as Leopold B.) [the inquisitive Leopold of Brabant] were your rooms on the journey and at Potsdam arranged according to English fashion? Then I see by the papers you wore a green dress at the Cologne concert. Was that the one with black lace?—You must not be impatient about all these details which I am so anxious to know.

To King Leopold   *9 February 1858*

The separation was awful, and the poor child was quite broken-hearted, particularly at parting from her dearest beloved papa, whom she idolises. How we miss her, I can't say, and never having been separated from her since thirteen years above a fortnight. I am in a constant fidget and impatience to know everything about everything. . . . The blank she has left behind is very great indeed . . .

To-morrow is the eighteenth anniversary of my blessed marriage, which has brought such universal blessings on this country and Europe! For what has not my beloved and perfect Albert done? Raised monarchy to the highest pinnacle of respect, and rendered it popular beyond what it ever was in this country!

To Princess Frederick William   11 February 1858

You cannot think what a delight for us it was to receive your dear, long, affectionate and interesting letter . . . But pray don't write on that enormous paper—for it will go into no box or book.

How glad I am that you are comfortably established but how are the rooms placed—and are there passages or must you (as in the greater part of those old places) go through all the rooms to get to the others? . . .

How do you like the houses and the diet? Lady Churchill says the rooms at night are so awfully hot.

15 February 1858

Only remember that the better you become acquainted with the family and court the more you must watch yourself and keep yourself under restraint. No familiarity—no loud laughing. You know, dearest, how necessary it is to have self control, tiresome as it may be. Kindness, friendliness and civility but no familiarity except with your parents (in-law) . . .

That you are so happy is a great happiness and great comfort to us and yet it gives me a pang, as I said once before to see and feel my own child so much happier than she ever was before, with another . . . You know, my dearest, that I never admit any other wife can be as happy as I am—so I can admit no comparison for I maintain Papa is unlike anyone who lives or ever lived and will live.

19 February 1858

The Duchess of Orleans kindly sent me a letter with an extract from a letter from the Grand Duchess Alexandrine of Mecklenburg-Schwerin about you which is very amiable and kind. It was naturally not written for us to see, so don't pray let her perceive any thing. Only one thing I can't understand she says: 'She is very small' which considering that you are a good deal taller than me, and I am not a dwarf, is rather hard.

22 February 1858

That you should have longings and yearnings for your own dear home and your parents and brothers and sisters is not only natural but it is right; if it were not so, we should be surprised and grieved.

Now I must tell you that you numbered the pages of your letter wrong, and then I must scold you a little bit for not answering some questions; but above all for not telling me what you do. My good dear child never liked matter of fact things—but Mama does, and when Lady Churchill leaves you I shall know nothing of what is going on which makes me sad. I tell you all that is going on, so that you may follow everything, daily.

Pray do answer my questions, my dearest child, else you will be as bad as Bertie used to be, and it keeps me in such a fidget.

I asked you several questions on a separate paper about your health, cold sponging—temperature of your rooms etc. and you have not answered one! You should just simply and shortly answer them one by one and then there could be no mistake about them. My good dear child is a little unmethodical and unpunctual still. Fritz always answers all questions. Just write them down on a bit of paper—when you have time—and put them into your letter; never mind if they are old—only pray do answer them ... Are you even happier than at Windsor? I thought that you could not be. Bertie is shocked at your liking every thing so much. But I have no fear of old England and home suffering by that. It is in one's own power to be happy—and to be contented. And with a husband whom one loves, as you do, one is sure to see every thing en couleur de rose and so one ought.

1 March 1858

One great drawback you all (our daughters I mean) will have, which I have not viz: the being exiled from your native land. This is a sad necessity. One great advantage however you all have over me, and that is that you are not in the anomalous position in which I am,—as Queen Regnant. Though dear Papa, God knows, does everything—it is a reversal of the right order of things which distresses me much and which no one, but such a perfection, such an angel as he is—could bear and carry through.

2 March 1858

You said in your long letter that the happiest time for you—was when you were alone with Fritz; you will now understand why I often grudged you children being always there, when I longed to be alone with dearest Papa! Those are always my happiest moments!

9 March 1858

Affie is going on admirably [with his studies preparatory to his joining the Royal Navy]; he comes to luncheon to-day (which is a real, brilliant Osborne day) and oh! when I see him and Arthur and look at ...! (You know what I mean!) [the Prince of Wales] I am in utter despair! The systematic idleness, laziness—disregard of everything is enough to break one's heart, and fills me with indignation. Alice behaved so admirably about it—and has much influence with him—but, to you I own, I am wretched about it. But don't mention this to a human being!

15 March 1858

That you should feel shy sometimes I can easily understand. I do so very often to this hour. But being married gives one one's position which nothing else can. Think however what it was for me, a girl of 18 all alone, not brought up at court as you were—but very humbly at Kensington Palace. No, no one knows what a life of difficulties mine was—and is! How thankful I am that none of you, please God! ever will have that anomalous and trying position. Now do enter into this in your letters, you so seldom do that, except to answer a question.

31 March 1858

Most certainly I shall give your message to poor Bertie. Alas! I feel very sad and anxious about him: he is so idle and so weak! But God grant that he may take things more to heart and more seriously for the future, and get more power. The heart is good, warm and affectionate—if there were but reflection and power, and self-control . . .

We must look out for princesses for Bertie—as his wife ought not to be above a year or 2 younger than him, therefore 14 or 15 now, pretty, quiet and clever and sensible. Oh! if you would find us one!

14 April 1858

And now I must again come with a little scold; you have not written me one single word, for more than a week!! I am vexed, for you could easily have managed—if you would but be a little more expert—to say: 'I am well—had a good journey and am delighted with it etc.' And this could have been done in 1 minute, and would have given me pleasure, and this you did even on your fatiguing journey from England every day! You seem to think that if you can't write to me a long letter you are not to write at all. And yet I (and also Papa) wrote volumes from Osborne . . . Now let this not happen again promise me and answer this . . .

You have not answered me about the little Princess of Hesse [as a possible bride for the Prince of Wales]—though I asked you nearly 5 weeks ago—and repeated it—now don't forget. Whoever has the happiness of marrying B must be nearly his age; this we all feel and Mr. Tarver [the Prince's classical tutor] said it to me the other day. I must own I feel greatly relieved at his absence; he is so insupportable with the younger ones. I really hope you are not getting fat again? Do avoid eating soft, pappy things or drinking much—you know how that fattens.

3 May 1858

I think people really marry far too much; it is such a lottery after all, and for a poor woman a very doubtful happiness.

8 December 1858

I think unmarried people are very often very happy—certainly more so than married people who don't live happily together of which there are so many instances.

16 May 1860

All marriage is such a lottery—the happiness is always an exchange—though it may be a very happy one—still the poor woman is bodily and morally the husband's slave. That always sticks in my throat. When I think of a merry, happy, free young

1 Princess Victoria in 1836

2 The young Queen in 1840

3 The Duchess of Kent in 1841. 'Talked of Mama and her humour being so variable . . . of her being touchy and jealous.'

4 Louise Lehzen, a drawing done by Queen Victoria in 1833. 'At five years old Miss Lehzen was placed about me, and though she was most kind, she was very firm and I had a proper respect for her.'

5 'Good faithful Stockmar', sent to England by King Leopold to be the Queen's adviser

6 A drawing by the Queen of 'my excellent Lord Melbourne' at her Coronation in 1838

7 Lord John Russell. 'The Queen must demand that respect which is due from a Minister to a Sovereign.'

8 'The Mayor was knighted on deck, like in times of old.' The Queen in Cork on her Irish tour, 1849

9 Prince Alfred and Edward, Prince of Wales, with their tutor Dr Becker in 1854

10 A family group at Osborne in
1854, 'our new house, which is a
constant interest and
amusement.' The portly lady on
the right is the Duchess of Kent

11 From the left, Princess Alice,
Prince Alfred, the Princess Royal
(Vicky) and Princess Helena in
Les Petits Savoyards, 1854

12 The Queen with the Emperor Napoleon III of
France at Windsor, 1855: 'Our Imperial
guests with whom I am much pleased.'

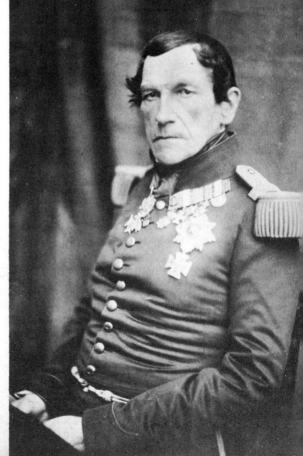

13 King Leopold of the Belgians in 1857, one of
the Queen's most regular correspondents:
'That dear loving Uncle, who has ever been to
me as a Father.'

14 Prince Alfred as 'Autumn' in 1854, propping up the weight of the wreath of grapes in his hair

15 Princess Beatrice taking her first ride on her second birthday in 1859

16 John Brown in 1860, whisky flask to hand

17 The Highland idyll: Prince Albert as seen by Landseer, 1860

girl—and look at the ailing, aching state a young wife generally is doomed to—which you can't deny is the penalty of marriage.

5 May 1858

How dreadfully vexed worried and fidgety I am at this untoward sprain I can't tell you! How could you do it? I am sure you had too high-heeled boots! I am haunted with your lying in a stuffy room in that dreadful old Schloss—without fresh air and alas! naturally without exercise and am beside myself. Only do take care and let some fresh air into your room and do get yourself carried out at least to get air!

26 May 1858

The horrid news [that the Princess was pregnant] contained in Fritz's letter to Papa upset us dreadfully. The more so as I feel certain almost it will all end in nothing . . .

The only one of all the children, who neither drew, wrote, played or did anything whatever to show his affection—beyond buying for me a table in Ireland—was Bertie. Oh! Bertie alas! alas! That is too sad a subject to enter on.

29 May 1858

I am so unhappy about you! It is well Fritz is not in sight just now or he would not be graciously received. Tell him that if he leaves you quite alone for a fortnight (he promised me never to do so without you were with his mother or sister or us) I shall not call him my son any more as I shall consider he has forfeited the claim! There is a threat! We tell everyone your foot is the cause of your not going to Coburg—and that the lying up has weakened you. I hope you do the same—and Fritz don't allow his own people and relations to enter into such subjects; it is so indelicate; Papa never allowed it and I should have been frantic.

11 June 1858

I hear that Chauvening [the Princess's hairdresser] is not at Babelsberg [the summer palace at Potsdam] and that your maids *coiffé* you—and that you don't find they do it well. Why don't you have him to do your hair in the middle of the day? Why can't he live nearby—if he can't be lodged in the house—and come daily to do your hair? You might have it early in the morning just twisted as I have—and then made later, for when one is unwell—to have one's hair pulled about is dreadful. I am terrified to hear you have such fearfully full sleeves—for God's sake take care or you will set yourself on fire, and now that might be the death of you; they are so horridly uncomfortable and ugly too.

15 June 1858

What you say of the pride of giving life to an immortal soul is very fine, dear, but I own I cannot enter into that; I think much more of our being like a cow or a dog at such moments; when our poor nature becomes so very animal and unecstatic—but for you, dear, if you are sensible and reasonable not in ecstasy nor spending your day with nurses and wet nurses, which is the ruin of many a refined and intellectual young lady, without adding to her real maternal duties, a child will be a great resource. Above all, dear, do remember never to lose the modesty of a young girl towards others (without being prude); though you are married don't become a matron at once to whom everything can be said, and who minds saying nothing herself—I remained particular to a degree (indeed feel so now) and often feel shocked at the confidences of other married ladies. I fear abroad they are very indelicate about these things.

22 June 1858

Do you know that you've got into a habit of writing so many words with a capital letter at the beginning? With nouns that would not signify so much, but you do it with verbs and adjectives, which is very incorrect, dear, and would shock other people if you wrote to them so. In your last to me for instance you wrote 'I fell Low' and 'women have not much to Live for', both of which are quite wrong. You must try to break yourself of what might get a habit at last and would be wrong in any language.

29 June 1858

But now I must give you a grand scold! You write to wish me joy of the 25th!! and write also to Bertie about my accession day! How could you make such an extraordinary mistake? It was the 20th, Sunday week, I was surprised at your saying nothing then and now understand why. Yesterday was my Coronation Day.

Promise me one thing, dear; don't stoop when you sit and write, it is very bad for you now, and later it will make you ill; remember how straight I always sit, which enables me to write without fatigue at all times. I always was distressed to see you bend so in drawing and writing, and now it is very, very bad for you, dear. I hope Fritz will remind you. It is a mere bad habit. Now pray don't do it.

30 June 1858

I delight in the idea of being a grandmama; to be that at 39 (D.V.) and to look and feel young is great fun, only I wish I could go through it for you, dear, and save you all the annoyance. But that can't he helped. I think of my next birthday being spent with my children and a grandchild. It will be a treat!

21 September 1858

Many thanks, dear, for your letter of the 18th which I received with heaps of letters on getting up. Papa says you write too much—he is sure you make yourself ill by it, and constantly declares (which I own offends me much) that your writing to me at such length is the cause of your often not writing fully to him.

5 October 1858

If you knew how Papa scolds me for (as he says) making you write! And he goes further, he says that I write far too often to you, and that it would be much better if I wrote only once a week! Therefore it is indeed a hard case and I know not what to say! I think however Papa is wrong and you do like to hear from home often. When you do write to Papa again just tell him what you feel and wish on that subject for I assure you—Papa has snubbed me several times very sharply on the subject and when one writes in spite of fatigues and trouble to be told it bores the person to whom you write, it is rather too much!

21 October 1858

The heartache I suffer each year, on leaving Balmoral and coming here [to Windsor] is most distressing. Then besides you know I have no feeling for Windsor—I admire it, I think it a grand, splendid place—but without a particle of anything which causes me to love it—none, I feel no interest in anything as if it were not my own; and that of course lessens all the enjoyment of one's existence.

27 October 1858

Dearest Affie is gone; and it will be 10 months probably before we shall see his dear face which shed sunshine over the whole house, from his amiable, happy, merry temper; again he was much upset at leaving and sobbed bitterly, and I fear the separation from dear Papa will have been equally trying. Still, sad as it is to part from dear Affie, it is nothing to parting with a daughter; she is gone, as your own child, for ever; she belongs to another, and that is so dreadful a feeling for a mother.

I hope Fritz is duly shocked at your sufferings, for those very selfish men would not bear for a minute what we poor slaves have to endure. But don't dread the *dénouement*; there is no need of it; and don't talk to ladies about it, as they will only alarm you, particularly abroad, where so much more fuss is made of a very natural and usual thing . . . I could not tell such a child as Lenchen [Princess Helena] about you; those things are not proper to be told to children, as it initiates them into things which they ought not to know of, till they are older. Affie knows nothing either . . . How you can call Windsor 'dear' I cannot understand. It is prison-like, so large and gloomy—and for me so dull after Balmoral too, it is like jumping from day into night—fine as it is!

17 November 1858

I know that the little being will be a great reward for all your trouble and suffering—but I know you will not forget, dear, your promise not to indulge in 'baby worship', or to neglect your other greater duties in becoming a nurse. You know how manifold your duties are, and as my dear child is a little disorderly in regulating her time, I fear you might lose a great deal of it, if you overdid the passion for the nursery. No lady, and still less a Princess, is fit for her husband or her position, if she does that . . . I can not bear to think Bertie is going to you and I can't—and when I look at the baby things, and feel I shall not be, where every other mother is—and I ought to be and can't—it makes me sick and almost frantic. Why in the world did you manage to choose a time when we could not be with you? In Nov: Dec: or the beginning of January we could have done it so easily . . . Well, it is no use complaining. Let us hope on another similar occasion to be more fortunate.

Poor Bertie! He vexes us much. There is not a particle of reflection, or even attention to anything but dress! Not the slightest desire to learn, on the contrary, il se bouche les oreilles, the moment anything of interest is being talked of! I only hope he will meet with some severe lesson to shame him out of his ignorance and dullness. Colonel Bruce [the Prince of Wales's Governor] is most anxious you should speak very openly to him about Bertie, and I told him I was sure you would. He is a very superior and a very charming person. Poor Mr. Gibbs [the Prince's tutor] certainly failed during the last 2 years entirely, incredibly—and did Bertie no good.

24 November 1858

As for Leopold [who was suffering from haemophilia] he still bruises as much as ever, but has *(unberufen)* not had accidents of late. He is tall, but holds himself worse than ever, and is a very common looking child, very plain in face, clever but an oddity—and not an engaging child though amusing. I hope the new governess will be able to make him more like other children. He has not the least forgotten you. Arthur is a precious love. Really the best child I ever saw. Louise very naughty and backward, though improved and very pretty, and affectionate.

27 November 1858

We are glad to hear so good an account of poor Bertie; I have no doubt his visit to you—and the mild but firm influence of Colonel Bruce will do him much good. But we always found that he appeared for the first week—much improved, then (as is always the case with him in everything) he gradually went down hill; not paying attention to what is said or read or what he sees is the real misfortune. His natural turn and taste is very trifling, and I think him a very dull companion. But he has been quite altered, for the last few months (in short since he lived at the White Lodge) as to manner, and he is no longer difficile à vivre. Handsome I cannot think him, with that painfully small and narrow head, those immense features and total want of chin.

4 December 1858

I am so glad that Bertie is amiable and companionable towards you, and occupies himself, as I am sure it will do him good. I own I think him very dull; his three other brothers are all so amusing and communicative.

8 December 1858

Alfred is ... quite well, but his letters of which he has given us only three specimens are too shockingly and disgracefully written. Strange that both the boys should write so ill—and that all the girls (at least you three) so well. But Affie's is very much worse than Bertie's.

22 December 1858

Dear Papa is still not quite well—he went yesterday evening with Bertie (who understood not a word of it) to see the Westminster boys act one of their (very improper) Latin plays.

Bertie talks constantly of Berlin and all he has seen—but particularly of the people, parties, theatres what people said etc. Of the fine works of art etc., he says nothing—unless asked.

29 January 1859

My precious darling, you suffered much more than I ever did—and how I wish I could have lightened them for you!* Poor dear Fritz—how he will have suffered for you! I think and feel much for him; the dear little boy if I could but see him for one minute, give you one kiss. . . . You will and must feel so thankful all is over! But don't be alarmed for the future, it never can be so bad again!

9 February 1859

Don't you feel such a weight off your mind, such a sense of returning freedom and thankfulness? I always felt that intense happiness on first waking, so different to the mornings of anxious expectation, of dread and anxiety. It is not a pleasant affair God knows, for any one, but you, my own darling, have had the very worst beginning possible from suffering so much! How I do wish I could go to you now and read to you . . . How I do long to see my little grandson! I own it seems very funny to me to be a grandmama, and so many people tell me they can't believe it!

To King Leopold   15 February 1859

Bertie's† interview with the Pope went off extremely well. He was extremely kind and gracious, and Colonel Bruce was present; it would never have done to have let Bertie go alone, as they might hereafter have pretended, God knows! what Bertie had said.

1 March 1859

It is rumoured that you are going to Berlin to the Christening, but I doubt it! Oh! dearest Uncle, it almost breaks my heart not to witness our first grandchild christened! I don't think I ever felt so bitterly disappointed about anything as about this! And then it is an occasion so gratifying to both Nations, which brings them so much together, that it is most peculiarly mortifying! It is a stupid law in Prussia, I must say, to be so particular about having the child christened so soon. However, it is now no use lamenting; please God! we shall be more fortunate another time!

To Princess Frederick William   2 March 1859

I have not read Barchester Towers all through, but I am told it is not meant to be so ill-natured. But I don't like reading it aloud to Papa as there was not enough romance in it. The people I could not interest myself in.

---

*After a difficult and dangerous labour the Princess Frederick William's first child, the future Kaiser Wilhelm II, was born on 27 January.
†The Prince of Wales had been sent to Rome with his governor, Colonel Bruce, to encourage his appreciation of art and to acquire 'knowledge and information'. He was taken to an audience with Pope Pius IX within a week of his arrival.

16 March 1859

In so many ways and things your case and mine are so different; and though I hated the thought of having children and have no adoration for very little babies, (particularly not in their baths till they are past 3 or 4 months, when they really become very lovely) still I know what a fuss and piece of work was made with you; far too much I think, for it was not good to dress you as often as you were, and to have you up so late. I used to have you in my dressing room—while I dressed for dinner ... dancing on Mrs. Pegley's [the royal nurse] knees—till you got so lively that you did not sleep at night. All that was very foolish, and I warn you against it—but one is very foolish with one's first child ... You, who dote on this little child, will understand now what a pang it is to the poor mother's heart to give that child entirely up to another, whose property she becomes—as I have done you.

19 March 1859

Air, air is what you want, and bracing and not hot stuffy rooms and theatres, or you will become sickly and old before you are 20!

20 March 1859

Now dear, you should positively get someone to be answerable that the rooms and still more the passages, (which ought to be cooler than the rooms), are never above a certain and given temperature, having (as we have everywhere) thermometers hanging up in them and the rooms, and by keeping to that—the stoves ought to be kept up or let out, according to whether it is warm out, or freezing. And then the windows should be opened regularly 3 times a day, or oftener, if it is warmer. Now if you would get someone of the servants to attend to this, your nerves would rapidly improve—and there would be no colds and for Fritz too it would be much better ... I rather dread your going to the theatre, intense hot rooms are the thing of all others to be avoided. But I envy your seeing Lohengrin for I delight in the music, at least in many parts of it, and am constantly playing it.

9 April 1859

Bertie continues such an anxiety, I tremble at the thought of only three years and a half being before us—when he will be of age and we can't hold him except by moral power! I try to shut my eyes to that terrible moment! He is improving very decidedly—but oh! it is the improvement of such a poor or still more idle intellect.

Oh! dear, what would happen if I were to die next winter! One shudders to think of it: it is too awful a contemplation. His journal is worse a great deal than Affie's letters. And all from laziness! Still we must hope for improvement in essentials. But the greatest improvement I fear, will never make him fit for his position. His only safety and the country's—is in his implicit reliance in every thing, on dearest Papa, that perfection of human beings!

My greatest of all anxieties is that dearest Papa works too hard, wears himself quite out by all he does. It makes me often miserable. If it were not for Osborne and Balmoral and then again Windsor at Easter—I don't know what we should do, though really London when it is warm, disagrees more with me, even more than with dearest Papa!

16 April 1859

I hear you model and even paint in oils; ᵗhis last I am sorry for; you remember what Papa always told you on the subject. Amateurs never can paint in oils like artists and what can one do with all one's productions? Whereas water colours always are nice and pleasant to keep in books or portfolios. I hope, dear, you will not take to the one and neglect the other!

I am shocked to hear baby leaves off his caps, so soon, I hope however only in the nursery, for they look so frightful to be seen without caps. In the nursery it is wholesome but it is not pretty.

By the by you went to see the 'Merry Wives'; you must have found it very coarse; even I have never had courage to go to see it—having always been told how very coarse it was—for your adored Shakespeare is dreadful in that respect and many things have to be left out in many of his plays.

20 April 1859

With regard to what you say about Shakespeare, I quite agree. You need not be afraid of seeing Faust; I am as bad and shy as anyone, matron as I am, about these things—and it is so beautiful that really one does not feel put out by it. I advise you to see it, dear. Also as regards the French plays—you should go; there are many—indeed quantities of charming little plays—and dear Papa—who you know is any thing but favourable to the French—used to delight in going to the French play—more than to any other, and we used for many years—when we had a good company (we have had none since 54) to go continually and enjoyed it excessively. It is such good practice for the language. So, I hope, dear, you will go. One's dislike to a nation need not prevent one's admiring and being amused by what is good, clever and amusing in it . . .

I am glad you bear out what I said about our dear correspondence. It is an immense pleasure and comfort to me, for it is dreadful to live so far off and always separated. I really think I shall never let your sisters marry—certainly not to be so constantly away and see so little of their parents—as till now, you have done, contrary to all that I was originally promised and told . . . That last night when we took you to your room, and you cried so much, I said to Papa as we came back 'after all, it is like taking a poor lamb to be sacrificed'. You now know—what I meant, dear. I know that God has willed it so and that these are the trials which we poor women must go through; no father, no man can feel this! Papa never would enter into it all! As in fact he seldom can in my very violent feelings. It really makes me shudder when I look around at all your sweet, happy, unconscious sisters—and think that I must give them up too—one by one!! Our dear Alice, has seen and heard more (of course not what no one ever can know before they marry and before they have had children) than you did, from your marriage—and quite enough to give her a horror rather of marrying.

To King Leopold   26 April 1859

It is a melancholy, sad Easter; but what grieves me the most (indeed, distracts me)—for I have had nothing but disappointments in that quarter since November—is that in all probability Vicky will be unable to come in May! It quite distracts me.

To Princess Frederick William   27 April 1859

I don't know what makes you speak of the English not being as good to their wives as the Germans, when England is the country of family life and good *ménages!* What makes you say that dear? In the higher classes amongst the fashionable, slang, disreputable young people—there are certainly some selfish, careless husbands, but they are the exception to the rule, and you must retract that assertion . . .

I must ask you a 1000 pardons. I have just read over your dear letter of today again (I generally read them 3 or 4 times) and find that I misread what you said about the German and English *ménages!* And that it was just the contrary. You must think Mama as bad in this respect, as poor dear Grandmama, which really is not the case, for I am very particular about my letters but I devour yours and with such eagerness—that I always read them slowly again the second time, and had not yet had time to do so.

I think as a rule—you are right not to go generally to the theatre on a Sunday, and that it even is better for you that you should not do so;—still to do so sometimes is equally right, if Fritz wishes it. You know I am not at all an admirer or approver of our very dull Sundays, for I think the absence of innocent amusement for the poor people, a misfortune and an encouragement of vice.

30 April 1859

On Monday poor dear Papa (who is very much fagged and has had toothache into the bargain) goes at 6 in the morning to Plymouth to open that great large bridge at Saltash over the Tamar and returns the same night at one!

2 May 1859

Your letter received today of the 30th ought to be numbered 84 instead of 78!! So that the next would be 85—or—if you have written since—the one you write after this, would be 86. If you numbered them down in your remembrancer as I do, and looked before you wrote, you would not make mistakes.

Abstractedly, I have no *tendre* for them [babies] till they have become a little human; an ugly baby is a very nasty object—and the prettiest is frightful when undressed—till about four months; in short as long as they have their big body and little limbs and that terrible frog-like action. But from four months, they become prettier and prettier. And I repeat it—your child would delight me at any age . . .

Now goodbye and God bless you, my dearest, and as Papa says (the policeman says it to Hawkesley in 'Still Waters Run Deep') 'Keep up your pecker; that's right' meaning keep up your spirits and don't be downhearted.

To KING LEOPOLD   3 May 1859

I venture to send you a letter I received some days ago from dear Vicky, and the religious tone of which I think will please you. May I beg you to return it me, as her letters are very valuable to me?

JOURNAL   21 May 1859

Such happiness to be at last together again* ... Vicky only began to cry when she talked of her poor little boy's left arm being so weak, which it has been since his birth, having been injured in being brought into the world.

To KING LEOPOLD   25 May 1859

Albert, who writes to you, will tell you how dreadfully our great, great happiness to have dearest Vicky, flourishing and so well and gay with us, was on Monday and a good deal too yesterday, clouded over and spoilt by the dreadful anxiety we were in about dearest Mamma. Thank God! to-day I feel another being—for we know she is 'in a satisfactory state,' and improving in every respect, but I am thoroughly shaken and upset by this awful shock; for it came on so suddenly—that it came like a thunderbolt upon us, and I think I never suffered as I did those four dreadful hours till we heard she was better! I hardly myself knew how I loved her, or how my whole existence seems bound up with her—till I saw looming in the distance the fearful possibility of what I will not mention.

To PRINCESS FREDERICK WILLIAM   15 June 1859

Now I must scold you a wee bit for an observation which really seems at variance with your own expressions. You say 'how glad' Ada [Princess of Schleswig-Holstein] 'must be' at being again in that most charming situation, which you yourself very frequently told me last year was so wretched. How can anyone, who has not been married above two years and three quarters, (like Ada) rejoice at being a third time in that condition? I positively think those ladies who are always *enceinte* quite disgusting; it is more like a rabbit or guinea-pig than anything else and really it is not very nice.

                                                       29 June 1859

Bertie is improved—I see it more now than I did at first—but still he does nothing, and they mean to make him work very hard at Holyrood—where he will go very soon. I think he will stop with us—a little more than a fortnight from the time he arrived. He is a little grown ... but his nose and mouth are much grown also; the nose is becoming the true Coburg nose and begins to hang a little, but there remains unfortunately the want of chin which with that large nose and very large lips is not so well in profile.

                                                       2 September 1859

Bertie has been doing better at Edinburgh [where he was continuing his education for three months] than he ever did before; he has worked hard and shown a desire to

*Princess Frederick William had come to visit her mother at Osborne.

learn, instead of resistance. He is also grown and spread; but not improved in looks; the mouth is becoming so very large and he will cut his hair away behind and divide it nearly in the middle in front, so that it makes him appear to have no head and all face. It is a frightful *coiffure*.

26 October 1859

I will not be angry with you for your feeling for Windsor in consequence of your short honeymoon; still I cannot share it, though my happiness began here too. Early impressions—the unpleasant and unhealthy climate, the restriction of the walks—the Court life and the impossibility of doing what one wishes here—without Court officials etc. all this makes it to me an undesirable and unenjoyable residence.

Dear Papa was a little indisposed with his old enemy, but it was not a very bad attack without sickness or shivering. Today he has gone to Oxford to see how Bertie is going on in that old monkish place, which I have a horror of.

29 February 1860

Darling Affie is back again. If possible the joy is greater than usual this time, as I was very anxious about him all Monday when it blew a perfect hurricane and knowing the ship was not in a good state, it kept me in a fever till in the evening the happy news of his safe arrival came to relieve our anxiety and gladden our hearts. Papa will have told you of his brilliant examination etc. etc. He looks well, though rather tired from his broken nights. Dear child, I feel so proud of the hardship he has endured—the way he has worked and when I think of ——! The very best there is wretched mediocrity. The joy of having Affie in the house is so great and alas! with —— it is such a contrary feeling! I dare not look forward! There is a dark cloud there—in spite of much good!

7 March 1860

Dear Affie is our great delight so full of fun and conversation and so full of anxiety to learn—always at something, never an instant idle—such steam power, such energy it is such a great pleasure to see this—but the contrast with someone else is sad.

31 March 1860

Bertie came yesterday evening and has passed his examination very well [he was now at Oxford]—the Dean of Christ Church finding a decided improvement since last December. He is likewise grown, but not handsomer I think. Affie is, I really think beautiful (excepting Papa who is much more so)—but it is such a darling, handsome, round face. Bless him, he is such a dear, dear boy, and I must say we have not had a single fault to find with him since he has been here. How can you ask, dear, which day he is to be confirmed? You know it can only be on Maundy Thursday—(5th of April) if it is at Easter; you have all been confirmed on that day, and I can't think why you should be uncertain about it.

7 April 1860

Bertie is delighted to see you [on a visit to Germany] which I am very jealous of; he is not at all in good looks; his nose and mouth are too enormous and as he pastes his hair down to his head, and wears his clothes frightfully—he really is anything but good looking. That *coiffure* is really too hideous with his small head and enormous features. He is grown however.

18 April 1860

You don't once enter into any of my observations upon Bertie? It is such a proof of my confidence in you when I speak to you so openly about your brothers—that your silence seems strange to me. Poor Bertie, I pity him; but I blame him too, for that idleness is really sinful.

TO KING LEOPOLD   25 April 1860

Bertie returned last night delighted with his tour [of Coburg and Gotha which he had not previously visited], and with our beloved old Coburg, in spite of snow. I will tell him to give you an account of it. He made a very favourable impression there. He gives a good account of dear Stockmar too.

TO PRINCESS FREDERICK WILLIAM   6 June 1860

Louis and Heinrich of Hesse dined with us on Sunday and again here yesterday, and stop till Friday. I like them extremely, so nice, natural, sensible; quiet and so unblasé—or foppish, and taking interest in everything. I think them the nicest young men I have seen for very long. You will imagine my agitation, not to do too much and yet not to neglect anything. Louis gets on extremely well with Alice, who is wonderfully composed and quite à son aise. I am quite proud of her. The difference between these young men and the Prince of Orange is very striking.

Alice is as amiable and quiet and cheerful as possible. The man who marries her will indeed have an enviable lot too, for she is so gentle and so very unselfish.

11 July 1860

The arrangements [for her second confinement], you mention are indeed too horrid—and quite like an execution. Oh! if those selfish men—who are the cause of all one's misery, only knew what their poor slaves go through! What suffering—what humiliation to the delicate feelings of a poor woman, above all a young one—especially with those nasty doctors.

JOURNAL   27 July 1860

Our darling grandchild was brought in [while the Queen and the Prince Consort were staying at Coburg]. Such a little love! He came walking in at Mrs Hobbs' [his nurse's] hand in a little white dress with black bows and was so good. He is a fine fat child, with a beautiful soft white skin . . . He has Fritz's eyes and Vicky's mouth and very fair curly hair. We felt so happy to see him at last!

To Princess Frederick William   1 August 1860

Many, many thanks for that charming, tasteful locket and the dear hair. The dear little nameless lady seems to have a great quantity of it! How I long to see her! I assure you I am not at all offended at hearing her called like me, for though I am no admirer of babies generally—there are exceptions (besides all of you were always thought like me when born)—for instance Alice, and Beatrice were very pretty from the very first—yourself also—rather so—Arthur too—though not so much so as the 2 first named. Bertie and Leopold—too frightful. Little girls are always prettier and nicer (Arthur alone making an exception).

20 October 1860

Beatrice is my darling, but she is fast, alas! growing out of the baby—is becoming long-legged and thin. She is however still most amusing and very dear.

17 November 1860

Bertie is then at last arrived [from America]—well—grown, and decidedly improved; he tells us a great deal of what he has seen. He looks a little yellow and sallow—and his hair so fair near Affie. Affie is very dark and very handsome I must say.

I have finished *The Mill on the Floss* and I must say it made a deep impression upon me. The writing and description of feelings is wonderful and painful!

28 November 1860

Our dear good Louis [of Hesse-Darmstadt] gets on extremely well, though I see that he is nervous and agitated; but he takes great pains to speak English to the people who are presented to him. Nothing has passed yet between any of us— but every day seems to bring him and Alice nearer. It is a little trying for them to be looked at by everybody—but it is well and right that everybody should see his marked attentions to her, and her bright and happy face when speaking to him. Alice behaves admirably; perfectly quiet and behaving just as usual and satisfied with everything that is done

I saw how very much agitated Louis was at dinner, and after it—while I was talking to some of the gentlemen and Alice happened to be standing alone at the chimney piece with Louis—he seized the opportunity (which dear Papa in his very quiet way thought he might wait for till today or tomorrow—as if people violently in love could wait for a stated time) and when I passed to go to the other room Alice and Louis whispered it to me. We had to sit quiet and crochet, till the evening was over and then Alice came to our room, much agitated and we told Papa.

8 December 1860

You are quite right in saying 'he [Prince Louis] has fascinated Alice'—but he has fascinated me too, and quite entwined himself round my heart. They are not at all sentimental but like two very happy children—adoring one another and full of fun and play. He is so intelligent and clever. Papa has begun talking a little upon German politics with him—which will I am sure be very useful—but from Papa's having last week had a great deal to do—and these last days been unwell and unable to go out, he has been much more with me than dearest Papa.

To KING LEOPOLD   11 December 1860

Your approval of this marriage of our dear Alice, which, I cannot deny, has been for long an ardent wish of mine, and just therefore I feared so much it never would come to pass, gives us the greatest pleasure. Now—that all has been so happily settled, and that I find the young man so very charming—my joy, and my deep gratitude to God are very great! He is so loveable, so very young, and like one of our own children—not the least in the way—but a dear, pleasant, bright companion, full of fun and spirits, and I am sure will be a great comfort to us, besides being an excellent husband to our dear, good Alice, who, though radiant with joy and much in love (which well she may be), is as quiet and sensible as possible.

The Empress [of the French who was making a tour of England and Scotland] is still here, and enjoys her liberty of all things. We went to town for the Smithfield Cattle Show yesterday, and visited her at Claridge's Hotel. She very civilly wanted us to avoid the trouble, but we felt that it would not be civil if we did not, and that hereafter even the French might say that she had not been treated with due respect. She looked very pretty, and was in very good spirits, but again carefully avoided any allusion to her husband and to politics, though she talked a great deal about all she was seeing!

To PRINCESS FREDERICK WILLIAM   13 February 1861

Poor dear Papa has been suffering badly with toothache since three days—which wears and worries him dreadfully, and seems particularly obstinate. I hope, however, it is a little better today, but dear Papa never allows he is any better or will try to get over it, but makes such a miserable face that people always think he's very ill. It is quite the contrary with me always; I can do anything before others and never show it, so people never believe I am ill or ever suffer. His nervous system is easily excited and irritated, and he's so completely overpowered by everything.

25 February 1861

As regards Princess Alexandra of D[enmark], you could surely for yourself get Wally [Walburga, Countess Hohenthal] to find out everything about her education and general character; whether she is clever, quiet, not frivolous or vain, fond of occupation etc. The looks and manners we know are excellent, and whether she seems very *outrée* Danish.

The subject is so important—the choice so circumscribed, that I am sure you will kindly set about at once finding out all these things. It is so very important—with the peculiar character we have to deal with. The Princess of Meiningen, he did not like, and she is not strong; Marie of the Netherlands is clever and ladylike, but too plain and not strong, and poor Addy [Princess Alexandrine, niece of the King of Prussia] not clever or pretty.

JOURNAL   15 March 1861

Oh, what agony what despair was this [to witness her mother's dying moments]. I knelt before her, kissed her dear hand and placed it next my cheek; but though she opened her eyes, she did not, I think, know me. She brushed my hand off, and the dreadful reality was before me, that for the first time she did not know the child she had ever received with such tender smiles! I went out to sob ... I asked the doctors if there was no hope. They said, they feared, none whatever ... As the night wore on into the morning I lay down on the sofa, at the foot of my bed ... I heard each hour strike ... At four I went down again. All still—nothing to be heard but the heavy breathing, and the striking, at every quarter, of the old repeater, a large watch in a tortoiseshell case, which had belonged to my poor father, the sound of which brought back all the recollections of my childhood ... Feeling faint and exhausted, I went upstairs again and lay down in silent misery ... The dreaded terrible calamity has befallen us, which seems like an awful dream ... Oh God! how awful! how mysterious! ... The constant crying was a comfort and relief ... but oh! the agony of it!

TO KING LEOPOLD   16 March 1861

On this, the most dreadful day of my life, does your poor broken-hearted child write one line of love and devotion. She is gone! That precious, dearly beloved tender Mother—whom I never was parted from but for a few months—without whom I can't imagine life—has been taken from us! It is too dreadful! But she is at peace—at rest—her fearful sufferings at an end! It was quite painless—though there was very distressing, heartrending breathing to witness. I held her dear, dear hand in mine to the very last, which I am truly thankful for! But the watching that precious life going out was fearful! ... Dearest Albert is dreadfully overcome—and well he may, for she adored him!

                                                          26 March 1861

On Sunday I took leave of those dearly beloved remains—a dreadful moment; I had never been near a coffin before, but dreadful and heartrending as it was, it was so beautifully arranged that it would have pleased her, and most probably she looked down and blessed us—as we poor sorrowing mortals knelt around, overwhelmed with grief! It was covered with wreaths, and the carpet strewed with sweet, white flowers. I and our daughters did not go yesterday—it would have been far too much for me—and Albert when he returned, with tearful eyes told me it was well I did not go—so affecting had been the sight—so universal the sympathy.

But oh! dearest Uncle—the loss—the truth of it— which I cannot, do not realise even when I go (as I do daily) to Frogmore—the blank becomes daily worse!

I try to be, and very often am, quite resigned—but dearest Uncle, this is a life sorrow. On all festive or mournful occasions, on all family events, her love and sympathy will be so fearfully wanting. Then again, except Albert (who I very often don't see but very little in the day), I have no human being except our children ... and besides, a woman requires woman's society and sympathy sometimes, as men do men's. All this, beloved Uncle, will show you that, without dwelling constantly upon it, or moping or becoming morbid, though the blank and the loss to me, in my isolated position especially, is such a dreadful, and such an irreparable one, the worst trials are yet to come. My poor birthday, I can hardly think of it!

30 March 1861

I think you may like to hear from your poor motherless child. It is to-day a fortnight already, and it seems but yesterday—all is before me, and at the same time all, all seems quite impossible ... Weeping, which day after day is my welcome friend, is my greatest relief ... To open her drawers and presses, and to look at all her dear jewels and trinkets in order to identify everything, is like a sacrilege, and I feel as if my heart was being torn asunder!

JOURNAL   9 April 1861

It is dreadful, dreadful to think we shall never see that dear kind loving face again, never hear that dear voice again! ... The talking of any ordinary things is quite unbearable to me ... The outbursts of grief are fearful and at times unbearable ... One of my great comforts is to go to Frogmore, to sit in her dear room ... dread as it is to feel the awful stillness of the house ... I had never been near a coffin before ... The dreadful thing as I told Albert yesterday is the certainty that the loss is irrevocable.*

TO PRINCESS FREDERICK WILLIAM   10 April 1861

As regards Bertie—I quite agree with you, dear child—that he must be a little more tender and affectionate in his manner—if he is to expect it from me—and take a little more interest in what interests us if he is to be at all pleasant in the house. And now, dearest child, I must say, without I hope making you angry—that you did not quite set about making matters better, for you kept telling me all his most stupid and silly remarks (said as he too often does—without thinking—partly to tease you and partly to give vent to his temper) and enraged me, low and wretched as I was—greatly. If one wishes to pour oil and not to 'keep the kettle boiling' one must not repeat everything another who irritates has said—else it of course makes matters much worse. He left on Monday. His voice made me so nervous I could hardly bear it. Altogether I never felt in such a state of nerves for noise or sound.

12 June 1861

You are, I know—perhaps a little inclined to be carried away if you are pleased with a person—like you were with the Empress E [Eugénie], but Fritz is not and as he so entirely coincides with what you say about Princess Alix [Alexandra of Denmark]

---

*The Queen's grief was exacerbated by remorse for what she now took to be her unfeeling attitude in the past towards her mother to whom she had recently become much closer. The Prince could not comfort her. When he suggested that she might feel less miserable if she went back to London, she turned on him in fury: how could she be expected to leave at such a time? How cruel were even the best of men when compared with so kind and tender a woman as her mother! It was not until six months later that the Prince could propose to her without fear of violent contradiction that a change of scene and outside interests might help her to control her feelings.

(why is she called so?) I feel quite sure she really must be charming in every sense of the word—and really a pearl not to be lost. The thought of having in Bertie's wife so charming a daughter—would be a great comfort for me and there is nothing I should not do to be quite a real mother to her; I shall else be dreadfully alone—when your sisters marry—one after the other—though I hope and think some at least, will be much more with me than you ever can be. But, dearest, we only look at one side of the question—have you at all thought if she will take him? I am sure I think it is not so certain. Alice (who is the only one consulted) declared she would strongly object to be selected without knowing for whom.

19 June 1861

Dear Papa and I are both so grateful to you about all the trouble you have taken about Princess Alix. May he only be worthy of such a jewel! There is the rub! When I look at Louis and —— at the charming—sweet, bright, lively expression of the one—and at the sallow—dull, blasé—and heavy look of the other I own I feel very sad.

To King Leopold   13 August 1861

On the 17th we shall visit that dear grave! Last year she was still so well, and so full of life; but it was a very sad birthday, two days after the loss of that dear beloved sister, whom she has joined so soon! . . . Beloved Mamma, how hourly she is in my mind!

20 August 1861

We parted from our dear children and grandchildren with heavy hearts at seven on the morning of the 16th, for their visit, excepting the blank which clouds over everything, has been most peaceful and satisfactory, and we have learnt to know and most highly appreciate the great excellence of dear Fritz's character; noble, high-principled, so anxious to do what is right, and to improve in every way, and so sweet-tempered and affectionate—so, beyond everything, devoted to Vicky.

We went that afternoon (16th) to Frogmore, where we slept. The first evening was terribly trying, and I must say quite overpowered me for a short time; all looked like life, and yet she was not there! But I got calmer; the very fact of being surrounded by all she liked, and of seeing the dear pretty house inhabited again, was a satisfaction, and the next morning was beautiful, and we went after breakfast with wreaths up to the Mausoleum, and into the vault which is *à plain-pied*, and so pretty—so airy—so grand and simple, that, affecting as it is, there was no anguish or bitterness of grief, but calm repose!

On Saturday we all went over to the camp, where there was a field-day [at the Curragh military camp near Dublin where the Prince of Wales was spending ten weeks attached to the Grenadier Guards]. It is a fine *emplacement* with beautiful turf. We had two cooling showers. Bertie marched past with his company, and did not look at all so very small.

Yesterday was again a very bad day. I have felt weak and very nervous, and so low at times; I think so much of dearest mamma, and miss her love and interest and solicitude dreadfully; I feel as if we were no longer cared for, and miss writing to her and telling her everything, dreadfully. At the Review they played one of her marches, which entirely upset me.

I hardly know how to write, for my head reels and swims, and my heart is very sore! What an awful misfortune this is [the death of King Pedro of Portugal on 11 November, following that of his brother Ferdinand, on the 6th]! How the hand of death seems bent on pursuing that poor, dear family! once so prosperous. Poor Ferdinand so proud of his children—of his five sons—now the eldest and most distinguished, the head of the family, gone, and also another of fifteen, and the youngest still ill! . . . Dear Pedro was so good, so clever, so distinguished! He was so attached to my beloved Albert.

To Princess Frederick William   1 October 1861

We can never sufficiently thank you and dear Fritz for all your love, affection and kindness in this important matter. Bertie is certainly much pleased with her—but as for being in love I don't think he can be, or that he is capable of enthusiasm about anything in the world. But he is shy and I dare say we shall hear more from Alice, to whom he is sure to open his heart. Poor boy—he does mean well—but he is so different to darling Affie!

You say no one is perfect but Papa. But he has his faults too. He is very often very trying—in his hastiness and over-love of business—and I think you would find it very trying if Fritz was as hasty and harsh (momentarily and unintentionally as it is) as he is!

6

MONARCHS AND MINISTERS
1852–1860

By the end of 1852 Lord Derby's Government was tottering. The Queen considered ways in which it might be strengthened, including the replacement of Disraeli, under whom the Peelites refused to serve, as Leader of the House of Commons. She discussed the problem with Derby who suggested Palmerston as Leader. The Queen quickly offered another name, that of Gladstone. But Derby objected, so the Queen noted in her journal on 27 November: 'Mr. G. was in his opinion quite unfit for it. He possessed none of that decision, boldness, readiness and clearness so necessary for leading a party.' Unstrengthened, Derby's Government was defeated in the House on 17 December. The Queen sent for Lord Aberdeen to try to form a Peelite ministry.

To King Leopold   24 December 1852

The success of our excellent Aberdeen's arduous task and the formation of so brilliant and strong a Cabinet [with Lord John Russell as Foreign Secretary and Gladstone as Chancellor of the Exchequer] would, I was sure, please you. It is the realisation of the country's and our most ardent wishes, and it deserves success, and will, I think, command great support.

Our Government is very satisfactorily settled. To have my faithful friend Aberdeen as Prime Minister is a great happiness and comfort for me personally. Lord Palmerston [Home Secretary] is terribly altered, and all his friends think him breaking. He walks with two sticks, and seemed in great suffering at the Council, I thought.

Ill as he seemed, Palmerston's career was far from over. And when war broke out between Russia and Turkey, he displayed all his old vigour as he energetically advocated England's going to the help of her Turkish allies in conjunction with France. The Queen at first hoped that war could be avoided. On 11 October 1853 she wrote to Lord Clarendon, Lord John Russell's successor as Foreign Secretary:

The Queen has received Lord Clarendon's letter. She had written to Lord Aberdeen that she felt it her duty to pause before giving her consent to the measures decided on in the Cabinet, until she should have received an explanation on the views which dictated that decision ... She has now received and read the Despatches, which have in the meantime been sent off to their points of destination without having received her sanction!

The instructions to Lord Stratford,* appear to her very vague, and entrusting him with enormous powers and a latitude of discretion which is hardly to be called safe. As matters have now been arranged, it appears to the Queen, moreover, that we have taken on ourselves in conjunction with France all the risks of a European war, without having bound Turkey to any conditions with respect to provoking it. The hundred and twenty fanatical Turks constituting the Divan at Constantinople are left sole judges of the line of policy to be pursued, and made cognisant at the same time of the fact that England and France have bound themselves to defend the Turkish Territory! This is entrusting them with a power which Parliament has been jealous to confide even to the hands of the British Crown. It may be a question whether England ought to go to war for the defence of so-called Turkish Independence; but there can be none that if she does so, she ought to be the sole judge of what constitutes a breach of that independence, and have the fullest power to prevent by negotiation the breaking out of the war.

To Aberdeen   5 November 1853

Although the Queen will have the pleasure of seeing Lord Aberdeen this evening, she wishes to make some observations on the subject of Lord Stratford's last private letters communicated to her yesterday by Lord Clarendon. They exhibit clearly on his part a desire for war, and to drag us into it. When he speaks of the sword which will not only have to be drawn, but the scabbard thrown away, and says, the war to be successful must be a 'very comprehensive one' on the part of England and France, the intention is unmistakable, and it becomes a serious question whether we are justified in allowing Lord Stratford any longer to remain in a situation which gives him the means of frustrating all our efforts for peace.

To Princess Augusta of Prussia   13 January 1854

For the last three weeks there have been vile attacks in the newspapers against my dear husband, who is accused of intriguing in the interests of Russia! They are quite mad, and although such nonsense gains no credit among sensible people who respect and love Albert, yet they have provided an occasion for many dreadful remarks, and the whole affair will probably continue until notice is taken of it in Parliament. It probably originated from a few envious and malicious people, and will be eagerly pursued and even believed by the ignorant. It has all arisen through the enormous excitement in the country over the Eastern question. You will easily understand how enraged and indignant I feel about it ...

---

*Lord Stratford de Redcliffe, British Ambassador in Constantinople, had been authorised to employ the British fleet as he thought fit in order to defend Turkish territory from aggression, and had been instructed that if the Russian navy left Sebastopol, British ships were to pass through the Bosphorus.

TO KING LEOPOLD   21 February 1854

War is, I fear, quite inevitable. You will have seen that the Emperor Nicholas has not given a favourable answer to our Brother Napoleon (which I hear has disappointed him extremely, as he expected very great results from it); and the last proposals or attempts made by Buol [Austrian Premier and Minister of Foreign Affairs] it is to be hoped will not be accepted by Russia, for France and England could not accept them; but if Austria and Prussia go with us—as we hope they will—the War will only be a local one. Our beautiful Guards sail to-morrow. Albert inspected them yesterday. George [Duke of Cambridge] is quite delighted to have a division.

TO ABERDEEN   24 February 1854

The Queen must write to Lord Aberdeen on a subject which at this moment appears to her of paramount importance—viz., the augmentation of the Army. The ten thousand men by which it has been ordered to be augmented can hardly be considered to have brought it up to more than an improved peace establishment, such as we have often had during profound peace in Europe; but even these ten thousand men are not yet obtained. We have nearly pledged ourselves to sending twenty-five thousand men to the East, and this pledge will have to be redeemed. To keep even such a force up in the field will require a strong, available reserve at home, of which we shall be quite denuded. But we are going to make war upon Russia!

TO KING LEOPOLD   24 February 1854

The last battalion of the Guards (Scots Fusiliers) embarked to-day. They passed through the courtyard here [Buckingham Palace] at seven o'clock this morning. We stood on the balcony to see them—the morning fine, the sun rising over the towers of old Westminster Abbey—and an immense crowd collected to see these fine men, and cheering them immensely as they with difficulty marched along. They formed line, presented arms, and then cheered us very heartily, and went off cheering. It was a touching and beautiful sight.

TO THE KING OF PRUSSIA   17 March 1854

The dreadful and incalculable consequences of a War weigh upon my heart not less than on your Majesty's. I also know that the Emperor of Russia does not wish for it. He, none the less, demands from the Porte things which all the Powers of Europe—among them, yourself—have solemnly declared to be incompatible with the independence of the Porte, and the European balance of power. In view of this declaration and of the presence of the Russian Army of invasion in the Principalities [Turkey's territories near the Danube], the Powers could not but be ready to confirm their word by action. You suppose that War may already have been declared; you express, however, at the same time, the hope that it may not already have actually broken out. I cannot unfortunately hold out any hope that the sentence will be followed by any stay of execution.

To Aberdeen   1 April 1854

The Queen is rather startled at seeing Lord Aberdeen's answer to Lord Roden upon the subject of a day of humiliation, as he has never mentioned the subject to her, and it is one upon which she feels strongly. The only thing the Queen ever heard about it was from the Duke of Newcastle [the Secretary for War], who suggested the possibility of an appropriate prayer being introduced into the Liturgy, in which the Queen quite agreed; but he was strongly against a day of humiliation, in which the Queen also entirely agreed, as she thinks we have recourse to them far too often, and they thereby lose their effect. The Queen therefore hopes that this will be reconsidered carefully, and a prayer substituted for the day of humiliation. Were the services selected for these days of a different kind to what they are—the Queen would feel less strongly about it; but they always select chapters from the Old Testament and Psalms which are so totally inapplicable that it does away with all the effect it ought to have. Moreover, really to say (as we probably should) that the great sinfulness of the nation has brought about this War, when it is the selfishness and ambition of one man and his servants who have brought this about, while our conduct has been throughout actuated by unselfishness and honesty, would be too manifestly repulsive to the feelings of every one, and would be a mere act of hypocrisy.

To Lord Clarendon, Foreign Secretary   24 September 1854

The French show their usual vivacity in pressing so hard for decision upon what is to be done with Sebastopol when taken. Surely we ought to have taken it first before we can dispose of it, and everything as to the decision about it must depend upon the state in which we receive it, and the opinion of the Military and Naval Commanders after they find themselves in possession of it. The Queen hopes, therefore, that Lord Clarendon will succeed in restraining French impatience as he has often done before.

To King Leopold   13 October 1854

We are, and indeed the whole country is, entirely engrossed with one idea, one anxious thought—the Crimea. We have received all the most interesting and gratifying details of the splendid and decisive victory of the Alma [on 20 September]; alas! it was a bloody one. Our loss was a heavy one—many have fallen and many are wounded, but my noble Troops behaved with a courage and desperation which was beautiful to behold. The Russians expected their position would hold out three weeks; their loss was immense—the whole garrison of Sebastopol was out. Since that, the Army has performed a wonderful march to Balaklava, and the bombardment of Sebastopol has begun. Lord Raglan's [the British Commander-in-Chief] behaviour was worthy of the old Duke's—such coolness in the midst of the hottest fire. We have had all the details from young Burghersh (a remarkably nice young man), one of Lord Raglan's Aides-de-camp whom he sent home with the Despatches, who was in the midst of it all. I feel so proud of my dear noble Troops, who, they say, bear their privations, and the sad disease which still haunts them, with such courage and good humour.

George did enormously well, and was not touched.*

TO PRINCESS AUGUSTA OF PRUSSIA   23 October 1854

You will understand it when I assure you that I regret exceedingly not to be a man and to be able to fight in the war. My heart bleeds for the many fallen, but I consider that there is no finer death for a man than on the battlefield!

TO CLARENDON   9 November 1854

No consideration on earth ought to stand in the way of our sending what ships we can lay hold of to transport French reinforcements to the Crimea, as the safety of our Army and the honour of the Country are at stake. The Queen is ready to give her own yacht for a transport which could carry 1,000 men. Every account received convinces the Queen more and more that numbers alone can ensure success in this instance, and that without them we are running serious risks.

TO KING LEOPOLD   14 November 1854

I am so bewildered and excited, and my mind so entirely taken up by the news from the Crimea, that I really forget, and what is worse, I get so confused about everything that I am a very unfit correspondent. My whole soul and heart are in the Crimea. The conduct of our dear noble Troops is beyond praise; it is quite heroic, and really I feel a pride to have such Troops, which is only equalled by my grief for their sufferings. We now know that there has been a pitched battle on the 6th [the Battle of Inkerman of 5 November], in which we have been victorious over much greater numbers, but with great loss on both sides—the greatest on the Russian. But we know nothing more, and now we must live in a suspense which is indeed dreadful. Then to think of the numbers of families who are living in such anxiety! It is terrible to think of all the wretched wives and mothers who are awaiting the fate of those nearest and dearest to them!

TO LORD RAGLAN   18 November 1854

The Queen has received with pride and joy the telegraphic news of the glorious, but alas! bloody victory of the 5th. These feelings of pride and satisfaction are, however, painfully alloyed by the grievous news of the loss of so many Generals, and in particular Sir George Cathcart—who was so distinguished and excellent an officer ... Both the Prince and Queen are anxious to express to Lord Raglan their

---

*The Duke of Cambridge who was not at his best in battle did not, in fact, distinguish himself as commander of the 1st Division at the Alma. He subsequently applied for sick leave. 'We were horrified,' the Queen recorded in her journal. 'I am sure this will have the very worst effect.' To the Duke she wrote on 30 December 1854: 'I hope you will be back in the Crimea by this time. Forgive my telling you frankly that I hope you will not let your low spirits and desponding feelings be known to others; you cannot think how ill natured people are here, and I can assure you that the Clubs have not been slow in circulating the most shameful lies about you.'

unbounded admiration of the heroic conduct of the Army, and their sincere sympathy in their sufferings and privations so nobly borne.

TO THE DUKE OF NEWCASTLE   30 November 1854

The Queen thinks that no time should be lost in announcing the intention of the Queen to confer a medal on all those who have been engaged in the arduous and brilliant campaign in the Crimea.

The medal should have the word 'Crimea' on it, with an appropriate device (for which it would be well to lose no time in having a design made) and clasps—like to the Peninsular Medal, with the names Alma and Inkerman inscribed on them, according to who had been in one or both battles. Sebastopol, should it fall, or any other name of a battle which Providence may permit our brave troops to gain, can be inscribed on other clasps hereafter to be added. The names Alma and Inkerman should likewise be borne on the colours of all the regiments who have been engaged in these bloody and glorious actions.

The Queen is sure that nothing will gratify and encourage our noble troops more than the knowledge that this is to be done.

TO ABERDEEN   10 January 1855

Before Parliament meets for probably a very stormy Session, the Queen wishes to give a public testimony of her continued confidence in Lord Aberdeen's administration, by offering him the vacant Blue Ribbon [the Order of the Garter]. The Queen need not add a word on her personal feelings of regard and friendship for Lord Aberdeen, which are known to him now for a long period of years.

TO NEWCASTLE   12 January 1855

The Queen returns the enclosed despatch to the Duke of Newcastle, which she has read with much pleasure, as bringing before Lord Raglan in an official manner—which will require official enquiry and answer—the various points so urgently requiring his attention and remedial effort. It is at the same time so delicately worded that it ought not to offend, although it cannot help, from its matter, being painful to Lord Raglan. The Queen has only one remark to make, viz. the entire omission of her name throughout the document. It speaks simply in the name of the People of England, and of their sympathy, whilst the Queen feels it to be one of her highest prerogatives and dearest duties to care for the welfare and success of her Army.

The Duke of Newcastle might with truth have added that, making every allowance for the difficulties before Sebastopol, it is difficult to imagine how the Army could ever be moved in the field, if the impossibility of keeping it alive is felt in a stationary camp only seven miles from its harbour, with the whole British Navy and hundreds of transports at its command.

MEMORANDUM BY THE QUEEN   30 January 1855

Lord Aberdeen arrived here at three. He came from the Cabinet, and tendered their unanimous resignation [following the Government's defeat on a motion calling for a committee of inquiry into their conduct of the war] ... We discussed future possibilities, and agreed that there remained nothing to be done but to offer the Government to Lord Derby, whose Party was numerically the strongest. He supposed Lord Derby would be prepared for it, although he must have great difficulties, unless he took in men from other Parties, about which, however, nothing could be known at present.

MEMORANDUM BY THE QUEEN   31 January 1855

We went up to Buckingham Palace and saw Lord Derby at half-past eleven. The Queen informed him of the resignation of the Government, and of her desire that he should try to form a new one. She addressed herself to him as the head of the largest Party in the House of Commons, and which had by its vote chiefly contributed to the overthrow of the Government. Lord Derby threw off this responsibility ... He owned that his Party was the most compact—mustering about two hundred and eighty men—but he had no men capable of governing the House of Commons, and he should not be able to present an Administration that would be accepted by the country unless it was strengthened by other combinations; he knew that the whole country cried out for Lord Palmerston as the only man fit for carrying on the war with success, and he owned the necessity of having him in the Government, were it even only to satisfy the French Government, the confidence of which was at this moment of the greatest importance; but he must say, speaking without reserve, that whatever the ignorant public might think, Lord Palmerston was totally unfit for the task. He had become very deaf as well as very blind, was seventy-one years old, and ... in fact, though he still kept up his sprightly manners of youth, it was evident that his day had gone by.

This led us to a long discussion upon the merits of the conduct of the war, upon which he seemed to share the general prejudices, but on being told some of the real facts and difficulties of the case, owned that these, from obvious reasons, could not be stated by the Government in their defence, and said that he was aware that the chief fault lay at headquarters in the Crimea. Lord Raglan ought to be recalled, as well as his whole staff, and perhaps he could render this less painful to him by asking him to join the Cabinet, where his military advice would be of great value.

To be able to meet the House of Commons, however, Lord Derby said he required the assistance of men like Mr Gladstone and Mr S. Herbert.

The Queen having laid great stress on a good selection for the office of Foreign Affairs, Lord Derby said he would have to return to Lord Malmesbury, who, he thought, had done well before, and had now additional experience ...

Lord Derby returned a little before two from Lord Palmerston, to whom he had gone in the first instance. Lord Palmerston was ready to accept the Lead of the House of Commons, and acknowledged that the man who undertook this could not manage the War Department besides. He undertook to sound Mr Gladstone and Mr S. Herbert, but had, evidently much to Lord Derby's surprise, said that it must be a coalition, and not only the taking in of one or two persons, which does not seem to suit Lord Derby at all—nor was he pleased at Lord Palmerston's suggestion that he ought to try, by all means, to retain Lord Clarendon at the Foreign Office.

To Russell   2 February 1855

The Queen has just seen Lord Lansdowne [to whom she had vainly turned on Lord Derby's failure to form a Government]. As what he could tell her has not enabled her to see her way out of the difficulties in which the late proceedings in Parliament have placed her, she wishes to see Lord John Russell in order to confer with him on the subject.

2 February 1855

As moments are precious, and the time is rolling on without the various consultations which Lord Lansdowne has had the kindness and patience to hold with the various persons composing the Queen's late Government having led to any positive result, she feels that she ought to entrust some one of them with the distinct commission to attempt the formation of a Government. The Queen addresses herself in this instance to Lord John Russell, as the person who may be considered to have contributed to the vote of the House of Commons, which displaced her late Government, and hopes that he will be able to present her such a Government as will give a fair promise successfully to overcome the great difficulties in which the country is placed.

Journal   3 February 1855

Lord John came at 6, much put out and disturbed at having 'nothing encouraging' to report. He had seen both Sir G. Grey and Lord Clarendon, who had strongly expressed their disinclination to taking office . . . Lord John had previously seen Mr. Gladstone and Mr. S. Herbert, who had both declined.

To Palmerston   4 February 1855

Lord John Russell having just informed the Queen that he was obliged to resign the task which the Queen confided to him, she addresses herself to Lord Palmerston to ask him whether he can undertake to form an Administration which will command the confidence of Parliament and efficiently conduct public affairs in this momentous crisis? Should he think that he is able to do so, the Queen commissions him to undertake the task. She does not send for him, having fully discussed with him yesterday the state of public affairs, and in order to save time. The Queen hopes to receive an answer from Lord Palmerston as soon as possible.

To King Leopold   6 February 1855

Van de Weyer will have informed you of the successive failures of Lord Derby and Lord John . . . and of Lord Palmerston being now charged with the formation of a Government! I had no other alternative . . . I am a good deal worried and knocked up by all that has passed; my nerves, which have suffered very severely this last year, have not been improved by what has passed during this trying fortnight—for it will be a fortnight to-morrow that the beginning of the mischief began . . .

Six o'clock p.m.—One word to say that Lord Palmerston has just kissed hands as Prime Minister.

TO ABERDEEN   7 February 1855

Though the Queen hopes to see Lord Aberdeen at six, she seizes the opportunity to say what she hardly dares to do verbally without fearing to give way to her feelings; she wishes to say what a pang it is for her to separate from so kind and dear and valued a friend as Lord Aberdeen has ever been to her since she has known him. The day he became Prime Minister was a very happy one for her; and throughout his Ministry he has ever been the kindest and wisest adviser—one to whom she could apply on all and trifling occasions even. This she is sure he will still ever be. But the thought of losing him as her First Adviser in her Government is very painful. The Queen is sure that the Prince and herself may ever rely on his valuable support and advice in all times of difficulty, and she now concludes with the expression of her warmest thanks for all his kindness and devotion, as well as of her unalterable friendship and esteem for him and with every wish for his health and happiness.

JOURNAL   22 February 1855

We saw 26 of the sick and wounded of the Coldstreams ... There were some sad cases;—one man who had lost his right arm at Inkermann, was also at the Alma, and looked deadly pale—one or two others had lost their arms, others had been shot in the shoulders and legs,—several, in the hip joint ... A private, Lanesbury, with a patch over his eye, and his face tied up, had had his head traversed by a bullet, penetrating through the eye, which was gone,—through the nose, and coming out at the neck! He looked dreadfully pale, but was recovering well. There were 2 other very touching and distressing cases, 2 poor boys. I cannot say how touched and impressed I have been by the sight of these noble brave, and so sadly wounded men and how anxious I feel to be of use to them, and to try and get some employment for those who are maimed for life. Those who are discharged will receive very small pensions but not sufficient to live upon.

TO KING LEOPOLD   27 February 1855

Altogether, affairs are very unsettled and very unsatisfactory. Lord John's return to office [as Secretary for the Colonies] under Lord Palmerston is very extraordinary!

The [French] Emperor's meditated voyage [to the Crimea where, he had announced, he would personally assume the conduct of the war]—though natural in him to wish—I think most alarming. I own it makes one tremble, for his life is of such immense importance. I still hope that he may be deterred from it, but Walewski was in a great state about it.

TO LORD PANMURE, SECRETARY OF WAR   22 March 1855

The other day, when the Queen spoke to Lord Panmure on the subject of the distribution of the Medal for the Crimean Campaign amongst the Officers, and those who are in this country, no decision was come to as to how this should be done. The Queen has since thought that the value of this Medal would be greatly enhanced if she were personally to deliver it to the officers and a certain number of men (selected for that purpose). The valour displayed by our troops, as well as the sufferings they have endured, have never been surpassed—perhaps hardly equalled; and as the Queen has been a witness of what they have gone through, having visited them in their hospitals, she would like to be able personally to give them the reward they have earned so well, and will value so much. It will likewise have a very beneficial effect, the Queen doubts not, on the recruiting.

To King Leopold   17 April 1855

The impression is very favourable.* There is great fascination in the quiet, frank manner of the Emperor, and she is very pleasing, very graceful, and very unaffected, but very delicate. She is certainly very pretty and very uncommon-looking. The Emperor spoke very amiably of you. The reception by the public was immensely enthusiastic.

                                                              19 April 1855

I have not a moment to myself, being of course entirely occupied with our Imperial guests, with whom I am much pleased, and who behave really with the greatest tact. The Investiture went off very well, and to-day (we came from Windsor) the enthusiasm of the thousands who received him in the City was immense. He is much pleased. Since the time of my Coronation, with the exception of the opening of the great Exhibition, I don't remember anything like it. To-night we go in state to the Opera.

                                                              24 April 1855

Our great visit is past, like a brilliant and most successful dream, but I think the effect on the visitors will be a good and lasting one; they saw in our reception, and in that of the whole Nation, nothing put on, but a warm, hearty welcome to a faithful and steady Ally.

Memorandum   2 May 1855

In reflecting on the character of the present Emperor Napoleon, and the impression I have conceived of it, the following thoughts present themselves to my mind:

That he is a very extraordinary man, with great qualities there can be no doubt—I might almost say a mysterious man. He is evidently possessed of indomitable courage, unflinching firmness of purpose, self-reliance, perseverance, and great secrecy; to this should be added, a great reliance on what he calls his Star, and a belief in omens and incidents as connected with his future destiny, which is almost romantic—and at the same time he is endowed with wonderful self-control, great calmness, even gentleness, and with a power of fascination, the effect of which upon all those who become more intimately acquainted with him is most sensibly felt.

How far he is actuated by a strong moral sense of right and wrong is difficult to say.

My impression is, that in all apparently inexcusable acts, he has invariably been guided by the belief that he is fulfilling a destiny which God has imposed upon him,

---

*The Emperor and Empress of the French had arrived on a visit to England the day before.

and that, though cruel or harsh in themselves, they were necessary to obtain the result which he considered himself as chosen to carry out, and not acts of wanton cruelty or injustice; for it is impossible to know him and not to see that there is much that is truly amiable, kind, and honest in his character. Another remarkable and important feature in his composition is, that everything he says or expresses is the result of deep reflection and of settled purpose, and not merely *des phrases de politesse*, consequently when we read words used in his speech made in the City, we may feel sure that he means what he says; and therefore I would rely with confidence on his behaving honestly and faithfully towards us ... He is very well read in German literature, to which he seems to be very partial. It is said, and I am inclined to think with truth, that he reads but little, even as regards despatches from his own foreign Ministers, he having expressed his surprise at my reading them daily. He seems to be singularly ignorant in matters not connected with the branch of his special studies, and to be ill informed upon them by those who surround him.

If we compare him with poor King Louis Philippe, I should say that the latter (Louis Philippe) was possessed of vast knowledge upon all and every subject, of immense experience in public affairs, and of great activity of mind; whereas the Emperor possesses greater judgment and much greater firmness of purpose, but no experience of public affairs, nor mental application; he is endowed, as was the late King, with much fertility of imagination.

Another great difference between King Louis Philippe and the Emperor is, that the poor King was thoroughly French in character, possessing all the liveliness and talkativeness of that people, whereas the Emperor is as unlike a Frenchman as possible, being much more German than French in character ... How could it be expected that the Emperor should have any experience in public affairs, considering that till six years ago he lived as a poor exile, for some years even in prison, and never having taken the slightest part in the public affairs of any country?

TO KING LEOPOLD   22 May 1855

Ernest will have told you what a beautiful and touching sight and ceremony (the first of the kind ever witnessed in England) the distribution of the Medals was. From the highest Prince of the Blood to the lowest Private, all received the same distinction for the bravest conduct in the severest actions, and the rough hand of the brave and honest private soldier came for the first time in contact with that of their Sovereign and their Queen! Noble fellows! I own I feel as if they were my own children; my heart beats for them as for my nearest and dearest. They were so touched, so pleased; many, I hear, cried—and they won't hear of giving up their Medals, to have their names engraved upon them, for fear they should not receive the identical one put into their hands by me, which is quite touching. Several came by in a sadly mutilated state.

                                                            23 August 1855

I am delighted, enchanted, amused, and interested, and think I never saw anything more beautiful and gay than Paris—or more splendid than all the Palaces. Our reception is most gratifying—for it is enthusiastic and really kind in the highest degree. Our entrance into Paris was a scene which was quite *feenhaft*, and which could hardly be seen anywhere else; was quite overpowering—splendidly decorated—illuminated—immensely crowded—and 60,000 troops out—from the Gare de Strasbourg to St Cloud, of which 20,000 Gardes Nationales, who had come great distances to see me.

The Emperor has done wonders for Paris, and for the Bois de Boulogne. Everything is beautifully *monté* at Court—very quiet, and in excellent order; I must say we are both much struck with the difference between this and the poor King's time, when the noise, confusion, and bustle were great . . . They have asked to call a new street, which we opened, after me!

The heat is very great, but the weather splendid, and though the sun may be hotter, the air is certainly lighter than ours—and I have no headache.

The Zouaves are on guard here [St. Cloud], and you can't see finer men . . .

The children are so fond of the Emperor, who is so very kind to them. He is very fascinating, with that great quiet and gentleness. He has certainly excellent manners, and both he and the dear and very charming Empress (whom Albert likes particularly) do the *honneurs* extremely well and very gracefully, and are full of every kind attention.

29 August 1855

Here we are again [at Osborne], after the pleasantest and most interesting and triumphant ten days that I think I ever passed. So complete a success, so very hearty and kind a reception with and from so *difficile* a people as the French is indeed most gratifying and most promising for the future. The Army were most friendly and amicable towards us also.

In short, the complete Union of the two countries is stamped and sealed in the most satisfactory and solid manner, for it is not only a Union of the two Governments—the two Sovereigns—it is that of the two Nations!

I have formed a great affection for the Emperor, and I believe it is very reciprocal, for he showed us a confidence which we must feel as very gratifying, and spoke to us on all subjects, even the most delicate. I find no great personal rancour towards the Orleans. . . . Nothing could exceed his tact and kindness.

To BARON STOCKMAR   1 September 1855

I know no one who puts me more at my ease, or to whom I felt more inclined to talk unreservedly, or in whom involuntarily I should be more inclined to confide, than the Emperor! He was entirely at his ease with us—spoke most openly and frankly with us on all subjects.

He is so simple, so *naïf*, never making *des phrases*, or paying compliments—so full of tact, good taste, high breeding; his attentions and respect towards us were so simple and unaffected, his kindness and friendship for the Prince so natural and so gratifying, because it is not forced, not *pour faire des compliments*. He is quite The Emperor, and yet in no way playing it . . . Wonderful it is that this man—whom certainly we were not over well-disposed to—should by force of circumstances be drawn into such close connection with us, and become personally our friend, and this entirely by his own personal qualities, in spite of so much that was and could be said against him!

TO KING LEOPOLD   11 September 1855

The great event has at length taken place—Sebastopol has fallen! We received the news here [at Balmoral] last night when we were sitting quietly round our table after dinner. We did what we could to celebrate it; but that was but little, for to my grief we have not one soldier, no band, nothing here to make any sort of demonstration. What we did do was in Highland fashion to light a bonfire on the top of a hill opposite the house, which had been built last year when the premature news of the fall of Sebastopol deceived every one, and which we had to leave unlit, and found here on our return!

JOURNAL   10 September 1855

Albert said they should go at once and light the bonfire ... In a few minutes, Albert and all the gentlemen, in every species of attire, sallied forth, followed by all the servants, and gradually by all the population of the village—keepers, gillies, workmen—up to the top of the cairn. We waited, and saw them light it; accompanied by general cheering. The bonfire blazed forth brilliantly, and we could see the numerous figures surrounding it—some dancing, all shouting ... About three-quarters of an hour after, Albert came down, and said the scene had been wild and exciting beyond everything. The people had been drinking healths in whisky, and were in great ecstasy. The whole house seemed in a wonderful state of excitement. The boys were with difficulty awakened, and when at last this was the case, they begged leave to go up to the top of the cairn.

We remained till a quarter to twelve; and, just as I was undressing, all the people came down under the windows, the pipes playing, the people singing, firing off guns, and cheering—first for me, then for Albert, the Emperor of the French, and the 'downfall of Sebastopol'.

TO KING LEOPOLD   5 December 1855

My time [has been] entirely taken up with my Royal brother, the King of Sardinia [King Victor Emmanuel, Britain's ally in the Crimea], and I had to make up for loss of time these last days. He leaves us to-morrow at an extraordinary hour—four o'clock in the morning (which you did once or twice)—wishing to be at Compiègne to-morrow night, and at Turin on Tuesday ... He is so frank, open, just, straightforward, liberal and tolerant, with much sound good sense. He never breaks his word, and you may rely on him, but wild and extravagant, courting adventures and dangers, and with a very strange, short, rough manner, an exaggeration of that short manner of speaking which his poor brother had. He is shy in society, which makes him still more brusque, and he does not know (never having been out of his own country or even out in Society) what to say to the number of people who are presented to him here, and which is, I know from experience, a most odious thing ... He is more like a Knight or King of the Middle Ages than anything one knows nowadays.

To Panmure   5 January 1856

The Queen returns the drawings for the 'Victoria Cross.' She has marked the one she approves with an X; she thinks, however, that it might be a trifle smaller. The motto would be better 'For Valour' than 'For the Brave', as this would lead to the inference that only those are deemed brave who have got the Victoria Cross.

To Miss Florence Nightingale   January 1856

You are, I know, well aware of the high sense I entertain of the Christian devotion which you have displayed during this great and bloody war, and I need hardly repeat to you how warm my admiration is for your services, which are fully equal to those of my dear and brave soldiers, whose sufferings you have had the privilege of alleviating in so merciful a manner. I am, however, anxious of marking my feelings in a manner which I trust will be agreeable to you, and therefore send you with this letter a brooch, the form and emblems of which commemorate your great and blessed work, and which, I hope, you will wear as a mark of the high approbation of your Sovereign!*

It will be a very great satisfaction to me, when you return at last to these shores, to make the acquaintance of one who has set so bright an example to our sex.

To Palmerston   11 April 1856

Now that the moment for the ratification of the Treaty of Peace is near at hand, the Queen wishes to delay no longer the expression of her satisfaction as to the manner in which both the War has been brought to a conclusion, and the honour and interests of this country have been maintained by the Treaty of Peace, under the zealous and able guidance of Lord Palmerston. She wishes as a public token of her approval to bestow the Order of the Garter upon him

To Panmure   Undated ? June 1857

The Queen thinks that the persons decorated with the Victoria Cross [which had just been instituted by Royal Warrant] might very properly be allowed to bear some distinctive mark after their name. The warrant instituting the decoration does not style it 'an Order,' but merely 'a Naval and Military Decoration' and a distinction ... V.C. would not do. K.G. means a Knight of the Garter, C.B. a Companion of the Bath, M.P. a Member of Parliament, M.D. a Doctor of Medicine, etc., etc., in all cases designating a person. No one could be called a Victoria Cross. V.C. moreover means Vice-Chancellor at present. D.V.C. (decorated with the Victoria Cross) or B.V.C. (Bearer of the Victoria Cross) might do. The Queen thinks the last the best.

To Palmerston   27 June 1857

The Queen has just received Lord Palmerston's letter and is likewise much alarmed

---

*The brooch resembled a badge. It bore a St. George's Cross in red enamel, and the royal cypher surmounted by a crown in diamonds. The inscription 'Blessed are the Merciful', encircled the badge which also bore the word 'Crimea'.

at the news from India.* She has for some time been very apprehensive of the state of affairs in the army there, and her fears are now fully realised. She trusts that Lord Palmerston and Lord Panmure will consult with the Duke of Cambridge [the Commander-in-Chief] without delay as to what measures should be taken to meet this great danger and that no time will be lost in carrying them out.

To Lady Canning, wife of the Governor-General of India and a former Lady-in-Waiting   5 July 1857

I had long intended writing ... when I received your last of 19th May with all the sad and alarming news of the insurrection at Meerut and Delhi. It is an anxious moment but we have great confidence in Lord Canning and in General Anson (the Commander-in-Chief in India who had, in fact, already died of cholera) and trust to hear soon of the fall of Delhi. Still I fear that there is a dangerous spirit amongst the Native Troops and that a fear of their religion being tampered with is at the bottom of it. I think that the greatest care ought to be taken not to interfere with their religion—as once a cry of that kind is raised amongst a fanatical people—very strictly attached to their religion—there is no knowing what it may lead to and where it may end.

To Palmerston   11 July 1857

The Queen has just received Lord Palmerston's letter and highly approves the proposed appointment of Sir Colin Campbell as commander-in-chief in India and thinks it very handsome of this distinguished, loyal and gallant general to be ready to start at once on so important and arduous a mission. The Queen likewise approves of ... the intention of sending out more troops forthwith.

                                                                     22 August 1857

The Queen is afraid from the telegram of this morning that affairs in India have not yet taken a favourable turn ... Delhi seems still to hold out ... The Queen must repeat to Lord Palmerston that the measures hitherto taken by the Government are not commensurate with the magnitude of the crisis.

To King Leopold   2 September 1857

We are in sad anxiety about India, which engrosses all our attention. Troops cannot be raised fast or largely enough. And the horrors committed [at Cawnpore] on the poor ladies—women and children—are unknown in these ages, and make one's blood run cold. Altogether, the whole is so much more distressing than the Crimea—where there was glory and honourable warfare, and where the poor women and children were safe. Then the distance and the difficulty of communication is such an additional suffering to us all. I know you will feel much for us all. There is not a family hardly who is not in sorrow and anxiety about their children, and in all ranks—India being the place where every one was anxious to place a son!

*In May 1857 sepoys in the Bengal army of the East India Company had mutinied at Meerut and had seized Delhi and other nearby towns; and by the middle of June the revolt had spread to the Ganges valley.

To Lady Canning   8 September 1857

I have to thank for several kind and interesting letters ... That our thoughts are almost solely occupied with India and with the fearful state in which everything there is—that we feel as we did during Crimean days and indeed far more anxiety, you will easily believe. That my heart bleeds for the horrors that have been committed by people once so gentle—(who seem to be seized with some awful mad fanaticism [for that is what] it is there cannot be a doubt) on my poor Country Women and their innocent little children—you, dearest Lady Canning who have shared my sorrows and anxieties for my beloved suffering Troops will comprehend. It haunts me day and night. You will let all who have escaped and suffered and all who have lost dear ones in so dreadful a manner know of my sympathy;—you cannot say too much. A Woman and above all a Wife and Mother can only too well enter into the agonies gone through of the massacres. I ask not for details, I could not bear to hear more, but of those who have escaped I should like to hear as much about as you can tell me.

I feel for you and Lord Canning most deeply! What a fearful time for you both, but what a comfort for Lord Canning to have such a wife as he has in you.

The deaths of Sir H Lawrence—Sir Hugh Wheeler and Sir H Barnard (the latter an old acquaintance of mine who seemed to be doing so well with his small force) are most grievous, and the loss of Sir H. Lawrence irreparable. The retribution will be a fearful one, but I hope and trust that our Officers and Men will show the difference between Christian and Mussulmen and Hindoo—by sparing the old men, women and children. Any retribution on these I should deeply deprecate for then indeed how could we expect any respect or esteem for us in future?

Those Troops (Native) who have remained faithful deserve every reward and praise for their position must be very trying and difficult. The accounts of faithfulness and devotion on the part of servants are also touching and gratifying. I cannot say how sad I am to think of all this blood shed in a country which seemed so prosperous—so improving and for which, as well as for its inhabitants, I felt so great an interest.

To Lady Canning   22 October 1857

Thank God—the accounts are much more cheering and those of Lucknow are a very great relief. The continued arrival of Troops will I trust be of great use, and that no further mutinies and atrocities will take place. As regards the latter I should be very thankful if you and Lord Canning could ascertain how far these are true. Of course the mere murdering—(I mean shooting or stabbing) innocent women and children is very shocking in itself—but in civil War this will happen, indeed I fear that many of the awful insults etc. to poor children and women are the inevitable accompaniments of such a state of things—and that the ordinary sacking of Towns by Christian soldiers presents spectacles and stories which if published in Newspapers would raise outbursts of horror and indignation: Badajoz and St Sebastian I fear were two examples which would equal much that has occurred in India and these the Duke of Wellington could not prevent—and they were the acts of British Soldiers, not of black blood. I mention this not as an excuse but as an explanation of what seems so dreadful to our feelings. Some of these stories certainly are untrue—as for instance that of Colonel and Mrs Farquarson who were said to be sawn asunder and has turned out be a sheer invention, no such people existing in India! What I wish to know is whether there is any reliable evidence of eye witnesses—of horrors, like people having to eat their children's flesh—and other unspeakable and dreadful atrocities which I could not write? Or do these not rest on Native intelligence and witnesses whom one cannot believe implicitly. So many fugitives have arrived at Calcutta that I'm sure you could find out to a great extent how this really is.*

I am delighted to hear that that most loyal excellent veteran Hero Sir Colin Campbell is well and that you like him; I was sure you would, for it is impossible not to do so—and we never for a moment credited the shameful lies of disagreement between him and Lord Canning. If he is still with you say everything most kind to him. I am glad to hear that he does not share that indiscriminate dislike of all brown skins which is very unjust—for the Inhabitants have, it appears, taken no part in this purely Military Revolution—and while summary punishment must alas! be dealt out to the mutinous sepoys—I trust he will see that great forbearance is shown towards the innocent and that women and children will not be touched by Christian soldiers. I hope also that some rule may be laid down as to Ladies in future living in such an unprotected way as they have done in many of those stations and that at the first alarm they will be sent away to places of security, for really they must be dreadfully in the way and it must be so paralysing to the Officers and Men if they have their wives and children in danger.

Now that the rebellion had been suppressed, the Queen continued to oppose harsh reprisals and to support Lord Canning whose clement policies had led to calls for his recall.

To Lady Canning   1 July 1858

I am quite shocked at my long and really unpardonable silence ... You know, dearest Lady Canning, what I always have felt about Lord Canning, and you will believe that those feelings are unaltered ... I only hope that Lord Canning will not think of leaving his post or mind what has passed for it has passed and there is but one feeling now about him here. People are very strange here, about six months ago the blood thirsting was too horrible and really quite shameful! ... All this came from judging of things from a distance, and not understanding them and not waiting for explanations. It is very melancholy but I hope that neither of you will mind it.

The troubles in India had not long been over when the Queen was faced with the anxiety of another change of Government, for Palmerston was forced to resign after being

*As the Queen suspected, nearly all these lurid stories of atrocities were quite without foundation.

defeated on the second reading of the Conspiracy to Murder Bill, and she had to send for Lord Derby.

To Derby   21 February 1858

The Queen has just received Lord Derby's letter, and would wish under all circumstances to see him at six this evening, in order to hear what progress he has made in his plans. The two offices the Queen is most anxious should not be prejudiced in any way, before the Queen has seen Lord Derby again, are the Foreign and the War Departments [appointments given to Lord Malmesbury and General Peel].

Derby did not last long. And after his defeat on the issue of Reform, the Queen, in her endeavour to keep out Lord Palmerston whose pro-Italian and anti-Austrian policies she deplored as strongly as ever, sent for the more amenable Lord Granville.

To Palmerston and Russell   11 June 1859

The Queen gives these lines to Lord Granville, whom she has entrusted with the task of forming an administration on the resignation of Lord Derby. She has selected him as the Leader of the Liberal Party in the House of Lords. She feels that it is of the greatest importance that both Lord Palmerston and Lord John Russell should lend their services to the Crown and country in the present anxious circumstances, and thought at the same time that they might do so most agreeably to their own feelings by acting under a third person.

To Derby   12 June 1859

The Queen writes to inform Lord Derby that after a fruitless attempt on the part of Lord Granville to form a Government comprising Lord Palmerston and Lord John Russell, she has now charged Lord Palmerston with the task, which she trusts may prove more successful.

To Granville   13 June 1859

The Queen is much shocked to find her whole conversation with Lord Granville yesterday and the day before detailed in this morning's leading article of the *Times*. What passes between her and a Minister in her own room in confidential intercourse ought to be sacred, and it will be evident to Lord Granville that if it were not so, the Queen would be precluded from treating her Ministers with that unreserved confidence which can alone render a thorough understanding possible; moreover, any Minister could state what he pleased, against which the Queen would have no protection, as she could not well insert contradictions or explanations in the newspapers herself.

To Palmerston   2 July 1859

The Queen has received Lord Palmerston's letter of to-day. She is sorry not to be able to give her assent to his proposal with regard to Mr Bright.* Privy Councillors have sometimes exceptionally been made without office, yet this has been as rewards, even in such cases, for services rendered to the State. It would be impossible to allege any service Mr Bright has rendered, and if the honour were looked upon as a reward for his systematic attacks upon the institutions of the country, a very erroneous impression might be produced as to the feeling which the Queen or her Government entertain towards these institutions. It is moreover very problematical whether such an honour conferred upon Mr Bright would, as suggested, wean him from his present line of policy, whilst, if he continued in it, he would only have obtained additional weight in the country by his propounding his views as one of the Queen's Privy Councillors.

To Russell   10 July 1859

The Queen has just received Lord John Russell's letter with the enclosure which she returns, and hastens to say in reply, that she does not consider the Emperor of the French or his Ambassador justified in asking the support of England to proposals he means to make to his antagonist to-morrow. He made war on Austria in order to wrest her two Italian kingdoms from her, which were assured to her by the treaties of 1815, to which England is a party; England declared her neutrality in the war. The Emperor succeeded in driving the Austrians out of one of these kingdoms after several bloody battles. He means to drive her out of the second by diplomacy, and neutral England is to join him with her moral support in this endeavour.

The Queen having declared her neutrality, to which her Parliament and people have given their unanimous assent, feels bound to adhere to it. She conceives Lord John Russell and Lord Palmerston ought not to ask her to give her 'moral support' to one of the belligerents. As for herself, she sees no distinction between moral and general support; the moral support of England is her support, and she ought to be prepared to follow it up.

The Queen wishes this letter to be communicated to the Cabinet.

To Princess Frederick William   24 August 1859

Poor dear Papa had one of his stomach attacks on Monday, which made him look fearfully ill, but he remained in the field [at Aldershot] in that broiling sun the whole time and said he was all the better for it. He is however not quite right yet. He is so fagged and worked and our 2 Italian Masters [Russell and Palmerston] almost drive us crazy. Really I never saw two such obdurate ... I won't use any expression because I can't trust what it would be.

To Russell   2 September 1859

The Queen was extremely sorry to find from Lord John Russell's letter of yesterday that he contemplates the possibility of our joining France in a fresh Italian war or demonstration of war against Austria, which the Queen had put entirely out of the

---

*John Bright, whose speeches were at that time unacceptable to the Queen, was later a welcome guest at Windsor as one of her favourite ministers.

question. If the Emperor of the French were allowed to believe in such a possibility, he would have it in his power to bring it about, or obtain a just cause of complaint against us, if we abandoned him. It would be just as dangerous and unfair towards the Emperor to mislead him in this respect as it would be for the Queen to conceal from Lord John that under no pretence will she depart from her position of neutrality in the Italian quarrel, and inflict upon her country and Europe the calamity of war on that account.

7 September 1859

The Queen is determined to hold to her neutrality in the Italian intrigues, revolutions, and wars. It is true, Lord John says, 'it becomes a great power like Great Britain to preserve the peace of Europe, by throwing her great weight into the scale which has justice on its side.' But where justice lies, admits of every variety of opinion.

TO PALMERSTON   10 February 1860

The Queen sends a letter to Lord Palmerston which she has received yesterday evening from Lord John Russell.* She is induced to do so from a feeling that it is to Lord Palmerston, as head of the Government, that she has to look, when she may have reason to take exception to the tone of communications she may receive from members of his Cabinet. Lord Palmerston will not fail to perceive that the enclosed is not the kind of communication which the Foreign Secretary ought to make, when asked by his Sovereign to explain the views of the Cabinet upon a question so important and momentous as the annexation of Savoy to France, and the steps which they propose to take with regard to it. She need not remind Lord Palmerston that in her letter communicated to the Cabinet she had given no opinion whatever upon Italian liberation from a foreign yoke, nor need she protest against a covert insinuation, such as is contained in Lord John's letter, that she is no well-wisher of mankind and indifferent to its freedom and happiness. But she must refer to the constitutional position of her Ministers towards herself. They are responsible for the advice they gave her, but they are bound fully, respectfully, and openly to place before her the grounds and reasons upon which their advice may be founded, to enable her to judge whether she can give her assent to that advice or not. The Government must come to a standstill if the Minister meets a demand for explanation with an answer like the following: 'I was asked by the Cabinet to give an answer, but as I do not agree with you, I think it useless to explain my views.'

   The Queen must demand that respect which is due from a Minister to his Sovereign. As the Queen must consider the enclosed letter as deficient in it, she thinks Lord John Russell might probably wish to reconsider it, and asks Lord Palmerston to return it to him with that view.

*This letter ran 'Lord John Russell unfortunately does not partake your Majesty's opinions in regard to Italy, and he is unwilling to obtrude on your Majesty unnecessary statements of his views. ... Whatever may be the consequence, the liberation of the Italian people from a foreign yoke is, in the eyes of Lord Palmerston and Lord John Russell, an increase of freedom and happiness at which as well-wishers to mankind they cannot but rejoice.'

TO PRINCESS FREDERICK WILLIAM   22 June 1860

We received yesterday the 2 Moorish Ambassadors and they also came to our Ball. You would be charmed with their appearance. They are entirely wrapped up in their white *burnouses*—and nothing can be more picturesque. The one, the 1st, who is not tall—is very handsome, the 2nd—the spokesman is also handsome—with a fair complexion and fine features, and a white beard. They are so like Horace Vernet's pictures and make one think of what the patriarchs of old must have been. I shall get their photographs.

10 November 1860

We are somewhat shocked at your speaking of 'those horrid Yankees'—when Bertie was received in the United States* as no one has ever been received anywhere, principally from the (to me incredible) liking they have for my unworthy self; the Duke of Newcastle's words were 'No sovereign or prince in any country or at any time ever received such an ovation' and that the order and good behaviour throughout was wonderful. He and all anticipate the most wonderful results from this visit. Don't therefore abuse the 'Yankees' for their natural defects—on this occasion at least; for their reception of Bertie has been something so marvellous and naturally so uncalled for and unexpected.

TO RUSSELL   12 February 1861

The Queen has received Lord John Russell's letter enclosing the draft of one to General Garibaldi, which she now returns. She had much doubt about its being altogether safe for the Government to get into correspondence, however unofficial, with the General, and thinks that it would be better for Lord John not to write to him. Lord Palmerston, who was here this afternoon on other business, has undertaken to explain the reasons in detail to Lord John—in which he fully concurs.

12 February 1861

The Queen has received Lord John Russell's reiterated request for her sanction to his writing to General Garibaldi. She still entertains the same objections to the step, as implying a recognition of the General's position as a European Power as enabling him to allow the impression to prevail, that he is in communication with the British Government and acts under its inspiration, as possibly leading to a prolonged and embarrassing correspondence, and as implying for the future that when the disapprobation of the Government is not expressed (as in the present instance), it gives its consent to his aggressive schemes. The Queen will not prevent, however, Lord John from taking a step which he considers gives a chance of averting a great European calamity. Should Lord John therefore adhere to his opinion, she asks him to let her see the letter again, upon the precise wording of which so much depends.

---

*The Prince of Wales had gone out to represent his parents in Canada. He had afterwards entered the United States in the character of a private student.

LIFE AT BALMORAL
1842–1861

Sir James Clark, the Queen's physician, who was born in Banffshire, urged upon her the healthiness of his native land. Both she and the Prince loved the Highlands from the beginning.

To Lord Melbourne   10 September 1842

The Highlands are so beautiful, and so new to me, that we are most anxious to return there again ... This house [Drummond Castle, seat of Lord Willoughby of Eresby] is quite a cottage, but the situation is fine, and the garden very beautiful. We leave this on Tuesday for Dalkeith [seat of the Duke of Buccleuch] where we sleep, and re-embark the next day for England. We greatly admire the extreme beauty of Edinburgh; the situation as well as the town is most striking; and the Prince, who has seen so much, says it is the finest town he ever saw.

To King Leopold   8 October 1844

I cannot reconcile myself to be here [at Windsor] again, and pine for my dear Highlands, the hills, the pure air, the quiet, the retirement, the liberty—all—more than is right.

                 13 September 1848

My letter to Louise will have informed you of our voyage and our arrival here.* This house is small but pretty, and though the hills seen from the windows are not so fine, the scenery all around is the finest almost I have seen anywhere. It is very wild and solitary, and yet cheerful and beautifully wooded, with the river Dee running between the two sides of the hills ...

 Then the soil is the driest and best known almost anywhere, and all the hills are as sound and hard as the road. The climate is also dry, and in general not very cold, though we had one or two very cold days. There is a deer forest—many roe deer, and on the opposite hill (which does not belong to us) grouse. There is also black

---

*Balmoral Castle. The Queen had bought the lease in 1847 and, after receiving a large and unexpected bequest from an eccentric miser, purchased the entire estate of over 17,000 acres. As at Osborne, the house was too small. So a new one was built to the designs of Prince Albert, this time assisted by William Smith of Aberdeen.

cock and ptarmigan. Albert has, however, no luck this year, and has in vain been after the deer, though they are continually seen, and often quite close by the house. The children are very well, and enjoying themselves much. The boys always wear their Highland dress.

JOURNAL   18 September 1848

At a quarter-past ten o'clock we set off in a postchaise with Bertie, and drove beyond the house of Mr. Farquharson's keeper in the Balloch Buie. We then mounted our ponies, Bertie riding Grant's [the head keeper] pony on the deer-saddle, and being led by a gillie, Grant walking by his side. Macdonald [the Prince's jäger] and several gillies were with us. They took us up a beautiful path winding through the trees and heather in the Balloch Buie; but when we had got about a mile or more they discovered deer. A 'council of war' was held in a whisper ...

We scrambled up an almost perpendicular place to where there was a little box, made of hurdles and interwoven with branches of fir and heather, about five feet in height. There we seated ourselves with Bertie, Macdonald lying in the heather near us, watching and quite concealed; some had gone round to beat, and others again were at a little distance. We sat quite still, and sketched a little; I doing the landscape and some trees, Albert drawing Macdonald as he lay there. This lasted for nearly an hour, when Albert fancied he heard a distant sound, and, in a few minutes, Macdonald whispered that he saw stags, and that Albert should wait and take a steady aim. We then heard them coming past. Albert did not look over the box, but through it, and fired through the branches, and then again over the box. The deer retreated; but Albert felt certain he had hit a stag. He ran up to the keepers, and at that moment they called from below that they 'had got him,' and Albert ran on to see. I waited for a bit; but soon scrambled on with Bertie and Macdonald's help; and Albert joined me directly, and we all went down and saw a magnificent stag, 'a royal,' which had dropped, soon after Albert had hit him, at one of the men's feet. The sport was successful, and every one was delighted.

                                                                    11 September 1849

The morning was very fine. I heard the children repeat some poetry in German, and then at ten o'clock we set off. The road got worse and worse. It was particularly bad when we had to pass the Burn of the Glassalt, which falls into the loch, and was very full. There had been so much rain, that the burns and rivers were very full, and the ground quite soft. We rode over the Strone Hill, the wind blowing dreadfully hard when we came to the top. Albert walked almost from the first, and shot a hare and a grouse; he put up a good many of them. We walked to a little hollow immediately above the Dhu Loch, and at half-past three seated ourselves there, and had some very welcome luncheon. The loch is only a mile in length and very wild; the hills, which are very rocky and precipitous, rising perpendicularly from it.

                                                                    13 September 1850

We walked with the boys, and Vicky to the river side above the bridge, where all our tenants were assembled with poles and spears, or rather 'leisters,' for catching salmon. They all went into the river, walking up it, and then back again, poking about under all the stones to bring fish up to where the men stood with the net. It had a very pretty effect; about one hundred men wading through the river, some in kilts with poles and spears, all very much excited. Not succeeding the first time, we went higher up, and moved to three or four different places, but did not get any salmon . . .

In one place there was a very deep pool, into which two men very foolishly went, and one could not swim; we suddenly saw them sink, and in one moment they seemed drowning, though surrounded by people. There was a cry for help, and a general rush, including Albert, towards the spot, which frightened me so much, that I grasped Lord Carlisle's arm in great agony. However, Dr. Robertson [factor at Balmoral] swam in and pulled the man out, and all was safely over; but it was a horrid moment.

A salmon was speared here by one of the men; after which we walked to the ford, or quarry, where we were very successful, seven salmon being caught, some in the net, and some speared.

TO KING LEOPOLD   30 September 1851

I write to you from our little bothy in the hills [Shiel of Allt-Na-Giuthasach], which is quite a wilderness—where we arrived yesterday evening after a long hill expedition to the Lake of Loch Nagar, which is one of the wildest spots imaginable. It was very cold. To-day it pours so that I hardly know if we shall be able to get out, or home even. We are not snowed, but rained up. Our little Shiel is very snug and comfortable and we have got a little piano in it.

                                                         6 October 1851

I love my peaceful, wild Highlands, the glorious scenery, the dear good people who are much attached to us . . . One of our Gillies, a young Highlander who generally went out with me, said, in answer to my observation that they must be very dull here when we left: 'It's just like death come all at once.'

JOURNAL   10 September 1852

We dined at a quarter-past six o'clock in morning gowns, (not ordinary ones, but such as are worn at a 'breakfast,') and at seven started for Corriemulzie, for a torch-light ball in the open air. I wore a white bonnet, a grey watered silk, and (according to Highland fashion) my plaid scarf over my shoulder; and Albert his Highland dress which he wears every evening.

                                                         28 September 1853

A fine morning early, but when we walked out at half-past ten o'clock it began raining, and soon poured down without ceasing. Most fortunately it cleared up before two, and the sun shone brightly for the ceremony of laying the foundation stone of the new house.

7 September 1855

At a quarter-past seven o'clock we arrived at dear Balmoral. Strange, very strange, it seemed to me to drive past, indeed through, the old house; the connecting part between it and the offices being broken through. The new house looks beautiful. The tower and the rooms in the connecting part are, however, only half finished, and the offices are still unbuilt: therefore the gentlemen (except the Minister [the Cabinet Minister always in attendance upon the Queen at Balmoral]) live in the old house, and so do most of the servants; there is a long wooden passage which connects the new house with the offices. An old shoe was thrown after us into the house, for good luck, when we entered the hall. The house is charming; the rooms delightful; the furniture, papers, everything perfection.

8 September 1855

The view from the windows of our rooms and from the library, drawing-room, etc. below them, of the valley of the Dee, with the mountains in the background,—which one never could see from the old house, is quite beautiful. We walked about, and alongside the river, and looked at all that has been done, and considered all that has to be done; and afterwards we went over to the poor dear old house, and to our rooms, which it was quite melancholy to see so deserted; and settled about things being brought over.

30 August 1856

On arriving at Balmoral at seven o'clock in the evening, we found the tower finished as well as the offices, and the poor old house gone! The effect of the whole is very fine.

13 October 1856

Every year my heart becomes more fixed in this dear Paradise, and so much more so now, that all has become my dearest Albert's own creation, own work, own building, own laying out, as at Osborne; and his great taste, and thè impress of his dear hand, have been stamped everywhere. He was very busy to-day, settling and arranging many things for next year.

To King Leopold   13 October 1856

Since the 6th we have the most beautiful weather—with the country in the most brilliant beauty—but not the bracing weather which did one so much good; yesterday and to-day it is quite warm and relaxing. Albert has continued to have wonderful sport; not only has he killed seven more stags since I wrote, but the finest, largest stags in the whole neighbourhood—or indeed killed in almost any forest!

To Princess Augusta of Prussia   14 October 1856

The last week, actually the last ten days have been glorious here. Every year we grow more and more attached to these magnificent mountains, to the great loneliness, to the pleasant quiet and freedom here, and truly it makes my heart bleed to have to tear myself away from it all. For now there is everything here—the house, the beautiful grounds, all the creation of my dear Albert, which we have seen take shape before our very eyes.

JOURNAL   26 September 1857

Albert went out with Alfred for the day, and I walked out with the two girls and Lady Churchill, stopped at the shop and made some purchases for poor people and others; drove a little way, got out and walked up the hill to Balnacroft, Mrs. P. Farquharson's, and she walked round with us to some of the cottages to show me where the poor people lived, and to tell them who I was. Before we went into any we met an old woman, who, Mrs. Farquharson said, was very poor, eighty-eight years old, and mother to the former distiller. I gave her a warm petticoat, and the tears rolled down her old cheeks, and she shook my hands, and prayed God to bless me: it was very touching.

TO PRINCESS FREDERICK WILLIAM   22 October 1859

I must say that (though I know the feeling for 'home' prompts you to say it) the appellation of 'dear, dear Windsor', coming at this moment, when I am struggling with my homesickness for my beloved Highlands, the air—the life, the liberty—cut off for so long—almost could make me angry. I cannot ever feel the slightest affection or tendre for this fine, old dull place, which please God shall never hold my bones! I think I dislike it more and more though I am quite aware of its splendour. I feel the change (as you know you always did) much, though I am well now and in good spirits and won't give way in the slightest degree. You don't say a word about all the affectionate speeches of those dear people at Balmoral, which I wrote to you about, but I am sure you were touched by them all the same.

JOURNAL   4 September 1860

Arrived this evening [at Hotel Grantown] after a most interesting tour; I will recount the events of the day. Breakfasted at Balmoral in our own room at half-past seven o'clock, and started at eight or a little past, with Lady Churchill and General Grey, in the sociable (Grant and [John] Brown on the box as usual) for Castleton, where we changed horses. . . . About a mile from [Loch Inch we came to] the ferry. There we parted from our ponies, only Grant and Brown coming on with us. Walker, the police inspector, met us, but did not keep with us. He had been sent to order everything in a quiet way, without letting people suspect who we were: in this he entirely succeeded. The ferry was a very rude affair; it was like a boat or cobble, but we could only stand on it, and it was moved at one end by two long oars, plied by the ferryman and Brown, and at the other end by a long sort of beam, which Grant took in hand. A few seconds brought us over to the road, where there were two shabby vehicles, one a kind of barouche, into which Albert and I got, Lady Churchill and General Grey into the other—a break; each with a pair of small and rather miserable horses, driven by a man from the box. Grant was on our carriage, and Brown on the other. We had gone so far 40 miles, at least 20 on horseback. We had decided to call ourselves Lord and Lady Churchill and party, Lady Churchill passing as Miss Spencer, and General Grey as Dr. Grey! Brown once forgot this, and called me 'Your Majesty' as I was getting into the carriage; and Grant on the box once called Albert 'Your Royal Highness;' which set us off laughing, but no one observed it . . .

The mountains gradually disappeared,—the evening was mild, with a few drops of rain. On and on we went, till at length we saw lights, and drove through a long and straggling 'toun,' and turned down a small court to the door of the inn. Here we got out quickly—Lady Churchill and General Grey not waiting for us. We went up a small staircase, and were shown to our bed-room at the top of it—very small, but clean—with a large four-post bed which nearly filled the whole room. Opposite was the drawing and dining-room in one—very tidy and well-sized. Then came the room where Albert dressed, which was very small. The two maids . . . had driven over by another road in the waggonette . . . Made ourselves 'clean and tidy,' and then sat down to our dinner. Grant and Brown were to have waited on us, but were 'bashful' [drunk] and did not. A ringletted woman did everything; and, when dinner was over, removed the cloth and placed the bottle of wine (our own which we had brought) on the table with the glasses, which was the old English fashion. The dinner was very fair, and all very clean:—soup, 'hodge-podge,' mutton-broth with vegetables, which I did not much relish, fowl with white sauce, good roast lamb, very good potatoes, besides one or two other dishes, which I did not taste, ending with a good tart of cranberries. After dinner, I tried to write part of this account (but the talking round me confused me), while Albert played at 'patience.' Then went away, to begin undressing, and it was about half-past eleven, when we got to bed.

5 September 1860

What a delightful, successful expedition! Dear Lady Churchill was, as usual, thoroughly amiable, cheerful, and ready to do everything. Both she and the General seemed entirely to enjoy it, and enter into it, and so I am sure did our people. To my dear Albert do we owe it, for he always thought it would be delightful, having gone on many similar expeditions in former days himself. He enjoyed it very much. We heard since that the secret came out through a man recognizing Albert in the street yesterday morning; then the crown on the dog-cart made them think that it was some one from Balmoral, though they never suspected that it could be ourselves! 'The lady must be terrible rich,' the woman observed, as I had so many gold rings on my fingers!—I told Lady Churchill she had on many more than I had. When they heard who it was, they were ready to drop with astonishment and fright.

20 September 1861

Looked anxiously at the weather at seven o'clock—there had been a little rain, there was still mist on the hills, and it looked doubtful. However, Albert said it would be best to keep to the original arrangements, and so we got up early, and by eight the sun shone, and the mist began to lift everywhere. We breakfasted at half-past eight, and at half-past nine we started in two sociables—Alice and Louis with us in the first, and Grant on the box; Lady Churchill and General Grey in the second, and Brown on the box . . .

At a quarter-past seven o'clock we reached the small quiet town, or rather village, of Fettercairn, for it was very small—not a creature stirring, and we got out at the quiet little inn, 'Ramsay Arms,' quite unobserved, and went at once upstairs. There was a very nice drawing-room, and next to it, a dining-room, both very clean and tidy—then to the left our bed-room, which was excessively small, but also very clean and neat, and much better furnished than at Grantown. Alice had a nice room, the same size as ours; then came a mere morsel of one, (with a 'press bed,') in which Albert dressed; and then came Lady Churchill's bed-room just beyond. Louis and General Grey had rooms in an hotel, called 'The Temperance Hotel,' opposite. We dined at eight, a very nice, clean, good dinner. Grant and Brown waited. They were rather nervous, but General Grey and Lady Churchill carved, and they had only to change the plates, which Brown soon got into the way of doing. A little girl of the house came in to help—but Grant turned her round to prevent her looking at us! The landlord and landlady knew who we were, but no one else except the coachman, and they kept the secret admirably.

The evening being bright and moonlight and very still, we all went out, and walked through the whole village, where not a creature moved;—through the principal little square, in the middle of which was a sort of pillar or Town Cross on steps, and Louis read, by the light of the moon, a proclamation for collections of charities which was stuck on it. We walked on along a lane a short way, hearing nothing whatever—not a leaf moving—but the distant barking of a dog! Suddenly we heard a drum and fifes! We were greatly alarmed, fearing we had been recognized; but Louis and General Grey, who went back, saw nothing whatever. Still, as we walked slowly back, we heard the noise from time to time,—and when we reached the inn door we stopped, and saw six men march up with fifes and a drum (not a creature taking any notice of them), go down the street, and back again. Grant and Brown were out; but had no idea what it could be. Albert asked the little maid, and the answer was. 'It's just a band,' and that it walked about in this way twice a week. How odd! It went on playing some time after we got home.

8 October 1861

At a quarter to nine, we reached the inn of Dalwhinnie,—29 miles from where we had left our ponies,— which stands by itself, away from any village. Here again, there were a few people assembled, and I thought they knew us; but it seems they did not, and it was only when we arrived that one of the maids recognized me. She had seen me at Aberdeen and Edinburgh. We went upstairs: the inn was much larger than at Fettercairn, but not nearly so nice and cheerful; there was a drawing-room and a dining-room; and we had a very good-sized bed-room. Albert had a dressing-room of equal size. Mary Andrews [a wardrobe-maid] (who was very useful and efficient) and Lady Churchill's maid had a room together, every one being in the house; but unfortunately there was hardly anything to eat, and there was only tea, and two miserable starved Highland chickens, without any potatoes! No pudding, and no fun; no little maid (the two there not wishing to come in), nor our two people—who were wet and drying our and their things—to wait on us! It was not a nice supper.

TO KING LEOPOLD   21 October 1861

I am glad to see that my account of our mountain expedition amused you, and that you remember all so well ... We have had a most beautiful week, which we have thoroughly enjoyed—I going out every day about twelve or half-past, taking luncheon with us, carried in a basket on the back of a Highlander, and served by an invaluable Highland servant I have [John Brown], who is my factotum here, and takes the most wonderful care of me, combining the offices of groom, footman, page, and maid, I might almost say, as he is so handy about cloaks and shawls, etc. He always leads my pony, and always attends me out of doors, and such a good, handy, faithful, attached servant I have nowhere; it is quite a sorrow for me to leave him behind.

BALMORAL CASTLE FROM THE NORTH-WEST.

'OUR MOST PRECIOUS INVALID'

1861

One winter's day in 1862 the Queen wrote a detailed description of her life with the Prince whose days started at seven o'clock when the wardrobe maid came into the bedroom to open the shutters.

He then went to his room—sitting room where in the winter a fire was made and his green German lamp lit. He brought the original one from Germany, and we always have 2 on our 2 tables which everywhere stand side by side in my room and wrote letters, read etc. and at a little after 8, sometimes a little sooner or later, he came in to tell me to get up ... and constantly he brought me in his letters (English ones) to read thro' ... Also drafts of answers and letters to the Ministers (all of which are preserved as most precious and invaluable documents, in those invaluable books of political and family events which he compiled so beautifully) ... He used to write quite a short diary in a little book, a 'remembrancer', with the days printed on the leaves into which he entered his letters ... Formerly he used to be ready frequently before me ... and he would either stop in my sitting room next door to read some of the endless numbers of despatches which I placed on his side, having either read them or looked into them before and turned the label ... If he was not ready—Baby [Princess Beatrice] generally went into his dressing room and stopped with him till he followed with her at his hand coming along the passage with his dear heavenly face ... Poor darling little Beatrice used to be so delighted to see him dress and when she arrived and he was dressed she made dearest Albert laugh so, by saying 'What a pity'! ... He went out shooting three or four times a week, hunted once a week, wore a green coat with gold buttons, white breeches and high black boots and a cord in his hat. Walked very fast, worked eating his luncheon. Dr Jenner suggested he should eat less lunch, a little cold meat, a little sweet or fruit and a little more claret ... He never went out or came home without coming thro' my room or into my dressing room—dear, dear Angel with a smile on his dearest beautiful face ... and I treasured up everything I heard, kept every letter in a box to tell and show him, and was always so vexed and nervous if I had any foolish draft or despatch to show him, as I knew it would distress and irritate him and affect his poor dear stomach ... My

Angel always wore the blue ribbon of the garter under his waistcoat which looked so nice. At breakfast and luncheon and also our family dinners he sat at the top of the table and kept us all enlivened by his interesting conversation, his charming anecdotes and funny stories of his childhood and people at Coburg—of our good people in Scotland and endless amusing stories which would make him laugh so heartily and which he repeated with the most wonderful mimicry ... The younger children he constantly kept in order if they ate badly or untidily ... He could not bear bad manners and always dealt out his dear reprimands to the juveniles and a word from him was instantly obeyed.

MEMORANDUM   May 1856

It is a strange omission in our Constitution that while the wife of a King has the highest rank and dignity in the realm after her husband assigned to her by law, the husband of a Queen regnant is entirely ignored by the law. This is the more extraordinary, as a husband has in this country such particular rights and such great power over his wife, and as the Queen is married just as any other woman is, and swears to obey her lord and master, as such, while by law he has no rank or defined position. This is a strange anomaly ...

When I first married, we had much difficulty on this subject; much bad feeling was shown, and several members of the Royal Family showed bad grace in giving precedence to the Prince, and the late King of Hanover positively resisted doing so. Naturally my own feeling would be to give the Prince the same title and rank as I have, but a Titular King is a complete novelty in this country, and might be productive of more inconveniences than advantages to the individual who bears it. Therefore, upon mature reflection, and after considering the question for nearly sixteen years, I have come to the conclusion that the title which is now by universal consent given him of 'Prince Consort,' with the highest rank in and out of Parliament immediately after the Queen, and before every other Prince of the Royal Family, should be the one assigned to the husband of the Queen regnant once and for all. This ought to be done before our children grow up, and it seems peculiarly easy to do so now that none of the old branches of the Royal Family [who had made objections in the past] are still alive.

The question has often been discussed by me with different Prime Ministers and Lord Chancellors, who have invariably entirely agreed with me; but the wish to wait for a good moment to bring the matter before Parliament has caused one year after another to elapse without anything being done. If I become now more anxious to have it settled, it is in order that it should be so before our children are grown up, that it might not appear to be done in order to guard their father's position against them personally, which could not fail to produce a painful impression upon their minds.

TO PALMERSTON   16 June 1856

The Queen would now wish the subject to be brought before the Cabinet and begs Lord Palmerston to read her memorandum to his colleagues. She hopes however that he will give them strict injunctions of secrecy, as it is of the greatest importance that the subject should be properly brought before the public and properly understood and not ooze out, so that misapprehensions might arise about it in the public mind or be created by the press—before it is explained. As Lord Palmerston has not given his opinion on the subject to the Queen, she concludes that he shares her views, but it would be satisfactory for her to hear this from him.

10 March 1857

Without wishing to appear impatient or indeed troublesome at a moment when so much business of importance presses upon Lord Palmerston, the Queen must nevertheless recall to his recollection that ten days have now elapsed since the minute of the Cabinet of the 'Prince Consort Bill' was to have been submitted.

15 March 1857

The Queen has to acknowledge the receipt of the minute of Cabinet upon 'The Prince Consort Bill'. Lord Palmerston will not be astonished when she tells him that the perusal of this document has caused her much surprise, so totally at variance is it with what had been expressed to her up to this moment as the opinion of the Cabinet on this question.*

TO KING LEOPOLD   12 February 1861

On Sunday we celebrated, with feelings of deep gratitude and love, the twenty-first anniversary of our blessed marriage, a day which had brought us, and I may say the world at large, such incalculable blessings! Very few can say with me that their husband at the end of twenty-one years is not only full of the friendship, kindness, and affection which a truly happy marriage brings with it, but the same tender love of the very first days of our marriage!

26 August 1861

This is the dearest of days, and one which fills my heart with love, gratitude, and emotion. God bless and protect for ever my beloved Albert—the purest and best of human beings! We miss our four little ones and baby sadly, but have our four eldest (except poor Vicky) with us [at Vice-Regal Lodge, Phoenix Park].

26 November 1861

Albert is a little rheumatic, which is a plague—but it is very difficult not to have something or other of this kind in this season.

TO THE CROWN PRINCESS OF PRUSSIA†   30 November 1861

I can begin by saying that dear Papa is in reality much better—only so much reduced and as usual desponding as men really only are—when unwell. Dr. Jenner [physician in ordinary to the Queen] said yesterday evening Papa was so much

*The Lord Chancellor had discovered legal reasons why the Prince could not be created Prince Consort by Act of Parliament. The Queen, therefore, conferred the title upon him by Letters Patent on 25 June 1857.

†Princess Frederick William's (Vicky's) new title.

better, he would be quite well in two or three days—but he is not inclined himself ever to admit he is better! I can only write a line to say beloved Papa is improving, and I hope now each day will make a decided difference. But he is so depressed and so low—that it is always very distressing and the amount of sleepless nights has lowered him, besides the impossibility to touch food is very vexatious. This time the bowels are perfectly right. He likes being read to constantly.

JOURNAL   2 December 1861

My dearest Albert did not dress, but lay on his sofa . . . Sir James [Clark] came over and found him much in the same state . . . sometimes lying on his sofa in his dressing gown, and then sitting in an armchair in his sitting room . . . he kept saying, it was very well he had no fever, as he should not recover!—which we all told him was too foolish and he must never speak of it. He took some soup with brown bread in it which unfortunately disagreed with him.

                                                            3 December 1861

Dreadfully annoyed at a letter from Lord Palmerston suggesting Dr Ferguson should be called in as he heard Albert could not sleep and eat. Very angry about it. In an agony of despair about my dearest Albert and crying much, for saw no improvement and my dearest Albert was so listless and took so little notice.

TO KING LEOPOLD

My poor dear Albert's rheumatism has turned out to be a regular influenza, which has pulled and lowered him very much. Since Monday he has been confined to his room. It affects his appetite and sleep, which is very disagreeable, and you know he is always so depressed when anything is the matter with him.

JOURNAL   7 December 1861

I went to my room and cried dreadfully and felt oh! as if my heart must break—oh! such agony as exceeded all my grief this year. Oh, God! help and protect him! . . . I seem to live in a dreadful dream. My angel lay on the bed in the bedroom and I sat by him watching him and the tears fell fast . . . saw Sir James and Dr Jenner talked over what could have caused this. Great worry and far too hard work for long! That must be stopped.

TO THE CROWN PRINCESS OF PRUSSIA   7 December 1861

I feel shaken—as for four nights I had not more than two or three hours sleep! And though I slept more last night and for Papa's own sake slept in the next room—I was for two hours in such a state of anxious suspense listening to every sound—that I feel very trembly myself. The doctors and I of course, would not have felt the least anxious had the dear invalid not been that most precious and perfect of human beings. Beloved Papa has never been confined to his bed, and is dressed and walks about his rooms. He is very irritable today. Dr. Jenner has been most attentive, and is excellent, very clever, very kind and very determined.

JOURNAL   8 December 1861

He wanders for moments a little, seldom smiles and is still very impatient ... so impatient because I tried to help in explaining something to Dr Jenner and quite slapped my hand, poor dear darling ... Went in again to see my dearest Albert. He was so pleased to see me—stroked my face and smiled and called me his '*Fräuchen*' [little woman] ... He was so dear and kind. Precious love!

TO THE CROWN PRINCESS OF PRUSSIA   8 December 1861

Dearest Papa is going on as favourably as we could wish ... But it is all like a bad dream! To see him prostrate and worn and weak, and unable to do any thing and never smiling hardly—is terrible.

   I am well but very tired and nervous for I am so constantly on my legs—in and out and near him. I sleep in my dressing-room, having given up our bedroom to dear Papa.

JOURNAL   9 December 1861

He wanders frequently and they say it is of no consequence tho' very distressing, for it is unlike my own Angel. He was so kind calling me '*gutes Weibchen*' (excellent little wife), and liking me to hold his dear hand. Oh! it is an anxious, anxious, time but God will help us thro' it.

TO KING LEOPOLD   9 December 1861

Our beloved invalid goes on well—but it must be tedious, and I need not tell you what a trial it is to me. Every day, however, is bringing us nearer the end of this tiresome illness.

TO THE CROWN PRINCESS OF PRUSSIA   10 December 1861

Thank God! beloved Papa had another excellent night and is going on quite satisfactorily. There is a decided gain since yesterday and several most satisfactory symptoms. He is now in bed—and only moves on the sofa made like a bed, for some hours. He takes a great deal of nourishment—and is really very patient. But it must be still some days before the fever leaves him, and it is very, very trying to watch and witness. I am constantly in and out—and a great deal with him. It is my greatest pleasure and comfort—but it is a life of intense anxiety and requires courage.

TO KING LEOPOLD   11 December 1861

I can report another good night, and no loss of strength, and continued satisfactory symptoms. But more we dare not expect for some days; not losing ground is a gain, now, of every day.

   It is very sad and trying for me, but I am well, and I think really very courageous; for it is the first time that I ever witnessed anything of this kind though I suffered from the same at Ramsgate [where the Queen had typhoid when she was sixteen] and was much worse. The trial in every way is so very trying, for I have lost my guide, my support, my all, for a time—as we can't ask or tell him anything.

12 December 1861

I can again report favourably of our most precious invalid. He maintains his ground well—had another very good night—takes plenty of nourishment, and shows surprising strength. I am constantly in and out of his room, but since the first four dreadful nights, last week, before they had declared it to be gastric fever—I do not sit up with him at night as I could be of no use; and there is nothing to cause alarm. I go out twice a day for about an hour ... I cannot sufficiently praise the skill, attention, and devotion of Dr Jenner, who is the first fever Doctor in Europe, one may say—and good old Clark is here every day; good Brown is also most useful ... We have got Dr Watson (who succeeded Dr Chambers) and Sir H. Holland has also been here. But I have kept clear of these two. Albert sleeps a good deal in the day. He is moved every day into the next room on a sofa which is made up as a bed. He has only kept his bed entirely since Monday. Many, many thanks for your dear, kind letter of the 11th. I knew how you would feel for and think of me. I am very wonderfully supported, and, excepting on three occasions, have borne up very well.

JOURNAL   13 December 1861

Found him very quiet and comfortably warm, and so dear and kind, called me '*gutes Fräuchen*' and kissed me so affectionately and so completely like himself, and I held his dear hands between mine ... They gave him brandy every half-hour.

14 December 1861

Went over at 7 as I usually did. It was a bright morning; the sun just rising and shining brightly ... Never can I forget how beautiful my darling looked lying there with his face lit up by the rising sun, his eyes unusually bright gazing as it were on unseen objects and not taking notice of me ... Sir James was very hopeful, so was Dr Jenner, and said it was a 'decided rally', but that they were all 'very, very, anxious' ... I asked if I might go out for a breath of fresh air. The doctors answered 'Yes, just close by, for half an hour!' ... I went out on the Terrace with Alice. The military band was playing at a distance and I burst out crying and came home again ... Sir James was very hopeful; he had seen much worse cases. But the breathing was the alarming thing—so rapid, I think 60 respirations in a minute ... I bent over him and said to him '*Es ist Kleines Fräuchen*' (it is your little wife) and he bowed his head; I asked him if he would give me '*ein Kuss*' (a kiss) and he did so. He seemed half dozing, quite quiet ... I left the room for a moment and sat down on the floor in utter despair. Attempts at consolation from others only made me worse ... Alice told me to come in ... and I took his dear left hand which was already cold, though the breathing was quite gentle and I knelt down by him ... Alice was on the other side, Bertie and Lenchen ... kneeling at the foot of the bed ... Two or three long but perfectly gentle breaths were drawn, the hand clasping mine and ... all, all, was over ... I stood up, kissed his dear heavenly forehead and called out in a bitter and agonising cry, 'Oh! my dear Darling!'

TO THE CROWN PRINCESS OF PRUSSIA   18 December 1861

What is to become of us all? Of the unhappy country, of Europe, of all? For you all, the loss of such a father is totally irreparable! I will do all I can to follow out all his wishes—to live for you all and for my duties. But how I, who leant on him for all and everything—without whom I did nothing, moved not a finger, arranged not a print or photograph, didn't put on a gown or bonnet if he didn't approve it shall be able to go on, to live, to move, to help myself in difficult moments? How I shall long to ask his advice! Oh! it is too, too weary! The day—the night (above all the night) is too sad and weary. The days never pass! I try to feel and think that I am living on with him, and that his pure and perfect spirit is guiding and leading me and inspiring me!

Sweet little Beatrice comes to lie in my bed every morning which is a comfort. I long so to cling to and clasp a loving being. Oh! how I admired Papa! How in love I was with him! How everything about him was beautiful and precious in my eyes! Oh! how, how I miss all, all! Oh! Oh! the bitterness of this—of this woe! I saw him twice on Sunday—beautiful as marble—and the features so perfect, though grown very thin. He was surrounded with flowers. I did not go again.

To King Leopold   20 December 1861

My own dearest, kindest Father,—For as such, have I ever loved you! The poor fatherless baby of eight months is now the utterly broken-hearted and crushed widow of forty-two! My life as a happy one is ended! the world is gone for me! If I must live on (and I will do nothing to make me worse than I am), it is henceforth for our poor fatherless children—for my unhappy country, which has lost all in losing him—and in only doing what I know and feel he would wish, for he is near me—his spirit will guide and inspire me! But oh! to be cut off in the prime of life—to see our pure, happy, quiet domestic life, which alone enabled me to bear my much disliked position, cut off at forty-two—when I had hoped with such instinctive certainty that God never would part us, and would let us grow old together (though he always talked of the shortness of life)—is too awful, too cruel!

                                                          24 December 1861

Though, please God! I am to see you so soon, I must write these few lines to prepare you for the trying, sad existence you will find it with your poor forlorn, desolate child—who drags on a weary, pleasureless existence! I am also anxious to repeat one thing, and that one is my firm resolve, my irrevocable decision, viz. that his wishes—his plans—about everything, his views about every thing are to be my law! And no human power will make me swerve from what he decided and wished—and I look to you to support and help me in this. I apply this particularly as regards our children—Bertie, etc.—for whose future he had traced everything so carefully. I am also determined that no one person, may he be ever so good, ever so devoted among my servants—is to lead or guide or dictate to me. I know how he would disapprove it. And I live on with him, for him; in fact I am only outwardly separated from him, and only for a time.

To Palmerston   26 December 1861

Business she can as yet hardly think of, for her whole soul, bruised and crushed as it is, and her utterly broken heart lives but in that future World which is now nearer than ever to her, as it contains he who was the life of her life, the sunshine of her existence, her guide, support,—her all, and who was too pure, too perfect for this world. The Queen feels her life ended, in a worldly point of view; her own wish is to leave it very soon to join him for whom she would have given hers a hundred times over, would have followed bare foot over the world!

TO THE CROWN PRINCESS OF PRUSSIA   27 December 1861

The wickedness of the world was too much for him to bear. Thank dear Fritz (who was a great comfort) for his dear letter and all he did and said to poor, unhappy Bertie.* Tell him that Bertie (oh! that boy—much as I pity I never can or shall look at him without a shudder as you may imagine) does not know, that I know all, (beloved Papa told him that I could not be told 'the disgusting details') that I try to employ and use him—but I am not hopeful. I believe firmly in all Papa foresaw. I am very fond of Lord Granville and Lord Clarendon—but I should not like them to be his moral guides, for dearest Papa said to me that neither of them would understand what we felt upon Bertie's 'fall'.

---

*While attached to the Grenadier Guards at the Curragh military camp near Dublin, the Prince of Wales had returned to his quarters to find a vivacious and cheerfully promiscuous young actress, Nellie Clifden, waiting for him in his bed. The subsequent scandal had horrified his father who, already ill, felt it his duty to go to Cambridge after the Prince's return there, to impress upon him the disgrace he had brought upon himself and his family. The emotional strain and the exacerbation of his illness were responsible, so the Queen became convinced, for her husband's death.

THE LONELY WIDOW

1862–1869

The Queen lived on as a widow for forty years. At first she thought she was 'going mad' with grief. She kept tapping her forehead and repeating, 'My reason! My reason!' She was only forty-two but it was as if she were already an old woman. She shut herself away with her family and Household, seeing her Cabinet Ministers as rarely as she could. She had been able to tolerate her duties as Queen, the ceremonial functions of monarchy, she explained, only because of the support she had received from her husband. Now that he was gone, they could not expect her to carry on as before. She abandoned herself to the past and to her memories of him with a passionate intensity. She could never forget him; no one else must. It was selfish and hysterical; but it was understandable. As she herself admitted, Prince Albert had not only been a beloved husband but a mother and father to her as well. He was, as she put it, her 'only assistant in her communications with the officers of the government, her private secretary and permanent minister'. He had been the mainstay of her private as well as of her official life; he had not only relieved her of the burden of work, given her the benefit of what she called his 'great mind', accompanied her in all her official engagements, but he had guided her taste in all things and had even chosen her hats and dresses for her. She felt utterly lost and alone without him and the loneliness was exacerbated by the isolation of her unique position as Queen. 'What a dreadful going to bed! What a contrast to that tender lover's love! All alone!' she lamented after one of those scarcely endurable nights of her early widowhood. 'Yet the blessings of 22 years cast their reflection.' And the precious memories of those years of their love were, indeed, an inspiration that was eventually to help her overcome the trials of the future.

JOURNAL   1 January 1862

Have been unable to write my Journal since the day my beloved one left us, and with what a heavy broken heart I enter on a new year [at Osborne] without him! My dreadful and overwhelming calamity gives me so much to do, that I must henceforth merely keep notes of my sad and solitary life. This day last year found us so perfectly happy, and now! Last year music woke us; little gifts, new year's wishes, brought in by maid, and then given to dearest Albert, the children waiting with their gifts in the next room—all these recollections were pouring in on my mind in an overpowering manner. Alice slept in my room, and dear baby [Princess Beatrice, the Queen's youngest child] came down early. Felt as if living in a dreadful dream. Dear Alice much affected when she got up and kissed me. Arthur gave me a nosegay, and the girls, drawings done by them of their dear father and me. Could hardly touch my breakfast.

When dressed saw Dr Jenner, Mr. Ruland, [Librarian at Windsor] and Augusta Bruce [Resident Bedchamber Woman to the Queen]. Went down to see the sketch for a statue of my beloved Albert in Highland dress, which promises to be good. Then out with Lenchen, Toward [Land Steward at Osborne] always following and pointing out trees and everything. When I came in, saw the Duke of Newcastle [Colonial Secretary] in dear Albert's room, where all remains the same. Talking for long of him, of his great goodness, and purity, quite unlike anyone else. Saw Sir J. Clark, Sir C. Phipps [Keeper of Her Majesty's Privy Purse], and then dear, kind Uncle Leopold.

6 January 1862

Held a Council, which was well and kindly arranged. Lord Granville [Lord President of the Council] and others, with Mr. Helps [Clerk of the Council], were in dear Albert's room, and I in mine, with the door open. The business was all summed up in two paragraphs, and Mr. Helps read 'approved' for me. This was unlike anything which had been done before. The Council after dear Mama died took place in the Red Room, and dearest Albert handed me the papers and was with me. But now!

TO THE CROWN PRINCESS OF PRUSSIA   8 January 1862

Oh! weary, weary is the poor head which has no longer the blessed precious shoulder to rest on in this wretched life! My misery—my despair increase daily, hourly.

I have no rest, no real rest or peace by day or by night; I sleep—but in such a way as to be more tired of a morning than at night and waken constantly with a dreamy, dreadful confusion of something having happened and crushed me! Oh! it is too awful, too dreadful! And a sickness and icy coldness bordering on the wildest despair comes over me—which is more than a human being can bear ... On Saturday it was three weeks that he left me. Oh—Oh! I can't bear to live on so! I never shall live on! God cannot mean to tear me alive into pieces! Forgive my grieving you, dearest, good child—but my misery is fearfully great!

TO EARL CANNING   10 January 1862

Lord Canning little thought when he wrote his kind and touching letter of the 22nd November, that it would only reach the Queen when she was smitten and bowed down to the earth by an event similar to the one which he describes—and, strange to say, by a disease greatly analogous to the one which took from him all that he loved best. [Lady Canning had died of jungle fever]. To lose one's partner in life is, as Lord Canning knows, like losing half of one's body and soul, torn forcibly away—and dear Lady Canning was such a dear, worthy, devoted wife! But to the Queen—to a poor helpless woman—it is not that only—it is the stay, support and comfort which is lost! To the Queen it is like death in life! Great and small—nothing was done without his loving advice and help—and she feels alone in the wide world, with many helpless children (except the Princess Royal) to look to her—and the whole nation to look to her—now when she can barely struggle with her wretched existence! Her misery—her utter despair—she cannot describe!

To the Crown Princess of Prussia   11 January 1862

On my knees do I implore God it may be very, very few, even though it may be some [years before she died]! But I have the feeling that I am getting on in my journey. I must work and work, and can't rest and the amount of work which comes upon me is more than I can bear! I who always hated business, have now nothing but that! Public and private, it falls upon me! He, my own darling, lightened all and every thing, spared every trouble and anxiety and now I must labour alone!

B[ertie]'s journey is all settled [a tour of the Near East]. Many wished to shake my resolution and to keep him here—to force a constant contact which is more than ever unbearable to me you can well imagine. And though the intentions are good, the tact, the head, the heart all are lamentably weak. The marriage is the thing and beloved Papa was most anxious for it.

To Earl Russell, Foreign Secretary   10 January 1862

The Queen leads the most utterly wretched and desolate life that can be imagined. Where all was peaceful sunshine and perfect happiness (which the troubles and worries of her position rendered very necessary) there is now utter desolation, darkness, and loneliness, and she feels daily more and more worn and wretched. The eternal future is her only comfort.

                    14 January 1862

The Queen approves of the drafts to Lord Lyons [British Ambassador in Paris]. She must, however, observe that she should have seen the despatch which he [Lord Russell] read to Mr. Adams [American Minister in London], before it was sent to its destination, and Lord Russell will perhaps take care that the rule should not be departed from, viz. that no drafts should be sent without the Queen's having first seen them.

To the Crown Princess of Prussia   15 January 1862

I hope the old Baron [Stockmar] knows the real truth about B[ertie]? If he does not would Fritz kindly ask Baron E. [Ernest, Stockmar's son] to put him in possession of the sad truth, and of the awful fact of its having made beloved Papa so ill—for there must be no illusion about that—it was so; he was struck down—and I never can see B[ertie]—without a shudder! Oh! that bitterness—oh! that cross!

Journal   21 January 1862

The expressions of universal admiration and appreciation of beloved Albert are most striking, and show how he was beloved and how his worth was recognised. Even the poor people in small villages, who don't know me, are shedding tears for me, as if it were their own private sorrow . . .

After luncheon saw Lord Palmerston [Prime Minister] (who has been very ill). It made me very nervous seeing him for the first time since my great misfortune, but I felt it was right not to put it off any longer. He seemed very nervous himself. Spoke of Uncle, whom he had been to see, and of my remaining here for the present. He could in fact hardly speak for emotion. It showed me how much he felt my terrible loss, and he said what a dreadful calamity it was. Then he spoke about Bertie, and the desirability for his travelling, which would be such a good thing for him. I repeated that it had been his Father's wish he should do so; and Lord Palmerston said it was most important he should marry. I observed that he was a very good and dutiful son, but that for him, just at his age, the loss of his Father was terrible, which Lord Palmerston thoroughly understands and feels keenly. Everything was quiet, he thought there would be no trouble, but 'the difficulty of the moment' was Bertie. I felt the same, and would hardly have given Lord Palmerston credit for entering so entirely into my anxieties. He alluded to Princess Alexandra [of Denmark, as a possible bride for the Prince of Wales] and thought the political objections must not be minded, as they did not affect this country. I did not speak as if there were any certainty, but praised the young lady.

TO THE CROWN PRINCESS OF PRUSSIA   24 January 1862

As regards B[ertie]'s affair, one thing I think very necessary and which has been strongly urged on me by those who wish my comfort—viz. that I should see the girl [Princess Alexandra] before B. sees her again so that I could judge, before it is too late, whether she will suit me. That whole affair is our 'forlorn hope' and if it were not to be successful, or she not to take to me, all would be lost! You will understand me.

TO LORD DERBY   17 February 1862

The Queen wishes herself to express her thanks to Lord Derby [Leader of the Conservative Opposition] for the copy of his speech, and her satisfaction at his serving on the Committee of the Memorial [to the Prince Consort], so full of interest to her poor broken heart. She hopes to see him some day at Windsor, to which living grave she intends to return for a short while next week.

To express what the Queen's desolation and utter misery is, is almost impossible; every feeling seems swallowed up in that one of unbounded grief! She feels as though her life had ended on that dreadful day when she lost that bright Angel who was her idol, the life of her life; and time seems to have passed like one long, dark day!

She sees the trees budding, the days lengthen, the primroses coming out, but she thinks herself still in the month of December! The Queen toils away from morning till night, goes out twice a day, does all she is desired to do by her physician, but she wastes and pines, and there is that within her inmost soul, which seems to be undermining her existence!

TO RUSSELL   5 March 1862

She could have wished to hear Lord Russell's opinion as to the probable effect of the change [of the Government in Italy] on the future course of events in Italy, as well as on the relations of this country with the Italian Government.

The Queen must also observe that she would be very glad of more assistance generally from Lord Russell in forming her opinion on the various important questions affecting the foreign policy of this country, which now engage the attention of her Government

It is very difficult for the Queen, when she is left without one word of explanation to assist her, to draw her own conclusions from the perusal of voluminous despatches from abroad (not always very regularly sent), when she receives drafts for her approval, and to judge, in her ignorance of the views of the Government, or of the reasons which have dictated them, whether she should approve them or not.

The assistance the Queen asks for is more than ever necessary now in her present desolation, when she has alas! alas! no one to look to for advice and help in these matters, or to prevent the duties she has to perform from becoming too much for her strength and health, which is now very far from what it was.

JOURNAL   19 March 1862

Saw Sir C. Phipps, who had been speaking with Mr. Gladstone [Chancellor of the Exchequer] about money matters. I feel very anxious about a provision for Bertie and his wife, in the event of his marrying; a provision for my younger sons on their coming of age and marrying; and a provision for the younger Children under age, in case of my death. He showed me a very satisfactory paper from Mr. Gladstone on the subject. Then saw Mr. Gladstone for a little while, who was very kind and feeling. We talked of the state of the country. He spoke with such unbounded admiration and appreciation of my beloved Albert, saying no one could ever replace him.

                                                          14 April 1862

I went down to see Tennyson, who is very peculiar-looking, tall, dark, with a fine head, long black flowing hair, and a beard; oddly dressed, but there is no affectation about him. I told him how much I admired his glorious lines to my precious Albert, and how much comfort I found in his In Memoriam. He was full of unbounded appreciation of beloved Albert. When he spoke of my own loss, of that to the nation, his eyes quite filled with tears.

TO THE CROWN PRINCESS OF PRUSSIA   16 April 1862

Has perhaps Princess Christian [Princess Alexandra's mother] heard of poor, wretched Bertie's miserable escapade—and thinks him a regular 'mauvais sujet'? The Aunt here [the Duchess of Cambridge, aunt of both the Queen and of Princess Christian] may have written in that way? I fear we can say no more. The meeting must be at Laeken [King Leopold's palace near Brussels], and can't be before the 2nd or 3rd Sept:. I will however let Bertie know that she is much sought after; but more we cannot do. Your account of the family is certainly as bad as possible, and that is the weak point in the whole affair, but dearest Papa said we could not help it. Oh! the whole thing is so disheartening to me! Alone! to do all this, and with B.! If he turns obstinate I will withdraw myself altogether and wash my hands of him, for I cannot educate him, and the country must make him feel what they think. Affie would be ready to take her at once, and really if B. refused I would recommend Affie's engaging to marry her in three years.

To Palmerston   30 April 1862

The Queen feels it necessary to call the attention of Lord Palmerston to the enclosed published Memorandum from the Admiralty.

Lord Palmerston will see that the Lords of the Admiralty state that they have caused additions to be made to the Queen's regulations. There is no mention of the Queen's sanction having been sought or obtained. The Queen is sure that the power thus assumed, of altering regulations issued under her authority, is one that Lord Palmerston will feel, equally with herself, cannot belong to any subject, whatever office he may hold.

                                                                    6 May 1862

[The Queen] can give but a very bad report of herself; the journey [to Balmoral] was no fatigue, but this place—this country and everything indoors and out-of-doors are perfectly overwhelming to the Queen and she feels herself still weaker and her nerves especially more shaken even than before. Her handwriting will show this. She feels as if she could not bear the torture of this existence without the idol of her poor life!

No one knows what her bitter anguish and sufferings are—or how that poor heart is pierced and bleeding.

Journal   14 June 1862

Heard of Bertie's landing, after a boisterous crossing. The good General [on his return from his tour in the Near East] none the worse for it. Bertie arrived at half past five, looking extremely well. I was much uspet at seeing him, and feeling his beloved father was not there to welcome him back. He would have been so pleased to see him so improved, and looking so bright and healthy. Dear Bertie was most affectionate, and the tears came into his eyes when he saw me.

To the Marchioness of Clanricarde   17 June 1862

To you as the sister of dear Lord Canning, I write to express my deep sorrow at his untimely end, and to say how my beloved and precious husband and I valued and esteemed him, and how we looked to him to be of the greatest use to his Sovereign and country! ...

Under these circumstances [the deaths of Lady Canning and the Prince Consort] Lord Canning returned to his native land [from India]. God has taken him, and he is again with her after only seven months' separation! Oh! for him how blessed! How enviable to follow so soon the partner of your life!

To the Crown Princess of Prussia   14 June 1862

A daughter-in-law never, in my present position, can be what a daughter is, and I must never, during the few (very few I think) years still remaining, be left without one of you—and with five daughters this will be quite easy. Dear Papa said so himself. Bertie could not cross yesterday from the dreadful gale—but has done so this morning and will be here very shortly. It makes me terribly nervous! I am so weak, so shattered, so terribly excitable that any new hint of anxiety alarms and agitates me.

I am so terribly nervously affected now; my pulse gets so high, it is constantly between 90 and 100 instead of being at 74! This wears me terribly. It exhausts me so and I am so weak, and then my poor memory fails me so terribly.

18 June 1862

I can give you a very good report of Bertie. He is much improved and is ready to do every thing I wish, and we get on very well. He is much less coarse looking and the expression of the eyes is so much better.

25 June 1862

It would be well if Wally [Walburga Paget, German wife of the British Minister in Copenhagen, once the Crown Princess's lady-in-waiting] could let Princess Christian know the truth; viz: that wicked wretches had led our poor, innocent boy into a scrape which had caused his beloved father and myself the deepest pain (the knowledge of which we only obtained just before the fatal illness) but that both of us had forgiven him this (one) sad mistake, that we had never disagreed, and that I was very confident he would make a steady husband; that quite the contrary I looked to his wife as being his salvation, for that he was very domestic and longed to be at home. That I was exceedingly satisfied and pleased with him since his return and thought him immensely improved.

JOURNAL   1 July 1862

Scarcely got any sleep. Towards morning heard all the preparations for to-day's ceremony [the marriage of Princess Alice to Prince Louis of Hesse] going on. It tried me terribly. Alice got up and came and kissed me, and I gave her my blessing and a Prayer Book, like one dear Mama gave me on our happy wedding morning. Went to look at the Dining-room, which was very prettily decorated, the altar being placed under our large family picture. All the furniture had been removed, and plants and flowers placed everywhere . . .

The time had come, and I, in my 'sad cap,' [her widow's cap] as baby calls it, most sad on such a day, went down with our four boys, Bertie and Affie leading me. It was a terrible moment for me . . .

I sat all the time in an armchair, Bertie and Affie close to me . . . After a short pause Louis came in, conducted by Lord Sydney [the Lord Chamberlain], and followed by his two brothers William and Henry. After another pause came the dear dear Bride on her Uncle's arm, followed by the Bridesmaids, a touching sight. The service then commenced, the Archbishop performing it beautifully. Alice answered so distinctly and was full of dignity and self-possession. Louis also answered very distinctly. I restrained my tears, and had a great struggle all through, but remained calm.

To the Crown Princess of Prussia   2 July 1862

Poor Alice's wedding (more like a funeral than a wedding) is over and she is a wife! I say God bless her—though a dagger is plunged in my bleeding, desolate heart when I hear from her this morning that she is 'proud and happy' to be Louis' wife! I feel what I had, what I hoped to have for at least 20 years more and what I can only have in another world again. All that has passed since December 14 seems gone—forgotten. What I shall not forget is Alice herself, and her wonderful bearing—such calmness, self-possession and dignity, and how really beautiful she looked, so tall, and graceful, and her voice so sweet. The Archbishop of York read that fine service (purified from its worst coarsenesses) admirably, and himself had tears running down his cheeks—for he too lost his dear partner not long ago. I sat the whole time in an armchair, with our four boys near me; Bertie and Affie led me down stairs. The latter sobbed all through and afterwards—dreadfully.

29 July 1862

Poor Bertie!—he is very affectionate and dutiful but he is very trying. The idleness is the same— and there is a great roughness of manner to his brothers and sisters which must be got the better of. Still he is most anxious to do what is right, that is every thing. But his idleness and 'désœuvrement', his listlessness and want of attention are great, and cause me much anxiety.

In September the Queen travelled to Laeken, where she met Princess Alexandra of Denmark and her parents, Prince and Princess Christian.

Journal   3 September 1862

Alexandra is lovely, such a beautiful refined profile, and quiet ladylike manner, which made a most favourable impression. Dagmar [Princess Alexandra's eldest sister] is quite different, with fine brown eyes. Princess Christian must have been quite good-looking. She is unfortunately very deaf. Uncle soon came in, and after a rather stiff visit they all (excepting myself) went to luncheon . . . Baby lunched with me.

Afterwards Marie B [Duchess of Brabant, King Leopold's daughter-in-law], brought Prince and Princess Christian upstairs, leaving them with me. Now came the terribly trying moment for me. I had alone to say and do what, under other, former happy circumstances, had devolved on us both together. It was not without much emotion that I was able to express what I did to the Princess: my belief that they knew what we wished and hoped, which was terrible for me to say alone. I said that I trusted their dear daughter would feel, should she accept our son, that she was doing so with her whole heart and will. They assured me that Bertie might hope she would do so, and that they trusted he also felt a real inclination . . . Dined as yesterday, and afterwards Prince and Princess Christian and Princess Alexandra came upstairs. She looked lovely, in a black dress, nothing in her hair, and curls on either side, which hung over her shoulders, her hair turned back off her beautiful forehead. Her whole appearance was one of the greatest charm, combined with simplicity and perfect dignity. I gave her a little piece of white heather, which Bertie gave me at Balmoral, and I told her I hoped it would bring her luck. Dear Uncle Leopold, who sat near me, was charmed with her. Very tired, and felt low and agitated.

Saw Lord Russell and talked of Bertie's marriage, of France, Germany, and the Schleswig-Holstein question. We discussed the importance of Bertie's marriage being in no sense considered a political one. Had a telegram from Bertie, which shortly afterwards General Grey [formerly the Prince's Private Secretary who now acted as the Queen's] sent back deciphered, to the effect that he 'had proposed and been accepted this day,' and asking 'for my consent and blessing.' So it is settled.

TO THE CROWN PRINCESS OF PRUSSIA   8 September 1862

You did not say too much about dear Alix or Alexandra. No, she is a dear, lovely being—whose bright image seems to float—mingled with darling Papa's—before my poor eyes—dimmed with tears! Dearest child! this very prospect of opening happiness of married life for our poor Bertie—while I thank God for it—yet wrings my poor heart, which seems transfixed with agonies of longing! I am alas! not old—and my feelings are strong and warm; my love is ardent.

TO PALMERSTON   22 October 1862

The Queen's nerves and strength and general health are just the same, and everything approaching to society or even having several of her own family together with her at her meals, is more than she can bear. But, trying and heartrending in many ways she found going to dear Coburg was, the many dear recollections of her beloved angel's childhood and youth (as well as of her dear mother), seeing the many scenes (so beautiful and peaceful in themselves), he so loved and the many kind old attached friends, high and low,—the hearing his native tongue—and the breathing of his native air—were soothing and sweet in their very sadness to her bruised spirit and her aching, bleeding heart!!

TO THE CROWN PRINCESS OF PRUSSIA   29 October 1862

In Affie's case—there is not a particle of excuse [for a sexual escapade at Malta], his conduct was both heartless and dishonourable. But he does feel it though he can't poor boy give utterance ever to any thing. But his great palour, thinness, his subdued tone, and his excessive anxiety by every little act to give me pleasure, to do what I like and wish show me he feels enough. I had wished not to see him and thought for himself it would have been better. But for the world it was necessary. So I saw him on the 28th and he stopped till today. It was very trying.

JOURNAL   5 November 1862

At last, at 9, dear Alexandra arrived with her father, looking very lovely and well. A gleam of satisfaction for a moment shone into my heart as I led 'our' future daughter upstairs to her room. The event I had so fondly, eagerly, looked forward to for years, feeling it would be such a joy in comparison to the weddings of our daughters, and the sorrow of parting with them, was now really coming to pass. How I realised this as I clasped dear Alix in my arms!

TO THE CROWN PRINCESS OF PRUSSIA   6 November 1862

Now to tell you that darling Alexandra arrived safely last night by moonlight, with the ships lit up by blue lights—a touching landing ... I waited downstairs, and received her in the hall [at Osborne]. She looked lovely and fresh and is very dear, but very shy when others are presented to her. We dined *à quatre* in the Council room and stayed a short while together. Dear sweet Alexandra looks lovely and clings with such affection to Lenchen—who is obliged to do every thing. It rains and I fear we can't show off the place. I must get a little mouthful of air—for in my weak state I require it ...

Tell Bertie I am grieved to find that in spite of all I said and all he promised he never writes to Alix in any thing but English! This I know is mere laziness and it grieves and pains me as the German element is the one I wish to be cherished and kept up in our beloved home—now more than ever; it is doubly necessary in this case, as Alix's parents are inclined to encourage the English and merge the German into Danish and English and this would be a dreadful sorrow to me; the very thing dear Papa and I disliked so much in the connexion is the Danish element.

8 November 1862

I can't say how I, and we all love her [Princess Alexandra]! She is so good, so simple, unaffected, frank, bright and cheerful, yet so quiet and gentle, that her presence soothes me. Then how lovely! She is quite at home, comes in and out of the room to me as the sisters, is most attentive and dear in her manner to me and quite at ease with us all. She and Lenchen adore one another, and seem to suit so well. Oh! may Bertie be worthy of such a sweet wife! Does he quite deserve it? But he will, if he does all he can to follow in beloved Papa's footsteps and to remember what a husband he was!

12 November 1862

We cannot thank you enough for all you have both done in securing for dear Bertie and us—this jewel. She is one of those sweet creatures who seem to come from the skies to help and bless poor mortals and brighten for a time their path! You couldn't know her, but you guessed what she was, and we love her. She lives in complete intimacy with us and she is so dear, so gentle, good, simple, unspoilt—so thoroughly honest and straightforward—so affectionate; she has been sitting for an hour with me this evening and I told her all about former happy times, our life, a great deal about dearest Papa, whom she seems to love quite dearly and to long so to see; all about his illness; she showed such feeling, laid her dear head on my shoulder and cried.

28 November 1862

Dear sweet Alix left us yesterday evening and I hardly could bear to let her go! I always tremble lest something should happen to her! She seems to be too charming.

She loves us all, and was much affected when she left us.

I hope you have Germanised Bertie as much as possible, for it is most necessary [he was staying with his sister in Germany]. You speak of time taking away the edge of sorrow. No doubt it does and perhaps ought to those who have still a dearly beloved husband to lean on—and to impart every thing to! Yes, it is easy for them to say that—but not to one, who every, every day meets with fresh cause to pine and long for him who was her guide and support and help. But I never, never shall be able to bear that dreadful, weary, chilling, unnatural life of a widow! It is too dreadful for any one to conceive who is still blessed with a loving husband.

Dear Alice and Louis are very happy—and he quietly, really so; I like him more and more. He is so thoroughly amiable and unpretending and very intelligent. May God bless them.

3 December 1862

Bertie arrived at 8 looking extremely well and really very much improved. It is such a blessing to hear him talk so openly, sensibly and nicely—with such horror of what is bad that I feel God has been merciful in listening to our prayers!

I have been for the first time to poor Buckingham Palace today and went through all our dear rooms! Though dismantled all was standing as in happy, happy, busy times, and I cannot say how overwhelmed I was!

10 December 1862

We are very much occupied (I wish I need not be) with the preparations and arrangements for the marriage. And it will be a tour de force to squeeze every one in.

JOURNAL   14 December 1862

Oh! this dreadful, dreadful day [the first anniversary of the Prince Consort's death]! At 10 we went into the dear room (all the children but Baby there) and Dr. Stanley [later Dean of Westminster who had accompanied the Prince of Wales on his tour in the Near East] most kindly held a little service for us, reading Prayers and some portions of the 14th and 16th Chapters of St. John, and spoke a few and most comforting and beautiful words. The room was full of flowers, and the sun shining in so brightly, emblems of his happiness and glory, which comforted me. I said it seemed like a birthday, and Dr. Stanley answered, 'It is a birthday in a new world.' Oh! to think of my beginning another year alone!

TO THE CROWN PRINCESS OF PRUSSIA   17 December 1862

I have to thank you for two letters from Rome . . . But I have had none for the sacred and heart-rending anniversary of the 14th, and only heard of your spending that day in the railroad and arriving that evening at Vienna—where the Emperor received you!!! Even now I can hardly believe it! I should have thought that you would have preferred remaining in the smallest wayside inn and going to pray to God to support your broken-hearted mother rather than do that! What I and we all here have felt I will not now say.

22 January 1863

You never will believe how unwell and how weak and nervous I am, but any talking or excitement is far too much for me. I must constantly dine alone, and any merriment or discussion are quite unbearable. Poor Bertie is far too noisy for me; he is very fond of disputing and this is what in my best times I never could bear, but now without beloved Papa to lead the conversation and to check the young people, it is quite too much for my very shattered nerves.

28 January 1863

Bertie returns today—and then my bad headaches will begin I fear. But he is very much pleased with his place and takes great interest in it [Sandringham which he had bought for £220,000].

4 February 1863

Bertie's marriage is finally to be on Tuesday the 10th; Alix etc. to arrive on the 7th. Oh! if it only was all over! I dread the whole thing awfully and wonder even how you can rejoice so much at witnessing what must I should think be to you, who loved Papa so dearly, so terribly sad a wedding! Dear child! your ecstasy at the whole thing is to me sometimes very incomprehensible! Think what it will be to see Windsor full of people, and both your parents absent; a marriage in state, also without them—that day to which we looked forward with such joy for many years and which now is to me far worse than a funeral to witness! Will you be able to rejoice when at every step you will miss that blessed guardian angel, that one calm great being that led all? And poor B.

JOURNAL   9 February 1863

Twenty-three years ago this day fell on a Sunday. There was a service in Buckingham Palace in the Bow Room, dearest Albert coming up afterwards and giving me the beautiful sapphire and diamond brooch ... After lunching together with Mama, my Father-in-law and Ernest, I saw dear Albert drive out through an immense crowd to pay his visits to the Royal Family. Then he came and sat with me and we read over the marriage service together and he was so dear and kind. I can still feel all the excitement at that last dinner and the joy and pride at knowing he was to be my very own and now, all is past and gone for ever like a dream. Oh! it is too dreadful!

TO LORD GRANVILLE   10 February 1863

[The Queen] quite approves of the advice [Lord Granville] gave the Prince of Wales, respecting the Literary Fund dinner, and she is at a loss to understand how any true friend of our son could advise him to preside at any such dinner; for he is far too young and inexperienced to take part in such Societies. Some years hence this might be different, but, till a few years have passed, the Queen thinks he should upon no account be put at the head of any of those Societies or Commissions, or preside at any of those scientific proceedings, in which his beloved great Father took so prominent a part. It would not be at all fair by the Prince of Wales.

With respect to his attendance in the House of Lords, the Queen thinks, that, whenever there is anything of interest or importance going on, and the Prince of Wales is in town, he should attend; but she is clearly of opinion that he should not do so regularly, for many reasons, which she can state verbally to Lord Granville.

The Queen thinks (and the Prince of Wales quite agrees in this) that with the exception of Lord Granville, Lord Palmerston, and possibly Lord Derby, and the three or four only great houses in London, Westminster House, Spencer House, Apsley House, the Prince and future Princess of Wales should not go out to dinners and parties, and not to all these the same year.

TO THE CROWN PRINCESS OF PRUSSIA   11 February 1863

I wish you would say in your affectionate letters that you will do all to help me in checking noise, and joyousness in my presence for I fear you always think that I am not ill, that I can bear it, and I cannot and it makes me wretched and miserable beforehand if I think I shall be excited. I cannot join you at dinner; I must keep very, very quiet.

TO KING LEOPOLD   14 February 1863

You say that work does me good; but the contrary is the fact with me, as I have to do it alone, and my Doctors are constantly urging upon me rest. My work and my worries are so totally different to any one else's: ordinary mechanical work may be good for people in great distress, but not constant anxiety, responsibility, and interruptions of every kind, where at every turn the heart is crushed and the wound is probed!

TO THE CROWN PRINCESS OF PRUSSIA   14 February 1863

The architect's sketches for the memorials [the Albert Memorial] are all up—in St. George's Hall—and will remain till the 28th, and no decision be taken till you arrive. There is only one that would come within the means, and which might be made very handsome. The idea is handsome. One by [George Gilbert] Scott is very handsome, but too much an imitation of W[alter]. Scott's [memorial in Edinburgh] and too like a market cross.

Have you read Kinglake's book [*The Invasion of the Crimea* which was highly critical of Napoleon III and far from complimentary about the British Government]? It is very scurrilous. I go daily to the beloved Mausoleum, and long to be there!

JOURNAL   24 February 1863

Vicky arrived [for the Prince of Wales's wedding] safely yesterday, and is looking really very pretty, so young and fresh and slim. I can but look at her, thinking how pleased her precious Papa would have been, and how happy, how proud we might have been to have our two married daughters and our little grandson [Prince William of Prussia, afterwards the Kaiser, William II] with us, and to have received this lovely Bride! Now all, all is spoilt; a heavy black cloud overhangs every thing now, and turns pleasure into woe.

7 March 1863

The bells began to ring, and at length, in pouring rain and when it was getting dark, the carriages and escort were seen coming.

I went down nearly to the bottom of the staircase, and Bertie appeared, leading dear Alix, looking like a rose. I embraced her warmly, and with her parents, Dagmar, the two sons, Thyra and Waldemar, went upstairs. Alix wore a grey dress, with a violet jacket, trimmed with fur, and a white bonnet. We all went into the White Drawing-room, where we remained a few minutes, and then Vicky took them over to their rooms. I went back to my room, desolate and sad. It seemed so dreadful that all this must take place, strangers arrive, and he, my beloved one, not be there! Vicky and Alice soon came to me and tried to cheer me, and kiss away my tears.

While I was waiting, Vicky returned and was sitting with me, dressed for dinner, when dear gentle Alix knocked at the door, peeped in, and came and knelt before me, with that sweet, loving expression which spoke volumes. I was much moved and kissed her again and again. She said the crowd in London had been quite fearful, and the enthusiasm very great, no end of decorations, etc., but the crush in the City had been quite alarming. Bertie came in for a moment whilst Alix was there. There was a family dinner, I dining alone.

To Palmerston   7 March 1863

The Queen is indeed most deeply touched and gratified at the extraordinary exhibition of loyalty and affection exhibited on the occasion of the arrival of our future daughter, a tribute which she well knows, and wishes all should know, is owing to her great and good husband, who led the Queen in the right path and to whom she owes everything and the country owes everything!

Journal   9 March 1863

Drove with Alix, Lenchen and Bertie to the Mausoleum where Vicky and Fritz met us. I opened the shrine and took them in. Alix was much moved and so was I. I said, 'He gives you his blessing', and joined Alix and Bertie's hands, taking them both in my arms and kissing them. It was a very touching moment and we all felt it . . . Saw General Grey who was very kind and feeling and had the tears in his eyes. He said there was but one feeling in all this, sympathy for me and regret that my beloved one was not spared to be with us! . . . Greatly agitated at the thought of tomorrow.

10 March 1863

All is over and this (to me) most trying day is past, as a dream, for all seems like a dream now and leaves hardly any impression upon my poor mind and broken heart! Here I sit lonely and desolate, who so need love and tenderness, while our two daughters have each their loving husbands, and Bertie has taken his lovely, pure, sweet Bride to Osborne, such a jewel whom he is indeed lucky to have obtained. How I pray God may ever bless them! Oh! what I suffered in the Chapel [St. George's Chapel, Windsor], where all that was joy, pride, and happiness on January 25th, '58 [when the Princess Royal was married at St. James's], was repeated without the principal figure of all, the guardian angel of the family, being there. It was indescribable. At one moment, when I first heard the flourish of trumpets, which brought back to my mind my whole life of twenty years at his dear side, safe, proud, secure, and happy, I felt as if I should faint. Only by a violent effort could I succeed in mastering my emotion!

But now I must return to the beginning of the day. Directly after breakfast went over to the State Rooms, to embrace darling Alix, and give her my blessing. Her mother was much affected. Went with her into Alix's bedroom, where she was in her dressing-gown, and very *émotionnée*. Then I went back to my room and could see from my windows all the crowds of people assembling and arriving. Cold from nervousness and agitation, I dressed, wearing my weeds, but a silk gown with crape, a long veil to my cap, and, for the first time since December '61, the ribbon, star, and badge of the Order of the Garter, the latter being one my beloved one had worn, also the Victoria and Albert Order, on which I have had dearest Albert's head put above mine, and a brooch containing a miniature of him set round with diamonds, which I have worn ever since '40.

We started from the usual door, going on to the North Terrace, where we got out and went through a covered way down the small stairs, quite quietly, up into the Deanery. A Guard of Honour was mounted in the Quadrangle. Before I had left I had seen Lenchen in her pretty dress and train, lilac and white, and Louise and sweet baby, the same colours. Louise wore the pearls belonging to dearest Albert's mother, which he had always intended to give her. To see them go alone was dreadful. We waited a short while in the Deanery, and then went along a covered way prepared over the leads, which brought us into the Royal Closet. The divisions had been removed, and, when I stepped up to the window, the Chapel full of smartly dressed people, the Knights of the Garter in their robes, the waving banners, the beautiful window, altar, and reredos to my beloved one's memory, with the bells ringing outside, quite had the effect of a scene in a play.

Sat down feeling strange and bewildered. When the procession entered to the playing of the March in *Athalie*, and after Aunt Cambridge [the Duchess of Cambridge], Mary [the Duchess of Cambridge's daughter, afterwards Duchess of Teck], and our five fatherless children (the three girls and two little boys) came into view, the latter without either parent (at Vicky's wedding they walked before, behind, and near me), I felt terribly overcome. I could not take my eyes off precious little baby, with her golden hair and large nosegay, and smiled at her as she made a beautiful curtsey. Everyone bowed to me. I quite overlooked Alice coming in, looking extremely well in a violet dress, covered with her wedding lace, and a violet velvet train, from the shoulders trimmed with the miniver beloved Mama had worn at Vicky's wedding, Louis in the Garter robes leading her. Last came dear Vicky (leading little William), in a white satin dress trimmed with ermine, etc. When she caught sight of me, coming up the Choir, she made a very low curtsey, with an inexpressible look of love and respect, which had a most touching effect. There was a pause, and then the trumpets sounded again, and our boy, supported by Ernest C[oburg] and Fritz [the Crown Prince of Prussia, his brother-in-law], all in Garter robes, entered; Bertie looking pale and nervous. He bowed to me, and during the long wait for his Bride kept constantly looking up at me, with an anxious, clinging look, which touched me much. At length she appeared, the band playing Handel's Processional March, with her eight Bridesmaids, looking very lovely. She was trembling and very pale. Dearest Albert's Chorale was sung, which affected me much, and then the service proceeded. When it was over, the young couple looked up at me, and I gave them an affectionate nod and kissed my hand to sweet Alix. They left together, immediately followed by all the others, Beethoven's Hallelujah Chorus (from *The Mount of Olives*) being played.

I went back to the Castle, getting out at the North Terrace, and went upstairs for a few minutes. Then hearing the couple were coming, I hastened down the Grand Staircase (the first time since my misfortune) where all the Beefeaters were drawn up. My only thought was that of welcoming our children, and I stepped out and embraced both dear Bertie and Alix most warmly, walking upstairs next to them and past several of the guests, who had already arrived. ... Went then with Alice over to the Dining-room, and afterwards to the White Drawing-room, where the young couple and all the others came, for the signing of the Register, which took a very long time. A family luncheon of thirty-eight followed, in the Dining-room ... I lunched alone with baby. The two luncheons over, I went back to the White Drawing-room, the whole assemblage of Royalties being in the Green Drawing-room.

Shortly afterwards went over with Lenchen, baby, etc., to the other side, where all the family were assembled, and Bertie soon appeared, then darling Alix, looking lovely in a white silk dress, lace shawl, and white bonnet with orange flowers. She was much agitated and affected, and was embraced by all her family, who were in tears; then I once more embraced her and Bertie, with feelings I cannot describe, and gave them my warmest blessing. When I saw them go down the crowded staircase, I hurried back to the Corridor, and from there saw the open carriage in which the young couple were seated, and they stopped for a moment under the window, Bertie standing up, and both looking up lovingly at me.

Then we hastened to my room, where I saw them drive off, through the enthusiastic crowds. It was so like our driving away twenty-three years ago to Windsor, amidst the same crowds and shouts of joy! Aunt Cambridge and Mary came in to wish me good-bye, and then I drove with Lenchen down to the Mausoleum, and prayed by that beloved resting-place, feeling soothed and calmed.

TO THE CROWN PRINCESS OF PRUSSIA   18 March 1863

That unbounded love for every thing English I own I can't share. Loving and admiring my own country as I do, I have seen so much of the cold, harsh cheerlessness of my countrymen and have seen with such grief the very bad effect it has had on your two elder brothers—in so many ways—that I cannot admire much of it as you do. I know the bad and mischievous sides of it more than you can do ...

The young couple returned [to Windsor after their honeymoon at Osborne] yesterday looking well and happy. To look at darling Alix and into those eyes is a satisfaction. And then she is so quiet, so placid, that it is soothing to one, and I am sure that must do Bertie good. He looks bright and happy and certainly totally different to what he used to be. Pray, dearest, when you write to Bertie and Affie don't write with frantic adoration of the Navy and all English feelings—for our sole object is to smooth that down and to Germanise them! Now, without beloved Papa, it is more than ever. To your sisters that is quite different.

21 March 1863

Dear Alix is not I fear reasonable or careful of her health and I must speak seriously to both else there will be mishaps and an end to good health and possibly to much of their happiness. It is amusing to see how Bertie keeps her in order (not in an improper way) and takes care of her; so, I hope to be able to get him to understand what is necessary. She will require care, that I am sure of.

TO PALMERSTON   21 March 1863

The Queen is glad to hear that Lord Palmerston proposes to give notice on Monday for a vote in aid of the national memorial to the beloved Prince. But she trusts that he will not propose any sum so utterly unworthy of the House of Commons and the country as £30,000. Lord Derby, to whom the Queen will now write on the subject, himself mentioned £50,000 as a fitting sum for Parliament to vote—and every opinion which the Queen had heard on the subject points to that as the very lowest sum which it would become the House to appropriate to such an object. The saving is really a paltry one, and the Queen trusts that her ministers will not place themselves in the position of having possibly to resist an amendment having for its object to make the vote more worthy of the country.*

TO THE CROWN PRINCESS OF PRUSSIA   25 March 1863

The dear young couple are here and I must again say that I am quite astonished at Bertie's improvement. Dear Alix felt the parting from her parents very much, but she is always calm and sweet and gentle and lovely. Very clever I don't think she is, but she is right-minded and sensible and straightforward. Dagmar [Princess Alexandra's sister] is cleverer, and would I am sure be very fit for the position in Russia; she is a very nice girl.

8 April 1863

Of course I shall take great interest in our dear little grand-daughter [Princess Victoria of Hesse], born at poor, sad, old Windsor in the very bed in which you all were born, and poor, dear Alice had the same night shift on which I had on when you all were born! I wish you could have worn it too. But I don't admire babies a bit the more or think them more attractive.

18 April 1863

Many, many thanks for your dear letter of the 14th and for your good wishes for our dear little darling Baby! She is the only thing I feel keeps me alive, for she alone wants me really. She, perhaps as well as poor Lenchen, are the only two who still

---

*Parliament voted £50,000 for the Albert Memorial which was designed by George Gilbert Scott. Its ultimate cost, including the statue of the Prince by John Foley, was £120,000.

love me the most of any thing—for all the others have other objects . . . I know how you all love me, but I see and feel with my terribly sensitive feelings that constantly I am *de trop* to the married children and that every thing I love I must give up!

Alice's child is to be called Victoria Alberta Elisabeth Matilde Marie, and will be called Victoria—the first of our grand-children that will be called after either of us. This I know was not your fault—but it grieved me.

22 April 1863

Poor, dear child! for you with your clever mind and love of all that beloved Papa loved, the loss of that intercourse is quite dreadful! How I feel that, words cannot say, but it is not the least of all my bitter sufferings to feel what you children have lost! I always feel when I write to you what stupid, uninteresting letters mine are—for Papa's were and are so beautiful. You possess a mine of wealth in those letters . . .

Bertie and Alix are here since Saturday. I do so wonder how she can be happy. He has let himself down to his bad manners again. She is dear and good but I think looks far from strong and will never be able to bear the London season unless she has but few late nights. She is but 18 and has gone through so much.

To Disraeli
24 April 1863

The Queen cannot resist from expressing, personally, to Mr. Disraeli her deep gratification at the tribute he paid to her adored, beloved, and great husband [in the House of Commons in a speech about the proposed Albert Memorial]. The perusal of it made her shed many tears, but it was very soothing to her broken heart to see such true appreciation of that spotless and unequalled character.

To the Crown Princess of Prussia
6 May 1863

I fear there is none [no signs of pregnancy] with Alix and though to be sure, unintellectual children which one might fear with B.'s children, would be a great misfortune, it would be very sad if they had none, and I sometimes fear they won't. Are you aware that Alix has the smallest head ever seen? I dread that—with his small empty brain—very much for future children. The doctor says that Alix's head goes in, in the most extraordinary way just beyond the forehead: I wonder what phrenologists would say.

To King Leopold
12 May 1863

To show how nervous and weak I am, I made the effort to go and visit the truly magnificent Military Hospital at Netley, in which my Angel took such immense interest and constantly went to see; I felt it a duty, and I don't regret it; it was the first time I had gone anywhere, where Officers, etc. (tho' it was as private as possible) accompanied me! I had never in my life gone to a Hospital before I visited them with him during the War, and to walk alone, and see those poor men, some dying—without Albert—was dreadful! I went through it all, but I have been ill ever since—bad headaches—restless nights, and an increase of despair! It shows how shattered I am!

18 May 1863

Alice's departure is a great loss and adds to my loneliness and desolation! She is a most dear, good child, and there is not a thing I cannot tell her; she knows everything and is the best element one can have in the family! Louis, too, is quite excellent ... A married daughter I must have living with me, and must not be left constantly to look about for help, and to have to make shift for the day, which is too dreadful! I intend (and she wishes it herself) to look out in a year or two (for till nineteen or twenty I don't intend she should marry) for a young, sensible Prince, for Lenchen to marry, who can during my lifetime make my house his principal home. Lenchen is so useful, and her whole character so well adapted to live in the house, that (unless Alice lived constantly with me, which she won't) I could not give her up, without sinking under the weight of my desolation. A sufficient fortune to live independently if I died, and plenty of good sense and high moral worth are the only necessary requisites. He need not belong to a reigning house.

TO THE CROWN PRINCESS OF PRUSSIA   19 May 1863

The Drawing-room was a fearful one ... Poor dear [Princess Alexandra], she looks so sallow and is losing her 'fraîcheur'. Alas! she is deaf and everyone observes it, which is a sad misfortune. Strong she is not, and they overtire her too much.

Alfred is well but not strong; he is quite wild about salmon fishing. In confidence I may tell you that we do all we can to keep him from Marlborough House as he is far too much 'épris' with Alix to be allowed to be much there without possibly ruining the happiness of all three and Affie has not the strength of mind (or rather more of principle and character) to resist the temptation, and it is like playing with fire. Beloved Papa always said the feelings of admiration and even love are not sinful—nor can you prevent the impulses of one's nature, but it is your duty to avoid the temptation in every way.

8 June 1863

Bertie and Alix left Frogmore today—both looking as ill as possible. We are all seriously alarmed about her—for though Bertie writes and says he is so anxious to take care of her, he goes on going out every night till she will become a skeleton, and hopes there cannot be!! I am quite unhappy about it. Oh! how different poor, foolish Bertie is to adored Papa whose gentle, loving, wise, motherly care of me when he was not 21 exceeded everything.

TO KING LEOPOLD   16 June 1863

You will, I trust, excuse me if my letter is short and ill-written, but since Dr. Jenner wrote to you in my name, I have been so unwell, the result of over-exertion this last week, that I can hardly hold my pen for shaking, and hardly know what I am about. I was so unwell on Sunday, from violent nervous headache and complete prostration, that I nearly fainted, and Clark and Jenner both say that, with the extreme state of weakness which I am in, if I did faint I might not come back to life. My weakness has increased to that extent within the last two months, as to make all my good doctors anxious. It is all the result of overwork, over-anxiety, and the weight of responsibility and constant sorrow and craving and yearning for the one absorbing object of my love, and the one only Being who could quiet and calm me; I feel like a poor hunted hare, like a child that has lost its mother, and so lost, so frightened and helpless.

I own, beloved Uncle, that I think my life will end more rapidly than any of you think; for myself this would be the greatest, greatest blessing.

TO THE CROWN PRINCESS OF PRUSSIA   27 June 1863

I fear she [Princess Alexandra] will never be what she would be had she a clever, sensible and well-informed husband, instead of a very weak and terribly frivolous one! Oh! what will become of the poor country when I die! I foresee, if B. succeeds, nothing but misery—for he never reflects or listens for a moment and he [would] do anything he was asked and spend his life in one whirl of amusements as he does now! It makes me very sad and angry.

TO KING LEOPOLD   9 July 1863

How deeply will you mourn for our dearest, wisest, best, and oldest friend—Stockmar [who had died at Coburg on 9 July]. Last night I felt sure that the end was near at hand, but the loss is totally irreparable! To him my Angel looked for advice and support, and his troubles and anxieties certainly greatly increased after Stockmar left! Again and again he longed for Stockmar.

TO THE CROWN PRINCESS OF PRUSSIA   11 July 1863

Our dearest, beloved, wisest friend! I knew he was ill, as I wrote to you, but never never did I dream of a sudden blow like this which would deprive us of this invaluable friend! I fully believe that if the dearest old Baron had continued his visits to England adored Papa and he would both be alive—for Stockmar's support and advice lightened Papa's burthen so much, and being here kept up the dearest Baron's spirits and nerves!

Bertie and Alix are at Frogmore since the 8th (having much over fatigued themselves in London, as every one has observed) and stay there till the 18th, when they come here for a fortnight, and I hope she may then bathe, and get a little fatter and stronger. There are no hopes and I sometimes have my fears and misgivings about it altogether ... I send you dear Alix's letter I received this morning to show you that, though very affectionate and dear, she does not write well. I fear the learning has been much neglected and she cannot either write or I fear speak French well.

TO PALMERSTON   11 August 1863

Before leaving on the Queen's sad journey [to Coburg] (which she dreads, as her nerves are so shattered that any exertion is great pain), she wishes to state once more her desire that no step is taken in foreign affairs without her previous sanction being obtained.

TO THE CROWN PRINCESS OF PRUSSIA   22 September 1863

Bertie and Alix seem going on extremely well; he is much improved and she has done a great deal for him, and is indeed a dear, noble excellent being whom one must love and respect. There is no doubt, I am sure about her condition, for, though she has not suffered from sickness and has been particularly well here—she has increased very much in size—her waist being quite broad and her clothes having all to be let out.

Dear Alice and Louis are a great comfort to me—so good and kind and so quiet; he is so improved, really so excellent, with much decision and firmness of character. I do love him dearly. Oh! if the boys (our sons) had his golden heart—especially Affie—who is a slippery youth, for I never feel sure (alas!) of what he says.

TO KING LEOPOLD   12 November 1863

You will be sorry to hear Lady Augusta [Bruce, General Bruce's sister and a favourite Lady-in-Waiting], at 41, without a previous long attachment, has, most unnecessarily, decided to marry (!!) that certainly most distinguished and excellent man, Dr. Stanley [soon to become Dean of Westminster]!! It has been my greatest sorrow and trial since my misfortune! I thought she never would leave me! She seems, however, to think that she can by his guidance be of more use than before even. She will remain in my service and be often with me, but it cannot be the same, for her first duty is now to another!

TO THE PRINCE OF WALES   13 January 1864

I wish now to say a few words again about the names, sponsors, and christening.* I will begin with the last, that is to say, that though I should for myself prefer its being at Windsor, which is now associated with so much that is precious to me, I quite agree to its being best for the people of London that they should not be deprived of the honour and gratification of having some event in town; and by having it, as all our christenings but two were, in the private chapel at Buckingham Palace, I think I shall be able to be present, and hold the dear baby myself, D.V., which, trying though it will be, I wish to do. Don't think of settling the time for it, till you have consulted the doctors . . . for with so small a child, who won't be at its full size for six weeks, the christening ought rather to be delayed; ours were generally nearly two months old, and I think you would find it would be safer for the baby, if that were the case.

As regards the names, if others besides Albert Victor are added (which I don't the least object to), you must take dear Uncle Leopold's also. You could not give King Christian's and the Landgrave's without also giving Uncle Leopold's. I would advise reserving Edward for a second or third son.

Respecting your own names, and the conversation we had, I wish to repeat, that it was beloved Papa's wish, as well as mine, that you should be called by both, when you became King, and it would be impossible for you to drop your Father's. It

*Prince Albert Victor Christian Edward, later Duke of Clarence, the Prince of Wales's first child had been born prematurely on 8 January at Frogmore. The Queen had gone to the Princess to offer her (not altogether welcome) help and had stayed with her for several days.

would be monstrous, and Albert alone, as you truly and amiably say, would not do, as there can be only one Albert! You will begin a new line, as much as the Tudors and Brunswicks, for it will be the Saxe-Coburg line united with the Brunswick, and the two united names will mark it, in the way we all wish, and your son will be known by the two others, as you are by Albert Edward.

For some time now the Queen had been much distressed by the crisis presented by the Duchies of Schleswig and Holstein which had been ruled for years by the Kings of Denmark but which the Germans considered they had a good right to annex.

To RUSSELL   4 October 1863

The Queen must repeat the expression of her determination not to consent to any measures which may involve her in the threatened rupture between Denmark and Germany, and she must ask that no step may be taken which may commit her Government, without the mature consideration of the Cabinet.

To KING LEOPOLD   9 November 1863

Fritz is very violent, Vicky sensible ... and I miserable, wretched, almost frantic without my Angel to stand by me, and put the others down, and in their right place! No respect is paid to my opinion now, and this helplessness almost drives me wild, and in the family his loss is more dreadfully felt than anywhere. It makes visits like Fritz and Vicky's very painful and trying. Oh! God, why, oh! why was all this permitted? and now this year everything that interested my Angel and that he understood takes place, and he is not here to help us, and to write those admirable memoranda which are gospel now. Oh! my fate is too too dreadful! If I could but go soon to him, and be at rest! Day and night I have no rest or peace.

                                                    3 December 1863

I am much worried by this S.-Holstein business, which is so strongly felt in Germany ... With him [the Prince Consort] by my side, I needed, so far less, all assistance and advice; but now? I hope and think that I shall not long require any, for I hope I am gradually nearing the end of my sad and wearisome journey.

To RUSSELL   24 December 1863

The news from Copenhagen and Germany are both very alarming.

To PALMERSTON   8 January 1864

The Queen has read with the greatest alarm and astonishment the draft of a despatch ... in which Lord Russell ... stated that, in the event of the occupation of Schleswig by Prussia ... Denmark would resist such an occupation and that Great Britain would aid her in that resistance. The Queen has never given her sanction to any such threat, nor does it appear to agree with the decision arrived at by the Cabinet upon this question ... England cannot be committed to assist Denmark in such a collision.

The Queen has declared that she will not sanction the infliction upon her subjects of all the horrors of war, for the purpose of becoming a partisan in a quarrel in which both parties are much in the wrong.

To the Crown Princess of Prussia 27 January 1864

With regard to this sad S. Holstein question, I can really speak with more thorough impartiality than anyone (and that the dear Crown Prince can bear witness to); my heart and sympathies are all German.

Where I do, however, blame Germany is in their wanting the two great Powers to break their engagements, and in not being contented with all the rights of the Duchies being obtained. They have mixed up the two questions, and gone so violently mad upon the subject, that they lose sight of the far greater evils which may be produced by provoking war . . .

That England is detested I know, alas! too well; but I must bear it, as many other trials and sorrows, with patience; and continue to do all I can to prevent further irritation and in future to avoid further complications. I am glad darling Papa is spared this worry and annoyance, for he could have done even less than I can.

To Palmerston 2 February 1864

The quarrel is now beyond our reach [because war had broken out and German forces had invaded Schleswig the day before], and we must wait to see the march of events. A time will very soon arrive, when our advice, and possibly our mediation, will be asked; and if it is given with perfect impartiality, and a due regard to the interests of all parties, and the wishes and rights of the peoples concerned, [it] may conduce to the peace and permanent security of Europe.

But till this time arrives, the Queen cannot but think that far the most dignified course for England will be to remain passive. The Queen would wish Lord Palmerston to let Lord Russell see this letter.

To the Crown Princess of Prussia 3 February 1864

Though I blame the haste and violence of the Germans my feelings and sympathies in the war can only be with them! . . . Poor Alix is in a terrible state of distress and Bertie frantic, thinking every one wishes to crush Denmark! This is not true.

13 February 1864

Oh! if Bertie's wife was only a good German and not a Dane! Not, as regards the influence of the politics but as regards the peace and harmony in the family! It is terrible to have the poor boy on the wrong side, and aggravates my sufferings greatly.

To Russell 13 February 1864

Lord Russell already knows that she will never, if she can prevent it, allow this country to be involved in a war in which no English interests are concerned.

To King Leopold   25 February 1864

I long for quiet and peace, and to be enabled to dwell on the blessed future!

I never really realised the power of prayer till now! When in an agony of loneliness, grief, and despair, I kneel by that bed, where he left us, decked with flowers, and pray earnestly to be enabled to be courageous, patient, and calm, and to be guided by my darling to do what he would wish; then, a calm seems to come over me, a certainty my anguish is seen and heard not in vain, and I feel lifted above this miserable earth of sorrows!

To the Crown Princess of Prussia   2 March 1864

The newspapers are very bad, and the feeling against Prussia very strong. I saw Mr. Oliphant [Laurence Oliphant who had been out to see something of the Schleswig-Holstein war] on Sunday (whom you never told me you had seen). He is such an agreeable clever man. I had seen him formerly. Augusta and the little Dean were here on Monday—looking as unsuited as possible. He runs after her like a little boy and looks at her whenever he speaks!! Both were rather embarrassed. He read to us after dinner part of a lecture of his on Isaiah—which is very interesting.

19 March 1864

Many thanks, for your dear letter of the 16th in which however you neither take notion of that dear, sad anniversary, nor of Louise's birthday which was yesterday—nor did you telegraph for either!

To Russell   13 April 1864

The Queen much regrets the extravagant excitement respecting Garibaldi [who had been most enthusiastically received on a visit to England], which shows little dignity and discrimination in the nation, and is not very flattering to others who are similarly received.

The Queen fears that the Government may find Garibaldi's views and convictions no little cause of inconvenience with foreign Governments hereafter, and trusts they will be cautious in what they do for him in their official capacity. Brave and honest though he is, he has ever been a revolutionist leader.

To Granville   21 April 1864

The Queen feels it a duty from which she must not shrink to call the attention of Lord Granville to the report, in the *Globe* of last night, of speeches made by Mazzini and Garibaldi on Sunday ...

It appears to the Queen that the object for which it was declared that the Government should receive and honour Garibaldi, namely that of keeping him out of dangerous hands, has hardly been attained when he boasts himself to have been the pupil of Mazzini ... She cannot but deeply regret that, whatever personal feeling it may have been right to testify towards a man of remarkable honesty and singleness of purpose, the members of her Government should have lavished honours usually reserved for Royalty upon one who openly declares his objects to be to lead the attack upon Venice, Rome, and Russia, with the Sovereigns of which countries the Government, in her name, profess sentiments of complete friendship and alliance.

The Queen thinks that the representatives of these countries might well remonstrate at the unusual adulation shown in official quarters to one professing objects so hostile to their Royal Masters.

TO THE CROWN PRINCESS OF PRUSSIA 27 April 1864

Garibaldi—thank God!—is gone. It has been a very absurd and humiliating exhibition and was becoming very dangerous by the connection with Mazzini and all the worst refugees etc. The whole crowned by the incredible folly and imprudence of your thoughtless eldest brother going to see him without my knowledge! It has shocked me much and his people are much to blame. For no foreigner can be presented to the Royal Family who has not been received by the Sovereign and none be presented to him or her except by his own Minister!

TO PALMERSTON 2 May 1864

As long as the Queen exercises her functions for the good of the country alone, and according to that Constitution which has through her reign been her sole guide, she must be content to see unjust remarks [about her alleged pro-German feelings] in obscure newspapers and must continue to disregard them. She quite agrees with Lord Palmerston that it ought to be put into the fire.

TO LORD CLARENDON 9 May 1864

The Queen is truly thankful to hear of the Armistice, though she wishes it was for longer. But she concludes it can be renewed. The Queen grieves to see such a disposition to put the worst interpretation upon the actions and motives of the Prussians. Most earnestly must she repeat her wish and indeed her hope and trust that her Government will enter upon the discussions with a true spirit of impartiality, and not with the decided Danish view which has, alas! all along characterised their conduct, and which has done so much harm, for it has encouraged the Danes all along, and has made Germany look upon us as her enemies.

TO KING LEOPOLD 12 May 1864

I have asked General Grey to write to you about Politics. Pilgerstein [Palmerston, called Pilgerstein by the Prince Consort, *Pilger* being German for palmer] is gouty, and extremely impertinent in his communications of different kinds to me.

TO CLARENDON 17 May 1864

It strikes the Queen that it would be very useful, if Lord Clarendon would take an opportunity of seeing the Prince of Wales,* and preparing him for what the Danish

---

*The Prince had made no secret of his pro-Danish sympathies and of his disapproval of the British Government's refusal to support his father-in-law. 'This horrible war,' he had written, 'will be a stain for ever on Prussian history, and I think it is very wrong of our Government not to have interfered.'

Government, in all probability, will have to consent to . . .

The Queen would not wish Lord Clarendon to enter into too many details, or to say anything of a very confidential nature; but to speak generally, and in doing so, to speak as he, and as Lord Russell did, to the Queen—in a true spirit of impartiality; showing that (as Lord Clarendon said to the Queen) 'you cannot ignore the strong and unanimous feeling of whole Germany'; and cautioning the Prince of Wales against violent abuse of Prussia; for [it] is fearfully dangerous for the Heir to her throne to take up one side violently; while he is bound by so many ties of blood to Germany, and only quite lately, by marriage, to Denmark . . .

In the interest of this country, above all, but also in the interest of Denmark, it is most essential that the Prince of Wales should understand this. And she feels sure that Lord Clarendon can do this better than anyone else.

To Russell   27 May 1864

The Queen hears this evening, from Lord Granville, what passed yesterday in the House of Lords,* and though she has not seen what was said, she wishes to thank Lord Russell for having spoken the truth.

It is not the first time that Lord Russell has had to defend the highest in the country from base calumny; and she never forgot what he did in '54 [when he defended the Prince Consort against charges of improper interference in affairs of state]; but she must own that she thought her terrible misfortunes, her unprotected position without a husband to stand by and protect her, her known character for fearless straightforwardness, her devotion now and ever to her country (a proof of which is her weakened health and strength), ought to have prevented such an attack—which to a lady she can only characterise as ungentlemanlike . . . The Queen hopes everyone will know how she resents Lord Ellenborough's conduct and how she despises him!

Much concerned at your telegram [about the objections to the peace proposals raised by the German plenipotentiaries]. Would it be of any use the Queen writing to the King of Prussia? The Queen trusts to your not allowing the Conference to break off abruptly, without trying every means of keeping Peace.

To the King of Prussia   28 May 1864

You know how anxious I was from the first to prevent this unhappy war; and how much I have deprecated the violent language which has on both sides embittered the feelings of our two nations . . .

I should deeply regret if the prospect of Peace were to be marred by too great demands of Prussia. Your arms have been victorious, and it now depends on the use you make of your victory, whether the public opinion which now inclines to the weaker side be rallied on your side or not.

To the Crown Princess of Prussia   31 May 1864

I am grieved and distressed to say that the feeling against Prussia has become most

---

*Lord Ellenborough had alluded to the German sympathies of the Queen and had suggested that these influenced Government policies.

violent in England, and quite ungovernable, as I heard from everyone. The people are carried away by imaginary fancies, and by the belief that Prussia wants to have the duchies for herself, and that she has (and this we can't get entirely contradicted) broken through the stipulations of the armistice, by her exactions in Jutland.

I don't share these ideas, and invariably say that I know it to be false; but the feeling is there, and at present no reason is listened to. I hope that my opinions and my actions will not be quoted in opposition to my Government, for that beloved Papa never permitted; and in the present instance the Government are perfectly impartial, and only anxious to come to a settlement, which can, once for all, be accepted by Denmark and Germany.

To Russell   12 June 1864

The Queen hears with much regret of the decision at which the Cabinet has arrived—for it amounts to this: that if the Danes accept and Germany refuses the offer of arbitration, England will be pledged to declare war against Germany! ...

The Queen shudders to think of the position in which this country would thus be placed—standing alone in support of Denmark against all Europe!

Under no circumstances should a threat of affording material aid be a one-sided measure. In order to take away its offensive character it should be so worded as to show to both belligerents, that whichever side objected to a proposal, which must be regarded as equitable in itself, would forfeit all claim to the support of the English Government ...

The Queen would wish her letter to be brought before the Cabinet before any further step is taken, and with as little delay as possible.

To Clarendon   25 June 1864

Lord Clarendon will easily believe with what feelings of relief and thankfulness she has received Lord Clarendon's and Lord Russell's letters this evening. The terrible anxiety she experienced, unsupported and uncheered by her beloved great husband (by whose side she went through so many trials and anxieties with a light heart) ... was almost wearing her out.

To King Leopold   30 June 1864

After a week of the greatest anxiety, the only wise and reasonable course has been pursued; and this country is safe. I feel that my darling has blessed and guided me, and that he works on for us all. As Kingsley (the celebrated author) said to me on Sunday: 'I think that God takes those who have finished their career on earth to another and greater sphere of usefulness.' ...

The third Court (which was quite unexpected) gave much satisfaction, but what did even more was my drive to the station through the full Park, in my open carriage and four; it was quite unexpected, and, though very painful, pleased people more than anything; and, if done occasionally in this way, will I believe go farther to satisfy them than anything else always. I was thanked for it, and told how kind they felt it was of me.

Everyone said that the difference shown, when I appeared, and [when] Bertie and

Alix drive, was not to be described. Naturally for them no one stops, or runs, as they always did, and do doubly now for me.

14 July 1864

Dear little William, Vicky's eldest boy [the future Kaiser Wilhelm II], a sweet, darling, promising child, on whom my own darling doted, and who has that misfortune with his poor little left arm, it is, who is come for sea bathing and change of air ... The dear child remembers his dear grandpapa!

To Queen Augusta of Prussia   8 November 1864

My sojourn here [at Windsor] during this winter season, where everything reminds me of last year and of my Angel's last weeks, is truly melancholy and painful, and yet I love being here and re-living everything in my thoughts! But alas, how desolate it all is, how hard my life will be without joy! Time only makes things emptier and more lonely, and my outbursts of passionate grief are always comforting.

Tomorrow is another bitter day: Bertie's birthday! Since 1861 he has no longer kept it here, and before that always with jollifications! My Angel was always so good and affectionate to his children, and always wanted them to be gay and happy. On the last occasion I was very sad, and had ordered music to be played at dinner for the first time since the loss of dear Mama, and my beloved Albert ordered Bertie to take my arm! Tomorrow everything will be quiet!

To Russell   8 December 1864

The Queen would wish to say with reference to what Lord Russell said to her this morning about the opening of Parliament, that she would be thankful if he would take any opportunity that might offer to undeceive people upon that head.

The Queen was always terribly nervous on all public occasions, but especially at the opening of Parliament, which was what she dreaded for days before, and hardly ever went through without suffering from headache before or after the ceremony; but then she had the support of her dear husband, whose presence alone seemed a tower of strength, and by whose dear side she felt safe and supported under every trial.

Now this is gone, and no child can feel more shrinking and nervous than the poor Queen does, when she has to do anything, which approaches to representation; she dreads a Council even.

Her nerves are so shattered that any emotion, any discussion, any exertion causes much disturbance and suffering to her whole frame. The constant anxieties inseparable from her difficult and unenviable position as Queen, and as mother of a large family (and that, a Royal family), without a husband to guide, assist, soothe, comfort, and cheer her, are so great that her nervous system has no power of recovery, but on the contrary becomes weaker and weaker.

This being the case, Lord Russell (whose kind' consideration she fully appreciated) will at once see that any great exertion which would entail a succession of moral shocks as well as very great fatigue, which the Queen must avoid as much as possible, would be totally out of the question.

She has no wish to shut herself up from her loyal people, and has and will at any time seize any occasion which might offer to appear amongst them (painful as it ever is now), provided she could do so without the fatigue or exertion of any State ceremony entailing full dress, etc.

Lord Russell may make what use he chooses of the substance of this letter.

JOURNAL   14 December 1864

I cannot believe I am writing again for the third time on this terrible anniversary. It seems but yesterday and there again so far off . . . Went in [the Mausoleum] to pray and gaze at that peaceful and beautiful face of the statue. What a day of harrowing memories.

TO THE CROWN PRINCESS OF PRUSSIA   11 January 1865

I hope you don't spoil Affie with your admiration of him. He has already a good amount of vanity and nothing is worse than that brother-worship of sisters. But he is indeed in many things wonderfully like adored Papa—and his figure is a miniature of that angel! Oh! that he were as pure!

27 January 1865

Accept on this day my warmest good wishes for our darling William. That beloved and promising child was adored Papa's great favourite; he took (and he takes I am sure) so deep an interest in him and in his physical and moral well-doing. He is so dear and so good that with care and God's blessing he will grow up to be a blessing to his country and a comfort to his parents! But bring him up simply, plainly, not with that terrible Prussian pride and ambition, which grieved dear Papa so much and which he always said would stand in the way of Prussia taking that lead in Germany which he ever wished her to do! Pride and ambition are not only very wrong in themselves, but they alienate affection and are in every way unworthy of really great minds and great nations! I hope my writing case will have arrived in time and will give the darling pleasure.

Little A.V. [Albert Victor, the Prince of Wales's eldest son] is a perfect *bijou*—very fairy-like but quite healthy, very wise-looking and good. He lets all the family carry him and play with him—and Alix likes him to be accustomed to it. He is very placid, almost melancholy-looking sometimes. What is not pretty is his very narrow chest (rather pigeon-breasted) which is like Alix's build and that of her family and unlike you all with your fine chests.

TO KING LEOPOLD   24 February 1865

We had snow and frost, and now complete thaw—and to-day, pouring rain, which has made sad old Windsor look gloomier than ever! But it is full of precious recollections—dear beyond measure. I continue to ride daily (I fear to-day I shall not be able) on my pony, and have now appointed that excellent Highland servant [John Brown] of mine to attend me always and everywhere out of doors, whether riding or driving or on foot; and it is a real comfort, for he is so devoted to me—so simple, so intelligent, so unlike an ordinary servant, and so cheerful and attentive . . .

Oh! life goes on; young people are happy, and I at 45½ look at life as ended! Last Friday was our dear Wedding Day ... Oh! and now? But I must and do ever look back with gratitude unbounded on that blessed time.

TO THE CROWN PRINCESS OF PRUSSIA   1 March 1865

I had a reception yesterday of the whole Corps Diplomatique at Buckingham Palace—a great bore. There were a hundred of them with attachés. The good Bernstorffs [the Prussian Ambassador and his wife] were, as usual, in a sort of porcupine condition which is so odious. It seems to me such a loss of time to be always offended and Brown's observation about a cross person seems to me very applicable here 'it can't be very pleasant for a person themselves to be always cross' which I think so true and so original. His observations upon everything he sees and hears here are excellent and many show how superior in feeling, sense and judgment he is to the servants here! The talking and indiscretion shocks him.

I have all along meant to ask you if you have ever read an extraordinary poem by a lady now dead a Mrs. Browning (Augusta Stanley knows her) called 'Aurora Leigh'? The Countess read it to me. It is very strange, very original full of talent and of some beautiful things—but at times dreadfully coarse—though very moral in its tendency—but an incredible book for a lady to have written.

18 March 1865

I am so fond of Burns's poems. They are so poetical—so simple in their dear Scotch tongue, which is so full of poetry.

TO PALMERSTON   13 March 1865

As it appears that a promise was made to Sir A. Cockburn, though unauthorised by her, the Queen will sanction his Peerage;* but her Majesty still retains her opinion of the absolute duty, which devolves upon her, of requiring that Peerages shall not be conferred upon any persons who do not in addition to other qualifications possess a good moral character.

TO THE CROWN PRINCESS OF PRUSSIA   5 April 1865

I have not, I think, told you that I have taken good J. Brown entirely and permanently as my personal servant for out of doors—besides cleaning my things and doing odd 'jobs'—as I found it so convenient and saving me so much trouble to have one and the same person always for going out, and to give my orders to, which are taken by him from me personally to the stables. He comes to my room after breakfast and luncheon to get his orders—and everything is always right; he is so quiet, has such an excellent head and memory, and is besides so devoted, and attached and clever and so wonderfully able to interpret one's wishes. He is a real treasure to me now, and I only wish higher people had his sense and discretion, and that I had as good a maid.

*Sir Alexander Cockburn, the Lord Chief Justice, did not, in fact, receive his peerage.

To King Leopold 27 April 1865

These American news [the assassination of President Lincoln] are most dreadful and awful! One never heard of such a thing! I only hope it will not be catching elsewhere.

To Mrs. Lincoln 29 April 1865

Though a stranger to you, I cannot remain silent when so terrible a calamity has fallen upon you and your country, and must express personally my deep and heartfelt sympathy with you under the shocking circumstances of your present dreadful misfortune.

No one can better appreciate than I can, who am myself utterly broken-hearted by the loss of my own beloved husband, who was the light of my life, my stay, my all, what your sufferings must be; and I earnestly pray that you may be supported by Him to Whom alone the sorely stricken can look for comfort, in this hour of heavy affliction!

To King Leopold 8 June 1865

Alix was again confined too soon, but this time only a month; and she is recovering extremely well, and the child is said to be much larger than little Albert Victor, and nice and plump. Bertie seems very much pleased with this second son [Prince George, later King George V]. I can't deny that I am glad that I am spared the anxiety and fatigue of being with Alix at the time, though I should never shun it, if I could be of use to anyone, high or low. I always feel drawn to the sick-bed of anyone, to be of use and comfort.

To the Prince of Wales 13 June 1865

I fear I cannot admire the names you propose to give the Baby. I had hoped for some fine old name. Frederic is, however, the best of the two, and I hope you will call him so; George only came over with the Hanoverian family. However, if the dear child grows up good and wise, I shall not mind what his name is. Of course you will add Albert at the end, like your brothers, as you know we settled long ago that all dearest Papa's male English descendants should bear that name, to mark our line, just as I wish all the girls to have Victoria at the end of theirs! I lay great stress on this; and it is done in a great many families.*

To King Leopold 17 June 1865

We returned here yesterday morning, and I feel painfully the change, though Windsor is hallowed to me in so many, many ways. How delightful it would be to show you our beloved Balmoral, with its glorious scenery and heavenly air, its solitude and absence of all contact with the mere miserable frivolities and worldlinesses of this wicked world! The mountains seem fresh from God's hand, nearer to Heaven, and the primitive people to have kept that chivalrous loyalty and devotion—seen hardly, indeed now nowhere, else! . . .

*The names given were George Frederick Ernest Albert

I have seen our new grandson; he is very small and not very pretty, but bigger than Albert Victor, who is a dear little fellow, was at that age.

TO THE CROWN PRINCESS OF PRUSSIA   21 June 1865

Do not say that your proposal about Christmas [to come to stay with her family] is disagreeable to me, my dearest child; that is not the word—but it would not be possible or feasible as my letter of Monday will explain. You never will believe how bad my poor nerves are, how more and more shattered they become, and how a large, merry family-party tries me; but on days of former festivity and joy—above all—I am quite and totally incapable of bearing joyousness and merriment. And you would not be happy I know.

7 July 1865

To-day the christening of the new baby takes place, but quietly and not *en grande tenue*. Still, these ceremonies and events are painful in the extreme to me, as you know.

3 August 1865

The meeting of so many relations at Coburg, without my darling Angel, will be most painful and trying. It is just twenty years that we had that very very happy meeting at dear Coburg, when you and dear Louise were there! Oh! how many are gone, and when one thinks that one of the youngest has vanished!

In Germany things look rather critical and threatening. Prussia seems inclined to behave as atrociously as possible, and as she always has done! Odious people the Prussians are, that I must say.

24 August 1865

I told General Grey to tell you how favourable the impression is, left by Prince Christian [of Schleswig-Holstein-Sonderburg-Augustenburg, the Duke of Augustenburg's younger brother who was to marry Princess Helena]. He is extremely pleasing, gentlemanlike, quiet, and distinguished. Lenchen (who knows nothing as yet) has of her own accord told me how amiable and pleasing and agreeable she thought him.

JOURNAL   9 September 1865

I received Queen Emma, the widowed Queen of the Sandwich Islands . . . Nothing could be nicer or more dignified than her manner. She is dark, but not more so than an Indian with fine features and splendid soft eyes. She was dressed in just the same widow's weeds as I wear . . . I asked her to sit near to me on the sofa. She was much moved when I spoke of her great misfortune in losing her husband and only child. She was very discreet and would only remain a few minutes.

JOURNAL   18 October 1865

Lenchen met me on the top of the stairs, when I came in, with the news of poor Lord Palmerston's death, which had taken place this morning at a quarter to 11. Strange, and solemn to think of that strong, determined man, with so much worldly ambition,—gone! He had often worried and distressed us, though as Prime Minister he had behaved very well. To think that he is removed from this world, and I alone, without dearest Albert to talk to or consult with!

To Russell   19 October 1865

The Queen can turn to no other but to Lord Russell, an old and tried friend of hers, to undertake the arduous duties of Prime Minister, and to carry on the Government.

To King Leopold   20 October 1865

You know now already long, of the death of poor Lord Palmerston, alias Pilgerstein! It is very striking, and is another link with the past—the happy past—which is gone, and in many ways he is a great loss. He had many valuable qualities, though many bad ones, and we had, God knows! terrible trouble with him about Foreign affairs. Still, as Prime Minister he managed affairs at home well, and behaved to me well. But I never liked him, or could ever the least respect him, nor could I forget his conduct on certain occasions to my Angel. He was very vindictive, and personal feelings influenced his political acts very much. Still, he is a loss! I shall have troubles and worries, and to face them alone without my Angel is dreadful! I feel no energy, no interest—nothing left—no one to talk to. I sometimes wish I could throw everything up and retire into private life.

25 October 1865

I feel for poor Lord Russell; to begin at his age afresh, after thirteen years, as Prime Minister, is very trying. I can't say I rejoice to have Clarendon [as Foreign Secretary]; I don't quite trust him; still he is conciliatory to other Powers, and is attached to me. I feel all these changes sadly, painfully! I feel more and more alone!

Memorandum   29 October 1865

Lord Russell came at 10 o'clock and kissed hands as Prime Minister. He seemed much impressed with the weight of the task he had undertaken, saying that it was different to do so at 73, to 54; but that, if he had been in the House of Commons, he could not have undertaken it at all.

Journal   21 November 1865

Dearest Vicky's birthday—our first born, my beloved one's great favourite of whom he was so proud. She, indeed, well deserves it, dear, warm hearted, highly gifted child! May God bless and long preserve her ... When dressed went to meet her with a nosegay and she and I were both much affected, thinking of former happy blessed times and of the inexpressible joy it would have been to dearest Albert to have her and her dear, promising children here with us ... Later, drove with her and Fritz down to the Mausoleum and placed a wreath over the pillow on which the beloved head rests. Oh! that this should be all he can do now ... By Vicky's wish a conjuror gave an entertainment for the servants [in St. George's Hall] in honour of the day. It was a great trial for me, but I [went to it] for Vicky's sake and to please my people. Remained only for a short while, and saw some very curious tricks, but of what interest were they to me without my beloved one whom I missed so dreadfully.

TO KING LEOPOLD   30 November 1865

As the secret about Lenchen's marriage had come out in the papers (most kindly received), we thought it best to ask him to come over at once ... Prince Christian will therefore arrive here to-morrow.

JOURNAL   4 December 1865

The news of beloved Uncle, which had improved, are again very anxious. There are fresh alarming symptoms and great increasing weakness ... If Uncle took the necessary nourishment prescribed, he might pull through. He was to take brandy every hour and broth every two hours. All this reminds me too painfully of beloved Albert's terrible illness!

10 December 1865

A sad, sad blow, which has long been impending, has at last fallen on us, and I can hardly believe what I write, and am stupefied and stunned. Dearly beloved Uncle Leopold is no more; that dear loving Uncle, who has ever been to me as a Father, has gone to that everlasting Home, where all is peace and rest.

TO THE CROWN PRINCESS OF PRUSSIA   23 December 1865

You have also never said one word about my poor little Highland book [the privately printed edition of *Leaves from the Journal of Our Life in the Highlands*]—my only book. I had hoped that you and Fritz would have liked it.

30 December 1865

I must speak of Leopold [of Hohenzollern-Sigmaringen] and Antoinette. I must now say how charmed with her we are. How lovely she is and what a look of dear Aunt Victoire she has. How like her she is—only handsomer perhaps. And so simple and unaffected—and dear Leopold I am so fond of. Oh! if B[ertie] and Al[ix] were like them! Oh if Antoinette was in Al.'s place! She is so much more *sympathique* and *grande dame*. Our good Al. is like a distinguished lady of society but nothing more!

I am so thankful for all you say about art, and hope that Mr. Robinson will be appointed [to succeed Sir Charles Eastlake as Director of the National Gallery. He was not appointed]. He has already been recommended by several people to Lord Russell. I hope you will always write to tell me of anything of this kind, because I do not understand these things well, having always placed implicit reliance with adored Papa, but you have inherited that from him, and he talked to you so much about it that I feel the greatest confidence in you.

TO RUSSELL   22 January 1866

To enable the Queen to go through what she can only compare to an execution, it is of importance to keep the thought of it as much from her mind as possible, and therefore the going to Windsor to wait two whole days for this dreadful ordeal would do her positive harm.

The Queen has never till now mentioned this painful subject to Lord Russell, but she wishes once for all to just express her own feelings. She must, however, premise her observations by saying that she entirely absolves Lord Russell and his colleagues from any attempt ever to press upon her what is so very painful an effort. The Queen must say that she does feel very bitterly the want of feeling of those who ask the Queen to go to open Parliament. That the public should wish to see her she fully understands, and has no wish to prevent—quite the contrary; but why this wish should be of so unreasonable and unfeeling a nature, as to long to witness the spectacle of a poor, broken-hearted widow, nervous and shrinking, dragged in deep mourning, alone in State as a Show, where she used to go supported by her husband, to be gazed at, without delicacy of feeling, is a thing she cannot understand, and she never could wish her bitterest foe to be exposed to!

She will do it this time—as she promised it, but she owns she resents the unfeelingness of those who have clamoured for it. Of the suffering which it will cause her—nervous as she now is—she can give no idea, but she owns she hardly knows how she will go through it.

TO THE CROWN PRINCESS OF PRUSSIA   2 February 1866

Your little ebullition about children and the exquisite delight of having them (certainly a thing very few share in) I will not reply to—except to say that you will find the pleasure less great when they grow up, and the sorrows and anxieties of one kind or another overwhelm you—and hardly counterbalance the pleasure.

JOURNAL   6 February 1866

Great crowds out, and so I had (for the first time since my great misfortune) an escort. Dressing after luncheon, which I could hardly touch. Wore my ordinary evening dress, only trimmed with miniver, and my cap with a long flowing tulle veil, a small diamond and sapphire coronet rather at the back, and diamonds outlining the front of my cap.

It was a fearful moment for me when I entered the carriage alone, and the band played; also when all the crowds cheered, and I had great difficulty in repressing my tears. But our two dear affectionate girls [Princesses Helena and Louise who faced the Queen in the carriage] were a true help and support to me, and they so thoroughly realised all I was going through. The crowds were most enthusiastic, and the people seemed to look at me with sympathy. We had both windows open, in spite of a very high wind.

When I entered the House, which was very full, I felt as if I should faint. All was silent and all eyes fixed upon me, and there I sat alone. I was greatly relieved when all was over, and I stepped down from the throne ...

So thankful that the great ordeal of to-day was well over, and that I was enabled to get through it.

TO RUSSELL   18 February 1866

The Queen will now return [from Osborne] on Wednesday in accordance with the wishes of her Ministers; but in return she hopes Lord Russell will in future trust to her doing what she believes to be necessary for the good of the Country, without her movements being dictated to her.

If the Queen seems to shrink at this moment from returning to Windsor, it is that, knowing the importance of her life, and that it is her duty in consequence to try and preserve her broken health and very shattered nerves (every day more and more shattered) from becoming seriously worse, she fears she could not now at Windsor find the pure air or be able to take the exercise so essential for this object.

The Government may rely on her always doing what she can without being pressed. It is only for the sake of the Country, of her people and her children that she has any desire to live; but to enable her to live, she must do what she believes to be indispensable for her health.

TO THE CROWN PRINCESS OF PRUSSIA   18 April 1866

Though it is very naughty of me to show dearest Fritz's English up to you I must tell you, as you will laugh so, he telegraphed you were 'Happily delivered from a strong and healthy daughter!' [Princess Victoria, the Crown Princess's fifth child]—and the telegraph clerk corrected it, putting an x and 'of' below.

JOURNAL   27 April 1866

Christian came to take leave. I have had very satisfactory conversations with him regarding all the arrangements for the marriage [which was to take place on 5 July] and the proposed date. Nothing can be nicer, more gentlemanlike, or more full of regards than he is. I feel this very much, and this visit has made me get to know him much better, and appreciate his excellent qualities and sterling worth.

TO RUSSELL   25 May 1866

The Queen thinks it right to send this memorandum to Lord Russell relative to her health.

She must say that she feels she could not go on working as she does, without any real relaxation (for she never is without her boxes and despatches, etc.; which her Ministers often are, for a few weeks at least) if she did not get that change of scene and that pure air, which always gives her a little strength, twice a year. Nine or ten days are very short, but still they will do her some good, and she will have more courage to struggle onwards, though every year, which adds to her age, finds her nervous system and general strength more and more shaken. She always fears some complete breakdown some day; and she is just now greatly in want of something to revive her after an autumn, winter, and spring of great anxiety, and many sorrows and annoyances of a domestic nature, which shake her very nervous temperament very severely.

The Queen's absence in Scotland has never caused any inconvenience hitherto, and the Queen would talk over with Lord Russell and Lord Clarendon every possible contingency which would have to be decided, with great promptitude, so that a mere reference by cypher (and she could establish one specially between herself and Lord Clarendon) would give an answer without a moment's delay.

MEMORANDUM   26 June 1866

Lord Russell came at a quarter to 1, and told the Queen that he had in fact come to tender the resignation of the Government.*

TO DERBY   27 June 1866

Her Ministers having placed their resignations in her hands, the Queen turns to Lord Derby, as to the only person whom she believes capable of forming such an Administration as will command the confidence either of the country or herself, and have the best chance of permanency.

The differences of opinion as to the principles on which the Government should be conducted, do not appear to the Queen to be such as ought to make it impossible for Lord Derby to obtain the assistance of some, at least, of those who have been supporters, or even Members, of the late Government.

TO GENERAL GREY   28 June 1866

The Queen wishes General Grey would tell Lord Derby that things are naturally greatly changed and altered since the terrible misfortune which laid low, for ever, the Queen's happiness; that she depends much on those who surround her; that her health and nerves are much shattered; and that therefore she cannot be expected to change those who are in constant attendance on her like her Equerries, and in this Lord Russell has concurred.

TO LORD CHARLES FITZROY, THE QUEEN'S EQUERRY   1 September 1866

The Queen felt too nervous to tell Lord Charles this evening what she meant, viz.: that unless it were a really rainy day, she had made up her mind to make the great effort of going for a short while this year to the Gathering [the Highland Gathering at Braemar]. It will be a great effort, and is a thing we always disliked, but she wishes to do what she can, to appear in public, and day occasions are positively the only ones, when she can do it, without completely knocking herself up. She hears that it would give great satisfaction in the country, and the Queen thinks Prince and Princess Christian's arrival a good occasion for her doing so.

TO THE CROWN PRINCESS OF PRUSSIA   9 October 1866

You speak of English maids not being grand in England. I assure you they are so—to an extent that is unbearable and that that is the bane of the present day. Pride, vulgar, unchristian pride in high and low.

TO THE PRINCE OF WALES   16 October 1866

I yesterday evening, after dinner, received your letter of the preceding day, on the subject of your visit to St. Petersburg [for the marriage of Princess Dagmar to the

*Russell's Government had been defeated on 18 June on the question of Parliamentary Reform.

future Tsar, Alexander III]. That you should like to see Russia, and, above all, to be present at the marriage of dear Alix's sister, and that Dagmar should wish to see her kind brother-in-law's face at so trying a time, I think perfectly natural. I own I do not much like the idea. First, I think it is a bad time of the year for you to go there. Secondly, that your visit to St. Petersburg ought to be for itself alone, and not on such an occasion; and thirdly, I think the Government over-rate the importance of it, in a political point of view. These are my reasons against it, and to that I may add another, which, dear Child, you know I have often already alluded to, viz.: your remaining so little quiet at home, and always running about. The country, and all of us, would like to see you a little more stationary, and therefore I was in hopes that this autumn and winter this would have been the case. However, if you are still very desirous to go now, I will not object to it.

JOURNAL   30 November 1866

With a sinking heart and trembling knees got out of the train [at Wolverhampton], amidst great cheering, bands playing, troops presenting arms, etc. . . . All along the three or four miles we drove, the town was beautifully decorated, with flags, wreaths of flowers, and endless kind inscriptions. There were also many arches. It seemed so strange being amongst so many, yet feeling so alone, without my beloved husband! Everything so like former great functions, and yet so unlike! I felt much moved, and nearly broke down when I saw the dear name and the following inscriptions— 'Honour to the memory of Albert the Good,' 'the good Prince,' 'His works follow him,' and so many quotations from Tennyson. There were barriers all along, so that there was no overcrowding, and many Volunteers with bands were stationed at different points.

The Mayor was completely taken by surprise when I knighted him, and seemed quite bewildered, and hardly to understand it when Lord Derby told him. There was some slight delay in the uncovering of the statue, but it [the covering sheet] fell well and slowly, amidst shouts and the playing of the dear old Coburg March by the band. How I could bear up, I hardly know, but I remained firm throughout. At the conclusion of the ceremony I walked round the statue followed by the children. I had seen it before at Thornycroft's studio, and it is upon the whole good . . .

We drove back through quite another, and the poorest, part of the town, which took half an hour. There was not a house that had not got its little decoration; and though we passed through some of the most wretched-looking slums, where the people were all in tatters, and many very Irish-looking, they were most loyal and demonstrative. There was not one unkind look or dissatisfied expression; everyone, without exception, being kind and friendly. Great as the enthusiasm used always to be wherever dearest Albert and I appeared, there was something peculiar and touching in the joy and even emotion with which the people greeted their poor widowed Queen!

TO DERBY   12 January 1867

After what Lord Derby has said of the importance which her Ministers attach to the moral support which would be afforded to them, particularly as regards this question [of Parliamentary Reform] by the Queen's opening Parliament in person, she will not hesitate, great, trying, and painful as the exertion will be to her, to comply with the wishes of her Government.

But, in doing so, under the peculiar circumstances of the time, the Queen must have it clearly understood that she is not to be expected to do it as a matter of course, year after year; and she must call upon Lord Derby to give her an assurance, that, except under a very pressing and self-evident necessity, she shall not be asked to make a similar exertion next year.

The shock to her nerves and the fatigue of the long journey the Queen cannot over-rate!

TO THE CROWN PRINCESS OF PRUSSIA   22 January 1867

I wish to say a few more words in confidence about Alice, but which I would beg you to make use of. I told you in my last letter that she had, from the time when she married and came back here, not been liked in the house from her ordering and commanding and from want of tact and discretion . . .

Well, when Alice came the last two times she grumbled about everything—and Louis also sometimes—the rooms, the hours, wanting to make me do this and that and preventing my being read to of a evening as Louis would come and he always fell asleep. Of course this is not kind or right and if Alice wishes to come, she should accommodate herself to my habits.

5 February 1867

Yesterday was a wretched day, and altogether I regret I went [to open Parliament]—for that stupid Reform agitation has excited and irritated people, and there was a good deal of hissing, some groans and calls for Reform, which I—in my present forlorn position—ought not to be exposed to. There were many, nasty faces—and I felt it painfully. At such times the Sovereign should not be there. Then the weather being very bad—the other people could not remain to drown all the bad signs. Of course it was only the bad people.

26 June 1867

The only objection I have to him [Prince Leopold's tutor] is that he is a clergyman. However he is enlightened and so free from the usual prejudices of his profession that I feel I must get over my dislike to that. Mr. Duckworth is an excellent preacher and good-looking besides.

TO DERBY   25 June 1867

The Queen has not troubled Lord Derby with many observations during the past discussions on the Reform Bill. He knows how anxious she is that this dangerous question should be set at rest, in a manner likely to prevent its being further agitated for many years at least to come!

But Lord Derby will not argue from the Queen's silence that she has not observed the progress of the measure with the deepest interest, and, she must acknowledge, not a little anxiety. He is also aware that she has had so little knowledge beforehand of what it was intended to propose, and the decisions of the House of Commons have been taken so suddenly and so unexpectedly, that there has been no opportunity for her to make any remarks at a time when they could have any effect. Nor does the Queen wish now to say anything that can embarrass the Government. She feels that it is impossible to recede from whatever concessions to popular feeling the Government has already made. All she would earnestly urge upon Lord Derby, is, to allow any amendments which may be proposed in the House of Lords, with a view to avert the danger which many people apprehend from the great increase of democratic power, to receive a fair consideration from the Government. She cannot believe that, even in the House of Commons, some modification of a measure, which even the most advanced Liberals regard with some degree of alarm, would not be willingly accepted.

The Queen earnestly prays that the result of the deliberations of the two Houses may yet be the adoption of a measure likely to settle the question on a safe and constitutional basis . . :

The Queen received Lord Derby's letter this morning. The word distasteful is hardly applicable to the subject;* it would be rather nearer the mark to say extremely inconvenient and disadvantageous for the Queen's health. As, however, a Sultan is not likely to come again, or a Paris Exhibition to take place soon again, the Queen does not object to delay her departure for Osborne, for three or four days, though it is very annoying to her. She thinks, however, that, if the Sultan knew how inconvenient it was to the Queen, and that she had purposely delayed her departure and changed all her arrangements, to receive him at Windsor, he might be induced to arrive a day sooner. The Queen would then receive him here on Friday afternoon, say about three or four, and would herself leave for Osborne quite early the following morning.

The Queen has moreover thought it absolutely necessary to send Dr. Jenner to Lord Derby to tell him of the real state of her health and nerves, as she is almost driven to desperation by the want of consideration shown by the public for her health and strength, and she foresees ere long a complete breakdown of her nervous system. She really believes that unless she were crippled by rheumatism or gout, or something which absolutely prevented her walking, people would worry her till she sunk under it! . . .

It is very wrong of the world to say that it is merely her distaste to go out and about as she could when she had her dear husband to support and protect her, when the fact is that her shattered nerves and health prevent her doing so. Still whatever the poor Queen can do she will; but she will not be dictated to, or teased by public clamour into doing what she physically cannot, and she expects her Ministers to protect her from such attempts. They cannot complain that she has not supported them, and she relies with perfect confidence on their well-known loyalty.

---

*Lord Derby had written to the Queen to urge her to postpone her journey to Osborne for three days so that she could receive the Sultan of Turkey earlier than she had intended. Derby suggested that the Sultan would otherwise be much offended, and unfavourable contrasts would be drawn between his reception in Paris and that in London.

To Lord Charles FitzRoy   26 June 1867

Lord Charles FitzRoy having always been so kind to the Queen in all that concerns her convenience and comfort, and having only lately informed her that the Duke of Beaufort [Master of the Horse] so completely understood her wishes and entered into her feelings respecting her faithful Brown, and having also told her last year that people quite understood his going as an upper servant with her carriage, and he (Lord Charles) thinking there should be no difference in London to the country, and moreover having taken him everywhere with her for two years on public as well as private occasions, she is much astonished and shocked at an attempt being made by some people to prevent her faithful servant going with her to the Review in Hyde Park, thereby making the poor, nervous, shaken Queen, who is so accustomed to his watchful care and intelligence, terribly nervous and uncomfortable.* Whatever can be done, the Queen does not know on this occasion, or what it all means she does not know; but she would be very glad if Lord Charles could come down to-morrow morning any time before luncheon, that she may have some conversation with him on this subject, not so much with a view as to what can be done on this occasion, but as to what can be done for the future to prevent her being teased and plagued with the interference of others, and moreover to make it completely understood once and for all that her Upper Highland servant (whether it be Brown or another, in case he should be ill, replaces him) belongs to her outdoor attendants on State as well as private occasions. The Queen will not be dictated to, or made to alter what she has found to answer for her comfort, and looks to her gentlemen and especially her Equerries setting this right for the future, whatever may be done on this single occasion.

To the Crown Princess of Prussia   13 July 1867

Alice is very amiable but she and Louis are no comfort. They are not quiet. Alice is very fond of amusing herself and of fine society and I think they do everyone harm. They ruined Affie.

16 July 1867

I often wonder how I shall ever be able to go on. Everything upsets me. Talking especially tries me. Reading in the open air quietly is what does me the most good. Sitting and reading and writing, working and drawing soothes me.

20 July 1867

You write so warmly and affectionately about your poor old Mama that I am greatly touched by it. I have been now 30 years in harness—and therefore ought to know what should be—but I am terribly shy and nervous and always was so.

Journal   17 July 1867

A very bright, but to our vexation a very windy morning. Still there was no news of a

---

*The Queen had agreed to attend this review on the supposition that John Brown would be allowed to attend her. The death of the Emperor Maximilian enabled the authorities to cancel the review which they did with profound relief.

postponement of the Review, so felt we were obliged to go, and at half past ten left with all the family for Trinity Pier, the ladies and gentlemen having preceded us. Embarked in the *Alberta*, and steamed close up to the *Victoria and Albert*. Had to go on board the yacht in barges, in a nasty swell. However, we managed well enough.

Went on till opposite Osborne, where we lay waiting one hour and a half for the Sultan in violent squalls with heavy rain. At length the *Osborne* appeared, followed by other vessels, and came close up to us. After waiting a little, on account of a heavy shower, the Sultan, with Bertie, George, the three Princes, and a good many others, came on board in our barge with a beautiful crimson and gold Turkish standard. I received them at the top of the accommodation ladder, and made the Sultan, who was in plain uniform, put on his cloak again. The Viceroy of Egypt, accompanied by his people, and flying his standard, followed and was received by me. He was shy and subdued in the presence of the Sultan. I sat outside the deck saloon with the Sultan, all the others beyond; and he made the Viceroy sit opposite to us and interpret, which he did, sitting at the edge of his chair, his short legs hardly reaching the ground. It must have been a curious sight.

The Sultan feels very uncomfortable at sea . . . He was continually retiring below, and can have seen very little. None of the ships could get under way. Still, it was a very fine sight. The men manned the rigging instead of the yards, and cheered. The saluting took place before we came close up. But, after we passed, they kept up a sort of sham fight, the ships firing at one another, whilst boats were to attack the shore. At length, it being near three, we lunched, I having, just before, given the Sultan the Garter, which he had set his heart upon, though I should have preferred the Star of India, which is more suited for those who are not Christians.

TO THE CROWN PRINCESS OF PRUSSIA   5 October 1867

If only she [the Princess of Wales] understood her duties better. That makes me terribly anxious. If only I had the comfort to feel that if I closed my eyes, things would go well—but alas! I feel just the contrary, and see it more and more every day. I feel how necessary my poor life is—and yet how uncertain is anyone's life!

                                                                                14 October 1867

Gen. Grey asked to see me when I came in, and said he was sorry to alarm me, but must show me a telegram . . . reporting . . . the news from a reliable source, that the Fenians had said they meant to try and seize me here [at Balmoral], and were starting to-day or to-morrow! Too foolish!!

                                                                                1 December 1867

Saw the Bishop of New Zealand [G. A. Selwyn], who came with the Dean. After having been pressed by me and the Archbishop of Canterbury, he has consented to accept the Bishopric of Lichfield. He is a very earnest man, devoted to his missionary work in New Zealand and to the natives, having been bishop there for 26 years. He says it is a sacrifice to take up this new post, but he feels he ought 'to obey orders', I having pressed him on account of the disturbed, distracted state of the

Church. He said, 'It will be a great change to me, after my wild life in the hills.' I do indeed pity him, for the exchange to the Black Country, where there will be no romance, no primitive races, but the worst kind of uncivilised civilisation, will be most trying.

TO LORD STANLEY, FOREIGN SECRETARY   16 December 1867

[The Queen] cannot doubt that it is Lord Stanley's wish to observe the constitutional practice of taking the Queen's pleasure before he forwards despatches of any importance to their destination. But she cannot approve of the irregular habit which has crept in, of sending off despatches without her sanction having been previously obtained, trusting to the power of cancelling them by telegraph, if disapproved. She would therefore certainly wish that the old custom of obtaining her approval before they are sent, should be resumed.

TO THE CROWN PRINCESS OF PRUSSIA   18 December 1867

I wished to answer what you said about the bar between high and low. What you say about it is most true [that the 'lower classes must rise to the upper—and not vice versa—or the consequences are dreadful, as the first French Revolution has proved] but alas! that is the great danger in England now, and one which alarms all right-minded and thinking people.

The higher classes—especially the aristocracy (with of course exceptions and honourable ones)—are so frivolous, pleasure-seeking, heartless, selfish, immoral and gambling that it makes one think (just as the Dean of Windsor said to me the other evening) of the days before the French Revolution. The young men are so ignorant, luxurious and self-indulgent—and the young women so fast, frivolous and imprudent that the danger really is very great, and they ought to be warned. The lower classes are becoming so well-informed, are so intelligent and earn their bread and riches so deservedly—that they cannot and ought not to be kept back—to be abused by the wretched, ignorant, high born beings who live only to kill time. They must be warned and frightened or some dreadful crash will take place. What I can, I do and will do—but Bertie ought to set a good example in these respects by not countenancing even any of these horrid people.

21 December 1867

It [a reference to Balmoral] makes me think of my little book [*Leaves from the Journal of Our Life in the Highlands*] which you know I gave you a private copy of. Well, it was so much liked that I was begged and asked to allow it to be published—the good Dean of Windsor amongst other wise and kind people saying it would, from its simplicity and the kindly feelings expressed to those below us, do so much good. I therefore consented—cutting out some of the more familiar descriptions and being subjected by Mr. Helps and others to a very severe scrutiny of style and grammar (the correspondence about which would have amused you very much) and adding our first journeys and visits to Scotland, yachting tours to the Channel Islands and visits to Ireland. I have likewise added a little allusion to your engagement to Fritz and what I wrote on the death of the Duke of Wellington.

The whole is edited by Mr. Helps, who has written a very pretty preface to it. It has given a great deal of trouble for one had so carefully to exclude even the slightest observation which might hurt anyone's feelings. But it has been an interest and an occupation—for no-one can conceive the trouble of printing a book, and the mistakes, which are endless.

11 January 1868

The country was never more loyal or sound. I would throw myself amongst my English and dear Scotch subjects alone (London excepted as it is so enormous and full of Irish) and I should be as safe as in my room.

To THEODORE MARTIN, THE PRINCE CONSORT'S OFFICIAL BIOGRAPHER
16 January 1868

The Queen was moved to tears on reading Mr. Martin's beautiful and too kind letter. Indeed it is not possible for her to say how touched she is by the kindness of everyone.* People are too kind. What has she done to be so loved and liked? She did suffer acutely last year; she will not deny it; and it made her ill, but it has vanished entirely and the very thought of it seems to have lost its sting . . .

To THE CROWN PRINCESS OF PRUSSIA
18 January 1868

I send you again several newspaper articles about the book, the effect produced by which is wonderful, and will I know gladden your heart from the extreme loyalty it displays. From all and every side, high and low, the feeling is the same, the letters flow in, saying how much more than ever I shall be loved, now that I am known and understood, and clamouring for the cheap edition for the poor—which will be ordered at once. 18,000 copies were sold in a week. It is very gratifying to see how people appreciate what is simple and right and how especially my truest friends—the people—feel it. They have (as a body) the truest feeling for family life.

22 January 1868

Newspapers shower in—the poorest, simplest full of the most touching and affectionate expressions. The kind and proper feeling towards the poor and the servants will I hope do good for it is very much needed in England among the higher classes.

29 January 1868

I have such quantities of beautiful and touching letters from people whom I don't know, or have ever heard of—all about my little book, but I send you none, and indeed have been doubtful of sending you the *Quarterly* with a review by the Bishop of Oxford, as you seem to take so little interest in it and only mentioned it once. Here everyone is so full of gratitude and loyal affection, saying it is not to be told the good it will do the Throne, and as an example to people in the higher classes.

---

*On the publication of her *Leaves from the Journal of Our Life in the Highlands*. Not everyone was kind. Her family in general disapproved of the book, particularly of its references to John Brown. The Prince of Wales considered it 'twaddle'.

To Derby   22 February 1868

The Queen has just received Lord Derby's (dictated) letter, and hastens to acknowledge it, though she purposes to answer him more fully to-morrow or Monday. Lord Derby's decision has not surprised the Queen, though she is deeply grieved at the necessity he feels himself under of resigning his office.* He is bound to take all the care of his health that he can, for the sake of his devoted wife and his children, and of his friends, amongst whom she considers herself. Though no longer in her service his advice will be still most valuable to his Sovereign.

24 February 1868

The Queen sent General Grey yesterday up to town to communicate with Mr. Disraeli, and inform him of Lord Derby's communication to her, and of her wish that he should undertake the arduous post of successor to Lord Derby. This he has accepted, relying on Lord Derby's kind promise of support, as well as on that of his colleagues.

To Disraeli   26 February 1868

The Queen thanks Mr. Disraeli very much for his kind letter received to-day, and can assure him of her cordial support in the arduous task which he has undertaken. It must be a proud moment for him to feel that his own talent and successful labours in the service of his country have earned for him the high and influential position in which he is now placed.

The Queen has ever found Mr. Disraeli most zealous in her service, and most ready to meet her wishes, and she only wishes her beloved husband were here now to assist him with his guidance.

To Queen Augusta of Prussia   26 February 1868

Mr. Disraeli has achieved his present high position entirely by his ability, his wonderful, happy disposition and the astounding way in which he carried through the Reform Bill, and I have nothing but praise for him. One thing which has for some time predisposed me in his favour is his great admiration for my beloved Albert and his recognition of and respect for his great character.

To the Crown Princess of Prussia   29 February 1868

I think the present man will do well, and will be particularly loyal and anxious to please me in every way. He is very peculiar, but very clever and sensible and very conciliatory.

4 March 1868

He is full of poetry, romance and chivalry. When he knelt down to kiss my hand which he took in both of his he said 'in loving loyalty and faith'.

---

*Derby had been suffering from severe attacks of gout and felt unable to continue as Prime Minister any longer.

To Disraeli   7 March 1868

The Queen has this moment received this letter from Mr. Disraeli. Lord Abercorn's letter is very proper, and the Queen would have no objection to the Prince of Wales's going.* But she entirely objects to the latter part of Mr. Disraeli's letter; and General Grey knows well enough her grounds for doing so.

In the first place, every other part of the Queen's dominions—Wales, and the Colonies even, might get up pretensions for residence, which are out of the question; and this she is most anxious General Grey should fully explain to Mr. Disraeli. And in the Prince of Wales's case, any encouragement of his constant love of running about, and not keeping at home, or near the Queen, is most earnestly and seriously to be deprecated.

But if the Irish behave properly, the Queen would readily send, from time to time, other Members of her family (and she particularly wishes the Prince of Wales not to be the only one)—for instance, Prince Arthur (who is called Patrick) and Prince and Princess Christian, etc. But with this understanding, that the expenses of these Royal visits should be borne by the Government, who press them constantly (and most annoyingly) on the Queen; and which are solely for political purposes.

For health and relaxation, no one would go to Ireland, and people only go who have their estates to attend to. But for health and relaxation thousands go to Scotland.

Pray let this question be thoroughly understood.

To the Prince of Wales   9 March 1868

I have heard from Mr. Disraeli on the subject of your going to Ireland, and, as the Government seem to wish it so much, and to think that it will do so much good, I will naturally sanction it. But I much regret that the occasion chosen should be 'Races,' [the Punchestown Races which the Prince attended] as it naturally strengthens the belief, already far too prevalent, that your chief object is amusement; and races have become so bad of late, and the connection with them has ruined so many young men, and broken the hearts thereby of so many fond and kind parents, that I am especially anxious you should not sanction or encourage them.

2 May 1868

It is not for herself [the Princess of Wales who was pregnant again] that I grieve but for the poor child! For they are such miserable, puny, little children (each weaker than the preceding one) that it is quite a misfortune. I can't tell you how these poor, frail, little fairies distress me for the honour of the family and the country. Darling Papa would have been in perfect despair.

To Theodore Martin   14 May 1868

Mr. Martin has been so kind and feeling, and knowing so well how the Queen's

---

*Lord Abercorn, Lord Lieutenant of Ireland, had suggested that the Prince of Wales should visit that country for a week or so. Disraeli had added that the Prince 'might make a longer visit later in the year, hunt . . . and occupy some suitable residence'.

health and nerves are shaken—therefore will understand how very great the efforts made this year have been—a week in London, three Drawing-rooms, and the great ceremony yesterday [laying the foundation stone of St. Thomas's Hospital], from which she is suffering much to-day; he will accordingly not be surprised at the indignation and pain with which she read the Article in the *Globe* to-night [criticising the Queen's continuing isolation], and her great anxiety therefore that he should try and prevent similar Articles appearing in *The Times* and *Daily Telegraph*. Every increased effort is rewarded by such shameless Articles; and the discouragement and pain they cause are very great. This therefore is the return for increased efforts made which cause her painful suffering. Her head is very painful, her nerves are so much shaken and her brain was feeling this evening quite confused and overtaxed! ... She really is feeling utterly worn out; and does wish some newspaper would point out how much she has done, and how necessary it is to keep her well enough to go on, for else she may be unable to do so.

To the Crown Princess of Prussia 8 July 1868

I am not as proud of Affie [who had been shot and wounded by a Fenian in Australia] as you might think, for he is so conceited himself and at the present moment receives ovations as if he had done something—instead of God's mercy having spared his life.

5 August 1868

Yes, Affie is a great, great grief—and I may say source of bitter anger for he is not led astray. His conduct is gratuitous! Oh! he is so different to dear Bertie, who is so loving and affectionate, and so anxious to do well, though he is some times imprudent—but that is all.

To Gathorne Hardy, Home Secretary 20 July 1868

The Queen cannot help directing Mr. Hardy's attention to an article in the *Daily Telegraph* of to-day, on the cruelty to animals, and to ask him to make enquiry on the subject.

Nothing brutalises people more than cruelty to dumb animals, and to dogs, who are the companions of man, it is especially revolting.

The Queen is sorry to say, that she thinks the English are inclined to be more cruel to animals than some other civilised nations are.

To the Crown Princess of Prussia 30 August 1868

What you told me of dear Willy interested me very much. I share your anxiety especially as regards pride and selfishness. In our days—when a Prince can only maintain his position by his character—pride is most dangerous. And then besides I do feel so strongly that we are before God all alike, and that in the twinkling of an eye, the highest may find themselves at the feet of the poorest and lowest. I have seen the noblest, most refined, high-bred feelings in the humblest and most unlearned, and this it is most necessary a Prince should feel. I am sure you, darling, who never had any pride will feel and understand this well.

6 September 1868

All you say of adored Papa is so true! But alas! he was too perfect for this world; it was impossible for him ever to have been really happy here. I saw how dreadfully the wickedness of this world grieved his pure, noble, heavenly spirit, how indeed, like our Saviour.

JOURNAL   20 September 1868

Saw Mr. Disraeli, who arrived yesterday [as Minister in attendance at Balmoral] and talked a good deal about affairs and about the Church appointments. He expressed himself most anxious to make good, liberal and moderate ones. Jane Churchill and Mr. Disraeli (who is extremely agreeable and original) dined.

28 September 1868

Took leave of Mr. Disraeli, who seemed delighted with his stay and was most grateful. He certainly shows more consideration for my comfort than any of the preceding Prime Ministers since Sir Robert Peel and Lord Aberdeen.

TO THE CROWN PRINCESS OF PRUSSIA   1 October 1868

We had Mr. Disraeli with us for ten days and he was most agreeable; he is so original and full of poetry and admiration for nature.

TO LORD CHARLES FITZROY   6 October 1868

The Queen writes to Lord Charles FitzRoy, who is acting as Master of the Household, to ask him to see that the Smoking room is always closed at 12. This was her original intention, as well as that 11 o'clock should be the time for leaving the Drawing room; but she hears that the smoking often goes on till very late, a thing which in her house she does not intend to allow. The servants (she has heard) feel these late hours very much (and they are not wrong), and she must say that it is a bad thing for them and a bad example, especially in these days. Perhaps Lord Charles would draw up a short Memorandum to be handed over from one Equerry to the other gentlemen who succeed him. Lord Charles would perhaps simply mention to Prince Christian without giving it as a direct order, that the Queen felt it necessary for the sake of the servants, who were kept up so late and who had to be up so early in the morning, to direct that the Smoking room should be closed and the lights put out by 12 o'clock—not later; that this had been her original intention, and that it was necessary for everyone that this should be the case.

Lord Charles will no doubt agree with the Queen, and he will see that it is done quite quietly but effectually; someone coming to remind the Prince and gentlemen that the hour has come when the lights must be put out.

TO DISRAELI   9 October 1868

The Queen has been greatly shocked to hear to-day (and only to-day) of the death of the worthy and amiable Archbishop of Canterbury [Dr. Longley], which came to her quite unexpectedly, though she knew he had been ill.

The position of the Primate of England is one of such importance, and he is brought into so much personal contact with the Sovereign, that his appointment is the most important as well as the highest in the Church, and the Queen therefore writes at once to Mr. Disraeli to say that she thinks there is no one so fit, (indeed she knows of no one who would be fit), than [as] the Bishop of London [Dr. Tait], an excellent, pious, liberal-minded, courageous man, who would be an immense support and strength to the Church in these times. His health, which is not good, would be benefited by the change.

The Queen hopes to hear without delay on this subject from Mr. Disraeli, as it would not be good to wait too long before coming to a decision.

31 October 1868

The Queen has read Mr. Disraeli's letter of this morning [in which Disraeli recommended the Bishop of Gloucester and Bristol for the Archbishopric of Canterbury]. She cannot alter her opinion, that the Bishop of London is the only fit man to succeed the Archbishop.

The Bishop of Gloucester and Bristol, though a very good man, has not the knowledge of the world, nor the reputation and general presence (which is of so great importance in a position of such very high rank, constantly called upon to perform all the highest functions in connection with the Sovereign and Royal Family).

6 November 1868

The Queen thanks Mr. Disraeli for his long letter which she found on arriving here yesterday. She has read with the most careful attention all the objections made by him to the promotion of the Bishop of London; but is still of opinion that he would be the proper person, indeed the only proper person, to succeed the late Archbishop, and she cannot agree in the opinion given by Mr. Disraeli of the failing estimation in which he is held.

The Queen herself would feel much more confidence in his dealing wisely and prudently with the existing difficulties of the Church, and at the same time with more firmness and decision, than any other Bishop on the Bench.*

At the General Election held on 17 November the Conservatives were soundly defeated, and the Liberals returned with a majority of about 120.

TO GLADSTONE   1 December 1868

Mr. Disraeli has tendered his resignation to the Queen. The result of the appeal to the country is too evident to require its being proved by a vote in parliament, and the Queen entirely agrees with Mr. Disraeli, and his colleagues in thinking that the most dignified course for them to pursue, as also the best for the public interests, was immediate resignation. Under these circumstances the Queen must ask Mr. Gladstone, as the acknowledged leader of the liberal party, to undertake the

*The Queen had her way; and Dr. Tait was appointed.

formation of a new administration. With one or two exceptions, the reasons for which she has desired General Grey (the bearer of this letter) to explain, the Queen would impose no restrictions on Mr. Gladstone as to the arrangement of the various offices in the manner which he believes to be best for the public service.

MEMORANDUM   3 December 1868

I saw Mr. Gladstone at twenty minutes past five, he having come from Hawarden with General Grey. I said I knew he had consented to form a Government. He was most cordial and kind in his manner, and nothing could be more satisfactory than the whole interview.

JOURNAL   4 March 1869

Drove to the Deanery at Westminster, where the Dean and Augusta had invited the following celebrities to meet me: Mr. Carlyle, the historian, a strange-looking eccentric old Scotchman, who holds forth, in a drawling melancholy voice, with a broad Scotch accent, upon Scotland and upon the utter degeneration of everything; Mr. and Mrs. Grote, old acquaintances of mine from Kensington, unaltered, she very peculiar, clever and masculine, he also an historian, of the old school; Sir C. and Lady Lyell, he [the geologist] an old acquaintance, most agreeable, and she very pleasing; Mr. Browning, the poet, a very agreeable man. It was, at first, very shy work speaking to them, when they were all drawn up; but afterwards, when tea was being drunk, Augusta got them to come and sit near me, and they were very agreeable and talked very entertainingly.

TO THE PRINCE OF WALES   4 May 1869

The dear little children are very well, and I shall be very sorry to lose them.* They have been very well with me and are very fond of Grandmama. You must let me see them often, and sometimes let one or other of them come and stay with me for a little while, as I should not like them to become strangers to me. The great thing which I have observed from watching them is to keep to as much regularity of hours as possible . . .

You will, I fear, have incurred immense expenses, and I don't think you will find any disposition (except perhaps as regards those which were forced upon you at Constantinople) to give you any more money.

JOURNAL   20 June 1869

My Accession Day, already thirty-two years ago. May God help me in my solitary path, for the good of my dear people, and the world at large. He has given me a very difficult task, one for which I feel myself in many ways unfit, from inclination and want of power. He gave me great happiness, and He took it away, no doubt for a wise purpose and for the happiness of my beloved one, leaving me alone to bear the heavy burden in very trying and troubled times. Help I have been given, and for this I humbly thank Him; but the trials are great and many.

*The Prince and Princess of Wales had been for a tour in the winter and early spring to Egypt and Constantinople, and their children had paid a visit to their grandmother.

TO THE CROWN PRINCESS OF PRUSSIA   25 September 1869

Mr. Gladstone left this morning. I cannot find him very agreeable, and he talks so very much. He looks dreadfully ill.

JOURNAL   6 November 1869

I am afraid I can only give a very imperfect account of this most successful and gratifying progress and ceremony [the opening of Blackfriars Bridge and Holborn Viaduct]. Drove [from Paddington Station] the same way we usually do, going to Buckingham Palace and from there down the Mall and up to Westminster Palace. Everywhere great crowds of people and many amongst them well-dressed, all cheering and bowing and in the best of humour. Crossed Westminster Bridge, going up Stamford Street, all the time at a gentle trot, the streets admirably kept. From here the crowds became more and more dense, the decorations commenced, and the enthusiasm was very great—not a window empty and people up to the very tops of the houses, flags and here and there inscriptions, everyone with most friendly faces. The day was cold, but quite fine, no fog and, though no real sunshine, there were occasional glimpses of it.

At 12 o'clock we reached Blackfriars Bridge, the first portion of which was entirely covered in; here on a platform with raised seats and many people, stood the Lord Mayor, who presented the sword, which I merely touched, and he introduced the engineer, Mr. Cubitt (son of the eminent engineer Sir William Cubitt, well known to my dearest Albert) and another gentleman. I was presented with an Address and a fine illuminated book describing the whole. The Bridge was then considered opened, but neither I nor the Lord Mayor said so. This however has not been found out.

At length we moved on, preceded by the Lord Mayor and Sheriffs and some of the Aldermen, which led to very frequent complete stoppages. The greatest and most enthusiastic crowds were in the City. From the Bridge we proceeded through New Bridge Street to the foot of Ludgate Hill and Fleet Street. The crowds were quite immense, up to the very roofs, and down every small street as far as the eye could reach ... I never saw more enthusiastic, loyal or friendly [crowds] and there were numbers of the very lowest. This, in the very heart of London, at a time when people were said to be intending to do something, and were full of all sorts of ideas, is very remarkable.

Felt so pleased and relieved that all had gone off so well. Nothing could have been more gratifying. But is was a hard trial for me all alone with my children in an open carriage amongst such thousands!

THE PRINCE AND PRINCESS
OF WALES

1866–1872

TO THE CROWN PRINCESS OF PRUSSIA   31 March 1866

Dear Alix I don't think improved. She is grown a little grand, I think, and we never get more intimate or nearer to each other.

11 August 1866

I am sorry for Bertie; I don't think she [Princess Alexandra] makes his home comfortable; she is never ready for breakfast—not being out of her room till 11 often, and poor Bertie breakfasts alone and then she alone. I think it gets much worse instead of better; it makes me unhappy and anxious.

4 September 1866

Affie makes me very unhappy; he hardly ever comes near me, is reserved, touchy, vague and wilful and I distrust him completely. All the good derived from his stay in Germany has disappeared. He is quite a stranger to me. Bertie, on the other hand, is really very amiable.

11 May 1867

I really don't known why they had such hosts of sponsors [at the baptism of the third child of the Prince and Princess of Wales]. The child ought to be called 'Victoria' [she was christened Louise Victoria Alexandra Dagmar]. But upon those subjects Bertie and Alix do not understand the right thing.

9 November 1867

Really Bertie is so full of good and amiable qualities that it makes one forget and overlook much that one would wish different. Dearest Alix walks about, and up and down stairs—everywhere with the help of one or two sticks—but of course very slowly.* She even gets in and out of a carriage, but is a sad sight to see her thus and to those who did not see her so ill as we did, when one really did not dare to hope she would get better, it is sad and touching to see. She is very thin and looks very frail

---

*The Princess had been suffering from rheumatic fever. She ever afterwards walked with a limp—the 'Alexandra limp' which some women considered so fetching that they adopted it themselves—and her deafness was exacerbated.

but very pretty, and is so good and patient under this heavy trial. The poor leg is completely stiff and it remains to be seen whether it will ever get quite right again. I much fear not.

22 April 1868

Bertie's and Alix's visit to Ireland has gone off well—as ours always did—but like ours, it will be of no real use.

10 July 1868

Alix continues to go on quite well, but I thought she looked pale and exhausted. The baby [Princess Victoria]—a mere little red lump was all I saw; and I fear the seventh grand-daughter and fourteenth grand-child becomes a very uninteresting thing—for it seems to me to go on like the rabbits in Windsor Park!

5 March 1870

Dear Alix has just left us, and with increased love and affection and regard on my part. We agree so well, and she is so good and honest and right-minded. She is looking wonderfully well—quite fat for her. She has felt everything, that passed lately, deeply—but she is I think quite easy as to Bertie's conduct; only regretting his being foolish and imprudent.* The dear children I do not think improved. They are not well trained or managed.

9 March 1870

You are wrong in thinking that I am not fond of children. I am. I admire pretty ones—especially peasant children—immensely but I can't bear their being idolised and made too great objects of—or having a number of them about me, making a great noise.

TO THE PRINCE OF WALES   1 June 1870

Now that Ascot Races are approaching, I wish to repeat earnestly and seriously, and with reference to my letters this spring, that I trust you will, ... as my Uncle William IV and Aunt, and we ourselves did, confine your visits to the Races, to the two days Tuesday and Thursday and not go on Wednesday and Friday, to which William IV never went, nor did we ...

If you are anxious to go on those two great days (though I should prefer your not going every year to both) there is no real objection to that, but to the other days there is. Your example can do much for good, and may do an immense deal for evil, in the present day.

I hear every true and attached friend of ours expressing such anxiety that you should gather round you the really good, steady, and distinguished people.

---

*This is a reference to the Mordaunt divorce case in which the Prince of Wales was required to go into the witness box where he loudly replied, 'No, never!' to counsel's question, 'Has there ever been any improper familiarity or criminal act between yourself and Lady Mordaunt?' The Queen and the Princess of Wales both loyally supported him.

MEMORANDUM   25 June 1871

Mr. Gladstone spoke to me on the subject of Ireland this afternoon, and on the wish expressed again and again that there should be a Royal residence there, and said that a motion on the subject was about to be brought on, to which an answer must be given. We went over the old ground, the pretensions of the Irish to have more done for them than the Welsh or English; the visits to Scotland being in no one way political or connected with the wishes of the people, but merely because the climate and scenery are so healthy and beautiful, and the people so charming, so loyal, and the residence there of the greatest possible advantage (to mind and body) to our family, myself, and everyone connected with me and my Household. That, therefore, to press and urge this was unreasonable . . .

Mr. Gladstone contended how important it was in these days to connect the Royal family with public functions and offices, and that it would be very important to do this in Ireland. I contended that it would be wasting time in spending this in Ireland, when Scotland and England deserved it much more.

We then talked over the possibility of the Prince of Wales doing this [becoming a non-political Lord-Lieutenant], though I doubted the wisdom of identifying the future King with Ireland, and depriving him of his own home, and of going to Scotland, both of which were important for his health, Ireland being a bad climate. In this case, Mr. Gladstone said, three or four months would be quite enough. If it could take him more away from the London season I said it would be a good thing, but not if it took him away from the country, and from Scotland, where I saw most of him. It would take him away more from London, Mr. Gladstone said, and he thinks it would give him something to do.

JOURNAL   3 July 1871

Bertie remained for the Investiture. After '61 I could hardly bear the thought of anyone helping me and standing where my dearest one had always stood, but as years go on I strongly feel that to lift up my son and heir and keep him in his place near me, is only what is right. Bertie offered that I should use his sword instead of the Equerry's, which I did.

                                                                  22 November 1871

Heard dear Bertie had 'mild typhoid fever' and I at once determined to send off Sir William Jenner to Sandringham. This was gratefully accepted by Alix. Felt very anxious. This fearful fever, and at this very time of the year! Everyone much distressed.

                                                                  27 November 1871

At eight a telegram arrived. The report not good, a restless night and incessant wandering. We are all in the greatest anxiety. The alarm in London great. Immense sympathy all over the country. Heard from Sir William Jenner, who had returned to Sandringham, that, though Bertie was certainly very ill, he had found him less alarmingly so than he had expected. Though quite delirious, he knew people quite well. The expression and appearance good and unchanged, and the voice strong . . .

Dear Alix was bearing up well, but looking quite wretched and the picture of sorrow ... Decided to go to Sandringham on Wednesday.

29 November 1871

At eleven I left ... Reached Wolferton after three ... A quarter of an hour's drive brought us to Sandringham. The road lay between commons, and plantations of fir trees, rather wild-looking, flat, bleak country. The house, rather near the high-road, a handsome, quite newly built Elizabethan building, was only completed last autumn. Dear Alix and Alice met me at the door, the former looking thin and anxious, and with tears in her eyes. She took me at once through the great hall upstairs to my rooms, three in number.

I took off my things and went over to Bertie's room, and was allowed to step in from behind a screen to see him sleeping or dozing. The room was dark and only one lamp burning, so that I could not see him well. He was lying rather flat on his back, breathing very rapidly and loudly. Of course the watching is constant, and dear Alix does a great deal herself... How all reminded me so vividly and sadly of my dearest Albert's illness! Went over to take tea in Alix's pretty room, with her, Alice, and Affie. Saw Sir William Jenner, who said that the breathing had all along been the one thing that caused anxiety. It was a far more violent attack than my beloved husband's was, and we could not now look for any improvement till after the 21st day. The temperature was higher than yesterday. Dined with Alix, Alice, and Affie, in a small room below. Afterwards, we went upstairs and into Bertie's dressing-room. Dr. Gull came in, saying they were a little more anxious, as the pulse and temperature were higher. I remained till about ten and then went to my room.

10 December 1871

The feeling shown by the whole nation is quite marvellous and most touching and striking, showing how really sound and truly loyal the people really are. Letters and telegrams pour in and no end of recommendations of remedies of the most mad kind. Receive the kindest letters full of sympathy from the Ministers, my own people and friends.

11 December 1871

This has been a terrible day. At half past five I was woke by a message from Sir William Jenner saying dear Bertie had had a very severe spasm, which had alarmed them very much, though it was over now. I had scarcely got the message, before Sir William returned saying there had been another. I saw him at once, and he told me the spasm had been so severe, that at any moment dear Bertie might go off, so that I had better come at once. I hurriedly got up, put on my dressing-gown, and went to the room, where I found Alix and Alice by the bedside, and Dr. Gull and the two devoted nurses. It was dark, the candles burning, and most dreary. Poor dear Bertie was lying there breathing heavily, and as if he must choke at any moment. I remained sitting behind the screen. Louise and her three brothers came into the dressing-room. Everything was done that could be thought of to give a little relief. After a little while he seemed easier, so the doctors advised us to go away, and I went back to my room, breakfasted, and dressed.

I went backwards and forwards continually. The talking was incessant, without a moment's sleep. Dr. Gull said he was much alarmed. Went away with a very heavy heart and dreading further trouble. Felt quite exhausted.

13 December 1871

This really has been the worst day of all, and coming as it has so close to the sad 14th [the anniversary of the Prince Consort's death], filled us and, I believe, the whole country with anxious forebodings and the greatest alarm. The first report early in the morning was that dear Bertie seemed very weak, and the breathing very imperfect and feeble. The strength, however, rallied again. There had been no rest all night, from the constant delirium. The pulse varied in quality of strength at intervals, from hour to hour. Got up and dressed quickly, taking a mouthful of breakfast before hurrying to Bertie's room. Sat near by on the sofa, but so that he could not see me. Remained a long time. It was very distressing to hear him calling out and talking incessantly quite incoherently. Strolled round the house and pleasure grounds for a short while. It was raw and damp, and thawing all day.

Returned to Bertie's room, and, whilst there, he had a most frightful fit of coughing, which seemed at one moment to threaten his life! Only Alix and one of the nurses were there, and the doctors were at once hastily summoned. But the dreadful moment had passed. Poor dear Alix was in the greatest alarm and despair, and I supported her as best I could. Alice and I said to one another in tears, 'There can be no hope.' . . . I went up to the bed and took hold of his poor hand, kissing it and stroking his arm. He turned round and looked wildly at me saying, 'Who are you?' and then, 'It's Mama.' 'Dear child,' I replied. Later he said, 'It is so kind of you to come,' which shows he knew me, which was most comforting to me. I sat next to the bed holding his hand, as he seemed dozing . . .

When I returned I found dear Bertie breathing very heavily and with great difficulty. Another symptom which frightened me dreadfully, was his clutching at his bed-clothes and seeming to feel for things which were not there. The gasping between each word was most distressing. We were getting nearer and nearer to the 14th, and it seemed more and more like ten years ago, and yet it was very different too.

14 December 1871

This dreadful anniversary, the 10th, returned again. It seems impossible to believe all that time has passed. Felt painfully having to spend the day away from Windsor, but the one great anxiety seems to absorb everything else. Instead of this date dawning upon another deathbed, which I had felt almost certain of, it brought the cheering news that dear Bertie had slept quietly at intervals, and really soundly from four to quarter to six; the respirations much easier, and food taken well . . .

Breakfasted with Beatrice and Leopold, and then went over to dear Bertie. When I stood near the screen, he asked the nurse if that was not the Queen, and she asked me to go up to him, which I did. He kissed my hand, smiling in his usual way, and said, 'So kind of you to come; it is the kindest thing you could do.' He wanted to talk more, but I would not allow him, and left.

To the Crown Princess of Prussia   16 December 1871

This morning all is much the same and we would like to say it was a marked improvement but it is not yet that.

I do not think this place wholesome and the drainage is defective—though he did not catch the fever here but at Scarborough.

20 December 1871

We all feel that if God has spared his life it is to enable him to lead a new life—and if this great warning is not taken, and the wonderful sympathy and devotion of the whole nation does not make a great change in him, it will be worse than before and his utter ruin. All the papers and the many sermons all tend to show the same feeling ... The loyalty shown, is most striking. Beloved Alix I can never praise enough. Her devotion, calmness and simple, strong religious faith were touching and beautiful to see. She is dearer to me than I can say—and so true, so discreet, so kind to all. It was a great lesson to all—to see the highest surrounded by every luxury which human mortal can wish for—lying low and as helpless and miserable as the poorest peasant.

23 December 1871

His progress is terribly slow. Broken sleep—wandering (yesterday morning a great deal) and great prostration. The nerves have received a terrible shock and will take long recovering. Another person, a woman who helped in the kitchen has now got the fever. I think the house very unhealthy—drainage and ventilation—bad; bad smells in some rooms—of gas and drains. It would never do to let the children go back there.

30 December 1871

I wish I could give you a really satisfactory report of dear Bertie—but I cannot. He has the most dreadful pain in his leg which comes on in violent spasms—and the temperature is very high which if it were merely pain would not be the case.

Journal   31 December 1871

Went over once more to Bertie's room. He was quiet, thank God. How I pray the New Year may see him safely on the road to recovery!

The New Year did bring the Prince safely to recovery.

To the Crown Princess of Prussia   14 February 1872

Dear Bertie looks very delicate, very pale and thin and drawn—walks slowly, still a little lame, but is very cheerful and quite himself, only gentler and kinder than ever, and there is something different which I can't exactly express. It is like a new life—all the trees and flowers give him pleasure as they never used to do—and he was quite pathetic over his small wheelbarrow and little tools at the Swiss cottage. He is constantly with Alix and they seem hardly ever apart!

A thanksgiving service—described by the Queen, who did not like 'religion being made a vehicle for a great show', as 'this dreadful affair at St. Paul's'—was arranged for 27 February.

JOURNAL   27 February 1872

Luckily a fine morning. After breakfast went over to the luncheon room and saw such crowds already collecting. Went to dress, and wore a black silk dress and jacket, trimmed with miniver, and a bonnet with white flowers and a white feather. Beatrice looked very nice in mauve, trimmed with swan's down. Awaited Bertie and Alix and they soon came with their two little boys. Bertie was very lame and did not look at all well, I grieved to see. My three other sons were there, and the poor Emperor Napoleon and Empress Eugénie, who were anxious to see the Procession quietly, and whom I had specially invited to come to the Palace. The boys with little George [the future King George V] went on and got into an open carriage and four, and in a few minutes I followed, taking poor Bertie's arm, for he could only walk very slowly, down to the Grand Entrance. We entered an open State landau with six horses, riden by three postilions. Alix (in blue velvet and sable) sat next to me, and Bertie opposite, with Beatrice, and little Eddy [Prince Albert Victor] between them. We had a Sovereign's escort, as on all State occasions. Seven open dress carriages with a pair of horses went in front of us, and immediately in front, the Lord Chancellor in his carriage, and the Speaker in his strange quaint old one ...

The deafening cheers never ceased the whole way, and the most wonderful order was preserved. We seemed to be passing through a sea of people, as we went along the Mall. Our course going to St. Paul's was down the Mall, by Pall Mall, Trafalgar Square, straight up the Strand, Fleet Street, and Temple Bar, which was handsomely decorated. There were stands and platforms in front of the Clubs, etc., full of well-dressed people, and no end of nice and touching inscriptions. At the corner of Marlborough House there was a stand on which stood Bertie's dear little girls, who waved their handkerchiefs. At Temple Bar the Lord Mayor, in a crimson velvet and ermine robe, came up to the carriage to present the sword, which I touched and returned to him, after which he got on horseback, bareheaded, and carrying the sword rode in front, preceded by the Mace-bearer, City Marshal, and three other Aldermen. Everywhere troops lined the streets, and there were fifteen military bands stationed at intervals along the whole route, who played 'God save the Queen' and 'God bless the Prince of Wales' as the carriages approached, which evoked fresh outbursts of cheering. I saw the tears in Bertie's eyes and took and pressed his hand! It was a most affecting day, and many a time I repressed my tears. Bertie was continually with his hat off.

Got back to the Grand Entrance at twenty minutes to four, and Bertie and Alix, with their boys, took leave in the hall, going straight home. I went upstairs and stepped out on the balcony with Beatrice and my three sons, being loudly cheered. Rested on the sofa after taking some tea. Could think and talk of little else, but to-day's wonderful demonstration of loyalty and affection, from the very highest to the lowest. Felt tired by all the emotion, but it is a day that can never be forgotten!

Gladstone assured the Queen that the joyful celebrations were perhaps the most satisfactory that the City had ever witnessed. They were 'a quite extraordinary manifestation of loyalty and affection'. That evening all over London thousands of people continued the celebrations in the streets, dancing and singing so happily that the poet, A. J. Munby, had never seen 'a sight so striking in England'. Republicanism as a significant force in British politics, already damaged by the excesses of the Paris Commune, had suffered a blow from which it was never completely to recover.

'FREQUENT DIFFICULTIES'

1870–1878

TO THE CROWN PRINCESS OF PRUSSIA   12 January 1870

I send you [Tennyson's] 'The Holy Grail' but must say—beautiful as are passages in it—it is still more unclear than any of his writings and leaves me quite bewildered.

JOURNAL   9 March 1870

I saw Mr. Helps [Clerk of the Privy Council] this evening at half past six, who brought and introduced Mr. Dickens, the celebrated author. He is very agreeable, with a pleasant voice and manner. He talked of his latest works, of America, the strangeness of the people there, of the division of classes in England, which he hoped would get better in time. He felt sure that it would come gradually.

23 March 1870

The newly cleaned monument of Henry VII [in Westminster Abbey] was what they [the Dean of Westminster and Lady Augusta Stanley] particularly wanted me to see, and it is beautiful, all bright gold. Went into the Deanery for tea, where, as last year, were assembled some celebrities: Lady Eastlake [widow of Sir Charles Eastlake and author of *Letters from the Baltic*], tall, large, rather ponderous and pompous; Mr. Froude [the historian], with fine eyes, but nothing very sympathetic; Professor Owen [the zoologist], charming as ever; Professor Tyndall [the physicist] (not very attractive), who has a great deal to say; Sir Henry Holland [Physician in Ordinary to the Queen], quite wonderful and unaltered; and Mr. Leikie [? Lecky the historian], young, pleasing, but very shy.

26 March 1870

At eleven, when the messenger arrived, there came a note from Sir William Jenner with the dreadful news that General Grey had had a seizure soon after seven this morning and three attacks of convulsions since! We were horrified, though I had foreseen some impending illness, as had Sir William. Saw Colonel Ponsonby and sent him up to town at once, to enquire, also telegraphed to Sir William Jenner to tell me the exact truth, and to poor Mrs. Grey. Most dreadful! Could think of nothing else. All greatly shocked.

1 April 1870

After talking for a little while [to Mrs. Grey in the General's house at St. James's Palace] she took me into the room where the dear General lay, looking so peaceful, nice and unaltered, without that dreadful pallor one generally sees after death. His bed was covered with flowers, of which he was so fond. Poor dear General, I could not bear to think I should never look again on his face in this world! He was most truly devoted and faithful and had such a kind heart.

Saw Sir T. Biddulph [Master of the Household]. Colonel [Henry] Ponsonby is to replace our dear General [as the Queen's Private Secretary] which he himself had recommended when he talked of retiring.

TO THE CROWN PRINCESS OF PRUSSIA   9 April 1870

Colonel Ponsonby is a very decided Liberal, but he never has mixed in politics—and is very discreet which our poor, dear General was not, I must own, though it may appear very strange to you.

JOURNAL   14 April 1870

Darling Beatrice's 13th birthday. It is to me quite sad that she is growing so fast out of the dear little engaging child and so far from the time when she was her precious father's little pet.

6 June 1870

Il ne faut pas disputer des goûts but I do not fancy him [Lord Lorne, who was to marry Princess Louise]. He has such a forward manner, and such a disagreeable way of speaking but I know he is very clever and very good.

In July 1870, in his endeavours to complete the unification of Germany under Prussia, Bismarck cleverly manoeuvred France into declaring war. The subsequent Franco-Prussian War caused the Queen intense distress.

TO THE CROWN PRINCESS OF PRUSSIA   16 July 1870

Beloved child, I cannot say what my feelings of horror and indignation are, or how frightfully iniquitous I think this declaration of war! My heart boils and bleeds at the thought of what misery and suffering will be caused by this act of mad folly!

20 July 1870

My poor, dear, beloved child, words are far too weak to say all I feel for you or what I think of my neighbours!!! We must be neutral as long as we can, but no one here conceals their opinion as to the extreme iniquity of the war—and the unjustifiable conduct of the French. Still, more publicly we cannot say but the feeling of the people and country here is all with you—which it was not before. And need I say what I feel? My whole heart and my fervent prayers are with beloved Germany! Say that to Fritz—but he must not say it again—and that I shall suffer cruelly for you all—thinking of beloved Papa too, who would have gone to fight if he could.

TO GRANVILLE   20 July 1870

[The Queen] is overwhelmed with letter-writing, telegrams and the terrible anxiety and sorrow which this horrible war will bring with it. The Queen hardly knows how she will bear it! Her children's home threatened, their husbands' lives in danger, and the country she loves best next to her own—as it is her second home, being her beloved husband's, and one to which she and all her family are bound by the closest ties—in peril of the gravest kind, insulted and attacked, and she unable to help them or to come to their assistance. Can there be a more cruel position than the unhappy Queen's? She knows what her duty is and will do what must be done, but she will suffer dreadfully.

TO QUEEN AUGUSTA OF PRUSSIA   20 July 1870

What can I say? This dreadful war is vile and unforgivable! May God protect our dear, beloved Germany! My heart is indeed heavy and bleeds for you! We have made every possible effort to preserve peace!

JOURNAL   9 August 1870

More telegrams. The losses of the unfortunate French seem to be greater and greater, 12,000 killed and wounded and 4,000 prisoners. They are quite disorganised. Dreadful excitement at Paris ...

Saw Mr. Gladstone, who was full of the extraordinary events which have taken place. Perhaps, he said, it might all be for the best for Europe, for, though he was always very fond of the French, he thought a Bonaparte on the throne had always an element of uncertainty and danger.

                                                      17 August 1870

After luncheon held a Council, before which I saw Lord Granville. Talked of the extraordinary reverses and defeats of the French and the great victories of the Prussians. Both in diplomacy and war the failure of the French had been so utterly complete. He thinks it is chiefly due to a loose unprincipled Government, and to everything having become so corrupt.

TO QUEEN AUGUSTA OF PRUSSIA   17 August 1870

This frightful bloodshed is really too horrible in Europe in the 19th century. With the weapons of today it is really too ghastly, and when this war is at an end, there ought to be some attempt made to find means of preventing such wars once and for all. Otherwise the peoples will become extinct!

TO THE CROWN PRINCESS OF PRUSSIA   22 August 1870

The position of the French seems to get hourly worse! Such a complete tumbling to pieces of their empire and its far famed army has really never been seen! It does seem like a judgement from heaven! Everything seems to fail! Odiously impertinent, insulting and boastful as the French have always been, one cannot help feeling for them.

JOURNAL   3 September 1870

Just as I was going out arrived a telegram, which nearly took one's breath away! The whole of MacMahon's Army have laid down their arms and capitulated [at Sedan], while the Emperor surrendered himself to the King of Prussia, who was to see him immediately. My first thought, and that of many, was that this might lead to peace!

                                                              5 September 1870

Heard that the mob at Paris had rushed into the Senate and proclaimed the downfall of the dynasty, proclaiming a Republic! This was received with acclamation and the proclamation was made from the Hôtel de Ville. Not one voice was raised in favour of the unfortunate Emperor! How ungrateful!

MEMORANDUM   9 September 1870

The French evidently wish for peace and are in the greatest want of it, but seem still to think they can dictate terms! This is madness! . . .

   A powerful Germany can never be dangerous to England, but the very reverse, and our great object should therefore be to have her friendly and cordial towards us.

JOURNAL   12 September 1870

My dearest kindest friend, dear old Lehzen, expired quite quietly and peacefully on the 9th. For two years she had been quite bedridden, from the results of breaking her hip. Though latterly her mind had not been clear, still there were days when she constantly spoke of me, whom she had known from the age of six months. She had devoted her life to me, from my fifth to my eighteenth year, with the most wonderful self-abnegation, never even taking one day's leave! After I came to the throne she got to be rather trying and especially so after my marriage, but never from any evil intention, only from a mistaken idea of duty and affection for me. She was an admirable governess, and I adored her, though I also feared her.

TO THE CROWN PRINCESS OF PRUSSIA   13 September 1870

In England I can assure you the feeling is far more German than French, and far the greater part of the press is in your favour. All reflecting people are.

                                                             17 September 1870

In one of your former long letters you said you thought now of many things which dear Papa had said—which showed that all was after all not so entirely unexpected in France!! The system of corruption, immorality and *gaspillage* was dreadful. Nothing annoyed dear Papa more than the abject court paid to the Emperor and the way in which we were forced to flatter and humour him, which was shortsighted policy, and spoilt him.

TO THE KING OF PRUSSIA   19 September 1870

[translated from the German]
The Queen asks the King of Prussia as a friend whether, in the interests of suffering humanity, he could so shape his demands as to enable the French to accept them.

The King and his splendid victorious Army stand so high that the Queen thinks they can afford, on obtaining necessary securities for preventing similar events or attacks, to be generous. The King's name will stand even higher if he make peace now.

To Granville   1 October 1870

The Queen is so glad to see how firmly and resolutely Lord Granville refuses to be dragged into mediation and interference, though it must be very difficult to avoid it.

The Queen feels so very strongly the danger to this country of giving advice which will not help the one party, and may turn the very powerful other party, already much (and unjustly) irritated against us, into an inveterate enemy of England, which would be very dangerous and serious.

To the Crown Princess of Prussia   3 October 1870

I must now tell you that I have changed my opinion of Lord Lorne [later 9th Duke of Argyll] since I have got to know him (he has been here since Thursday) and I think him very pleasing, amiable, clever—his voice being only a little against him. And he is in fact very good looking [he married Princess Louise in 1871].

To Queen Augusta of Prussia   17 November 1870

The very groundless and unjust feeling against England in Germany is beginning to arouse great indignation here, especially as from the very first, with the exception of a few of the upper classes who are fond of going to Paris, all sympathies were with Germany, and because people have done so much for the wounded. I write this to you quite frankly, for I consider the danger great and serious that the 2 great nations should become so far irritated against one another as to be unable to put things right again and allow the feelings of hostility to grow. Please warn the dear King and Fritz and everyone.

Journal   30 November 1870

Dull, raw, and cold. At quarter to eleven, started . . . for Chislehurst, in Kent, where the poor Empress Eugénie is staying.

At the door [of Camden Place] stood the poor Empress, in black, the Prince Imperial, and, a little behind, the Ladies and Gentlemen. The Empress at once led me through a sort of corridor or vestibule and an ante-room into a drawing-room with a bow window. Everything was like a French house and many pretty things about. The Empress and Prince Imperial alone came in, and she asked me to sit down near her on the sofa. She looks very thin and pale, but still very handsome. There is an expression of deep sadness in her face, and she frequently had tears in her eyes. She was dressed in the plainest possible way, without any jewels or ornaments, and her hair simply done, in a net, at the back. She showed the greatest tact in avoiding everything which might be awkward, and enquired after Vicky and Alice, asked if I had had any news, saying, 'Oh! si seulement l'on pouvait avoir la paix.' Then she said how much had happened since we had met at Paris and that she could not forget the dreadful impressions of her departure from there . . . The night before she had lain down fully dressed, on her bed. The crossing had been fearful. Afterwards she talked of other things. The Prince Imperial is a nice little boy, but rather short and stumpy.

31 December 1870

I ended this dreadful year of bloody conflict in no cheerful mood. It is always sad to me to be getting further and further off from the time when my beloved husband was still with me, and the thought of this dreadful war, indeed there has perhaps never been a worse one, or such loss of life . . . lies like a heavy weight on one's heart and mind.

To the Crown Princess of Prussia 14 January 1871

The bombardment [of Paris] is a sad thing and I cannot say how I pray for the ending of this dreadful slaughter, which seems alas! so useless, for the feeling in England is becoming sadly hostile to Germany. Everything will be done to calm this, and Parliament in this respect will do good they say, though things will be said which are painful and may have a bad effect. The fact is people are so fond of Paris—so accustomed to go there that the threatened ruin of it makes them furious and unreasonable.

To Queen Augusta of Prussia [whose husband was proclaimed German Emperor that day] 18 January 1871

I hate having to tell you about the increase of bad feeling in this country towards Germany, and in particular against Prussia, and how unhappy it makes me.

To the Crown Princess of Prussia 1 February 1871

God be praised truly and really for this blessed capitulation and armistice which will soon dissipate the sentimentality here. It came so suddenly at last but the wretched people must be in a terrible state.

4 February 1871

Here the terms have created a bad feeling [the French were required to pay £200,000,000 within a fortnight; and agreed to do so]. But I should think they will be abated—at any rate the money part, which would be quite impossible for the wretched French to pay.

10 February 1871

I am glad to see that you feel prouder than of anything to be an Englishwoman or rather more a Briton, for you may be as proud of the Scotch blood in your veins as of any other. One of the last walks I took with darling Papa, he said to me 'England does not know what she owes to Scotland'. She is the brightest jewel in my crown—energy, courage, worth, inimitable perseverance, determination and self-respect.

1 March 1871

The preliminaries of peace are declared but they are very hard. That was to be expected, and I fear that they may not be accepted. This march through Paris alarms us all very much. If only nothing untoward happens. The feeling here towards Prussia is as bitter as it can be. It is a great grief to me—and I can do nothing! ... To see the enmity growing up between two nations—which I am bound to say began first in Prussia, and was most unjust and was fomented and encouraged by Bismarck—is a great sorrow and anxiety to me—and I cannot separate myself or allow myself to be separated from my own people. For it is alas! the people, who from being very German up to three months ago are now very French! I tell you this with a heavy heart but it is the fact.

TO THE GERMAN EMPEROR   20 March 1871

From my heart I hope that our two countries may draw nearer to one another, and that the passing ill-humour which sprang from misunderstandings and mistaken judgments on both sides may disappear!

JOURNAL   27 March 1871

At a little before three, went down with our children and Ladies and Gentlemen to receive the Emperor Napoleon [at Windsor Castle]. I went to the door with Louise and embraced the Emperor 'comme de rigueur.' It was a moving moment, when I thought of the last time he came here in '55, in perfect triumph, dearest Albert bringing him from Dover, the whole country mad to receive him, and now! He seemed much depressed and had tears in his eyes, but he controlled himself and said, 'Il y a bien longtemps que je n'ai vu votre Majesté.' He led me upstairs and we went into the Audience Room. He is grown very stout and grey and his moustaches are no longer curled or waxed as formerly, but otherwise there was the same pleasing, gentle, and gracious manner. My children came in with us. The Emperor at once spoke of the dreadful and disgraceful state of France, and how all that had passed during the last few months had greatly lowered the French character.

29 March 1871

At a little after twelve, started in nine dress closed carriages (mine with a pair of creams) for the Albert Hall, for its opening ... Immense and very loyal crowds out. Bertie received us at the door and then we walked up the centre of the immensely crowded Hall (8,000 people were there), which made me feel quite giddy. Bertie read the address from the dais, to which we had been conducted, very well, and I handed to him the answer, saying : 'In handing you this answer I wish to express my great admiration of this beautiful Hall, and my earnest wishes for its complete success.' This was greatly applauded. The National Anthem was sung, after which Bertie declared the Hall open. Good Mr. Cole [Henry Cole, Chairman of the Society of Arts, who had worked hard for the success of the Great Exhibition of 1851] was quite crying with emotion and delight. It is to Colonel Scott of the Engineers, who built the Hall, that the success of the whole is due. We then went upstairs to my Box, which is not quite in the centre, and heard Costa's cantata performed, which is very fine. I had never been at such a big function since beloved Albert's time, and it was naturally trying and *émotionnant* for me. I thought of poor dear General Grey, who had been so enthusiastic and anxious about this undertaking and who was not permitted to see the building completed!

8 April 1871

Still dreadful news from Paris. The Commune have everything their own way, and they go on quite as in the days of the old Revolution in the last century, though they have not yet proceeded to commit all the same horrors. They have, however, thrown priests into prison, etc. They have burnt the guillotine and shoot people instead. I am so glad I saw Paris once more, though I should not care to do so again.

TO GLADSTONE   23 April 1871

With respect to the Budget, it is difficult not to feel considerable doubt as to the wisdom of the proposed tax on matches, which is a direct tax and will be at once felt by all classes, to whom matches have become a necessity of life. Their greatly increased price will in all probability make no difference in the consumption by the rich; but the poorer classes will be constantly irritated by this increased expense and reminded of the tax by the Government stamp on the box.

Above all it seems certain that the tax will seriously affect the manufacture and sale of matches, which is said to be the sole means of support of a vast number of the very poorest people and little children, especially in London, so that this tax, which it is intended should press on all equally, will in fact be only severely felt by the poor, which would be very wrong and most impolitic at the present moment.

The Queen trusts that the Government will reconsider this proposal and try and substitute some other which will not press upon the poor.*

JOURNAL   27 May 1871

Most dreadful news from Paris. The wretched Archbishop, another Bishop, a Curé, and sixty-four other prisoners have been shot by these horrid Communists, before the prison could be taken.

31 July 1871

A very fine day. Breakfast in the tent [at Osborne]. Afterwards met good Fritz and talked with him of the war. He is so fair, kind, and good and has the intensest horror of Bismarck, says he is no doubt energetic and clever, but bad, unprincipled, and all-powerful; he is in fact the Emperor, which Fritz's father does not like, but still cannot help . . . That he felt they were living on a volcano, and that he should not be surprised if Bismarck some day tried to make war on England. This corroborates and justifies what many people here have said.

4 September 1871

In the afternoon took a little turn in the garden chair [at Balmoral, where the Queen

*The Government did reconsider the proposal and, after a procession of thousands of poor match makers to the Houses of Parliament, the tax was given up.

was feeling dreadfully ill with rheumatism in her leg and an abscess on her arm]. It was so fine. On coming in heard Mr. Lister had arrived. Sir William Jenner explained everything about my arm to him, but he naturally said he could do nothing or give any opinion till he had made an examination. I had to wait nearly half an hour before Mr. Lister and Dr. Marshall appeared! In a few minutes he had ascertained all and went out again with the others. Sir William Jenner returned saying Mr. Lister thought the swelling ought to be cut; he could wait twenty-four hours, but it would be better not. I felt dreadfully nervous, as I bear pain so badly. I shall be given chloroform, but not very much, as I am so far from well otherwise, so I begged the part might be frozen, which was agreed on. Everything was got ready and the three doctors came in. Sir William Jenner gave me some whiffs of chloroform, whilst Mr. Lister froze the place, Dr. Marshall holding my arm. The abscess, which was six inches in diameter, was very quickly cut and I hardly felt anything excepting the last touch, when I was given a little more chloroform. In an instant there was relief.

11 September 1871

To-day I have been very miserable from a violent attack of rheumatism or even rheumatic gout, which has settled in my left ankle, completely crippling me and causing me dreadful pain.

18 September 1871

My foot much swollen, and I could hardly walk a step. The doctors, after looking at it, pronounced it to be severe rheumatic gout, and I was not to walk, indeed I could not. How distressing and disappointing! Was rolled into my sitting-room, where Alice came to see me, much shocked and grieved. Was carried downstairs and took a little drive. The rest of the day I remained in my room. By degrees agonies of pain came on which continued almost without intermission, the foot swelling tremendously.

18 October 1871

A most dreadful night of agonising pain. No sedative did any good. I only got some sleep between five and eight this morning. Felt much exhausted on awaking, but there was no fever, and the pain was much less. Had my feet and hands bandaged. My utter helplessness is a bitter trial, not even being able to feed myself ... Was unable all day hardly to eat anything. Dictated my Journal to Beatrice, which I have done most days lately.

23 October 1871

My hands much better. Was able to sign, which is a great thing.

22 November 1871

Breakfasted for the first time again with my children, and felt it was a step forward and I was returning to ordinary life.

To the Crown Princess of Prussia   17 February 1872

Our Government here does not get on very well. They have contrived to get so very unpopular. Mr. Gladstone is a very dangerous Minister—and so wonderfully unsympathetic. I have felt this very much, but find his own followers and colleagues complain fully as much.

Journal   29 February 1872

At half past four drove in open landau and four with Arthur, Leopold, and Jane C[hurchill], the Equerries riding. We drove round Hyde and Regent's Parks, returning by Constitution Hill, and when at the Garden Entrance a dreadful thing happened ... It is difficult for me to describe, as my impression was a great fright, and all was over in a minute. How it all happened I knew nothing of. The Equerries had dismounted, Brown had got down to let down the steps, and Jane C. was just getting out, when suddenly someone appeared at my side, whom I at first imagined was a footman, going to lift off the wrapper. Then I perceived that it was someone unknown, peering above the carriage door, with an uplifted hand and a strange voice, at the same time the boys calling out and moving forward. Involuntarily, in a terrible fright, I threw myself over Jane C., calling out, 'Save me,' and heard a scuffle and voices! I soon recovered myself sufficiently to stand up and turn round, when I saw Brown holding a young man tightly, who was struggling. They laid the man on the ground and Brown kept hold of him till several of the police came in. All turned and asked if I was hurt, and I said, 'Not at all.' Then Lord Charles [FitzRoy], General Hardinge, and Arthur came up, saying they thought the man had dropped something. We looked, but could find nothing, when Cannon, the postilion, called out, 'There it is,' and looking down I then did see shining on the ground a small pistol! This filled us with horror. All were as white as sheets, Jane C. almost crying, and Leopold looked as if he were going to faint.

It is to good Brown and to his wonderful presence of mind that I greatly owe my safety, for he alone saw the boy rush round and followed him! When I was standing in the hall, General Hardinge came in, bringing an extraordinary document which this boy had intended making me sign! It was in connection with the Fenian prisoners!

                                                    1 March 1872

Drove to Marlborough House, where Alix and the little boys received me, and where I saw dear Bertie, who is still not allowed to walk, suffers a good deal of pain, and does not know if he will be able to get away on the 4th, which is very tiresome. He has appointed General Probyn as his Equerry, which is a very good appointment. Took leave of dear Bertie and Alix with great regret. On returning saw Mr. Gladstone, who was dreadfully shocked at what [had] happened, and to whom I recounted the whole thing.

To the Crown Princess of Prussia   8 May 1872

I am most thankful to hear you are going on so satisfactorily [the Princess had given birth to her fourth daughter, Margaret Beatrice Feodore]. I never thought you cared (having 3 of each) whether it was a son or a daughter; indeed I think many Princes a great misfortune—for they are in one another's and almost everybody's way. I am sure it is the case here—and dear Papa felt this so much that he was always talking of establishing if possible one or two of your brothers and eventual grandchildren (of which I fear there is the prospect of a legion with but little money) in the colonies. I don't dislike babies, though I think very young ones rather disgusting, and I take interest in those of my children when there are two or three—and of people who are dear to me and whom I am fond of—but when they come at the rate of three a year it becomes a cause of mere anxiety for my own children and of no great interest.

5 June 1872

It gave me much pleasure to receive your dear letter, as it showed me that you can understand what I meant about the relations of children and parents. The higher the position the more difficult it is.—And for a woman alone to be head of so large a family and at the same time reigning Sovereign is I can assure you almost more than human strength can bear. I assure you I feel so disheartened. I should like to retire quietly to a cottage in the hills and rest and see almost no one. As long as my health and strength will bear it—I will go on—but I often fear I shall not be able for many years (if I live). If only our dear Bertie was fit to replace me! Alas! Alas! I feel very anxious for the future ... And so is everyone.

26 June 1872

I come now to this very important subject of the position of the working classes. You know that I have a very strong feeling on that subject. I think the conduct of the higher classes of the present day very alarming—for it is amusement and frivolity from morning till night— which engenders selfishness, and there is a toleration of every sort of vice with impunity in them. Whereas the poorer and working classes who have far less education and are much more exposed—are abused for the tenth part less evil than their betters commit without the slightest blame. The so called immorality of the lower classes is not to be named on the same day with that of the higher and highest. This is a thing which makes my blood boil, and they will pay for it.

JOURNAL   4 July 1872

Had some music in the Red Drawing-room, to which my three children, Lenchen, and the ladies and some of the gentlemen came. Adelina Patti, the famous prima donna, now the favourite, since the last six years ... I was charmed with Patti, who has a very sweet voice and wonderful facility and execution. She sings very quietly and is a very pretty ladylike little thing.

TO THE CROWN PRINCESS OF PRUSSIA   3 September 1872

In your dear letter you speak of beloved Papa and his irreparable loss and how few if any were like him. But I feel and see that if he had lived he would have suffered cruelly from many inevitable things which have taken place and which he never would have approved ... The style of life of your elder brothers which he could not have prevented would have shocked and angered him. He foresaw this and often was greatly depressed in speaking of it to me, and it is this which makes me often sad when I see things go on so exactly contrary to what he would have wished and liked—but which cannot be avoided. I often say 'thank God! he is spared this'; I rather bear the burthen alone and submit to what I cannot prevent, doing all I can to prevent serious evil. He could not have borne it.

9 September 1872

I have this evening seen a Mr. Stanley, who discovered Livingstone, a determined, ugly, little man—with a strong American twang.

JOURNAL   23 September 1872

Can I write it? My own darling, only sister, my dear excellent, noble Feodore is no more! ... This was to have been and is still a day of rejoicing for all the good Balmoral people, on account of dear Bertie's first return after his illness; and I am here in sorrow and grief, unable to join in the welcome. God's will be done, but the loss to me is too dreadful! I stand so alone now, no near and dear one near my own age, or older, to whom I could look up to, left! All, all gone! She was my last near relative on an equality with me, the last link with my childhood and youth. My dear children, so kind and affectionate, but no one can really help me.

TO GLADSTONE   22 October 1872

A subject which demands the most serious attention of the Government is the very alarming and increasing insecurity of the Railroads. The Queen has repeatedly spoken and written about this, but she thinks that nothing has yet been done by Government which tends to remedy this most alarming subject. Legislation is applied to every possible subject, but the one fully as important as Education, viz. the safety of human life, seems to be much less thought of than any other. In no country except ours are there so many dreadful accidents, and for the poor people who have to travel constantly by rail, and who cannot even have the comparative security which those who travel in first-class carriages can have, to be in perpetual danger of their lives is monstrous. Independent of this, the Queen's own family, not to speak of her servants and visitors, are in perpetual danger, and are put to the most serious inconvenience by the inexactitude of the trains.

TO DISRAELI   15 December 1872

The Queen well knows that Mr. Disraeli will not consider the expression of her heart-felt sympathy an intrusion in this his first hour of desolation and overwhelming grief [at the death of his wife that day], and therefore she at once attempts to express what she feels. The Queen knew and admired as well as appreciated the unbounded devotion and affection which united him to the dear partner of his life, whose only thought was him.

JOURNAL 29 February 1873

A very foggy, raw day. At quarter past ten, left Windsor for Chislehurst [to commiserate with the Empress Eugénie on the death of the Emperor], by the South-Western. We passed through London, which was wrapped in a thick yellow fog ... drove to Camden House, where at the door, instead of his poor father, who had always received me so kindly, was the Prince Imperial, looking very pale and sad. A few steps further on, in the deepest mourning, looking very ill, very handsome, and the picture of sorrow, was the poor dear Empress, who had insisted on coming down to receive me. Silently we embraced each other and she took my arm in hers, but could not speak for emotion.

TO THE CROWN PRINCESS OF PRUSSIA 6 March 1873

The Prince Imperial came to luncheon here on Tuesday. He is a very dear, nice boy—with such charming manners—reminding me much of his father and yet he is like her too. The tears come to his eyes whenever he speaks of the Emperor. His nose is getting like the Emperor's and the shape of his head as well as the colour of his eyes—but the shape of his eyelids, and eyebrows and his smile—a very sweet one—are the Empress's. His hair is very dark.

2 April 1873

I asked Lady Caledon [Lady of the Bedchamber] to write you an account of my visit to the Victoria or People's Park today—which is beyond Bethnal Green and we had to drive through the poorest and worst parts of London—but nothing could go off better or the enthusiasm be greater. We had no escort and I am sure there were as many people out as on the Thanksgiving Day! ... It was a splendid day. In few countries could such a sight have been seen.

16 April 1873

My beloved Baby [Princess Beatrice]—who is really the apple of my eye—and who is very much with me—though not of an evening as I keep her as young and child-like as I can—and who I pray God may remain with me as long as I live for she is the last I have and I could not live without her—was very happy [on her 16th birthday].

25 May 1873

It gives me so much pleasure to hear you speak so lovingly of dear Bertie for he deserves it. He is such a kind, good brother a very loving son and a very true friend, and so kind to all below him for which he is universally beloved; poor A. [Prince Alfred] is not at all, either by high or low.

In fulfilment of a promise made to Gladstone, the Queen agreed, in the summer of this year to receive the Shah of Persia, his friendship being highly important in the affairs of the Middle East.

JOURNAL   2 June 1873

Felt nervous and agitated at the great event of the day, the Shah's visit. All great bustle and excitement. The guns were fired and bells ringing for my Accession Day, and the latter also for the Shah. The Beefeaters were taking up their places, pages walking about, in full dress, etc. Arthur arrived, crowds appeared near the gates, the Guard of Honour and Band marched into the Quadrangle, and then I dressed in a smart morning dress, with my large pearls, and the Star and Ribbon of the Garter, the Victoria and Albert Order, etc. Was much surprised at seeing no troops lining the hill, as was done when the Sultan came here [to Windsor]. Sent for Colonel Ponsonby, who could not understand it, as he knew the order had been given. He ran down to give some directions, in hopes of getting them still, and some makeshift was arrived at, just as we heard the Shah had arrived at the station. Arthur and Leopold had gone down to meet him, and Lenchen, Louise, Beatrice, and Christian were with me in my room, watching the gradual approach, heralded by cheering.

The carriage was quite near, followed by eleven others!! and we hastened down. The great Officers of State, the ladies and gentlemen, Lord Granville, etc., had all preceded us below. The band struck up the new Persian March, and in another moment the carriage drove up to the door. The Grand Vizier, who, with my sons, was in the same carriage as the Shah, got out first, and then the Shah. I stepped forward and gave him my hand. Then took his arm and walked slowly upstairs, and along the Corridor, the Grand Vizier close behind, and the Princes and Princesses, including all the Persian ones, the ladies, etc., following, to the White Drawing-room. The Shah is fairly tall, and not fat, has a fine countenance and is very animated. He wore a plain coat (a tunic) full in the skirt and covered with very fine jewels, enormous rubies as buttons, and diamond ornaments, the sword belt and epaulettes made entirely of diamonds, with an enormous emerald in the centre of each. The sword-hilt and scabbard were richly adorned with jewels, and in the high black astrakan cap was a aigrette of diamonds. I asked various questions through the Grand Vizier, but the Shah understands French perfectly and speaks short, detached sentences . . .

We entered the White Drawing-room . . . I asked him to sit down, which we did on two chairs in the middle of the room (very absurd it must have looked, and I felt very shy), my daughters sitting on the sofa. Lord Granville handed me the Garter and diamond Star and Badge, and helped by Arthur and Leopold I put it over the Shah's shoulder. He then took my hand and put it to his lips, and I saluted him . . .

The doors were then opened into the Green Drawing-room, where everyone was assembled, and we proceeded slowly to luncheon, the Shah giving me his arm. We lunched in the Oak Room and sat down twenty. The Shah sat on my right, and the Grand Vizier on my left, Lenchen next the Shah, Louise next the Grand Vizier, Prince Abdul between Lenchen and Beatrice, and one of the old uncles on her other side. The band played during luncheon, and the Pipers at dessert, walking round the table, which seemed to delight the Shah. I talked a good deal to the Grand Vizier, and through him to the Shah, but also directly to the latter, in French. He takes great interest in everything, spoke of Vicky and her children, and said she was well; that he should so much like to see Scotland, had had my book translated into

Persian, and had read it. The Shah ate fruit all through luncheon, helping himself from the dish in front of him, and drank quantities of iced water.

To the Crown Princess of Prussia 3 July 1873

We [the Shah and the Queen upon his departure] took some refreshments in the White Room [at Windsor Castle] (only fruit—ice and tea) during which he spoke most kindly of his visit—his earnest wish that the closest alliance should be maintained between the two countries and from this time a new era should commence; that he hoped I would keep him and his Country in my remembrance and not forget it, which I replied was very reciprocal on my part. I gave him a nosegay and my photograph which he kissed (I hear) as he was leaving the station! I took him again down and he kissed my hand!

Journal 11 July 1873

Out to tea with Beatrice, near the pines and ilexes. Just as we got out a telegram was brought, which came from Affie ... He said the following: 'Marie [the Grand Duchess Marie Alexandrovna, only daughter of the Tsar Alexander II] and I were engaged this morning. Cannot say how happy I am. Hope your blessing rests on us.' Was greatly astonished at the great rapidity with which the matter has been settled and announced ... Felt quite bewildered. Not knowing Marie, and realising that there may still be many difficulties, my thoughts and feelings are rather mixed, but I said from my heart 'God bless them,' and I hope and pray it may turn out for Affie's happiness.

To the Crown Princess of Prussia 16 July 1873

Affie and Marie seem very happy and I pray she may continue so, for she really seems a very sweet girl, who marries him entirely for his sake (!!)—I wonder—but never mind that. She has written me such a pretty letter in English of which I will send you a copy another day. Difficulties there will be and delays and troubles but if she is so amiable and dear, much will be got over.

To Gladstone 25 July 1873

The Queen is glad to see the Bishop of Winchester's own family declined the proposal of Westminster Abbey as his last resting-place, than which nothing more gloomy and doleful exists.

The Queen thinks it would have been open to grave objections, for, while all concurred in thinking poor Bishop Wilberforce most agreeable, talented, and eloquent, many entertained grave doubts as to his conduct and views as a Churchman, which she must own was her own case, while others extolled him and rated him very high in that capacity. And such controversies would have been very painful.

To the Crown Princess of Prussia 26 July 1873

I cannot entirely agree with your religious ideas as I feel much more strongly and deeply than you do ... Does not every true Protestant feel the errors of a superstitious religion—full of strange observances repugnant to all the simplicity of our Saviour's teaching? Do you really think that the exclusion of Roman Catholics from the possibility of marrying any of our family merely a political necessity? If you love your children and then they marry, is it not a terrible bar to be unable to talk to them on religious subjects except with totally different feelings and to have no sympathy with them? If you don't feel this, I fear you cannot feel really deeply and earnestly on these subjects. Then you say one ought to be very tolerant; certainly I would never persecute others for their religion and would always respect it, but we Protestants are not aggressive and when I was a child, our Church was not in danger of the alarming innovations which its unfortunately too Catholic nature and forms have exposed it to and which are most alarming. In Germany you can afford not to dread these terrible High Church and ritualistic attempts and movements—which are mere aping of Catholic forms and an undermining of Protestantism, because your Church is really Protestant and all Catholic forms are expunged from it. But here flowers, crosses, vestments, all mean something most dangerous! Thank God the Scotch Church is a stronghold of Protestantism, most precious in these realms . . .

I think I ought seriously to condole with you at there being no baby coming. What a misfortune when you have only seven children—and only three boys!! Is Mama very naughty?

2 August 1873

Affie seems very well satisfied so far [with the Grand Duchess Marie of Russia]—but I find no improvement in him as yet—otherwise. There is the same ungracious, reserved manner which makes him so little liked.

3 September 1873

It is strange that you should dream so often of beloved Papa, while I do so seldom—and so much oftener of dear Grandmama, and as if I lived with her. You see, your life has not changed, whereas mine has entirely. Married life has totally ceased and I suppose that is why I feel as though I were again living with her.

14 September 1873

I hope the cold at St. Petersburg [where Prince Alfred was to be married] will not be too much for you. I shall feel not being present for the first time at the marriage of one of our children, but at the same time I dislike now witnessing marriages very much, and think them sad and painful, especially a daughter's marriage.

JOURNAL   1 October 1873

After luncheon heard that the great artist and kind old friend [Sir Edwin Landseer] had died peacefully at eleven. A merciful release, as for the last three years he had been in a most distressing state, half out of his mind, yet not entirely so. The last time I ever saw him was at Chiswick, at Bertie's garden party, two years ago, when he was hardly fit to be about, and looked quite dreadful. He was a great genius in his day, and one of the most popular of English artists. It is strange that both he and

Winterhalter, our personal, attached friends of more than thirty years' standing, should have gone within three or four months of each other! I cannot at all realise it. How many an incident do I remember, connected with Landseer! He kindly had shown me how to draw stags' heads, and how to draw in chalks, but I never could manage that well.

To the Crown Princess of Prussia   20 October 1873

As to her [Princess Beatrice] appearing at state parties;—as she is my constant companion and I hope and trust will never leave me while I live, I do not intend she should ever go out as her sisters did (which was a mistake) but let her see (except of course occasionally going to theatres) as much as she can with me. I may truly and honestly say I never saw so amiable, gentle, and thoroughly contented a child as she is. She has the sweetest temper imaginable and is very useful and handy and is unselfish and kind to everyone ... Thank God too she is not touchy and offended like several of her brothers and sisters are. That has increased with poor Lenchen (partly from health and partly from Christian's inordinate spoiling and the absence of all actual troubles and duties) to a degree that it makes it very difficult to live with her. But pray keep this to yourself and say nothing to anyone about it—but it grieves me to see it and to see what poor health she has. She won't either do anything to get better and says she don't care if she is ill or well!!

To Dean Stanley   13 November 1873

The Queen now turns with much anxiety to the very pressing question of the state of the English Church; its Romanising tendencies which she fears are on the increase, its relations with other Protestant Churches, and the universal struggle, which has begun between the Roman Catholic Church and Protestant Governments in general ... She thinks a complete Reformation is what we want. But if that is impossible, the Archbishop should have the power given him, by Parliament, to stop all these Ritualistic practices, dressings, bowings, etc., and everything of that kind, and, above all, all attempts at confession ... Her mind is greatly occupied with the state of the Church in England, and with the terrible amount of bigotry and self-sufficiency and contempt of all other Protestant Churches, of which she had some incredible instances the other day. The English Church should bethink itself of its dangers from Papacy, instead of trying to widen the breach with all other Protestant Churches, and to magnify small differences of form. The English Church should stretch out its arms to other Protestant Churches.

To the Crown Princess of Prussia   27 December 1873

Affie left my yesterday morning and was a good deal upset in taking leave and is, I think, very nervous about the whole thing—I mean the fearful ordeal he will have to go through! I wrote to him that I hoped and prayed he felt the very solemn and serious step he was going to take, how I prayed he would make the dear, amiable, young girl—who is leaving all for him—happy and that she alone must have his heart and love—and all old habits must be given up. But he has said nothing in return! Oh if he only does break with old habits! It would be awful if he did not.

4 January 1874

If only he has principle and shows heart. He can be so hard—and so sharp and unkind in speaking of and to others when he disagrees and he always knows best. This makes him not a pleasant inmate in a house and I am always on thorns and *gêne* when he is at dinner. Alix feels just as I do.

I am sorry it [the wedding] is on the 23rd for that is the anniversary of my poor father's death.

20 January 1874

Pray be photographed by Bergamasco [the St. Petersburg photographer] who has done all those lovely photos of Marie. Pray ask Alix to be done too. I think Alix does not dress her hair to advantage just now. Too high and pointed and close at the sides for her small head. The present fashion with a frizzle and fringe in front is frightful.

To Gladstone   20 January 1874

The progress of these alarming Romanising tendencies has become so serious of late, the young clergy seem so tainted with these totally anti-protestant doctrines, and are so self-willed and defiant, that the Queen thinks it absolutely necessary to point out the importance of avoiding any important appointments and preferments in the Church, which have any leaning that way.

The Queen must speak openly, and therefore wishes to say that she thinks this especially necessary on the part of Mr. Gladstone, who is supposed to have rather a bias towards High Church views himself, but the danger of which she feels sure he cannot fail to recognise.

To the Crown Princess of Prussia   10 February 1874

We have a large Conservative majority and the change of ministry will take place very shortly!! Mr. Gladstone has contrived to alienate and frighten the country. Since '46, under the great, good and wise Sir Robert Peel—there has not been a Conservative majority!! It shows a healthy state of the country.

Memorandum   17 February 1874

I saw Mr. Gladstone at quarter to three to-day. I began by saying what extraordinary things had occurred since I had seen him, and how very unexpected the result of the elections was. He then went on to say that he had come to the decision with his colleagues to tender their resignation to me at once...

We then talked of the causes of the great defeat of the Government in the elections ... I could, of course, not tell him that it was greatly owing to his own unpopularity, and to the want of confidence people had in him. He said that he thought it was the greatest expression of public disapprobation of a Government that he ever remembered, though he did not think it was quite just. He then asked whether I would approve of the various honours for people connected with the Government, which he had submitted to me this morning. I said my only objection to them was their great number, which he, however, said he thought, when taken altogether, would not prove to be so. After agreeing to approve them and discussing the individual claims, I asked him what I could do for him? to which he replied 'Oh! nothing.'

JOURNAL   18 February 1874

Mr. Disraeli came at half past 12. He expressed great surprise at the result of the elections. He had thought there might have been a very small majority for them—but nothing like this had been anticipated, and no party organisation could have caused this result of a majority of nearly 64!!

20 February 1874

I saw Mr. Disraeli at quarter to three to-day. He reported good progress ... He knelt down and kissed hands, saying: 'I plight my troth to the kindest of Mistresses'!
   Mr. Gladstone came at 6, and delivered up his Seals. He was very grave, and little disposed to talk.

TO THE CROWN PRINCESS OF PRUSSIA   24 February 1874

You speak of the Liberals being more in accordance with the views of the times;—but many real Liberals would tell you that they all looked with fear and trembling upon 'what next?' Everything was being altered and in many cases ruined—and Lord Palmerston was quite right when he said to me 'Mr. Gladstone is a very dangerous man.' And so very arrogant, tyrannical and obstinate, with no knowledge of the world or human nature. Papa felt this very strongly. Then he is a fanatic in religion. All this and much want of *égard* towards my feelings (though since I was so ill that was better) led to make him a very dangerous and unsatisfactory Premier. He was a bad leader in the House of Commons—too.

JOURNAL   7 March 1874

Could hardly believe the long expected day had come [the arrival of the Grand Duchess Marie]. All bustle and excitement. Bells ringing, troops arriving, bands playing, etc. At half past 11 we started for the South-Western Station, all the gentlemen in full dress. I had my jacket trimmed with miniver, my cloak the same, and lined with ermine. The town [Windsor], which was lined with troops, was very full and completely decked out with flags, flowers, festoons, and inscriptions, some of which were in Russian. There was a very pretty triumphal arch. Lenchen and Christian met us at the station. The train drew up, Affie and Arthur stepped out, and then dear Marie, whom I took in my arms and kissed warmly several times. I was quite nervous and trembling, so long had I been in expectation.

TO THE CROWN PRINCESS OF PRUSSIA   9 March 1874

And now about Marie. She is dear, and most pleasingly natural, unaffected and civil; very sensible and frank and unaffected not pretty (excepting *fraîcheur*) and not at all graceful. At first in her white bonnet I thought her prettier than I expected, but without it—and since—I think her less pretty even than I expected. The chin is so short and runs into the throat and the neck and waist are too long for the dear little child's face though the bust is very pretty and then she holds herself badly and walks badly. She is however quite at her ease with me and we get on very well—and she is very sensible. She is not a bit afraid of Affie and I hope will have the very best influence upon him.

2 June 1874

Bertie's dear little boys left yesterday;—they are dear, intelligent and most thoroughly unpretending children, who never are allowed to be 'great Princes' than which there is no greater mistake. It is already so difficult to prevent little Princes from becoming spoiled as everybody does what they wish.

TO DISRAELI　　　　　　　　　　　　　　　　　　　10 July 1874

[The Queen] is deeply grieved to see the want of Protestant feeling in the Cabinet [on the subject of the Public Worship Regulation Bill which sought to purge the Anglican Church of Romish practices]. Mr. Gladstone's conduct is much to be regretted though it is not surprising: but she wrote to him in the strongest terms of the danger to the Church and of the intention of the Archbishop to bring forward a measure to try and regulate the shameful practices of the Ritualists.

He [Disraeli] should state to the Cabinet how strongly the Queen feels and how faithful she is to the Protestant faith, to defend and maintain which, her family was placed upon the Throne! She owns she often asks herself what has become of the Protestant feeling of Englishmen.

TO THE CROWN PRINCESS OF PRUSSIA　　　　　　23 September 1874

Affie and Marie left us on Monday. I have formed a high opinion of her; her wonderfully even, cheerful satisfied temper—her kind and indulgent disposition, free from bigotry and intolerance, and her serious, intelligent mind—so entirely free from everything fast—and so full of occupation and interest in everything makes her a most agreeable companion. Everyone must like her. But alas! not one likes him! I fear that will never get better.

20 October 1874

In a former letter of yours you said that there was an excellent article about Bertie's supposed debts in *The Times*; now I think you must have overlooked the great untruths it contained, of a dangerous nature, in which it tried to make out that B.'s expenses were caused by me, which is an abominable falsehood.

25 October 1874

There is a most indiscreet book of Mr. C. Greville's (the Duchess of Richmond's uncle, former Clerk of the Council) published.* It is Mr. Reeve's intense indiscretion to publish it—and shows a nasty, and most ill-conditioned disloyal disposition towards my two Uncles in whose service he was and whose hospitality he enjoyed. And I am most indignant that I should be praised at the expense of my

---

*This was the first series of Greville's memoirs, covering the years 1818–1830, edited by Henry Reeve who had been a colleague of Greville at the Privy Council and a leader writer on the staff of *The Times*.

poor old Uncle and predecessor, who though not dignified or very clever—was very honest—most anxious to do what he thought was right and always very kind to me. But the accounts in many ways are very full of truth—and the one of my first Council wonderfully exact.

TO THEODORE MARTIN   26 October 1874

What does [Mr. Martin] say to the dreadful indiscretion and disgracefully bad taste of Mr. Reeve in publishing Mr. C. Greville's scurrilous Journal, without eliminating what is very offensive and most disloyal towards the Sovereigns he served, and the Sovereigns and Princes whose hospitality and even intimacy he enjoyed! And to leave the names in full when the children and near relatives of those he abuses are alive, is unheard of!

The Queen hopes and wishes Mr. Reeve will and should know what she thinks of such conduct. It is especially revolting to her, as she is put in comparison with her uncle and predecessor, who, though undignified and peculiar and not highly gifted, was very honest, most extremely conscientious and anxious to do his duty, and most kind to herself, though not always in a judicious manner. The Queen is determined that on some occasion or other she will make known what she knows of his character. Of George IV he speaks in such shocking language; language not fit for any gentleman to use of any other gentleman or human being, still less of his Sovereign.

JOURNAL   14 January 1875

Saw Mr. Martin, after taking tea with Leopold. He is much gratified at the success of the dear *Life* [the first volume of the *Life of the Prince Consort*]. Mrs. Oliphant's review in *Blackwood* is extremely good. Talked about it, and how pleased he is at the letters he has received from my children.

TO GLADSTONE   31 January 1875

The Queen was sure that Mr. Gladstone would be shocked at that horrible book [Greville's *Memoirs*] to which he alludes. Her dear husband's *Life* so pure and bright, presents a favourable and useful contrast to this most scandalous publication.

TO THE CROWN PRINCESS OF PRUSSIA   6 January 1875

The Emperor of Austria has kindly sent me a very fine copy of that beautiful picture by our unequalled and ever to be lamented Winterhalter, of the lovely Empress with her hair down! I never saw a lovelier picture and so like. All these great artists Angeli, Richter etc. cannot throw that life and lightness and animation into a portrait that dear old Winterhalter could.

16 April 1875

You ask if I don't approve of your trying to improve your stock of knowledge? To a certain amount yes;—but I think that when one has seen so much as you, one does not require it—especially should not carry the artistic culture too far, and not make it a chief object in life.

1 June 1875

I can imagine that you regret your pleasant tour in your beloved Italy—though to me dirt, insects and absence of many comforts (not luxuries which are the bane of the present day) would destroy all pleasure! You say pleasant things seem to come to an end much sooner than disagreeable ones. It certainly often does seem so, but don't you think it is because one longs for the ending of the latter and vice versa with the former?

JOURNAL   6 May 1875

Saw Mr. Disraeli and talked about the very alarming rumours from Germany, as to war. This began by dictatorial and offensive language to Belgium, then by reports of the Germans saying they must attack the French, as these threatened to attack them, and a war of revenge was imminent, which the increase in their armaments proved. I said this was intolerable, that France could not for years make war, and that I thought we ought, in concert with the other Powers, to hold the strongest language to both Powers, declaring they must not fight, for that Europe would not stand another war! In fact, as Mr. Disraeli said, Bismarck is becoming like the first Napoleon, against whom all Europe had to ally itself.

TO THE CROWN PRINCESS OF PRUSSIA   8 June 1875

I have just received your dear long letter with the enclosure which I have not had time to read properly, but I wish just to answer those principal points in your letter, though of course you know how absurd these ideas and notions of Bismarck's are . . .

No one wishes more, as you know, than I do for England and Germany to go well together; but Bismarck is so overbearing, violent, grasping and unprincipled that no one can stand it, and all agreed that he was becoming like the first Napoleon whom Europe had to join in putting down . . .

As for anyone working upon me in the sense Bismarck thinks, it is too absurd. I am not worked upon by anyone; and though I am very intimate with the dear Empress, her letters hardly ever contain any allusion to politics, certainly never anything which could be turned against her or me, and she sends her letters either by messenger or in indirect ways, and I mine the same.

You know how I dislike political letters and politics in general, and therefore that it is not very likely that I should write to her on them! and the Empress Eugénie I only see once or twice a year and she never writes to me!!

26 June 1875

How are all your dogs? I feel so much for animals—poor, confiding, faithful, kind things and do all I can to prevent cruelty to them which is one of the worst signs of wickedness in human nature!

14 July 1875

I am trying all I can to get some better and more eminent persons added to this list [of those who were to accompany the Prince of Wales on his tour of India] which I send you, as I promised, but the difficulty is very great and I fear dear B. has a number of stupid, *soi-disant* friends who put all sorts of ideas into his head. The whole thing is very full of difficulties. You say that Bertie's breakfast must have been charming. I myself think them dreadful and very fatiguing bores, walking and standing about and seeing fresh faces in every direction—but it don't last long and pleases people and so there it is and easily done.

4 August 1875

[Bertie] is grown so large—and nearly quite bald. Dear Alix is very thin but seems well and very dear as ever.

31 August 1875

Let me now answer your dear question about your coming to see me in November; you know how dear you are to me, how impossible it is that you could ever become a stranger to me and how gladly I should see you. But this winter I fear it is impossible. Dear Alix will be here with me a good deal, and I believe her parents will come to her and stay with her, and would come and see me—and this would prevent my seeing you at all quietly. D.V. next year that would do—and we might besides meet for a day or two in the spring at Coburg—if I can get there—which I hope to do. So I fear we cannot think of it this year, which I grieve to say, darling child!*

7 September 1875

You are quite wrong in saying you are 'unwelcome'; that is a very wrong word, and I am extremely sorry to have to refuse you—but this year I cannot help myself, as I told you— for reasons which I explained.

JOURNAL   21 October 1875

Much grieved at its being a worse day than ever for the funeral of Brown's father [who had died on 18 October aged 86] which sad ceremony was to take place to-day. The rain is hopeless—the ninth day! Quite unheard of! I saw good Brown a moment before breakfast; he was low and sad ... Brown and his four brothers, took us to the kitchen [of their father's house], where was poor dear old Mrs. Brown sitting near the fire and much upset, but still calm and dignified ... Mr. Campbell, the minister of Crathie, stood in the passage at the door, every one else standing close outside. As soon as he began his prayer, poor dear old Mrs. Brown got up and came and stood near me—able to hear, though, alas! not to see—and leant on a chair during the very impressive prayers, which Mr. Campbell gave admirably. When it was over, Brown came and begged her to go and sit down while they took the coffin away, the brothers bearing it. Every one went out and followed, and we also hurried out and just saw them place the coffin in the hearse, and then we moved on to a hillock, whence we saw the sad procession wending its way sadly down ... I went back to the house, and tried to soothe and comfort dear old Mrs. Brown, and gave her a mourning brooch with a little bit of her husband's hair which had been cut off

---

*The Crown Princess grew accustomed to receiving such answers to her proposals of visits to England.

yesterday, and I shall give a locket to each of the sons.

When the coffin was being taken away, she sobbed bitterly.

We took some whisky and water and cheese, according to the universal Highland custom, and then left, begging the dear old lady to bear up.

TO THE CROWN PRINCESS OF PRUSSIA   24 November 1875

Captain Montagu's* accident is serious. I pity any one who has an accident to their eye—else I could not feel very sorry if the great O. had a good lesson and was shaken a good bit—for he is an odious individual who annoys B. very much—often.

Angeli's picture of me is hung up in the Oak Room and looks very well, only we must arrange lights to show it off at night—ugly as the old lady is to behold.

JOURNAL   24 November 1875

Received a box from Mr. Disraeli, with the very important news that the Government has purchased the Viceroy of Egypt's shares in the Suez Canal for four millions, which gives us complete security for India, and altogether places us in a very safe position! An immense thing. It is entirely Mr. Disraeli's doing. Only three or four days ago I heard of the offer and at once supported and encouraged him, when at that moment it seemed doubtful, and then to-day all has been satisfactorily settled.

TO THE CROWN PRINCESS OF PRUSSIA   29 December 1875

Many loving thanks for dear letters of the dear children, Charlotte's and Vicky's are very well written. Willy's is a little peculiar as to the English which Henry's is not—but what I grieve over most is the handwriting of both which is bad like their uncle's, whereas the girls' is very good. Do watch over that with the dear boys.

5 January 1876

You will find as the children grow up that as a rule children are a bitter disappointment—their greatest object being to do precisely what their parents do not wish and have anxiously tried to prevent.

19 January 1876

Most extraordinary it is to see that the more care has been taken in everyway the less they often succeed! And often when children have been less watched and less taken care of—the better they turn out!! This is inexplicable and very annoying.

2 February 1876

Bertie's progresses [in India, as reported in his boring letters] lose a little interest and are very wearing—as there is such a constant repetition of elephants—trappings—jewels—illuminations and fireworks.

*Captain the Hon. Oliver Montagu, an officer in the Household Cavalry, who nursed a romantic, idealistic passion for the Princess of Wales.

9 February 1876

I am terribly pressed for time and can only write very hurriedly but I must just observe that it is very strange and not right in you to take all my observations about your meeting me or coming to see me always amiss. You I think also hardly know with what a suite you always move about—which makes everything difficult. This is not said to offend but as the truth. To see you both is always a pleasure, but you hardly know how tired and fagged I am, overpowered with work, and how easily I am knocked up and tired, I never get to bed till one—and with the greatest wish to see those I love I must have time to rest—not because you are in my way.

JOURNAL   26 February 1876

After luncheon saw Mr. Disraeli, who talked of the Titles Bill [the Royal Titles Bill which, instigated by the Queen, sought to give her the title of Empress of India] causing trouble and annoyance, he could not tell why. I spoke of the feeling about the Colonies and gave him full power to add anything to the title. He thought a plan of his to give to two of my sons the titles of Duke of Canada and Duke of Australia might be a good way of solving the difficulty, and I saw no objection to it if he found it would be of use. He was greatly pleased at the Suez Debate going off so well.

TO THEODORE MARTIN   14 March 1876

The Queen is sure that Mr. Martin (though he has not mentioned it) is as shocked and surprised at the conduct of the Opposition, and the sort of disgraceful agitation caused thereby, on the subject of her additional title . . . The reason the Queen now writes to Mr. Martin is to ask whether he cannot get inserted into some papers a small paragraph to this effect, only worded by himself: 'There seems a very strange misapprehension on the part of some people, which is producing a mischievous effect; viz. that there is to be an alteration in the Queen's and Royal family's ordinary appellation. Now this is utterly false. The Queen will be always called 'the Queen,' and her children 'their Royal Highnesses,' and no difference whatever is to be made except officially adding after Queen of Great Britain, 'Empress of India,' the name which is best understood in the East, but which Great Britain (which is an Empire) never has acknowledged to be higher than Queen or King.'

JOURNAL   14 March 1876

Heard, on getting up, that the second reading of the Titles Bill had been carried by 105!—an immense majority. It is to be hoped now no more stupid things will be said, and that the matter will be dropped. I cannot understand how the quite incorrect rumour can have got about, that I did not care for it; it is really too bad. But all sensible people know that this Bill will make no difference here, and that I am all for it, as it is so important for India. There is no feeling whatever in the country against it, but the Press took it up, having at first been all the other way.

TO THE CROWN PRINCESS OF PRUSSIA   26 April 1876

I know that you have many great difficulties—and that your position is no easy one, but so is mine full of trials and difficulties and of overwhelming work—requiring that rest which I cannot get. The very large family with their increasing families and interests is an immense difficulty and I must add burthen for me. Without a husband and father, the labour of satisfying all (which is impossible) and of being just and fair and kind—and yet keeping often quiet which is what I require so much—is quite fearful. You will one day have to encounter this though never like me; for you will not be the Sovereign and please God will always have your dear husband to guide and help you. Dear Willy seems a dear, amiable, good and natural boy. May he ever remain so! I shall always take the warmest interest in him.

16 May 1876

Bertie's arrival [from India] and the hearty reception he met with and I also met with, which was very striking and of which I sent you an account in the *Daily Telegraph* as *The Times* did not condescend to notice—was a proof of the immense loyalty of the country in spite of the attempts of the Opposition (not all) and of their very radical supporters as well as of the Press to agitate and rouse them against the Throne in which they themselves say they entirely failed, is very remarkable and very gratifying. When I appeared at the window—though Bertie with Alix and the boys were just driving away—the whole immense crowd turned round and cheered and waved their handkerchiefs without ceasing.

18 May 1876

You speak only of the enthusiasm for Bertie! That for your own Mama was I thought much greater.

In the summer of 1876 the Daily News *published a horrifying story of fearful atrocities committed by Turkish irregular troops on Bulgarian peasants, 25,000 of whom were believed to have been murdered. Predisposed to prefer the Turks to their subject peoples, and encouraged in his prejudice by the Turcophil British Ambassador at Constantinople, Disraeli made light of the reports; and even when it became clear that, while the* Daily News's *accounts were exaggerated, terrible slaughter had indeed taken place, he continued to talk dismissively of 'coffee-house babble' and to refer to the 'atrocities' in inverted commas, as though they were figments of some journalist's imagination.*

The country in general took a different view; and Gladstone, outraged by the reports and sensing the time had come to emerge from his premature retirement, gave voice to this dissent in his famous, fiercely condemnatory pamphlet, The Bulgarian Horrors and the Question of the East, *which sold 200,000 copies within a month.*

The scene was now set for one of the bitterest political arguments that has ever erupted in England. People ranged themselves on the side of the Turks or on that of the Russians, protectors of the Slavs and the Turks' traditional enemies, decrying those who stood behind the opposing barricades with the most ferocious animosity. To Disraeli, Gladstone's pamphlet, which denounced the Turkish race as 'the one great anti-human specimen of humanity', was contemptible, 'vindictive and ill-written', 'perhaps the greatest . . . of all the Bulgarian horrors', the product of an 'unprincipled maniac'. To the Queen, its author, 'that half madman', was a 'mischief maker and firebrand', his conduct 'shameful' and 'most reprehensible'. The Queen was also highly critical of what she took to be the weak responses of Lord Derby, the Foreign Secretary, Lord Carnarvon, Colonial Secretary, and Lord Salisbury, Secretary for India.

JOURNAL   23 August 1876

More news of the horrors committed by the Turks, which seem to be more and more verified, and are causing dreadful excitement and indignation in England, or indeed in Great Britain. Constant telegrams arriving, giving most conflicting accounts. A mediation is most anxiously hoped for.

TO THE CROWN PRINCESS OF PRUSSIA   16 September 1876

Pray send me what music is published of the new opera of Wagner—as I admire his operas very much. I should be very grateful if you would send it me—as Beatrice would play it. *Lohengrin* is our great favourite but I delight in Gounod too. His *Faust, Romeo and Juliet*, and *Mireille* and *Joan of Arc* are so lovely.

TO DISRAELI, NOW THE EARL OF BEACONSFIELD   28 September 1876

The Queen understands Lord Beaconsfield's motive for not expressing 'horror' at the 'Bulgarian atrocities.' She had only suggested a word of sympathy if an occasion offered, and she now leaves this entirely to Lord Beaconsfield's judgment.

MEMORANDUM   17 October 1876

The difficulties to be solved are very great. On the one hand we have Russia, who under the pretext of wishing to protect the Christians in the principalities wishes to obtain possession of a portion of Turkey, if not of Constantinople. On the other hand Turkey is in great difficulty from her Mussulman subjects, who will try to prevent her yielding to pressure in favour of the Christians.

It seems to me that the great object in view ought to be to remove from Russia the pretext for constantly threatening the peace of Europe on the Eastern or Oriental question. The only way to do this seems to me to free the principalities from Turkish rule and to unite them under an independent Prince, to make that united principality a neutral State. Nothing short of this will, I think, ever prevent a frequent recurrence of difficulties and alarming complications like the present.

TO THE CROWN PRINCESS OF PRUSSIA   7 November 1876

Only think my horror that Bertie without even saying a word to me has invited the Prince of Orange [renowned for his dissipation] to Sandringham!! Oh what a contrast to the 'noble life' [of the Prince Consort] which is now being universally admired and looked upon as one of the purest and best! I often pray he may never survive me, for I know not what would happen.

21 November 1876

I have excellent accounts of Arthur. He is so universally respected and liked. He is called 'the model Prince' for his wonderfully steady and perfect conduct. He at least follows in his beloved father's footsteps as regards character and sense of duty.

TO BEACONSFIELD   21 March 1877

The Queen ... trusts the Cabinet will be very firm, and Lord Derby seemed so yesterday. She is prepared to speak or write to good but nervous and somewhat weak and sentimental Lord Carnarvon, if necessary, as well as to Lord Salisbury. This mawkish sentimentality for people who hardly deserve the name of real Christians, as if they were more God's creatures and our fellow-creatures than every other nation abroad, and forgetting the great interests of this great country—is really incomprehensible.

In April 1877 the Russo-Turkish War broke out; and Russian troops were soon advancing on Sofia.

TO THE CROWN PRINCESS OF PRUSSIA   3 May 1877

It is a very anxious time and we must and will take some marked line to show that Russia is not to have all her own way—which thanks to the most unfortunate and ill-judged agitation of last autumn and this spring has led Russia on to think she may do anything, and they are the cause of what is happening now! You never answer when I constantly tell you this—as if you thought the Liberals and that madman Gladstone must be right and the Government wrong! If you only knew how I have but one object—the honour and dignity of this country (of course the stopping of all acts of cruelty is as great an object to us as to anyone else) and it is the neglect of this which distresses and grieves me so much. The other party seem wilfully blind to the fact that Russia never would have dared to go as far as she has—had not they thought England not only would not fight but was with them.

19 June 1877

You say you hope we shall keep out of the war and God knows I hope and pray and think we shall—as to fighting. But I am sure you would not wish Great Britain to eat humble pie to these horrible, deceitful, cruel Russians? I will not be the Sovereign to submit to that!

TO BEACONSFIELD   27 June 1877

The Queen must write to Lord Beaconsfield again and with the greatest earnestness on the very critical state of affairs. From so many does she hear of the great anxiety evinced that the Government should take a firm, bold line. This delay—this uncertainty, by which, abroad, we are losing our prestige and our position, while Russia is advancing and will be before Constantinople in no time! Then the Government will be fearfully blamed and the Queen so humiliated that she thinks she would abdicate at once. Be bold!

TO THE CROWN PRINCESS OF PRUSSIA   11 July 1877

Affie, I am grieved to say, has become most imprudent in his language and I only hope he does not make mischief. It is very awkward with this Russian relationship just now. This is what I always feared and dreaded. Lord Beaconsfield is well again and the mainstay of everything.

25 July 1877

I cannot help smiling at your complaints about Charlotte.* I—and dear Papa even if possible more than me—so very much disapproved of that system of complete intimacy before marriage, and in that respect I am bound to say that you never gave us the slightest trouble or annoyance, but Fritz did, and made me very impatient. You were in other ways very difficult to manage, but not in that. I think there is a great want of propriety and delicacy as well as dutifulness in at once treating your bridegroom as though (except in one point) he were your husband. Papa felt this so strongly and it applies still more strongly to very long engagements like yours, your sisters' and Charlotte's. You, as time goes on, I am sure will change your great passion for marriage—and will understand the great change it is to a mother especially, though to a father too! Here now they have lost all modesty for not only do they go about driving, walking, and visiting—everywhere alone, they have also now taken to go out everywhere together in society—which till a year or so no young lady just engaged, ever did, and make a regular show of themselves—and are laughed at and stared at! In short young people are getting very American, I fear in their views and ways.

TO BEACONSFIELD   1 August 1877 (B52 17)

The Queen ... [urges] so strongly ... the importance of the Czar knowing that we will not let him have Constantinople! Lord Derby and his wife most likely say the reverse right and left and Russia goes on.† It maddens the Queen ... She must say she can't stand it.

TO THE CROWN PRINCESS OF PRUSSIA   11 August 1877

Now let me answer about Bernard. I would naturally like very much to see him and know him before he marries dear Charlotte. But I own I fear (much as it really vexes and grieves me to say it) that I cannot manage it now. I am overwhelmed with business just now!

18 August 1877

To make Bernard's acquaintance is a pleasure and he seems very amiable and intelligent—reminding me much of his grandmother's family. But I must say I was astonished and annoyed at the way in which you received my expression of deep

---

*The Princess had written: How differently the younger generation expects to be treated from what we were. Fancy that Charlotte never tells me when she writes to Bernard [the Hereditary Prince of Saxe-Meiningen, whom Princess Charlotte was to marry] or when he writes to her—they correspond daily almost, I believe, but he would be quite furious if I were only to ask, and she consider herself highly offended and very indignant if her letters were interfered with. Fritz thinks this all right for a German engaged couple and says it ought to be so, but considering how young and how immature she is, I have my little doubts sometimes, and find it rather difficult to know what to do. They resent the slightest restraint put upon them and Bernard thinks they ought to do just as they like—so I am obliged to let it all alone.
†In his anxiety to prevent war breaking out, Lord Derby went so far as to reveal Cabinet secrets to the Russian ambassador.

regret at being unable on account of the overwhelming work I have had to do . . . and it was only in order to show my affection for you and Charlotte that I made the proposal to Bertie to bring him here. But I must protest against these visits in future at the very end of my stay here [Balmoral] when I want a little rest. I can't enjoy seeing them and can't receive them as I should wish. So pray dear child let this be understood in future.

4 September 1877

What you say about Willy [his neglect of his mother] grieves me and is not right. I can't help smiling sometimes, though I am truly sorry too for you, that you should have such experiences—that you should now learn what I—without a dear husband to share all—have had and have to go through. But excepting Affie who is very wanting in attention and consideration and Leopold occasionally, I have not had to suffer in the way you speak of. Bertie and dear Arthur are always most attentive. It is really a thing to be most carefully watched in education viz the neglect of parents—the total want of gratitude and thought for their feelings—and how much one feels it—how it cuts one to the heart you can now understand.

MEMORANDUM   7 September 1877

The Turco-Russian War is unexampled in its savageness, while its commencement was most iniquitous, being merely the result of the Emperor of Russia's declaring he could not accept a slap in the face ('un soufflet') from the Turks, when they refused the proposals of the Conference. For this he has plunged his own nation as well as the Turkish Empire into one of the most bloody wars ever known, and which no one thought possible in this century. Under the cloak of religion and under the pretence of obtaining just treatment for the so-called 'Christians' of the principalities, but who are far worse than the Mussulmans, and who moreover had been excited to revolt by General Ignatieff, who prevented regular troops being sent out to quell the revolt, leading thereby to the so-called 'Bulgarian atrocities' as the irregular troops were sent out, this war of extermination (for that it is) has been iniquitously commenced!

The question now arises whether in the interests of humanity, justice, and of the British Empire, this is to be allowed to go on to the bitter end; merely to remain neutral, and to avoid all interference?

The Queen is most decidedly of opinion that this should not be. When it is clear that Russia is not inclined to make even an offer for peace, but to press on for two campaigns, the Queen thinks we ought to declare that, having taken part in all the negotiations previous to the war, we feel determined to put an end to so horrible a slaughter, which the longer it lasts the more savage it will become, and the more difficult to stop. We should then propose certain terms recapitulating the disinterested protestations of Russia, and should at the same time say that, if these are rejected, we shall support Turkey in defence of her capital, and in preventing her extermination.

To Beaconsfield   5 November 1877

('Inkerman Day').—The Queen has to thank Lord Beaconsfield for a most interesting and important letter received yesterday. Lord Carnarvon will evidently resign.

She must own that she is shocked at his views, for how can he think true religion and civilisation can be advanced by Russians who are more barbarous and cruel almost than the Turks, though they may not kill or murder in the same way, but the slow killing by imprisonment and exile to Siberia, and ill-usage of every kind which no one hears of, is as bad if not worse.

13 November 1877

Pray do insist on action, or the Russians will crow over us, and our uncertain and weak policy! Weak, because it is delayed.

Lord Beaconsfield has a good majority in his Cabinet, but we want energy like his own. Pray do beware of delay, for the state of Turkey is most alarming!

The Queen wishes to repeat most emphatically that she hopes Lord Beaconsfield will be very firm and decided to-morrow, and not give way to any one, even if Lord Derby [Foreign Secretary] should wish to resign. No time is to be lost in deciding what is to be done, and if Lord Beaconsfield will be very decided, supported as he will be, and has been, all along by herself, the other Ministers will surely yield.

Make what use of the Queen's name Lord Beaconsfield wishes.

The Queen knows that the greater part of the country is strongly anti-Russian, even if it does not entirely sympathise with Turkey; and now that Plevna is taken, people are getting alarmed and will feel, as the people of this country always do feel, for the poor country which is getting the worst of the conflict, and for the heroic army who are defending their home and hearth. England will never stand (not to speak of her Sovereign) to become subservient to Russia, for she would then fall down from her high position and become a second-rate Power!!

On 15 December the Queen decided to pay a visit to Lord Beaconsfield at Hughenden, his country house near High Wycombe.

To Beaconsfield   15 December 1877

The Queen is anxious to express her concern at having inadvertently fixed this day [the anniversary of his wife's death] of such sad recollections to Lord Beaconsfield for her visit to Hughenden; and she wishes he should know that she only found out what she had done, when it was too late to alter it. But it has annoyed her very much.

Journal   15 December 1877

It took us hardly quarter of an hour to reach Hughenden, which stands in a park, rather high, and has a fine view. Lord Beaconsfield met me at the door, and led me into the library, which opens on to the terrace and a pretty Italian garden, laid out by himself. We went out at once, and Beatrice and I planted each a tree, then I went

18 The Queen with some of her children, Helena, Leopold, Alice and Louise, at Balmoral in 1860. 'Every year my heart becomes more fixed in this dear Paradise.'

19 'Dear Albert's room, where all remains the same.' The blue room at Windsor, where the Prince Consort died.

20 'The beloved Mausoleum' at Frogmore in Windsor Great Park where the Prince Consort was buried

21 The Queen and Princess Louise at Windsor in 1862, looking at a photograph of the Prince Consort with his marble bust just behind them

22 Group taken in 1863 the day before the wedding of the
Prince of Wales and Princess Alexandra of Denmark.
Left to right: Prince and Princess Christian of Denmark;
Prince Frederick of Denmark; Alice, Princess Louis of
Hesse; Crown Prince Frederick William of Prussia (back
row): Prince Louis of Hesse; Princess Helena; Princess
Alexandra of Denmark; the Prince of Wales; Victoria
(Vicky), Crown Princess of Prussia; Princess Dagmar of
Denmark; Princess Louise; Prince William of Denmark

23 The wedding cake

24 The Queen in 1867. 'Whatever the poor Queen can do she will; but she will not be dictated to, or teased by public clamour into doing what she physically cannot.'

25 The Queen in 1872, dressed for the thanksgiving service for the recovery of the Prince of Wales, 'wore a black silk dress and jacket, trimmed with miniver, and a bonnet with white flowers and a white feather.'

26 The Prince of Wales with the Shah of Persia in 1873

27 The Queen's son, Arthur, Duke of Connaught, ready for the Egyptian Expedition in 1882. 'Arthur told us what his dress would be, serge, quite loose, flannel shirt, high boots over breeches and white helmet with a puggaree.'

28 Lord Granville in 1880: 'has not the courage of his convictions and is therefore of not the slightest use to the Queen.'

29 The Jubilee Procession, 1887, going round Hyde Park Corner. 'This never-to-be-forgotten day will always leave the most gratifying and heart-stirring memories behind.'

30 The Queen with her Indian secretary or Munshi in 1895. 'Abdul is most handy in helping when she signs by drying signatures.'

31 The Diamond Jubilee Procession, 1897, on Westminster Bridge, just below Big Ben. The figure on the white horse is Field Marshal Lord Roberts. 'No one ever, I believe, has met with such an ovation as was given to me.'

32 The Queen in Dublin, 1900. 'Even when I used to go round the grounds in my pony chair and the people outside caught sight of me they would at once cheer and sing *God Save the Queen*.'

back into the library and he gave me an account of yesterday's Cabinet, which had been very stormy. Lord Beaconsfield is determined to bring things to an issue on the 17th, in which I strongly encouraged him.

MEMORANDUM   11 January 1878 (B54 65)

We must stand by what we have always declared viz that any advance on Constantinople would free us from our position of neutrality ... There is not a moment to be lost or the whole of our policy of centuries of our honour as a great European power will have received an irreparable blow.

TO BEACONSFIELD   14 January 1878 (B55 5)

The Queen had a great deal of conversation with Colonel Wellesley of whom she thinks very highly and only wishes he was at Lord Derby's elbow. He said truly if only Lord Beaconsfield was Foreign Secretary as well as Prime Minister! Why not? ... He could do with ease what he has now to do with a constant drag and perpetual difficulties ... Nothing can be worse than it is at present. Lord Derby will do nothing, originate nothing and besides is indiscreet.

20 January 1878

The Queen has received Lord Beaconsfield's letter of yesterday, which has much distressed her, as he seems so much out of heart. But he must not be that, and he must not give way an inch. All that has been foreseen for months is taking place; and we shall be for ever disgraced if we submit to the seizure of Gallipoli, or the attack on Constantinople! It simply must not be. War with Russia is, the Queen believes, inevitable now or later. Let Lord Derby and Lord Carnarvon go, and be very firm. A divided Cabinet is of no use.

The Queen would wish to confer the vacant Garter on Lord Beaconsfield, as a mark of her confidence and support. She and the country at large, have the greatest confidence in him.

TO THE CROWN PRINCESS OF PRUSSIA   21 January 1878

One hurried line before going to bed to thank you so very, very much for the darling little Dackels [dachshunds]. I could not believe it, when on coming in this afternoon Brown came in and said 'here are two dackels come—bonnie dogs' and I at once guessed who sent them and we are so delighted with the sweet dackel babies. You know how I adore doggies and these are two darlings.

TO BEACONSFIELD   27 January 1878 (B55 58)

I think the countermanding of the Fleet when it was just at the entrance to Constantinople was most unfortunate.

On 31 January an armistice was signed at Adrianople. Soon afterwards an unfounded rumour reached London that the Russians in flagrant disregard of it, had crossed the demarcation line.

To BEACONSFIELD   6 February 1878 (B56 12)

Since beginning this, the Queen has heard the dreadful news of Russia's monstrous treachery!

7 February 1878

The Queen could not go on last night. No words are strong enough to express the Queen's indignation. The Powers have now received a far worse 'soufflet' than the Emperor of Russia declared he had done, when Turkey refused the advice of the Conference.

We must, cannot submit to this and we must at once show what we feel at being duped and led by the nose ... The error, the fatal error of recalling the Fleet has no doubt encouraged the monstrous Russians to set every rule of war, every principle of good faith at defiance ... You have the country with you, only act quickly and firmly and show that Great Britain will not be trampled upon.

7 February 1878 (B56 18)

The Queen ... cannot enough express in the very strongest terms, her extreme indignation at hearing that we cannot prevent the Russians from entering Constantinople. Again and again have we told the Russians that we would not stand this ... The Queen expects that we shall use force to drive them out ... The Queen has worked and laboured and supported Lord Beaconsfield with all her might to maintain the honour and interests of the country, but if these are to be abandoned, she would be disgraced and humiliated in the disgrace of her country. She can and will not abide it.

To DERBY   10 February 1878

The Queen acknowledges Lord Derby's two letters of the 7th and 8th, and is very glad to see that a portion of the fleet, in company with French and Italian vessels, is to go up to-day to Constantinople; but she regrets the explanation to Russia about it. There is, however, an observation in Lord Derby's letter of yesterday which she cannot leave unanswered. He says this step will in his belief 'do much to satisfy the feeling of those who are complaining of inaction on the part of the Government'. The Queen for one does not feel this satisfaction, and never has felt satisfied at our inaction, which has brought about, what the Queen feels, and so do many others, a painful humiliation for this country, which no action now can remedy; for it ought to have been taken long ago—and we ought to have acted up to our repeated declarations with regard to Constantinople.

To THE CROWN PRINCESS OF PRUSSIA   15 February 1878

Mr. Gladstone goes on like a madman. I never saw anything to equal the want of patriotism and the want of proper decency in Members of Parliament. It is a miserable thing to be a constitutional Queen and to be unable to do what is right. I would gladly throw all up and retire into quiet.

To Beaconsfield   5 March 1878

Remember vacillation and delay will be ruinous to the country, not to speak of the Government. Lord Derby must go, for he is believed abroad to be the person who acts and no one trusts him! What use is there in keeping him, and yesterday's speech, so different to good Sir Stafford's [Northcote, Chancellor of the Exchequer], shows the great danger of such discrepancies. It besides makes Sir Stafford's position untenable!

Peace was eventually achieved by the Treaty of San Stefano by which the independence of Rumania, Serbia, Montenegro and an enlarged Bulgaria was recognized; and Russia obtained some towns in the Caucasus. The Treaty was severely criticised throughout Europe. In England it was feared that Russia would now be placed in a much stronger position from which to dominate the Balkans. The Russians were persuaded to submit the settlement to a congress in Berlin.

To Beaconsfield   27 March 1878

The Queen must own, that she feels Lord Derby's resignation an unmixed blessing ... His name had suffered and was doing great harm to us abroad: and the very fact of his becoming a mere cypher and putting his name to things he disapproved, was very anomalous and damaging ... The Queen, therefore, without a moment's hesitation, sanctions Lord Beaconsfield's acceptance of his resignation.

To Ponsonby   2 May 1878

The Queen cannot deny that she does not rejoice so much at the event [the forthcoming marriage of Prince Arthur to Princess Louise Margaret, daughter of Prince Frederick Charles of Prussia]—she thinks that so few marriages are really happy now and they are such a lottery. Besides Arthur is so dear a son to her that she dreads any alteration.

But it is entirely his own doing and as she, the Princess, is so much praised and said to be so good, unassuming and unspoilt, serious minded and very English we must hope for the best and that one so good as he is being very happy.

To the Crown Princess of Prussia   undated, ? 19 May 1878

The dear three weeks' visit is over like everything pleasant and everything evil in this uncertain world. I enjoyed it much, found you so sympathetic and improved, and understanding and sharing the load of anxiety which weighs on me especially about that wayward, undutiful Leopold!

                                                                    21 May 1878

I admire your knowledge and great talents and your great energy and perseverance but I would venture to warn against too great intimacy with artists as it is very seductive and a little dangerous ... All you say of my darling Beatrice pleases and touches me, but it is only the truth, for she is like a sunbeam in the house and also like a dove, an angel of peace who brings it wherever she goes and who is my greatest comfort.

TO BEACONSFIELD   23 May 1878

I have the greatest suspicions of Russian proposals. There must be no half measures. The conduct of ... Mr. Gladstone, and others, which is shameful, must not deter you from acting boldly.

TO THE PRINCE OF WALES   30 May 1878 (B57 69)

The subject of Lord Beaconsfield attending the Conference [in Berlin] has been before me, and if it were to be at Brussels, The Hague or Paris ... I should—and I have done so—urge it, but you know that Lord Beaconsfield is 72½, is far from strong, and that he is the firm and wise head that rules the Government, and who is my great support and comfort, for you cannot think how kind he is to me, how attached! His health and life are of immense value to me and the country and should on no account be risked. Berlin is decidedly too far and this is what I have said. I wrote to him on the subject two days ago, and have not had an answer yet. I don't believe that without fighting and giving those detestable Russians a good beating any arrangement will be lasting, or that we shall ever be friends! They will always hate us and we never can trust them.

TO BEACONSFIELD   31 May 1878

The Queen again cyphered about Lord Beaconsfield's going to the Congress if it takes place. There is no doubt that no one could carry out our views, proposals, etc., except him, for no one has such weight and such power of conciliating men and no one such firmness or has a stronger sense of the honour and interests of his Sovereign and country. If only the place of meeting could be brought nearer!

Beaconsfield left for Berlin on 7 June 1878. At the Congress, to which he was accompanied by Derby's successor, Lord Salisbury, he had to agree to Turkey's loss of more territory than he would have liked; but he returned to England having gained Cyprus for the Queen. He thus secured what he termed a place d'armes *from which Russian designs on the crumbling Turkish Empire could be resisted. He returned home to a hero's welcome, declaring that he had gained 'peace with honour'. 'High and low are delighted,' the Queen assured him, 'excepting Mr. Gladstone who is frantic.'*

TO BEACONSFIELD   16 July 1878

The Queen ... sends these lines with some Windsor flowers to welcome [Lord Beaconsfield] back in triumph! He has gained a wreath of laurels which she would willingly herself offer him, but hopes that the Blue Ribbon [the Order of the Garter] she may greet him with [*sic*] at Osborne.

                                                                                                    17 July 1878

The Queen was much touched by Lord Beaconsfield's very kind letter. Would he not accept a Marquisate or Dukedom in addition to the Blue Ribbon? And will he not allow the Queen to settle a Barony or Viscounty on his Brother and Nephew? Such a name should be perpetuated!

TO THE CROWN PRINCESS OF PRUSSIA   27 July 1878

Lord Beaconsfield and Mr. Corry [his Secretary] spent three nights here [at Osborne] and I think the former really very well. He never coughed once—walked and drove out and was in the best spirits. I am so glad you have learnt to know and appreciate him. He is unlike other people and unless you know him well you cannot entirely appreciate him. He has a large mind. One other great quality which Lord Beaconsfield possesses—which Mr. Gladstone lacks entirely—and that is a great deal of chivalry and a large, great view of his Sovereign's and country's position.

Relations between England and Russia continued to be tense after the Congress of Berlin. Russian troops had advanced to the borders of Afghanistan and in July a Russian mission entered Kabul and signed a convention with the Ameer who had declined to receive a British mission. Lord Lytton, the Viceroy of India, therefore sent a delegation, insisting upon its reception. The Ameer stopped it and war with the Afghans began.

JOURNAL   6 October 1878

The Cabinet had been much occupied with this alarming Afghan affair. Lord Lytton should not have sent the mission, having been forbidden to do so by the Cabinet. Now, of course, we must punish the insult, and support Lord Lytton. Care must be taken that we are quite sure of success, and that there should be no repetition of the misfortunes at Cabul in 1840 [where, in the winter of 1841–2, disaster overtook a large British force which withdrew from the town and was later overwhelmingly defeated]. This time the Kyberins, and other Hill tribes are with us. The Indian reliefs are being stopped coming over. Sir Stafford said he felt very anxious, till we heard more. All depended on whether the Ameer was assisted by Russia or not. That she is at the bottom of it all, there is little doubt.

TO BEACONSFIELD   23 October 1878

Any doubt, want of firmness or delay now may be fatal to us. The whole of India will watch our conduct, and the assistance we may expect will depend on our energy.

TO LORD LYTTON   6 December 1878

The Queen must begin by her earnest congratulations and the expression of her pride and satisfaction at the brilliant successes [the victories of General (afterwards Lord) Roberts at Peiwar Kotel on 2 and 3 December] of her brave, noble soldiers, which is of the greatest importance in every way; but in no way surprises her, for British soldiers always do their duty and almost always are victorious. The loss of brave officers and men is always a source of deep sorrow to the Queen, but it is unavoidable; and to die for one's country and Sovereign in the discharge of duty is a worthy and noble end to this earthly life for a soldier.

TO THE CROWN PRINCESS OF PRUSSIA   9 December 1878

How distressed and alarmed you will all be when you hear of our poor dear Alice being ill with diphtheria since yesterday! It is indeed dreadful! I fear she must have given up all precautions too soon and got it from Ernie or Louis [her sons]—or else from the house. She has had great trouble with the house which is not well built. However *unberufen* it seems not to be violent or malignant as yet with her.

12 December 1878

Alas! It is a severe attack that our poor darling Alice has got ... I fear it was from poor Ernie she got it when she told him of little May's death [his four-year-old sister. His mother kissed him to comfort him in his distress]. The greatest sympathy is shown. People know how nobly she behaved and stood by me when darling Papa died, and how bravely and devotedly she has watched husband and children.

JOURNAL   13 December 1878

Terribly anxious day, just like in '61 and '71. At a little after 11, came a telegram from Louis, which gave me an awful shock: 'Jenner has just seen Alice, is consulting with doctors. He does not despair, but I see no hope; my prayers are exhausted.' This upset me too dreadfully ... Dear Beatrice and I felt nearly hopeless. My distress great. Walked down to the dear peaceful Mausoleum. Just beyond Frogmore, we met a footman with a telegram. Stopped and read it. It was from Sir William [Jenner] and bore bad tidings: 'Disease in wind-pipe extended, difficulty of breathing at times considerable; gravity of condition increased since I last telegraphed. Restlessness very great.' Too dreadful! Could settle down to nothing, agony great. Lenchen came to luncheon. All so terribly anxious, hoping—fearing.

14 December 1878

This terrible day come round again [the anniversary of the Prince Consort's death]! Slept tolerably, but awoke very often, constantly seeing darling Alice before me. When I woke in the morning, was not for a moment aware of all our terrible anxiety. And then it all burst upon me. I asked for news, but nothing had come. Then got up and went, as I always do on this day, to the Blue Room [where the Prince Consort had died], and prayed there. When dressed, I went into my sitting-room for breakfast, and met Brown coming in with two bad telegrams: I looked first at one from Louis, which I did not at first take in, saying: 'Poor Mama, poor me, my happiness gone, dear, dear Alice. God's will be done.' (I can hardly write it!) The other from Sir Wm. Jenner, saying: 'Grand Duchess became suddenly worse soon after midnight, since then could no longer take any food.' Directly after, came another with the dreadful tidings that darling Alice sank gradually and passed away at half past 7 this morning! It was too awful! I had so hoped against hope. Went to Bertie's sitting-room. His despair was great. As I kissed him, he said, 'It is the good who are always taken.' That this dear, talented, distinguished, tender-hearted, noble-minded, sweet child, who behaved so admirably during her father's illness, and afterwards, in supporting me, and helping me in every possible way, should be called back to her father on this very anniversary, seems almost incredible, and most mysterious! To me there seems something touching in the union which this brings, their names being for ever united on this day of their birth into another better world!

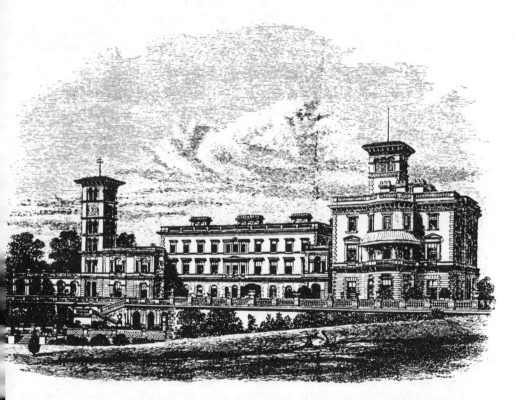

OSBORNE HOUSE.

12

'THE QUEEN WILL SOONER ABDICATE'

1879–1885

JOURNAL   8 January 1879

Heard from Lord Beaconsfield, that there would be nothing to prevent my going to the North of Italy in the latter part of March, which I am very anxious to do, as I feel it an absolute necessity for my nerves and health, to have a complete change of scene. Germany I could not bear to go to this spring, Switzerland was too cold, and Italy I have long desired to see.

TO THE CROWN PRINCESS OF PRUSSIA   19 March 1879

This is a letter to wish you goodbye [after the Crown Princess's visit to England for the marriage of the Duke of Connaught]. It will I know be a very sad day to you, as you cling so much to your own country and I fear dislike Berlin more and more! It grieves me so to think that you have so many trials and difficulties and that life is so little congenial. It must however be borne with courage and patience!

JOURNAL   27 March 1879

Received a telegram, which on opening I found, to my unbounded grief and horror, to contain the terrible words: 'Have just taken a last look at the beloved child, [her eleven-year-old grandson, Prince Waldemar of Prussia]. He expired at half-past three this morning, from paralysis of the heart. Your broken-hearted daughter Victoria.' How heart-rending! My poor darling Vicky!

TO THE CROWN PRINCESS OF PRUSSIA   29 March 1879

My poor dear darling child, my heart bleeds and aches for you to that extent I cannot describe it. I am so miserable about you. You have, thank God, dear kind Fritz near you who will share all with you and that is a blessing. Could I only do something for you to comfort you? I wish you could come here but you cannot leave Charlotte [the Princess's eldest daughter, wife of the Hereditary Prince of Saxe-Meiningen, who was expecting her first child] and if later on you should like one of the cottages at Osborne or Abergeldie I would too gladly offer them. I would wish to do anything for you to help and soothe you at this terrible affliction.

JOURNAL   1 April 1879

In the afternoon drove with Beatrice and Jane C[hurchill] beyond Gravellona, and back. The mist was low down on the hills, but one could just see that there was much more snow on them. The children on the roads know me quite well, and call out 'La Regina d'Inghilterra.' There are such dreadful, queer-looking pigs here, as thin as greyhounds, and with quite long legs. Two mounted Carabinieri generally follow the carriage at some little distance, and ones on foot patrol the roads. They look very smart and well turned out.

                                                                           7 May 1879

Saw Lord Beaconsfield at 1. Talked of the loss of the Bill, permitting the marriage with a sister-in-law, in favour of which Bertie presented a petition, and which we are most anxious should pass.* It has passed the Commons but is thrown out in the Lords, the Bishops being so much against it. Lord Beaconsfield is in favour of it, but the whole Cabinet against it!! Incredible! Spoke also of another, for opening Museums on Sundays, which likewise has been lost, but only by 10. It ought to be granted, for it is the only way to improve the masses and check drink.

                                                                           12 May 1879

Received the news that Charlotte had been safely delivered of a little girl [Princess Feodora of Saxe-Meiningen who was to marry Prince Henry XXX of Reuss], all doing quite well, and I have thus become a great-grandmother! Quite an event.

TO PRINCESS VICTORIA OF HESSE   29 May 1879

Your dear letter for my poor old birthday touched me very much and I thank you so much for it as well as for the Head which is really beautifully done! Oh! how sad was that morning!! How I missed darling Mama, her letter, her gift—her telegram. From all my other children I heard but not from her! It quite overwhelmed me and when I got dear Papa's and your letters how my heart sank within me! It was too terrible. How I do miss her and you poor dear children must feel it more and more! To try and be like her—unselfish and courageous, loving and good—that is how best to fulfil her dear wishes.

JOURNAL   19 June 1879

Just before 11, a telegram was given me with the message that it contained bad news. When I, in alarm, asked what, I was told it was that the Prince Imperial had been killed. I feel a thrill of horror in even writing it. I kept on saying 'No, no, it can't be!' To die in such an awful way is too shocking! Poor dear Empress! her only child, her all, gone! I am really in despair. He was such an amiable, good young man, who would have made such a good Emperor for France one day. It is a real misfortune.

---

*The Queen and the Prince had a personal interest in the Bill: they hoped that the Grand Duke of Hesse-Darmstadt, a widower since the death of Princess Alice, might marry her youngest sister, Princess Beatrice. When the Bill was passed by the Commons the Queen professed her delight that 'the bigots' had been defeated; but the bishops defeated the Bill in the Lords.

The more one thinks of it the worse it becomes. Got to bed very late, it was just dawning! and little sleep did I get.

20 June 1879

Had a bad, restless night, haunted by this awful event, seeing those horrid Zulus constantly before me, and thinking of the poor Empress, who did not yet know it.

TO THE CROWN PRINCESS OF PRUSSIA   21 June 1879

Your sore and wounded heart will bleed for the poor, poor Empress—who has lost her all, her only child and the only hope she had left. And in such a horrible way.*
Good, exemplary, brave but alas! far too daring young man, and to think of his being murdered in such a way—though I am sure it was the affair but of a few seconds—is enough for ever to haunt a mother's heart.

23 June 1879

She stood pale and bowed down with grief—I clasped her in my arms and she sobbed, but quietly, and she then took me into her little boudoir which was very dark all the blinds being down where I had been with her after the Emperor's death and sat an hour with her. She is so uncomplaining, so gentle so resigned not accusing anyone but utterly broken-hearted in her terrible grief. 'Pauvré petit, seulement 23 ans' that he has never caused her a moment's sorrow 'il était si droit, si loyal', but that she felt it was to be so 'Sa destinée l'a entrâiné'.

TO BEACONSFIELD   26 June 1879

The Queen wishes to say, as she cannot see Lord Beaconsfield, that she cannot consent to send Leopold to Australia. Since the loss of her beloved child [Princess Alice who had died of diphtheria the year before], the separation from another, and this dreadful event of the dear Prince Imperial's death, she cannot bring herself to consent to send her very delicate son, who has been four or five times at death's door, who is never hardly a few months without being laid up, to a great distance, to a climate to which he is a stranger, and to expose him to dangers which he may not be able to avert.

TO THE CROWN PRINCESS OF PRUSSIA   28 June 1879

The questions you ask are in everyone's mouth and there is a terrible feeling in which we all share at the thought that this precious young life should have been sacrificed without being defended and fought for by British officers. I own it is quite dreadful to me! As regards the war itself there is no doubt the right thing has not been done—which is distracting. No head at the head and the want of organisation etc.

---

*The Prince Imperial had joined Lord Chelmsford's staff in South Africa. After the defeat of the British forces by the Zulus at Isandhlwana the Prince had been killed in a skirmish with Zulu warriors. The Queen's distress was exacerbated by the belief that the Prince had not been properly protected by those responsible for looking after him. On 23 June the Queen visited the Empress at Chislehurst.

TO BEACONSFIELD   28 July 1879

One great lesson is again taught us, but it is never followed: never let the Army and Navy down so low as to be obliged to go to great expense in a hurry.

This was the case in the Crimean war. We were not prepared. We had but small forces at the Cape: hence the great amount having to be sent out in a hurry ... If we are to maintain our position as a first-rate Power ... we must, with our Indian Empire and large Colonies, be prepared for attacks and wars, somewhere or other, continually. And the true economy will be to be always ready. Lord Beaconsfield can do his country the greatest service by repeating this again and again, and by seeing it carried out. It will prevent war.

TO THE CROWN PRINCESS OF PRUSSIA   18 October 1879

I am afraid I am as bad a sightseer as Willie, not from want of interest but that it fatigues me most dreadfully and finally bores me too. One or two fine things at a time is all I like or am able for ... Visiting is (at least in England) the worst thing I know and such a bore. The gentlemen go out shooting and the ladies spend the whole day idling and gossiping together. Alix never hardly goes now—she hates it so. And Marie dislikes it very much indeed.

Willie's want of interest or taste—and mere love of Potsdam is quite tiresome and provoking—but I think young people have such an *esprit de contradiction* that it very likely will get better if no notice is taken of that.

TO PONSONBY   12 March 1880

The Queen is anxious to write once more and more decidedly her very strong objections—indeed her determination not to accept Sir C. Dilke as a minister of any future Liberal Government. It is well known that he is a democrat—a disguised republican, who is in communication with the extreme French republicans. He has been personally most offensive in his language respecting the Court—the expenses, etc.—and to place him in the Government, not to speak of the Cabinet would be a sign to the whole world that England was sliding down into democracy and a republic. If the Liberals intend to lean to the extreme radicals, they can never expect any support from the Queen.

                                                          13 March 1880

The Queen is no partizan and never has been since the first three or four years of her reign when she was so from her inexperience and great friendship with Lord Melbourne. But she has, in common with many sound Liberals or Whigs, most deeply grieved over and been indignant at the blind and destructive course pursued by the Opposition which would ruin the country and her great anxiety is to warn them not to go on committing themselves to such a very dangerous and reckless course.

By now Disraeli's government was beset by agricultural and economic depression at home and by successive troubles abroad. The unnecessary war against the Afghans, and another war in South Africa in which the Zulus won their victory at Isandhlwana, aroused fierce criticism from his opponents and gave Gladstone just the kind of ammunition he needed to attack the policies of his rival. In the general election of 1880 the Tory organization proved no match for the well-run machine of their opponents.

To Beaconsfield   3 April 1880

This is a terrible telegram [announcing the defeat of the Government in the elections in which 349 Liberals were elected against 243 Conservatives]. The Queen cannot deny she ... thinks it a great calamity for the country and the peace of Europe.

To Ponsonby   4 April 1880

The great alarm in the country is Mr. Gladstone, the Queen perceives and she will sooner abdicate than send for or have any communication with that half-mad fire-brand who would soon ruin everything and be a Dictator.

Others but herself may submit to his democratic rule, but not the Queen.

She thinks he himself don't wish for or expect it.

To the Crown Princess of Prussia   5 April 1880

Mr. Gladstone's mad, unpatriotic ravings and the sad want of patriotism of the Opposition have done this harm which is very grievous to me; it is a serious sorrow and trouble—it is that strange, ungrateful love of change which I do believe is the chief if not sole cause of this extraordinary result. People ignorant and unreasoning think a change of Government will give them a good harvest and restore commerce.

To Beaconsfield   7 April 1880

What your loss to me as a Minister would be, it is impossible to estimate. But I trust you will always remain my friend, to whom I can turn and on whom I can rely.

Hope you will come to Windsor in the forenoon on Sunday, and stop all day, and dine and sleep.

To Ponsonby   8 April 1880

What the Queen is especially anxious to have impressed on Lords Hartington and Granville [two leading Liberals] is, firstly, that Mr. Gladstone she could have nothing to do with, for she considers his whole conduct since '76 to have been one series of violent, passionate invective against and abuse of Lord Beaconsfield ... The Queen does feel the Opposition to have been unusually and very factious, and to have caused her great annoyance and anxiety, and deep regret. She wishes, however, to support the new Government and to show them confidence, as she has hitherto done all her Governments, but that this must entirely depend on their conduct. There must be no democratic leaning, no attempt to change the Foreign policy (and the Continent are terribly alarmed), no change in India, no hasty retreat from Afghanistan, and no cutting down of estimates.

16 April 1880

The Queen ... cannot leave 2 expressions of his [Ponsonby's] without a remark.

He says 'Mr. Gladstone is loyal and devoted to the Queen'!!!

He is neither; for no one can be, who spares no means—contrary to anything the Queen and she thinks her Predecessors ever witnessed or experienced—to vilify—attack—accuse of every species of iniquity a Minister who had most difficult times and questions to deal with—and who showed a most unpardonable and disgraceful spite and personal hatred to Lord Beaconsfield who has restored England to the position she had lost under Mr. Gladstone's Government.

Is this patriotism and devotion to the sovereign? And what has he brought upon the Queen and country?

Such conduct is unheard of and the only excuse is—that he is not quite sane.

MEMORANDUM   18 April 1880

I saw Lord Beaconsfield this morning at ½ pt. 12. After remarks on the sad and startling result of the elections which no one was in the least prepared for, I asked him what he advised me to do for the real good of the country, which we both agreed was inseparable from my own; and he replied that, irrespective of any personal feeling which I might have respecting Mr. Gladstone, the right and constitutional course for me to take was to send for Lord Hartington. He was in his heart a conservative, a gentleman, and very straightforward in his conduct. Lord Granville was less disinterested and looked more for his own objects.

JOURNAL   22 April 1880

Saw Lord Beaconsfield, and thought him very low. Wrote last night to Lord Hartington, asking him to come to-day at 3, as I wished to charge him with forming a Government. I saw him at that hour, and he spoke quite frankly. I was equally frank with him. But the result was unsatisfactory. When I stated that I looked to him to form a Government, he said that, though they had not consulted Mr. Gladstone, both he (Lord H.) and Lord Granville feared they would have no chance of success if Mr. Gladstone was not in the Government, and that they feared he would not take a subordinate position. If he were to remain a sort of irresponsible adviser outside the Government, that would be unconstitutional and untenable, and that, if he were quite independent, he might make it impossible for any Liberal Government to go on. They therefore thought it would be best and wisest if I at once sent for Mr. Gladstone.

I said there was one great difficulty, which was, that I could not give Mr. Gladstone my confidence.

MEMORANDUM   23 April 1880

I saw Mr. Gladstone at quarter to 7, and told him [that] ... I had applied to Lord Hartington as the Leader (which he said was quite correct), but as Lord H. and Lord Granville said they could not act without him, I wished now to know if he could form a Government? He replied that, considering the part he had taken, he felt he must not shrink from the responsibility, and that he felt he would be prepared to form a Government . . .

I then said I wished to be frank and say something; which was that I hoped he would be conciliatory, as it had been a cause of pain to me to see such asperity and such strong expressions used, and I thought 'peace was blessed.' He replied that he considered all violence and bitterness 'to belong to the past'.

MEMORANDUM   27 April 1880

Saw Lord Beaconsfield at 3 and gave him a parting gift, my statuette in bronze, and the plaster casts of the group of Brown, the pony, and 'Sharp,' and the statuettes of Sharp and William Brown, all of which he had never seen and with which he expressed himself much delighted. We then talked of the new Government, which he thought very moderate, but which I told him I heard the Radicals were very indignant at! His intention, he said, was to impress upon his party, of whom he should have a large meeting before the opening of Parliament, not to attack the Government excepting when extreme measures were proposed, or any change in foreign policy. Otherwise they should let them alone . . .

He would not come to town or to the drawing-room, and wished to 'keep out of sight,' only coming up when it was necessary for him to be in the House of Lords.

I then took leave of him, shaking hands, when he kissed mine. I would not consider this as a leave-taking, as I said I was sure to see him again before we left for Scotland, and that I begged he would always let me know his whereabouts so that I could always give him news of myself.

TO THE CROWN PRINCESS OF PRUSSIA   27 April 1880

These are very trying, worrying, painful days. It is a terrible change. Dear, kind, wise Lord Beaconsfield so dignified and worthy is 'overwhelmed' as he says to leave me, for whom he had really the most wonderful devotion and attachment and I don't think he was at all pleased to be relieved. He was so anxious to see things settled and England more and more raised and strengthened—and the result was so totally unexpected on both sides, the ingratitude of the country so great that he feels the disappointment much. Mr. Gladstone, too, as Prime Minister seems hardly possible to believe. I had felt so sure he could not return and it is a bitter trial for there is no more disagreeable Minister to have to deal with. It is a weak Cabinet. Lords Hartington and Granville said they had no chance without him.

2 May 1880

I am very tired and have much to do. Tomorrow is another Council. I send you here a list of the Government. All these Radicals are a great trial—but they may not prove dangerous when in office. Still it alarms people. The first Council was a great trial. To take people I cannot trust and whose object was to drive the late Government, which had done so well, out merely to put themselves in, and who will have to pursue much the same policy or there will [be] war and every sort of disaster—is dreadful—is a dreadful trial. To me 'the people's William' is a most disagreeable person—half crazy, and so excited—(though he has been respectful and proper in his manner and professes devotion) to have to deal with. I insisted on receiving assurances on the subject of the principles and languages of Mr. Chamberlain [President of the Board of Trade] and especially Sir C. Dilke. The last named has made confessions of sins, and promises not to repeat them. He is only an Under Secretary [at the Foreign Office].

TO LORD GRANVILLE, FOREIGN SECRETARY   5 June 1880

The Queen cannot help feeling uneasy at the state of the House of Commons. There is such an amount of interference and meddling in everything, that, unless it is firmly resisted, Government will soon become impossible. It would be grievous indeed and very serious if this democratic tendency were not checked, and the Queen thinks Mr. Gladstone has it in his power, by his experience and influence as well as by his large majority, to raise the tone, and not let the House of Commons become, as it were, the executive power, which is what this constant interference and constant questioning increasingly leads to.

If Mr. Gladstone would refuse shortly and firmly to answer questions of a totally unfit character for Parliament, and would desire his colleagues to do the same, he would be doing immense good to the Monarchy and Constitution. A Constitutional Sovereign at best has a most difficult task, and it may become almost an impossible one, if things are allowed to go on as they have done of late years.

TO THE CROWN PRINCESS OF PRUSSIA   7 July 1880

Mr. Gladstone seems to me very infirm of purpose—weak, irresolute and ready to give way to extreme opinions. There is no support in the Government. I feel it is a reed—where I used to have a rock to lean on.

TO PRINCESS VICTORIA OF HESSE   4 August 1880

I was, darling Child, rather shocked to hear of your shooting at a mark but far more so at your idea of going out shooting with dear Papa. To look on is harmless but it is not lady like to kill animals and go out shooting—and I hope you will never do that. It might do you great harm if that was known as only fast ladies do such things.

TO GRANVILLE   4 September 1880

The House of Commons is becoming like one of the Assemblies in a Republic, and the Ministers ought in the interest of the much vaunted and admired British Constitution at least to stand up for it. It is becoming very serious, and the Ministers only weaken their own authority and their own power by yielding to it.

TO THE CROWN PRINCESS OF PRUSSIA   11 August 1880

And now let me say how horrified and how distressed I am about your cat!* It is monstrous—and the man ought to be hung on the tree. I could cry with you as I adore my pets. When they belonged to a loved and lost object it must be quite a

---

*The cat had been shot by 'a stupid jäger', who had then hanged the dead animal on a tree and cut off her nose. The Princess confessed she was 'very silly perhaps' but could not stop crying: 'The man was neither punished nor reproved.'

grief! We always put a collar with V.R. on our pet cats and that preserves them. Our keeper once shot a pet one of Beatrice's. Keepers are very stupid but none would dream of mutilating an animal here! I think it right and only due to the affection of dumb animals, who (the very intelligent and highly developed ones) I believe to have souls, to mourn for them truly and deeply.

TO PRINCESS VICTORIA OF HESSE   8 December 1880

I cannot tell you what pleasure your dear affectionate letter gave me! It shows so much good sense and right, proper feeling—as well as such confidence in your old loving Grandmama who loves you as a Mother and wishes to be one to you as much as she possibly can. God bless you for it sweet Child! Beloved Mama, excepting Auntie Beatrice (and she was and is too young to understand many things which only a wife and Mother can) was the one of my daughters who felt with me so much, and agreed with me in her views about my Children and I think I can therefore tell you what she would have wished—

You are right to be civil and friendly to the young girls you may occasionally meet, and to see them sometimes—but never make friendships; girls friendships and intimacies are very bad and often lead to great mischief—Grandpapa and I never allowed it, and dear Mama was quite of the same opinion. Besides, as you so truly say, you are so many of yourselves that you want no one else. I think also that you are quite right not to have large parties for you are both, and Ella [her younger sister], decidedly too young for them. And at the dinners remember not to talk too much and especially not too loud and not across the table . . .

There is another most important thing which you are quite old enough for me to speak or write to you about. Dear Papa will, I know, be teazed and pressed to make you marry, and I have told him you were far too young [she was seventeen] to think of it, and that your first duty was to stay with him, and to be as it were the Mistress of the House, as so many eldest daughters are to their Fathers, when God has taken their beloved Mother away—I know full well that you have no ideas of this sort and that you (unlike, I am sorry to say, so many Princesses abroad)—don't wish to be married for marrying's sake and to have a position. I know darling Child that you would never do this, and dear Mama had a horror of it; but it is a very German view of things and I would wish you to be prepared and on your guard when such things are brought before Papa, and possibly to your Grandmama.

TO THE CROWN PRINCESS OF PRUSSIA   14 September 1880

Mr G. is not what he was—he is *très baissé* and really a little crazy. He has not recovered his illness yet and I doubt (and fervently hope) he won't be able to go through another session.

I am very proud of Sir F[rederick] Roberts' wonderful march [to Kandahar where he defeated the Afghans] and brilliant victory. No one knows what our brave troops have to go through in our constant Indian and Colonial warfare. It is fearful; very different to Europe.

12 October 1880

You say I am worried! God knows I am—and disgusted. I have warned and tried to do all I can but in vain and I now feel disheartened and disgusted. The mad passion of one half-crazy enthusiast is ruining all the good of six years peaceful, wise government and I often wish I could retire quietly and let people work out their own policy and reap its fruits.

JOURNAL   23 November 1880

Finished *Jane Eyre*, which is really a wonderful book very peculiar in parts, but so powerfully and admirably written, such a fine tone in it, such fine religious feeling, and such beautiful writing. The description of the mysterious maniac's nightly appearances awfully thrilling, Mr. Rochester's character a very remarkable one, and Jane Eyre's herself a beautiful one. The end is very touching, when Jane Eyre returns to him and finds him blind, with one hand gone from injuries during the fire in his house, which was caused by his mad wife.

TO THE CROWN PRINCESS OF PRUSSIA   8 December 1880

Sir F. and Lady Roberts also are here tonight. He is a great man, and so small in appearance but a very keen, eagle eye. It is so interesting to hear about his wonderful march and he is so modest. He was so ill the day of the great battle that they had to give him champagne—which one of his A.D.C.s found—every half hour. He managed to get through it but was dreadfully exhausted afterwards. On the march he had to be carried some days. His wife is a very nice person and must have been very pretty. She is an excellent soldier's wife and devoted to him and proud of him.

During the winter of 1880–81 the Queen was extremely worried about conditions in Ireland. Never well advised about Irish affairs, she considered the people 'impossible' and vehemently condemned the placatory measures proposed by Gladstone and supported by Bright, the Chancellor of the Duchy of Lancaster, and Joseph Chamberlain, the President of the Board of Trade. Before concessions were made she wanted the Government to renew the Peace Preservation Act, in other words to pursue a policy of 'coercion' in order to re-establish law and order.

TO LORD HARTINGTON, INDIAN SECRETARY   12 December 1880

The Queen is so very anxious about the present very alarming state of affairs that she cannot refrain from appealing to Lord Hartington in the very strongest manner possible to use all his influence with Mr. Gladstone and his Whig Colleagues to act very strongly and firmly at the present very anxious moment. The Queen gathered from Lord Hartington that Mr. Bright and Mr. Chamberlain would have resigned if strong measures of coercion had been proposed sooner, and that it was considered of importance to prevent their doing so—for fear of the few Radicals who would follow them, and therefore delayed the meeting of Parliament. The danger of this is not to be compared to the danger, hourly increasing, of allowing a state of affairs like the present in Ireland to go on. The law is openly defied, disobeyed, and such an example may spread to England, if it prove successful in Ireland. It must be put down and nothing but boldness and firmness will succeed. You moderate Ministers must be firm, and insist on means being used to put an end to this dreadful state of affairs. Don't yield to satisfy Messrs. Bright and Chamberlain; let them go: declare you will not be parties to a weak and vacillating policy, which is ruining the country and bringing great discredit on the Government. The Queen does not here speak of what she feels herself, as Sovereign of the country. It is most painful to her, and she has a right to appeal to her Ministers to uphold her authority and to expect them to do so.

To W. E. Forster, Chief Secretary for Ireland   25 December 1880

The Queen is as sincerely liberal in her views for the improvement of her Empire as anyone can be, but she is as sincerely and determinedly opposed to those advanced, and what she must call destructive, views entertained by so many who unfortunately are in the Government. If these prevail instead of those of the moderate, far-seeing, and loyal ones, the Queen will not remain where she is; she cannot and will not be the Queen of a democratic monarchy; and those who have spoken and agitated, for the sake of party and to injure their opponents, in a very radical sense must look for another monarch; and she doubts [if] they will find one. The Queen has spoken very strongly, but she thinks the present Government are running a very dangerous course.

To Gladstone   26 December 1880

As the time for the reassembling of the Cabinet approaches, the Queen wishes to repeat her decided opinion that no measure of any kind should be brought forward till those of coercion and those to give power to put down the state of lawlessness in Ireland have been passed ... The state of affairs is very serious, and the language used by Mr. Bright and Mr. Chamberlain on some recent occasions is totally at variance with their position in the Cabinet, and calculated to encourage the Irish ...

The Queen cannot help sometimes remembering the day when Mr. Gladstone was sworn in at Claremont in 1841 (40 years ago) on the formation of Sir R. Peel's Conservative Government! That was an admirable Government!

To Ponsonby   1 January 1881

Would Sir Henry prepare a letter for her to write to Mr. Gladstone. She does expect him to warn both Mr. Bright and Mr. Chamberlain not to attack the House of Lords—for they cannot remain in the Cabinet if they do that again.

Journal   3 January 1881

Only received the Speech [to be delivered at the Opening of Parliament] this morning, which is very wrong. In it found the announcement of the abandonment of Kandahar, without my having heard a word about it ... Directly after breakfast I telegraphed to Mr. Gladstone to have the Speech altered, or the part about Kandahar left out. All was ready for my Council, and I waiting, when Sir H.

Ponsonby came to say that the Ministers, Lord Spencer [Lord President of the Council] and Sir Wm. Harcourt [Home Secretary], declared they would wait till Mr. Gladstone's answer came. In vain I assured them (through Sir H. Ponsonby) that I would approve the Speech, leaving out that paragraph; they insisted on waiting. Suggested that if no answer came in time, or not a favourable one, I would have the Speech approved, but send a protest. Sir H. Ponsonby sent in a proposal which I altered and wrote. 3 o'clock passed, and still no answer came, but it at length did so at half past 3. It was not favourable, saying the matter had been agreed upon yesterday.

So I had my Memorandum given to the Ministers, and settled to hold my Council at once. After waiting 10 minutes in the drawing-room, Sir H. Ponsonby came in, saying the Ministers objected to the word 'disapproval,' which rather amused me. Called in Leopold, and after some difficulty, suggested altering 'disapproval' to 'much regrets.' This seemed to settle the matter, but 20 minutes elapsed before Sir H. Ponsonby again returned, saying they objected to the last part, in which I ask for an assurance. So I said, Very well, I would not send my Memorandum through them, but straight to Mr. Gladstone, and would hold the Council. Dreadfully put out, they at length came in, after 4 ... The business was hurriedly gone through, and the Speech approved. I spoke to no one, and the Ministers nearly tumbled over each other going out. My headache had got very bad.

To Hartington   16 January 1881

The Queen trusts that means will be found to prevent these dreadful Irish people from succeeding in their attempts to delay the passing of the important measures of Coercion.

As well as trouble in Ireland, there was now war in South Africa where the Transvaal Boers, who had proclaimed a Republic in December 1880, invaded Natal.

To the Prince of Wales   20 January 1881

I most earnestly protest against the Princes serving with a Naval Brigade on shore at the Cape. I strongly objected to their both going to sea, but consented on the suggestion that it was necessary for their education. The proposal to send them on active service destroys the cause for my consent, and there is no reason for, but many against, their incurring danger in the South African war.

To Lord Northbrook, 1st Lord of the Admiralty   7 February 1881

The Queen ... is glad to hear that the Princes are not to be attached to a Naval Brigade. They are so young, and the only two sons of the Prince of Wales ... so that to expose their lives needlessly, and moreover in a civil war, would have been wrong in every way.

The Queen trusts that in future Lord Northbrook will not give any fresh orders without letting the Queen first know, so that the disagreement of opinions may not again occur.

TO THE PRINCESS OF WALES   18 February 1881

I am very sorry Bertie should have been sore about the boys; but I think he must have forgotten the arrangements and conditions and instructions respecting their going to sea. I, and even Bertie and you, only consented to their both going to sea for their education and moral training.

JOURNAL   28 February 1881

Dreadful news reached me when I got up. Another fearful defeat [Majuba Hill]. When I opened the telegram, I hoped it might be the news of a victory. It is too dreadful!

TO LORD KIMBERLEY, COLONIAL SECRETARY   9 March 1881

I find that an impression prevails that we are about to make peace with the Boers on their own terms. I am sure you will agree with me that even the semblance of any concessions after our recent defeats would have a deplorable effect.

TO THE CROWN PRINCESS OF PRUSSIA   16 March 1881

I share your horror, condemnation and sorrow at the death (unparalleled) of the poor, kind Emperor Alexander [who had been assassinated in St. Petersburg]. A sense of horror thrills me through and through! Where such a criminal succeeds the effect is dreadful. The details are too terrible. No punishment is bad enough for the murderers who planned it; hanging is too good. That he, the mildest and best sovereign Russia had, should be the victim of such fiends is too grievous. Poor darling Marie on whom her poor father doted, it is too much almost to bear. But she is very courageous.

JOURNAL   28 March 1881

The accounts of good Lord Beaconsfield are not very satisfactory. He is in bed and very weak.

                                                          19 April 1881

Received the sad news that dear Lord Beaconsfield had passed away. I am most terribly shocked and grieved, for dear Lord Beaconsfield was one of my best, most devoted, and kindest of friends, as well as wisest of counsellors. His loss is irreparable to me and the country. To lose such a pillar of strength, at such a moment, is dreadful!

                                                          20 April 1881

As regards the funeral, Lord Beaconsfield expresses his wish for it to be private, and that he should be laid near his wife at Hughenden. He was asked, what if the Queen should wish Westminster Abbey? and he was silent. But I at once said, I did not wish it, but thought he should be buried at Hughenden, as he had desired.

TO THE CROWN PRINCESS OF PRUSSIA   20 April 1881

A heavy cloud has fallen over this Empire; a great statesman who in six years raised its name again and placed it where it ought to be, has passed away at the very time when Kandahar is given up—a wretched peace in Africa—and disorder is rampant in Ireland—while murder is openly preached in every country. To me the blow is terrible and I was quite ill yesterday and am still feeling poorly and shaken. Dear Lord Beaconsfield was the truest, kindest friend and wisest counsellor and he too is gone! I feel much crushed by it.

23 April 1881

You say his loyalty ought not to be put foremost of his merits as I am blessed by loyalty in general. Yes, as a general term, this may be true but I know too well for discussion what a difference there was in his loyalty. My good, my comfort were his first objects—not the party's or his own.

And now with the Liberal Party alas! The creed is party and self first and country and sovereign last. I feel this most sadly, and more and more each succeeding Liberal or I fear radical Government. They have to be told and taught what they owe to me—and I suffer much from this. It is the one thing which is so totally different. You would see this at once if you were here.

Few people understand me so well as dear Lord Beaconsfield, or what gentle, true tenderness there was in his nature combined with such great firmness and courage, and how he struggled with a feeble frame and bad health with his indomitable will and self control . . .

I agree with you in thinking the executions in Russia will have no effect but they were absolutely necessary—only they should have been strictly private. Public executions are horrid.

JOURNAL   30 April 1881

After an early luncheon, started . . . in an open landau, for Hughenden. We drove past Cliveden, and changed horses at a small inn, just outside the gates . . . We went into the Church . . . The flowers still remained, as at the funeral. Then we walked round to the vault, which had been opened purposely for me to see. There, in a small space, is dear Lord Beaconsfield's coffin, covered with wreaths and flowers, next to his wife's, and there are others of his family also buried there. Could hardly realise it all, it seemed too sad, and so cheerless. I placed a wreath of china flowers. Now the vault is to be finally closed, and not used again.

We got into the carriage and drove up to the house. All was just the same as when, two and a half years ago, dear Lord Beaconsfield received us there, such a sad contrast! Went into the library and drawing-room, where hangs my picture, all, all the same, only he not there! Took tea in the library, where I had sat with my kind friend, and where he had given me a long account of a very stormy Cabinet he had had, when he had expected several of the Ministers would resign. I seem to hear his voice, and the impassioned, eager way he described everything.

17 May 1881

Saw Lord Granville, and talked of the topics of the day, but with no satisfactory result. He lamented over things, shrugged his shoulders, but is weak as water.

TO GLADSTONE   25 May 1881

The Queen has to thank Mr. Gladstone for a very kind letter on the occasion of her now somewhat ancient birthday. The affectionate loyalty of her subjects is very gratifying to her. Her constant object, which only increases with years, is the welfare, prosperity, honour, and glory of her dear country.

But the work and anxiety weigh heavily on her, unsustained by the strong arm and loving advice of him who now nineteen and a half years ago was taken to a higher and better world!

JOURNAL   29 May 1881

Letters from Lord Kimberley and Mr. Gladstone. After considering my remonstrance [against the Government's decision to recognize the independence of the whole of the Transvaal], they still determined to disregard my opinion. Feel very indignant. Telegraphed and wrote strongly. I can do no more, but feel utterly disgusted and disheartened.

TO THE CROWN PRINCESS OF PRUSSIA   16 July 1881

Mrs. Gladstone told me you had been to tea with her and Mr. G. [on the Gladstones' visit to Germany]. Was that necessary?

TO GLADSTONE   7 August 1881

The Queen has to thank Mr. Gladstone for his regular and interesting reports of the proceedings in the House of Commons, many of which are of a most disgraceful character. But how can you expect better from so many Members of such low and revolutionary views who are now in the House of Commons?

TO PONSONBY   13 August 1881

The H. of Lords cannot be totally set aside or a republic with one House had better be proposed. Mr. Gladstone is dragged along by his dreadful Radical following and is ruining the country.

TO GLADSTONE   21 August 1881

The Queen was grieved and shocked to perceive, as Mr. Gladstone so strongly and justly remarks in the letter she received from him yesterday, with what bitterness, violence, and pertinacity these rebel Irishmen continue their attacks and opposition to the very last.

JOURNAL   21 September 1881

Saw Lord Hartington, who arrived to-day. The state of Ireland was one of the subjects touched on in our conversation. He feared it was very bad.

TO GLADSTONE   1 October 1881

The news from South Africa are serious. The Queen relies confidently on the firmness of the Government and on their not yielding further ... We have gone as far as we could already, but yielding now would only be weakness and injure us seriously.

20 October 1881

I see you are to attend a great Banquet at Leeds. Let me express a hope that you will be very cautious not to say anything which may bind you to any particular measures. Every word is looked for and criticised, and the times are serious.

TO THE CROWN PRINCESS OF PRUSSIA   8 November 1881

I quite agree in what you say about that dreadful man [Bismarck]—it is deplorable. No wonder you dislike Conservatives; but you would not if you lived here. Here they are the only security.

JOURNAL   18 November 1881

Very shortly after tea, received a telegram in cypher from Leopold saying: 'I have proposed to Princess Helen of Waldeck, and been accepted. May I receive your consent to the engagement?' This was hardly a surprise to me, as since last autumn, when he had met the young lady, I knew he had taken a liking to her, and now they had met again at Frankfort. But the news rather upset me, and I cannot help feeling as if I were losing dear Leopold, but as Hélène Waldeck [Princess Helen of Waldeck-Pyrmont] is said to be so good and nice, it may be a blessing to us all. Telegraphed my consent and best wishes, and then wrote to my other children.

TO THE CROWN PRINCESS OF PRUSSIA   26 November 1881

You say you can't be enthusiastic about Leopold's bride! I never can be about any marriage and can naturally be less about Leopold's than about any other for I think his marrying such a risk and experiment. But everyone (from different sides) spoke most highly of Helen Waldeck and her charming character, excellent education, solid, sterling qualities!

TO PRINCESS VICTORIA OF HESSE   10 December 1881

I hope you work regularly. Some knitting and plain work for the poor is such a nice occupation. Only reading is not good and work is such a good occupation for girls and women and it is such a pleasure to be able to be of use to others.

TO GLADSTONE   31 December 1881

The state of Ireland causes the Queen much painful thought and must she thinks do so to Mr. Gladstone.

Notwithstanding the official assurances that matters are improving in Ireland, the Queen continues to read in the newspapers and to hear from various sources the most distressing accounts of the disorder and anarchy that seem to prevail ...

If there are not sufficient soldiers to perform the duties required of them, let more regiments be sent. If the law is powerless to punish wrong-doers, let increased powers be sought for and at any rate let no effort be spared for putting an end to a state of affairs which is a disgrace to any civilised country.

JOURNAL   2 March 1882

At 4.30 left Buckingham Palace for Windsor. Just as we were driving off from the station there, the people, or rather the Eton boys, cheered, and at the same time there was the sound of what I thought was an explosion from the engine, but in another moment I saw people rushing about and a man being violently hustled, people rushing down the street. I then realised that it was a shot, which must have been meant for me, though I was not sure. No one gave me a sign to lead me to believe anything amiss had happened. Brown however, when he opened the carriage, said, with a greatly perturbed face, though quite calm: 'That man fired at Your Majesty's carriage.'

Took tea with Beatrice, and telegraphed to all my children and near relations. Brown came in to say that the revolver had been found loaded, and one chamber discharged. Superintendent Hayes, of the Police here, seized the man [Roderick (sometimes called Frederick) Maclean], who was wretchedly dressed, and had a very bad countenance. Sir H. Ponsonby came in to tell me more. The man will be examined to-morrow. He is well spoken, and evidently an educated man ... An Eton boy had rushed up, and beaten him with an umbrella. Great excitement prevails. Nothing can exceed dearest Beatrice's courage and calmness, for she saw the whole thing, the man take aim, and fire straight into the carriage, but she never said a word, observing that I was not frightened. Telegrams began arriving in numbers, in answer to mine, and one or two sent before, to enquire if the report, which spread instantly to London and all over the world, was true. Was really not shaken or frightened.

                                                          7 April 1882

Dear Leopold's birthday, his 29th. How thankful we must be that he has been preserved to us! How often has his poor young life hung on a thread, and how many bad and wearisome illnesses has he not recovered from! Though the idea of his marrying makes me anxious, still, as he has found a girl, so charming, ready to accept and love him, in spite of his ailments, I hope he may be happy and carefully watched over.

TO THE CROWN PRINCESS OF PRUSSIA   19 April 1882

This day week we were at lovely Mentone which we miss sadly—the bright sunshine and the sea, mountains, vegetation and lightness of the air and the brightness and gaiety of everything. Still it is at least a month more forward here than usual. The lilacs are out in blossom and chestnuts blooming away, all the thorns in leaf; many oaks out even—and the primroses still out. I sent a wreath of them to place on the grave of my dear and even more missed and regretted friend Lord Beaconsfield today—the first anniversary of his death. I sent them to him in his dear lifetime as they were his favourite flowers.

Oh! were he only here now to try and put order into the frightful misrule of the present people which is really appalling. Murder upon murder in Ireland and no one getting any rents.

JOURNAL   27 April 1882

(Leopold's wedding day).—This exciting day is all over, and past, like a dream, and the last, but one, of my children is married, and has left the paternal home, but not entirely, as he still keeps his rooms. It was very trying to see the dear boy, on this important day of his life, still lame and shaky, but I am thankful it is well over [he had slipped on some orange peel at Mentone]. I feel so much for dear Helen, but she showed unmistakably how devoted she is to him. It is a great blessing. God bless them both!

                                                                       6 May 1882

Drove through enormous crowds, who lined the whole way, nearly 3 miles, to High Beech, where an Address was received, and read, and I declared the Park [Epping Forest which the Corporation of the City of London had been mainly instrumental in securing for the City of London] open. The sight was very brilliant ...

The enthusiasm was very great, and many quite poor people were out. The Park has been given to the poor of the East End, as a sort of recreation ground. Nothing but loyal expressions and kind faces did I hear and see; it was most gratifying. Got home shortly before dinner, and only had the ladies to dine with us.

Directly afterwards a telegram was received by Sir H. Ponsonby, from Lord Spencer [Viceroy of Ireland], conveying the following shocking news: 'I grieve to say the Under-Secretary Mr. Burke has been murdered, and Lord Frederick Cavendish [Forster's successor as Chief Secretary for Ireland] most dangerously wounded, in the Pheonix Park.' This is too terrible ... At midnight Janie Ely came in with a telegram, saying 'All is over with Lord Frederick. Both he and Mr. Burke were stabbed with knives ...'

How could Mr. Gladstone and his violent Radical advisers proceed with such a policy, which inevitably has led to all this? Surely his eyes must be opened now.

TO GRANVILLE   7 May 1882

While the Queen does not wish to pain Lord Granville at this moment when he must himself be deeply grieving ... she cannot withhold from him that she considers this horrible event the direct result of what she has always considered and has stated to Mr. Gladstone and to Lord Spencer as a most fatal and hazardous step.

TO GLADSTONE   9 May 1882

The Queen has not wished to write to Mr. Gladstone and make any observations on the state of Ireland, while she knew he was still stunned and overwhelmed by the terrible blow which has shocked all the world.

She wishes now to express her earnest hope that he will make no concessions to those whose actions, speeches, and writings have produced the present state of affairs in Ireland, and who will be encouraged by weak and vacillating action to make further demands.

TO THE CROWN PRINCESS OF PRUSSIA   10 May 1882

Oh! but I can think of nothing but this unexampled horror of what has just happened in Ireland, which fills everyone's mind and one can think of nothing, nothing else. Horrible, awful beyond belief!

To Gladstone   11 May 1882

While this horrible and daring murder of two valuable and excellent men and public servants, perpetrated in a most unexampled manner, has brought the danger of the present state of Ireland and the necessity for action home to everyone, we must not forget how many lives have already been sacrificed, though their names were hardly known, and how many humble homes have been rendered desolate, like those we have so deeply to sympathise with now.

To the Crown Princess of Prussia   13 May 1882

I saw Mr Gladstone yesterday; greatly shaken—and seemingly despondent and as if his energy was gone; very pale. I of course spoke of the necessity of strengthening measures—but also how fatal I thought the step of letting the suspects out was, before a stringent measure had been passed. He defended himself but, when I said I knew Mr Burke (one of the victims) had been of the same opinion and greatly deprecated it, it seemed to shake him and he said the catastrophe had certainly followed very rapidly.

To the Prince of Wales   27 May 1882

The state of affairs—this dreadfully Radical Government which contains many thinly-veiled Republicans—and the way in which they have truckled to the Home Rulers—as well as the utter disregard of all my opinions which after 45 years of experience ought to be considered, all make me very miserable, and disgust me with the hard, ungrateful task I have to go through.

To Ponsonby   27 May 1882

The Queen must complain bitterly of the want of respect and consideration of her views (which with her experience of 45 years might and ought to be regarded on the part of the Government, especially Lord Granville who she has known so long and who also lately ignores all her remarks! She feels hurt and indignant, as he is the only friend (though he has never really proved to be that) or at least the only person she has been in the habit of speaking out to in the Cabinet.

Sir Henry she fears (being so much inclined to the Liberal party himself) may not express strongly enough her views and fears. He should defend her as much as he can. Instead of resisting and trying to stem the downward and alarming course of radicalism and indeed but thinly veiled republicanism, the Cabinet weakly yields to Mr. Gladstone and Mr. Chamberlain who head and lead this dangerous policy instead of trying as they perfectly could to check it and rally all the good Whigs and moderate liberals whom they have entirely alienated toward them.

31 May 1882

With respect to Ireland and the Ministers, the Queen regrets to see, that she has no one real independent friend in the Cabinet, never hearing exactly what passes, and finding that no one will see the danger of going so much ahead with the radicals. The truth is, that, like so many people, Lord Granville has not the courage of his opinions and therefore is of not the slightest use to the Queen.

Sir Henry Ponsonby must know well, that the Queen cannot communicate frequently and openly with Mr. Gladstone, as he does not possess her confidence; and that was one of the reasons, why she so strongly objected to taking him as her Prime Minister, as she felt and feels how false and painful her position is with regard to him.

JOURNAL   25 June 1882

I told him [Lord Hartington] I wished he were at the head of the Government, instead of Mr. Gladstone. He admitted that Mr. Gladstone had a leaning towards the Irish, as he is always excusing them to me!

20 July 1882

When I read that my darling, precious Arthur, was really to go [as commander of the 1st Guards Brigade in Egypt where there had been a nationalist revolt], I quite broke down. It seemed like a dreadful dream. Telegraphed to him. Still, I would not on any account have him shirk his duty. Went with a heavy heart to bed.

28 July 1882

A little after 5, dear Arthur and Louischen arrived, come for the terrible leave-taking! We took tea all together, and sat talking for some time. Arthur told us of all his preparations, that he took only one horse, and was selling the others, what his dress would be, serge, quite loose, flannel shirt, high boots over breeches, and white helmet with a puggaree. His canteen and spy glass I give him, and Louischen his *tente d'abri*. Felt very low. We four dined together. Arthur and Louischen both looked sad, but we tried to talk as cheerily as we could of all the preparations. Went out a little on the Terrace; the moonlight was beautiful.

14 August 1882

Cetewayo* is a very fine man, in his native costume, or rather no costume. He is tall, immensely broad, and stout, with a good-humoured countenance, and an intelligent face. Unfortunately he appeared in a hideous black frock coat and trousers, but still wearing the ring round his head, denoting that he was a married man. His companions were very black, but quite different to the ordinary negro. I said, through Mr. Shepstone [British Administrator in Zululand], that I was glad to see him here, and that I recognised in him a great warrior, who had fought against us, but rejoiced we were now friends. He answered much the same, gesticulating a good deal as he spoke, mentioned having seen my picture, and said he was glad to see me in person. I asked about his voyage, and what he had seen and then named my three daughters, at which he said 'Ah!' After further commonplace observations, the interview terminated. Both in coming in, and going out, they gave me the royal Zulu

*King of the Zulus. He had been allowed to come to England to present his case after the defeat of his people, and was to be restored as ruler of Central Zululand in January 1883.

salute, saying something all together, and raising their right hands above their heads. Capt. Bigge says it is very striking, when thousands do this at the same time. Cetewayo walked about on the Terrace, and came close enough to Leopold's window for him to see him. They had lunch by themselves, and we watched them afterwards from the Colonnade, leaving. As they drove away, Cetewayo caught sight of me, and got up in the carriage, and remained standing till they were out of sight.

The Queen was now as concerned about events in Egypt as about Ireland. The Egyptian Foreign Minister, Arabi Pasha, revolted against the authority of the Khedive; and, after some hesitation and unsatisfactory talks with the French, who were reluctant to act against a nationalist leader, Gladstone decided upon naval and military action. The battle of Tel-el-Kebir on 13 September 1882, in which Sir Garnet Wolseley defeated the nationalists, brought the short Egyptian campaign to a successful conclusion.

JOURNAL   13 September 1882

Had a telegram saying that fighting was going on and that the enemy had been routed with heavy loss at Tel-el-Kebir. Much agitated. On coming in got a telegram saying, 'A great victory, Duke* safe and well.' The excitement very great. Felt unbounded joy and gratitude.

TO PONSONBY   17 September 1882

The Queen does not like the words 'early withdrawal' [as applied to British troops from Egypt] and would wish Sir Henry to cypher as follows:

The Queen to Earl Granville

Think you should be very cautious in speaking of early withdrawal of troops. We must bind ourselves to nothing. We have not fought and shed precious blood and gone to great expense for nothing.

JOURNAL   18 September 1882

Took my short walk, and while I did so, a telegram was brought out to me, saying that all was quietly over, and our dear Dean [Gerald Valerian Wellesley, Dean of Windsor] gone to his rest. Could not believe it, and felt stunned. Such a kind devoted friend for 33 years! By degrees and imperceptibly, he had grown to be our best friend. We consulted him so much, and since my great misfortune in '61, he was quite invaluable to me, and helped me in so many difficulties. He was truly sympathetic, loved our children, and was of the greatest use to me in many ways. The last of my four intimate and confidential friends has now been taken, and I feel it very deeply.

*The Duke of Connaught had commanded the Guards in the battle.

To Sir Garnet Wolseley   18 September 1882

The Queen begins her letter to Sir Garnet Wolseley, by congratulating him again on the very brilliant victory of her brave troops, which he may well be proud of having commanded. The praise of her beloved son and his safety are subjects of deep thankfulness and rejoicing . . . The Queen trusts that there will be no undue haste in withdrawing the troops from Egypt, where the disorganisation of everything must be very great.

To Gladstone   21 September 1882

The Queen is especially anxious that no troops should move in a hurry, as she feels convinced no reliance can be placed yet on the Egyptians, who would, if they saw a chance of success, again rise; that is, the army.

If Arabi and the other principal rebels, who are the cause of the death of thousands, are not severely punished, revolution and rebellion will be greatly encouraged and we may have to do all over again. The whole state of Egypt and its future are full of grave difficulties, and we must take good care that, short of annexation, our position is firmly established there and that we shall not have had to shed precious blood and expended much money for nothing.

To Ponsonby   26 September 1882

The Queen is glad to hear that Mr. Gladstone understands that the appointment of Dean of Windsor is a personal, and not a political appointment; she will therefore not expect Mr. Gladstone to suggest names to her. For obvious reasons the Queen thinks that it will be best to associate again the office of Domestic Chaplain with that of Dean of Windsor. It is therefore of more importance that the future Dean should be a person with whom she is pretty well acquainted, and whom she can confide in, than that he should be a distinguished Churchman or a brilliant scholar . . .

What the Queen wants is a tolerant, liberal-minded, broad Church clergyman, who at the same time is pleasant socially, and is popular with all members and classes of her Household; who understands her feelings not only in ecclesiastical but also in social matters; a good, kind man, without pride. The Queen, after much thought and consideration, has thought of Canon Connor, who unites the different qualifications which the Queen has enumerated. She only regrets that he is not of higher social and ecclesiastical rank. But he is of a good family and a thorough gentleman, and universally respected.

To Wolseley   30 September 1882

The Queen need not say *how immensely* gratified and proud she is to hear his high praise of her darling son Arthur. This dear soldier son has never given her a day's sorrow or anxiety except on the score of his health occasionally . . .

The Queen would wish to thank Sir Garnet very particularly for what he has said in that letter about Arthur. He has proved himself worthy of his own dear father who she wishes could have lived to see his child distinguish himself as he has done, and of the name he bears, and of his great godfather [the Duke of Wellington].

TO GLADSTONE   30 September 1882

The Queen feels strongly that by retaining our present force in Egypt we shall be in a far better and more dignified position to take that leading part amongst the other Powers in arranging for the future of the country, to which, after all the precious blood and treasure we have spent, we are so justly entitled: and it seems to her that, if once any troops are withdrawn, we shall have no pretext for replacing them.

TO GRANVILLE   11 October 1882

The Queen's opinion is that, short of annexation, our power in Egypt and control over it ought to be great and firm, and we ought to show to other Powers that we shall maintain this position, though without detriment to them. We should maintain a large force there for a long time. As regards the punishment of the principal rebels, there seems to be but one feeling, except amongst people of morbid sentimentality, in this country!!

JOURNAL   30 October 1882

Went down to the Drawing-room to receive him [Sir Garnet Wolseley]. He was looking uncommonly well, and said he was not in the least tired. He gave very good accounts of Arthur, but there was a great deal of sickness amongst the troops, though he hoped it was but slight. He supposed I had heard what alarming news had arrived from the Soudan, where a most serious insurrection, in favour of the so-called false prophet, was going on. A number of the Egyptian troops had been killed and defeated, and the false prophet was said to be besieging Khartoum. Should this be taken, the whole country would go with him, and the consequences might be disastrous.

TO PONSONBY   15 December 1882

Would Sir Henry tell Mr. Gladstone how unpleasant it is for her to have Lord Derby as a Minister [Lord Hartington's successor as Colonial Secretary] for she utterly despises him; he has no feeling for the honour of England and the language in the French press and the alarm felt in Germany (she had a letter from her daughter hoping it was not true) show what a very bad effect his name will have on the Government. All will believe that a 'cotton-spinning', 'peace-at-all-prices' policy is now to be favoured! Tell this all to Mr. Gladstone. It is too bad of Lord Granville not to give her a hint of this before it was too late. He had not once (since he came into office) been of the slightest help to her at all.

                                                         16 December 1882

The Queen was greatly shocked when Mr. Gladstone proposed Mr. Chamberlain in these uncomfortable words 'Sir C. Dilke says his friend Mr. Chamberlain is quite ready to exchange with him'! The Queen greatly objected and was told Mr. Chamberlain had never said anything like [it to] Sir C. Dilke. But she maintained her objections and she said that there might be other arrangements. The Queen is determined not to have a man like Mr. Chamberlain to hold such a personal appointment and Mr. Gladstone should be told the Queen will not have him. If Sir

C. Dilke and Mr. Chamberlain are such 'friends' their power for mischief may be very great.

Lord Granville's silence in all this shocks the Queen greatly. Lord Hartington is the most straight forward and reliable and far less radical.

To GLADSTONE   5 January 1883

The Queen is sure that Mr. Gladstone will not misunderstand her, when she expresses her earnest hopes that he will be very guarded in his language when he goes to Scotland shortly, and that he will remember the immense importance attached to every word falling from him. Words spoken are often the cause of difficulties hereafter . . .

Is it not rather venturesome for Mr. Gladstone to undertake such a visit at this time of year, and with so short a time of rest before him—not having been well?

10 January 1883

The Queen thanks Mr. Gladstone for three letters and is very sorry to hear of his indisposition and sleeplessness which is but too common a result of overwork and very exhausting. But he must be really quiet and not occupy himself at all with affairs and not write long letters like the one he did yesterday. Perfect quiet is ordered and the prescription ought to be thoroughly obeyed and followed.

When the Queen wrote she was not aware of this new feature in Mr. Gladstone's health, but thought he had only been suffering from a chill.

To THE CROWN PRINCESS OF PRUSSIA   10 January 1883

Now let me soothe your feelings of injustice about my 'little hit' at your being a Radical. I will yield to none in true liberalism, but republicanism and destructiveness are no true liberalism.

Where is there greater tyranny and greater oppression of religious thought and sects than in France? You say you are a progressist. Right and well—when fundamental landmarks are not abolished and swept away, and every good thing changed for change's sake, and that unfortunately has become the doctrine of people like Mr. Chamberlain, Sir C. Dilke . . . and Mr. Gladstone himself as well as a good many others. Mr. Gladstone has helped to roll the stone downhill—instead of guiding and checking it. There is the danger. If you lived here and saw and understood all that goes on you would see that these so-called, but not really 'liberal' ideas are very mischievous.

To GRANVILLE   20 January 1883

The Queen has been expecting to hear from Lord Granville upon the subject of Mr. Gladstone's somewhat sudden breakdown and indisposition, the more so as she knows that Lord Granville is acting for him during his absence. But he may be expecting to hear from her, and she therefore writes to him to say she has had a long conversation with Sir William Harcourt, who thinks very seriously of his state. The brain and nervous system have clearly been overtaxed, as indeed must every Public man's be now, unless they husband their strength.

To the Crown Princess of Prussia   5 March 1883

Though I am not an admirer of babies—I must say this [Princess Alice, daughter of Prince Leopold] is a beautiful child, so plump and so big with such neat little features and such a complete head of dark hair.

To Princess Victoria of Hesse   7 March 1883

Oh! dear! How very unfortunate it is of Ella to refuse good Fritz of Baden [Prince Frederick of Baden] so good and steady, with such a safe, happy position, and for a Russian [the Grand Duke Sergius, a reactionary who was to be assassinated in 1905]. I do deeply regret it. Ella's health will *never* stand the climate which killed poor Aunt [Princess Marie of Hesse, wife of the Emperor Alexander II] and has ruined the healths of almost all the German Princesses who went there.

Journal   29 March 1883

Leopold came to my dressing-room, and broke the dreadful news to me that my good, faithful Brown had passed away early this morning. Am terribly upset by this loss, which removes one who was so devoted and attached to my service and who did so much for my personal comfort. It is the loss not only of a servant, but of a real friend.

To Ponsonby   3 April 1883

The Queen is trying hard to occupy herself but she is utterly crushed and her life has again sustained one of those shocks like in 1861 when every link has been shaken and torn and at every turn and every moment the loss of the strong arm and wise advice, warm heart and cheery original way of saying things and the sympathy in any large and small circumstances—is most cruelly missed. She hopes to see Sir Henry for a moment this evening but he must be prepared to find her very nervous and very much shaken.

The Queen can't walk the least and the shock she has sustained has made her very weak—so that she can't stand.

To the Crown Princess of Prussia   4 April 1883

The terrible blow which has fallen so unexpectedly on me—and has crushed me—by tearing away from me not only the most devoted, faithful, intelligent and confidential servant and attendant who lived and, I may say (as he overworked himself) died for me—but my dearest best friend has so shaken me—still quite helpless as I am—prevented my writing on Wednesday last. The shock—the blow, the blank, the constant missing at every turn of the one strong, powerful reliable arm and head almost stunned me and I am truly overwhelmed.

The sympathy is universal—the appreciation of his noble, grand and yet simple nature—true and great—which is soothing. To be tied to my chair*, not able to go

---

*The Queen had fallen downstairs at Windsor and this had caused a series of rheumatic attacks.

and see after him while he was ill or go about now from one chair to another even, has greatly added to the cruel suffering and sorrow.

8 April 1883

I am crushed by the violence of this unexpected blow which was such a shock—the reopening of old wounds and the infliction of a new very deep one. There is no rebound left to recover from it and the one who since 1864 had helped to cheer me, to smooth, ease and facilitate everything for my daily comfort and who was my dearest best friend to whom I could speak quite openly is not here to help me out of it! I feel so stunned and bewildered and this anguish that comes over me like a wave every now and then through the day or at night is terrible! He protected me so, was so powerful and strong—that I felt so safe! And now all, all is gone in this world and all seems unhinged again in thousands of ways!—I feel so discouraged that it requires a terrible effort to bear up at all against it.

To Princess Victoria of Hesse   19 June 1883

You will I know be anxious to know what I think [about the Princess's engagement to Prince Louis of Battenberg]? I think that you have done well to choose only a Husband who is quite of your way of thinking and who in many respects is as English as you are—whose interests must be the same as yours and who dear Mama liked ... One only drawback I see—and that is 'the fortune'. I don't think riches make happiness, or that they are necessary, but I do think a certain amount is a necessity so as to be independent. And this I hope you will be able to reassure me upon—and have well thought of before you take the irrevocable step ...   I am still very lame and suffer from neuralgic pains—My spirits are the same—and it is in short the constant missing of that merry buoyant 'nature' of my dearest Brown which depresses me so terribly and which makes everything so sad and joyless!—And this I carry everywhere about with me. He helped in so many things—more perhaps here [at Balmoral] than any where.

Journal   7 August 1883

After luncheon saw the great Poet Tennyson, who remained nearly an hour, and most interesting it was. He is grown very old, his eyesight much impaired, and he is very shaky on his legs. But he was very kind, and his conversation was most agreeable. He spoke of the many friends he had lost, and what it would be if he did not feel and know that there was another world where there would be no partings, of his horror of unbelievers and philosophers, who would try to make one believe there was no other world, no immortality, who tried to explain everything away in a miserable manner ... He spoke of the state of Ireland with abhorrence, and the wickedness of ill-using and maiming poor animals. 'I am afraid I think the world is darkened; but I dare say it will be brighter again.' I told him what a comfort In Memoriam had always been to me, which seemed to please him; but he said, I could not believe the numbers of shameful letters of abuse he had received about it. Incredible! When I took leave of him I thanked him for his kindness, and said how much I appreciated it, for I had gone through much, to which he replied, 'You are so alone on that terrible height. I have only a year or two more to live, but I am happy to do anything for you I can.'

TO PRINCESS VICTORIA OF HESSE   22 August 1883

Your dear little Note from Havre on board the *Osborne* gave me great pleasure. To take leave of you at night, and never to see you again (as I fear it will not be possible for me to go to your marriage)—as a dear girl,—as you are now, was very painful to me and I felt it terribly.

Every thing upsets me now and the heavy cloud which overhangs every thing causes partings to be more deeply felt than ever.

Its being at night too was particularly sad. I always hate 'goodbyes' at night or leaving any place at night. I saw the *Osborne* lit up gliding like a meteor over the Solent!

You are so good and sensible that I am sure you will be a steady good wife and not run after amusements, but find your happiness chiefly in your own home. Beware of London and Marlborough House [where the Prince of Wales lived].

There is one thing which I had wished to speak to you about but had no opportunity of doing so, and that is: that I would earnestly warn you against trying to find out the reason for and explanation of everything.

Science can explain many things, but there is a spiritual as well as a material World and this former cannot be explained. We must have faith and trust, and believe in all ruling, all wise and benificent Providence which orders all things. To try and find out the reason for everything is very dangerous and leads to nothing but disappointment and dissatisfaction, unsettling your mind and in the end making you miserable.

TO GLADSTONE   23 August 1883

The Queen feels very strongly that the withdrawal of our troops from Cairo and Egypt must be put off, she believes *sine die*; for the state of utter corruption and helplessness there is quite dreadful.

JOURNAL   29 August 1883

Went in the pony chair to the Kirkyard where I visited good Brown's grave and looked at the granite stone I have had placed over it which is simple and nice, and I laid a wreath there. It always upsets and makes me sad to realise that that excellent and faithful servant is really gone, never to return.

TO PONSONBY   16 September 1883

The Queen is a good deal surprised and she must say annoyed at Mr. Gladstone's 'Progress'. He 'was not to cross the border' and yet he has been landing and receiving Addresses from many places in Scotland and now is off to Norway.

The Queen thinks it most unusual and not she thinks respectful towards herself that the Prime Minister should go to a foreign country without mentioning it to the Sovereign. And she thinks considering the extraordinary and tactless publicity given to every single movement and trifling act of his, his presence in Norway when affairs are very critical seems indiscreet and ill-judged.

Sir H. Ponsonby is in such frequent correspondence with Mr. Gladstone's Private Secretaries that she wishes him to tell them what she feels (and she does so very strongly) on this subject. Why does Mr. Gladstone not try to stop this sort of Court Circular which the Queen believes never appeared for any other Prime Minister and which is not approved of by many.

TO THE CROWN PRINCESS OF PRUSSIA   2 October 1883

Mr Gladstone's journey well deserves blame. But can you imagine that he is so ignorant of what is going on that he never knew till the pilot told him when they were on the way to Copenhagen that the Emperor of Russia was there? He trusts Russia, hates Austria and don't like Germany. Republican France and Italy are all he cares for.

TO GRANVILLE   22 October 1883

The Queen has no doubt that Lord Granville feels as Lord Palmerston did; who with all his many faults, had the honour and power of his country strongly at heart, and so had Lord Beaconsfield. But she does not feel that Mr. Gladstone has. Or at least he puts the House of Commons and party first; thinking no doubt that he is doing what is best by keeping this country out of everything and swallowing offences.

TO GLADSTONE   30 October 1883

The Queen has been much distressed by all she has heard and read lately of the deplorable condition of the Homes of the Poor in our great towns ... She cannot but think that there are questions of less importance than this, which are under discussion, and might wait till one involving the very existence of thousands—nay millions—had been fully considered by the Government.

TO THE CROWN PRINCESS OF PRUSSIA   6 November 1883

It is very trying to be able to walk so very little out of doors; ten minutes or a quarter of an hour is the outside I can do but the colder weather does me good. You never will take in, how infirm and shaken I still am. I think as you don't wish to believe it you won't, and never allude to it!*

JOURNAL   9 November 1883

Dear Bertie's 42nd birthday. May God bless and long preserve him for the good of his country! Warm-hearted, kind, and amiable, he is always a very good son to me.

                                                    16 January 1884

Saw some soldiers dressed in Khaki or even a rather darker, very ugly *café au lait* colour, which is proposed for a service and fighting dress, but which is very ugly and

---

*The Princess had not been able to bring herself to express much more than formal regrets about John Brown's death.

I do not want to hear of, excepting for *foreign* service in hot climates. The whole dress, including helmet, is of the same colour.

After the proclamation of a holy war, the Mahdi's forces had overrun the Sudan and had laid siege to Khartoum. General Gordon was sent out to evacuate Egyptian forces there.

To Sir Evelyn Wood, C in C of the Egyptian Army
1 February 1884

Much has taken place since the last letter she received, and what Sir Evelyn suggested in one of his letters as the best solution of the Soudan difficulty, has taken place, viz.: the employment of General (China) Gordon. Why this was not done long ago and why the right thing is never done till it is absolutely extorted from those who are in authority, is inexplicable to the Queen.

To the Crown Princess of Prussia
6 February 1884

Mr Gladstone cares little for and understands still less foreign affairs. He is a great optimist and thinks all is doing well. Lord Granville is absolutely *passé* and *baissé* and neglects things (not answering them even) in a dreadful way. Lord Hartington is very idle and hates business! And so on. God knows what is to happen next.

To Gladstone
9 February 1884

The Queen has to thank Mr. Gladstone for several letters. She feels very strongly about the Soudan and Egypt, and she must say she thinks a blow must be struck, or we shall never be able to convince the Mohammedans that they have not beaten us. These are wild Arabs and they would not stand against regular good troops at all.

We must make a demonstration of strength and show determination, and we must not let this fine and fruitful country, with its peaceable inhabitants, be left a prey to murder and rapine and utter confusion. It would be a disgrace to the British name, and the country will not stand it.

The Queen trembles for General Gordon's safety. If anything befalls him, the result will be awful.

12 February 1884

I am glad that my Government are prepared to act with energy at last.

May it not be too late to save other lives!

To the Crown Princess of Prussia
20 February 1884

Though I know it don't interest you much, it will still gratify you to hear that the enthusiasm and affectionate sympathy with which my book* has been received is

*More Leaves from the Journal of A Life in the Highlands From 1862 to 1882. It was dedicated to 'My Loyal Highlanders and especially to the memory of my devoted personal attendant and faithful friend John Brown.'

most touching and gratifying. There have been some 30 or 40 different articles—really beautifully written. So many from the country—county papers—such as appreciation of all I have gone through.

JOURNAL   21 February 1884

Leopold started for Cannes to stay at the Villa Nevada there, as he thinks he requires a little change and warmth, but he is going alone, as Helen's health does not allow her to travel just now. I think it rather a pity that he should leave her.

TO HARTINGTON   25 March 1884

Gordon is in danger: you are bound to try and save him. Surely Indian troops might go from Aden: they could bear the climate. You have incurred fearful responsibility.

TO PONSONBY   27 March 1884

The Queen merely wished to cypher as follows to Mr. Gladstone: 'Am greatly distressed at the news about Gordon ... You told me, when I last saw you, Gordon must be trusted and supported; and yet what he asked for repeatedly nearly 5 weeks ago has been refused. If not only for humanity's sake, for the honour of the Government and the nation, he must not be abandoned!' ...

The Queen has no confidence in Lord Granville; he is as weak as water.

JOURNAL   28 March 1884

Another awful blow has fallen upon me and all of us to-day. My beloved Leopold, that bright, clever son, who had so many times recovered from such fearful illnesses, and from various small accidents, has been taken from us!* To lose another dear child, far from me, and one who was so gifted, and such a help to me, is too dreadful!

Am utterly crushed. How dear he was to me, how I had watched over him! Oh! what grief, and that poor loving young wife, who has been kept on her sofa, more or less since the middle of January, for fear of any accident, how may this news affect her! Too, too dreadful!

The poor dear boy's life had been a very tried one, from early childhood [he suffered from haemophilia]! He was such a dear charming companion, so entirely the 'Child of the House.' ...

I went back to rest a little, feeling stunned, bewildered, and wretched. I am a poor desolate old woman, and my cup of sorrow overflows! Oh! God, in His mercy, spare my other dear children!

                                                             29 March 1884

Had a very kind letter from Bertie, who asked me to let him go to Cannes to fetch the precious remains, to which I consented. When I awoke, I cried very much, feeling all pleasure has gone for ever for me.

Lenchen came up and told me that a letter from dear Leopold to Helen had been

*A fall in a club-house in Cannes had led to a brain haemorrhage of which Prince Leopold died at the Villa Nevada.

found, with his last directions, expressing a wish to be buried in St. George's Chapel, as he had been married there, and because there would always be singing over him! Of course I could say nothing against this, as I consider a wish of that kind as sacred. But personally, I should have liked the Mausoleum.

To Ponsonby   17 May 1884

The conduct of the Government in this Egyptian business is perfectly miserable; it is universally condemned; and this weakness and vacillation have made us despised everywhere. The Queen must make an effort to save her country from disgrace; if they must make a declaration it must be with this provision, after the five years are mentioned, 'Provided the state of Egypt is such as to enable us to leave the country without the danger of anarchy and confusion.' . . .

The Queen feels much aggrieved and annoyed. She was never listened to, or her advice followed, and all she foretold invariably happened and what she urged was done when too late! It is dreadful for her to see how we are going downhill, and to be unable to prevent the humiliation of this country.

To Gladstone   5 June 1884

The Queen cannot alter her decided opinion that to put any limit to our occupation of Egypt is a very fatal mistake.

To the Crown Princess of Prussia   10 June 1884

I am so powerless—the Ministers are so weak and obstinate that I have no hopes of any kind. To be a sovereign and to be unable to prevent grievous mistakes is a very hard and ungrateful task. This Government is the worst I have ever had to do with. They never listen to anything I say and commit grievous errors.

To Gladstone   15 July 1884

The Queen thanks Mr. Gladstone for his letter received this morning*. She is sorry that she cannot agree with Mr. Gladstone in his opinion of the House of Lords, which has rendered such important services to the nation, and which at this moment is believed by many to represent the true feeling of the country. The House of Lords is in no way opposed to the people.

The existence of an independent body of men acting solely for the good of the country, and free from the terror which forces so many Commoners to vote against their consciences, is an element of strength in the State, and a guarantee for its welfare and freedom.

To Gladstone   22 October 1884

The Queen must again call Mr. Gladstone's attention to Mr. Chamberlain's

---

*Gladstone had written: 'Mr. Gladstone knows, like the rest of the world, how formidable an opponent the House of Lords has habitually been, and especially for the last thirty years, to the Liberal policy, which has had the nearly uniform assent of the nation. He perceives with pain that the tendency of the Lords to separate from the people becomes more marked with the lapse of years, indicated as it is by the increase of the Tory majority in that House.'

speeches. He approves of the disgraceful riot at Birmingham [where Liberal demonstrators had broken up a Conservative meeting]! If a Cabinet Minister makes use of such language, and sets the Prime Minister's injunctions at defiance, he ought not to remain in the Cabinet. His language if not disavowed justifies the worst apprehensions of the Opposition.

TO THE CROWN PRINCESS OF PRUSSIA   21 November 1884

I must write you a line from this dreary, gloomy old place [Windsor Castle]—to wish you again joy of your dear birthday, which makes me think much of former days and of your birth on a dark, dull, windy, rainy day with smoking chimneys, and dear Papa's great kindness and anxiety—though he was disappointed you were not a boy.

27 November 1884

Dear Helen is wonderfully well and so good such an example to all, I do love respect and admire her, poor darling—always a kind, sweet smile on her poor, sad face and cheerful and always thinking of others and not of herself. But it is heartbreaking to see her and still more almost, the dear innocent pretty merry little children. Little Alice is very intelligent and such a healthy child.

30 December 1884

Lenchen and Beatrice have both written to you, and the former has told you of the pain it has caused me that my darling Beatrice should wish (which she never did till she had lost her dear brother) to marry, as I hate marriages especially of my daughters, but as I like Liko [Prince Henry of Battenberg] very much and as they are both so very much devoted to each other, and she remains always with me, I cannot refuse my consent.

3 January 1885

Your second letter pleased me as I saw you take the right view of darling Beatrice's engagement with Liko. But you who are so fond of marriages which I (on the other hand) detest beyond words, cannot imagine what agonies, what despair it caused me and what a fearful shock it was to me when I first heard of her wish! It made me quite ill.

7 January 1885

I am surprised at myself—considering the horror and dislike of the most violent kind I had for the idea of my precious Baby's marrying at all (never personally against dear Liko) how I should have been so much reconciled to it now that it is settled. But it is really Liko himself who has so completely won my heart. He is so modest, so full of consideration for me and so is she, and both are quietly and really sensibly happy. There is no kissing, etc. (which Beatrice dislikes) which used to try me so with dear Fritz. But the wedding day is like a great trial and I hope and pray there may be no results! That would aggravate everything and besides make me terribly anxious.

Now I must tell you how very unamiably the Empress and even dear Fritz have written to me [about Princess Beatrice's engagement to Prince Henry of Battenberg]. I think really the Empress has no right to write to me in that tone. I send you copies of her letter and my rather stern answer but I cannot swallow affronts.

Please return them to me and ask Fritz to let you see what I wrote to him, for his letter was very cold and not what it should be to me.

Dear Fritz speaks of Liko as not being of the blood—a little like about animals. Don't mind my saying this . . .*

Lord Derby [now Colonial Secretary] (as I foresaw) has made dreadful messes. He is a terrible Minister, for he won't do anything to prevent wars and complications.

JOURNAL   8 January 1885

Eddy's 21st birthday. It seems quite like a dream, and but so short a while ago, that I hurried across from Osborne to Windsor, or rather Frogmore, to find that poor little bit of a thing, wrapped in cotton! May God bless him and may he remain good and unspoilt, as he is!

TO THE CROWN PRINCESS OF PRUSSIA   10 January 1885

I have to thank you for your letter of the 6th in which you tell me of the extraordinary impertinence and insolence, and I must add, great unkindness of Willie, that foolish Dona [Prince William's wife]—and of Henry [Prince William's younger brother]. It is most impertinent and I shall not write to either. As for Dona, a poor, little, insignificant Princess raised entirely by your kindness to the position she is in, I have no words [she had been Princess Augusta of Schleswig-Holstein-Sonderburg-Augustenburg before her marriage to Prince William].**

Lord Granville very truly said with respect to the extraordinary behaviour of the Emperor and Empress (but much more of the latter) and your naughty, foolish sons that if the Queen of England thinks a person good enough for her daughter what have other people got to say? Think if I was to do such a thing! No other sovereign in the world would do such a thing. All have telegraphed so kindly. I own I do resent it. And I do feel angry with dear Fritz too. I think Liko the handsomest of the three handsome brothers.

                                                                                14 January 1885

Don't fear that I should be angry with dear Fritz—for I know his attachment to all of you. But your children or rather sons' behaviour and Dona's impertinence I do resent and shall do so. I had intended writing to both but shall not do so and wish that they should know why I do not do so. I had got a present for Willie's birthday. Shall I send it to him or not? I shall certainly not write to him.

*Prince Henry's father had morganatically married a Countess and thus deprived his sons of their Hessian rank. The Queen bestowed upon him the title of His Royal Highness for use in England, but her German relations refused to recognise it.

**They had all expressed their disapproval of the marriage of Princess Beatrice to Prince Henry of Battenberg.

To Ponsonby   24 January 1885

The Queen always has telegraphed direct to her Generals, and always will do so, as they value that and don't care near so much for a mere official message . . .

She thinks Lord Hartington's letter very officious and impertinent in tone.* The Queen has the right to telegraph congratulations and enquiries to any one, and won't stand dictation. She won't be a machine. But the Liberals always wish to make her feel that, and she won't accept it.

Journal   5 February 1885

Dreadful news after breakfast. Khartoum fallen, Gordon's fate uncertain! All greatly distressed. Sent for Sir H. Ponsonby, who was horrified. It is too fearful. The Government is alone to blame, by refusing to send the expedition till it was too late. Telegraphed *en clair* to Mr. Gladstone, Lord Granville, and Lord Hartington, expressing how dreadfully shocked I was at the news, all the more so when one felt it might have been prevented. Saw Sir H. Ponsonby, who is to go to London to speak to Mr. Gladstone, Lord Hartington and others, about the alarming state of affairs.

To Lord Hartington   5 February 1885

These news from Khartoum are frightful, and to think that all this might have been prevented and many precious lives saved by earlier action is too frightful.

Express to Lord Wolseley my great sorrow and anxiety at these news, and my sympathy with Lord Wolseley in this great anxiety; pray, but have little hope, brave Gordon may yet be alive.

To Ponsonby   5 February 1885

It seems to me these are the vital questions to be considered.

First. Absolutely necessary to leave no stone unturned to ascertain Gordon's fate. We are bound in honour and respect to him to do that.

Secondly. We must not retire without making our power felt.

Thirdly. Some means must be found to try and place some sort of Government at Khartoum, or try to treat with the rebels.

If we merely turn straight back again, our object having been defeated by the vacillation and delays of the Government, our position in sending out the Expedition, and our power in the East will be ruined; and we shall never be able to hold our heads up again! The country will be furious! and we are bound to show a bold front. Tame submission would oblige us very likely to fight in some other direction in Egypt soon again. Such an ending as this would be fatal.

Something strong must be written to the Cabinet.

To Gladstone   6 February 1885

It is absolutely necessary that we must ascertain Gordon's fate.

Trust the Cabinet will promptly agree to a bold and decided course. Hesitation

*Hartington had complained to Ponsonby of the Queen's telegraphing direct to Lord Wolseley: 'I cannot help thinking,' he had written, 'that it would on the whole be most convenient that any message from the Queen should be sent through the Secretary of State.'

and half measures would be disastrous.

What is Lord Wolseley's and Sir E. Baring's [Consul-General in Egypt] advice? Have you consulted the Duke of Cambridge and Lord Napier on the military situation?

TO PONSONBY   17 February 1885

Mr. Gladstone and the Government have—the Queen feels it dreadfully— Gordon's innocent, noble, heroic blood on their consciences. No one who reflects on how he was sent out, how he was refused, can deny it! It is awful ... May they feel it, and may they be made to do so! ...

Pray read this last letter of Lord Wolseley's, and what he says about delay. If Mr. Gladstone tries to throw it on him, the Queen herself will remind Mr. Gladstone, and hopes Lord Wolseley will be indignant!

TO HARCOURT   23 February 1885

I am horrified at the disgraceful scene at Exeter, at Lee's execution. Surely he cannot now be executed? It would be too cruel. Imprisonment for life seems the only alternative. But since this new executioner has taken it in hand there have been several accidents. Surely some safe and certain means could be devised which would make it quite sure. It should be of iron not wood, and such scenes must not recur.*

TO LADY WOLSELEY   3 March 1885

In strict confidence I must tell you I think the Government are more incorrigible than ever, and I do think that your husband should hold strong language to them, and even threaten to resign if he does not receive strong support and liberty of action.

I have written very strongly to the Prime Minister and others, and I tell you this; but it must never appear, or Lord Wolseley ever let out the hint I give you. But I really think they must be frightened.

TO LORD WOLSELEY   31 March 1885

Whatever happens, the Queen hopes and trusts Lord Wolseley will resist and strongly oppose all idea of retreat! His words on that subject, both in one of his telegraphs when asked whether he could do so, and also in his despatch of the 1st of March, have plainly spoken out on this point. But she fears some of the Government are very unpatriotic, and do not feel what is a necessity. This and the absolute necessity of having a good Government at Khartoum the Queen trusts Lord Wolseley will insist on. But then comes the health of the troops. Those at Suakin (whom she fears have been pretty harassed) as well as those at Dongola must if possible have change and be moved, for it would be too dreadful to lose many by sickness! Altogether the Queen's heart is sorely troubled for her brave soldiers ... Our soldiers fight and have on every single occasion in this exceptionally trying

*John Lee had been sentenced to death for murder. Three attempts to hang him failed, owing to a defect in the mechanism. His sentence was commuted to penal servitude for life.

campaign fought like heroes individually, and she hopes he will tell them so from the highest to the lowest from her.

The Queen would ask Lord Wolseley to destroy this letter as it is so very confidential, though it contains nothing which she has not said to her Ministers and over and over again.

To GLADSTONE   14 April 1885

Think it would be fatal to our reputation and honour if we were to abandon active operations in and still more to withdraw from the Soudan. It would be such an exhibition of weakness, and of the triumph of savages over British arms, that it would seriously affect our position in India and elsewhere.

To PONSONBY   15 April 1885

With respect to Mr. Gladstone the Queen does feel she is always kept in the dark.

In Lord Melbourne's time she knew everything that passed in the Cabinet and the different views that were entertained by the different Ministers and there was no concealment. Sir R. Peel who was completely master of his Cabinet (and the Prime Minister ought to be) was, after the first strangeness for her [who] hardly knew him, also very open. Lord Russell less communicative but still far more so than Mr. Gladstone and Lord Palmerston too. They mentioned the names of the Ministers and their views. Lord Palmerston again kept his Cabinet in great order. Lord Derby was also entirely master of his Cabinet. Lord Aberdeen most confidential and open and kind—Lord Beaconsfield was like Lord Melbourne. He told the Queen everything (he often did not see her for months) and said: 'I wish you to know everything so that you may be able to judge.' Mr. Gladstone never once has told her the different views of his colleagues. She is kept completely in the dark.

JOURNAL   19 April 1885

The anniversary of dear Lord Beaconsfield's death. Oh! were he but still alive!

To GRANVILLE   28 April 1885

The last fortnight has given [the Queen] terrible trouble and anxiety, which, if Lord Granville saw any of her very distinct telegrams to Mr. Gladstone (not one of which were in the slightest degree listened to by Mr. Gladstone) he will easily understand. The way in which the expedition to Egypt, or rather the Soudan, has been given up, the way in which public fancy and whims (which are known to be quite unreliable) seem to govern Mr. Gladstone's actions, make all maintenance of principle, all dignity and prudence, impossible, and the Queen must say that she is quite at a loss how to go on and communicate with a Minister who never will consider the effect of all his constant shiftings and changes on the country and on the whole world.

Moreover, he is so reserved and writes such unsatisfying letters, that the Queen never knows where she is. She does not know who takes his or other views (which all his predecessors kept her informed of) and she is left powerless to judge of the state of affairs!

To Hartington   17 May 1885

Though the Queen has little if any hope of bringing the Cabinet to see what is right and what is due to the honour of England, and may be even to the safety of our Eastern possessions, she cannot satisfy her conscience without making one more appeal to Lord Hartington ... Does Parliament know that you are acting against the earnest advice of the civil as well as military authorities and of those who can alone be fit judges of the state of affairs? Can you go on persevering blindly in your own opinion without letting Parliament know the truth?

As the present Government seem to act entirely according to the dictates of Parliament irrespective of any settled policy, Parliament should be told the truth and at once ...

To see her brave soldiers as the Queen did yesterday gashed and mutilated for nothing is dreadful! And to see for the second time our troops recalled and retreating before savages—probably and most probably only to have to send them out again in a little while—is to make us the laughing-stock of the world! For military reasons the strongly expressed opinion of the Generals should be listened to!

The Queen writes strongly, but she cannot resign if matters go ill, and her heart bleeds to see such short-sighted humiliating policy pursued, which lowers her country before the whole world.

To Ponsonby   18 May 1885

We are becoming the laughing-stock of the world!

The Queen is quite miserable ... Can the G.O.M. [Grand Old Man, Gladstone] not be roused to some sense of honour?

To the Crown Princess of Prussia   31 May 1885

I am so thankful that you do see the blind folly and weakness of my present Government and that you saw and were pleased with Lord Rosebery, for I like him, and dear Leopold liked him very much. He is quite the rising politician of the day and is always very respectful and anxious to please me.

Journal   9 June 1885

Heard by telegram that the Government were defeated by 12 on the Budget. When we were at breakfast heard from Mr. Gladstone that a Cabinet was summoned. This sounded serious. After luncheon received a telegram from Mr. Gladstone resigning.

                                                                10 June 1885

Mr. Gladstone is to be finally asked whether he considers his resignation as definite, in which case I should accept it.

To Gladstone   11 June 1885

While fully appreciating the reason of Mr. Gladstone's state of health precluding his coming here [Balmoral], which she certainly expected he would have done, she much regrets his not having at once offered to send a Minister up here to communicate personally with her ... The Queen, however, has now sent for Lord Salisbury, having accepted Mr. Gladstone's resignation ...

With respect to the Queen's return south, she must observe first, that the Railway authorities, unless previously warned, do not consider it safe for her to start without some days' notice. Secondly, that the Queen is a lady nearer 70 than 60, whose health and strength have been most severely taxed during the 48 years of her arduous reign, and that she is quite unable to rush about as a younger person and a man could do. And lastly it is extremely inconvenient and unpleasant from the noise and great crowds at Windsor during the Ascot week for her to be there, and for 24 years the Queen has carefully avoided being there at that time. However, if she finds it necessary, the Queen will return early next week to Windsor. She is not feeling strong, and must husband her strength for the fatigues she has before her.

JOURNAL   12 June 1885

After luncheon Lord Salisbury arrived, and I saw him very soon afterwards. I said I hoped he was not tired and began speaking at once about the resignation of the Government, and my wish that he should form one. He replied that he thought the Government ought to have remained on, but of course that he would not refuse to come to my assistance in the present difficulty.

TO GLADSTONE   13 June 1885

Mr. Gladstone mentioned in his last letter but one, his intention of proposing some honours. But before she considers these, she wishes to offer him an Earldom, as a mark of her recognition of his long and distinguished services, and she believes and thinks he will thereby be enabled still to render great service to his sovereign and country—which if he retired, as he has repeatedly told her of late he intended to do shortly,—he could not. The country would doubtless be pleased at any signal mark of recognition of Mr. Gladstone's long and eminent services, and the Queen believes that it would be beneficial to his health,—no longer exposing him to the pressure from without, for more active work than he ought to undertake.

JOURNAL   17 June 1885

At 4, saw Lord Salisbury, who said that it was 'hard work,' but that he had got over the principal difficulties.

After luncheon saw Capt. Bigge about the honours Mr. Gladstone wishes for, which are very numerous. At quarter past 4 received Lord Salisbury, who seemed pleased that things were at last settled, and thanked me. But he feared he had given me a great deal of trouble, which I did not deny.

TO PONSONBY   12 July 1885

The Queen must ask that both Sir Henry and Major Edwards [Assistant Private Secretary] should not be out on Sunday morning or any other at the same time. Not 5 minutes after the service in the Chapel was over she sent to say she wished to see Sir Henry in a ¼ of an hour but was told he was gone to church. She then sent for Major Edwards and was then told he was out too. This is extremely inconvenient ...

She must ask Sir Henry to take care that this does not happen again and that by 12 one or other of the gentlemen . . . should be at hand.

There is now especially so much of a public and private nature wanting arrangement and constant attention [she] must have the necessary assistance she requires, being quite done up with all her work and anxieties for the last 6 or 7 weeks.

JOURNAL   23 July 1885

A happier-looking couple [Princess Beatrice and Prince Henry of Battenberg] could seldom be seen kneeling at the altar together. It was very touching. I stood very close to my dear child, who looked very sweet, pure, and calm. Though I stood for the ninth time near a child and for the fifth time near a daughter, at the altar, I think I never felt more deeply than I did on this occasion, though full of confidence. When the Blessing had been given, I tenderly embraced my darling 'Baby'.

TO THE CROWN PRINCESS OF PRUSSIA   25 July 1885

I bore up bravely till the departure and then fairly gave way. I remained quietly upstairs and when I heard the cheering and 'God Save The Queen' I stopped my ears and cried bitterly.

TO VICTORIA, PRINCESS LOUIS OF BATTENBERG   21 August 1885

Let me again ask you to remember that your first duty is to your dear and most devoted Husband to whom you can never be kind enough and to whom I think a little more tenderness is due sometimes.

Lastly let me add one word which as your Godmother as well as Grandmama I may. It is not to neglect going to Church or to read some good and serious religious book, not materialistic and controversial ones—for they are very bad for everyone—but especially for young people.

And now darling Victoria, I will end my long lecture with a kiss to you and Baby and many loves to dear Papa and sisters

Ever your devoted loving Grandmama

TO GLADSTONE   2 October 1885

The Queen trusts Mr. Gladstone is recovering from the hoarseness with which he has been troubled for so many months, and takes this opportunity of expressing a hope that he will spare himself from speaking at public meetings for some time to come.

TO MRS. GLADSTONE   4 January 1886

You must both rejoice at Mr. Gladstone's rest—which he so often spoke of as his great wish and which is essential at his time of life, when overwork and excitement are always detrimental to health.

To G. J. GOSCHEN [a leading opponent of Home Rule for Ireland]
24 January 1886

Mr. Gladstone's speech was very unsatisfactory. While speaking for the union under the Crown of the two countries, he did not speak out, or retract any of his ambiguous utterances, which have caused so much alarm; and the way in which the Irish cheered him shows what his real leanings are supposed to be.

13

IRISH REBELS
AND GERMAN RELATIONS
1886–1891

Lord Salisbury's government survived for only a few months; and in 1886 the Queen had once more to accept Mr. Gladstone as her Prime Minister.

MEMORANDUM   28 January 1886

Saw Lord Salisbury at half-past seven this evening ... I lamented greatly what had occurred [the defeat of the Government] ...

Said if Mr. Gladstone came in I would refuse objectionable people. Sir C. Dilke, of course, I must and would never accept on account of his dreadful private character [he was then being cited as co-respondent in a divorce case]; Mr. Chamberlain Lord S. advised me not to refuse, for it would not be understood and make him a martyr; nor Mr. Morley, a clever and extreme man; but, on observing I did not wish, and meant to object to, Lord Granville as Foreign Secretary, he said I should be 'perfectly justified' in doing so, and thought Lord Rosebery would do very well ... I omitted to state that he said I naturally would object to Mr. Labouchere [a prominent Radical] as a Minister (if Mr. Gladstone dared propose him) and to any Separatist [Home Ruler].

What a dreadful thing to lose such a man as Lord Salisbury for the country, the world, and me!

JOURNAL   30 January 1886

Received a telegram early (at 9.30) from Sir H. Ponsonby saying Mr. Gladstone had 'accepted' (alas!), and soon after came another telegram saying that Mr. Gladstone thought he could form a Government.

TO GOSCHEN   31 January 1886

Mr. Gladstone really intends to bring forward Home Rule, and I trust that the line you and Lord Hartington and other influential people will take will soon bring a good many Liberals to their senses and open their eyes.

In fact, Mr. Gladstone is in the hands of Mr. Parnell without his knowing it. You must be strong and courageous and show publicly what your views are, and you will be supported.

JOURNAL   1 February 1886

Saw Mr. Gladstone before luncheon. He looked very pale when he first came in, and there was a momentary pause, and he sighed deeply.

To GLADSTONE   11 February 1886

The Queen cannot sufficiently express her indignation at the monstrous riot [in Trafalgar Square]* which took place the other day in London, and which risked people's lives and was a momentary triumph of socialism and disgrace to the capital. If steps, and very strong ones, are not speedily taken to put these proceedings down with a high hand, to punish severely the real ringleaders, and to 'probe to the bottom,' as Mr. Gladstone has promised, the whole affair, the Government will suffer severely. The effect abroad is already very humiliating to this country.

To LORD ROSEBERY   13 February 1886

The Queen is delighted to be able to assist Lord Rosebery in his very difficult task [as Foreign Secretary in Gladstone's Cabinet], which he has begun so well. She has nearly fifty years' experience, and has always watched particularly and personally over foreign affairs, and therefore knows them well . . .

All the Powers are very suspicious of a Government at the head of which Mr. Gladstone's name appears. But Lord Rosebery has done much to dispel this already.

To PONSONBY   26 February 1886

The Queen has been very much hurt by an article in the *St. James's* imputing to her the bad state of Society and making out there had been no Courts for 20 years!! Whereas excepting Balls and parties and going to Theatres and living in Town the Queen neglected nothing and it is most disheartening and ungrateful to almost threaten her if she does not do much more, when morally and physically she is doing much more than she can bear with her overwhelming work. It is very wrong. She never went out into general Society before 1861 and never would have done it.

Since her illness in 71—these attacks had ceased but now when she is older and far less fit to do things, they have begun again.

JOURNAL   19 April 1886

Primrose Day! Already five long years since good Lord Beaconsfield was taken.

                                                          4 May 1886

Got out at the entrance to the Exhibition [the Colonial and Indian Exhibition at South Kensington], amidst great acclamations, the flourish of trumpets and the playing of God Save the Queen. Bertie and Arthur met us here . . . Of the Exhibition itself, and the things in it, it was impossible to judge, for excepting some very high towering trophies one could see nothing, on account of the masses of people standing on either side, as we walked along. Everything seemed beautifully

*The rioters were not the 'deserving unemployed', the Queen told her daughter but 'horrid thieves and a few socialists'.

arranged and the people all looked much pleased. Bands stationed at different points played as we walked along. How pleased my darling husband would have been at the whole thing, and who knows but that his pure bright spirit looks down upon his poor little wife, his children and children's children, with pleasure, on the development of his work! The walk was very long and fatiguing, though very interesting. Bertie kindly helped me up and down the steps, whenever we came to any . . .

Bertie read a very long Address, to which I read an Answer. Dear Bertie, who was most kind throughout, then kissed my hand. What thoughts of my darling husband came into my mind, who was the originator of the idea of an exhibition, an idea fraught with such fearful difficulties and carried through against such fearful odds! There were many allusions in the Address, as well as Answer, which were full of the dearest and saddest memories, and very agitating to me . . .

We drove to Buckingham Palace, as we had driven from the station, and got there at 1.30. The crowds enormous and most good-humoured and enthusiastic; the heat both inside and out very great. I felt very tired, but much gratified and pleased that all had gone off so well.

To Gladstone   6 May 1886

The Queen is anxious, before leaving for Windsor, to repeat to Mr. Gladstone what she tried to express—but which she thinks perhaps she did not do very clearly—viz.: that her silence on the momentous Irish measures, which he thinks it his duty to bring forward, does not imply her approval, or acquiescence in them . . .

The Queen writes this with pain, as she always wishes to be able to give her Prime Minister her full support, but it is impossible for her to do so, when the union of the Empire is in danger of disintegration and serious disturbance.

Journal   12 May 1886

Alas! the weather [in Liverpool where the Queen was to open an International Exhibition of Navigation and Commerce] had not cleared by luncheon time, and it blew and rained hopelessly. But there was nothing for it but resolutely to brave the elements, put on waterproofs and hold up umbrellas, as the carriages could not be shut . . . The streets along which we drove were beautifully decorated, and there was not a house which had not some motto or flags up. Some of these mottoes were very touching. The crowds were enormous, and every shop window was full of people . . .

Then we moved on again, the rain coming down worse than ever. But it did not seem to reduce the numbers of people or mar their enthusiasm. I never saw anything like it . . . In spite of the weather, the whole thing was a great success, and the wonderful loyalty and enthusiasm displayed, most touching and gratifying. We got home at seven, quite bewildered and my head aching from the incessant perfect roar of cheering.

8 June 1886

Did not sleep well, as I felt so worried and anxious. When I got up a telegram was brought in to me, which gave the news that the Government had been defeated by a majority of thirty! Cannot help feeling relieved, and think it is the best for the interests of the country.

Saw Lord Rosebery after luncheon . . . Lord Rosebery admires everything here [at Balmoral] very much. He is certainly a very clever, pleasant man, and very kind.

20 June 1886

Have entered the fiftieth year of my reign and my Jubilee year. I was upset at the thought of those no longer with me, who would have been so pleased and happy, in particular my beloved husband, to whom I owe everything, who are gone to a happier world.

There were beautiful and most kind articles in *The Times*, *Standard* and *St. James's*. I don't want or like flattery, but I am very thankful and encouraged by these marks of affection and appreciation of my efforts.

TO VICTORIA, PRINCESS LOUIS OF BATTENBERG   14 July 1886

The Elections are going quite wonderfully—No losses these last 2 days at all and nothing but gains! Today 8—the Conservatives will have above 300 and the Unionist Liberals 72. Nearly 130 majority against the G.O.M. who writes more [?] dreadful letters setting class against class—and behaves abominably. I really think he is cracked.

TO GLADSTONE   20 July 1886

The Queen has received Mr. Gladstone's report of the proceedings of the Cabinet held this day, and the tender of his resignation of office consequent on the result of the recent elections.

The Queen will accept this resignation, and has at once sent to Lord Salisbury; but, as he is abroad, there may be a little delay before any arrangement can be finally decided.

JOURNAL   24 July 1886

Saw Lord Salisbury, who is looking remarkably well. We talked over the whole situation. Though Lord Salisbury had expected their party to gain a good deal, they had not been prepared for such successes.

30 July 1886

After luncheon saw Mr. Gladstone, who looked pale and nervous . . . Spoke of education, it being carried too far, and he entirely agreed that it ruined the health of the higher classes uselessly, and rendered the working classes unfitted for good servants and labourers. I then wished him good-bye, shaking hands with him, and he kissed mine.

TO GLADSTONE   31 July 1886

On the occasion of Mr. Gladstone's visit yesterday the Queen did not like to allude to the circumstances which led to his resignation; but she would wish to say a few words in writing. Whatever Mr. Gladstone's personal opinion may be as to the best means of promoting contentment in and restoring order to Ireland, the country has unequivocally decided against his plan, and the new Government will have to devise some other course in due time.

She trusts that his sense of patriotism may make him feel that the kindest and wisest thing he can do for Ireland is to abstain from encouraging agitation by public speeches; which, though not so intended by Mr. Gladstone, may nevertheless increase the excitement and be considered as supporting the violent proceedings of those who do not hesitate to defy the law.

MEMORANDUM   14 August 1886

The Queen wishes to restate in writing the points which she is determined to see carried out, 1st, respecting the dogs, and 2nd, respecting cruelty to animals in general.

I. As regards her poor dear friends the dogs, she would repeat that no dogs should ever be killed by police unless the veterinary surgeon declared they were mad. That dogs who were close to their masters or mistresses or their house door, poor quiet dogs, should be left alone and not molested. A faithful dog will often snap and snarl and bite if interfered with by strangers.

2. The Dogs' Homes should be augmented and enlarged, and the time for keeping them considerably lengthened.

3. Muzzles, except in the cases of very savage dogs, should not be used, nor should dogs be run after and hunted to be caught.

4. The best veterinary surgeon should be consulted as to whatever is best to be done. But no dog should be killed till it is certain he is mad. Fits are no proof of this.

II. As regards cruelty to animals in general, the Queen thinks a commissioner should be appointed to enquire into the abattoirs . . .

Cats should be likewise well cared for in Homes.

To SALISBURY [once more Prime Minister]   26 October 1886

Conclude that we shall not receive the monstrous declaration of the Russian Government of the illegality of the Bulgarian elections without expressing our dissent. Austria's expression of wish to please Russia is contemptible and unwise. A firm front and declaration of disapproval of Russia's pretensions would not have provoked war but done great good. To allow her to bully Europe and trample on Bulgaria is as impolitic as it is wrong. We at least must not do that.

To VICTORIA, PRINCESS LOUIS OF BATTENBERG   23 November 1886

As a rule I like girls best but I did wish it should be a boy [Princess Beatrice's son, Prince Alexander, who had been born at Windsor on 5 November], to be like his dear father and Uncles, and as I thought it would be a pleasure to poor dear Sandro [Prince Alexander of Battenberg] in all his trials and troubles.

To GOSCHEN   24 December 1886

I have so often written to you of late on events of great importance and difficulty that I feel bound to do so again now, and you will doubtless guess why. This resignation of Lord Randolph Churchill [who had hoped it would provoke a popular outcry in his favour] has placed Lord Salisbury in considerable difficulty; and its abruptness and, I am bound to add, the want of respect shown to me and to his colleagues have added to the bad effect which it has produced. Lord Randolph dined at my table on Monday evening, and talked with me about the Session about to commence, and about the procedure, offering to send me the proposed rules for me to see! And that very night at the Castle, he wrote to Lord Salisbury resigning his office! It is unprecedented! . . .

I believe that one of your chief objections to join Lord Salisbury's Government was Lord Randolph Churchill; consequently the difficulty of doing so would no longer be so great.

JOURNAL   28 December 1886

Saw Lord Salisbury, and talked of course a great deal about Lord Randolph's extraordinary conduct. He had evidently made a great mistake, for he had no following, and the excitement at his resignation was beginning to subside.

1 January 1887

A small family party for New Year, quite unusually so. Only Beatrice and Liko. but the children were there. Gave each other cards; even the sweet little baby, who was brought in, had one tied to a string round his little arm. Telegrams pouring in and being sent out, and letters and cards innumerable. Bertie most kindly sent me a Jubilee inkstand, which is the first that has yet been sold. It is the crown, which opens, and on the inside there is a head of me.

TO SALISBURY   2 January 1887

Delighted at good news [Goschen's acceptance of office as Chancellor of Exchequer in Churchill's place] which telegram from Mr. Goschen last night led me to expect. Quite agree as to great importance of this; and, though I should regret seeing any change, I will make only one exception, and that is, that one Member of your Cabinet, who is a real personal friend of mine [Lord Cross, Secretary for India] and has been of great use to me, who was twice nearly sacrificed to please Lord Randolph and was made a Peer which he did not wish, should on no account be moved. On this I must insist.

JOURNAL   12 January 1887

Dreadfully grieved that dear excellent Lord Iddesleigh [until recently Foreign Secretary] has died. It was quite sudden. He fainted at the top of the stairs, at Downing Street, and died in twenty minutes! . . . I felt quite bewildered and stunned, at first, for I could not believe or realise it. People are dreadfully shocked, as Lord Iddesleigh was immensely beloved and respected by all shades of politicians. For Lord Salisbury it is a very severe blow.

TO VICTORIA, PRINCESS LOUIS OF BATTENBERG   2 March 1887

I feel very deeply that my opinion and my advice are never listened to and that it is almost useless to give any.

JOURNAL   4 March 1887

The behaviour of the Irish in the House of Commons is simply dreadful; such
language, such delay... The questions and amendments made, and the
determination to worry the Government to death, are really quite wicked. The Irish
hope to force Home Rule by making themselves as disagreeable as possible.

6 March 1887

Feel very tired and exhausted, being really much overdone; and fell asleep in my
chair, after tea—a very rare thing for me.

23 March 1877

The crowds [in Birmingham] were tremendous, and the greater number were of the
poorest class, excepting those who filled the windows from top to bottom. Though
the crowd were a very rough lot, they were most friendly, and cheered a great deal.

Before entering the train, I expressed to the Mayor in warm terms the
gratification I had experienced at the splendid reception I had received. It was as
good as at Liverpool, but I think that, while there were not such ragged people as we
saw there, there were at Birmingham more generally poor and rough ones. Here the
weather was very different from that terrible rain at Liverpool. What was very
remarkable was that in that very Radical place, amongst such a very rough
population, the enthusiasm and loyalty should have been so great.

*In April 1887, so as to have a holiday before the strain of the Jubilee celebrations, the
Queen went to Cannes and Aix-les-Bains. While on the Continent she paid a visit to the
Grande Chartreuse.*

We were shown where the cells were, and told I should see a young *compatriote*, an
Englishman who had been there for some time. The Grand Prieur unlocked the cell,
which is composed of two small rooms, and the young inmate immediately
appeared, kneeling down and kissing my hand, and saying, 'I am proud to be a
subject of your Majesty.' The first little room looked comfortable enough, and he
had flowers in it. The other contained his bed and two little recesses, in one of which
stood a small altar, where he said he performed his devotions and said his prayers. In
the other deeper recess, with a small window, is the study, containing his books. I
remarked how young he looked, and he answered, 'I am 23,' and that he had been
five years in the Grande Chartreuse, having entered at 18!! I asked if he was
contented, and he replied without hesitation, 'I am very happy.' He is very good-
looking and tall, with rather a delicate complexion and a beautiful, saintly, almost
rapt expression. When we left the cell and were going along the corridor, the
Général said I had seen that the young man was quite content, to which I replied
that it was a pleasure to see people contented, as it was so often not the case.

To Ponsonby   30 April 1887

The Queen approves of these answers, but always wishes the words 'my dear Mother' to be inserted [into speeches made by the Queen's sons and daughters on public occasions]. Not only on this occasion, but always. If Sir Henry thinks that it could come in in any other place better, he can alter it. But the Queen wishes it should never be omitted when her children represent her.

Journal   11 May 1887

We went at once into the Abbey [to discuss the arrangements for the Jubilee Service], and into the Choir, which is one mass of boarding and lumber, so that it looks dreadful. The dais has been put in the same place where the Throne was when the homage took place at my Coronation, and the old Coronation Chair, with the stone from Scone, is placed on it for me to sit in. Various discussions took place, and I hope there will be room for everybody and everything.

                                                              14 May 1887

At half-past three left Windsor for London for the opening of the People's Palace [in Mile End Road, now part of Queen Mary College], accompanied by Beatrice and Liko . . .

From the moment we emerged from the station the crowds were immense and very enthusiastic with a great deal of cheering; in the City especially, it was quite deafening.

Still, what rather damped the effect of the really general and very enthusiastic reception, to me, was the booing and hooting, of perhaps only two or three, now and again, all along the route, evidently sent there on purpose, and frequently the same people, probably Socialists and the worst Irish.

                                                              19 May 1887

Got a cypher telegram from Vicky begging me to send Dr. Morell Mackenzie at once for consultation. We therefore concluded it was Fritz's throat, which has been causing him a good deal of trouble lately. Dr. M. Mackenzie is a celebrated specialist for the throat, and I sent Dr. Reid at once to him to try and get him to start as soon as possible. In the evening got another urgent telegram, asking for the doctor to come without delay, and a distracted letter from poor Vicky, saying that two eminent professors at Berlin had examined dear Fritz's throat in which there is a very small growth, which they had declared to be suspicious, and possibly malignant. They consider that the only safe thing would be to remove it from the outside, a most alarming remedy; that, however, they would not do this without having the first European opinion, which they considered Dr. Mackenzie to be. Greatly distressed, and cannot bear to think of poor darling Vicky's anguish and sorrow. We could hardly believe it; Fritz was otherwise well, but depressed at his loss of voice, and was to know nothing.

                                                              24 May 1887

Could hardly realise the fact of its being my 68th birthday. Beatrice came in quite early to wish me joy, as she generally does, and brought in the sweet little boy, who held a bunch of lilies of the valley tightly in his hand. I thought so much of my former happy birthdays at Osborne. Found a pile of letters on my table, amongst them a very sad and anxious one from poor dear Vicky, which made me very sad.

25 May 1887

Was immensely relieved by a telegram yesterday from Dr. Mackenzie, saying that a second very careful microscopic examination of the small piece he had removed from dear Fritz's throat had proved entirely satifactory.

5 June 1887

A sad and very anxious letter from poor dear Vicky, as the German doctors will insist that Fritz's throat is in a dangerous state, which Dr. M. Mackenzie says it is not. The latter was to arrive to-day or to-morrow, and they hoped soon to leave for England. It is terrible for poor dear Vicky.

8 June 1887

A very satisfactory telegram from Dr. Mackenzie. He had again made a slight operation in the throat, and a further small piece had been detached, and proved on examination to be perfectly healthy.

19 June 1887

Saw Lord Salisbury. He said the state of excitement and preparation in London was quite marvellous; the only anxiety one felt was about the enormous number of people, half a million being expected to come into London.

20 June 1887

The day has come, and I am alone, though surrounded by many dear children. I am writing, after a very fatiguing day, in the garden at Buckingham Palace, where I used to sit so often in former happy days. Fifty years to-day since I came to the Throne! God has mercifully sustained me through many great trials and sorrows.

Had a large family dinner. All the Royalties assembled in the Bow Room, and we dined in the Supper-room, which looked splendid with the buffet covered with the gold plate. The table was a large horseshoe one, with many lights on it. The King of Denmark took me in, and Willy of Greece sat on my other side. The Princes were all in uniform, and the Princesses were all beautifully dressed. Afterwards we went into the Ball-room, where my band played.

21 June 1887

This very eventful day has come and is passed. It will be very difficult to describe it, but all went off admirably. This day, fifty years ago, I had to go with a full Sovereign's escort to St. James's Palace, to appear at my proclamation, which was very painful to me, and is no longer to take place.

The morning was beautiful and bright with a fresh air. Troops began passing early with bands playing, and one heard constant cheering. Breakfasted with

Beatrice, Arthur, Helen, and Liko, in the Chinese room. The scene outside was most animated, and reminded me of the opening of the Great Exhibition, which also took place on a very fine day. Received many beautiful nosegays and presents. As I left the breakfast-room, met the Connaught children and little Willy of Prussia, who is a dear little boy. Then dressed, wearing a dress and bonnet trimmed with white point d'Alençon, diamond ornaments in my bonnet, and pearls round my neck, with all my orders.

At half-past eleven we left the Palace, I driving in a handsomely gilt landau drawn by six of the Creams, with dear Vicky and Alix, who sat on the back seat. Just in front of my carriage rode the 12 Indian officers, and in front of them my 3 sons, 5 sons-in-law, 9 grandsons, and grandsons-in-law. Then came the carriages containing my 3 other daughters, 3 daughters-in-law, granddaughters, one granddaughter-in-law, and some of the suite. All the other Royalties went in a separate procession ... At the door [of Westminster Abbey] I was received by the clergy, with the Archbishop of Canterbury and Dean at their head, in the copes of rich velvet and gold, which had been worn at the Coronation ... The crowds from the Palace gates up to the Abbey were enormous, and there was such an extraordinary outburst of enthusiasm as I had hardly ever seen in London before; all the people seemed to be in such good humour. The old Chelsea Pensioners were in a stand near the Arch. The decorations along Piccadilly were quite beautiful, and there were most touching inscriptions. Seats and platforms were arranged up to the tops of the houses, and such waving of hands. Piccadilly, Regent Street, and Pall Mall were all alike most festively decorated.

I sat alone (oh! without my beloved husband, for whom this would have been such a proud day!) where I sat forty-nine years ago and received the homage of the Princes and Peers, but in the old Coronation Chair of Edward III, with the old stone brought from Scotland, on which the old Kings of Scotland used to be crowned. My robes were beautifully draped on the chair. The service was very well done and arranged. The *Te Deum*, by my darling Albert, sounded beautiful. When the service was concluded, each of my sons, sons-in-law, grandsons (including little Alfred), and grandsons-in-law, stepped forward, bowed, and in succession kissed my hand, I kissing each; and the same with the daughters, daughters-in-law ... granddaughters, and granddaughter-in-law. They curtsied as they came up and I embraced them warmly. It was a very moving moment, and tears were in some of their eyes.

The procession then reformed, and we went out as we came in, resting a moment in the waiting-room, whilst the Princes were all getting on their horses. Came back another way until we got into Piccadilly. The heat of the sun was very great, but there was a good deal of wind, which was a great relief.

We only got back at a quarter to three. Went at once to my room to take off my bonnet and put on my cap. Gave Jubilee brooches to all my daughters, daughters-in-law, granddaughters, granddaughter-in-law, and pins to all my sons, sons-in-law, grandsons, and grandsons-in-law and George Cambridge.

Only at four did we sit down to luncheon, to which all came ... After luncheon, I stood on the small balcony of the Blue Room, which looks out on the garden, and saw the Bluejackets march past. After this we went into the small Ball-room, where the present given me by all my children was placed. It is a very handsome piece of plate ... I felt quite exhausted by this time and ready to faint, so I got into my rolling chair and was rolled back to my room. Here I lay down on the sofa and rested, doing nothing but opening telegrams, coming from every part of the country, so that they could no longer be acknowledged, and this will have to be done through the papers.

I was half dead with fatigue, and after sitting down a moment with Marie of Belgium, slipped away and was rolled back to my room, and to the Chinese room to try and see something of the very general illuminations, but could not see much. The noise of the crowd, which began yesterday, went on till late. Felt truly grateful that all had passed off so admirably, and this never-to-be-forgotten day will always leave the most gratifying and heart-stirring memories behind.

22 June 1887

The illuminations last night are said to have been splendid. Thousands thronged the streets, but there was no disorder. They shouted and sang till quite late, and passed the Palace singing *God Save the Queen* and *Rule Britannia*. Went into the garden for a little while, and on coming home rested. Again a big luncheon in the Dining-room ... I went round and spoke to as many as I could ...

This over, I went through the Blue Drawing-room and Bow room, full of ladies, to the White Drawing-room, equally full. This was a Deputation from the 'Women of England,' who brought me the signatures of the millions who have subscribed to a gift ... From here I passed into the Picture Gallery, where were assembled all the people who came with other presents, which extended down the whole length of the Gallery. Was really greatly touched and gratified.

Rested on the sofa for some time, and took a cup of tea before leaving Buckingham Palace at half-past five ... Enormous and enthusiastic crowds on Constitution Hill and in Hyde Park ... We drove past the statue of Achilles, right on to the grass in the middle of the park, where 30,000 poor children, boys and girls with their schoolmasters and mistresses, were assembled. Tents had been pitched for them to dine in, and all sorts of amusements had been provided for them ... The children sang *God Save the Queen* somewhat out of tune, and then we drove on to Paddington station. All the Princesses came into my saloon.

The train stopped at Slough, and we got out there ... All along the road there were decorations and crowds of people. Before coming to Eton, there was a beautiful triumphal arch, made to look exactly like part of the old College, and boys dressed like old Templars stood on the top of it, playing a regular fanfare. The whole effect was beautiful, lit up by the sun of a bright summer's evening. Stopped at the College to receive two Addresses and the Eton Volunteers were drawn up there. The town was one mass of flags and decorations. We went under the Castle walls up the hill, slowly, amidst great cheering, and stopped at the bottom of Castle Hill, where there was a stand crowded with people and every window and balcony were full of people, Chinese lanterns and preparations for illuminations making a very pretty effect ... Amidst cheering, the ringing of bells, and bands playing, we drove up to the Castle ...

These two days will ever remain indelibly impressed on my mind, with great gratitude to that all-merciful Providence, Who has protected me so long, and to my devoted and loyal people. But how painfully do I miss the dear ones I have lost!

                                                                    29 June 1887

Drove to Buckingham Palace through great crowds, who were as enthusiastic as ever, getting there a little after four. People were beginning to arrive. I went up to my Sitting-room and rested a little, Beatrice having met me at the door, and before five I joined all my family (which is legion!) and the enormous number of foreign guests ... People were spread all over the garden, and there were a number of tents, and a large one for me, in front of which were placed the Indian escort. I walked right round the lawn in front of the Palace with Bertie, and I bowed right and left, talking to as many as I could, but I was dreadfully done up by it and could not speak to, or see, all those I wished.

TO ROSEBERY   21 July 1887

It is impossible for me to say how deeply, immensely touched and gratified I have been and am by the wonderful and so universal enthusiasm displayed by my people, and by high and low, rich and poor, on this remarkable occasion, as well as by the respect shown by Foreign Rulers and their peoples. It is very gratifying and very encouraging for the future, and it shows that fifty years' hard work, anxiety, and care have been appreciated, and that my sympathy with the sorrowing, suffering, and humble is acknowledged ...

Alone I did feel, in the midst of so many, for I could not but miss sadly those who were so near and dear, and who would have so rejoiced in those rejoicings, above all him to whom the nation and I owe so much!

Yesterday afternoon, I was most agreeably surprised by your kind and most valuable present, accompanied by such flattering words.

It is the beautiful little miniature [of Queen Elizabeth] in its quaint setting which you once sent for me to see, and which I shall greatly value, though I fear I have no sympathy with my great predecessor, descended as I am from her rival Queen, whom she so cruelly sacrificed.

In 1887 the Queen for the first time engaged some Indian servants, one of whom in particular, Abdul Karim, was to cause a great deal of trouble by his pretensions. Originally a domestic servant, he was eventually promoted to become Munshi Hafiz Abdul Karim, the Queen's Indian Secretary, a post for which the members of the Household, with some justification, considered him peculiarly ill-qualified.

JOURNAL   23 June 1887

Felt very tired. Drove down to Frogmore with Beatrice to breakfast and met Vicky and young Vicky there. My two Indian servants were there and began to wait. The one, Mohamed Buxsh, very dark with a very smiling expression ... and the other, much younger, called Abdul Karim, is much lighter, tall, and with a fine serious countenance. His father is a native doctor at Agra. They both kissed my feet.

3 August 1887

Am learning a few words of Hindustani to speak to my servants. It is a great interest to me, for both the language and the people. I have naturally never come into real contact with it before.

TO PONSONBY   12 September 1887

The Queen cannot say what a comfort she finds [her Indian servants]. Abdul is most handy in helping when she signs by drying the signatures. He learns with extraordinary assiduity and Mahomet is wonderfully quick and intelligent and understands everything.

JOURNAL   26 August 1887

Dear Fritz came over from Braemar to luncheon, where he has been staying at the Fife Arms. It has done him so much good and he is wonderfully better, still hoarse, but not without any voice, as when he arrived in England. He seemed in excellent spirits.

10 October 1887

An early luncheon, after which dear Bertie left, having had a most pleasant visit, which I think he enjoyed and said so repeatedly. He had not stayed alone with me, excepting for a couple of days in May in 1868, at Balmoral, since he married! He is so kind and affectionate that it is a pleasure to be a little quietly together.

7 November 1887

Dr. Reid brought in a Reuter's telegram, with rather an alarming account of dear Fritz. Sir M. Mackenzie's answer to my telegram of yesterday has come, and is the following: 'Fresh development lower down, exact nature uncertain, but looks unfavourable; have advised Schrotter of Vienna and Krause of Berlin should be consulted; no immediate danger.' This alarms me dreadfully, and we feel it is very anxious. So troubled and anxious about dear Fritz, could hardly think of anything else. God grant we may soon have more cheering news!

9 November 1887

Received a very distressed letter from poor dear Vicky. Sir M. Mackenzie had arrived, and declared the new growth to be of a malignant character. It is too sad, for dear Fritz had really seemed better.

12 November 1887

Received such a heart-broken letter from poor dear Vicky. It is too dreadfully sad, and makes me quite miserable. She says: 'The doctors read to me their protocol,—cruel indeed it sounded. I hardly expected much else, still when the crude facts of one's doom are read out to one, it gives one an awful blow! ... My darling has got a fate before him which I hardly dare think of. How I shall have the strength to bear it, I do not know! . .'

Dr. Reid brought me a letter from Sir M. Mackenzie, which alas! is not very satisfactory, though not quite devoid of hope! The doctors seem to agree that the growth is of a malignant character. As regards the treatment too, there is a difference of opinion. Sir M. Mackenzie advocates simply palliatives to prolong life, but the German surgeons, on the other hand, are in favour of an operation, which they say is not dangerous, and which offers a prospect, or at least a chance, of recovery, with an impaired voice. It is dreadful, but there must still be some hope! Poor darling Vicky, the thought of all her trouble makes my heart ache.

14 November 1887

Dr. Reid showed me his answer to Sir M. Mackenzie, which is excellent. He is so afraid that both sides of the question may not have been duly weighed, for it must be borne in mind that palliatives cannot eradicate the disease, whereas the operation of opening the throat and removing all the growths might do so. The German doctors, as well as some in England, do not consider this operation as very dangerous. It is a terrible state of affairs, and I am haunted by the thought of it.

31 December 1887

Never, never can I forget this brilliant year, so full of the marvellous kindness, loyalty, and devotion of so many millions, which really I could hardly have expected. I felt sadly the absence of those dear ones, who would so entirely have rejoiced in this eventful time. Then, how thankful I must be for darling Beatrice coming safely through her severe confinement, and now again in the great improvement in dear Fritz's condition! We had been in such terrible anxiety about him in November. May God help me further!

TO THE PRINCE OF WALES   3 January 1888

I must just answer your observations about Lord Randolph Churchill [whom the Prince wished to have reinstated in Salisbury's Cabinet]. I cannot, I own, quite understand your high opinion of a man who is clever undoubtedly, but who is devoid of all principle, who holds the most insular and dangerous doctrines on foreign affairs, who is very impulsive and utterly unreliable. If you knew how infamously he behaved towards his colleagues (Lord Iddesleigh he treated atrociously), holding views which were utterly impossible to be listened to and which he holds now, you would see that to have him again in the Government would be to break it up at once; and I shall do all I can to prevent such a catastrophe . . .

Pray don't correspond with him, for he really is not to be trusted and is very indiscreet, and his power and talents are greatly overrated.

JOURNAL   10 February 1888

Rather anxious reports again about dear Fritz, and the necessity for tracheotomy having to be performed, so telegraphed to enquire. Heard from Sir M. Mackenzie that it was quite true that tracheotomy was found necessary, and was to be performed at once! Greatly troubled. Heard that the operation was well over, and that dear Fritz was doing quite well.

3 March 1888

A sad, distracted letter from poor dear Vicky about the trouble with the doctors, the German ones disagreeing with Sir M. Mackenzie. The latter gives a good report.

Have a new Indian servant, called Ahmed Hussain, a fine soldier-like looking man, very tall and thin.

10 March 1888

Already twenty-five years since dear Bertie married sweet Alix! May God give them many more happy years together! To me it was not permitted to celebrate this happy anniversary with my husband Albert.

5 April 1888

At eleven received [at the Villa Palmieri, Florence] the King and Queen of Italy, who had arrived at Florence yesterday evening. The King is aged and grown grey, the Queen is as charming as ever. To my astonishment Signor Crispi, the present very Radical Prime Minister, came into the room, and remained there, which was very embarrassing . . . They were most kind and amiable, making many excuses for Crispi's behaviour this morning—the King saying that he was a very clever man, but had no manners.

6 April 1888

I drove to the Palazzo Pitti to lunch with the King and Queen. The King talked very pleasantly and sensibly. Signor Crispi came in, with whom I had some conversation. He has no *savoir faire* whatever.

On her way home the Queen passed through Berlin.

JOURNAL   *24 April 1888*

After I had tidied myself up a bit, dear Vicky came and asked me to go and see dear Fritz. He was lying in bed, his dear face unaltered; and he raised up both his hands with pleasure at seeing me and gave me a nosegay. It was very touching and sad to see him thus in bed. Vicky then took me through his rooms, into a very pretty little green one with rococo decorations in silver. Here I breakfasted with her, her three girls, Beatrice and Liko. Afterwards saw Sir M. Mackenzie with Vicky. He seemed to think Fritz was better.

We visited the Empress Augusta at the Schloss, going in at a side entrance, where Fritz and Louise of Baden met us at the door. I went up in a lift alone, and there was the Empress, in deep mourning, with a long veil, seated in a chair, quite crumpled up and deathly pale, really rather a ghastly sight. Her voice was so weak, it was hardly audible. One hand is paralysed, and the other shakes very much. She seemed much pleased at seeing me again, after nine years.

25 April 1888

A little after twelve, Vicky brought Prince Bismarck* to my room and left him there. I had a most interesting conversation with him, and was agreeably surprised to find him so amiable and gentle. I spoke of William's inexperience and his not having travelled at all. Prince Bismarck replied that [William] knew nothing at all about civil affairs, that he could however say 'should he be thrown into the water, he would be able to swim,' for that he was certainly clever. We talked of other personal affairs, and I asked the Prince to tell Princess Bismarck to come to the English Embassy, where I was going in the afternoon, and this seemed to give him much pleasure. He remained with me over half an hour.

At half-past four drove with Vicky in a phaeton, the others following as yesterday. Almost the whole way we passed through double lines of carriages, and when we got into the town there were great crowds, who were most enthusiastic, cheering and throwing flowers into the carriage ... We went first to the British Embassy in the Wilhelmstrasse, which is a pretty house. Sir Edward Malet [the British Ambassador] and Lady Ermyntrude received me at the door, and he led me into the drawing-room, where we found Princess Bismarck, an elderly, rather masculine and not very sympathique lady, and her son Count Herbert, both very civil ...

Got home at seven, and I went directly with Vicky to see dear Fritz, bringing him a bouquet which had been given me at the church. Vicky took me back to my room and talked some time very sadly about the future, breaking down completely. Her despair at what she seems to look on as the certain end is terrible. I saw Sir M. Mackenzie, and he said he thought the fever, which was less, though always increasing at night, would never leave dear Fritz, and that he would not live above a few weeks, possibly two months, but hardly three!! We talked so long that I forgot the time, and had a terrible scramble to get ready for dinner.

26 April 1888

Went over to dear Fritz with Vicky, before our early dinner, and gave him my photograph, which he kissed, but, a fit of coughing coming on, we left him. Went back to him after dinner, and, after a few minutes' talk, took leave of him, which fortunately passed off without either of us being upset. I kissed him as I did every day, and said I hoped he would come to us when he was stronger ...

Vicky struggled hard not to give way, but finally broke down, and it was terrible to see her standing there in tears, while the train moved slowly off, and to think of all she was suffering and might have to go through. My poor poor child, what would I not do to help her in her hard lot!

11 June 1888

Much distressed at this morning's telegram from Sir M. Mackenzie; 'A festulona communication has occurred between larynx and oesophagus. I consider the condition extremely serious.' How terrible!

*According to Bigge, Bismarck was 'unmistakably nervous and ill at ease'. He came out of the room mopping his brow. 'That was a woman,' he said, 'one could do business with her.'

15 June 1888

A terrible day! No news, and we could hardly understand. While we were at breakfast a telegram arrived saying it could only last hours. Out for a little, going along the river, and then sat at the Cottage, but Beatrice had hardly left me before she returned with a telegram in her hand, saying all was over! at 11.15. The telegram was from William. I cannot, cannot realise the dreadful truth—the awful misfortune! It is too, too dreadful! My poor dear Vicky, God help her! I went in directly and sent endless telegrams. One universal feeling of grief and alarm!

Feel very miserable and upset. None of my own sons could be a greater loss. He was so good, so wise, and so fond of me! And now? To think of it all is such pain. Received a most touching, distracted letter from darling Vicky. Oh! it is all too terrible for words! How well he was at Balmoral last year! How delighted to come to us here, and see everything here, how full of hopes for the future!

Greatly relieved to hear that dear Alix would go with Bertie to Berlin, as I begged her to. Everyone from almost every country telegraphed. The misfortune is awful. My poor child's whole future gone, ruined, which they had prepared themselves for for nearly thirty years! Heard poor Vicky was not ill but wretched.

To the German Emperor, Wilhelm II   15 June 1888

I am broken-hearted. Help and do all you can for your poor dear Mother and try to follow in your best, noblest, and kindest of father's footsteps.

Journal   26 June 1888

Saw Lord Salisbury and we talked a great deal about this all-engrossing misfortune of poor darling Fritz's death, which is such an untold tragedy; of the symptoms, in William's opening speech, of a leaning towards Russia, and there having been no mention of England; of Prince Bismarck's violent language, when talking to Bertie, which showed how untrue and heartless he is, after all he seemed to promise me, and after poor Fritz had placed Vicky's hand in his, as if to recommend her to him! It is incredible and disgraceful.

To the German Emperor   3 July 1888

I have been waiting in hopes to have an answer to my letter written just before darling, beloved Papa was taken, but as you have not written I will just send you a few lines by the messenger. That my thoughts are very very much with you all you will easily understand. I am naturally very much occupied with poor dear Mama's future home. She feels probably a certain awkwardness, amounting to pain, to ask for anything, where so lately all was her own; but, as Uncle Bertie told me you were only too anxious to do what she wished in that respect, would you not (if you are not going to live there yourself) offer her to stay, at any rate for the present let her have Friedrichskron, or else Sans Souci? Uncle Bertie told me you had mentioned the Villa Liegnitz, but that is far too small, and would not do I think for your mother, who is the first after you, and who is the first Princess after Aunt Alix in Great Britain. An Empress could not well live in a little villa where Charlotte and afterwards Henry lived.

Mama does not know I am writing to you on this subject, nor has she ever mentioned it to me, but after talking it over with Uncle Bertie he advised me to write direct to you. Let me also ask you to bear with poor Mama if she is sometimes irritated and excited. She does not mean it so; think what months of agony and suspense and watching with broken and sleepless nights she has gone through, and don't mind it. I am so anxious that all should go smoothly, that I write thus openly in the interests of both.

There are many rumours of your going and paying visits to Sovereigns. I hope that at least you will let some months pass before anything of this kind takes place, as it is not three weeks yet since dear beloved Papa was taken, and we are all still in such deep mourning for him.

TO THE PRINCE OF WALES   24 July 1888

I will let you have a copy of Willy's answer as soon as possible. How sickening it is to see Willy, not two months after his beloved and noble father's death, going to banquets and reviews! It is very indecent and very unfeeling!

JOURNAL   11 August 1888

Am making arrangements to appoint Abdul a *munshi*, as I think it was a mistake to bring him over as a servant to wait at table, a thing he had never done, having been a clerk or *munshi* in his own country and being of rather a different class to the others. I had made this change, as he was anxious to return to India, not feeling happy under the existing circumstances. On the other hand, I particularly wish to retain his services, as he helps me in studying Hindustani, which interests me very much, and he is very intelligent and useful.

TO SALISBURY   15 October 1888

As regarding the Prince's [the Prince of Wales's] not treating his nephew as Emperor [as the Emperor had complained]; this is really too vulgar and too absurd, as well as untrue, almost to be believed.

We have always been very intimate with our grandson and nephew, and to pretend that he is to be treated in private as well as in public as 'his Imperial Majesty' is perfect madness! He has been treated just as we should have treated his beloved father and even grandfather, and as the Queen herself was always treated by her dear uncle King Leopold. If he has such notions, he [had] better never come here.

The Queen will not swallow this affront.

JOURNAL   2 November 1888

Had my last Hindustani lesson, as good Abdul goes home to India to-morrow on leave, which I regret, as it will be very difficult to study alone, and he is very handy and useful in many ways.

6 November 1888

Received many letters. Had a long heartbroken one from poor dear Vicky, who feels her helplessness so much, and has to put up with the most monstrous behaviour from Prince Bismarck and his son. It makes my blood boil!

To Salisbury   10 November 1888

This new most ghastly murder [by 'Jack the Ripper'] shows the absolute necessity for some very decided action.

All these courts must be lit, and our detectives improved. They are not what they should be. You promised, when the first murder took place, to consult with your colleagues about it.

To Henry Matthews, Home Secretary   13 November 1888

The Queen fears that the detective department is not so efficient as it might be. No doubt the recent murders in Whitechapel were committed in circumstances which made detection very difficult; still, the Queen thinks that, in the small area where these horrible crimes have been perpetrated, a great number of detectives might be employed, and that every possible suggestion might be carefully examined and, if practicable, followed.

Have the cattle boats and passenger boats been examined?

Has any investigation been made as to the number of single men occupying rooms to themselves?

The murderer's clothes must be saturated with blood and must be kept somewhere.

Is there sufficient surveillance at night?

These are some of the questions that occur to the Queen on reading the accounts of this horrible crime.

Journal   19 November 1888

At length the anxiously expected day had arrived [the bereaved Empress Frederick's visit to England] . . . We stepped on board [the *Victoria and Albert*] and went at once to the deck saloon, young Vicky, Sophy, and Mossy being outside, as well as the ladies and gentlemen. I went in and found my poor darling child in her deep widow's mourning. She was very much upset when she first saw me. Many tears were shed, but then she became calmer . . .

Reached Windsor about two. A Guard of Honour at the station, and there were many people. I made darling Vicky go first everywhere. There was a Sovereign's Captain's escort. Great crowds, and flags hung out.

To John Bright   4 December 1888

The Queen hears with much concern of the severe indisposition of Mr. Bright, from which however she trusts he is recovering. She wishes to repeat in writing what she said to him last year, when he came with the Address, which is how much she thanks him for the very loyal and patriotic manner in which he had spoken and written on all occasions in support of the maintenance of the union, and of law and order.

The Queen would also wish to thank Mr. Bright for the kind manner in which a year or two after her great sorrow he had taken her part* when ignorant and unfeeling people attacked her for not going out into the world, and at that early time not taking part in Jubilee ceremonies which he overwhelming grief rendered at that time impossible. The Queen has never forgotten his kindness on that occasion.

TO VICTORIA, PRINCESS LOUIS OF BATTENBERG   5 January 1889

Christmas was to all of us a very sad time, and how strange and mysterious that with only a few hours difference, your dear Father-in-law and Great Uncle should have been taken from us all at the same time of year,—that dreadful month of December which took away your beloved Mama, and 17 years before, dear Grandpapa and which so nearly carried off Uncle Bertie in 71! It seems as though there was in this, the bond of sorrow (which is far stronger than that of joy) a new tie between our families which are already so much linked together!

31 March 1889

And now let me say a word about Alicky. Is there no hope about E[ddy]†? She is not 19—and she should be made to reflect seriously on the folly of throwing away the chance of a very good husband, kind, affectionate and steady and of entering a united happy family and a very good position which is second to none in the world! Dear Uncle and Aunt wish it so much and poor E. is so unhappy at the thought of losing her also! Can you and Ernie not do any good? What fancy has she got in her head?

JOURNAL   26 April 1889

We went down into the Ballroom [at Sandringham], which was converted into a theatre, after talking till ten. There were nearly 300 people in the room, including all the neighbours, tenants, and servants. We sat in the front row, I between Bertie and Alix. The stage was beautifully arranged and with great scenic effects, and the pieces were splendidly mounted and with numbers of people taking part. I believe there were between sixty and seventy, as well as the orchestra. The piece, *The Bells*, is a melodrama, translated from the French *Le Juif Polonais* by Erckmann-Chatrian, and is very thrilling. The hero (Irving), though a mannerist of the Macready type, acted wonderfully.

24 May 1889

My seventieth birthday. Beatrice sent in the dear little children to wish me joy the

---

*In 1866, after criticisms of the Queen's infrequent public appearances had been made at a Reform meeting, Bright had declared: 'I am not accustomed to stand up in defence of those who are possessors of crowns; but I could not sit here and hear that observation without a sensation of wonder and of pain. I think there has been by many persons a great injustice done to the Queen in reference to her desolate and widowed position. And I venture to say this, that a woman—be she the Queen of a great realm, or be she the wife of one of your labouring men—who can keep alive in her heart a great sorrow for the lost object of her life and affection, is not at all likely to be wanting in a great and generous sympathy with you.'

†The Queen was most anxious that her granddaughter, Princess Alix of Hesse, Princess Victoria's beautiful sister, should marry her grandson, Prince Albert Victor (Eddy), of whom she was extremely fond. But the Princess would not have him and was to marry Tsar Nicholas II instead.

first thing in the morning, and they sat on my bed, and were very good, Drino, with a nosegay, saying over and over again, 'Many happy returns, Gangan.' Then I got up and dressed, and went over to Beatrice, who was looking so well, and who also gave me flowers. How far away did this birthday seem from those bright happy ones from 1840 to 1861! Went into the Audience Room [at Windsor] with all the others, and here the table with my presents was arranged.

TO THE EARL OF FIFE   27 June 1889

I have received the announcement of your intended engagement to my dear granddaughter, Louise of Wales, with the greatest pleasure, and I most readily and gladly give my consent to it.

I love my dear granddaughters dearly and they are like my own children; their happiness is very near my heart.

TO SALISBURY   9 July 1889

I am quite horrified to see the name of that horrible lying Labouchere and of that rebel Parnell on the Committee for the Royal Grants. I protest vehemently against both. It is quite indecent to have such people on such a Committee.

JOURNAL   2 August 1889

All on the *qui vive* for William's arrival, which had been expected at five. The Guard of Honour with the band was drawn up and waited and waited. At length, at near half-past seven, he appeared. I received him at the door with Lenchen and Beatrice, and with all the ladies and gentlemen ... We took William, who was in British Admiral's uniform, into the Drawing-room, and then he went to his rooms downstairs. He was very amiable, and kissed me very affectionately on both cheeks on arriving.

5 August 1889

Just before dinner I received in the Drawing-room a deputation of the 1st regiment of Garde Dragoner, of which William has just nominated me Chief ... William introduced them and made a very pretty speech, in which he said that his father had served in the regiment, which was one of the reasons why he chose it for me. I answered as well as I could.

TO VICTORIA, PRINCESS LOUIS OF BATTENBERG   7 August 1889

The visit here is going off quite well—though it is very hard to have to swallow that horrid Herbert B[ismarck, Prussian Foreign Minister] who everyone dislikes ... Willy is quite amiable; he is grown very large and puffed in the face. He seems pleased with everything.

TO THE GERMAN EMPEROR   10 August 1889

I was just intending to inquire if there were no news when I received your kind telegram, for which many thanks. It gives me great pleasure to hear that you liked your stay and were happy here.

To Ponsonby   22 August 1889

Would Sir Henry thank Mr. Matthews, and say the only regret she feels about the decision is that so wicked a woman should escape by a mere legal quibble!*

The law is not a moral profession she must say. But her sentence must never be further commuted.

To Victoria, Princess Louis of Battenberg   30 October 1889

We have just a faint lingering hope that Alicky might in time look to see what a pleasant home, and what a useful position she will lose if she ultimately persists in not yielding to Eddy's really earnest wishes. He wrote to me he should not give up the idea (though it is considered for the present at an end)—and: 'I don't think she knows how I love her, or she could not be so cruel.'

There are so few for her to marry and Eddie is very good! However if after he returns [from India]—improved and developped [*sic*] as we must hope he will be—who could he marry? Do tell me.

To Salisbury   21 December 1889

The Queen has taken pains to enquire into some of the people who Lord Salisbury mentioned as candidates for bishoprics; and she finds that Canon Fleming would not be acceptable, though a very popular preacher and a kind-hearted man ... He has no intellectual powers of a high order; is very unpopular with the Clergy, especially the liberal-minded ones, and takes no interest in questions outside his parish. To this, the Queen ought to have mentioned the other day, that he has not a very good or high-bred manner.

Journal   27 June 1890

After luncheon started for London, and from Paddington station drove in a closed carriage to Kensal Green Cemetery. There were crowds out, we could not understand why, and thought something must be going [on], but it turned out it was only to see me. Got out and walked a short way along a path, where the vault is in which dear Janie Ely [Marchioness of Ely] rests. Placed our wreaths there. Unfortunately, there were such crowds that the privacy of my visit was quite spoilt; still, I felt glad so many bore witness to this act of regard and love paid to my beloved friend.

                                                      11 October 1890

After dinner, the other ladies and gentlemen joined us in the Drawing-room, and we pushed the furniture back and had a nice little impromptu dance, Curtis's band being so *entraînant*. We had a quadrille, in which I danced with Eddy!! It did quite well, then followed some waltzes and polkas.

*Mrs. Maybrick had been found guilty of administering poison to her husband with intent to murder. She was sentenced to death but the sentence was commuted to penal servitude for life on the grounds that there were reasonable grounds to doubt that the arsenic administered was the cause of death.

TO DR. DAVIDSON, DEAN OF WINDSOR   17 October 1890

The Queen has been much touched by the kindness of the Dean of Windsor's letter. She is naturally much grieved that he should leave Windsor, where she hoped he would and could have remained and been of such use to herself and others. But, when she saw and heard how useful he would be to the Church in another position, she felt she had no right to be selfish; and therefore gave her consent to a bishopric being offered to him. The Queen must honestly confess that she has (excepting in one case, the Bishop of Ripon) never found people promoted to the Episcopate remain what they were before. She hopes and thinks this will not be the case with the Dean. Many who preached so well before did no longer as bishops, excepting the Bishop of Ripon. The whole atmosphere of a cathedral and its surroundings, the very dignity itself which accompanies a bishopric, seems to hamper their freedom of speech.

TO VICTORIA, PRINCESS LOUIS OF BATTENBERG   29 December 1890

In my last letter I said I must write to you about a subject which I had no time for, then. It is about Alicky and N[icholas]. I had your assurance that nothing was to be feared in that quarter, but I know it for certain, that in spite of all your (Papa's Ernie's and your) objections and still more contrary to the positive wish of his Parents who do not wish him to marry A. as they feel, as everyone must do, for the youngest Sister to marry the son of the Emperor—would never answer, and lead to no happiness,—well in spite of all this behind all your backs, Ella [Elizabeth, Princess of Hesse] and Serge [her husband, the Grand Duke Sergius] do all they can to bring it about, encouraging and even urging the Boy to do it!—I promised not to mention who told me—I must do to you, it is Aunt Alix who heard it from Aunt Minny [the former Princess Dagmar of Denmark, the Princess of Wales's sister, whose husband had become Tsar in 1881] herself who is very much annoyed about it. You must never mention Aunt Alix's name, but this must not be allowed to go on. Papa must put his foot down and there must be no more visits of Alicky to Russia—and he must and you and Ernie must insist on a stop being put to the whole affair.

The state of Russia is so bad, so rotten that at any moment something dreadful might happen and though it may not signify to Ella, the wife of the Thronfolger [heir to the throne] is in a most difficult and precarious position.

I have written all to Papa, who must be strong and firm and I am so afraid he may not be. It would have the very worst effect here and in Germany (where Russia is not liked) and would produce a great separation between our families.

JOURNAL   2 July 1891

Went to the Green drawing-room [Windsor Castle] and heard M. Paderewski play on the piano. He does so quite marvellously, such power and such tender feeling. I really think he is quite equal to Rubinstein. He is young, about twenty-eight, very pale, with a sort of aureole of red hair standing out.

4 July 1891

We got ourselves ready for William and Dona's arrival [the arrival of the German Emperor and the Empress on their first State visit to England]. Alix, Beatrice, and Louischen went down to the station to meet them.

11 July 1891

Heard everything had gone off admirably at the lunch in the City, where William and Dona had an enthusiastic reception. The Lord Mayor made a touching allusion to my beloved Albert and to dear Vicky and Fritz, dwelling on William's being my grandson, which is the reason for their receiving him so well.

To Salisbury   20 July 1891

I am extremely anxious to show civility to the French Fleet, the French having been so very civil to me; but it is impossible for me to delay my departure for Scotland beyond Monday, 24th August, and I hope the French Fleet will make it possible to arrive on the 20th or latest 21st August.

Journal   20 August 1891

The great dinner took place in the Indian room at a quarter to nine [at Osborne for the French Ambassador, W. H. Waddington]. Arthur took me in, and on my other side sat M. Waddington. At dessert Arthur proposed the health of the President of the French Republic . . . Then the *Marseillaise* was played, and I remained standing till it was over, which gratified them greatly. After dinner I spoke first to the Admiral, who is very pleasing and full of conversation, and then M. Waddington called up the principal Captains [and the officers of the French fleet], who were all very civil and some particularly so . . . M. Waddington said it was necessary to speak to what he called the small fry, and that it was a great day for them all, which they would never forget, especially my standing while the *Marseillaise* was played. I did not say that I had not particularly liked doing it.

To Lord George Hamilton, First Lord of the Admiralty
5 September 1891

When the Queen saw Lord George Hamilton at Osborne, she had not time to say to him she hopes and expects that Prince Louis of Battenberg, to whose merits everyone who knows the service well can testify, will get his promotion at the end of the year, for several junior to him have been promoted over his head. There is a belief that the Admiralty are afraid of promoting officers who are Princes on account of the Radical attacks of low papers and scurrilous ones; but the Queen cannot credit this. She will always maintain that if a Prince is unfit for promotion in the Navy or Army he should not be promoted because he is a Prince. But if he is fit, he should be treated as any other officer, who can rise to the highest post in the service. This the Queen will insist on, whoever the Prince may be. But she knows how excellent an officer her grandson-in-law is, and she therefore trusts there will be no further delay in giving him what he deserves.

JOURNAL   26 November 1891

At four, we all went to the Waterloo Gallery [Windsor Castle] . . . where the opera of *Cavalleria Rusticana*, by a young Italian composer, of the name of Mascagni, was performed. I had not heard an Italian opera for thirty-one years. The story was most pathetic and touching beyond words. The whole performance was a great success and I loved the music, which is so melodious, and characteristically Italian.

5 December 1891

Heard that Eddy wished to see me. I suspected something at once. He came in and said, 'I have some good news to tell you; I am engaged to May Teck.' This had taken place at a ball at Luton . . . I was quite delighted. God bless them both! He seemed very pleased and satisfied, and I am so thankful, as I had much wished for this marriage, thinking her so suitable. Sat some time talking. Wrote to Bertie.

26 December 1891

Bertie admired the seat I put up [at Balmoral] to dear Leopold's memory, which he had not yet seen. Everything interests him here. After luncheon Bertie sat talking in my room, and said how happy he had been here, that he had liked to come here alone and stay, it reminded him of old times. He was most dear and affectionate. Then he and Georgie took leave, both very sorry to go, and I to lose them.

14

'THE QUEEN IS DEEPLY CONCERNED'

1892–1895

JOURNAL   10 January 1892

Was startled and rather troubled by a telegram from Bertie, saying dear Eddy had a 'very sharp attack of influenza and had now developed some pneumonia in left lung, the night restless'.

13 January 1892

Whilst I was dressing, Beatrice asked to see me, and brought a bad telegram from Dr. Broadbent to Dr. Reid to the following effect: 'Condition very dangerous.' How terrible! Felt I ought to fly to Sandringham, and yet I feared I should be in the way. I telegraphed to Bertie, and got the following most sad answer: 'Our darling Eddy is in God's hands. Human skill seems unavailing. There could not be a question of your coming here.' Can think of nothing else but dear Eddy and the terrible distress at Sandringham.

14 January 1892

A never-to-be-forgotten day! Whilst I was dressing, Lenchen came in, bringing the following heartrending telegram from poor dear Bertie: 'Our darling Eddy has been taken from us. We are broken-hearted.' Words are far too poor to express one's feelings of grief, horror, and distress! Poor, poor parents; poor May to have her whole bright future to be merely a dream! Poor me, in my old age, to see this young promising life cut short! I, who loved him so dearly, and to whom he was so devoted! God help us! This is an awful blow to the country too!

15 January 1892

I received a most affecting [letter] from dear Bertie, from which I copy some extracts: 'We always say God's Will be done, and it is right to say and think so, but it does seem hard to rob us of our eldest son, on the eve of his marriage. Gladly would I have given my life for his, as I put no value on mine ... You were, dearest Mama, always very kind to him and fond of him, which he greatly appreciated, as we did.' Such a heart-broken letter! I truly did love and understand the darling boy and how I shall miss him!

TO VICTORIA, PRINCESS LOUIS OF BATTENBERG   3 February 1892

Was there ever anything so sad, so tragic? It is really an overwhelming misfortune, and I believe Auntie has written you some of the sad, sad details! The real and actual illness only lasted from the Saturday 9th to the 14th (that fatal date) [the anniversary of the Prince Consort's death]—and the last 48 hours had flown to the brain and caused fearful delirium ...

When you think that his poor young Bride who had come to spend his birthday with him, came to see him die—it is one of the most fearful tragedies one can imagine. It would sound unnatural and overdrawn if it was put into a Novel.

Uncle Bertie, Aunt Alix, Georgie and the girls came here yesterday, and I hope will be the better for the change. Uncle B. looks very pale and sad and aged, but is not ill; Aunt Alix is grown very thin and looks so sad and delicate. It is too sad to see her so. Georgie is thin, his hair cut quite short and the merriment gone out of him at present. But he does not look ill ... Poor darling Eddy, he was so good and gentle, and I shall miss him greatly.

TO PONSONBY   30 May 1892

First of all, however, she must say how dreadfully disappointed and shocked she is at Lord Rosebery's speech,* which is radical to a degree to be almost communistic. Hitherto he always said he had nothing whatever to do with Home Rule, and only with foreign affairs; and now he is as violent as anyone. Poor Lady Rosebery is not there to keep him back. Sir Henry must try and get at him through some one, so that he may know how grieved and shocked the Queen is at what he has said. In case of the Government's defeat the Queen meant to send for him first, but after this violent attack on Lord Salisbury, this attempt to stir up Ireland, it will be impossible; and the G.O.M. at eighty-two is a very alarming look-out.

JOURNAL   11 June 1892

Georgie sat and talked with me after breakfast. He is a dear boy, with much character and most affectionate to me.

23 June 1892

Saw Lord Salisbury, who said no one could tell how things would go, it might be one way or the other. There was a marked improvement in favour of the Government within the last two months. There might be a small majority either way.

17 July 1892

Saw Lord Salisbury, who talked of the elections, and Mr. Gladstone having a majority of 40, which was, however, small, in comparison with what he had in '80, when he had 100, and in '85, when he had 160. He had only the Irish vote to depend on, which was insecure. No one knew what Lord Rosebery would do, but we agreed he must be Foreign Secretary.

*Since the death of his wife, Rosebery had taken no active part in politics. But his return to political life was marked by a speech in Edinburgh in which he had attacked Lord Salisbury in the strongest terms.

To GLADSTONE   12 August 1892

Lord Salisbury having placed his resignation in the Queen's hands, which she has accepted with much regret, she now desires to ask Mr. Gladstone if he is prepared to try and form a Ministry to carry on the Government of the country.

The Queen need scarcely add that she trusts that Mr. Gladstone and his friends will continue to maintain and promote the honour and welfare of her great Empire.

To THE PRINCE OF WALES   13 August 1892

Though I shall see you to-night, I just write to say that it would not do for me to press Lord Rosebery to join this Government, as I heard he had expressed to a person who repeated it to Lord Salisbury that he did not wish to throw in his lot with these people, as if he did so now he could never free himself from them and it would naturally ruin his career. If I tried to press him and he did it merely to oblige me against his own wish and convictions, it would put me under obligations to him, and I might find myself in a very awkward position. I must not interfere in the formation of this iniquitous Government. I am thankful to hear the G.O.M. is not going to make any offer to Mr. Labouchere.

JOURNAL   15 August 1892

Sir H. Ponsonby had a telegram informing him that Lord Rosebery had accepted [as Foreign Secretary], which is a great thing, but we could not think what had made him change his mind at the last. After luncheon went down to see Mr. Gladstone. I thought him greatly altered and changed, not only much aged and walking rather bent with a stick, but altogether; his face shrunk, deadly pale with a weird look in his eyes, a feeble expression about the mouth, and the voice altered . . .

The audience did not last very long. Mr. Gladstone sat close up to me, as he said he had grown deaf, but that I need not raise my voice, as it was clear. It is rather trying and anxious work to have to take as Prime Minister a man of eighty-two and a half, who really seems no longer quite fitted to be at the head of a Government, and whose views and principles are somewhat dangerous.

18 August 1892

Next came Lord Rosebery, who looked low and could hardly speak when I expressed my satisfaction at his accepting office, observing that he thought they would have done better to leave him where he was. I said his name would be of such importance, to which he replied: 'It is but the name,' which I would not admit and told him he knew I had helped him and would do so again, which made him smile. I said work would do him good; but he replied that he had no one to look after his children.

26 August 1892

Had some conversation with Mr. Asquith [Home Secretary], whom I thought pleasant, straightforward, and sensible. He is a very clever lawyer.

TO PONSONBY   9 November 1892

The Queen wonders if this principle [of allowing the Prince of Wales to receive information about Cabinet proceedings] was ever adopted by Lord Salisbury.

Would Sir Henry ask Lord Salisbury and ask Mr. Gladstone to pause before pursuing this course regularly? She thinks it can only have been on very particular occasions.

JOURNAL   17 January 1893

Talked to the Lord Chancellor about Ireland and the many Cabinets they were having about the Home Rule Bill. He said it was a difficult question, and did not seem at all reassured about it; but that one must look at the other side of the question, as things could not go on in the future as they had done. I replied, that I feared I did not agree.

                                                    14 February 1893

Mr. Gladstone brought in his Home Rule Bill yesterday, and spoke for two hours, his voice becoming very feeble at last. Bertie was present, and telegraphed that the speech was impressive.

                                                    15 February 1893

Mr. Balfour made a splendid speech on the Home Rule Bill last night. It is such an impracticable measure, which he pulled quite to pieces.

                                                    19 February 1893

This unfortunate Home Rule Bill was read a first time on Friday, after a splendid speech of Mr. Chamberlain's, but it is sad to think it will be [read] a second time! Please God, in committee it will be much altered. But I am much disturbed about this and other measures I cannot approve of.

TO GLADSTONE   20 February 1893

The Queen thanks Mr. Gladstone for his letters reporting the debates on the Irish Bill. She cannot conceal from him her feelings of anxiety and apprehension with reference to the provisions of this measure, which tend towards the disruption of her Empire and the establishment of an impracticable form of Government.

JOURNAL   18 March 1893

Bertie arrived; he and Lorne dined with us and afterwards Tennyson's play *Thomas à Becket* was performed in the Waterloo Gallery. Irving acted well and with much dignity, but his enunciation is not very distinct, especially when he gets excited. Ellen Terry as 'Rosamund' was perfect, so graceful and full of feeling and so young-looking in her lovely light dress, quite wonderfully so, for she is forty-six!!

                                                    3 May 1893

Received a telegram from Georgie . . . to say he was engaged to May Teck, and asked for my consent. I answered that I gladly did so.

1 July 1893

Just before two, the young Cesarewitch arrived, and I received him at the top of the staircase. All were in uniform to do him honour, and to show him every possible civility. He is charming and wonderfully like Georgie [his cousin]. He always speaks English, and almost without a fault, having had an English tutor who is still with him. He is very simple and unaffected.

5 July 1893

Dressed for the great dinner ... in the Supper-room, which, with all the plate and flowers, looked splendid. I sat between the King of Denmark, who led me in, and the Cesarewitch, who is charming. His great likeness to Georgie leads to no end of funny mistakes, the one being taken for the other!

6 July 1893

The great day [of the Duke of York's wedding], so anxiously looked forward to, was very bright and fine, but overpoweringly hot. To describe this day fully would be impossible. It was really (on a smaller scale) like the Jubilee; and the crowds, the loyalty and enthusiasm were immense. Telegrams began pouring in from an early hour. Was rolled to our usual dining-room, to see from the window all that was going on. Troops, Infantry, Cavalry, Volunteers, crowds, bands, all presented a most brilliant animated appearance. Already, whilst I was still in bed, I heard the distant hum of the people. I breakfasted alone with Beatrice. Began to dress soon after eleven. I wore my wedding lace over a light black stuff, and my wedding veil surmounted by a small coronet. While I was dressing, Mary (herself very handsome) brought in May, who looked very sweet. Her dress was very simple, of white satin with a silver design of roses, shamrocks, thistles and orange flowers, interwoven. On her head she had a small wreath of orange flowers, myrtle, and white heather surmounted by the diamond necklace I gave her, which can also be worn as a diadem, and her mother's wedding veil.

All the different Princes and Princesses came to the Palace and left in different processions from here, going up Constitution Hill, Piccadilly, and St. James's Street. At a quarter to twelve I left, driving with Mary Teck, in the new State glass coach with four creams, amidst a flourish of trumpets ... All along the route, which was wonderfully kept, there were very fine decorations and enormous crowds, who cheered tremendously and were in the best of humour ...   I was the first to arrive and enter the Chapel, which was not intended, but which I was glad of, as I saw all the processions, which were very striking and dignified. There was a flourish of trumpets, followed by a march played outside, and then taken up by the organ, as the Royalties slowly entered ... May was supported by her father and her brother Dolly. Georgie gave his answers very distinctly, while May, though quite self-possessed, spoke very low. The music was well played and sung, but sounded weak and inferior to that in St. George's Chapel. I could not but remember that I had stood, where May did, fifty-three years ago, and dear Vicky thirty-five years ago, and that the dear ones, who stood where Georgie did, were gone from us! May these dear children's happiness last longer!

We drove back amidst the same tremendous cheering. Mary had been a little upset, but was very brave.

We got home before everyone else. The heat was very great, quite overwhelming. Went to the middle room, with the balcony, overlooking the Mall, and stepped out amidst much cheering. Very soon the Bride and Bridegroom arrived, and I stepped out on the balcony with them, taking her by the hand, which produced another great outburst of cheering.

To GLADSTONE   12 August 1893

The Queen would wish, before approving of Lord Elgin as Viceroy of India, to observe that, though she has a regard for him as the son and nephew of that most distinguished family of Bruce, of which his father was the head, and with most of whom she was on terms of sincere friendship, she hardly thinks him well suited for this important post. He is very shy and most painfully silent, has no presence, no experience whatever in administration. He would not command the respect which is necessary in that office [to which he was, nevertheless, appointed].

JOURNAL   9 September 1893

This morning came the news of the Division: for the Home Rule Bill 41, and against it 419! A crushing majority indeed; and, what is most remarkable, the crowd outside cheered very much, and cheered Lord Salisbury.

To GLADSTONE   10 February 1894

The Queen has approved Canon Basil Wilberforce's appointment [to a canonry of Westminster], but wishes to add a condition to it, *viz.*, that he should not, when preaching at Westminster, use the very strong total abstinence language which he has carried to such an extreme hitherto.

Total abstinence is an impossibility; and, though it may be necessary in individual cases, it will not do to insist on it as a general practice; and the Queen relies on Mr. Gladstone's speaking strongly to him in this sense.

JOURNAL   27 February 1894

Saw Sir H. Ponsonby after breakfast, who had been sent for by Mr. Gladstone. It was about nothing more nor less than his resignation [on what he described as 'physical grounds'] ... He is growing blind, and is already very deaf, so that his decision is not to be wondered at. Meantime, the secret has come out, and it is placarded all over London!

28 February 1894

Saw Sir Henry Ponsonby, and then Mr. Gladstone, who was looking very old and was very deaf. I made him sit down, and said I had received his letter, and was sorry for the cause of his resignation. He said very little about it, only that he found his blindness had greatly increased since he had been at Biarritz. Then he talked of some of the honours for his friends, but not many.

3 March 1894

I wrote a few lines to Lord Rosebery, urging him to accept the Premiership, if even only for a short time, for the good of the country. These lines I sent up by Sir H. Ponsonby. Mrs. Gladstone I had seen after breakfast. She was very much upset, poor thing, and asked to be allowed to speak, as her husband 'could not speak.' This was to say, which she did with many tears, that, whatever his errors might have been, 'his devotion to your Majesty and the Crown were very great'. She repeated this twice, and begged me to allow her to tell him that I believed it, which I did; for I am convinced it is the case, though at times his actions might have made it difficult to believe. She spoke of former days, and how long she had known me and dearest Albert. I kissed her when she left.

To GLADSTONE   3 March 1894

Though the Queen has already accepted Mr. Gladstone's resignation and has taken leave of him, she does not like to leave his letter tendering his resignation unanswered. She therefore writes these few lines to say that she thinks, after so many years of arduous labour and responsibility, he is right in wishing to be relieved at his age of these arduous duties, and she trusts [will] be able to enjoy peace and quiet, with his excellent and devoted wife, in health and happiness, and that his eyesight may improve.

The Queen would gladly have conferred a peerage on Mr. Gladstone, but she knows that he would not accept it.

Lord Rosebery was Gladstone's natural successor as Prime Minister; but he was, so he told the Queen, reluctant to assume the responsibility. He advanced 'one main reason' for his 'wishing to avoid this heavy and thankless succession'. It was his apprehension that by advocating policies 'already laid down' to which he was bound by honour, he 'was likely to find himself in acute conflict' with Her Majesty's views.

To ROSEBERY   4 March 1894

[The Queen] is sorry to hear that he apprehends any trouble which might alienate him from her. The Queen can hardly think this possible or at any rate probable.

She does not object to Liberal measures which are not revolutionary, and she does not think it possible that Lord Rosebery will destroy well-tried, valued, and necessary institutions for the sole purpose of flattering useless Radicals or pandering to the pride of those whose only desire is their own self-gratification.

JOURNAL   5 March 1894

Lord Rosebery kissed hands on his appointment as Prime Minister. He said the task I had entrusted to him was very difficult, and not what he would have wished to undertake, but I repeated that he was the only person in the Government I considered suited to the post, and in whom I had absolute confidence, for which he thanked me.

TO ROSEBERY   11 March 1894

The Queen cannot refrain from writing a few lines of friendly advice to Lord Rosebery on the subject of the important speeches she hears he has to make tomorrow.

It is that she hopes he will not commit himself too strongly in what he says, but be as general, and to a certain extent as vague, as he can be. He need not say anything against what he feels himself pledged to, but he can do this and say nothing to discourage his party and yet not commit himself too strongly.

What the Queen says here, she does in the name of one who can no longer be a comfort and support to him [Rosebery's late wife], and who she [knows] felt very anxious on this subject.

17 March 1894

The House of Lords might possibly be improved, but it is part and parcel of the much vaunted and admired British Constitution and cannot be abolished. It is the only really independent House, for it is not bound as the House of Commons is, where they are constantly made to say what they would not otherwise do by their constituents, whom they try to please in order to be elected.

9 April 1894

The Queen cannot but think that some day even Lord Rosebery may be thankful for the power and independence of the Peers. There are some who whisper that many Gladstonians thanked God that Home Rule was destroyed by the much-abused Peers. The Queen in conclusion would most earnestly and solemnly conjure her Prime Minister not to let her Ministers join in any attempt to excite the passions of the people on this important subject, but rather to strive to restrain them (if they really exist, which she doubts), for if once a stone is set rolling they may not be able to stop it!

By now the members of the Queen's Household were finding the pretensions of Her Majesty's Indian Secretary, the Munshi Hafiz Abdul Karim, insufferable. They were also concerned that he might pass on confidential information to undesirable friends in India. The Queen declined to listen to their advice and protested angrily.

TO PONSONBY   10 April 1894

The Queen ... would wish to observe that to make out ... the poor good Munshi ... is so low is really outrageous and in a country like England quite out of place as anyone can [see] this. She has known two Archbishops who were the sons respectively of a Butcher and a Grocer, a Chancellor whose father was a poor sort of Scotch Minister, Sir D. Stewart and Lord Mount Stephen both who ran about barefoot as children and whose parents were very humble and the tradesmen M. and J. P. were made baronets.

Although it was discovered that the Munshi's father was a prison apothecary rather than a surgeon-general as his son had claimed, the Queen declined to believe the intelligence; and it was not for some years, after various members of the Household had threatened to resign and Lord Salisbury had intervened, that the Munshi was put more into his 'proper place'.

JOURNAL   23 June 1894

Very soon after I got to my room, I received the joyful news that dear May had been safely delivered of a son [the future Duke of Windsor], a fine strong child! What joy! What a blessing! How relieved and thankful to God I feel for this great mercy!

TO VICTORIA, PRINCESS LOUIS OF BATTENBERG   21 October 1894

All my fears about her future marriage [Princess Alix's to the Tzar Nicholas II] now show themselves so strongly and my blood runs cold when I think of her so young most likely placed on that very unsafe throne, her dear life and above all her husband's constantly threatened and unable to see her but rarely. It is a great additional anxiety in my declining years! Oh! how I wish it was not to be that I should lose my sweet Alicky. All I most earnestly ask now is that nothing should be settled for her future without my being told before. She has no Parents and I am her only Grandparent and feel I have a claim on her! She is like my own Child as you all are my dear Children but she and he are orphans.

TO ROSEBERY   21 November 1894

The Queen did not wish yesterday to speak to Lord Rosebery on the subject of his last speech at Glasgow, as she did not wish to enter into a discussion on a subject on which she feels so very strongly, but she cannot refrain from writing to him about it.*

It was with deep concern and surprise that the Queen read the expressions used by Lord Rosebery with regard to the Scotch Church. Lord Rosebery seems to forget, or perhaps he never knew, that at her accession, on the Throne in the House of Lords, she solemnly swore to 'maintain the Protestant Reformed Religion established by law,' and one of the principal provisions of the Act of Union in 1707 is that 'the Established Presbyterian Church of Scotland shall be maintained,' and the Queen will do all that lies in her power to be true to this promise.

It is indeed deeply to be deplored that the Prime Minister should have spoken as he did, and she thinks he will find that the Scotch will not be conciliated by it.

TO PONSONBY   21 December 1894

Many years ago Lord Palmerston laid down that, as a rule, all drafts as well as despatches for abroad should be written in a good round distinct hand, and insisted on it, that all the clerks should be forced to write in that way. This went on till within quite the last years, when indistinct and very small handwriting, like this draft, in

*In this speech, delivered at Glasgow before a large audience on 14 November, Rosebery had advocated reform of the House of Lords as well as disestablishment of both the Welsh and Scottish Churches.

very pale ink, has become the habit, and the Queen must quote and try to enforce Lord Palmerston's practice.

Would Sir Henry try to do this?

Unfortunately, Lord Rosebery himself is the very worst offender. The Queen (whose eyesight has become very faulty, and has not yet got glasses to suit) can hardly read them at all.

JOURNAL   7 January 1895

Just as we had finished breakfast and Beatrice had left the room, she quickly came back to say that poor Sir Henry Ponsonby was very ill. Dr. Reid, whom she had met in the passage, had been sent for, and he had spent half the night at Osborne Cottage. He was returning directly again. Sir Henry had had a stroke. He was not absolutely unconscious; but his speech was affected, and his right arm and leg were paralysed!

Ponsonby did not recover. He was succeeded by Lieutenant-Colonel Sir Arthur Bigge, later Lord Stamfordham.

JOURNAL   24 May 1895

My poor old birthday, my seventy-sixth!! May God spare me to do my duty for my people and country, children and friends!

MEMORANDUM   22 June 1895

Saw Lord Rosebery at a little after eight. After talking of my journey and the weather, he said they had had two Cabinets, and after a long discussion and much difference of opinion, they had agreed to resign after last night's vote. I asked whether I must accept the resignation, and he said, Yes. There were two alternatives, resignation or dissolution, and he thought decidedly the former the best. They had had a very bad week with various defeats and very small majorities; and he thought it would be very humiliating to go on with the certainty of being defeated sooner or later; and that it was very bad for the country, as well as for our foreign relations, to have such a small majority. Other nations would not trust us; and, if it were for that alone, he should be glad if the Government were to resign. There had been considerable difference of opinion, and Sir Wm. Harcourt [Chancellor of the Exchequer] had been very disagreeable. I said that I should be sorry to take leave of him (Lord R.), to which he replied that it was an immense relief to give up his office, and the unfortunate inheritance of Mr. Gladstone ...

After dinner I saw Sir Arthur Bigge, and told him what had passed; and he thought also it was much better that the Government should resign, and that all their friends and followers seemed to think it would be a great mistake if Lord Rosebery tried to hang on. I said I would write to Lord Salisbury to-night, and send him with the letter to Hatfield.

To Salisbury   11 August 1895

I very much regret Lord Wolseley's answer; but as soon as I found Lord Lansdowne had made him the double offer [of becoming either British Ambassador in Berlin or Commander-in-Chief], which he was not authorised to do by me, I felt afraid that the temptation of the C.-in-C.-ship would at once outweigh the other.

But Lord Lansdowne never asked my permission to offer him the last-named, and he ought not to have done that. I dislike the appointment of Lord Wolseley as C.-in-C., as he is very imprudent, full of new fancies, and has a clique of his own.

Journal   19 November 1895

After luncheon I went to the White drawing-room to receive three Chiefs from Bechuanaland, who are Christians. The Chiefs are very tall and very black, but their hair is not woolly. One of the Chiefs is said to be a very remarkable and intelligent man. One of their chief objects in coming was to obtain a permit from the Government to suppress strong drink, which demoralises and kills the poor natives. Alas! everywhere this terrible evil, which has such a fatal effect on the population, seems to follow civilisation!

                                                         14 December 1895

This terrible anniversary returned for the thirty-fourth time. When I went to my dressing-room found telegrams saying that dear May had been safely delivered of a son [the future King George VI] at three this morning. Georgie's first feeling was regret that this dear child should be born on such a sad day. I have a feeling it may be a blessing for the dear little boy, and may be looked upon as a gift from God!

                                                         16 December 1895

Received a dear letter from Georgie, which has given me the greatest pleasure, saying that they intended the baby to have the name of Albert.

15

THE FINAL YEARS
1896–1901

JOURNAL   2 January 1896

Beatrice read me telegrams after tea, as my sight is so bad, and I have not yet succeeded in getting spectacles to suit.

3 January 1896

More telegrams kept coming in, and one from President Kruger to Mr. Chamberlain apparently referring to those in the Raid who had been taken prisoners, saying they were to be treated with kindness.*

The papers are full of very strong articles against William, who sent a most unwarranted telegram to President Kruger, congratulating him, which is outrageous, and very unfriendly towards us.

TO THE GERMAN EMPEROR   5 January 1896

As your Grandmother to whom you have always shown so much affection and of whose example you have always spoken with so much respect, I feel I cannot refrain from expressing my deep regret at the telegram you sent President Kruger. It is considered very unfriendly towards this country, which I feel sure it is not intended to be, and has, I grieve to say, made a very painful impression here. The action of Dr. Jameson was of course very wrong and totally unwarranted; but considering the very peculiar position in which the Transvaal stands towards Great Britain, I think it would have been far better to have said nothing.

TO CHAMBERLAIN   7 January 1896

I cannot say how shocked I am at the terrible loss of life, and I am struck by the excess of killed over wounded, which is the reverse of the usual proportion.

TO SALISBURY   8 January 1896

While thoroughly appreciating and approving Mr. Chamberlain's prompt and firm action, I think that I should be consulted in such very important questions as

*Dr. Jameson, the Administrator of the British South Africa Company's territory (what is now Zimbabwe), had crossed the frontier into the Transvaal at the head of 400 to 500 troops in order to assist the Uitlanders to obtain the civil rights denied them by President Kruger. He had been defeated by a Boer force near Krugersdorp and his surviving followers had been taken prisoner.

sending troops to the Cape or mobilising a Flying Squadron, even though time may press.

To the Prince of Wales   11 January 1896

I send you here the answer I received yesterday to my letter from William, which please return when done with ... It would not do to have given him 'a good snub.' Those sharp, cutting answers and remarks only irritate and do harm, and in Sovereigns and Princes should be most carefully guarded against.

Journal   22 January 1896

A terrible blow has fallen on us all, especially on my poor darling Beatrice. Our dearly loved Liko has been taken from us! Can I write it? He was so much better, and we were anxiously awaiting the news of his arrival at Madeira. What will become of my poor child? All she said in a trembling voice, apparently quite stunned, was, 'The life is gone out of me.' ...

To Salisbury   22 March 1896

Every day I feel the blessing of a strong Government in such safe and strong hands as yours.

Journal   20 June 1896

Fifty-nine years since I came to the throne! What a long time to bear so heavy a burden! God has guided me in the midst of terrible trials, sorrows, and anxieties, and has wonderfully protected me. I have lived to see my dear country and vast Empire prosper and expand, and be wonderfully loyal!

                                                             11 September 1896

After luncheon saw Mr. Balfour [First Lord of the Treasury]. We talked over many important topics.

I am much struck, as is everyone, by Mr. Balfour's extreme fairness, impartiality, and large-mindedness. He sees all sides of a question, is wonderfully generous in his feelings towards others, and very gentle and sweet-tempered.

                                                             23 September 1896

To-day is the day on which I have reigned longer, by a day, than any English sovereign, and the people wished to make all sorts of demonstrations, which I asked them not to do until I had completed the sixty years next June. But notwithstanding that this was made public in the papers, people of all kinds and ranks, from every part of the kingdom, sent congratulatory telegrams, and they kept coming in all day.

                                                             28 September 1896

Dear little David [the Duke of York's son, Prince Edward, later Duke of Windsor] with the baby came in at the end of luncheon to say good-bye. David is a most attractive little boy, and so forward and clever. He always tries at luncheon time to pull me up out of my chair, saying 'Get up, Gangan,' and then to one of the Indian servants, 'Man pull it,' which makes us laugh very much.

3 October 1896

At twelve went down to below the terrace, near the ballroom, and we were all photographed by Downey by the new cinematograph process, which makes moving pictures by winding off a reel of films. We were walking up and down.

23 November 1896

After tea went to the Red drawing-room, where so-called 'animated pictures' were shown off, including the groups taken in September at Balmoral. It is a very wonderful process, representing people, their movements and actions, as if they were alive.

To Bigge   30 January 1897

Sir A. Bigge may tell the Prince of Wales that there is not the slightest fear of the Queen's giving way about the Emperor William's coming here in June. It would never do for many reasons, and the Queen is surprised that the Empress should urge it.

Journal   22 April 1897

At half-past six the celebrated and famous actress Sarah Bernhardt, who has been acting at Nice and is staying in this hotel [at Cimiez], performed a little piece for me in the drawing-room at her own request. The play was called *Jean Marie*, by Adrien Fleuriet, quite short, only lasting half an hour. It is extremely touching, and Sarah Bernhardt's acting was quite marvellous, so pathetic and full of feeling. She appeared much affected herself, tears rolling down her cheeks. She has a most beautiful voice, and is very graceful in all her movements ... When the play was over, Edith Lytton presented Sarah Bernhardt to me, and I spoke to her for a few moments. Her manner was most pleasing and gentle. She said it had been such a pleasure and honour to act for me. When I expressed the hope that she was not tired, she answered, 'Cela m'a reposée.'

24 May 1897

My poor old birthday again came round, and it seems sadder each year, though I have such cause for thankfulness, and to be as well as I am, but fresh sorrow and trials still come upon me. My great lameness, etc., makes me feel how age is creeping on. Seventy-eight is a good age, but I pray yet to be spared a little longer for the sake of my country and dear ones.

Before breakfast the little children, Lenchen and Beatrice, gave me flowers and took me to my birthday table, which was covered with presents.

20 June 1897

This eventful day, 1897 [the Diamond Jubilee], has opened, and I pray God to help and protect me as He has hitherto done during these sixty long eventful years!

At eleven I, with all my family, went to St. George's Chapel, where a short touching service took place . . . The service began with the hymn, 'Now thank we all our God,' followed by some of the usual morning prayers. Dear Albert's beautiful *Te Deum* was sung, and the special prayer for Accession Day followed, with a few others. Felt rather nervous about the coming days, and that all should go off well.

21 June 1897

The 10th anniversary of the celebration of my fifty years Jubilee . . . Passed through dense crowds [between Paddington Station and Buckingham Palace], who gave me a most enthusiastic reception. It was like a triumphal entry . . . The windows, the roofs of the houses, were one mass of beaming faces, and the cheers never ceased. On entering the park, through the Marble Arch, the crowd was even greater, carriages were drawn up amongst the people on foot, even on the pretty little lodges well-dressed people were perched. Hyde Park Corner and Constitution Hill were densely crowded. All vied with one another to give me a heartfelt, loyal, and affectionate welcome. I was deeply touched and gratified . . .

Dressed for dinner. I wore a dress of which the whole front was embroidered in gold, which had been specially worked in India, diamonds in my cap, and a diamond necklace, etc. The dinner was in the Supper-room at little tables of twelve each. All the family, foreign royalties, special Ambassadors and Envoys were invited.

22 June 1897

A never-to-be-forgotten day. No one ever, I believe, has met with such an ovation as was given to me, passing through those six miles of streets . . . The crowds were quite indescribable, and their enthusiasm truly marvellous and deeply touching. The cheering was quite deafening, and every face seemed to be filled with real joy.

At a quarter-past eleven, the others being seated in their carriages long before, and having preceded me a short distance, I started from the State entrance in an open State landau, drawn by eight creams, dear Alix, looking very pretty in lilac, and Lenchen sitting opposite me. I felt a good deal agitated, and had been so all these days, for fear anything might be forgotten or go wrong . . .

Before leaving I touched an electric button, by which I started a message which was telegraphed throughout the whole Empire. It was the following: 'From my heart I thank my beloved people, May God bless them!' At this time the sun burst out. Vicky was in the carriage nearest me, not being able to go in mine, as her rank as Empress prevented her sitting with her back to the horses, for I had to sit alone. Her carriage was drawn by four blacks, richly caparisoned in red . . . The denseness of the crowds was immense, but the order maintained wonderful. The streets in the Strand are now quite wide, but one misses Temple Bar. Here the Lord Mayor received me and presented the sword, which I touched. He then immediately mounted his horse in his robes, and galloped past bare-headed, carrying the sword, preceding my carriage, accompanied by his Sheriffs. As we neared St. Paul's the procession was often stopped, and the crowds broke out into singing *God Save the Queen*. In one house were assembled the survivors of the Charge of Balaclava.

In front of the Cathedral the scene was most impressive. All the Colonial troops, on foot, were drawn up round the Square. My carriage, surrounded by all the Royal Princes, was drawn up close to the steps, where the Clergy were assembled, the Bishops in rich copes, with their croziers, the Archbishop of Canterbury and the Bishop of London each holding a very fine one. A *Te Deum* was sung . . .

[On returning to Buckingham Palace] I stopped in front of the Mansion House, where the Lady Mayoress presented me with a beautiful silver basket full of orchids. Here I took leave of the Lord Mayor. Both he and the Lady Mayoress were quite *émus*.

6 July 1897

Mr. Whitelaw Reid [Special United States Envoy for the Diamond Jubilee, afterwards American Ambassador in London] was full of the kindest expressions to me personally, and said that the people in America were so much attached to me, and spoke of me as 'the good Queen,' and that there was in fact a very friendly feeling towards this country, the various disputes and disagreements being really entirely superficial.

30 September 1897

Took leave with much regret of Georgie and May, who are leaving the first thing to-morrow morning. Every time I see them I love and like them more and respect them greatly. Thank God! Georgie has got such an excellent, useful, and good wife!

30 January 1898

Spoke to [Cosmo Gordon Lang, then Vicar of Portsea, afterwards Archbishop of Canterbury] for some time after dinner. He is a very interesting and clever man, a Scotchman, and was at Oxford. He has a very hard time at Portsea, having 40,000 parishioners, and the population is not very pleasant, particularly the artizans, who are very difficult, sceptical, and full of prejudices. The sailors are true and warm-hearted, but, as well as the soldiers, somewhat difficult to manage. Mr. Lang has thirteen curates to assist him, and they all live together.

To Chamberlain   6 February 1898

The Queen thanks Mr. Chamberlain for his letter enclosing the despatch from New Zealand inviting the Duke and Duchess of York to visit that Colony.

The Queen duly appreciates the loyal and kind wish of the New Zealanders to see her grandchildren. But there are very strong reasons against it, which she feels cannot be disregarded. The Duke of York is the only surviving son of the Prince of Wales, and the only available Prince in this country, besides the Prince of Wales himself and the Duke of Connaught (both very much overworked) able to perform all that is expected of them, and to help the Queen, now in her seventy-ninth year, who has lost the able and affectionate help of her dear son-in-law Prince Henry of Battenberg. But this is not all. Life is so uncertain, that the risk of sending the Duke of York so far away and exposing him to the innumerable dangers of fatigue, climate, etc., are too great; and it would indeed be tempting providence were we to send him so far away.

The Queen cannot but think that Mr. Chamberlain will understand her strong reasons for declining this proposal.

JOURNAL   13 April 1898

At half-past three M. Faure, the President of the Republic, who has been spending some days at the Riviera Palace, came to see me [at Cimiez]. Bertie received him below, and brought him up, and the three Princesses with the ladies were at the top of the stairs. I stood at the door of the drawing-room and asked him to sit down. He was very courteous and amiable, with a charming manner, so *grand seigneur* and not at all *parvenu*. He avoided all politics, but said most kindly how I was *aimée par la population*, that he hoped I was comfortably lodged, etc.

19 April 1898

Already seventeen years ago that good Lord Beaconsfield died.

10 May 1898

A dull dark morning, very warm. Lenchen breakfasted with us three, and the dear little York children came, looking very well. David is a delightful child, so intelligent, nice, and friendly. The baby [Princess Mary] is a sweet, pretty little thing.

19 May 1898

Heard at breakfast time that poor Mr. Gladstone, who has been hopelessly ill for some time and had suffered severely, had passed away quite peacefully this morning at five. He was very clever and full of ideas for the bettering and advancement of the country, always most loyal to me personally, and ready to do anything for the Royal Family; but alas! I am sure involuntarily, he did at times a good deal of harm. He had a wonderful power of speaking and carrying the masses with him.

TO MRS GLADSTONE   28 May 1898

My thoughts are much with you to-day when your dear husband is laid to rest. To-day's ceremony will be most trying and painful for you, but it will be, at the same time, gratifying to you to see the respect and regret envinced by the nation for the memory of one whose character and intellectual abilities marked him as one of the most distinguished statesmen of my reign. I shall ever gratefully remember his devotion and zeal in all that concerned my personal welfare and that of my family.

JOURNAL   27 January 1899

William's fortieth birthday. I wish he were more prudent and less impulsive at such an age!

TO THE EMPEROR OF RUSSIA   March 1899

I feel I must write and tell you something which you ought to know and perhaps do not. It is, I am sorry to say, that William takes every opportunity of impressing upon Sir F. Lascelles [British Ambassador in Berlin] that Russia is doing all in her power to work against us; that she offers alliances to other Powers, and has made one with the Ameer of Afghanistan against us.

I need not say that I do not believe a word of this, neither do Lord Salisbury nor Sir F. Lascelles.

But I am afraid William may go and tell things against us to you, just as he does about you to us. If so, pray tell me openly and confidentially. It is so important that we should understand each other, and that such mischievous and unstraightforward proceedings should be put a stop to. You are so true yourself, that I am sure you will be shocked at this.

This year the Queen paid her last visit to the French Riviera. She was asked to open a bridge at the end of the Boulevard Carabacel in Nice.

JOURNAL   27 April 1899

The bridge was beautifully decorated with flags and garlands. The Maire and his Adjunct met us on arrival, the band playing *God Save the Queen*, and he addressed a few words to me, thanking me for the honour I had conferred upon the town, also presenting a most enormous and lovely bouquet. I answered in flattering words, 'Je suis bien touchée que vous m'avez demandé d'inaugurer votre nouveau pont, et je fais des voeux bien sincères pour la prospérité de la ville de Nice et de ses environs.' Flowers were given to the Princesses. We then drove over the bridge, the band playing the *Marseillaise*. There were great crowds, who were all most enthusiastic.

Drove to Beaulieu. Had our tea at St. Jean, where Lenchen and Beatrice joined us. Alas! my last charming drive in this paradise of nature, which I grieve to leave, as I get more attached to it every year. I shall mind returning to the sunless north, but I am so grateful for all I have enjoyed here.

24 May 1899

Quite overpowered with letters and, above all, telegrams [on the Queen's 80th birthday], of which between two and three thousand have been received, many more than at the Jubilee . . .

Drove out about five, with Alix and Beatrice, down the hill [at Windsor], which was densely crowded, through a beautiful arch covered with flowers. The enthusiasm of the people was tremendous and most touching . . .

A little after nine we all went to the Waterloo Gallery, which was arranged as a theatre, and had the first, third, and last acts of Wagner's *Lohengrin* performed . . . I was simply enchanted. It is the most glorious composition, so poetic, so dramatic, and one might almost say, religious in feeling and full of sadness, pathos, and tenderness.

TO THE GERMAN EMPEROR   12 June 1899

Your other letter, I must say, has greatly astonished me. The tone in which you write about Lord Salisbury I can only attribute to a temporary irritation on your part, as I do not think you would otherwise have written in such a manner, and I doubt whether any Sovereign ever wrote in such terms to another Sovereign, and that Sovereign his own Grandmother, about their Prime Minister.* I never should do such a thing, and I never personally attacked or complained of Prince Bismarck, though I knew well what a bitter enemy he was to England.

JOURNAL   6 July 1899

I sat [at dinner] between Count Deym and M. Cambon [French Ambassador in London], who is a most agreeable well-informed man with large views. He spoke of the 'misérable affaire de Dreyfus,' the feeling about which had greatly changed. That he would probably be acquitted, and that the whole affair had arisen from the fact of his being a Jew and being rather a miserable creature' That he was very intelligent and clever, but had been so much disliked in the War Office, that when a succession of betrayals of secrets had taken place, his colleagues in the office had fixed suspicion on him. 'Il était la victime de son caractère.'

To SALISBURY   9 September 1899

I am too horrified for words at this monstrous horrible sentence against this poor martyr Dreyfus. If only all Europe would express its horror and indignation!

To THE QUEEN OF THE NETHERLANDS   ?12 September 1899

I sympathise most deeply with your expressions of the horrors of war, than which no one can feel more strongly than I do; and earnestly hope that it may be averted. But I cannot abandon my own subjects who have appealed to me for protection. If President Kruger is reasonable, there will be no war, but the issue is in his hands.

War with the Boers was not to be averted, however; and on 19 October the Queen inspected the Gordon Highlanders before their embarkation for South Africa.

JOURNAL   19 October 1899

I addressed them a few parting words as follows: 'I desire to wish you Godspeed. May God protect you! I am confident that you will always do your duty, and will ever maintain the high reputation of the Gordon Highlanders.' The men then gave three cheers, and I called up Captain Kerr, who seemed much moved, and could hardly speak. I shook hands with him, and wished him a safe return, and also spoke to the two Lieutenants. It was very touching, and I felt quite a lump in my throat as we drove away, and I thought of how these remarkably fine men might not all return.

To SALISBURY   21 October 1899

I sincerely hope that the increased taxation, necessary to meet the expenses of the

*The Kaiser had complained that Lord Salisbury's Government despised Germany 'and this has stung my subjects to the quick'.

war, will not fall upon the working classes; but I fear they will be most affected by the extra sixpence on beer.

TO LORD LANSDOWNE, SECRETARY FOR WAR   22 October 1899

My heart bleeds for these dreadful losses [at Glencoe, the first engagement of the war]. Again to-day a great success, but I fear very dearly bought. Would you try and convey my warmest, heartfelt sympathy with the near relations of the fallen and wounded, and admiration of the conduct of those they have lost?

JOURNAL   11 November 1899

Every moment telegrams keep coming, announcing the arrival of troops at the Cape. It is very encouraging. After tea I saw Lord Wolseley for some little time. We went over everything concerning the war. He said he felt much easier now, but had been very anxious ten days ago. He was sure Ladysmith would be able to hold out. He was delighted at the way in which the Reserves had come up, as well as the way in which the employers had helped in furnishing the men. We lamented bitterly over the loss of so many horse, and I made him promise to see that everything possible was done in the ships for their safety and comfort. But it is at best a great risk transporting so many such a great distance by sea.

19 November 1899

On coming home [from Clifton where she had opened a convalescent hospital] saw Lady White, wife of Sir George White, who is besieged in Ladysmith. She said my sympathy had been her greatest comfort, and she knew it would be the same to her husband. After tea had a long talk with the Bishop of Winchester on Church affairs.

22 November 1899

William came to me after tea, which Bertie and Louise had taken with me. I had a long interesting conversation with him on all subjects. We first spoke about his dear Mama's health, which is not satisfactory, then of the shocking tone of the German Press and the shameful attacks on England, as well as monstrous misrepresentations and lies about the war, which he greatly deplores. But he says it is due to the 'poison' which Bismarck poured into the ears of the people; that the latter had hated England, and wished for an alliance with Russia. If he had not sent him away, he does not know what would have happened, and he became even worse latterly in his abuse, which his son continued. William himself wishes for a better understanding with us.

11 December 1899

Saw Sir A. Bigge on his return from London, whither he has gone by my desire to see Lord Wolseley and Lord Lansdowne, and found the latter depressed and the former much annoyed with the Generals, who he considers have not done what they were advised to do, but have attacked difficult and inaccessible positions, instead of trying to outflank the enemy. Lord Wolseley wanted to know what I thought about it all.

We also discussed whether Bertie should resign the Presidency of the French Exhibition, on account of the atrocious personal attacks on me in the French Press. He is himself most indignant, and only wishes to do what I desire. But I have not yet decided what it would be advisable to do in the matter.

14 December 1899

Already thirty-eight years since that dreadful catastrophe which crushed and changed my life, and deprived me of my guardian angel, the best of husbands and most noble of men! The news in the papers is very sad, and there is a confirmation of the report of Lord Winchester's and Colonel Downman's deaths, the latter a very nice man, who commanded the Gordon Highlanders, and had dined with me at Balmoral in September, after Bertie had given colours to the regiment. Received a list of the casualties. The Highlanders lost awfully, but I am glad young Freddie Kerr, who commanded my guard, and of whom I took leave at Ballater, escaped. Feel very low and anxious about the war.

11 January 1899

After tea saw Lord Rowton. Had much talk with him about the war and our want of preparedness, which has existed for a long time, and which is very culpable. Also asked him to see Lord Salisbury, to try and impress upon him the importance of having no official enquiry into the conduct of the war until it is over. It would only be repeated back to the Boers and to foreign countries, and would do us a great deal of harm.

TO SALISBURY   27 January 1900

Am much surprised to see by your telegram and letter that the very serious state of the war was not considered by the Cabinet yesterday. The feeling abroad, except in America and Italy, is so inimical that we ought to take further steps to protect this country, and to raise more troops if we can. There is not a doubt that the attempted relief of Ladysmith would have succeeded had we had more troops. All Militia must be called out. Red-tapings and useless difficulties must not be regarded at such a very serious moment. The loss of so many valuable lives for nothing is terrible. What do Lord Roberts and Lord Kitchener say? I have such faith in the latter, but we hear nothing from either.

TO BIGGE   27 January 1900

The Queen is determined to press any available measure to put us in a safe position, and to put an end to these terrible failures.

TO LANSDOWNE   30 January 1900

Thanks for your letter of yesterday. Am glad to hear of the measures determined on. Hope all Cavalry at home will be for ever kept at war footing. Trust that 8th Division will be sent as desired by Lord Roberts and as soon as possible. Still think the whole Militia should be embodied. Could not corrugated-iron huts be used instead of tents?

To Lady Roberts   30 January 1900

I send you with these lines what I know you and Lord Roberts will value, but which I forward with a very sad heart. It is the Victoria Cross, which I should have been so proud and pleased to decorate your darling son with myself. I would not let anyone send it you but myself; it is a melancholy pleasure to me to do so.

To Lansdowne   30 January 1900

I am horrified at the terrible list of casualties, twenty-two officers killed and twenty-one wounded. It is quite imperative that Lord Roberts should not move till he has plenty of troops, a really large force. Pray impress this on him. And we must hurry out more troops. You must call out the Militia at once. Would it be possible to warn the young officers not to expose themselves more than is absolutely necessary?

To Balfour   4 February 1900

The Queen must urge again on Mr. Balfour very strongly the necessity of resisting these unpatriotic and unjust criticisms of our Generals and of the conduct of the war. If the Government are firm and courageous the country will support them. If not, the number of Boer spies will telegraph back to South Africa, and great harm will be done. You must all show a firm front, and not let it be for a moment supposed that we vacillate in the least. An enquiry after the war itself is over can be held out, but not now. No doubt the War Office is greatly at fault, but it is the whole system which must be changed, and that cannot be just now.

To Chamberlain   14 February 1900

Please let me know what steps you intend to take to protect the Zulus from being attacked by the Boers. Feel certain you agree with me that we are bound in honour to stand by my native subjects.

To Sir Redvers Buller   27 February 1900

I have heard with the deepest concern of the heavy losses sustained by my brave Irish soldiers. I desire to express my sympathy and my admiration of the splendid fighting qualities which they have exhibited throughout these trying operations.

To Lady White   1 March 1900

Accept my warmest congratulations on this delightful news, of the relief of Ladysmith, and the safety of your most gallant husband, and all under him. His heroism was splendid.

To the German Emperor   7 March 1900

I wish to thank you for your last kind letter and at the same time for your kind and friendly expressions and views of the war in South Africa, as expressed both to Uncle Bertie and Sir Frank Lascelles.

I sincerely hope that your example may at last be followed by other Powers, and that the German Press may cease abusing and reviling us and telling lies about our army. Under these circumstances, while affairs, including the war, are so uncertain,

I have given up going abroad, but intend paying a visit to Ireland quite early next month.

JOURNAL 8 March 1900

At a quarter to four started with Lenchen and Beatrice, who sat opposite to me ... Everywhere the same enormous crowds and incessant demonstrations of enthusiasm; if possible, even beyond that of the two Jubilees.

9 March 1900

A little before four drove out as yesterday, only Alix went instead of Lenchen, I again sitting alone. We went by Buckingham Palace Road, Victoria Street, Parliament Street, Regent Street, Marylebone Road, Edgware Road, Hyde Park, Exhibition Road, Brompton Road, and home by Constitution Hill. Everywhere the greatest enthusiasm and enormous crowds of people, quite as great as yesterday. There were many decorations, and all looked most festive. It was a truly thrilling and touching manifestation of the deep devotion and loyalty of my people, and I feel it very much. Dear Alix was delighted to have been with me.

22 March 1900

Lunched early, and left at two with Lenchen and Thora [her daughter, Helena Victoria] for Woolwich, going by train via Waterloo. The whole line from Windsor was crowded with workmen and people, all cheering. At Waterloo Station there was an immense crowd, which was very enthusiastic. We three drove straight through the [Woolwich] Arsenal, where 20,000 workmen were assembled, who are working day and night on munitions, and had been given a half-holiday in honour of my visit. They cheered so tremendously that it quite drowned the band playing *God Save the Queen*.

In April the Queen paid her last visit to Ireland.

JOURNAL 4 April 1900

We landed at the Victoria Wharf at half-past eleven ... We three [Princesses Helena and Beatrice were with the Queen on this visit to Ireland] wore bunches of real shamrocks, and my bonnet and parasol were embroidered with silver shamrocks ...

The drive lasted two hours and a half. We went all along the quays in the poorer parts of the town, where thousands had gathered together and gave me a wildly enthusiastic greeting. At Trinity College the students sang *God Save the Queen*, and shouted themselves hoarse. The cheers and often almost screams were quite deafening. Even the Nationalists in front of the City Hall seemed to forget their politics and cheered and waved their hats. It was really a wonderful reception I got and most gratifying ... The whole route from Kingstown to Dublin was much crowded, all the people cheering loudly, and the decorations were beautiful. For some distance the road was kept by bluejackets, but in many parts of the more country roads there was scarcely a policeman or soldier ...

Lenchen came in with some startling news from Bertie, who had been shot at as their train was leaving Brussels. A man [Sipido, an Italian] jumped on to the step of the railway carriage in which he and Alix were sitting, and fired straight at them. Was greatly shocked and upset.

7 April 1900

Drove with Lenchen and Beatrice, the ladies in a second carriage, to the public part of the Phoenix Park, where 52,000 school-children from all parts of Ireland were assembled with their masters and mistresses. It was a wonderful sight, and the noise of the children cheering was quite overpowering. I drove down the line so that they could all see me.

9 April 1900

I drove through all the principal streets of Dublin, where all the decorations were still up. The crowds were just as large as on Wednesday, and the enthusiasm immense.

17 April 1900

We drove into Dublin to the Adelaide Hospital, situated in the very poorest part of the town. The street in which it stands is a very narrow one, and the people literally thronged round the carriage, giving me the most enthusiastic welcome, as indeed I receive everywhere.

26 April 1900

I can never forget the really wild enthusiasm and affectionate loyalty displayed by all in Ireland, and shall ever retain a most grateful remembrance of this warm-hearted sympathetic people. Even when I used to go round the grounds in my pony chair and the people outside caught sight of me they would at once cheer and sing *God Save the Queen*.

TO THE PRINCE OF WALES   ? 8 May 1900

I wish to express to you my earnest hope that you will not go to Paris for the Exhibition. All kinds of people of every sort will be there, and you would have to be perpetually watched and followed about—which would be very disagreeable to you—and even then would run great risk. We are all most anxious that your precious life should not be jeopardised.

JOURNAL   19 May 1900

Fine day. Went with Beatrice to the kennels. The following telegram was received from Major-General Baden-Powell, dated 17th May: 'Happy to report Mafeking successfully relieved to-day ...'

The people are quite mad with delight, and London is said to be indescribable.

TO LORD GEORGE HAMILTON   20 May 1900

I approve the honours proposed for my birthday, though still finding the same fault in not having enough natives amongst them.

But I am surprised and disappointed at not seeing the name of one of the devoted nurses for whom specially in fact I and you also wanted this medal to be instituted. It seems to me that very few of them who exposed their lives are mentioned.

JOURNAL   24 May 1900

Again my old birthday returns, my eighty-first! God has been very merciful and supported me, but my trials and anxieties have been manifold, and I feel tired and upset by all I have gone through this winter and spring . . .

The number of telegrams to be opened and read was quite enormous, and obliged six men to be sent for to help the two telegraphists in the house [Balmoral]. The answering of them was an interminable task, but it was most gratifying to receive so many marks of loyalty and affection.

TO SALISBURY   9 June 1900

Feel anxious for personal safety of Sir C. MacDonald [British Minister in Peking where the foreign legations were being besieged by the Boxers, in a nationalist uprising]. Have you considered possibility of removal of Foreign Ministers from Pekin? If one of them were killed war would be inevitable.

16 June 1900

Should be glad to hear your views on the state of affairs in China, which seem to me most serious; also please say what you propose to do.

JOURNAL   1 July 1900

Distressing news has come from China, that the German Minister was murdered at Pekin, already on the 18th of last month, by the Chinese regular troops, and that the other Legations are in the greatest danger. Feel very grieved and anxious about our good Minister Sir Claude MacDonald.

5 July 1900

Very bad news from Pekin, through Reuter. All the foreigners, including 400 soldiers, women and children, who held out at British Legation, till ammunition and food exhausted, reported killed, but this is not yet officially confirmed, so that one lives in hopes it may not all be true. Feel quite miserable, horror-struck.

TO SALISBURY   5 July 1900

I am horror-struck at the dreadful news from China, which Reuter gives this morning; is there not a hope it may not be entirely true? Such a thing would be quite unparalleled, and arc we to stand and bear this worse than insult without some strong action? I feel quite ill at the thought of the poor MacDonalds and all the ladies and children, it haunts me day and night. Ought not the Chinese Minister to receive his passports if these news be true?

JOURNAL   24 July 1900

Received a telegram from Mr. Goschen saying following just received from Admiral Seymour: 'By latest reports Legations at Pekin still holding out, prospect more hopeful.' The news from South Africa much the same. Fighting continues and results in nothing very definite.

31 July 1900

A terrible day! When I had hardly finished dressing Lenchen and Beatrice knocked at the door and came in. I at once asked whether there were any news, and Lenchen replied, 'Yes, bad news, very bad news; he has slept away!' Oh, God! my poor darling Affie gone too! My third grown-up child, besides three very dear sons-in-law. It is hard at eighty-one! It is so merciful that dearest Affie died in his sleep without any struggle, but it is heartrending. Poor darling Marie, who knew of no real danger when she left, such a short time ago, without a fear. It is too terrible also for the poor daughters, who adored their father! . . .

It is a horrible year, nothing but sadness and horrors of one kind and another. I think they should never have withheld the truth from me, as long as they did. It has come such an awful shock. I pray God to help me to be patient and have trust in Him, who has never failed me! Everyone is quite stunned . . . Felt terribly shaken and broken, and could not realise the dreadful fact. Recollections of dear Affie's childhood and youth, and nowhere more vivid than here, crowded in upon me. People are so dreadfully shocked, and the Navy feels it deeply, for he was much beloved in the service, and greatly admired, having been such an excellent officer . . . I asked Bertie to come here, but he said he was too unnerved to come to-day, but would do so to-morrow.

Bertie arrived with Tino. Bertie came at once to my room, and was a good deal upset, as he feels the loss of his dear brother terribly.

Quantities of telegrams kept pouring in, and the day was spent in answering them. We dined again alone, and later Beatrice read to me in my room out of some of my favourite religious books, which was soothing.

19 August 1900

Received the most welcome news, for which I thank God most earnestly, that the Allies had entered Pekin, and found all well at the Legations. General Gaselee, with my Indian troops, seems to have done admirably.

26 August 1900

This ever dear day has returned again without my beloved Albert being with me, who on this day, eighty-one years ago, came into the world as a blessing to so many, leaving an imperishable name behind him! How I remember the happy day it used to be, and preparing presents for him, which he would like! I thought much of the birthday spent at the dear lovely Rosenau in '45, when I so enjoyed being there, and where now his poor dear son, of whom he was so proud, has breathed his last.

17 September 1900

I have not been feeling very well these last days, and can eat very little. This has been a great trouble for some time past.

5 November 1900

Felt very poorly and wretched, as I have done all the last days. My appetite is completely gone, and I have great difficulty in eating anything.

9 November 1900

Had felt better through the day and free from pain, but I still have a disgust for all food.

11 November 1900

Had a shocking night, and no draught could make me sleep, as pain kept me awake. Felt very tired and unwell when I got up, and was not able to go to church, to my great disappointment . . . Could do nothing for the whole morning. Rested and slept a little.

19 November 1900

Had a very fair night, but my appetite much about the same. The sitting through meals, unable to eat anything, is most trying.

21 November 1900

Darling Vicky's sixtieth birthday. To think of her, who was so wonderfully active and strong, now so ill and suffering is heartbreaking.

28 November 1900

Had a very bad restless night, with a good deal of pain. Got up very late, and when I did felt so tired I could do nothing, and slept on the sofa.

Saw Lord Salisbury after tea, who talked a good deal about my health, the necessity for my going abroad and my getting a thorough change and rest . . .

Felt so exhausted and uncomfortable that I did not go to dinner, but I went for a short while into the drawing-room later.

29 November 1900

Had a good night and felt rather better. A wet morning, so I did not go out. In the afternoon, directly after luncheon, saw the contingent of the 1st Life Guards who landed this morning with the rest of the composite regiment. I inspected them just in front of the archway I always go out by, on their way from the station to the barracks.

2 December 1900

After a very wretched night, I passed a very miserable day, and could neither go out nor leave my room. Missed being able to go to church and hearing a very fine sermon from Dean Farrar, which annoyed me very much. Slept a good deal and, as my repulsion for food was very great, went to neither luncheon nor dinner. Beatrice read and played a little to me.

11 December 1900

Saw Sir Francis Laking for some time after tea. He encouraged me by saying he thought I should in time get over this unpleasant dislike for food and squeamishness, as well as the great discomfort I suffer from, and recommended my taking a little milk and whisky several times a day.

18 December 1900

Had a very bad night, and scarcely slept at all. Breakfasted late and left [Windsor] for Osborne at 11.40 with Beatrice and Ena, Thora meeting us at the station. Drino had arrived early from Wellington College. Slept for an hour in the train, and then I had a little broth, but I could not take much. Embarked on the *Alberta* at two. It was rather rough just outside Portsmouth, but became quite calm afterwards. Arrived up at the house about 3.30. Felt too tired to do anything, and dozed for a short while ... Had some dinner with Beatrice and Thora, in the room in which we generally breakfast. There was very unsatisfactory news from South Africa, the Boers being terribly active all over the country.

22 December 1900

I slept a little at first, and then was rather disturbed, after which I slept on again till quarter to twelve, at which I was very annoyed. I got up and had some breakfast, which I really liked, then drove out with Thora. I am rather better but still see very badly.

25 December 1900

Did not have a good night, was very restless, and every remedy that was tried failed in making me sleep. Then when I wished to get up I fell asleep again, which was too provoking.

27 December 1900

Had only a pretty good night, as I was much disturbed by the wind. I took several draughts, and then some milk, and fell asleep towards morning, so did not get up till nearly one. Felt very low and sad.

31 December 1900

A terribly stormy night. The same unfortunate alternations of sleep and restlessness, so that I again did not get up when I wished to, which spoilt my morning and day.

1 January 1901

Another year begun, and I am feeling so weak and unwell that I enter upon it sadly. The same sort of night as I have been having lately, but I did get rather more sleep and was up earlier.

2 January 1901

On Lord Roberts's arrival here [at Osborne], Arthur took him to the Council-room, where the family were assembled, and then took him to see Lenchen in her own room, after which he brought him to the drawing-room, where I was. I received him most warmly, shaking hands with him, and he knelt down and kissed my hand ... Lord Roberts spoke of several officers who had not done well, and of others who had done excellently; also of all the difficulties our army had had to content with. We deeply deplored the loss of so many valuable lives. He still wears his arm in a sling, the result of a fall from his horse.

After about half an hour Arthur came back with Louischen, Daisy, young Arthur, and Thora. I then gave Lord Roberts the Garter, which quite overcame him, and he said it was too much. I also told him I was going to confer an earldom on him, with the remainder to his daughter.

4 January 1901

From not having been well, I see so badly, which is very tiresome.

12 January 1901

Had a good night and could take some breakfast better. Took an hour's drive at half-past two with Lenchen. It was very foggy, but the air was pleasant.

The Queen died at half past six in the evening on 22 January 1901 surrounded by her children and grandchildren.

BALMORAL CASTLE—EVENING.

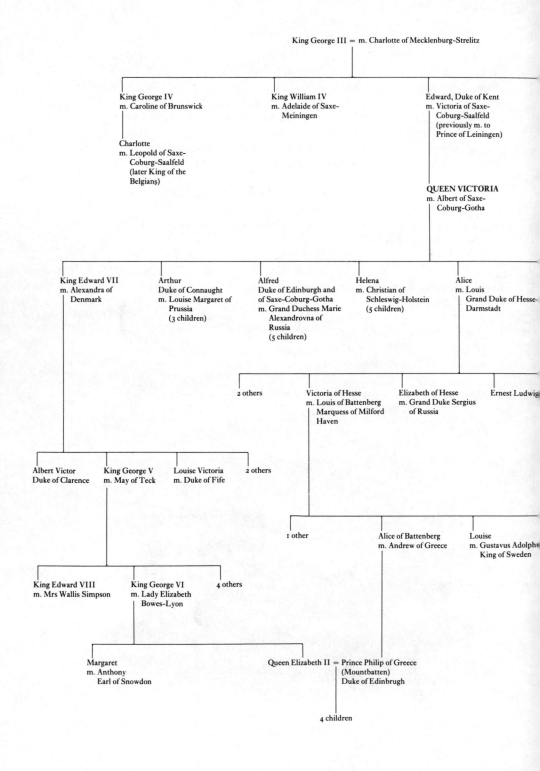

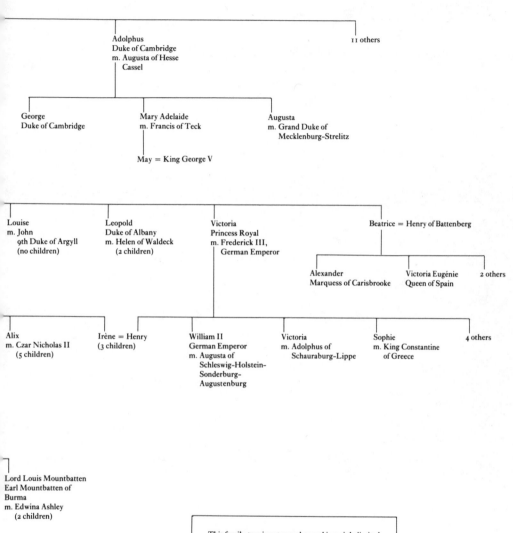

Adolphus
Duke of Cambridge
m. Augusta of Hesse
   Cassel

11 others

George
Duke of Cambridge

Mary Adelaide
m. Francis of Teck

Augusta
m. Grand Duke of
   Mecklenburg-Strelitz

May = King George V

Louise
m. John
   9th Duke of Argyll
   (no children)

Leopold
Duke of Albany
m. Helen of Waldeck
   (2 children)

Victoria
Princess Royal
m. Frederick III,
   German Emperor

Beatrice = Henry of Battenberg

Alexander
Marquess of Carisbrooke

Victoria Eugénie
Queen of Spain

2 others

Alix
m. Czar Nicholas II
   (5 children)

Irène = Henry
(3 children)

William II
German Emperor
m. Augusta of
   Schleswig-Holstein-
   Sonderburg-
   Augustenburg

Victoria
m. Adolphus of
   Schauraburg-Lippe

Sophie
m. King Constantine
   of Greece

4 others

Lord Louis Mountbatten
Earl Mountbatten of
Burma
m. Edwina Ashley
   (2 children)

This family tree is not complete and is mainly limited
to persons mentioned in the text. Children are not
always shown in order of birth. For dates of birth and
death, full names and titles, and nicknames see Index.

INDEX

Albert, Prince Consort is referred to as A, the Prince and Princess of Wales as P and Pss of W, and Queen Victoria as QV.